Advanced Structured Materials

Volume 241

Common engineering materials are reaching their limits in many applications, and new developments are required to meet the increasing demands on engineering materials. The performance of materials can be improved by combining different materials to achieve better properties than with a single constituent, or by shaping the material or constituents into a specific structure. The interaction between material and structure can occur at different length scales, such as the micro, meso, or macro scale, and offers potential applications in very different fields.

This book series addresses the fundamental relationships between materials and their structure on overall properties (e.g., mechanical, thermal, chemical, electrical, or magnetic properties, etc.). Experimental data and procedures are presented, as well as methods for modeling structures and materials using numerical and analytical approaches. In addition, the series shows how these materials engineering and design processes are implemented and how new technologies can be used to optimize materials and processes.

Advanced Structured Materials is indexed in Google Scholar and Scopus.

Soney C. George · Benjamin Tawiah
Editors

Advanced Bionanocomposite Materials

Innovations for Sustainable Development

Springer

Editors
Soney C. George
Center for Nanoscience and Technology
Amal Jyothi College of Engineering
(Autonomous)
Kottayam, Kerala, India

Benjamin Tawiah
Department of Industrial Art (Textiles
and Fashion)
College of Art and Built Environment
(CABE)
Kwame Nkrumah University of Science
and Technology (KNUST)
Kumasi, Ghana

ISSN 1869-8433 ISSN 1869-8441 (electronic)
Advanced Structured Materials
ISBN 978-981-96-8224-9 ISBN 978-981-96-8225-6 (eBook)
https://doi.org/10.1007/978-981-96-8225-6

This Springer imprint is published by the registered company Springer Nature Singapore Pte Ltd.
The registered company address is: 152 Beach Road, #21-01/04 Gateway East, Singapore 189721, Singapore

To the visionary leaders and innovators dedicated to a sustainable future—may your passion and commitment to advancing science and technology inspire generations to come. This book is dedicated to those working tirelessly to achieve the United Nations Sustainable Development Goals (SDGs), as we strive together to create a better world through the transformative power of bionanocomposites. Together, we can drive meaningful change and foster a more sustainable and equitable future for all.

Prof. Soney C. George
Dr. Benjamin Tawiah

Foreword

In the face of mounting global challenges, the intersection of science, technology, and sustainability has never been more critical. *Advanced Bionanocomposite Materials* emerges at this pivotal juncture, offering a profound exploration of how innovative materials can address some of the most pressing issues of our time. This work serves as not only a technical resource but also a manifesto for a future that prioritizes ecological balance and human well-being.

Bionanocomposites, which seamlessly integrate biological elements with nanoscale materials, represent a frontier in materials science that holds immense promise. Their applications span a diverse range of fields, from healthcare to energy and environmental management, making them a vital component in the quest to achieve the United Nations Sustainable Development Goals (SDGs). This book meticulously details how these materials can contribute to goals such as ensuring clean water and sanitation, promoting sustainable energy solutions, and advancing public health. Drawing from the insights of leading experts, each chapter offers a nuanced understanding of the principles underlying bionanocomposites, while also providing practical examples of their application. Readers will find themselves engaged with cutting-edge research that not only elucidates the scientific principles but also contextualizes them within the broader societal implications.

What sets this work apart is its commitment to fostering interdisciplinary dialogue. The challenges we face today are complex and multifaceted, requiring collaborative efforts that transcend traditional boundaries. The contributors to this volume exemplify this spirit of collaboration, bringing together expertise from materials science, engineering, biology, and social sciences to address the interconnected issues of sustainability.

As you delve into the pages of this book, I encourage you to consider the potential of bionanocomposites not just as materials, but as catalysts for change. They embody the promise of innovation; an opportunity to rethink our approaches to production, consumption, and environmental stewardship. In doing so, we can aspire to create solutions that not only meet the needs of the present but also safeguard the future of our planet.

This book is a vital contribution to the ongoing discourse on sustainable development. It challenges us to envision a world where scientific advancement and ecological responsibility go hand in hand, inspiring a collective commitment to the transformative power of bionano innovation. May it serve as a source of inspiration and a guide for all who seek to contribute to a more sustainable and equitable future.

Prof. Ebenezer Kof Howard
Kwame Nkrumah University
of Science and Technology (KNUST)
Kumasi, Ghana

Preface

In an era marked by unprecedented environmental challenges and societal demands, the quest for sustainable solutions has become a global imperative. *Advanced Bionanocomposite Materials* is not merely a compilation of scientific knowledge; it is a call to action for researchers, industry leaders, and policymakers alike to harness the transformative potential of bionanocomposites.

As we face the realities of climate change, resource scarcity, and public health crises, the innovative integration of biological materials with nanoscale technologies offers a promising path forward. This book delves into the multifaceted applications of bionanocomposites, revealing how they can contribute significantly to achieving the United Nations Sustainable Development Goals (SDGs).

Among the SDGs that this work addresses, Goal 3: Good Health and Well-Being stand out, as bionanocomposites have the potential to revolutionize healthcare through advanced drug delivery systems and biocompatible materials that enhance patient outcomes. Additionally, Goal 6: Clean Water and Sanitation is critical; bionanocomposites can be employed in water purification technologies, providing efficient solutions for clean water access in underserved communities.

Goal 7: Affordable and Clean Energy also benefits from this research, as these materials can be used in the development of more efficient solar cells and energy storage systems, promoting renewable energy solutions. Furthermore, Goal 12: Responsible Consumption and Production is addressed through the creation of biodegradable composites that reduce plastic waste and promote circular economy principles. Lastly, Goal 13: Climate Action is intertwined with the ability of bionanocomposites to contribute to carbon capture technologies and sustainable agricultural practices that enhance food security while minimizing environmental impact.

Each chapter is crafted to provide a deep understanding of the principles and practices surrounding these advanced materials. From their fabrication methods to their applications in areas such as energy, environmental remediation, healthcare, and agriculture, we explore how bionanocomposites can address pressing global issues while promoting ecological sustainability. The collective expertise of our contributors reflects a commitment to interdisciplinary collaboration, underscoring the importance of diverse perspectives in tackling complex challenges. We aim to inspire a new generation of innovators who are not only equipped with knowledge but are also driven by the urgency of creating impactful solutions.

As you engage with this book, we invite you to envision the possibilities that lie ahead. Together, we can pioneer a sustainable future, leveraging the power of bionano innovation to create materials and technologies that meet the needs of today while safeguarding the planet for generations to come. Your role in this journey is crucial, and we hope that this work ignites your passion for change and encourages you to be an active participant in this vital endeavor.

Kottayam, India — Soney C. George
Kumasi, Ghana — Benjamin Tawiah

Acknowledgements

The journey of editing *Advanced Bionanocomposite Materials* has been both challenging and rewarding, and it would not have been possible without the support and contributions of many individuals and institutions.

First and foremost, we would like to express our heartfelt gratitude to the contributors, whose expertise and insights have been invaluable throughout the writing process. Your dedication to advancing the field of composites and your willingness to collaborate have enriched this work immeasurably.

We are also deeply thankful to the researchers and practitioners in the field who have shared their knowledge and experiences. Your pioneering work has inspired us and provided the foundation upon which this book is built. Special thanks to those who participated in discussions, workshops, and conferences, where ideas were exchanged, and new perspectives were gained.

We would like to acknowledge the support of our academic institutions and organizations, which have fostered an environment of innovation and inquiry. Your encouragement and resources have been instrumental in bringing this project to fruition.

To our families and friends, thank you for your understanding and encouragement during the long hours spent researching, writing, and editing this book. Your unwavering support has been a source of strength and motivation, reminding us of the importance of our work.

Finally, we extend our appreciation to the editorial and publishing team who have worked diligently to bring this book to life. Your professionalism and commitment to excellence have made this process smooth and enjoyable.

Together, we hope this book serves not only as a valuable resource but also as a catalyst for further research and collaboration in the field of bionanocomposites. May it inspire future innovations that contribute to a more sustainable and equitable world.

Acknowledgements

About This Book

Advanced Bionanocomposite Materials offers an in-depth exploration of the innovative field of bionanocomposites, emphasizing their potential to promote sustainability across diverse sectors. This book serves as a comprehensive resource for researchers, engineers, and policymakers interested in the intersection of advanced materials science and sustainable development.

The book begins by establishing the foundational concepts of bionanocomposites, elucidating their significance in addressing contemporary environmental challenges. It highlights the urgent need for materials that not only meet performance demands but also minimize ecological footprints. This introductory chapter sets the stage for understanding the multifaceted roles that bionanocomposites play in sustainable development. Subsequent chapters delve into the fabrication of biomaterials, emphasizing sustainable approaches that prioritize renewable resources and eco-friendly processes. The discussion includes various techniques for synthesizing biomaterials, focusing on their compatibility with bionanocomposites. This section highlights advancements in material science that enable the creation of composites with enhanced mechanical properties and biodegradability.

Furthermore, the exploration of bioprepregs and hybrid fiber reinforcement techniques reveals how combining different materials can optimize performance while maintaining sustainability. This chapter discusses the unique advantages these composites offer, including improved strength-to-weight ratios and adaptability to various applications. A significant portion of the book is dedicated to smart and adaptive processing techniques. These methods not only enhance the functionality of bionanocomposites but also allow for dynamic responses to environmental factors. The integration of sensors and responsive materials is discussed, showcasing how these innovations can lead to breakthroughs in various industries. Characterization methodologies are crucial for understanding the properties of bionanocomposites. This section covers advanced characterization techniques that reveal the intricate relationships between structure and performance, enabling researchers to tailor materials for specific applications. The discussion includes insights into mechanical, thermal, and electrical properties, highlighting the versatility of bionanocomposites.

The applications of bionanocomposites in biotechnology and biomedical engineering are explored in detail. This chapter examines their role in drug delivery systems, tissue engineering scaffolds, and biosensors, illustrating how these materials can revolutionize healthcare by providing safer, more effective solutions. In the realm of energy, the book discusses the potential of bionanocomposites in energy storage and conversion technologies, such as batteries and fuel cells, and highlights the importance of sustainable materials in the quest for renewable energy sources.

Environmental applications, particularly in remediation and water purification, are addressed with a focus on how bionanocomposites can effectively address pollution and resource scarcity. The book further discusses various case studies that demonstrate their efficacy in removing contaminants and improving water quality. The agricultural sector receives attention as well, with discussions on how bionanocomposites can enhance food safety, extend shelf life, and improve packaging solutions. This chapter underscores the importance of sustainable practices in food processing and packaging, reflecting on the role of materials in promoting health and safety. The health and pharmaceutical sciences are explored through the lens of bionanocomposites, which offer innovative solutions for drug delivery and diagnostics. The potential for these materials to improve patient outcomes is a key focus, illustrating their impact on modern medicine. A dedicated chapter examines conductive bionanocomposites and their applications in green electronics. This section discusses the development of sustainable electronic devices, emphasizing the role of these materials in reducing environmental impact while maintaining functionality.

The intersection of artistry and cultural heritage preservation is explored, highlighting how bionanomaterials can contribute to the conservation of art and artifacts through sustainable practices. This unique perspective adds depth to the discussion of material applications and their societal implications. Transportation applications are addressed with a focus on lightweight bionanocomposites, which are critical for improving fuel efficiency and reducing emissions. This chapter discusses recent advancements in materials that support sustainable transportation solutions. The versatility of bionanocomposites is further illustrated through discussions on nanostructured coatings and wearable textiles. The book examines how these materials enhance functionality while promoting sustainability in everyday products. In structural engineering, the application of bionanocomposites in building and construction is analyzed, showcasing their potential to create resilient and sustainable infrastructures. This chapter emphasizes the need for innovative materials in addressing global challenges such as urbanization and climate change.

The book concludes with a thorough examination of lifecycle assessments of bionanocomposites, providing insights into their ecological footprints and sustainability metrics. This critical analysis is complemented by discussions on the regulatory frameworks and ethical considerations surrounding the manufacture and application of these materials, highlighting the importance of responsible innovation. The critical examination of the diverse themes makes *Advanced Bionanocomposite Materials* an important material in understanding the crucial role that advanced materials can play in creating a sustainable future. This book serves as a vital resource for those committed to advancing knowledge and practices that align technological innovation with environmental stewardship.

Contents

Contributors

Benjamin K. Asinyo Department of Industrial Art (Textiles), Kwame Nkrumah University of Science and Technology, Kumasi, Ghana

Festus Ben Centre for Nanoengineering and Advanced Materials, Department of Metallurgy, University of Johannesburg, Johannesburg, South Africa;
Centre for Advanced Materials Research and Development, Department of Physics, Federal Polytechnic Ede, Ede, Nigeria

Bhawana Chand Research Lab for Energy Systems, Department of Physics, Netaji Subhas University of Technology, Dwarka, New Delhi, India

Sujata Deb Associate Professor Department of Physics, University of Science and Technology Meghalaya (USTM), Ri-Bhoi, Techno City, Killing Road, Baridua, Meghalaya, India

Atiyeh Derakhshandehpour Department of Chemistry, University of Isfahan, Isfahan, Iran

Alessandro Distante Istituto Scientifico Biomedico Euro Mediterraneo (ISBEM), Brindisi, Italy

Benjamin Eghan Faculty of Art, Design and Humanities, De Montfort University, Leicester, UK

Alireza Fatahi Department of Chemistry, University of Isfahan, Isfahan, Iran

Bin Fei School of Fashion and Textiles (SFT), The Hong Kong Polytechnic University, Hung Hom, Kowloon, Hong Kong SAR, China

George Kwame Fobiri Department of Fashion Design and Textiles Studies, Kumasi Technical University, Kumasi, Ghana;
Department of Industrial Art (Textiles), Kwame Nkrumah University of Science and Technology, Kumasi, Ghana

Soney C. George Centre for Nanoscience and Technology, Amal Jyothi College of Engineering, Koovappally, Kottayam, India

Ahmed Fathy Mostafa Ghanem Packaging Materials Department, Chemical Industries Research Institute, National Research Centre, Dokki, Giza, Egypt; Korea Institute of Energy Technology, KENTECH, Naju, Republic of Korea

Priya Gupta Manyawar Kanshiram Government Polytechnic, Tirwa, Kannauj, Uttar Pradesh, India;
Department of Physics, University of Lucknow, Lucknow, Uttar Pradesh, India

Ranjana Jha Research Lab for Energy Systems, Department of Physics, Netaji Subhas University of Technology, Dwarka, New Delhi, India

Shouxiang Jiang School of Fashion and Textiles, The Hong Kong Polytechnic University, Kowloon, Hong Kong SAR, China;
Research Centre of Textiles for Future Fashion, The Hong Kong Polytechnic University, Kowloon, Hong Kong SAR, China

Jeny Jose Bar-Ilan University, Ramat Gan, Israel

Jerin Jose University of Catania, Sicily, Italy;
Amal Jyothi College of Engineering, Kanjirappally, Kerala, India

Juwel Jose Caritas Hospital, Kottayam, Kerala, India

Manpreet Kaur Department of Punjabi, Govt. Ranbir College, Sangrur, Punjab, India

Ashok Kumar Malik Department of Chemistry, Punjabi University, Patiala, Punjab, India

Abbas Mohammadi Department of Chemistry, University of Isfahan, Isfahan, Iran

Lixiao Ni Key Laboratory of Integrated Regulation and Resource Development on Shallow Lakes, MOE, School of Environment, Hohai University, Nanjing, China

Emmanuel Abankwah Ofori Department of Materials Science and Engineering, Hubei University of Automotive Technology, Shiyan, Hubei, China

Narendra Kumar Pandey Manyawar Kanshiram Government Polytechnic, Tirwa, Kannauj, Uttar Pradesh, India;
Department of Physics, University of Lucknow, Lucknow, Uttar Pradesh, India

Peeyush Phogat Research Lab for Energy Systems, Department of Physics, Netaji Subhas University of Technology, Dwarka, New Delhi, India;
Council of Scientific and Industrial Research–National Institute of Science Communication and Policy Research (CSIR–NIScPR), Pusa, New Delhi, India

Rituka Rani Department of English, GSSS Talwan, Talwan, Punjab, India

Amirhossein Raouffard Department of Chemistry, University of Isfahan, Isfahan, Iran

Karishma Rawat Research Lab for Energy Systems, Department of Physics, Netaji Subhas University of Technology, Dwarka, New Delhi, India

Satyam Rawat Research Lab for Energy Systems, Department of Physics, Netaji Subhas University of Technology, Dwarka, New Delhi, India

Bismark Sarkodie Department of Industrial Art (Textiles), Kwame Nkrumah University of Science and Technology, Kumasi, Ghana;
School of Environment, Key Laboratory of Yellow River and Huaihe River Water Environment and Pollution Control, Ministry of Education, Henan Key Laboratory for Environmental Pollution Control, Henan Normal University, Xinxiang, China

Raphael Kanyire Seidu School of Fashion and Textiles, The Hong Kong Polytechnic University, Hung Hom, Kowloon, Hong Kong SAR, China

Shana Shaji Istituto Scientifico Biomedico Euro Mediterraneo (ISBEM), Brindisi, Italy

Promila Sharma Department of Chemistry, Punjabi University, Patiala, Punjab, India;
Department of Chemistry, RIMT University, Mandigobindgarh, Punjab, India

Shreya Sharma Research Lab for Energy Systems, Department of Physics, Netaji Subhas University of Technology, Dwarka, New Delhi, India;
Council of Scientific and Industrial Research–National Institute of Science Communication and Policy Research (CSIR–NIScPR), Pusa, New Delhi, India

Shreya Research Lab for Energy Systems, Department of Physics, Netaji Subhas University of Technology, Dwarka, New Delhi, India;
Council of Scientific and Industrial Research—National Institute of Science Communication and Policy Research (CSIR–NIScPR), Pusa, New Delhi, India

Peramjeet Singh Department of Physics, University of Lucknow, Lucknow, Uttar Pradesh, India

Sukhveer Singh Research Lab for Energy Systems, Department of Physics, Netaji Subhas University of Technology, Dwarka, New Delhi, India

Sukhvir Singh Department of Nanotechnology, Delhi Skill and Entrepreneurship University, Dwarka, New Delhi, India

Vijay Singh Department of Botany, Mata Gujri College, Fatehgarh Sahib, Punjab, India

Benjamin Tawiah School of Fashion and Textiles (SFT), The Hong Kong Polytechnic University, Hung Hom, Kowloon, Hong Kong SAR, China;
Department of Textiles, Department of Industrial Art (Textiles and Fashion), Kwame Nkrumah University of Science and Technology, Kumasi, Ghana

Anita Thakur Department of Chemistry, RIMT University, Mandigobindgarh, Punjab, India

Jahanvi Thakur Research Lab for Energy Systems, Department of Physics, Netaji Subhas University of Technology, Dwarka, New Delhi, India

Vernica Verma Department of Physics, University of Lucknow, Lucknow, Uttar Pradesh, India

Jakpa Wizi Key Laboratory of Integrated Regulation and Resource Development on Shallow Lakes, MOE, School of Environment, Hohai University, Nanjing, China

Neetu Yadav Department of Physics, University of Lucknow, Lucknow, Uttar Pradesh, India

Chapter 1
Bionanocomposite Materials for Sustainable Development

Benjamin Tawiah, Emmanuel Abankwah Ofori, Bin Fei, and Soney C. George

Abstract The exploration of bionanocomposites for sustainable development offers a vital insight into innovative biodegradable polymer materials and their integration with nanomaterials. This approach is particularly relevant in tackling the pressing sustainability challenges we face globally. In this introductory chapter, we provide a comprehensive overview of bionanocomposites, emphasizing their distinctive properties and the diverse applications they hold across various sectors such as packaging, construction, and biomedical fields. We delve into the fundamental principles that underpin these composites and discuss their importance in mitigating environmental impact. A key focus is placed on how nanomaterials enhance the mechanical, thermal, and barrier properties of bionanocomposites, thereby broadening their functionality and potential uses. We also categorize different types of bionanocomposites and examine their environmentally friendly production processes, aligning them with sustainable development goals, particularly in terms of waste reduction, resource efficiency, and the advancement of circular economies. Moreover, this chapter addresses the challenges related to the commercialization of bionanocomposites, including regulatory hurdles and market acceptance. This foundational knowledge equips readers with the understanding necessary to explore bionanocomposite technologies in the rest of the chapters.

B. Tawiah · B. Fei (✉)
School of Fashion and Textiles, The Hong Kong Polytechnic University, Hung Hom, Kowloon, Hong Kong SAR, China
e-mail: bin.fei@polyu.edu.hk

B. Tawiah
Department of Industrial Art (Textiles), Kwame Nkrumah University of Science and Technology, Kumasi, Ghana

E. A. Ofori
Department of Materials Science and Engineering, Hubei University of Automotive Technology, Hubei, China

S. C. George (✉)
Centre for Nanoscience and Technology, Amal Jyothi College of Engineering, Koovappally, Kottayam, India
e-mail: soneycgeorge@amaljyothi.ac.in

S. C. George and B. Tawiah (eds.), *Advanced Bionanocomposite Materials*, Advanced Structured Materials 241, https://doi.org/10.1007/978-981-96-8225-6_1

Keywords Bionanocomposites · Biodegradable polymers · Sustainable development · Nanomaterials · Eco-friendly materials

1.1 Introduction to Bionanocomposites

In recent years, the pressing need for sustainable development has led industries to seek innovative materials that reduce environmental impact while enhancing performance (Nwokediegwu et al. 2024; Paidari et al. 2023). Among the promising options are bionanocomposite materials, which blend the principles of bio-based materials and nanomaterials to create advanced materials with unique properties. The concept of sustainable development emphasizes balancing economic growth, environmental care, and social equity (Sen et al. 2024). With the global population expected to reach nearly 10 billion by 2050, the pressure on natural resources is increasing (O'Sullivan 2023; Kenny and Gehan 2023). As a result, there is a strong demand for materials that are not only effective but also environmentally friendly. Bionanocomposites emerge as a viable solution, bridging the gap between traditional materials and sustainability (Singh et al. 2024).

Bionanocomposites consist of a matrix derived from fauna and flora combined with nano-sized fillers. The biological matrices can be sourced from renewable materials like plant fibers, proteins, or biodegradable polymers, while the nanomaterials may include nanoparticles, nanofibers, or nanotubes that enhance the mechanical, thermal, and barrier properties of the composites (Atodariya et al. 2024; Zankana et al. 2023). This unique combination results in materials that are lightweight, strong, and capable of performing well in various environments. The foundation of bionanocomposites often lies in biopolymers such as polylactic acid (PLA), polyhydroxyalkanoates (PHAs), and cellulose (Paidari et al. 2023; Agbakoba et al. 2023). These materials come from renewable sources and are biodegradable, making them suitable for sustainable applications. Incorporating nano-fillers like carbon nanotubes, graphene, or silica nanoparticles significantly enhances the properties of the biopolymer matrix. These fillers improve mechanical strength, thermal stability, and barrier properties, making the composites ideal for demanding applications. To further boost the performance of bionanocomposites, various additives can be included, such as plasticizers, stabilizers, or antimicrobial agents, which enhance the functionality and longevity of the material (Kishore et al. 2023; Dubey et al. 2023).

The development of bionanocomposites involves several manufacturing processes, including solution casting, extrusion, injection molding, and 3D printing. Solution casting entails dissolving the biopolymer in a solvent, mixing it with nanofillers, and casting the solution into molds. Once the solvent evaporates, a solid composite forms. Extrusion is a continuous process where the biopolymer and nanofillers are fed into an extruder, melted, and shaped into pellets or films, making this method particularly effective for large-scale production. Injection molding is commonly used to create complex shapes and structures from bionanocomposites, where the molten polymer is injected into molds and allowed to cool and solidify

(Mehdinia et al. 2024; Lecoublet et al. 2024). The rise of additive manufacturing has opened new avenues for bionanocomposites, enabling the creation of intricate designs and structures tailored to specific applications. Bionanocomposites have a wide range of applications across various sectors. One of the most significant applications is in the packaging industry. Traditional plastic packaging poses environmental challenges due to its non-biodegradable nature. Utilizing bionanocomposites in packaging materials can help reduce reliance on fossil fuels and decrease plastic waste. These composites can effectively block moisture, gases, and UV light, ensuring the preservation of food products while being biodegradable (Paidari et al. 2023; Nandhini et al. 2024a). In the construction industry, bionanocomposites are gaining traction for their lightweight yet strong characteristics. They can be used in insulation materials, panels, and even structural components. The incorporation of nanomaterials enhances the durability and thermal efficiency of building materials, contributing to energy savings and a reduced environmental impact (Mehdinia et al. 2024). The biomedical field is another area where bionanocomposites show great promise. They can be employed in drug delivery systems, tissue engineering, and wound healing applications. The biocompatibility of biopolymers combined with the functionality of nanomaterials can lead to innovative solutions for medical challenges, promoting sustainable healthcare practices (Sharma et al. 2023b; Nandhini et al. 2024b). Additionally, bionanocomposites are finding applications in the textile industry. The use of sustainable fibers combined with nanotechnology can result in fabrics that are not only durable but also possess unique properties, such as water resistance, antimicrobial effects, and UV protection.

The development and utilization of bionanocomposites can have profound environmental and economic implications. By replacing conventional petroleum-based materials with renewable resources, we can significantly reduce the carbon footprint associated with material production. Furthermore, the biodegradability of bionanocomposites ensures that they do not contribute to long-term pollution, addressing one of the critical challenges of the current linear economy (Nasir et al. 2023; Tripathy and Gupta 2024). From an economic perspective, adopting bionanocomposites can stimulate growth in the green economy, creating new job opportunities in research, production, and recycling. Additionally, the versatility of these materials opens up avenues for innovation, leading to the development of new products and markets (Kishore et al. 2023).

This foundational chapter sheds light on the multifaceted aspects of bionanocomposites, providing the basis for further exploration and application in the quest for sustainable solutions. This chapter provides an in-depth examination of the principles underlying bionanocomposites, their manufacturing principles, classifications, applications, and the challenges and opportunities for reducing environmental impact.

1.2 Fundamentals Principles of Bionanocomposites

The fundamental principles underlying bionanocomposites involve several key aspects, including material characteristics, interactions at the nanoscale, and the integrated performance of the composite structure (Farooq et al. 2024). The primary principles of bionanocomposites are:

(i) *Biopolymer Matrix*

The foundation of bionanocomposites usually starts with a biopolymer matrix, which comes from renewable biological sources. Some common examples of these biopolymers are starch, cellulose, chitosan, polylactic acid (PLA), and polyhydroxyalkanoates (PHA). One of the primary advantages of these materials is their biodegradable nature, which allows them to decompose naturally over time, thereby minimizing their environmental impact. Furthermore, these materials typically exhibit exceptional mechanical and thermal properties, making them suitable for a wide range of applications. Selecting the appropriate biopolymer is crucial, as it significantly influences the overall sustainability, compatibility, and performance of the composite (Sharma et al. 2023a).

(ii) *Nano-fillers and Their Characteristics*

Nano-fillers, such as carbon nanotubes, silica nanoparticles, graphene, and nanofibers, are incorporated into the biopolymer matrix to enhance its properties. The term "nano" refers to exceptionally small materials, typically measuring between 1 to 100 nm. At this nanoscale, materials exhibit distinct properties that can differ markedly from those of their larger counterparts. Understanding these principles is crucial for leveraging the advantages of nano-sized materials in bionanocomposites, which integrate biological matrices with nano-fillers. One of the most notable benefits of nano-sized materials is their increased surface area. Due to their diminutive size, they possess a significantly higher surface area-to-volume ratio, which enhances their effectiveness in various applications (Anwer et al. 2023). This means they can interact more effectively with the surrounding biopolymer matrix, leading to better bonding and performance. Additionally, many nanomaterials, such as carbon nanotubes and graphene, possess remarkable strength. When incorporated into biopolymers, they can significantly enhance the tensile strength and durability of the composite. Another important aspect of nanomaterials is their ability to improve thermal and electrical conductivity. For instance, graphene is known for its excellent electrical properties, making bionanocomposites suitable for applications in electronics and sensors. The small size of nano-fillers also allows for better compatibility within the biopolymer matrix, ensuring a more uniform dispersion. This uniformity is crucial for achieving consistent performance across various applications. These tiny materials have some remarkable features, such as a high surface area, increased strength, and better thermal stability. Due to their small size, these fillers can interact effectively at the molecular level, leading to substantial enhancements in the composite's mechanical strength, thermal resistance, and barrier properties. This integration fosters the development of materials that exhibit outstanding performance across a variety of applications.

(iii) *Interfacial Interactions*

The performance of bionanocomposites heavily relies on the interactions between the biopolymer matrix and the nano-fillers. These interactions are fundamental to how well the composite performs in various applications, making them a critical area of focus in material science. At the heart of these interactions is the concept of interfacial bonding. Strong bonding between the biopolymer matrix and the nano-fillers is essential for effective load transfer. When forces are applied to the composite, this strong bond allows the matrix to effectively transmit those forces to the fillers, which in turn enhances the overall mechanical strength and stability of the material. Without robust interfacial bonding, the composite may fail to perform as intended, leading to reduced durability and reliability.

To improve these interactions, several techniques can be employed. One common approach is surface modification of the fillers. By altering the surface characteristics of the nano-fillers, researchers can enhance their compatibility with the biopolymer matrix. This modification may involve changing the chemical properties of the filler surfaces to promote better adhesion or introducing functional groups that facilitate stronger bonding with the matrix. Another effective strategy is the use of coupling agents. These agents act as intermediaries that improve the interaction between the biopolymer and the nano-fillers. They can help bridge the gap between the two materials, ensuring that they work together more effectively. By enhancing the compatibility of the components, coupling agents play a vital role in achieving the desired properties of the composite. Ultimately, the success of bionanocomposites hinges on the quality of the interactions between the biopolymer matrix and the nano-fillers. Strong interfacial bonding not only enhances mechanical strength but also contributes to the stability and longevity of the composite. As researchers continue to explore innovative methods for improving these interactions, the potential applications for bionanocomposites will undoubtedly expand, paving the way for more sustainable and high-performance materials in various fields.

(iv) *Processing Techniques*

The manufacturing processes employed in the creation of bionanocomposites are pivotal in shaping their properties and overall performance. Techniques such as solution casting, extrusion, and injection molding are commonly utilized, and each method influences how well the nano-fillers are dispersed within the biopolymer matrix. This dispersion is crucial, as it directly impacts the effectiveness of the composite in various applications. In all these manufacturing processes, the goal is to achieve a homogeneous distribution of nano-fillers within the biopolymer matrix. This uniformity is vital for optimizing the performance of the composite, as uneven dispersion can lead to weak points and reduce the material's overall strength and durability.

(v) *Structure–Property Relationship*

The performance of bionanocomposites is deeply influenced by their microstructure and the arrangement of nano-fillers within the biopolymer matrix. Understanding the complex structure–property relationship is crucial for customizing these materials to fulfill specific application requirements. This relationship highlights how the physical arrangement of components can significantly dictate the overall functionality of the composite. One of the key factors in this relationship is the ***orientation and alignment of the nano-fillers***. When nano-fillers are strategically aligned within the biopolymer matrix, they can enhance mechanical strength by providing better load distribution during stress. For example, aligned fillers can create a more robust framework that resists deformation, thus improving the composite's durability. Similarly, the arrangement of these fillers can also impact thermal conductivity. Properly aligned nano-fillers can facilitate heat transfer more efficiently, making the composite suitable for applications requiring effective thermal management. Moreover, the barrier properties of bionanocomposites are influenced by the microstructure as well. An optimal arrangement of nano-fillers can enhance the material's ability to resist permeation by gases or liquids, which is particularly important in packaging and protective coatings. By understanding how the microstructure affects these properties, researchers can design composites that are not only strong but also functional in a variety of environments.

(vi) *Sustainability and Biodegradability*

The manufacture of bionanocomposites is fueled by the desire to contribute positively to sustainable development. In a world increasingly aware of environmental challenges, these innovative materials stand out for their potential to make a real difference. At the heart of bionanocomposites is their reliance on renewable resources. Materials made from plants and agricultural by-products not only help reduce our dependence on fossil fuels but also support local economies. This shift toward bio-based materials opens new doors, encouraging sustainable practices that can benefit both the planet and communities. Another remarkable aspect of bionanocomposites is their biodegradability. Unlike conventional plastics that linger in our environment for centuries, biopolymers can break down naturally. This means that once these materials have served their purpose, they will not contribute to the growing problems of pollution. Instead, they decompose, enriching the soil and leaving no harmful trace behind. This natural lifecycle is a step toward reducing the burden on landfills and promoting a cleaner, greener planet. Moreover, bionanocomposites bring more than just environmental benefits; they also offer enhanced performance. By combining the strength of nano-fillers with biodegradable materials, these composites can achieve impressive mechanical properties and thermal stability. This versatility makes them suitable for various applications.

(vii) *Functionalization and Tailoring Properties*

Another fundamental principle of bionanocomposites is that they can be functionalized to impart specific properties, such as antimicrobial activity, UV resistance, or enhanced barrier functions. This functionalization can be achieved through the incorporation of bioactive agents, the use of specific fillers, or chemical modifications of the biopolymer matrix. Also, it is possible to create composites that not only possess improved mechanical and thermal properties but also exhibit additional characteristics, such as electrical conductivity, optical properties, or self-healing capabilities. This multifunctionality expands the potential applications of bionanocomposites across various industries. Tailoring these properties allows the development of composites suited for a wide range of applications, from packaging to biomedical uses.

1.3 Types of Bionanocomposites

Bionanocomposites are categorized according to the types of reinforcement, matrix material, and production techniques. As illustrated in Fig. 1.1, bionanocomposites can be classified based on nanoparticle geometry into three categories: isodimensional (particulate nanoparticles utilized as fillers), acicular (nanotubes and cellulose nanofibrils employed as fillers), and layered (including flocculated/phase-separated nanocomposites, intercalated nanocomposites, and exfoliated nanocomposites). This classification depends on the size and shape of the reinforcement used (Saini and Bajpai 2017; Puiggalí and Katsarava 2017).

The final properties of bionanocomposites are significantly influenced by the shape of the incorporated nanoparticles. Biodegradable polymers, also known as biopolymers are naturally occurring macromolecules based on carbohydrates and protein molecules. Biodegradable and bio-based polymers can be categorized based on their sources into the following groups (Viacava et al. 2024):

(i) Polymers derived from bio-derived monomers (e.g., poly(lactic acid), poly(glycolic acid), hydroxypropyl methylcellulose, methylcellulose, and poly(vinyl alcohol) [PVA]).

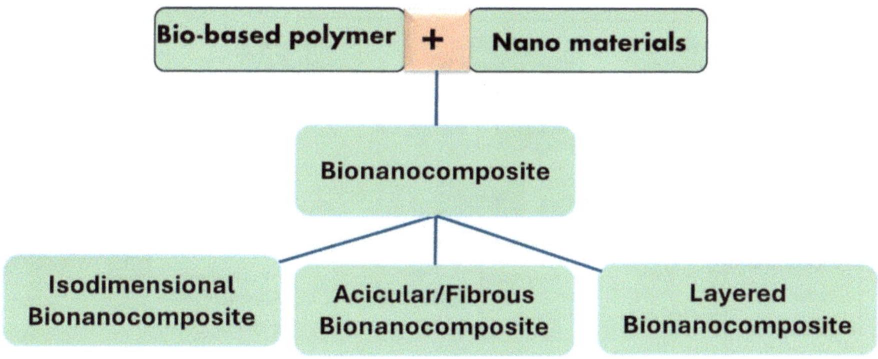

Fig. 1.1 Types of bionanocomposites

(ii) Polymers are sourced from living organisms, such as bacteria or microorganisms (e.g., poly(hydroxybutyrate), poly(hydroxybutyrate-co-valerate), poly(hydroxyl alkenoate), bacterial cellulose, and bacterial polyesters).
(iii) Polymers are derived from natural materials, including carbohydrates like starch, cellulose, chitosan, and alginic acid.

Biodegradable polymers can be classified further based on source as shown in Fig. 1.2. These materials have garnered research interest due to their biocompatibility, biodegradability, and versatility. Biodegradable materials can be recycled by microbes to provide valuable monomers and oligomers for product manufacture (Samir et al. 2022). Biodegradation of polymers typically undergo enzymatic decomposition by microorganisms, which results in changes in the material properties and the formation of environmentally friendly metabolic products such as methane, water, biomass, and CO_2.

The recyclable biological process of biodegradable polymers is shown in Fig. 1.3a (Zhong et al. 2020). The polymer's form, structure, chemical constituent, and molecular weight are the primary factors that influence the rate of the biodegradation process (Ghanbarzadeh and Almasi 2013; Künkel et al. 2016). Additionally, the melting temperature (T_m) of polymers affects the rate of enzymatic breakdown of polymers; the higher the melting point, the less degradation occurs (Saini and Bajpai 2017). This is represented by the equation: $T_m = \frac{\Delta H}{\Delta S}$, where entropy changes upon melting are represented by ΔS and enthalpy changes upon melting by ΔH.

(a) *Biopolymers*

Biopolymers obtained from biomass can further be classified based on renewable sources including polysaccharides, proteins, lipids, and microbes are further

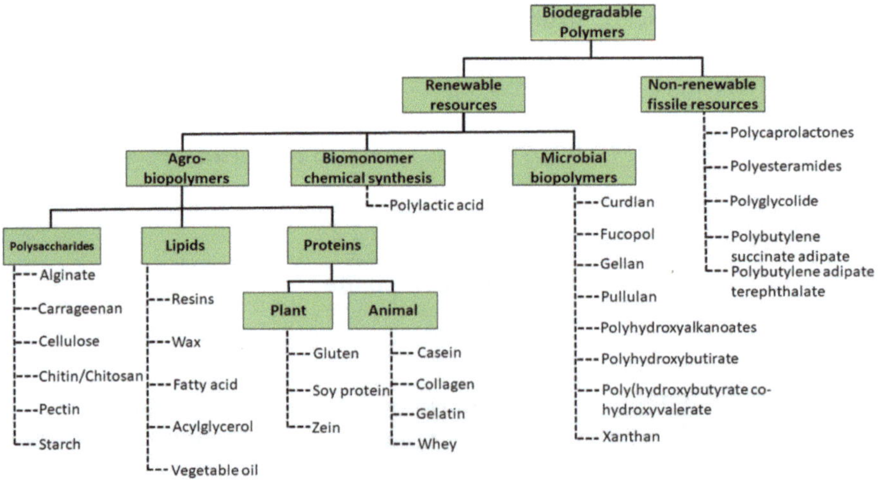

Fig. 1.2 Classification of biodegradable polymers. Reprinted from Mujtaba et al. (2022), an open-access article by Elsevier distributed under CC BY license

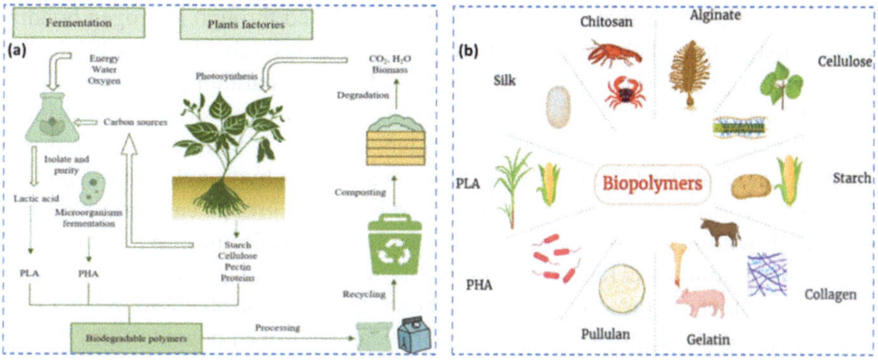

Fig. 1.3 **a** Ecological cycle of biodegradable polymers. Reprinted from Zhong et al. (2020), an open-access article by MDPI distributed under CC BY license. **b** Types of biopolymers. Reprinted from Nandhini et al. (2024b), an open-access article by MDPI distributed under CC BY 4.0 license

distinguished as shown in Fig. 1.3b (Nandhini et al. 2024b; Liu 2006; Stloukal et al. 2015). PHA (polyhydroxyalkanoates) and poly(hydroxybutyrate) (PHB) are examples of those produced by microbial fermentation; synthetic biopolymers include poly-e-caprolactone (PCL), polyvinyl alcohol (PVA), polyglycolic acid (PGA), poly(butylene succinate-co-adipate) (PBSA), and poly(lactic-co-glycolic acid) (PLA) (Dominic et al. 2023; Ghosal et al. 2024). Other synthetic biodegradable polymers, notably those made of physiological and non-toxic building blocks such as a-amino acids, fatty diols, and carbonic or dicarboxylic acids, have been produced in recent decades (Puiggalí and Katsarava 2017).

The biodegradable polymers from renewable resources are also called biopolymers.

This concept is introduced in 1980s.

(b) *Biomass-Based*

Under biomass-based biodegradable polymers, the primary polysaccharides of interest are cellulose and starch; however, there is a growing focus on more complex carbohydrate polymers. Notably, polysaccharides such as xanthan, curdlan, pullulan, and hyaluronic acid are receiving increased attention. These latter polymers typically consist of multiple types of carbohydrate units and often feature consistently organized branching structures in various contexts (Ghanbarzadeh and Almasi 2013). Polysaccharides are essential biopolymers for all living creatures. They are composed of homo or hetero monosaccharides, uronic acids, and β-(1,4)-glycosidic connections (see Fig. 1.4) (Li et al. 2016a).

(c) *Cellulose*

In 1838, Anselme Payen made the first attempt to separate cellulose from plants (Magalhães and Almeida 2023). In today's rapidly expanding economy, cellulose stands out as the most abundant polymer worldwide, with an estimated annual natural production of approximately 1.5×10^{22} tons. It is considered a virtually limitless source of raw material. Microfibrils are created through hydrogen bonding between long, straight chains of cellulose, which is an insoluble linear polymer. The strength and rigidity of the plant cell wall arise from the parallel alignment of these microfibrils, which combine to form microfibrils (Liu et al. 2021). The primary cell wall consists of approximately 6000 glucose units. It features three hydroxyl groups located at the C2 and C3 positions (secondary hydroxyl groups) and the C6 position (primary hydroxyl group). These hydroxyl groups can form hydrogen bonds both within the same molecule and with other molecules (Wu et al. 2023). Despite its simplicity, cellulose is remarkably resistant to enzymatic and chemical degradation thanks to the strong hydrogen bonds that hold the chains together. Cellulose is a homopolysaccharide $(C_6H_{12}O_5)_n$ that can be broken down to form glucose $(C_6H_{12}O_6)$. The source of the material influences the number of repeating units in the chain length. Its long molecular chains and dense packing, combined with numerous hydrogen bonds, both intermolecular and intramolecular contribute to its insolubility in polar solvents. These solvents include ionic liquids, LiCl/dimethylacetamide, 7–9% NaOH/water, NH_3/SO_2/DMSO, molten salt hydrates, and Cu/ethylenediamine. Cellulose is highly infusible and hence can be transformed into derivatives such as cellulose esters and cellulose ethers to make it more processable by decreasing cellulose's hydrophilicity (Shah and Vasava 2019). Apart from being a renewable and cheap material, cellulose composites offer a lot of other advantages, including high modulus and specific strength (Nishino et al. 2004). However, moisture absorption deteriorates the material's mechanical qualities because cellulose is hydrophilic. There are several uses for cellulose and its derivatives in the pharmaceutical and textile industries.

(d) *Starch*

Plants primarily utilize starch as an energy reserve, storing it for extended periods in their roots, tubers, seeds, and stems. A starch granule, which is a type of polysaccharide, consists of approximately 10–30% amylase-containing polymer and 70–90% amylopectin-containing polymer (Olobayo 2019; Kong 2020). The polymerization degree of amylose, which is a linear series of α(1–4)-linked D-glucose units, ranges from 300 to 10,000 (Arora et al. 2018). However, amylopectin is a high molecular weight polymer that has α(1–6) bonds connecting its backbone structure to that of amylase (Xie et al. 2013). The different forms of amylose and amylopectin are depicted in Fig. 1.5. Water sensitivity, low impact resistance, low mechanical capabilities, and brittleness are all characteristics of starch (Soykeabkaew et al. 2015). However, the properties of starch can be enhanced by incorporating fibers into the starch matrix or by chemically or physically modifying the starch with additional biodegradable polymers. Starch biopolymers offer numerous advantages, such as high biodegradability, renewability, and excellent oxygen barrier properties. As a result, they can be utilized in a wide range of commercial applications, including bags, packaging, adhesives, cosmetics, medical products, and agricultural mulch films (Gamage et al. 2022).

(e) *Alginate*

The primary source of alginate, a naturally occurring polymer is from marine plants, thus brown algae (Angra et al. 2021). Alginate exists in different varieties—namely, chlorophyceae, cyanophyceae, phaeophyceae, and rhodophyceae (Augst et al. 2006). The most widely used commercially available alginate products at the moment are sodium alginate, calcium alginate, and ammonium alginate (Lee and Mooney 2012). Alginate is a linear polymer made up of monomers of α-L-guluronic acid and β (1 → 4)-linked mannuronic acid. While alginate chains typically range from 100 to 1000 monomers in length, this can vary significantly based on the source and manufacturing process. Alginate is believed to have a persistence length of approximately 1 nm. Importantly, the composition and relative molecular mass of alginates determine their classification as low, moderate, or high viscosity. The low cost, non-toxicity, biocompatibility, biodegradability, good moisture absorption, ease of removal, high oxygen permeability, gelation properties, and ability to adsorb metal ions make alginates valuable in various industries, including food, medicine, and cosmetics (Lone et al. 2016).

(f) *Synthetic Biodegradable-Based*

Aliphatic polyesters, polylactide, and aliphatic copolymers are examples of synthetic biodegradable polymers designed by traditional polymerization techniques (Mangaraj et al. 2019). Compared to traditional plastics, biodegradable polyesters are seen to be the most promising materials for application because of their appropriate breakdown time and large-scale production. The two categories of synthetic biodegradable polymers are oil-based monomers like polycaprolactone (PCL) and bio-based polymers like polylactic acid (PLA) (Mangaraj et al. 2019), with low ecological contamination (Tian and Bilal 2020). However, among the various synthetic biodegradable-based materials, PLA is the most significant (Luckachan and Pillai 2011). The class of synthetic biodegradable-based materials is treated free-standing below.

(i) *Polylactic Acid (PLA)*

PLA is a linear polymer composed of repeating units of lactic acid monomers (LA) (Puiggalí and Katsarava 2017). PLA is derived from renewable resources, such as corn, wheat, rice, potatoes, starch, and sugar cane. Lactic acid exists in two enantiomeric forms, the L-optical isomer (levorotatory) and the D-optical isomer (dextrorotatory) (Trivedi et al. 2023) as shown in Fig. 1.6. Thus, commercial PLA is composed of copolymers of PDLLA and PLLA (de França et al. 2022).

The primary route to PLA production is the ionic polymerization of lactide, which is formed through the cyclic condensation of two lactic acid molecules, resulting in the formation of PLA (Deshmukh et al. 2017). Similarly, PLA can be synthesized via a polycondensation reaction (Sneha et al. 2016). PLA's characteristics are principally governed by the stereochemical structure of L- and D-lactic acids. These include increased elongation and poorer tensile strength. By managing the stereochemical structure, it is possible to finely tune material properties such as crystallinity, processing temperature, and mechanical strength (Mujtaba et al. 2022). PLA is more

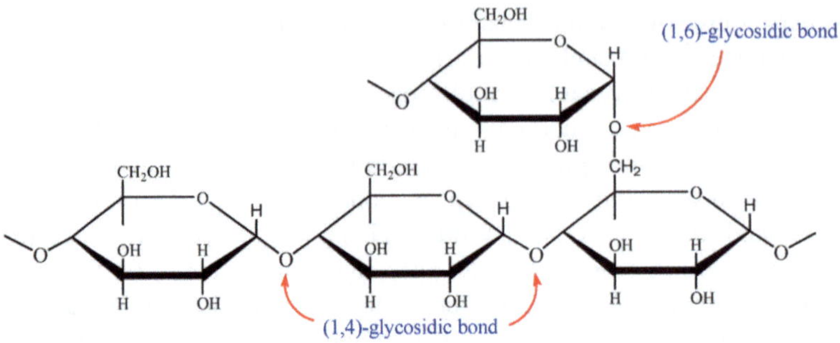

Fig. 1.4 Glycosidic linkage in polysaccharides

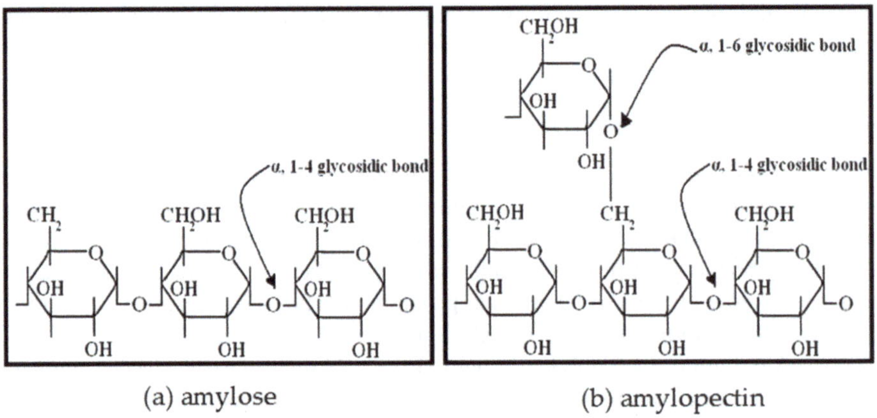

Fig. 1.5 Chemical structures of **a** amylose and **b** amylopectin. Reprinted from Amira et al. (2012), an open-access article by Intech distributed under CC BY license

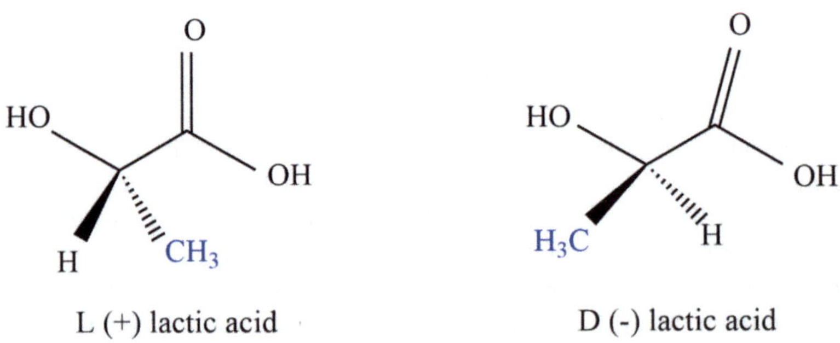

Fig. 1.6 Stereo-isomers of lactic acid. Reprinted from Yadav et al. (2023), an open-access article by MDPI distributed under CC BY license

environmentally friendly, transparent, and practical than other biodegradable polymers (Trivedi et al. 2023; Ranakoti et al. 2022). PLA is used in numerous applications such as interference screws, sutures, dentistry, ligating clips, bone pins, and rods (Luckachan and Pillai 2011; Middleton and Tipton 2000).

(ii) *Poly(Vinyl Alcohol) (PVOH or PVA)*

poly(vinyl alcohol) (PVOH or PVA), represented by the molecular formula $[CH_2CH(OH)]$, is a synthetic polymer distinguished by its white color, tastelessness, and odorlessness. It is commonly available as water-based solutions or beads. PVA's protective colloid properties enhance the process of emulsion polymerization. Initially, polyvinyl acetate is created through the polymerization of vinyl acetate, which is subsequently converted into PVA. This polymer holds significant promise in various applications, including membrane technology, pharmaceuticals, medical fields, and optics. Its solubility in water allows for the development of environmentally friendly material processes (Clémenson et al. 2007).

(iii) *Polyhydroxyalkanoates (PHA)*

Polyhydroxyalkanoates (PHAs) are linear biopolymers that are naturally hydrophobic, biocompatible, and biodegradable. Under conditions of unbalanced growth, microorganisms, including many Gram-positive and Gram-negative bacteria, are capable of synthesizing polyhydroxyalkanoates (PHAs) as a means to store energy (Reddy et al. 2003). Nonetheless, the most prevalent microbe utilized in the production of PHA is *Alcaligenes eutrophus* (Lenz and Marchessault 2005). PHA can be categorized into three groups according to chain length: short-chain (C3–C5), medium-chain (C6–C14), and long-chain (more than C14) (Lenz and Marchessault 2005). The two most common polymer types of PHA are poly(hydroxybutyrate-cohydroxyvalerate) (PHBV) and polyhydroxybutyrate (PHB). In addition to having a good resistance to UV light, this class of materials has decreased toughness, increased stiffness, and decreased solvent resistance. PHA's biodegradability and biocompatibility make it appealing for a variety of applications, including tissue engineering, packaging, medication delivery, and medical bio-implants (Ranakoti et al. 2022).

(g) *Animal/Protein-Based*

(i) *Chitin and Chitosan*

Chitin is a nitrogen-rich polymer classified as a polysaccharide, akin to cellulose. It is essential for the exoskeletons of insects, arachnids, and crustaceans (Sneha et al. 2016). Chitin assembles similarly to cellulose, a polysaccharide composed of β-(1 $\rightarrow$ 4) linked N-acetylglucosamine that crystallizes into whiskers/nanofibrils. Chitosan, one of the most promising biopolymers for constructing advanced materials (Gómez-Estaca et al. 2011; Xu et al. 2015), is a deacetylated derivative of chitin (refer to Fig. 1.7), a cationic polysaccharide that forms polyelectrolyte complexes (Kong et al. 2010). Chitosan's amine and hydroxyl groups allow it to be structurally modified with suitable crosslinking agents. As a result, chitosan's physical characteristics are strengthened, whereas swelling in its aqueous system is minimized (Xu et al. 2015). The antibacterial action of chitosan is determined by its type, molecular weight, and degree of acetylation. Chitosan's antibacterial activity comes from its electrostatic contact with the negatively charged bacterial cell membrane (Yilmaz Atay et al. 2019). Its antibacterial, non-toxic, biodegradable, affordable, and biocompatible qualities enable a wide range of uses (Aizat and Aziz 2019; Dave and Gor 2018). This allows its application across various fields including medical fields for wound dressing, tissue engineering, drug delivery, and absorbable sutures (Ahmad et al. 2017; Mousa et al. 2016; Shchipunov et al. 2010). Chitin and chitosan are of commercial interest due to their high amount of nitrogen (6.89%) as opposed to synthetically replaced cellulose (1.25%) (Ravi Kumar 2000).

(ii) *Gelatin*

Gelatin is a protein derived from animals, produced through the partial hydrolysis of collagen sourced from animal tissues, skeletal frameworks, and skin. The primary source of gelatin is type I collagen, which does not contain cysteine residues. Gelatin can be categorized based on its specific production methods and composition. Type A gelatin is obtained via acid hydrolysis of collagen, containing 18.5% nitrogen, while type B gelatin is derived from alkaline hydrolysis, with a nitrogen content of 18%. When gelatin is heated to high temperatures, it dissolves and takes on a coiled shape. When it cools, it becomes a helical fragment of the gelatin α-chain, featuring the triplet (Gly-X-Y)n repeated, where X and Y are typically proline and hydroxyproline according to the chemical structure shown in Fig. 1.8.

Unlike collagen, which is resistant to proteolytic enzymes, gelatin is susceptible to proteases. The versatility of gelatin arises from its many active groups, including

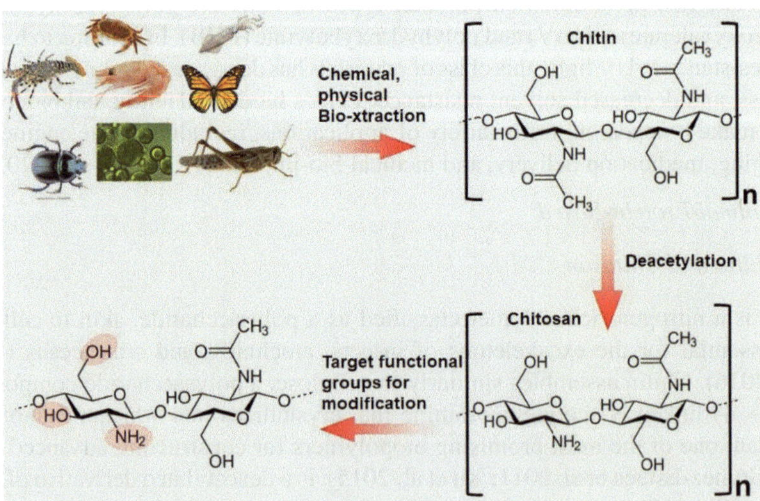

Fig. 1.7 Deacetylation of chitin by alkali treatment (concentrated NaOH) to produce chitosan. Reprinted from Pellis et al. (2022), an open-access article by MDPI distributed under CC BY license

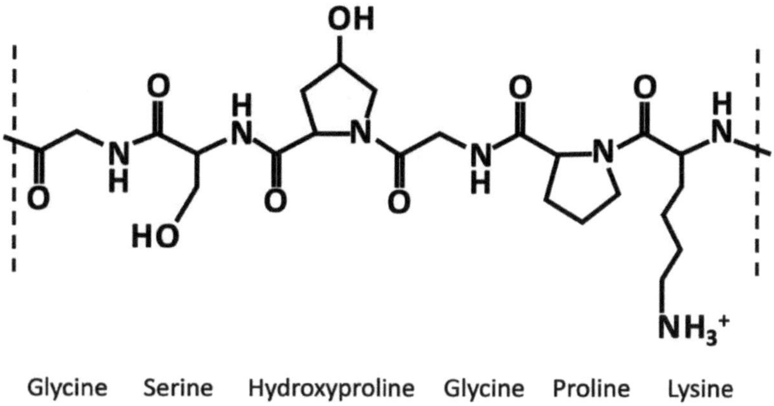

Fig. 1.8 Chemical structure of gelatin

amide, thiosulfate, and acetic groups, allowing for modifications using biomolecular frameworks and nanoparticles for a variety of applications (Magalhães and Almeida 2023).

(iii) *Collagen*

Collagen, the most abundant protein in the human body, provides strength and structural support and is found in tendons, skin, muscles, bones, and various other tissues. Approximately 30% of the body's total protein consists of collagen, with types I, II, and III accounting for 80–90% of this composition. Collagen is characterized by three intertwined polypeptide chains, which are connected by intermolecular hydrogen bonds, forming a rod-like structure measuring about 300 nm in length and 1.5 nm in diameter as shown in Fig. 1.9. This structure can self-assemble into fibrils both in vitro and in vivo, leading to the formation of three-dimensional networks.

Due to its remarkable biocompatibility, biodegradability, suitable mechanical strength, and flexibility, as well as its ability to absorb fluids and release nutrients, collagen is considered an ideal biomaterial for biomedical applications (Giraud Guille et al. 2010; Parenteau-Bareil et al. 2010), hence, its prevalent use in wound dressing (Sneha et al. 2016).

(iv) *Pullulan*

The polymorphic fungus Aureobasidium produces the commercially valuable biopolymer pullulan, which is an extracellular polysaccharide soluble in water. This linear polysaccharide is composed of repeating maltotriose (or trimer) units linked together by α-(1 → 6) glycosidic bonds. As shown in Fig. 1.10, maltotriose consists of α-(1 → 6) linkages to (1 → 4)-α-D-tri glucosides (Mhd Haniffa et al. 2016). Furthermore, only 1–7% of maltotetraose (tetramer) subunits may be found in pullulan (Le et al. 2022). Pullulan's unusual α-(1 → 6) connections between maltotriose subunits give it particular physical features, such as high water solubility and structural flexibility for shape-making. Pullulan is synthesized using a variety

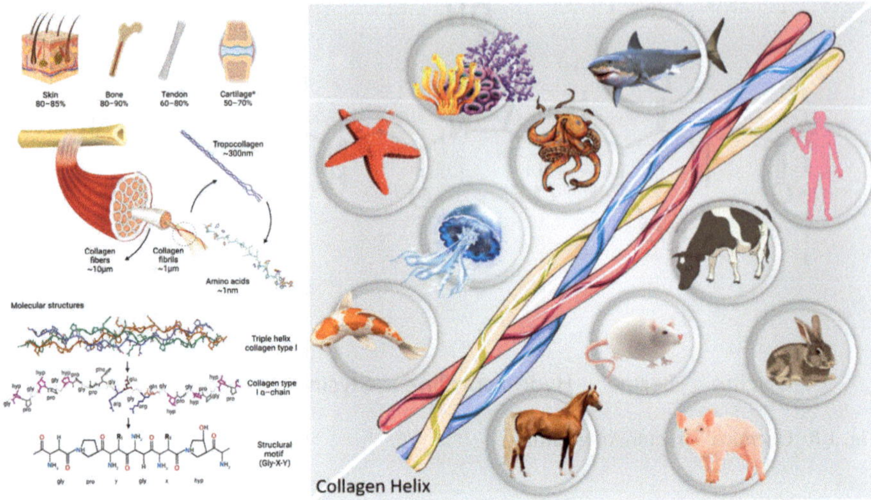

Fig. 1.9 Typical chemical structure of collagen and some sources. Reprinted from Feng et al. (2023), an open-access article by MDPI distributed under CC BY license

of ingredients, including sucrose, glucose, fructose, maltose, starch, and oligosaccharides. However, sucrose is widely preferred (Maltsev et al. 1994). Pullulan has a wide range of industrial applications, including pharmaceuticals, chemicals, energy generation, agriculture, and food.

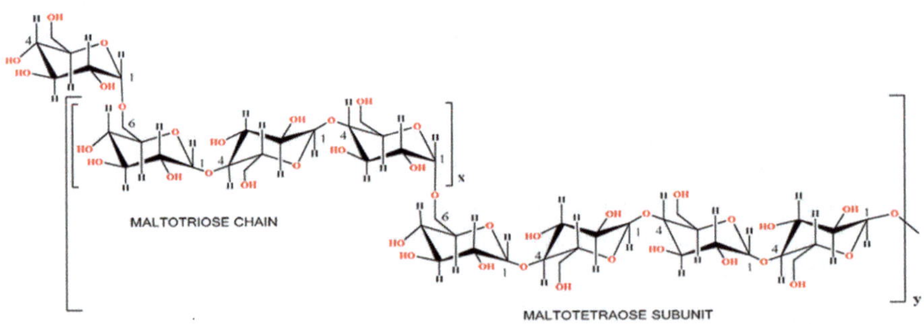

Fig. 1.10 Molecular structure of pullulan with maltotriose and maltotetraose subunits. Reprinted from Mhd Haniffa et al. (2016), an open-access article by MDPI distributed under CC BY license

1.4 Nanomaterials in Bionanocomposites

Nanomaterials are integral components of bionanocomposites. Their intimate combination with bio-based polymers results in reinforcements and unique functional composites. These materials produce protective barriers that secure the active substances from degradation, ensuring a controlled release on specified targets while preventing the loss of functionality (Giaconia et al. 2020; Shishir et al. 2018). Bionanocomposite materials are fabricated in such a way that two or more material phases can be insolubly blended at the molecular level. One component is defined as the matrix, which serves as the continuous phase, while the other material or components are identified as the filler, strengthening the composite (Huang et al. 2023; Bustamante-Torres et al. 2021) for enhanced performance (Fura et al. 2024). Interestingly, a variety of materials, including polymers, carbons, metals, and metal oxides, can be used to create reinforcing nanoparticles (Ghoshal 2018).

1.4.1 Types of Nanomaterials

Fundamentally, nanomaterials are classified based on their dimensionality beyond their nanoscale into 0D (e.g., nanoparticles), 1D (e.g., nanorods, nanotubes), 2D (e.g., graphene), and 3D (e.g., nano prisms) as shown in Fig. 1.11a (Chouhan et al. 2021; Saha et al. 2022). Some researchers have also classified them into five based on their *potential toxicity, pore size, origin, structural configuration, and dimension or size* (Mekuye and Abera 2023). However, for simplicity and convenience, they can be categorized broadly into organic, inorganic, and hybrid nanomaterials as shown in Fig. 1.11b (Dosawada et al. 2023).

(i) *Carbon-Based Nanomaterials*

Carbon-based nanomaterials are recognized for their remarkable properties and hence, their extensive applications across various fields. Among the most notable are carbon nanotubes (CNTs), graphene, and carbon quantum dots (CQDs) (Sajid 2022).

Carbon Nanotubes

CNTs have nanometer-sized interior diameters and micrometer-long lengths. CNTs are one of the most fascinating carbon allotropes in the shape of cylinders organized by rolling graphite sheets. This results in the distortion of their sp^2 hybrid orbitals and the production of a σ-p rehybridization structure (Elhissi et al. 2012; Karthik et al. 2014). CNTs are produced using a variety of processes, including laser ablation, carbon arc discharge, chemical vapor deposition, and electrolysis. CNTs are separated into two categories: single-walled and multi-walled. In contrast to multi-walled carbon nanotubes (MWCNTs), which come in three different configurations such as chiral, armchair, and zigzag, single-walled CNTs (SWCNTs) are composed

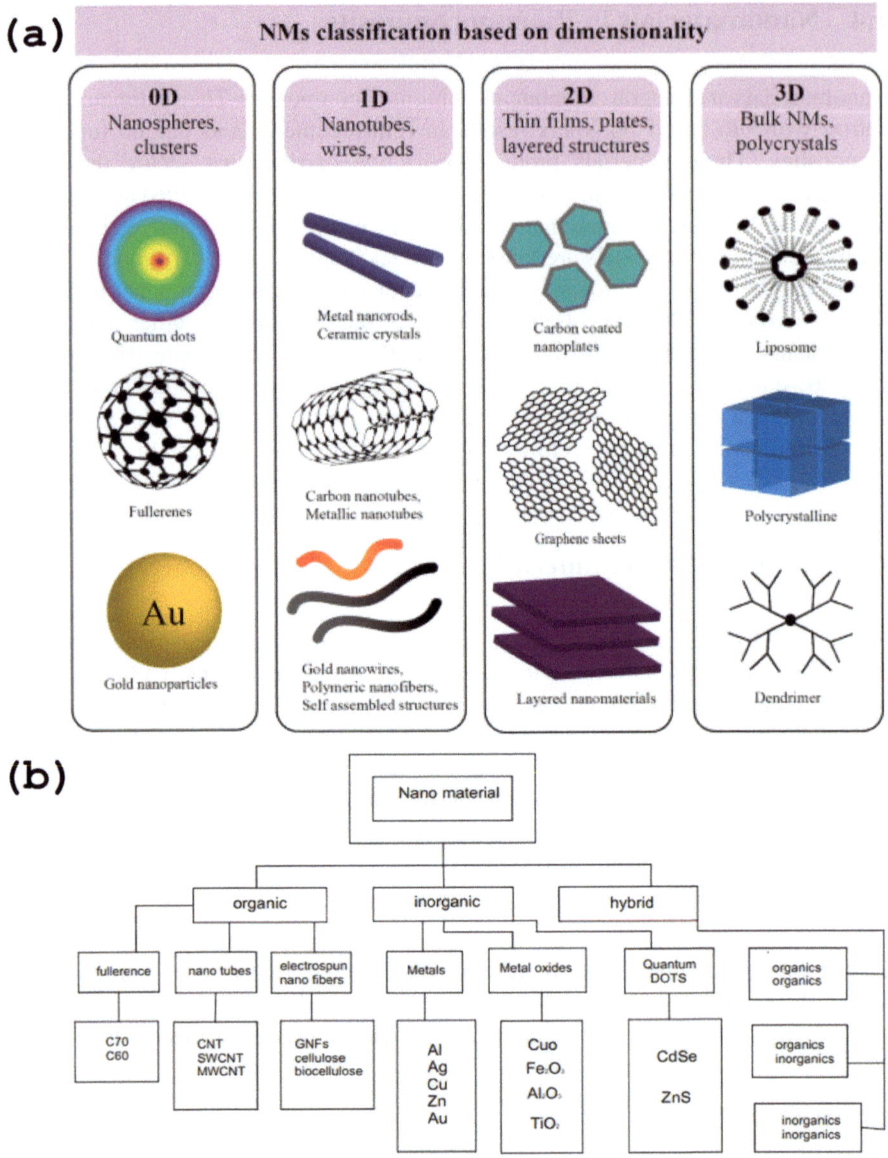

Fig. 1.11 **a** Classification of nanomaterials based on dimensions. Reprinted from Saha et al. (2022) with permission from Elsevier BV. **b** Classification of carbon. Reprinted from Dosawada et al. (2023) with permission from Elsevier BV

of two or more cylindrically rolled concentric graphene sheets (Karthik et al. 2014). The individual sheets are held together by a weak Van der Waals force with an adjacent gap of 0.34 nm. Depending on the diameter and helicity of the graphitic rings, SWCNTs can exhibit either metallic or semiconducting behavior (Rahman et al. 2019). Additionally, carbon nanotubes (CNTs) demonstrate significant electrical conductivities, with single-walled carbon nanotubes (SWCNTs) ranging from 10^2 to 10^6 S cm^{-1} and multi-walled carbon nanotubes (MWCNTs) also exhibiting conductivities in the same range. They possess exceptional thermal conductivities as well, with SWCNTs at 6000 W m^{-1} K^{-1} and MWCNTs at 2000 W m^{-1} K^{-1} (Jia et al. 2024; Khoswan et al. 2024). Due to their high tensile strength, carbon nanotubes (CNTs) are utilized as reinforcing agents in various composites and products. However, chemical treatment processes can compromise the properties of single-walled carbon nanotubes (SWCNTs) by causing defects from the breakage of carbon–carbon bonds. In contrast, multi-walled carbon nanotubes (MWCNTs) tend to retain their properties since only their outer surfaces are impacted (Li et al. 2024). MWCNTs offer numerous benefits, including low manufacturing costs and catalyst-free synthesis, while having a lesser pore volume and specific surface area than SWCNTs because of their complicated structure.

Graphene

Since its initial synthesis by mechanical exfoliation in 2004, graphene—a sheet of sp^2 hybridized carbon atoms arranged in a honeycomb-like crystal lattice has garnered a great deal of attention. Graphene exhibits exceptional carrier mobility (200,000 cm^2 V^{-1} s^{-1}), great theoretical surface area (2630 m^2 g^{-1}), outstanding electrical conductivity (2000 S m^{-1}), excellent transparency (97.7%), and very high thermal conductivity (5000 W m^{-1} K^{-1}) (Agbolade et al. 2024). Graphene, a one-atom-thick sheet of sp^2 hybridized carbon atoms packed in a honeycomb-like crystal lattice (Sur 2012; Moosa and Abed 2021), has received significant attention since its first production by mechanical exfoliation in 2004 (Li et al. 2016b). It has great chemical and mechanical strength, a remarkable quantum hall effect, and a low cost of production (Xin Jiat et al. 2018; Aafreen et al. 2024). Graphene derivatives include reduced graphene oxide (RGO), a byproduct of graphene due to the existence of oxygen-based functional groups, and graphene oxide (GO), which is produced from graphite powder (Đorđević et al. 2019), is produced via chemical, electrochemical, or thermal methods. Crucially, graphene exhibits favorable electrochemical performance and can be synthesized by a range of approaches, including chemical procedures, thermal degradation of SiC and other substrates, exfoliation and cleavage, thermal or plasma-based chemical vapor deposition, and unzipping of carbon nanotubes (Tarcan et al. 2020).

Carbon Nanodots (CNDs)

Carbon dots are zero-dimensional nanostructures with fluorescent capabilities composed of nanoparticles that were accidentally discovered during the purification of single-walled carbon nanotubes, with a size of less than 10 nm (Maholiya et al. 2023; Mehta et al. 2021). There are three types of carbon dots: graphene quantum dots (GQDs), carbon nanodots (CNDs), and carbon quantum dots (CQDs) (Li et al. 2023; Liu et al. 2020). CNDs are amorphous carbon particles that lack quantum confinement. However, both CQDs and GQDs exhibit crystal structure and quantum effects. Carbon dots are essentially a general term for all quasi-spherical carbon nanoparticles (Li et al. 2012). CQDs have oxygen and nitrogen functional groups such as amines carboxyl and hydroxyl, providing hydrophilic properties, most likely due to the formation of H bonds (Singh et al. 2023). Furthermore, CDs feature several surface groups that can attach to ions, chemical molecules, polymers, DNA, and proteins, altering the characteristics of CDs, and these functionalized CDs can be optimized to satisfy specific requirements in various sectors. There are several methods for synthesizing CND including arc discharge, laser ablation, microwave/ultrasonic synthesis, combustion thermal/hydrothermal/acidic oxidation, electrochemical synthesis, and plasma treatment (Liu et al. 2020; Geetha Bai et al. 2019). The superior photostability, higher quantum yield, lower toxicity levels, and excellent biocompatibility of CDs over traditional nanomaterials like cadmium/lead, rare earth, and metal oxides, some of which come from plentiful, inexpensive sources, allow them to have promising uses in biomedicine, optronics, sensors, and catalysis (Shen et al. 2012).

(ii) *Metallic Nanomaterials*

Metal-based nanomaterials fall into one of four categories: trimetallic, bimetallic, metal oxide, or metal nanoparticles (Sajid 2021; Roy et al. 2023). The most commonly employed metal oxide NPs are Fe_2O_3, Al_2O_3, TiO_2, SiO_2, CuO, SnO_2, and WO_3. Their high density of corner or edge surface sites and compact size give them their unique physicochemical characteristics. Metal oxide nanoparticles are particularly useful in optoelectronic, antibacterial, pharmaceutical, energy storage, sensing, and catalytic applications (Boutry and Muller 2018; Fazio et al. 2021). They are mostly synthesized using the hydrothermal or sol–gel method. Compared to monometallic nanoparticles, bimetallic nanoparticles have become materials of scientific and technological interest in recent years due to their multivalent effect (Sharma et al. 2019; Zaleska-Medynska et al. 2016). They are produced by mixing several metallic nanoparticle designs and are characterized by size-dependent optical, electrical, thermal, and catalytic capabilities (Pan et al. 2016). Bimetallization improves the catalytic properties of nanoparticles, and the electronic effect explains charge transfer in bimetallic catalysts. Trimetallic nanoparticles, composed of three metals, are superior to bimetallic catalysts due to their better tuning of catalytic properties (Ma et al. 2021). Trimetallic nanoparticles are synthesized using microwaves, selective catalytic reduction, microemulsions, coprecipitation, and hydrothermal methods. However, trimetallic nanoparticles have an unstable surface area and are rapidly precipitated out of liquids, limiting their catalytic activity (Sharma et al. 2017). Thus, stabilizers for trimetallic nanoparticles include block copolymers, organic ligands, surfactants, and dendrimers (Sharma et al. 2017; Molakkalu Padre et al. 2024). Nanomaterials like AuNPs are widely used due to their size and shape-dependent properties, with bright red hues for smaller particles and dirty yellowish hues for larger ones (Zakaria et al. 2013; Ahmad et al. 2016). Au NPs, with their small particle size, increased surface area, and reactivity, possess exceptional optical, electrical, and chemical properties, making them popular in optical sensors and imaging applications.

(iii) *Ceramic-Based Nanomaterials*

Ceramic nanomaterials typically consist of various compounds, including oxides, carbides, phosphates, and carbonates derived from metals or metalloids like calcium, titanium, and silica. These materials are notable for their unique properties and versatility, making them valuable in a range of applications (Thomas et al. 2015; Szczyglewska et al. 2023). They are resistant to degranulation and degradation due to their nanoscale porosity. There are three types of nanoceramics: nanoparticles, nano scaffolds, and nanoclays (Singh et al. 2016). These materials, known for their strong heat resistance and chemical inertness, are extensively used in biomedical applications due to their controllable size and ease of surface functionalization (Moreno-Vega et al. 2012; Kapil et al. 2023). These materials exhibit exceptional mechanical strength, pH tolerance, high load capacity, and ease of integration into hydrophobic and hydrophilic systems.

(iv) *Polymeric Nanomaterials*

Polymeric nanomaterials with particle sizes of at least 1–1000 nm in one dimension have garnered a lot of interest in a variety of scientific applications, mainly due to a notable rise in the ratio of surface area to volume (Zielińska et al. 2020; Rao and Geckeler 2011). Their optical, thermal, and electrical properties are superior to those of bulk polymers. Their main applications include drug delivery, separation, catalysis, nanoreactors, and surface coatings. They are ideal materials because of their controlled biodegradability and biological safety (Han et al. 2018). Polymeric nanoparticles (NPs), biodegradable and biocompatible polymers, are utilized in drug delivery applications as drug carriers by encasing or trapping medications (Pridgen et al. 2014; Gagliardi et al. 2021; Dang and Guan 2020). Polymers like PLA, PLG, and poly-î-caprolactone are safe and biodegradable, with some approved by the FDA and EMA for pharmaceutical applications (Palma et al. 2018). Polymeric nanoparticles (NPs) are excellent drug carriers due to their small size, water solubility, non-toxicity, high shelf life, and stability (Gagliardi et al. 2021).

(v) *Biomolecule-Derived Nanomaterials*

The readily available biological raw materials cellulose, keratin, collagen, and silk have intriguing material properties and play a major role in biomaterial research because of their advantageous chemical, biological, and mechanical characteristics, which are necessary for the processing and creation of biomaterials (refer to Fig. 1.12) (Datta et al. 2020). Thus, biomolecules such as proteins, lipids, polysaccharides, nucleic acids, and others have been successfully employed in the synthesis of nanomaterials (Niemeyer 2001; Yuan et al. 2023; Jebali et al. 2017). The growing applications of biocompatible and biodegradable nanomaterials have drawn a lot of attention to these nanoparticles, despite their previous lack of recognition compared to inorganic nanomaterials (Yaqoob et al. 2020). These materials created from either

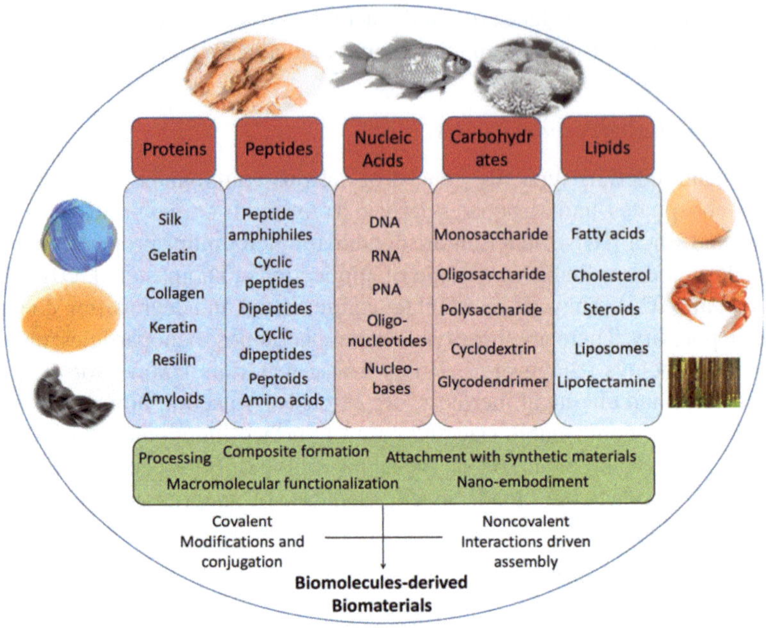

Fig. 1.12 Various classes of plant and animal-derived biomolecules (monomers, oligomers, and macromolecules) and their natural sources are utilized in the creation of biomaterials. Reprinted from Datta et al. (2020) with permission from Elsevier BV

biomolecules or their mixtures with synthetic materials typically find application in biomedical sectors for medication delivery, cosmetics, therapies, textiles, biofuels, pulp and paper industries, and wastewater treatment (Datta et al. 2020; Ding et al. 2020; Gade et al. 2023). Biomolecules like proteins and DNA can self-assemble, which underpins their role in creating a vast array of nanostructures (Lombardi et al. 2024). For instance, in biosensing, nucleic acid-conjugated nanomaterials enhance molecular identification and incorporate the signal-amplifying capabilities of DNA probes (Pan et al. 2024; Wang et al. 2021).

1.4.2 Mechanisms of Interaction

Nanoparticles have attracted lots of attention in many scientific designs and developments over the years (Sanchez and Sobolev 2010). However, while designing nanomaterials and bionanocomposites, the nano-interface construction is an important factor to consider. This is the interaction of nanoparticles with their ecosystem, primarily with other nanomaterials and biomolecules (Auria et al. 2019; Tian et al. 2020). Nanomaterials have been extensively studied for their antibacterial, anticancer, inflammatory, immune system regulation, antidiabetic, antioxidant, membrane penetration, and biocompatible properties (Wei et al. 2021).

However, the realization and knowledge of nano-bio interactions serve as a guide to rationally develop engineered nanomaterials (ENMs) and tailored nano-bio interfaces with adequate safety and efficacy (refer to Fig. 1.13a) (Zhao et al. 2022). Crucially, these interactions are driven by extremely complex factors, such as the size, shape, and surface characteristics of engineered nanomaterials (ENMs), as shown in Fig. 1.13b (e.g., hydrophobicity, hydrogen-bonding capabilities, π-bonds, and stereochemical interactions) (Mu et al. 2014; Walczyk et al. 2010). As a result, the mechanisms underlying these interactions can be studied at multiple levels, from molecular to macroscopic, and have an impact on the composite material's performance. The processes are influenced by metal–ligand coordination, ligand-receptor contact, hydrophobic interaction, hydrogen bonding, and electrostatic interaction (Zhao et al. 2022).

1.5 Fabrication and Characterization of Bionanocomposites

Generally speaking, there are two types of methods used to synthesize nanomaterials: top-down and bottom-up approaches. Among these, chemical, physical, or biological processes are provided as pathways (Mekuye and Abera 2023; Sajid 2022), as illustrated in Fig. 1.14. Compared to their larger counterparts, nanomaterials have different physical, mechanical, catalytic, optical, and biological properties that are influenced by the manufacturing process (Rai et al. 2021). When nanomaterials interact with other structures, like cells, and function, crucial elements include size, surface area-to-volume ratio, shape, morphology, coat, dispersity, reactivity, and compatibility. As a result, the production and characterization of bionanocomposites are key steps in developing these sophisticated materials that mix biological components with nanoscale materials. This field comprises a variety of approaches and procedures required for developing and comprehending these complex systems.

1.5.1 Fabrication of Bionanocomposites

Achieving multifunctional capabilities in a single, ideal, and distinct polymer matrix is essential for addressing the shortcomings of current nano or biomaterials. This requires a close collaboration between nanomaterials and biopolymers. When nanomaterials like nanoparticles, nanotubes, or nanoclays are added to bio-based polymers, the composite's mechanical, thermal, and functional qualities are enhanced. Bionanocomposites are typically made by selecting nanomaterials based on their

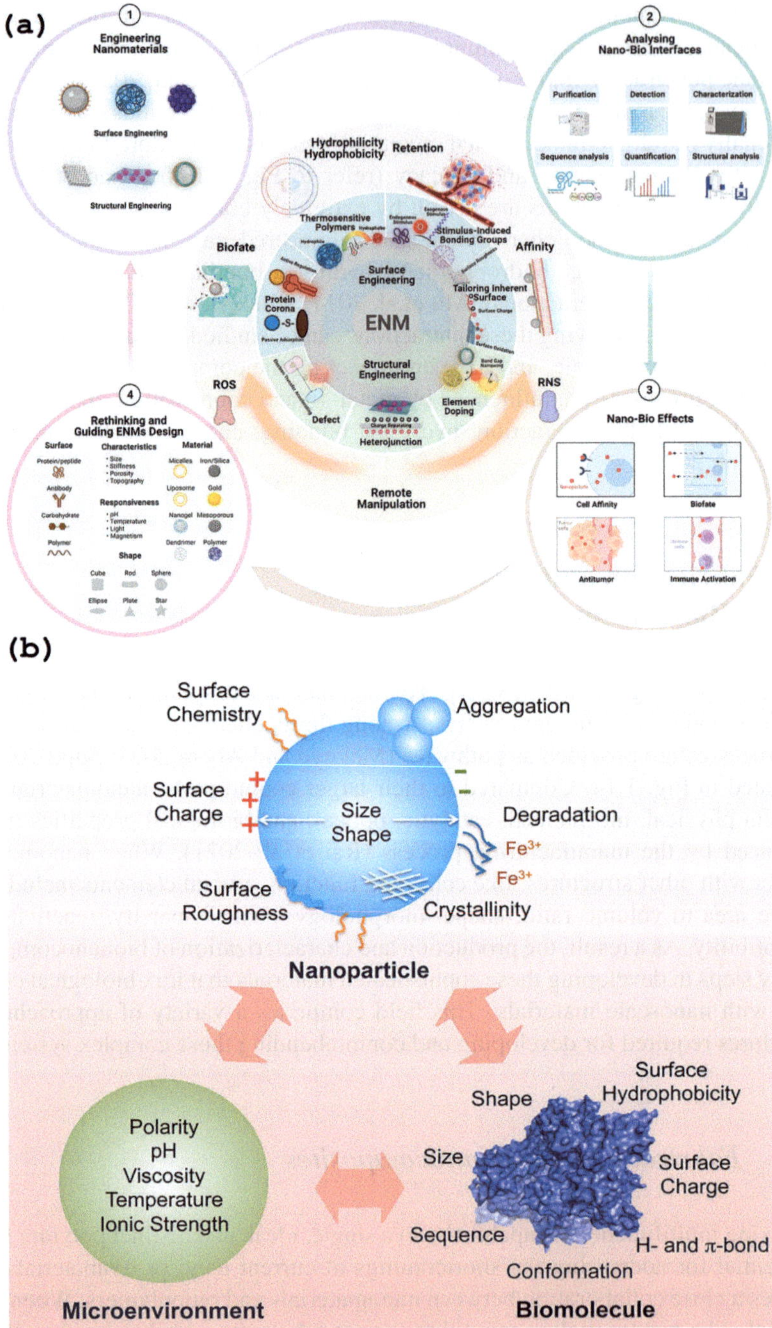

Fig. 1.13 **a** Understanding the nano-bio interfaces between engineered nanomaterials and biological systems. Adapted from Zhao et al. (2022), with permission from the American Chemical Society. **b** Factors influencing nanoparticle-biomolecule interactions. Adapted from Mu et al. (2014), with permission from the American Chemical Society

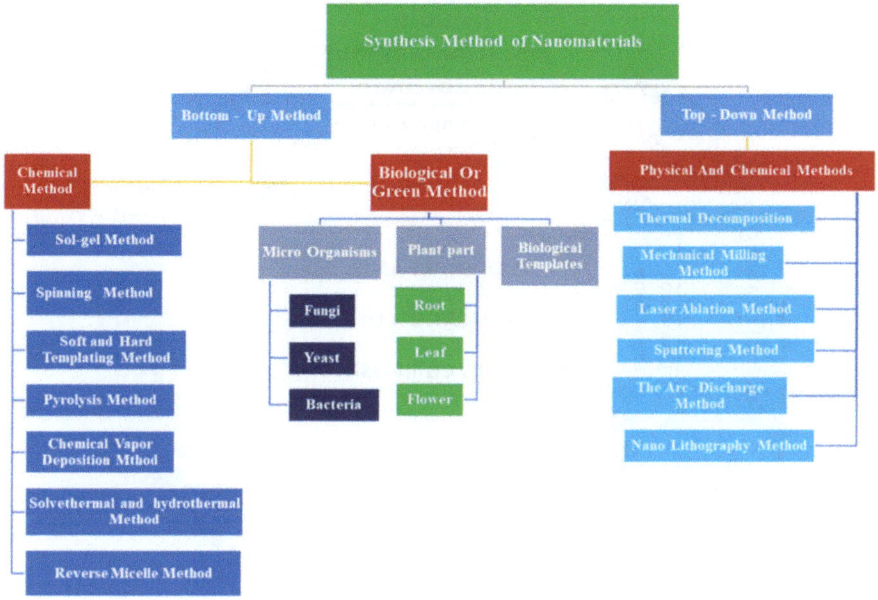

Fig. 1.14 Synthesis methods of nanomaterials. Reprinted from Mekuye and Abera (2023), an open-access article by Wiley under CC BY 4.0 license

compatibility with the biopolymer and the required qualities of the final composite, dispersing them uniformly with a suitable solvent, and molding the composite material into the appropriate shape. Nonetheless, bionanocomposites are often made utilizing three major techniques: (1) in situ polymerization, (2) solution casting, and (3) melt processing (Saini and Bajpai 2017). Additional preparation techniques shown in Fig. 1.15 (Nandhini et al. 2024b) have received significant research interest although their commercial viability is limited.

(i) *Solution Intercalation*

Intercalation from solution, or solution casting, is a three-step technique. The solution intercalation process starts with the dispersion of the nanoparticles in an appropriate solvent followed by the dissolution of the polymer matrix using the same solvent or solvent mixtures. Examples of biopolymers include proteins and carbohydrates, which are mixed with a soluble solvent in this system. Inorganic micro filler silicate platelets expand when exposed to solvents such as water, chloroform, and toluene. Depending on how the solvent and nanoparticles interact, the conjugated structure may stay intact while the nanoparticle aggregates may disintegrate in a good solvent because of the rise in entropy brought on by the weak van der Waals forces that stack the layers together (Darder et al. 2021; Camargo et al. 2009; Attaran et al. 2015). The type of polymer that can be utilized to synthesize nanocomposites is ultimately determined by the solvent used. However, because the majority of water-soluble polymers are polar and hydrophilic enough to interact with most nanomaterials without

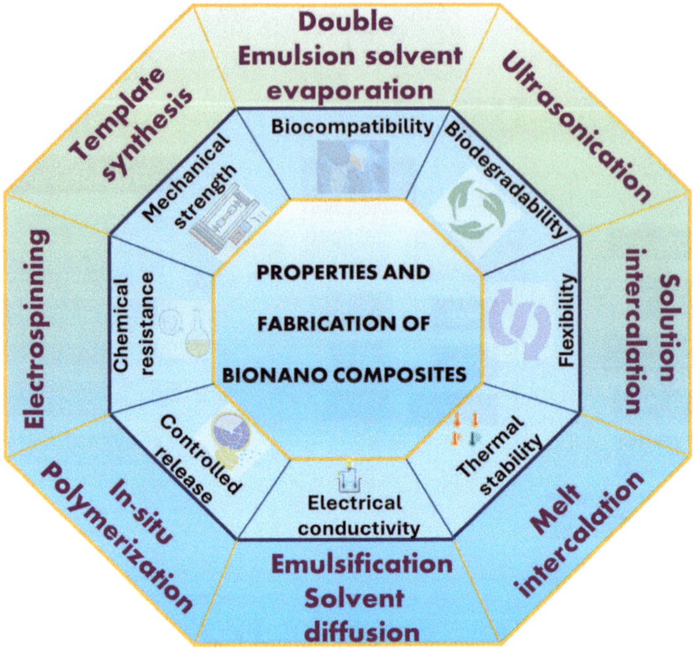

Fig. 1.15 Properties and fabrication techniques of bionanocomposite

requiring cation exchange modification, this method remains quite attractive for creating nanocomposites.

Advantages of Solution Intercalation

Solution intercalation can significantly improve the mechanical properties of bionanocomposites, such as tensile strength and modulus, due to the uniform dispersion of nanoparticles within the polymer matrix. Moreover, the intimate mixing of nanomaterials with biopolymers can enhance the thermal stability of the composite material, since the composite may not undergo initial thermal processing which could compromise its properties, thus making it more resistant to degradation at high temperatures. Also, the properties of bionanocomposites can be tailored more easily by varying the type, concentration, and distribution of nanoparticles, allowing for customization to specific application requirements.

Disadvantages

One of the main challenges is the tendency of nanoparticles to agglomerate, which can lead to uneven distribution and compromise the mechanical and thermal properties of the composite. Also, the solution intercalation process can be complex and requires precise control over parameters such as concentration, temperature, and mixing techniques to achieve the desired properties. Moreover, the use of certain nanoparticles and solvents in the intercalation process may pose environmental and health risks, necessitating careful handling and disposal procedures. Lastly, the commercial production of bionanocomposites using solution intercalation can be quite expensive compared to the other methods.

(ii) *Melt Intercalation*

Melt intercalation is used for creating polymer bionanocomposites, offering solvent elimination and an economically viable method to the other techniques (Attaran et al. 2015). The process is excellent for creating smaller nanoparticles with better impact strength, thermal stability, and tensile strength (Wani and Mundada 2023). The procedure's main drawback is that certain biopolymers may be destroyed by mechanical shearing force or temperature, leading to chain cleavage and molecular weight loss. The method involves annealing a biopolymer-layered silicate mixture, resulting in intercalations and exfoliations of polymer strands. The dispersion of nanoparticles is influenced by processing parameters and enthalpic interaction, useful for in situ polymerization (Kaffashi et al. 2016).

Advantages

Melt intercalation improves the mechanical, thermal, and barrier properties of bionanocomposites. The dispersion of nanoparticles within the polymer matrix leads to improved strength and stability. Also, the process is relatively straightforward, requiring fewer steps compared to other methods such as solvent intercalation. This simplicity often translates to lower production costs. Also, utilizing biodegradable polymers in conjunction with natural nano-fillers aligns with sustainable practices, making the resulting bionanocomposites environmentally friendly. Moreover, melt intercalation can be easily scaled up for industrial applications, allowing for the production of larger quantities of bionanocomposites without significant alterations to the process. This method can also be applied to a wide range of polymer matrices, enhancing the versatility of the bionanocomposites produced.

Disadvantages

The high temperatures required for melt intercalation can lead to the thermal degradation of sensitive biopolymers, potentially compromising the integrity of the final product. Another important setback of this process is the difficulty in achieving a uniform dispersion of nanoparticles can be challenging, often resulting in agglomeration. This can adversely affect the performance of the bionanocomposite. While the method is generally simple, achieving optimal processing conditions (temperature, time, and shear) can require extensive experimentation, which may complicate the formulation process. Although the method uses biodegradable materials, the energy consumption associated with high-temperature processing can raise concerns regarding its overall environmental impact. It is worth mentioning that not all types of nanoparticles are compatible with the melt intercalation process, hence, limiting the choice of fillers that can be used to enhance the properties of the bionanocomposite. That notwithstanding, this method remains the most commercially viable method of nanocomposite processing.

(iii) *In Situ Intercalative Polymerization*

In situ polymerization is a fundamental approach in nanocomposite creation, although it is currently hard to do with biopolymers. In this technique, the nanoparticle is dispersed in a monomer solution or liquid, which permits polymer synthesis between the intercalated sheets. This is followed by monomer polymerization by external stimuli like radiation, temperature, or suitable initiators. A high-shear mixer can be used to mix the monomer's particles evenly. Additionally, the rate at which monomers diffuse into the interlayer region is accelerated by the low viscosity and high diffusivity.

(iv) *Template Synthesis*

It is simple, quick, and adaptable to synthesize silicates in an aqueous solution with biopolymers. The fundamental constituents of hydroxide nanocomposites are these silicates. Multilayer silicates are rarely produced with this technique, though. The precursor serves as a template for producing water-soluble biomolecules, microorganisms, and whole-cell nanoparticles. Host crystals form and grow due to self-assembly. Finally, the bioorganic nano-template is captured by the mesoporous network. Polymer intercalation and sol–gel are the most common processes for producing bionanocomposites, whether they comprise organic or inorganic components (Postnova and Shchipunov 2022).

(v) *Double Emulsion Solvent Evaporation*

In double the emulsion solvent evaporation, the dispersed phase droplets encapsulate the smaller dispersed phase in double emulsion solvent evaporation. The two most prevalent combinations are water–oil–water (w/o/w) and oil–water–oil (o/w/o) DE. Following the formation of the double emulsion, the organic solvent is removed from the dispersed phase via evaporation. Because the active ingredient produces insolubility, it is wrapped in a stiff polymer. Depending on the boiling point of the organic solvent, it can be evaporated under pressure with a rotary evaporator or stirred at ambient temperature. For example, Shen et al. (2017) designed a technique that combines cancer treatment and diagnosis using nanocarriers like doxorubicin, Fe_3O_4 nanocrystals, and quantum dots. For tumor-specific targeting, cancer imaging, drug administration, and gene delivery, they designed biodegradable poly(D,L-lactic-co-glycolic acid) polymeric nanocomposites using the double emulsion solvent evaporation process. Bionanocomposites demonstrated significant anticancer effects in vitro and vivo, decreasing tumor volume by 81% and enhancing cell viability by about 14% compared to saline.

(vi) *Electrospinning*

Electrospinning is an electrohydrodynamic process and a versatile approach for producing bionanocomposites using biodegradable polymers with nano-fillers nanocrystals or clay nanoparticles (Singh et al. 2020). Nanofibers are electrostatically spun from a polymer solution that contains the desired spinning dope (spinning solution) (Nandhini et al. 2024c). The hemispherical droplets create a conical shape (Taylor cone) as a result of these forces (Bhushani and Anandharamakrishnan

2014). When a constant electric potential is applied, the electrostatic force counteracts the surface tension of the droplet, resulting in jet production and collection by the collector plate attached to the ground.

In this production process, nanofibrous mats are formed and nano-fillers are inserted into the polymer matrix to create bionanocomposites with superior structural and mechanical capabilities. Due to their enormous surface area, precise control over fiber diameter, and biocompatibility, these materials are useful for many applications, such as wound dressing, medication administration, and tissue engineering (Yoosefi et al. 2022; Eghbalifam et al. 2023). Samanta et al. (2022) assessed how the properties of the nanofibrous system and drug release were affected by reinforcing nano-fillers. They recovered nano collagen (NCG) from leftover fish scales and cellulose nanofibrils (CNF) from jute fibers. The nanofibrous structures filled with CNF-NCG bionanocomposites demonstrated excellent sustained release of ketorolac tromethamine for up to 16 h, making them ideal transdermal DDSs. Besides melt intercalation, electrospinning has also been used for the commercial fabrication of bionanocomposites. Table 1.1 shows studies that have used electrospinning to fabricate bionanocomposites for specific applications.

(vii) *Ultrasonication*

Ultrasonication is a technique that uses high-frequency ultrasonic vibrations to effectively produce bionanocomposites for biomedical applications (Müller et al. 2022). This method provides precise control over the size and location of nano-filler particles, leading to uniform dispersion and dosage across the polymer matrix. The resultant bionanocomposites have enhanced mechanical and bioactive characteristics and can be applied to a range of biological applications, including wound healing, drug delivery, and tissue scaffolding (Surendran and Sherje 2022).

Table 1.1 Bionanocomposite material fabricated by electrospinning and applications

Biomaterial type	Nanomaterial	Product fabricated	Application	Challenges	References
Polycaprolactone (PCL)	Carbon nanotubes (CNTs)	PCL/CNT nanofibers	Tissue engineering	Uniform dispersion of CNTs	Mirmusavi et al. (2022)
Poly(lactic-co-glycolic acid) (PLGA)	Hydroxyapatite (HA)	PLGA/HA nanofibers	Bone regeneration	Control of fiber morphology	Song et al. (2013)
Collagen	Silver nanoparticles (AgNPs)	Collagen/AgNP nanofibers	Wound healing	AgNP leaching	Iurilli et al. (2025)
Chitosan	Graphene oxide (GO)	Chitosan/GO nanofibers	Drug delivery	GO aggregation	Moradi et al. (2021)
Polyvinyl alcohol (PVA)	Silica nanoparticles (SiO_2)	PVA/SiO_2 nanofibers	Filtration membranes	Achieving high porosity	Wu et al. (2010)

1.5.2 *Properties and Characterization Techniques*

Characterizing bionanocomposites is critical for determining the effect of nanomaterial integration on the material's characteristics. In general, numerous aspects can be evaluated in the physical characterization of nanomaterials, such as various aspects can be investigated in the physical characterization of nanomaterials, including size, shape, surface area, surface charge, surface hydrophobicity, surface adsorbates, density, and interior structure (Patra and Baek 2014; Ingale and Chaudhari 2013). Particle size can be measured using a variety of techniques, and the best strategy is often chosen based on the measurement's goal and the average size distribution of NPs.

The most useful and applied tools for the physicochemical characterization of nanomaterials regarding their morphology, size, chemical composition, forces, and surface charge are "UV–Visible Spectroscopy (UV–Vis), Dynamic Light Scattering (DLS), Fourier Transform Infrared Spectroscopy (FTIR), Transmission Electron Microscopy (TEM), High-Resolution Transmission Electron Microscopy (HRTEM), Scanning Electron Microscopy (SEM), X-Ray Diffraction (XRD), and Energy Dispersive" (Mourdikoudis et al. 2018; Souza 2023). However, other significant techniques are accessible and can be used to supplement the characterization per the required specifications. "Nanoparticle Tracking Analysis (NTA), Atomic Force Microscopy (AFM), High-Resolution Transmission Electron Microscopy (HRTEM), Energy Dispersive Spectroscopy (EDS), Selected Area Electron Diffraction (SAED), X-Ray Photoelectron Spectroscopy (XPS), Diffuse Reflectance Spectrometer (DRS), inductively coupled plasma optical emission spectrometry (ICP-OES), Inductively Coupled Plasma Mass Spectrometry (ICP-MS), Inductively Raman Spectroscopy (IRS), Scanning" (Mourdikoudis et al. 2018; Choi et al. 2007; Gupta et al. 2013). Table 1.2 provides a brief description of the characterization methods for bionanocomposites. Detailed description and methods application are described in Chap. 4.

1.6 Applications of Bionanocomposites

Bionanocomposites, which combine biopolymers with nanoscale reinforcements, have emerged as multifunctional materials with superior performance and sustainability over traditional materials. As a result, the use of bionanocomposites is quickly rising, with a global market valued at many billions of dollars per year (Samir et al. 2022). Hence, bionanocomposites have been utilized in a wide range of applications across various industries, including biomedical devices (Nandhini et al. 2024b; Fura et al. 2024; Akbar et al. 2020), tissue engineering (Sharma et al. 2021; Ahmed and Tomer 2021), drug delivery systems (Mumtaz et al. 2023; Ali et al. 2023; Nandhini et al. 2024b), wound healing (Bal-Öztürk et al. 2024; Wazed Ali and Bairagi 2021), food packaging (Kishore et al. 2023; Iderawumi et al. 2024; Pires et al. 2023;

Table 1.2 Summary of the various characterization techniques for nanomaterials

Type	Technique	Specific purpose	References
Formation of nanoparticle	Ultraviolet–visible spectrophotometry	Gives details about the size, composition, stability, and aggregation of nanoparticles	Gupta et al. (2013)
Morphology and particle size	Transmission Electron Microscopy	Ascertain the nanoparticles' morphology, size (10^{-10} m), shape, and allographic structure	Pal et al. (2011), Kowalczyk et al. (2011) and Khan et al. (2019)
	High-Resolution Transmission Electron Microscopy	Ascertain the atoms' configuration and local microstructures, including the screw axis, lattice vacancies and defects, glide plane, lattice fringe, and surface atomic arrangement of crystalline nanoparticles	Faraji et al. (2010)
	Scanning Electron Microscopy	Visualize the morphology directly to ascertain it	Khan et al. (2019) and Buhr et al. (2009)
	Atomic Force Microscopy	Determine the size information (length, width, and height) and other physical properties (such as morphology and surface texture)	Sulania et al. (2018) and Mišić Radić et al. (2023)
	Dynamic Light Scattering	Determine the particle size distribution	Naiim et al. (2015) and Karmakar (2019)
	Nanoparticle Tracking Analysis	Classifies the many kinds of nanoparticles according to their size, which ranges from 30 to 1000 nm, and their refractive index determines the lower detection limit	Patois et al. (2012)
	Energy Dispersive X-Ray Spectra (EDX)	Detects the number of X-rays which are emitted to balance the energy difference between two electrons	Pathak et al. (2011) and MubarakAli et al. (2011)
Surface charge	Zeta potential	Assess colloidal nanoparticles' stability and surface charge in addition to the kind of components that are coated or encapsulated within them	Honary and Zahir (2013) and Shao et al. (2015)
	Fourier Transform Infrared Spectroscopy	For characterizing the functional groups	Baraton and Merhari (2007) and Irshad et al. (2021)

(continued)

Table 1.2 (continued)

Type	Technique	Specific purpose	References
	X-Ray Photoelectron Spectroscopy	Confirm the structure and speciation of the various elements present in the magnetic nanoparticles' chemical composition, as well as the mechanism of the reaction that takes place on their surface and the properties involved in the bonding of the various elements	Scimeca et al. (2018), Korin et al. (2017) and Geißler et al. (2021)
Thermal analysis	Thermal gravimetric analysis	Confirm the formation of coatings such as surfactants or polymers to estimate the binding efficiency on the surface of magnetic nanoparticles	Ali (2021) and Biehl et al. (2018)
Crystallinity	X-ray diffraction	Identify and quantify various crystalline forms or elemental compositions of nanoparticles	Holder and Schaak (2019) and Kumar et al. (2022)

Youssef and El-Sayed 2018), automotive industry (Ali and Hoque 2022; Ali et al. 2021a), cosmetics (Angra et al. 2021; Zia et al. 2020), and environmental remediation (Baruah et al. 2022; Varghese and Das 2015). Figure 1.16 depicts several applications of biomaterials (Yin and Yang 2020) which have been investigated.

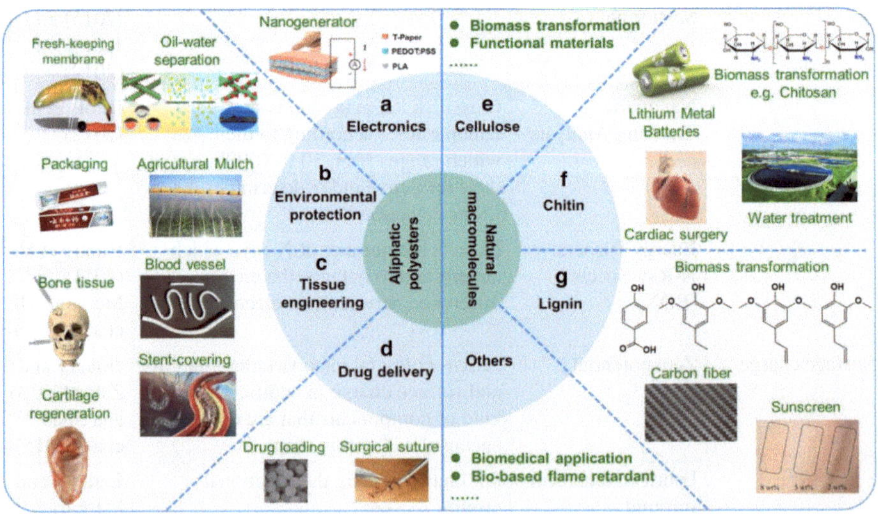

Fig. 1.16 Primary areas of application for biodegradable materials. Reprinted with permission from Yin and Yang (2020), Springer Nature

1.6.1 Packaging Solutions

Polymeric materials are utilized extensively in everyday life, particularly in packaging. Biodegradable polymers are frequently utilized to reduce trash generation addition to being biodegradable, biopolymers also have low-temperature sealability and air permeability (Isabelle and Lan 2009). Depending on the product to be packaged and the storage conditions, different biodegradable polymers have different physical properties. Thus, bionanocomposites have emerged as a viable solution for sustainable and efficient packaging, exhibiting better barrier qualities, thermal stability, antimicrobial capabilities, mechanical strength, and biodegradability when compared to standard materials (see Fig. 1.17) (Jayakumar et al. 2022; Ahari et al. 2022; Bharathi and Jayas 2024; Pires et al. 2023; Youssef and El-Sayed 2018). The two main areas of the food industry where NPs can be used are food processing and packaging (Shukla 2012; Espitia et al. 2012). Depending on the nanomaterials' qualities, bionanocomposites can also provide additional functionality, such as active and intelligent packaging via Radio-Frequency Identification (RFID) tags, allowing significant benefits in food preservation and monitoring. For example, Espitia et al. (2012) combined zinc oxide nanoparticles with polymeric materials to create packaging material. The freshly synthesized packing tissue demonstrated better antibacterial properties. Similarly, Rajamanickam et al. (2012) zinc oxide nanoparticles could be utilized to make food packaging materials containers and nutritious coatings. Food is kept fresh and contamination is decreased when nanoparticles are included in packaging materials. AgNPs can pierce bacterial cells' biofilm, which makes them more resilient to stress. AgNPs can therefore be used to clean and disinfect food as it is being packed (Huang et al. 2015).

Alexandre et al. (2016) examined the fabrication of gelatin-based films with a nanoemulsion of ginger essential oil and Montmorillonite (MMT) reinforcement.

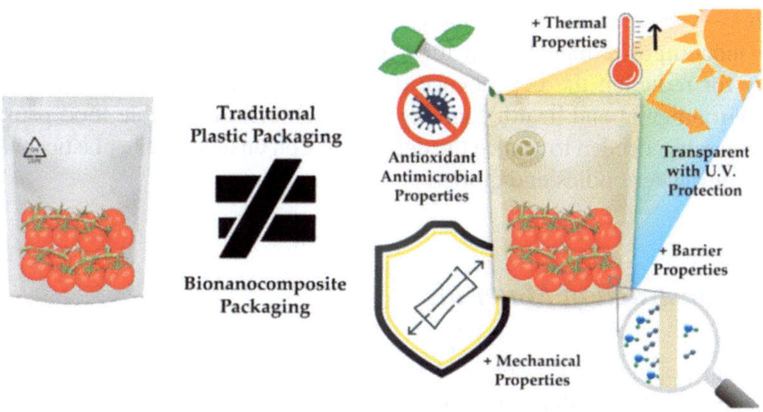

Fig. 1.17 Benefits of bionanocomposites over conventional packaging made of petroleum. Reprinted from Pires et al. (2023), an open-access article by MDPI

When MMT and GEO were added to the films, their thickness increased while their moisture content and water solubility decreased. MMT only improved the films' roughness and decreased surface hydrophobicity. Similarly, Souza et al. (2019) developed chitosan/MMT bionanocomposites with rosemary essential oil (REO) as a bionanocomposite film and evaluated its efficacy on fresh chicken meat. MMT reduced the permeability of the film to oxygen and water vapor. The bioactivity of the films and the release of the bioactive compounds into the fowl flesh both declined when MMT was added to the bionanocomposite. For food packaging applications, the use of cellulose nanocrystals (CNC) made from sugarcane bagasse as reinforcement in a PVA/CMC blend has been studied (Achaby et al. 2017). The adhesion capabilities of the CNC's surface and the abundant functional groups present in the macromolecular chains of the PVA/CMC blend improved the interfacial interactions of the ensuing bionanocomposite. It was observed that the use of CNC boosted the bio-based films' tensile modulus and strength with no effect on their transparency. However, the 5 wt.% CNC loading reduced the water vapor permeability of the bionanocomposite.

In a recent study, Souza et al. (2020) synthesized a ZnO NP to be evaluated in an active chitosan bionanocomposite via a more environmentally friendly way as a food package for poultry meat. The samples covered with the films demonstrated a slower rate of degradation when used as primary packaging for meat. This was demonstrated by the preservation of the meat's original reddish color as well as a decrease in oxidation and microbiological development. The films' antioxidant properties were enhanced by the nanoparticles, which also demonstrated their viability as food preservation agents for application in active food packaging. In another investigation by Bang and Kim (2012), Ag NPs were used in hydroxypropyl methylcellulose (HPMC) polymer for the fabrication of bactericidal food packaging films. Excellent mechanical and barrier qualities were displayed by the HPMC-AgNP nanocomposites. The study concluded that the nanocomposite films might be used as an active antimicrobial coating in the food packaging industry since the nanofillers improved the films' tensile strength. These studies underscore the significance of bionanocomposite materials in food packaging and related applications paving the way for reducing the dependence on the petrochemical plastics notorious for their environmental and health challenges. To achieve sustainable food packaging, more research is needed to reduce the dependence on unsustainable plastics that have polluted our water bodies and drains.

1.6.2 Construction Materials

The human population is estimated to grow to 9.3 billion by 2050 (Satdive and Muthukumar 2022). With this growth comes the desire for building and construction expansion and the associated energy demand is estimated to increase by ~ 40%. Currently, 35–60% of the remaining carbon budget needed to keep global warming to 2 °C may come from emissions from current civil works which are forecasted to worsen if no changes are made in the current material culture of the building industry (Churkina et al. 2020; Müller et al. 2013). As the world shifts toward a more sustainable future, bionanocomposite materials applications when properly leveraged are the game changers in the resolve to reduce the environmental problems and emissions from new infrastructure, architectural projects, and commercial housing (Satdive and Muthukumar 2022). Beyond reducing the environmental effect of traditional building materials, bionanocomposites could be designed to have multi-purpose functions beyond the architectural effect. For instance, bionanocomposites are designed with photocatalytic and antimicrobial properties for construction which could function with a self-cleaning effect and the deactivation/destruction of organic pollutants and impurities present on architectural surfaces (Bharathi et al. 2019). Moreover, TiO_2-based BNCs could be used for a variety of exterior and interior building materials, including glasses, plastic films, window blinds, metallic panels, paints and coatings, claddings, and roofs, thanks to the self-cleaning and depolluting properties of TiO_2 (Munafò et al. 2015; Ocak et al. 2014).

1.6.3 Biomedical Applications

Among the many uses for bionanocomposites in biomedicine are antimicrobial and anticancer treatments, drug delivery methods, wound dressings, scaffolds for tissue engineering, anti-anemia therapies, dental applications, and bioimaging technologies (Nandhini et al. 2024b; Akbar et al. 2020). Some typical applications of the bionanocomposite, current drawbacks, and areas for further improvement are shown in Table 1.3.

Due to the significant advancement in biomedical engineering in recent years, natural tissue architectures can be effectively restored using bionanocomposites. The form of tissues designed using bionanocomposites is normally identical to that of extracellular matrices made of inorganic materials such as proteins, and polysaccharides (Bal-Öztürk et al. 2024; Bramhill et al. 2017). To be able to achieve the identical properties of natural tissues using bionanocomposites, selecting the right biomaterial with biodegradable and bioresorbable structures is essential, especially for applications such as polymer scaffolds and HAPs. The right combination of biomaterial with nanomaterial is essential for bone regeneration in clinical orthopedics to promote bone formation at implant sites (Battafarano et al. 2021). More specifically, certain bionanocomposites need to have a structure that is sufficiently porous to permit the development of new tissues. For example, porous scaffolds made of HAP/collagen nanocomposites are utilized to treat bone abnormalities in animal models. Tissue engineering has specifically employed bionanocomposites reinforced with HNTs, such as PVA/HNT bionanocomposite that is compatible with fibroblast and osteoblast cells (Zhou et al. 2010). Electrospun PLGA/HNT bionanocomposites are also highly biocompatible, and the addition of HNTs does not impair cell proliferation (Qi et al. 2012). Similar to native extracellular matrices, PLGA/HNT nanofibrous scaffolds can provide 3D structures with linked pores to facilitate cell adhesion

Table 1.3 Current applications of bionanocomposite materials

Application	Description	Biocomposite type used	Current drawbacks	Areas for further research
Drug delivery systems	Targeted delivery of therapeutic agents	Polymer-based composites	Limited control over release rates	Enhanced release mechanisms
Tissue engineering scaffolds	Support for cell growth and tissue repair	Hydrogel composites	Mechanical instability in some cases	Improved mechanical properties
Antimicrobial coatings	Surface protection against microbial growth	Metal-nanoparticle composites	Potential toxicity of metals	Safer alternatives to metals
Diagnostic imaging agents	Enhancing visibility in imaging techniques	Nanoparticle-based composites	Stability issues in biological environments	Long-term stability in vivo
Biosensors	Detection of biological substances	Conductive polymer composites	Sensitivity and specificity challenges	Development of more sensitive detection methods
Wound healing materials	Promoting faster healing of injuries	Natural polymer composites	Variability in healing efficacy	Standardizing efficacy across populations
Bone regeneration	Supporting bone growth and healing	Bioceramic composites	Limited integration with host tissue	Strategies for better integration
Cancer therapeutics	Targeted treatment options for tumors	Lipid-based nanocomposites	Drug resistance and side effects	Overcoming drug resistance mechanisms
Gene delivery systems	Transporting genetic material for therapy	Viral and non-viral composites	Immunogenicity and low transfection efficiency	Improving transfection efficiency
Implants and prosthetics	Enhancing biocompatibility and functionality	Composite metal-ceramic materials	Long-term stability and wear resistance issues	Biodegradable materials for long-term use
Cartilage repair	Regeneration of cartilage tissue	Polymer-ceramic composites	Limited mechanical strength	Strengthening mechanical properties
Dental applications	Restoration and repair of dental structures	Resin-based composites	Durability issues under stress	Improving longevity and wear resistance

(continued)

Table 1.3 (continued)

Application	Description	Biocomposite type used	Current drawbacks	Areas for further research
Nerve regeneration	Supporting the repair of nerve tissues	Conductive polymer composites	Limited electrical conductivity in some cases	Enhancing electrical properties
Anti-inflammatory treatments	Reducing inflammation in tissues	Biopolymer composites	Variable efficacy among patients	Personalizing treatment approaches

and migration, according to the SEM results for cell morphology (Qi et al. 2010). In like manner, PLLA-based nanocomposites, such as PLLA/MMT bionanocomposites have been claimed to be ideal for bone replacement applications due to their excellent mechanical and thermal properties. In practice, bone-fixing models for animal models have achieved the use of PLLA self-reinforced bionanocomposites (Naseem et al. 2021). Similarly, the potential use of bionanocomposites in medication delivery has been discussed extensively in prior literature (Mumtaz et al. 2023; Ali et al. 2023; Carazo et al. 2018). For example, the bone morphogenetic protein was loaded into HAP/collagen-alginate bionanocomposites and used as a drug carrier (Sotome et al. 2004). Similarly, bionanocomposites reinforced with HNTs have been investigated extensively as potent drug carriers and practically loaded powders, solutions, and fibrous scaffolds (Liu et al. 2014). A delayed release rate is produced, for example, when diltiazem hydrochloride, 5 aminosalicylic acid, and propanol hydrochloride are injected into the lumen and exterior surfaces of HNTs (Aafreen et al. 2024). Despite its promise in practical application, HNT loading capacity is limited since the lumen volume percentage of HNTs cannot surpass 10 vol.% [327]. To enhance the loading volume of HNTs, alumina must be etched from their inner surfaces. Table 1.4 provides a summary of some bionanocomposite and their applications as drug carriers.

1.6.4 Environmental Remediation

The increasing release of dangerous and lethal chemicals and compounds into the environment as a result of human activities has resulted in the degradation and pollution of water, soil, and atmosphere. Most of the time, these organic pollutants are constantly present in the air, as pesticides, aerosols, polychlorinated biphenyls (PCBs), polycyclic aromatic hydrocarbons (PAHs), and volatile industrial waste which cannot be perceived by the human eye (Ali et al. 2021b; Rodriguez-Narvaez et al. 2017; Richardson and Ternes 2018). This has necessitated the need for effective technologies that can help clean the ecosystem to reduce the adverse health

Table 1.4 Specific applications of bionanocomposites and drugs delivered

Biopolymers	Nanomaterial used as reinforcement	Drugs were supplied	Key findings	References
Cellulose/pectin	Nanoclays	5-methylindole derivative	Neurodegenerative illnesses were treated with the controlled release with toxicity observed	Bravo et al. (2023)
Crosslinked nano collagen-cellulose nanofibrils	Electrospun polyvinyl alcohol/ methylcellulose/ polyethylene glycol	Ketorolac tromethamine	Continuous sustained release for the entire duration of treatment	Samanta et al. (2022)
Chitosan and starch	Chitosan and starch films	Cefixime	Efficient antibacterial effectiveness with the possibility of regulated release	Alotaibi et al. (2022)
Chitosan	Nanocellulose	Silver sulfadiazine and betamethasone	Improved wound dressings with specific qualities for antibacterial activity and controlled drug release	Riccio et al. (2021)

effects of these pollutants. Nanotechnology offers a sustainable solution for critical environmental challenges, from CO_2 collection to water purification using various technologies (Wang et al. 2019). To ensure sustainable access to quality drinking water, magnetic nanoparticles, and nano filters have been used to remove pollutants from water, while metal certain oxides have been employed to absorb flue gas and CO_2 emissions. In the application of bionanocomposites for environmental purification, Eldeeb et al. (2021) developed and characterized a bionanocomposite comprising alginate multiwall carbon nanotubes and dialdehyde cellulose. The SEM had an average size of 85–100 nm and was capable of adsorbing copper (II) ions from the aquatic environment in 2.5 h with a removal percentage of 100%. Similarly, bionanocomposites are used as filtration members, leveraging their unique qualities of enhanced filtration, greater strength, and increased resistance to fouling to eliminate residues of contaminants from water (Tomer et al. 2024). In another study, Gopal et al. (2023) used clay nanosheets supported by a Fe–Cu nanocomposite to build a nanocomposite for eliminating antibiotics from water. This innovative method involves immobilizing the composite in an alginate-carboxymethyl chitosan matrix covered with biodegradable chitosan, producing nanocomposite beads suitable for column reactors. The study demonstrated successful ciprofloxacin (CIPRO) removal, with a maximum removal capacity of 485.58 mg/g based on the Langmuir isotherm and an optimal batch mode removal rate of over 90%. Ehsan et al. (2023) synthesized an alginate-based nanocomposite containing a graphene oxide (GO) carbon network for oil–water separation. The use of biopolymeric materials in conjunction with nanotechnology offers a viable approach to resolving the environmental challenge of antibiotic contamination in water. Detailed description and applications of bionanocomposite for environmental re-engineering has discussed in Chap. 8 of this book.

1.7 Challenges and Opportunities

Bionanocomposites represent a unique class of advanced green composites utilizing biomass and nanotechnology. These composites have produced intriguing results for various applications which have piqued the interest of scientists as long-term replacements to conventional composite materials. These materials show great promise for addressing environmental issues while providing high-performance qualities. However, their development and implementation face many important problems that require short- to long-term attention and commercial solutions.

1.7.1 Market Acceptance and Commercialization

Manufacturing is a key driver of innovation and higher-value employment development in both industrialized and emerging economies, ensuring a globally sustainable economy (Diallo et al. 2013). The biocomposite market is expanding due to increased demand in various industries, including construction, transportation, and consumer products. Key players include Akrema, Mitsubishi Plastics, Huntsman Corporation, and Stemergy. The Asia Pacific market for bionanocomposites is rapidly expanding due to government regulations in China, India, South Korea, Thailand, and Indonesia, with manufacturers like FlexForm Technologies, Tecnaro GmbH, Trex Company, and Green Bay Decking (Angra et al. 2021). Bionanocomposites are increasingly used in food packaging due to their low permeability and government regulations, boosting their expansion in the food and beverage industries (Pires et al. 2023; Bharathi and Jayas 2024). In the biomedical field, bionanocomposites are being investigated for drug delivery systems, tissue engineering, and diagnostic instruments (Wani and Mundada 2021). Their biocompatibility and biodegradability make them appropriate for these applications. Bionanocomposites are increasingly being used in cosmetics because they promise to improve skin penetration and active ingredient delivery.

1.7.2 Regulatory Considerations

Nanotechnology advancements have improved risk management, leading to international harmonized testing frameworks and responsible development. Intergovernmental efforts promote responsible nanotechnology growth through science-based safety assessment, risk mitigation, ethical guidelines, capacity-building modules, and adaptive governance policies (Borrás and Edquist 2013). Major global organizations like ISO, OECD, UNESCO, and science academies are launching multi-stakeholder initiatives to tackle challenges in standardization (Soltani and Pouypouy 2019). The ISO established a Technical Committee on Nanotechnologies in 2005 to develop standards and technical reports on terminology, metrology, calibration, reference materials, test methodologies, modeling, and simulations for nanomaterial characterization (Benko 2017). The committee, comprising advisory groups under ISO/TC 229, offers protocols for evaluating essential physicochemical properties for toxicity assessment, including particle size distribution, zeta potential, water solubility, and agglomeration degree (Benko 2017). Similarly, the International Organization for Standardization (ISO) has established various standards relevant to nanotechnology, notably ISO/TS 80004-1, which gives vocabulary and definitions of nanomaterials (Kumari et al. 2023).

Bionanocomposites as FCM face concerns about potential nanoparticle migration into food, but their release in nanosize form is not fully established (Souza and Fernando 2016; Bott et al. 2014). Nanoparticles possess unique properties and behaviors compared to larger particles, which could potentially impact their safety (Bradley et al. 2011). Their migration into food can potentially cause unintended health effects on consumers. For example, Metak et al. (2015) discovered that orange juice had the maximum silver migration from nano-silver-coated films when compared to other food items such as bread, apple, cheese, milk powder, carrot, minced beef, butter, and water. Therefore, bionanocomposites must undergo rigorous safety evaluation before they can be utilized as food-contact materials to prevent potential health risks (More et al. 2021; DeLoid et al. 2017; Enescu et al. 2019). To address these concerns, the BioNanoPolys project aims to create safe additives and food-contact materials. This pledge is founded on rigorous conformity to European legislation, specifically Regulation (EC) No. 1935/2004 (Ballesteros et al. n.d.). The European Food Safety Authority (EFSA) released guidelines in 2021 on the risk evaluation of nanomaterials in the food and feed chain for human and animal health (Pires et al. 2023). The guidelines include the safe use of nanotechnology in packaging through comprehensive characterization and standard testing, to measure the nanoparticle migration and ensure that it does not exceed the regulatory limitations (Trotta et al. 2023). In the European Union, bionanocomposites intended for food contact must conform with Regulation (EU) No. 10/2011, which establishes composition and safety regulations for plastic materials. These laws ensure that materials do not constitute a health concern for consumers by controlling the migration of chemicals into food (Ballesteros et al. n.d.). This rule outlines the materials suitable for active packaging and the maximum discharge into food material and headspace (Melini and Melini 2018). The

USA, UK, EU, China, Japan, and Korea have implemented country-specific rules to address environmental, health, and safety concerns related to nanotechnology (Soltani and Pouypouy 2019; Dong et al. 2016). These include required reporting standards before commercializing items, worker protection equipment, acceptable exposure limits, and disposal and recycling methods in streamlining and enabling balanced factory monitoring.

1.7.3 Interdisciplinary Collaboration

The growing demand for sustainability has prompted the development of novel materials that can reduce environmental impact while yet providing great performance. One such class of materials that have received a lot of attention is bionanocomposites, a type of nanotechnology that combines biopolymers with nanomaterials to generate composite materials with unique features that are suited for a variety of applications, notably in sustainable development. For instance, the European Union (EU) allocated over €896 million in research related to nanotechnology from 2007 to 2011 (Tawiah et al. 2024).

Furthermore, academics, businesses, and NGOs are working together to establish legislation on the environmental, health, and safety aspects of engineered nanomaterials, demonstrating a shared interest in responsible nanotechnology development (Morris et al. 2011). Global funding increases have influenced the amount of scholarly papers relating to nanotechnology. For example, Bharathi and Jayas (2024) report almost 17,000 publications published on bionanocomposites in the Web of Science database. Similarly, as of 2016, national nanotechnology projects had been started in over 60 countries (Kargozar and Mozafari 2018). Nanotechnology advances are predicted to continue to increase with greater assistance from governments and the commercial sector (Rambaran and Schirhagl 2022). As a result, Yasri and Wiwanitkit (2017) and Allon et al. (2017) have done reviews on recent ethical issues in nanomedicine and nanomedical technology. This strategy has the potential to dramatically increase innovation and value creation in the private sector. Based on this, research groups may collaborate with relevant organizations to develop cutting-edge solutions through data-driven innovations (Morris et al. 2011; Bozeman et al. 2013). This data is then strategically used to construct or enhance products and services based on existing consumer gaps or desires. Similarly, recent studies have emphasized the importance of collaboration among chemists, biologists, and engineers in the development of bionanocomposites as biosensors in the field of electroanalysis (Deffo et al. 2024). With research-driven innovation, scientific or technological skills originating from labs are employed as a seed of new ideation (Pokrajac et al. 2021). The industry's commitment to eco-friendly practices can be communicated through updated regulatory guidelines, industry standards, and collaborations with environmental organizations, fostering a positive outcome for a sustainable future (Trotta et al. 2023).

1.8 Conclusion

As the world faces the dual challenges of environmental degradation and resource depletion, the development of advanced materials that are both high-performing and environmentally friendly is becoming increasingly critical. Hence, the integration of nanomaterials with biodegradable polymers, forming bionanocomposites, represents a transformative approach in the pursuit of sustainable materials. Bionanocomposites, which combine the strengths of renewable, and biodegradable biopolymers including polylactic acid (PLA), polyhydroxyalkanoates (PHA), and cellulose, with the enhanced properties of nanomaterials, offer a promising pathway to address these challenges while promoting sustainable development. Moreover, the incorporation of nanomaterials like nanoclays or carbon nanotubes can provide enhanced properties such as increased strength, thermal stability, and UV resistance, enabling biopolymers to replace traditional materials in more demanding applications. In the food packaging industry, bionanocomposites are addressing the growing demand for sustainable solutions by offering enhanced barrier properties, oxygen impermeability, extending food shelf life, and antimicrobial characteristics. From a technological perspective, the development of bionanocomposites aligns with the principles of circular economy, wherein materials are designed for reuse, recycling, and minimal waste. Furthermore, bionanocomposites are not limited to environmental benefits; they also hold promise in advancing other sustainability goals, including energy efficiency and resource conservation. For example, lightweight bionanocomposites can enhance fuel efficiency in the automotive industry and reduce energy consumption in construction materials, thereby reducing the environmental impact of buildings. The combination of nanotechnology and bio-based materials is not without its challenges. Issues such as the dispersion of nanomaterials, the optimization of matrix-filler interactions, and the scalability of production need to be addressed to fully realize the potential of bionanocomposites. Furthermore, lifecycle assessments (LCAs) are essential to understanding the true environmental benefits of these materials, as they must be considered in terms of energy use, resource extraction, and disposal. Importantly, continued research and innovation in the fabrication, characterization, and application of bionanocomposite materials serve a key purpose in unlocking their full potential contributing to a more sustainable and circular future in industries ranging from packaging to construction to biomedicine. While initial production costs may be higher than traditional materials, the long-term economic benefits of bionanocomposites are becoming increasingly apparent. Their durability, multifunctionality, and reduced environmental impact often result in lower lifecycle costs. Additionally, the growing market demand for sustainable materials is driving innovation and scale-up efforts, potentially leading to more cost-effective production methods and a more sustainable world.

References

Aafreen, Priyanka V, Haris S (2024) Recent advances in the synthesis of graphene and its derivative materials, chap 1. In: Enos WW (ed) Chemistry of graphene. IntechOpen, Rijeka
Agbakoba V et al (2023) PLA bio-nanocomposites reinforced with cellulose nanofibrils (CNFs) for 3D printing applications. Cellulose 30(18):11537–11559
Agbolade LO et al (2024) Revisiting the optoelectronic properties of graphene: a DFT approach. Int J Nanoelectron Mater (IJNeaM) 17(1):76–85
Ahari H et al (2022) Bio-nanocomposites as food packaging materials; the main production techniques and analytical parameters. Adv Coll Interface Sci 310:102806
Ahmad T et al (2016) Effect of reaction time on green synthesis of gold nanoparticles by using aqueous extract of elaise guineensis (oil palm leaves). Procedia Eng 148:467–472
Ahmad M et al (2017) Chitosan centered bionanocomposites for medical specialty and curative applications: a review. Int J Pharm 529(1):200–217
Ahmed S, Tomer A (2021) Bionanocomposites in tissue engineering and regenerative medicine
Aizat MA, Aziz F (2019) Chitosan nanocomposite application in wastewater treatments, chap 12. In: Ahsan A, Ismail AF (eds) Nanotechnology in water and wastewater treatment. Elsevier, pp 243–265
Akbar MU et al (2020) Biomedical applications of bionanocomposites, chap 17. In: Mahmood Zia K et al (eds) Bionanocomposites. Elsevier, pp 457–483
Alexandre EMC et al (2016) Gelatin-based films reinforced with montmorillonite and activated with nanoemulsion of ginger essential oil for food packaging applications. Food Packag Shelf Life 10:87–96
Ali N, Hoque M (2022) Bionanocomposites in the automotive and aerospace applications, pp 237–253
Ali M et al (2021a) Nano and bio-composites and their applications: a review. IOP Conf Ser Mater Sci Eng 1067:012093
Ali A et al (2021b) Review on recent progress in magnetic nanoparticles: synthesis, characterization, and diverse applications. Front Chem 9
Ali M et al (2023) Applications of chitosan based bionanocomposites in drug-delivery and anticancer treatment—a review. Eur Polym J 201:112576
Allon I et al (2017) Ethical issues in nanomedicine: tempest in a teapot? Med Health Care Philos 20(1):3–11
Alotaibi BS et al (2022) A new approach for the management of *Escherichia coli* and *Klebsiella pneumonia* by using cefixime-based bionanocomposite films. J Exp Nanosci 17(1):389–419
Amira E-F et al (2012) Starch and microbial α-amylases: from concepts to biotechnological applications, chap 21. In: Chuan-Fa C (ed) Carbohydrates. IntechOpen, Rijeka
Angra V et al (2021) Commercialization of bionanocomposites, pp 587–610
Anwer AH et al (2023) State-of-the-art advances in nanocomposite and bio-nanocomposite polymeric materials: a comprehensive review. Adv Colloid Interface Sci 102955
Arora B, Bhatia R, Attri P (2018) Bionanocomposites: green materials for a sustainable future. In: Hussain CM, Mishra AK (eds) New polymer nanocomposites for environmental remediation. Elsevier, pp 699–712
Atodariya J et al (2024) Advancements in green nanocomposites: a comprehensive review on cellulose-based materials in biocomposites and bionanocomposites. In: Rubber based bionanocomposites: characterisation, pp 213–239
Attaran SA, Hassan A, Wahit MU (2015) Materials for food packaging applications based on bio-based polymer nanocomposites: a review. J Thermoplast Compos Mater 30(2):143–173
Augst AD, Kong HJ, Mooney DJ (2006) Alginate hydrogels as biomaterials. Macromol Biosci 6(8):623–633
Auria C et al (2019) Interactions of nanoparticles and biosystems: microenvironment of nanoparticles and biomolecules in nanomedicine. Nanomaterials 9:1365

Ballesteros A, Díaz R, ITENE (n.d.) Ensuring safety and compliance: the cornerstone of innovation in BioNanoPolys. In: BioNanoPolys
Bal-Öztürk A et al (2024) Innovative approaches in skin therapy: bionanocomposites for skin tissue repair and regeneration. Mater Adv 5(12):4996–5024
Bang G, Kim SW (2012) Biodegradable poly(lactic acid)-based hybrid coating materials for food packaging films with gas barrier properties. J Ind Eng Chem 18(3):1063–1068
Baraton M-I, Merhari L (2007) Dual contribution of ftir spectroscopy to nanoparticles characterization: surface chemistry and electrical properties. Proc SPIE Int Soc Opt Eng 6768
Baruah R, Yadav A, Moni Das A (2022) Evaluation of the multifunctional activity of silver bionanocomposites in environmental remediation and inhibition of the growth of multidrug-resistant pathogens. New J Chem 46(21):10128–10153
Battafarano G et al (2021) Strategies for bone regeneration: from graft to tissue engineering. Int J Mol Sci 22(3)
Benko H (2017) ISO technical committee 229 nanotechnologies. In: Metrology and standardization of nanotechnology, pp 259–268
Bharathi VSK, Jayas DS (2024) Evolution of bionanocomposites: innovations and applications in food packaging. Foods 13(23). https://doi.org/10.3390/foods13233787
Bharathi D et al (2019) Preparation of chitosan coated zinc oxide nanocomposite for enhanced antibacterial and photocatalytic activity: as a bionanocomposite. Int J Biol Macromol 129:989–996
Bhushani JA, Anandharamakrishnan C (2014) Electrospinning and electrospraying techniques: potential food based applications. Trends Food Sci Technol 38(1):21–33
Biehl P et al (2018) Synthesis, characterization, and applications of magnetic nanoparticles featuring polyzwitterionic coatings. Polymers 10. https://doi.org/10.3390/polym10010091
Borrás S, Edquist C (2013) The choice of innovation policy instruments. Technol Forecast Soc Change 80(8):1513–1522
Bott J, Störmer A, Franz R (2014) A model study into the migration potential of nanoparticles from plastics nanocomposites for food contact. Food Packag Shelf Life 2(2):73–80
Boutry S, Muller RN (2018) Metal oxide particles and their prospects for applications, pp 3–42
Bozeman B, Fay D, Slade CP (2013) Research collaboration in universities and academic entrepreneurship: the-state-of-the-art. J Technol Transf 38(1):1–67
Bradley EL, Castle L, Chaudhry Q (2011) Applications of nanomaterials in food packaging with a consideration of opportunities for developing countries. Trends Food Sci Technol 22(11):604–610
Bramhill J, Ross S, Ross G (2017) Bioactive nanocomposites for tissue repair and regeneration: a review. Int J Environ Res Public Health 14(1)
Bravo I et al (2023) Cellulose/pectin-based materials incorporating Laponite-indole derivative hybrid for oral administration and controlled delivery of the neuroprotective drug. Int J Biol Macromol 234:123765
Buhr E et al (2009) Characterization of nanoparticles by scanning electron microscopy in transmission mode. Meas Sci Technol 20:084025
Bustamante-Torres M et al (2021) Interaction between filler and polymeric matrix in nanocomposites: magnetic approach and applications. Polymers 13. https://doi.org/10.3390/polym13172998
Camargo P, Satyanarayana KG, Wypych F (2009) Nanocomposites: synthesis, structure, properties and new application opportunities. Mater Res-Ibero-Am J Mater 12
Carazo E et al (2018) Advanced inorganic nanosystems for skin drug delivery. Chem Rec 18(7–8):891–899
Choi Y, Ho NH, Tung CH (2007) Sensing phosphatase activity by using gold nanoparticles. Angew Chem 119(5):721–723
Chouhan RS et al (2021) Emerging tri-s-triazine-based graphitic carbon nitride: a potential signal-transducing nanostructured material for sensor applications. Nano Select 2(4):712–743
Churkina G et al (2020) Buildings as a global carbon sink. Nat Sustain 3(4):269–276

Clémenson S, David L, Espuche E (2007) Structure and morphology of nanocomposite films prepared from polyvinyl alcohol and silver nitrate: influence of thermal treatment. J Polym Sci Part A Polym Chem 45(13):2657–2672
Dang Y, Guan J (2020) Nanoparticle-based drug delivery systems for cancer therapy. Smart Mater Med 1:10–19
Darder M et al (2021) Gentamicin-montmorillonite intercalation compounds as an active component of hydroxypropylmethylcellulose bionanocomposite films with antimicrobial properties. Clays Clay Miner 69(5):576–588
Datta LP, Manchineella S, Govindaraju T (2020) Biomolecules-derived biomaterials. Biomaterials 230:119633
Dave PN, Gor A (2018) Natural polysaccharide-based hydrogels and nanomaterials: recent trends and their applications, chap 3. In: Mustansar Hussain C (ed) Handbook of nanomaterials for industrial applications. Elsevier, pp 36–66
Deffo G et al (2024) Bionanocomposite materials for electroanalytical applications: current status and future challenges. Nanoscale Adv 6(19):4736–4750
de França JOC et al (2022) Polymers based on PLA from synthesis using D,L-lactic acid (or racemic lactide) and some biomedical applications: a short review. Polymers (Basel) 14(12)
DeLoid GM et al (2017) An integrated methodology for assessing the impact of food matrix and gastrointestinal effects on the biokinetics and cellular toxicity of ingested engineered nanomaterials. Part Fibre Toxicol 14(1):40
Deshmukh K et al (2017) Biopolymer composites with high dielectric performance: interface engineering, chap 3. In: Sadasivuni KK et al (eds) Biopolymer composites in electronics. Elsevier, pp 27–128
Diallo MS, Fromer NA, Jhon MS (2013) Nanotechnology for sustainable development: retrospective and outlook. J Nanopart Res 15(11):2044
Ding S et al (2020) Protein-based nanomaterials and nanosystems for biomedical applications: a review. Mater Today 43
Dominic S, Sabjan K, Vinoji S (2023) In vitro study of biocompatible hybrid scaffold of polyvinyl alcohol–polyaniline–nanocellulose for tissue engineering applications. Chem Afr 6
Dong H et al (2016) The nanotechnology race between China and the United States. Nano Today 11(1):7–12
Đorđević L, Arcudi F, Prato M (2019) Preparation, functionalization and characterization of engineered carbon nanodots. Nat Protoc 14(10):2931–2953
Dosawada PS et al (2023) Review on influence of nanomaterials on thermal energy storage methods. Mater Today Proc 78:580–593
Dubey A, Irudhayaraj S, Jaiswal A (2023) Bio-nanocomposites: a next generation food packaging materials. Synthesis and applications of nanomaterials and nanocomposites. Springer, pp 197–231
Eghbalifam N, Shojaosadati SA, Hashemi-Najafabadi S (2023) Role of bioactive magnetic nanoparticles in the prevention of wound pathogenic biofilm formation using smart nanocomposites. J Nanobiotechnol 21(1):161
Ehsan M et al (2023) Engineering nanocomposite membranes of sodium alginate-graphene oxide for efficient separation of oil-water and antifouling performance. J Environ Chem Eng 11(1):109185
El Achaby M et al (2017) Processing and properties of eco-friendly bio-nanocomposite films filled with cellulose nanocrystals from sugarcane bagasse. Int J Biol Macromol 96:340–352
Eldeeb TM et al (2021) Novel bio-nanocomposite for efficient copper removal. Egypt J Aquat Res 47(3):261–267
Elhissi A et al (2012) Carbon nanotubes in cancer therapy and drug delivery. J Drug Deliv 2012:837327
Enescu D et al (2019) Recent advances and challenges on applications of nanotechnology in food packaging. A literature review. Food Chem Toxicol 134:110814
Espitia P et al (2012) Zinc oxide nanoparticles: synthesis, antimicrobial activity and food packaging applications. Food Bioprocess Technol 5

Faraji M, Yamini Y, Rezaee M (2010) Magnetic nanoparticles: synthesis, stabilization, functionalization, characterization, and applications. J Iran Chem Soc 7:1–37
Farooq K et al (2024) Bio-nanocomposites: fundamentals and recent advances. In: Novel bio-nanocomposites for biomedical applications, pp 1–35
Fazio E et al (2021) Metal-oxide based nanomaterials: synthesis, characterization and their applications in electrical and electrochemical sensors. Sensors 21. https://doi.org/10.3390/s21072494
Feng Y et al (2023) The collagen-based scaffolds for bone regeneration: a journey through electrospun composites integrated with organic and inorganic additives. Processes 11(7):2105
Fura S et al (2024) Biomedical applications of bionanocomposites, chap 15. In: Sharma B et al (eds) Advances in bionanocomposites. Elsevier, pp 351–364
Gade JV, Singh A, Jain B (2023) Current industrial- and commercial-scale applications of biomolecules, chap 28. In: Verma C, Verma DK (eds) Handbook of biomolecules. Elsevier, pp 551–574
Gagliardi A et al (2021) Biodegradable polymeric nanoparticles for drug delivery to solid tumors. Front Pharmacol 12
Gamage A et al (2022) Applications of starch biopolymers for a sustainable modern agriculture. Sustainability 14:6085. https://doi.org/10.3390/su14106085
Geetha Bai R et al (2019) Graphene-based 3D scaffolds in tissue engineering: fabrication, applications, and future scope in liver tissue engineering. Int J Nanomed 14:5753–5783
Geißler D et al (2021) Analyzing the surface of functional nanomaterials—how to quantify the total and derivatizable number of functional groups and ligands. Microchim Acta 188(10):321
Ghanbarzadeh B, Almasi H (2013) Biodegradable polymers. In: Biodegradation-life of science, pp 141–186
Ghosal K et al (2024) Polyvinyl alcohol-based bionanocomposites: synthesis, properties, and applications, chap 5. In: Sharma B et al (eds) Advances in bionanocomposites. Elsevier, pp 117–132
Ghoshal G (2018) Recent trends in active, smart, and intelligent packaging for food products, chap 10. In: Grumezescu AM, Holban AM (eds) Food packaging and preservation. Academic Press, pp 343–374
Giaconia MA et al (2020) Overcoming restrictions of bioactive compounds biological effects in food using nanometer-sized structures. Food Hydrocolloids 107:105939
Giraud Guille MM et al (2010) Dense fibrillar collagen matrices for tissue repair. Soft Matter 6(20):4963–4967
Gómez-Estaca J et al (2011) Effects of gelatin origin, bovine-hide and tuna-skin, on the properties of compound gelatin–chitosan films. Food Hydrocolloids 25(6):1461–1469
Gopal G, Nirmala MJ, Mukherjee A (2023) A novel chitosan-coated Fe–Cu CNS loaded with CMC–alginate composite for adsorptive removal of ciprofloxacin from water. Surf Interfaces 39:102981
Gupta V, Gupta AR, Kant V (2013) Synthesis, characterization and biomedical application of nanoparticles. Sci Int 1(5):167–174
Han J et al (2018) Polymer-based nanomaterials and applications for vaccines and drugs. Polymers 10. https://doi.org/10.3390/polym10010031
Holder CF, Schaak RE (2019) Tutorial on powder X-ray diffraction for characterizing nanoscale materials. ACS Nano 13(7):7359–7365
Honary S, Zahir F (2013) Effect of zeta potential on the properties of nano-drug delivery systems—a review (part 2). Trop J Pharm Res 12
Huang J et al (2015) Bio-inspired synthesis of metal nanomaterials and applications. Chem Soc Rev 44(17):6330–6374
Huang X et al (2023) Advanced composite materials for structure strengthening and resilience improvement. Buildings 13. https://doi.org/10.3390/buildings13102406
Iderawumi A et al (2024) Applications of bionanocomposites in food packaging, pp 245–273

Ingale AG, Chaudhari A (2013) Biogenic synthesis of nanoparticles and potential applications: an eco-friendly approach. J Nanomed Nanotechol 4(165):1–7
Irshad MA et al (2021) Synthesis, characterization and advanced sustainable applications of titanium dioxide nanoparticles: a review. Ecotoxicol Environ Saf 212:111978
Isabelle V, Lan T (2009) Biodegradable polymers. Materials 2
Iurilli M et al (2025) Electrospun collagen-coated nanofiber membranes functionalized with silver nanoparticles for advanced wound healing applications. Membranes 15(2):39
Jayakumar A et al (2022) Recent innovations in bionanocomposites-based food packaging films—a comprehensive review. Food Packag Shelf Life 33:100877
Jebali A et al (2017) Nano-carbohydrates: synthesis and application in genetics, biotechnology, and medicine. Adv Coll Interface Sci 240:1–14
Jia Y et al (2024) Enhancement mechanisms of mechanical, electrical and thermal properties of carbon nanotube-copper composites: a review. J Mater Res Technol 32:1395–1415
Kaffashi B, Davoodi S, Oliaei E (2016) Poly(ε-caprolactone)/triclosan loaded polylactic acid nanoparticles composite: a long-term antibacterial bionanocomposite with sustained release. Int J Pharm 508(1–2):10–21
Kapil N, Mayani SV, Gopal Bhattacharyya K (2023) Environmental implications of nanoceramic applications. Results Chem 5:100724
Kargozar S, Mozafari M (2018) Nanotechnology and nanomedicine: start small, think big. Mater Today Proc 5(7, Part 3):15492–15500
Karmakar S (2019) Particle size distribution and zeta potential based on dynamic light scattering: techniques to characterise stability and surface distribution of charged colloids, pp 117–159
Karthik PS, Himaja A, Singh S (2014) Carbon-allotropes: synthesis methods, applications and future perspectives. Carbon Lett 15:219–237
Kenny C, Gehan Z (2023) Scenarios for future global growth to 2050. Center for Global Development, New York, London
Khan S, Khan S, Asiri AM (2019) Scanning electron microscopy: principle and applications in nanomaterials characterization
Khoswan I et al (2024) Why carbon nanotubes improve aqueous nanofluid thermal conductivity: a qualitative model critical review. Processes 12. https://doi.org/10.3390/pr12040834
Kishore A, Mithul Aravind S, Singh A (2023) Bionanocomposites for active and smart food packaging: a review on its application, safety, and health aspects. J Food Process Eng 46(5):e14320
Kong X (2020) Fine structure of amylose and amylopectin, pp 29–39
Kong M et al (2010) Antimicrobial properties of chitosan and mode of action: a state of the art review. Int J Food Microbiol 144(1):51–63
Korin E, Froumin N, Cohen S (2017) Surface analysis of nanocomplexes by X-ray photoelectron spectroscopy (XPS). ACS Biomater Sci Eng 3(6):882–889
Kowalczyk B, Lagzi I, Grzybowski BA (2011) Nanoseparations: strategies for size and/or shape-selective purification of nanoparticles. Curr Opin Colloid Interface Sci 16(2):135–148
Kumar M et al (2022) XRD analysis for characterization of green nanoparticles: a mini review. GJPPS 10
Kumari R et al (2023) Regulation and safety measures for nanotechnology-based agri-products. Front Genome Ed 5
Künkel A et al (2016) Polymers, biodegradable. In: Ullmann's encyclopedia of industrial chemistry, pp 1–29
Le N-MN et al (2022) Polyaminated pullulan, a new biodegradable and cationic pullulan derivative for mucosal drug delivery. Carbohydr Polym 282:119143
Lecoublet M et al (2024) Sustainable 3D-printed cellulose-based biocomposites and bio-nano-composites: analysis of dielectric performances. Ind Crops Prod 221:119332
Lee KY, Mooney DJ (2012) Alginate: properties and biomedical applications. Prog Polym Sci 37(1):106–126

Lenz RW, Marchessault RH (2005) Bacterial polyesters: biosynthesis, biodegradable plastics and biotechnology. Biomacromolecules 6(1):1–8

Li H et al (2012) Carbon nanodots: synthesis, properties and applications. J Mater Chem 22(46):24230–24253

Li S et al (2016a) Molecular modification of polysaccharides and resulting bioactivities. Compr Rev Food Sci Food Saf 15(2):237–250

Li X et al (2016b) Graphene in photocatalysis: a review. Small 12(48):6640–6696

Li G, Xu J, Xu K (2023) Physiological functions of carbon dots and their applications in agriculture: a review. Nanomaterials 13. https://doi.org/10.3390/nano13192684

Li H et al (2024) Functionalization of carbon nanotubes in polystyrene and properties of their composites: a review. Polymers 16. https://doi.org/10.3390/polym16060770

Liu L (2006) Bioplastics in food packaging: innovative technologies for biodegradable packaging. San Jose State Univ Packag Eng 13:1348–1368

Liu M et al (2014) Recent advance in research on halloysite nanotubes-polymer nanocomposite. Prog Polym Sci 39(8):1498–1525

Liu J, Li R, Yang B (2020) Carbon dots: a new type of carbon-based nanomaterial with wide applications. ACS Cent Sci 6(12):2179–2195

Liu K et al (2021) Recent advances in cellulose and its derivatives for oilfield applications. Carbohydr Polym 259:117740

Lombardi L, Li J, Williams DR (2024) Peptide-based biomaterials for combatting infections and improving drug delivery. Pharmaceutics 16. https://doi.org/10.3390/pharmaceutics16111468

Lone IH, Kowsalya E, Rebecca LJ (2016) Alginate fiber from brown algae. Pharm Lett 8(8):68–71

Luckachan GE, Pillai CKS (2011) Biodegradable polymers—a review on recent trends and emerging perspectives. J Polym Environ 19(3):637–676

Ma H et al (2021) Trimetallic PtNiCo branched nanocages as efficient and durable bifunctional electrocatalysts towards oxygen reduction and methanol oxidation reactions. J Mater Chem A 9

Magalhães M, Almeida A (2023) Nature-inspired cellulose-based active materials: from 2D to 4D. Appl Biosci 2:94–114

Maholiya A et al (2023) An insight into the role of carbon dots in the agriculture system: a review. Environ Sci Nano 10(4):959–995

Maltsev VA et al (1994) Cardiomyocytes differentiated in vitro from embryonic stem cells developmentally express cardiac-specific genes and ionic currents. Circ Res 75(2):233–244

Mangaraj S et al (2019) Application of biodegradable polymers in food packaging industry: a comprehensive review. J Packag Technol Res 3(1):77–96

Mehdinia M et al (2024) Developing bio-nano composites using cellulose-nanofiber-reinforced epoxy. J Compos Sci 8(7):250

Mehta VN et al (2021) Recent developments on fluorescent hybrid nanomaterials for metal ions sensing and bioimaging applications: a review. J Mol Liq 333:115950

Mekuye B, Abera B (2023) Nanomaterials: an overview of synthesis, classification, characterization, and applications. Nano Select 4(8):486–501

Melini V, Melini F (2018) Strategies to extend bread and gf bread shelf-life: from sourdough to antimicrobial active packaging and nanotechnology. Fermentation 4. https://doi.org/10.3390/fermentation4010009

Metak AM, Nabhani F, Connolly SN (2015) Migration of engineered nanoparticles from packaging into food products. LWT Food Sci Technol 64(2):781–787

Mhd Haniffa MA et al (2016) Review of bionanocomposite coating films and their applications. Polymers 8. https://doi.org/10.3390/polym8070246

Middleton JC, Tipton AJ (2000) Synthetic biodegradable polymers as orthopedic devices. Biomaterials 21(23):2335–2346

Mirmusavi MH, Ahmadian M, Karbasi S (2022) Polycaprolactone-chitosan/multi-walled carbon nanotube: a highly strengthened electrospun nanocomposite scaffold for cartilage tissue engineering. Int J Biol Macromol 209:1801–1814

Mišić Radić T et al (2023) Insights into the morphology and surface properties of microalgae at the nanoscale by atomic force microscopy (AFM): a review. Water 15. https://doi.org/10.3390/w15111983

Molakkalu Padre S et al (2024) Polyol-assisted synthesis of Ni/Cu/Ag trimetallic nanoparticles for nonlinear optical applications. ACS Omega 9(47):46773–46783

Moosa AA, Abed MS (2021) Graphene preparation and graphite exfoliation. Turk J Chem 45(3):493–519

Moradi S et al (2021) Chitosan/graphene oxide composite films and their biomedical and drug delivery applications: a review. Appl Sci 11(17):7776

More S et al (2021) Guidance on risk assessment of nanomaterials to be applied in the food and feed chain: human and animal health. EFSA J 19(8):e06768

Moreno-Vega A-I et al (2012) Polymeric and ceramic nanoparticles in biomedical applications. J Nanotechnol 2012(1):936041

Morris J et al (2011) Science policy considerations for responsible nanotechnology decisions. Nat Nanotechnol 6(2):73–77

Mourdikoudis S, Pallares RM, Thanh NTK (2018) Characterization techniques for nanoparticles: comparison and complementarity upon studying nanoparticle properties. Nanoscale 10(27):12871–12934

Mousa MH, Dong Y, Davies IJ (2016) Recent advances in bionanocomposites: preparation, properties, and applications. Int J Polym Mater Polym Biomater 65(5):225–254

Mu Q et al (2014) Chemical basis of interactions between engineered nanoparticles and biological systems. Chem Rev 114(15):7740–7781

MubarakAli D et al (2011) Biosynthesis and characterization of silver nanoparticles using marine cyanobacterium, *Oscillatoria willei* NTDM01. Dig J Nanomater Biostruct 6:385–390

Mujtaba M et al (2022) Trends and challenges in the development of bio-based barrier coating materials for paper/cardboard food packaging; a review. Sci Total Environ 851:158328

Müller DB et al (2013) Carbon emissions of infrastructure development. Environ Sci Technol 47(20):11739–11746

Müller R, Kuchinka J, Heinze T (2022) Studies about the design of magnetic bionanocomposite. Phys Sci Rev 7(12):1397–1420

Mumtaz S et al (2023) Bionanocomposites as a new platform for drug delivery systems, chap 13. In: Das S, Thomas S, Das PP (eds) Novel platforms for drug delivery applications. Woodhead Publishing, pp 289–315

Munafò P, Goffredo GB, Quagliarini E (2015) TiO_2-based nanocoatings for preserving architectural stone surfaces: an overview. Constr Build Mater 84

Naiim M et al (2015) Multiangle dynamic light scattering for the improvement of multimodal particle size distribution measurements. Soft Matter 11(1):28–32

Nandhini J et al (2024a) Advancements in metal oxide bio-nanocomposites for sustainable food packaging: fabrication, applications, and future prospectives. Food Bioeng 2024

Nandhini J, Karthikeyan E, Rajeshkumar S (2024b) Eco-friendly bio-nanocomposites: pioneering sustainable biomedical advancements in engineering. Discov Nano 19(1):86

Nandhini J et al (2024c) Pitolisant nanofibers: a promising frontier in drug delivery for narcolepsy—formulation, optimisation, and characterization insights. Biomed Technol 6:9–16

Naseem R et al (2021) Strategies for enhancing polyester-based materials for bone fixation applications. Molecules 26. https://doi.org/10.3390/molecules26040992

Nasir Z, Fatima A, Ali RM (2023) Chemistry and applications of nano composites: an update about recent developments. Int J Technol Educ Res 1(04):14–63

Niemeyer C (2001) Nanoparticles, proteins, and nucleic acids: biotechnology meets materials science. Angew Chem Int Ed 40:4128–4158

Nishino T, Matsuda I, Hirao K (2004) All-cellulose composite. Macromolecules 37(20):7683–7687

Nwokediegwu ZQS et al (2024) Advanced materials for sustainable construction: a review of innovations and environmental benefits. Eng Sci Technol J 5(1):201–218

Ocak Y et al (2014) Sustainable bio-nano composite coatings for the protection of marble surfaces. J Cult Herit 16
Olobayo OK (2019) Starch source and its impact on pharmaceutical applications, chap 3. In: Martins E (ed) Chemical properties of starch. IntechOpen, Rijeka
O'Sullivan JN (2023) Demographic delusions: world population growth is exceeding most projections and jeopardising scenarios for sustainable futures. World 4(3):545–568
Paidari S et al (2023) Bio-nanocomposites and their potential applications in physiochemical properties of cheese: an updated review. J Food Meas Character 17(3):2595–2606
Pal SL et al (2011) Nanoparticle: an overview of preparation and characterization. J Appl Pharm Sci 228–234
Palma E et al (2018) Antileishmanial activity of amphotericin B-loaded-PLGA nanoparticles: an overview. Materials 11. https://doi.org/10.3390/ma11071167
Pan C-J, Sarma LS, Hwang B-J (2016) Formation and characterization of bimetallic nanoparticles in electrochemistry. In: Aliofkhazraei M, Makhlouf ASH (eds) Handbook of nanoelectrochemistry: electrochemical synthesis methods, properties and characterization techniques. Springer International Publishing, Cham, pp 1–60
Pan X et al (2024) Harnessing nanomaterials for enhanced DNA-based biosensing and therapeutic performance. ChemBioChem e202400936
Parenteau-Bareil R, Gauvin R, Berthod F (2010) Collagen-based biomaterials for tissue engineering applications. Materials 3:1863–1887. https://doi.org/10.3390/ma3031863
Pathak K, Prasad R, Selvaraj K (2011) Biogenic synthesis of silver nanoparticles using *Nicotiana tabacum* leaf extract and study of their antibacterial effect. Afr J Biotechnol 10:8122–8130
Patois E et al (2012) Evaluation of nanoparticle tracking analysis (NTA) in the characterization of therapeutic antibodies and seasonal influenza vaccines: pros and cons. J Drug Deliv Sci Technol 22:427–433
Patra JK, Baek K-H (2014) Green nanobiotechnology: factors affecting synthesis and characterization techniques. J Nanomater 2014(1):417305
Pellis A, Guebitz GM, Nyanhongo GS (2022) Chitosan: sources, processing and modification techniques. Gels 8(7):393
Pires JRA et al (2023) Current applications of bionanocomposites in food processing and packaging. Polymers 15(10). https://doi.org/10.3390/polym15102336
Pokrajac L et al (2021) Nanotechnology for a sustainable future: addressing global challenges with the international network4sustainable nanotechnology. ACS Nano 15(12):18608–18623
Postnova I, Shchipunov Y (2022) Tannic acid as a versatile template for silica monoliths engineering with catalytic gold and silver nanoparticles. Nanomaterials 12. https://doi.org/10.3390/nano12234320
Pridgen EM, Alexis F, Farokhzad OC (2014) Polymeric nanoparticle technologies for oral drug delivery. Clin Gastroenterol Hepatol 12(10):1605–1610
Puiggalí J, Katsarava R (2017) Bionanocomposites, chap 7. In: Jlassi K, Chehimi MM, Thomas S (eds) Clay-polymer nanocomposites. Elsevier, pp 239–272
Qi R et al (2010) Electrospun poly(lactic-co-glycolic acid)/halloysite nanotube composite nanofibers for drug encapsulation and sustained release. J Mater Chem 20(47):10622–10629
Qi R et al (2012) Biocompatibility of electrospun halloysite nanotube-doped poly(lactic-co-glycolic acid) composite nanofibers. J Biomater Sci Polym Ed 23(1–4):299–313
Rahman G et al (2019) An overview of the recent progress in the synthesis and applications of carbon nanotubes. C 5. https://doi.org/10.3390/c5010003
Rai M et al (2021) Biogenic silver nanoparticles: what we know and what do we need to know? Nanomaterials 11. https://doi.org/10.3390/nano11112901
Rajamanickam U et al (2012) Biosynthesis of zinc nanoparticles using actinomycetes for antibacterial food packaging. In: International conference on nutrition and food sciences IPCBEE
Rambaran T, Schirhagl R (2022) Nanotechnology from lab to industry—a look at current trends. Nanoscale Adv 4(18):3664–3675

Ranakoti L et al (2022) Critical review on polylactic acid: properties, structure, processing, biocomposites, and nanocomposites. Materials 15. https://doi.org/10.3390/ma15124312
Rao J, Geckeler K (2011) Polymer nanoparticles: preparation techniques and size-control parameters. Prog Polym Sci. In Press, Accepted Manuscript
Ravi Kumar MNV (2000) A review of chitin and chitosan applications. React Funct Polym 46(1):1–27
Reddy CSK et al (2003) Polyhydroxyalkanoates: an overview. Bioresour Technol 87(2):137–146
Riccio BVF et al (2021) Chitosan/nanocellulose-based bionanocomposite films for controlled betamethasone and silver sulfadiazine delivery. J Appl Polym Sci 138(21):50468
Richardson SD, Ternes TA (2018) Water analysis: emerging contaminants and current issues. Anal Chem 90(1):398–428
Rodriguez-Narvaez OM et al (2017) Treatment technologies for emerging contaminants in water: a review. Chem Eng J 323:361–380
Roy A et al (2023) Bio-fabrication of trimetallic nanoparticles and their applications. Catalysts 13. https://doi.org/10.3390/catal13020321
Saha S, Bansal S, Khanuja M (2022) Classification of nanomaterials and their physical and chemical nature, chap 2. In: Ghorbanpour M, Shahid MA (eds) Nano-enabled agrochemicals in agriculture. Academic Press, pp 7–34
Saini R, Bajpai AK, Jain E (2017) Fundamentals of bionanocomposites
Sajid M (2021) Nanomaterials: types, properties, synthesis, emerging materials, and toxicity concerns. Curr Opin Environ Sci Health 25:100319
Sajid M (2022) Nanomaterials: types, properties, recent advances, and toxicity concerns. Curr Opin Environ Sci Health 25:100319
Samanta AP et al (2022) Crosslinked nanocollagen-cellulose nanofibrils reinforced electrospun polyvinyl alcohol/methylcellulose/polyethylene glycol bionanocomposites: study of material properties and sustained release of ketorolac tromethamine. Carbohydr Polym Technol Appl 3:100195
Samir A et al (2022) Recent advances in biodegradable polymers for sustainable applications. npj Mater Degrad 6(1):68
Sanchez F, Sobolev K (2010) Nanotechnology in concrete—a review. Constr Build Mater 24(11):2060–2071
Satdive A et al (2022) Bionanocomposites in the construction and building applications. In: Muthukumar C et al (eds) Polymer based bio-nanocomposites: properties, durability and applications. Springer Singapore, Singapore, pp 293–310
Scimeca M et al (2018) Energy dispersive X-ray (EDX) microanalysis: a powerful tool in biomedical research and diagnosis. Eur J Histochem 62(1):2841
Sen A et al (2024) Bio-nanocomposites: innovative solutions for addressing issues in health, agriculture, energy and environmental domains. Nano-Struct Nano-Obj 39:101270
Shah T, Vasava D (2019) A glimpse of biodegradable polymers and their biomedical applications. e-Polymers 19:385–410
Shao XR et al (2015) Independent effect of polymeric nanoparticle zeta potential/surface charge, on their cytotoxicity and affinity to cells. Cell Prolif 48(4):465–474
Sharma G et al (2017) Revolution from monometallic to trimetallic nanoparticle composites, various synthesis methods and their applications: a review. Mater Sci Eng C 71:1216–1230
Sharma G et al (2019) Novel development of nanoparticles to bimetallic nanoparticles and their composites: a review. J King Saud Univ Sci 31(2):257–269
Sharma S, Malik A, Gupta P (2021) Bionanocomposites in tissue engineering and regenerative medicine, chap 27. In: Ahmed S, Tomer A (eds) Bionanocomposites in tissue engineering and regenerative medicine. Woodhead Publishing, pp 507–532
Sharma B et al (2023a) Advances in bionanocomposites: materials, applications, and life cycle. Elsevier
Sharma J, Gole VL, Siddiqui MA (2023b) Bionanocomposites: biological aspects and biomedical applications. Polymer nanocomposites. CRC Press, pp 149–171

Shchipunov Y et al (2010) Hydrogels formed through regulated self-organization of gradually charging chitosan in solution of xanthan. Green Chem 12(7):1187–1195

Shen J et al (2012) Graphene quantum dots: emergent nanolights for bioimaging, sensors, catalysis and photovoltaic devices. Chem Commun 48(31):3686–3699

Shen X et al (2017) Luminescent/magnetic PLGA-based hybrid nanocomposites: a smart nanocarrier system for targeted codelivery and dual-modality imaging in cancer theranostics. Int J Nanomed 12:4299–4322

Shishir MRI et al (2018) Advances in micro and nano-encapsulation of bioactive compounds using biopolymer and lipid-based transporters. Trends Food Sci Technol 78:34–60

Shukla K (2012) Nanotechnology and emerging trends in dairy foods: the inside story to food additives and ingredients. Int J Nanosci Nanotechnol 1(1):41–58

Singh D et al (2016) Ceramic nanoparticles: recompense, cellular uptake and toxicity concerns. Artif Cells Nanomed Biotechnol 44(1):401–409

Singh YP et al (2020) Optimization of electrospinning process & parameters for producing defect-free chitosan/polyethylene oxide nanofibers for bone tissue engineering. J Biomater Sci Polym Ed 31(6):781–803

Singh P et al (2023) Assessing the biomass-based carbon dots and their composites for photocatalytic treatment of wastewater. J Clean Prod 413:137474

Singh S, Bala N, Kaur A (2024) Green sustainable bionanocomposites materials: short review. J Manag 1(3):98–106

Sneha M et al (2016) Biopolymers—application in nanoscience and nanotechnology, chap 3. In: Farzana Khan P (ed) Recent advances in biopolymers. IntechOpen, Rijeka

Soltani AM, Pouypouy H (2019) Standardization and regulations of nanotechnology and recent government policies across the world on nanomaterials, chap 19. In: Ghorbanpour M, Wani SH (eds) Advances in phytonanotechnology. Academic Press, pp 419–446

Song X et al (2013) Electrospun hydroxyapatite grafted poly(L-lactide)/poly(lactic-co-glycolic acid) nanofibers for guided bone regeneration membrane. Compos Sci Technol 79:8–14

Sotome S et al (2004) Synthesis and in vivo evaluation of a novel hydroxyapatite/collagen–alginate as a bone filler and a drug delivery carrier of bone morphogenetic protein. Mater Sci Eng C 24:341–347

Souza AOD (2023) Overview of nanomaterials and cellular interactions. Biointerface Res Appl Chem 13(4)

Souza VGL, Fernando AL (2016) Nanoparticles in food packaging: biodegradability and potential migration to food—a review. Food Packag Shelf Life 8:63–70

Souza VGL et al (2019) Activity of chitosan-montmorillonite bionanocomposites incorporated with rosemary essential oil: from in vitro assays to application in fresh poultry meat. Food Hydrocolloids 89:241–252

Souza VG et al (2020) Eco-friendly ZnO/chitosan bionanocomposites films for packaging of fresh poultry meat. Coatings 10. https://doi.org/10.3390/coatings10020110

Soykeabkaew N, Thanomsilp C, Suwantong O (2015) A review: starch-based composite foams. Compos A Appl Sci Manuf 78:246–263

Stloukal P et al (2015) The influence of a hydrolysis-inhibiting additive on the degradation and biodegradation of PLA and its nanocomposites. Polym Test 41:124–132

Sulania I, Yadav R, Karn R (2018) Atomic and magnetic force studies of co thin films and nanoparticles: understanding the surface correlation using fractal studies, pp 263–291

Sur UK (2012) Graphene: a rising star on the horizon of materials science. Int J Electrochem 2012(1):237689

Surendran G, Sherje AP (2022) Cellulose nanofibers and composites: an insight into basics and biomedical applications. J Drug Deliv Sci Technol 75:103601

Szczyglewska P, Feliczak-Guzik A, Nowak I (2023) Nanotechnology–general aspects: a chemical reduction approach to the synthesis of nanoparticles. Molecules 28. https://doi.org/10.3390/molecules28134932

Tarcan R et al (2020) Reduced graphene oxide today. J Mater Chem C 8(4):1198–1224

Tawiah B, Ofori EA, George SC (2024) Nanotechnology in societal development. In: George SC, Tawiah B (eds) Nanotechnology in societal development. Springer Nature Singapore, Singapore, pp 1–64
Thomas SC et al (2015) Ceramic nanoparticles: fabrication methods and applications in drug delivery. Curr Pharm Des 21(42):6165–6188
Tian K, Bilal M (2020) Research progress of biodegradable materials in reducing environmental pollution, chap 15. In: Singh P, Kumar A, Borthakur A (eds) Abatement of environmental pollutants. Elsevier, pp 313–330
Tian X, Chong Y, Ge C (2020) Understanding the nano–bio interactions and the corresponding biological responses. Front Chem 8
Tomer A et al (2024) Biopolymeric nanocomposites for wastewater remediation: an overview on recent progress and challenges. Polymers 16. https://doi.org/10.3390/polym16020294
Tripathy DB, Gupta A (2024) Nanocomposites as sustainable smart materials: a review. J Reinf Plast Compos 07316844241233162
Trivedi AK, Gupta MK, Singh H (2023) PLA based biocomposites for sustainable products: a review. Adv Ind Eng Polym Res 6(4):382–395
Trotta F et al (2023) Silver bionanocomposites as active food packaging: recent advances & future trends tackling the food waste crisis. Polymers 15. https://doi.org/10.3390/polym15214243
Varghese LR, Das N (2015) Application of nano-biocomposites for remediation of heavy metals from aqueous environment: an overview. Int J ChemTech Res 8:566–571
Viacava G, Marcovich N, Ansorena M (2024) Studies on nanostructure functionalized materials for active and intelligent food packaging, pp 477–511
Walczyk D et al (2010) What the cell "sees" in bionanoscience. J Am Chem Soc 132(16):5761–5768
Wang D et al (2019) Next-generation multifunctional carbon-metal nanohybrids for energy and environmental applications. Environ Sci Technol 53(13):7265–7287
Wang D-X et al (2021) DNA nanostructure-based nucleic acid probes: construction and biological applications. Chem Sci 12(22):7602–7622
Wani SD, Mundada AS (2021) A review: emerging trends in bionanocomposites. Int J Pharm Res Technol (IJPRT) 11(1):1–8
Wani SD, Mundada AS (2023) A review: emerging trends in bionanocomposites. Int J Pharm Res Technol (IJPRT) 11(1):1–8
Wazed Ali S, Bairagi S (2021) Chitosan-based bionanocomposites in wound healing, chap 7. In: Ahmed S, Tomer A (eds) Bionanocomposites in tissue engineering and regenerative medicine. Woodhead Publishing, pp 149–168
Wei Y et al (2021) Induced autophagy of macrophages and the regulation of inflammatory effects by perovskite nanomaterial $LaNiO_3$. Front Immunol 12
Wu S et al (2010) Effects of poly(vinyl alcohol) (PVA) content on preparation of novel thiol-functionalized mesoporous PVA/SiO_2 composite nanofiber membranes and their application for adsorption of heavy metal ions from aqueous solution. Polymer 51(26):6203–6211
Wu C et al (2023) Cellulose dissolution, modification, and the derived hydrogel: a review. ChemSusChem 16(21):e202300518
Xie F et al (2013) Starch-based nano-biocomposites. Prog Polym Sci 38:1590–1628
Xin Jiat L et al (2018) Review on graphene and its derivatives: synthesis methods and potential industrial implementation. J Taiwan Inst Chem Eng 98
Xu L et al (2015) Chitosan in molecularly-imprinted polymers: current and future prospects. Int J Mol Sci 16:18328–18347. https://doi.org/10.3390/ijms160818328
Yadav N et al (2023) Towards self-reinforced PLA composites for fused filament fabrication. Appl Sci 13(4):2637
Yaqoob AA et al (2020) Recent advances in metal decorated nanomaterials and their various biological applications: a review. Front Chem 8
Yasri S, Wiwanitkit V (2017) Important ethical issues for nanomedicine. J Res Med Sci 22(1)

Yilmaz Atay H (2019) Antibacterial activity of chitosan-based systems. In: Jana S, Jana S (eds) Functional chitosan: drug delivery and biomedical applications. Springer Singapore, Singapore, pp 457–489
Yin G-Z, Yang X-M (2020) Biodegradable polymers: a cure for the planet, but a long way to go. J Polym Res 27
Yoosefi S et al (2022) Dual drug delivery system based on layered double hydroxides/carboxymethyl cellulose-poly ethylene oxide bionanocomposite electrospun fibrous mats: fabrication, characterization, in-vitro and in-vivo studies. Int J Biol Macromol 222:3142–3154
Youssef AM, El-Sayed SM (2018) Bionanocomposites materials for food packaging applications: concepts and future outlook. Carbohydr Polym 193:19–27
Yuan X et al (2023) Application of nanomaterials in the production of biomolecules in microalgae: a review. Mar Drugs 21. https://doi.org/10.3390/md21110594
Zakaria NS et al (2013) Imaging of colloidal gold nanoparticle using atomic force microscope. Nano Hybrids 4:47–60
Zaleska-Medynska A et al (2016) Noble metal-based bimetallic nanoparticles: the effect of the structure on the optical, catalytic and photocatalytic properties. Adv Coll Interface Sci 229:80–107
Zankana MM, Al-Dalawy SM, Barzinjy AA (2023) Synthesis and characterization of bionanocomposites: functionalization of graphene oxide with a biocompatible amino acid. Hybrid Adv 3:100070
Zhao H et al (2022) Engineering nano-bio interfaces from nanomaterials to nanomedicines. Acc Mater Res 3(8):812–829
Zhong Y et al (2020) Biodegradable polymers and green-based antimicrobial packaging materials: a mini-review. Adv Ind Eng Polym Res 3(1):27–35
Zhou WY et al (2010) Poly(vinyl alcohol)/halloysite nanotubes bionanocomposite films: properties and in vitro osteoblasts and fibroblasts response. J Biomed Mater Res A 93(4):1574–1587
Zia MA et al (2020) Bionanocomposites: uses in applied sciences and their benefits, chap 21. In: Mahmood Zia K et al (eds) Bionanocomposites. Elsevier, pp 535–545
Zielińska A et al (2020) Polymeric nanoparticles: production, characterization, toxicology and ecotoxicology. Molecules 25(16)

Chapter 2
Biomaterials and Their Fabrication for Advanced Bionanocomposites: A Sustainable Approach

Vernica Verma, Priya Gupta, Peramjeet Singh, and Narendra Kumar Pandey

Abstract The escalating interest in bio-based nanocomposites and innovative process technologies has spurred the study of bionanocomposites—promising advanced materials that could lead us away from fossil-fuel dependence toward a more sustainable material future. Bionanocomposites, comprising biological components in their nano-dimension, hold remarkable potential in diverse applications ranging from artificial tissue culture to drug delivery systems. The increasing demand for sustainable materials has prompted significant advancements in the field of biomaterials and their integration into bionanocomposites. Various fabrication techniques, such as solution mixing, electrospinning, and 3D printing, have enabled the development of bionanocomposites with tailored properties. By incorporating nanoscale agents, such as nanoparticles and nanofibers, the mechanical, thermal, and barrier properties of biomaterials can be significantly enhanced, paving the way for high-performance materials. This article explores the latest developments in biomaterials, focusing on their fabrication methods and applications for creating advanced bionanocomposites that align with sustainable practices, emphasizing the reduction of reliance on fossil fuels, and the promotion of biodegradability. The integration of biomaterials and nanotechnology offers a promising pathway toward the development of sustainable materials, addressing critical challenges in health, energy, and environmental sectors, underscoring the importance of continued research and innovation in biomaterials' fabrication contributing to a more sustainable future. Conclusively, the creation of sustainable materials becomes feasible when biomaterials and nanoscience are combined. These materials are critically invaluable to the health, energy, and environment sectors. The ongoing extensive research on biomaterial fabrication can provide for an environment-friendly and imperishable future.

V. Verma · P. Singh
Department of Physics, University of Lucknow, Lucknow, Uttar Pradesh, India
e-mail: vernicaverma20@gmail.com

P. Gupta · N. K. Pandey (✉)
Manyawar Kanshiram Government Polytechnic, Tirwa, Kannauj, Uttar Pradesh, India
e-mail: profnarendrapandey137@gmail.com

S. C. George and B. Tawiah (eds.), *Advanced Bionanocomposite Materials*, Advanced Structured Materials 241, https://doi.org/10.1007/978-981-96-8225-6_2

Keywords Biomaterials · Nanotechnology · Bionanocomposites · Fabrication techniques

2.1 Introduction

Biomaterials are referred to as biologically derived or synthesized materials that are designed to efficiently interact with biological systems owing to their biomimetic and microstructural compatibility with their biological counterpart. They comprise a revolutionizing group of materials that numerous scientific and industrial fields are trying to develop through sustainable bioengineering. Biomaterials are pivotal in addressing challenges in healthcare, environmental sustainability, and advanced materials science as substances specifically engineered to interact seamlessly with biological systems (Kerman et al. 2008). Biomaterials, whether natural or synthetically engineered, are tailored to perform specific functions such as supporting tissue regeneration, delivering drugs with precision, or serving as integral components in bionanocomposites. These materials are indispensable in the pursuit of sustainable and innovative solutions for concerning global challenges due to their adaptability and multifunctionality (Adekoya et al. 2023). Biomaterials prove crucial for sustainability and advanced medical technologies, modern scientific efforts increasingly focus on developing materials that not only encounter functional and mechanical demands but also align with environmental goals. This is particularly relevant in the realm of bionanocomposites, where the integration of nanoscale features into biomaterials enables unparalleled achievement in performance, biocompatibility, and eco-friendliness. These advanced composites are useful for innovations in regenerative medicine, biodegradable implants, electronics, automotive, a food packaging, and environmentally sustainable industrial applications (El Mir et al. 2007). Biomaterials are important for such advancements, providing the structural and functional foundation upon which novel bionanocomposites are designed and materialized. By surveying their diverse types and properties, researchers can harness their full potential to create materials that are not only efficient but also sustainable.

The biomaterials extracted from natural sources have become an important component in the fabrication of bionanocomposites that is a viable way to generate sustainable materials with improved qualities. Polysaccharides, including chitosan, cellulose, starch, and chitin are widely used biomaterials that are biodegradable, non-toxic, and have high mechanical qualities employing their uses in packaging, biomedical applications, and textiles. Animal proteins, such as collagen, hydrogel, gelatin, and keratin have excellent mechanical properties, biocompatibility, and biodegradability, making them useful in biomedical, pharmaceutical, and cosmetic applications (Hedenqvist and Backman 2006). Triglycerides and waxes are examples of lipids with strong barrier properties, biocompatibility, and biodegradability, making them useful in packaging, biomedical, and pharmaceutical applications. Nucleic acids, i.e., DNA and RNA derived from genetic material, exhibit high biocompatibility, biodegradability, and potential for gene transfer, with applications in the biomedical, pharmacological, and gene delivery fields. These biomaterials integrated into bionanocomposites have accelerated the development of superior materials for applications in the packaging, biomedical, textile, and pharmaceutical industries, offering a promising approach to address environmental and sustainability concerns.

Biomaterials are essential element in the development of bionanocomposites and the potential production of sustainable materials with improved properties. With the rising demand for eco-friendly materials, natural biopolymers are poised to play a critical role in revolutionizing the way we live and interact with the environment. These natural biopolymers find wide applications in various industries, including packaging, medicine, agriculture, and food, offering a sustainable solution to reduce environmental pollution. Proteins and peptides are versatile biomolecules that possess great potential for biomedical applications, including tissue regeneration, wound healing, biomedical textiles as well as cosmetic applications. Natural proteins including silk fibroin, collagen, and keratin have shown promise in wound healing as they promote tissue regeneration by cell proliferation, migration, and differentiation. These proteins and peptides rigorously used on industrial scale in cosmetic applications in nail, skin, and hair care products as they are potent in promoting skin hydration, elasticity, and overall well-being. Additionally, these biomolecules offer excellent mechanical properties, biocompatibility, non-toxicity, and sustainability and can be incorporated into biomedical textiles such as sutures, implants, and scaffolds for tissue engineering. Moreover, with escalating research ventures, proteins and peptides are expected to play a central role in the development of innovative biomaterials in biomedical technologies, thereby enhancing human health and well-being.

Lipids constitute another class of diverse biomolecules that are revolutionizing the field of drug delivery and tissue engineering by offering a platform for inventive and therapeutic strategies. Certain lipids including solid lipid nanoparticles, liposomes, and lipid-polymer hybrids, are extensively explored for their potential in gene therapies and delivering therapeutics and vaccines. Lipids are also used as drug delivery vehicles for targeted and effective treatments owing to their mechanical flexibility, biocompatibility, and biodegradability. Lipids support cell growth by creating biomimetic scaffolds that mimic natural tissues and are thus seamlessly employed in tissue engineering (Alessandrini and Facci 2011). The potential of lipids and lipid-based materials in drug delivery and tissue engineering is extensive. Despite the advantages of lipid-based materials, challenges such as stability, scalability, and regulatory approval need to be resolved.

A separate spectrum of biomaterials involves bioceramics and bioglasses that have revolutionized the field of bone tissue engineering and implants, offering a unique combination of biocompatibility, bioactivity, and exceptional mechanical strength. These biomaterials promote cell growth and tissue regeneration as they are engineered to mimic the structure and function of natural bone. As bone grafts, they can repair or replace damaged or diseased bone tissue and have transformed the field of orthopedics and dentistry through their usage in implants for hip and knee replacements, spinal implants, and dental implants. Furthermore, bioceramics and bioglasses are being used as tissue engineering scaffolds, providing a three-dimensional framework for cell growth and tissue regeneration. With ongoing research focused on enhancing their mechanical properties, scalability, and regulatory approval, bioceramics and bioglasses are poised to play a significant role in the development of innovative bone repair and regeneration strategies, improving the lives of millions of people worldwide suffering from bone-related disorders and diseases.

The recent advancements in bioceramics and bioglasses have opened a transformative era in bone tissue engineering and implants, offering an unparalleled combination of biocompatibility, bioactivity, and mechanical properties. These materials demonstrate excellent mechanical properties, such as high compressive strength and stiffness, making them suitable for load-bearing applications. Bioceramics and bioglasses exhibit a unique blend of properties, including non-toxicity, biocompatibility, and bioactivity, which enables them to promote bone growth and regeneration by providing a scaffold for cell attachment and proliferation. Despite the challenges associated with scalability, mechanical properties, and regulatory approval, ongoing research is poised to overcome these hurdles, unlocking the full potential of bioceramics and bioglasses in the development of innovative bone repair and regeneration strategies, and transforming the lives of millions of people worldwide suffering from bone-related disorders and diseases.

2.1.1 Importance of Biomaterials in Modern Science and Engineering

Biomaterials have transformed several scientific and industrial disciplines by bridging the gap between biology and engineering. These compounds are designed to work seamlessly with biological systems. These materials, whether natural or synthetically created, are designed to serve specific roles such as promoting tissue regeneration, precisely delivering medications, or acting as integral components in bionanocomposites (Mohan et al. 2016). Biomaterials have an important role in tackling problems in healthcare, environmental sustainability, and advanced materials science. Their adaptability and multifunctionality make them crucial in the quest of innovative and long-term solutions to major global issues. Biomaterials offer advantages such as biocompatibility, remodeling, good adhesion, high absorption, permeability, and biodegradability that are highly beneficial for modern science, engineering, and technology driving innovation across a wide spectrum of applications. Their ability to interface with biological systems has opened new avenues in healthcare, enabling breakthroughs in areas such as regenerative medicine, prosthetics, and drug delivery systems. For instance, the use of biocompatible materials in implants and artificial organs has greatly improved patient outcomes, while enhanced drug delivery systems that incorporate biomaterials enable regulated and targeted therapy, reducing adverse effects and increasing efficacy (Macha et al. 2017). Biomaterials are crucial in engineering owing to their contribution to sustainable and high-performance materials. Moreover, biomaterials are essential to bionanocomposites, where their inclusion at the nanoscale results in materials with enhanced mechanical, thermal, and biological properties. This combination between biomaterials and nanotechnology allows for the fabrication of multifunctional composites that are lightweight, durable, and environmentally friendly, meeting the simultaneous criteria of performance and sustainability (Idrees et al. 1970).

Furthermore, biomaterials provide major contributions to environmental sustainability by providing biodegradable and renewable alternatives to traditional materials. Natural biomaterials including cellulose, alginate, and chitosan are utilized to create environmentally friendly food packaging, water purification systems, wound repairing, and bio-based sensors. Similarly, breakthroughs in biologically derived materials, such as proteins and peptides, are spurring innovation in green chemistry and sustainable industrial processes. The significance of biomaterials extends to their role in tackling global concerns such as rising demand for healthcare solutions, the need for sustainable resource utilization, and environmental impact reduction. Biomaterials alter established industries while also creating the groundwork for new materials and will continue to be used due to their versatility and possibilities for customization.

2.2 Types of Biomaterials

Biomaterials are classified into various categories based on their origin, chemical composition, and specific functionalities. Based on their origin biomaterials are categorized into natural and synthetic biomaterials, while based on their functionality, biomaterials are bioinert, bioactive, and bioresorbable materials. A deeper understanding of their types is discussed in the section below that is essential to infer their role in the sustainable fabrication of advanced bionanocomposites. Figure 2.1 gives an overview on the classification of biomaterials.

2.2.1 Biomaterials Based on Origin Sources

Natural Biomaterials

The materials derived from renewable sources including gelatin, silk, cellulose, collagen, and marine-derived materials fall into this category. These biomaterials exhibit inherent biodegradability and biocompatibility. They are readily used in wound healing, drug delivery, and environmentally sustainable applications. This class of biomaterials also includes materials such as extracellular matrix (ECM) components, peptides, and proteins. These biomaterials are tailored for specific biological interactions, making them ideal for advanced tissue engineering and regenerative medicine. Biologically derived materials offer unparalleled potential for

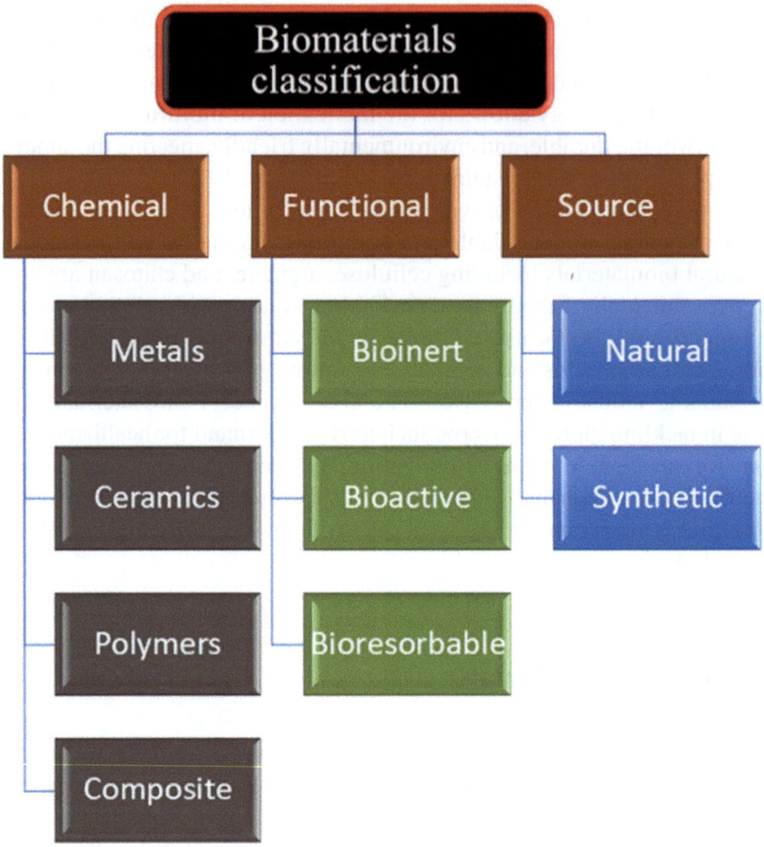

Fig. 2.1 Classification of biomaterials

creating bionanocomposites that closely mimic natural systems, thereby enhancing sustainability and efficiency.

Synthetic Biomaterials

Synthetic biomaterials are not naturally occurring materials, but their applications have become indispensable in many medical fields such as tissue engineering and drug delivery. This class of biomaterials includes materials that are biocompatible, bioresorbable, and have reproducible properties that can be adapted to a given application. Further, these biomaterials are synthesized and modified in a controlled manner to achieve specific properties or stability. Certain examples of these materials are polyethene, polyurethane, bioglass, ceramics, and alloys. Unlike natural materials, these substances' physical, chemical, mechanical, and kinetic characteristics can be modified to suit the intended applications. The use of synthetic polymers is highly restricted cost-wise, and in some respects, they are structurally far removed from the extracellular matrix. Bioactivity can be conferred to material by immobilizing biomolecules, such as peptide sequences, adhesion proteins, or polysaccharides, on its surface or within its structure. Another way is by interpolymer blending, which is achieved by introducing natural polymer or biomolecules into synthetic polymers.

2.2.2 Biomaterials Based on Functionality

Bioinert Biomaterials

Bioinert materials are referred to as the biomaterials that do not interact with their biological surroundings and do not cause immunological reactions when used as implants meant to be placed inside the body for an extended period of time. Stainless steel, titanium, zirconia, alumina, and ultra-high molecular weight polyethylene are among the materials that once inserted into the body do not cause immunological reactions and so they remain harmless inside the body. It is necessary to comprehend and evaluate before implants, as to how the immune system reacts to biomaterials in order to evaluate biocompatibility that is a major factor in the longevity and efficacy of medical implants. Furthermore, because biomaterials interact with the body's immune system, it is crucial to take into account physicochemical factors like size, bioutility, and biodegradability in order to alter immune responses and produce favorable treatment results particularly when it comes to ocular diseases.

Bioactive Biomaterials

A material that causes a response when bonded to a biological tissue is called a bioactive biomaterial. In order to improve tissue growth and adhesion to bone or other tissues bioactive materials such as soft biomaterials bioactive glass and hydroxyapatite are crucial. These materials' benefits include targeted immune responses, controlled release of bioactive compounds and environmental responsiveness to temperature and pH changes. According to research, bioactive glasses for instance can accelerate skin regeneration, encourage angiogenesis and aid in the formation of a calcium phosphate-like layer on hard tissues all of which support bone mending (Kaou et al. 2023). Moreover, zinc-based composites that incorporate bioactive particles have been shown to exhibit enhanced strength and good dispersion of second-phase particles allowing for customization of the rate of zinc corrosion.

Biosorbable Biomaterials

Once implanted inside the body, bioresorbable materials get dissolved and eventually replaced by fresh tissues through a process called resorption (bone for example). Bioresorbable materials, which the body breaks down, gradually are used in a variety of industries including drug delivery systems, orthopedics dentistry, and biodegradable electronics. Since these materials may break down and be absorbed by the body over time, they offer a significant advantage in drug administration and temporary implant applications. This class of materials includes metals, polymers, ceramics, and intelligent micro-nanomaterials. In the quest to enhance medical interventions and reduce their adverse effects, a variety of bioresorbable materials have been investigated for their biocompatibility and degradability (Deng et al. 2022). Among these are calcium phosphate, gelatin, chitosan, iron magnesium alloys and zinc alloys. The field of temporary implants and drug delivery systems may benefit from the development of bioresorbable materials and their fabrication technologies such as 3D and 4D printing. This could make it easier to introduce new medical treatments and body implants that will eventually dissolve or degrade naturally from the system. The selection of biomaterial for an implant application is based on its functionality and long-term effects on the body. When stability over an extended period is required, bioinert materials are frequently preferable to alternative materials such as bioactive or bioresorbable ones which offer benefits in promoting bone formation and tissue regeneration.

2.2.3 Biomaterials Based on Chemical Composition

Metallic Biomaterials

Metals and their alloys are extensively used as biomaterials because of their excellent electrical and thermal conductivities and mechanical properties. Their excellent mechanical strength, ductility, stiffness, corrosion resistance, wear resistance, and biocompatibility make them a durable and reliable choice for load-bearing applications. Metals and alloys like stainless steel, titanium, and cobalt-chromium alloys are integral to cardiovascular stents, orthopedic replacements (hip replacement, knee replacement), and dental devices. These biomaterials are not just used as a replacement for failed hard tissue but are also beneficial in the reconstruction of soft tissues such as blood vessels. Some typically used metals and alloys used as biomaterials include gold, steel, cobalt, titanium, and nickel–chromium alloys. These are mostly used materials for tension-bearing applications like hip or knee implants, skeletal system to replace joints like sockets, hip ball, and fixators such as plates and screws.

Three metallic groups that are majorly used in biomedical applications comprise of 316 stainless steel, cobalt (Co) and its alloy, and titanium and its alloys (Niinomi 2002). 316 stainless composed of 16–18% chromium, 10–14% nickel, 2–3% molybdenum, up to 2% manganese, up to 0.75% silicon, up to 0.10% nitrogen, up to 0.08%

carbon, up to 0.045% phosphorus, up to 0.03% sulfur; iron making up the remainder. 316 stainless steel is used to make implants for biomedical use, but 316 stainless steel may corrode under high-stress oxygen-depleted surroundings inside the body. Co-based alloys are durable and versatile in orthopedic implants. A number of cobalt-based biocompatible metallic biomaterials like cobalt-chromium (Co Cr), nickel-free cobalt-chromium tungsten (Co Cr W), and cobalt-chromium molybdenum (Co Cr Mo) exist for biomedical application. The alloys of nickel and titanium Ni–Ti are widely utilized in the biomedical industry as stents, bone plates, and orthodontics wires. Also, antibacterial agents are synthesized using the silver products and gold alloys, and mercury-silver-tin amalgam are used for dental restoration. However, challenges such as long-term degradation and sustainability concerns drive research into greener alternatives and recycling techniques.

Ceramic Biomaterials

Ceramic biomaterials are biocompatible materials that are employed in a wide range of medicinal applications. Although some bioceramics are flexible, they are usually utilized as stiff materials in surgical implants. They include bioactive glasses and glass ceramics, bioinert ceramics based on oxides, sulfides, and nitride, and bioresorbable calcium phosphate compounds. These materials' high surface area, resistance to corrosion, and enhanced mechanical qualities make them especially helpful for bone replacement and prosthetic applications. Compared to metal-based biomaterials, bioceramic materials have superior mechanical qualities, weak plasticity, a high melting temperature, non-degradability, and excellent biocompatibility. Bioceramics, whether synthetic or natural, are widely used in orthopedics, dentistry, and medical implants due to their excellent biocompatibility, high melting points, and non-corrosive nature. However, they exhibit poor plasticity and low fracture toughness compared to metals. Bioglasses, a subset of bioceramics, are gaining popularity due to their versatility and increasing market demand. The addition of dopants can enhance properties like biocompatibility and antibacterial activity, but excessive doping may lead to toxicity.

Ceramics including alumina, hydroxyapatite, and zirconia are primarily used in bone and dental applications owing to their hardness, wear resistance, and bioactivity. As they closely mimic the mineral composition of bones, ceramics are invaluable for orthopedic and dental implants. Their ability to be combined with polymers or other materials makes them ideal for creating sophisticated bionanocomposites, leading to their enhanced bio-functionality and mechanical strength. Sustainable bioceramics derived from agro-food wastes (e.g., rice husk ash, egg shells, and banana peels) offer a cost-effective alternative to synthetic bioglasses. These natural sources provide essential minerals like silica, calcium, and potassium, which can be used to produce bioactive materials such as hydroxyapatite and β-tricalcium phosphate (β-TCP). Combining synthetic minerals with agro-waste-derived materials enhances biomedical applications while reducing costs.

Polymeric Biomaterials

Polymers are highly versatile materials, available in synthetic and natural forms, and they are designed for interfacing with the biological environment when performing their dedicated functions within the human body. These biomaterials offer structural flexibility, biodegradability, biocompatibility and hence can be functionalized with desired biomolecules. Polymeric biomaterials are categorized on the basis of their origin, natural, or synthetic (Kohane and Langer 2008). Structurally, they are composed of covalently bonded long-range crosslinking chains. Their mechanical properties are influenced by crystallinity, entanglement, and molecular weight.

The advancement of polymeric biomaterials with outstanding therapeutic features is important for applications in diagnostic systems, regenerative medicine, pharmaceutical formulations, drug delivery, and tissue engineering. The physical, chemical, and mechanical properties of these biomaterials can be altered by chemical and structural modifications, functionalizing the relevant features with excellent therapeutic potential in tissue repair applications like repairing of cartilage, skin, bladder, muscle, neurite, bone, and blood vessels. Polymeric biomaterials are also extensively used cardiovascular, orthopedic, ophthalmologic, and dental prostheses.

Polyvinyl chloride (PVC), polyethylene (PE), polypropylene (PP), polymethyl methacrylate (PMMA), polystyrene (PS), and polyamides are examples of synthetic polymers with varying characteristics that are useful for a variety of applications. Each of these polymers possesses unique characteristics that render them suitable for diverse biomedical applications. PVC is commonly utilized in blood storage bags, intravenous (IV) administration hoses, dialysis equipment, and catheters. PE is found in pharmaceutical bottles, orthopaedic implants, as well as non-implantable medical compounds. PMMA, known for its optical clarity and biocompatibility, is used in intraocular lenses, blood pumps, and bone cement. Fluoropolymers like polytetrafluorethylene (PTFE), offer high resistance tolerance and are employed in vascular grids and catheters.

Biodegradable polymers, including glycolide-based polyesters, polylactic acid (PLA), polycaprolactone (PCL), and polycarbonates, are essential in the development of absorbable sutures, drug delivery systems, and temporary scaffolds in tissue engineering. These materials undergo degradation through hydrolysis or enzymatic action, which regulates their absorption within the body. Natural polymer biomaterials, particularly those derived from collagen, play a crucial role in wound healing, soft tissue augmentation, and bioengineered implants. Advances in collagen biotechnology facilitate the creation of absorbable medical implants that exhibit enhanced mechanical strength and controlled degradation rates, alongside tissue scaffolds composed of collagen and biodegradable polyesters. Ongoing progress in polymer chemistry and manufacturing techniques is paving the way for the next generation of biomaterials, which promise improved performance and a pivotal role in contemporary innovations in medical regeneration and device development. Synthetic polymers such as PLA, PCL, and polyethylene glycol (PEG) are extensively utilized in sutures, scaffolding, and systems for controlled drug release. Contrarily, natural polymers such as collagen, chitosan, and hyaluronic acid offer excellent biocompatibility and biodegradability. Polymeric biomaterials also play a crucial role in the development of bionanocomposites, where their properties are enhanced by incorporating nanomaterials, providing a sustainable alternative to conventional and biocompatible designs.

The development of biomaterials with enhanced performance is pivotal in driving modern advancements in medical, regenerative, and medical device technologies. Synthetic polymers, including polylactic acid (PLA), polycaprolactone (PCL), and polyethylene glycol (PEG), are extensively utilized in applications such as sutures, scaffolds, and controlled drug delivery systems. In contrast, natural polymers like collagen, chitosan, and hyaluronic acid provide remarkable biocompatibility and biodegradability. Furthermore, polymeric biomaterials are integral to the creation of bionanocomposites, where their properties are augmented through the integration of nanomaterials, offering a sustainable alternative to traditional and biocompatible designs. The significance of polymeric biomaterials in contemporary biomedical applications stems from their structural adaptability, biocompatibility, and the capacity for customization to meet specific medical requirements. These materials, which encompass both synthetic and natural polymers, engage with biological systems to support therapeutic and diagnostic processes. Structurally, polymeric biomaterials are composed of long-chain macromolecules linked by covalent bonds, with their mechanical properties influenced by factors such as crystallinity, molecular weight, and polymer chain entanglement. The degree of polymerization (DP) and molecular weight (MW) are critical determinants of their strength and functionality, with analytical methods like gel permeation chromatography (GPC) utilized to evaluate their properties. Figure 2.2 gives a classification of polymeric biomaterials.

Composite Biomaterials

The word biocomposites describe composite materials that include at least two different constituent materials or phases at the macroscopic level and can be used in bioengineering. Bones, cartilage, tendons, skin, ligaments, and teeth are examples of composite structures found in nature. These structures exhibit hierarchical and anisotropic structural characteristics.

Composite biomaterials, commonly referred to as biocomposites, are engineered materials synthesized by incorporating two or more distinct materials to harness their synergistic properties. One of the components of a biocomposite is naturally derived, constituting the reinforcing component and the other being the matrix. The reinforcement phase in these materials may include natural fibers, hydroxyapatite, or cellulose extracts which enhance the strength of the otherwise weaker biopolymer matrix typically composed of polymers, preferably sourced from renewable materials such as vegetable oils, collagen, chitosan, alginate, or starches. These unique combinations provide biocomposites with exceptional biocompatibility, mechanical strength, stiffness, and adaptable properties, making them highly suitable for diverse medical applications. Furthermore, biocomposites are essential in the advancement of medical implants, where their biodegradability and biocompatibility with biological

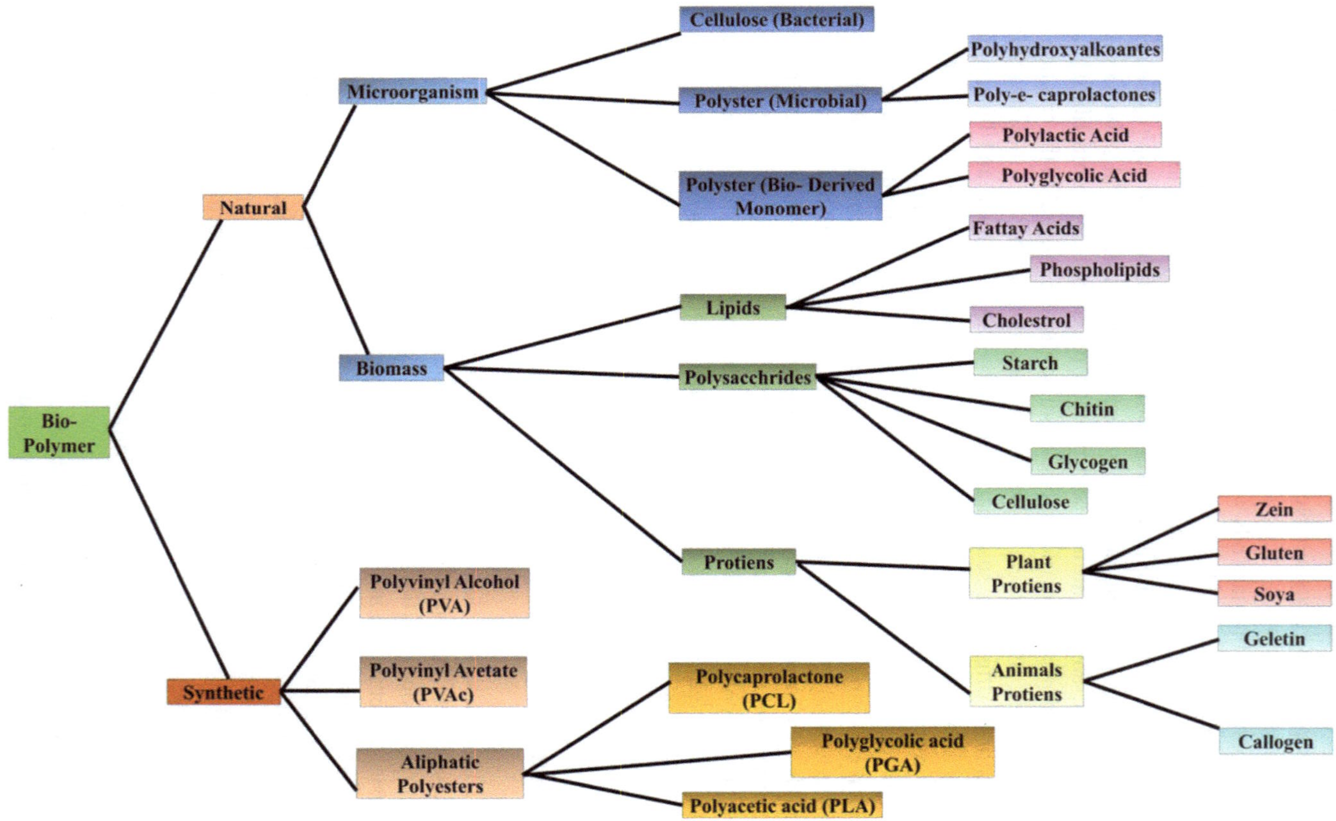

Fig. 2.2 Classification of polymeric biomaterials

systems are crucial. The hemocompatibility of these materials makes them useful in cardiovascular applications such as stents and grafts. Their adaptability also extends to drug delivery systems, allowing regulated pharmaceutical release.

In the fields of tissue engineering and regenerative medicine, biocomposites function as scaffolds that mimic the extracellular matrix, thereby facilitating cell proliferation and tissue repair. They play a crucial role in the creation of implantable medical devices, where their biodegradability and compatibility with bodily fluids enhance both performance and durability. Additionally, their adaptability extends to drug delivery systems, allowing for the controlled release of pharmaceuticals, while their hemocompatibility renders them essential in cardiovascular applications, such as stents and grafts. Despite their significant potential, challenges persist, particularly regarding the necessity for standardized testing and regulatory approvals. Ongoing research focused on improving the properties of biocomposites and broadening their applications underscores their transformative impact in the medical sector, capitalizing on their unique combination of biocompatibility and mechanical strength. The performance and mechanical characteristics of biocomposites are shaped by various factors, with the intrinsic properties of their constituent materials primarily determining their overall attributes. Key parameters influencing reinforcement include fiber architecture, which encompasses fiber geometry (size and shape), fiber orientation (alignment within the matrix), and fiber volume fraction (percentage within the composite). Natural composites display anisotropic properties, indicating that their mechanical characteristics differ based on direction, a vital consideration for biomedical applications such as implants. Furthermore, hierarchical structural features, including particulate, porous, and fibrous microstructures, significantly influence the ultimate behavior of tissues and organs. Synthetic biocomposites can be designed to emulate the mechanical properties of biological tissues, making them particularly valuable in prosthetics aimed at restoring functionality. Polymeric biocomposites are optimized to maintain a balance between mechanical strength and bioactivity by optimizing their mechanical properties by combining low modulus with high strength for orthopedic uses. In dentistry, they provide enhanced aesthetics and cost-effectiveness compared to metal-based restorations, making them a preferred choice for various biomedical applications. Biocomposites offer numerous benefits, particularly in medical and dental applications, due to their desirable mechanical and biological properties. The key features include their good durability in small to moderate restorations and offer high biocompatibility, minimized risk of adverse reactions with biological tissues. They demonstrate moderate wear and are corrosion resistant, ensuring longevity in biomedical applications, and their inert nature prevents any reactions with tissues, reducing degradation over time. These composites exhibit high fracture toughness, are thus well-suited for load-bearing applications, and exhibit resistance to fatigue failure, allowing them to withstand cyclic loading without significant deterioration. Additionally, biocomposites are highly compatible with diagnostic imaging techniques such as MRI and possess low density and weight, making them ideal for orthopedic implants and prosthetics. The development of sustainable composite biomaterials focuses on reducing waste and improving recyclability, aligning with the goals of eco-friendly innovation.

2.3 Nanomaterials in Bionanocomposites

Nanomaterials have revolutionized the field of bionanocomposites, offering a plethora of enhanced properties, including mechanical, electrical, optical, and thermal characteristics, which have far-reaching implications for various applications. Bionanocomposites, a synergistic combination of biomaterials, such as polymers, ceramics, or metals, with nanomaterials, have emerged as a fascinating research domain, where the unique attributes of nanomaterials are harnessed to create innovative materials with unprecedented performance. The nanomaterials employed in bionanocomposites encompass a broad spectrum of materials, including inorganic nanoparticles like carbon nanotubes, renowned for their exceptional mechanical strength and electrical conductivity, graphene, a highly conductive and flexible 2D material, and silver, celebrated for its antimicrobial properties, as well as organic nanoparticles like cellulose nanocrystals, derived from renewable biomass sources, and chitosan nanoparticles, exhibiting biocompatibility and biodegradability. Furthermore, hybrid nanomaterials, such as metal-organic frameworks and carbon-based hybrids, which combine the benefits of different materials, are also being explored for their potential in bionanocomposites (Niinomi 2002). The size and shape of these nanomaterials affect their properties, which are important in deciding how well they work in bionanocomposites. Researchers are always looking for ways to improve these properties. Compatibility and dispersion problems are among the major obstacles to integrating nanomaterials into biomaterials; these must be resolved by careful surface modification, compatibilizer use, and processing condition optimization. Notwithstanding these obstacles, nanomaterials in bionanocomposites have enormous promise for novel uses in energy, medicine, and the environment, which motivates more study and development to fully realize their advantages and open up new avenues.

Because they provide a special combination of physical, chemical, and biological qualities that can be used to produce novel materials with previously unheard of performance, nanomaterials are easily incorporated into bionanocomposites. Inorganic nanoparticles with remarkable mechanical strength, thermal conductivity, and electrical properties, such as carbon nanotubes, graphene, silver, silica, and clay, are among the wide range of nanomaterials being used. Carbon nanotubes typically have exceptional mechanical strength, stiffness, and toughness, which makes them perfect for creating sophisticated composites for energy storage and biomedical applications (Furukawa et al. 2013). Graphene, a two-dimensional nanomaterial, that is extensively explored for its exceptional electrical conductivity, thermal stability, and mechanical strength, rendering it suitable for applications in energy storage, sensing, and catalysis (Geim and Novoselov 2007).

Chitosan nanoparticles, cellulose nanocrystals, and liposomes are examples of organic nanomaterials that are used because they are biocompatible, biodegradable, and versatile in terms of surface modification and functionalization. Crustacean shell-derived chitosan nanoparticles have been studied for their biocompatibility, biodegradability, and capacity to encapsulate and transport therapeutic compounds, making them appropriate for use in tissue engineering and drug delivery (Katas et al. 2013). The primary source of cellulose nanocrystals is renewable biomass, which has remarkable mechanical strength, thermal stability, and optical qualities that make it appropriate for use in energy storage, biomedicine, and sustainable materials (Kumar et al. 2017).

Furthermore, the advantages of several nanomaterials are combined to generate novel properties and functionalities in hybrid nanomaterials, such as carbon-based hybrids and metal-organic frameworks. Apart from their chemical and thermal stability, metal-organic frameworks made of metal and organic linkers exhibit exceptional porosity which is suitable for their gas storage, catalysis and gas separation applications (Moon et al. 2011). Nanomaterials have numerous commercial applications, including the creation of multifunctional water- and oil-repellent, thermochromic, photochromic, and antibacterial materials. Additionally, the innovation of portable energy storage devices, high-performance electrode and hybrid materials for solar cells is solely led by nanomaterials.

The amalgamation of nanofillers in biocomposites offers the potential to transform the fields of energy storage, advance pharmaceutical, biomedical engineering, and sustainable materials. The bionanocomposites are specifically tailored for targeted applications for food packaging, drug delivery, tissue engineering, prosthetics, and biomedical implants. Hence, the incorporation of nanomaterials into bionanocomposites is promising for major advances in various sectors, enabling the development of advanced materials with targeted properties and functionalities.

The mechanical, thermal, electrical, optical, electrochemical, catalytic properties of the nanomaterials are different from the bulk materials, owing to their sizes of the order of a billionth of a meter. Hence, these materials finds a wide variety of application fields including sensitive sensors, batteries with high energy density, powerful magnets, pollutants removal, fuel-efficient automobiles, excellent weapons' platforms, satellites with longer-lasting and better insulation materials, next-generation computer chips, food additives, agricultural production, food processing, animal feed, large electro-chromic display devices.

2.4 Role of Nanoparticle Size and Shape in Enhancing Performance

The capacity to manipulate the shape and dimensions of nanostructures allows for the construction of materials whose macroscopic properties are determined by the nature of the nanoparticle. Although particle morphology is recognized as an important characteristic, the potential shape-dependent properties have only been partially exploited so far. Nanoparticle form is an important factor in determining nanocomposite hydrogel characteristics. Calcium-alginate hydrogels, poly(L-lactide) nanoparticles based adhesive, with platelet morphology exhibit improved adhesion over nanoparticle glues based on spherical or cylindrical micelles. Furthermore, gel nanocomposites containing platelets demonstrate greater resilience to breaking under tension than their spherical and cylindrical counterparts. This study opens the doors to a change in direction in the field of gel nanocomposites, where nanoparticle shape plays an important role in tuning mechanical properties (Niinomi 2002).

2.5 Challenges in Integrating Nanomaterials into Biomaterials

Integration of nanomaterials with biomaterials poses significant challenges that hinder their potential applications. Compatibility is one of the major obstacles due to the inherent differences in physical and chemical properties between nanomaterials and biomaterials. For instance, the hydrophobic nature of many nanomaterials, such as carbon nanotubes (CNTs) and graphene, can lead to inadequate interactions with hydrophilic biomaterials, resulting in aggregation, phase separation, or compromised mechanical properties. Moreover, the high surface energy of nanomaterials can also lead to non-specific binding with biomolecules, causing unwanted biological responses.

One of the challenges in the dispersion of matrix materials also afflicts the integration process, uniform dispersion of nanomaterials within biomaterials is crucial to achieve optimized constitutional and performance parameters, but achieving this uniformity is difficult due to the tendency of nanomaterials to agglomerate. This can lead to uneven distribution, reduced interfacial interactions, and diminished overall performance of the nanocomposites (Kim et al. 2019). For instance, studies have shown that the dispersion of CNTs in poly lactic acid (PLA) matrix can significantly affect the mechanical properties of the resulting nanocomposite, with uniform dispersion leading to improved tensile strength and modulus (Niinomi 2002; Giner-Grau et al. 2024). Additionally, the high aspect ratio of nanomaterials can also lead to complications in achieving uniform dispersion. For example, the high aspect ratio of graphene nanosheets can cause them to reaggregate during the dispersion process, resulting in non-uniform distribution within the biomaterial matrix. Various strategies have been employed to overcome these challenges, including modifying the surface of the nanomaterials, using surfactants or dispersants, and optimizing the processing conditions. In conclusion, addressing the compatibility and dispersion challenges is essential to harness the full potential of nanomaterials in biomaterials. Further research is needed to develop effective strategies for achieving uniform dispersion and optimizing interfacial interactions between nanomaterials and biomaterials.

2.6 Overview of Sustainability in Materials Science

A new generation of hybrid nanostructures aligns with an emerging field in the frontier between materials science, life science, and nanotechnology. Bionanocomposites may be said to typically refer to the nanocomposites formed by encapsulating naturally occurring biomolecules with conventional nanomaterials. These are often referred to as green composites or biohybrids whose properties are governed by their dimensionality occurring in nanoscale. The terms bionanocomposites, biocomposites, nanobiocomposites, green composites, and biohybrids are occasionally used interchangeably. Bionanocomposites exhibit good optical and mechanical properties along with decreased water sensitivity and good barrier properties against environmental microorganisms and are easily biodegradable. The nanosize of the biomaterials is the key factor in the advancement of novel properties and in controlling their structure at the nanoscale. The integration of nanotechnology with biomaterials produces composites that meet the functional requirements along with the principles of sustainability. The properties inherent to the biopolymers including large surface area, easy processing, cost-effectiveness, sustainable drug release, cancer treatment, drug delivery, biosensing, biocompatibility, and biodegradability revolutionizing the landscape of medical interventions (Mousa et al. 2016). Bionanocomposites is new interdisciplinary field ranging from biomineralization processes, bioinspired materials to biomimetic systems, this multifunctionality opens a potential research area that exploits the synergistic assembly of biopolymers with inorganic nanomaterials.

The composition and functionalization of the bionanocomposites are governed by concentration, type, angle of orientation, shape and size/aspect ratio, affinity/compatibility with the matrix polymer, use of external additives other than the nanoreinforcement in the synthesis and degree of exfoliation. Different nanofillers show variable thermal and optical characteristics, as well as diverse compatibility with the matrix material, biodegradability, and toxicity. These characteristics are important in determining the bionanocomposite's features. In terms of preparation technique, characteristics, functions, biocompatibility, biodegradability, and applications, they differ from nanocomposites. Because of its nanoscale dispersion with sizes smaller than 100 nm, bionanocomposites are important. Submicron-sized particles containing one or more therapeutic drugs that are distributed, adsorbed, or covalently bound in encapsulated vesicles, capsules, or polymer matrices are known as nano drug delivery systems. Their outstanding qualities, including biocompatibility, antibacterial activity, biodegradability, and excellent mechanical, optical, and barrier properties, make them widely used in a range of fields. The recent advances in biomedical applications using nanomaterials such as graphene are used in wound healing, drug or gene delivery, tissue engineering, bioimaging, and biosensing.

Additionally to excellent characteristics, bionanocomposites show the remarkable advantage of exhibiting biocompatibility, biodegradability, and functional properties provided by either the biological or inorganic moieties. Each type of biomaterial offers distinct advantages and limitations, influencing their selection for specific applications. In the context of this chapter, the focus lies on how these materials contribute to the fabrication of advanced bionanocomposites. The properties inherent to the biopolymers including large surface area, easy processing, cost-effectiveness, sustainable drug release, cancer treatment, drug delivery, biosensing, biocompatibility, and biodegradability are volutionizing the landscape of medical interventions (Mousa et al. 2016). Understanding the types and characteristics of biomaterials provides the necessary foundation for exploring their role in sustainable fabrication practices, which will be discussed in detail in the following sections.

The incorporation of organic/inorganic components into polymer matrices has several advantages, including outstanding optoelectrical, enhanced mechanical, magnetic, catalytic, and biological properties. Biodegradable polymers can be classified as (a) agro-polymers based on the biomass of agro-resources as raw materials (e.g., starch and cellulose), (b) polymers based on microbial production (e.g., polyhydroxyalkanoates), (c) polymers chemically synthesized using biomass-based monomers (e.g., poly(lactic acid) [PLA]), and (d) polymers produced from the chemical synthesis of petroleum resources.

Each type of biomaterial has various advantages and limits, which influence their suitability for specific applications. This chapter focuses on how these materials help to fabricate sophisticated bionanocomposites. Researchers are designing composites that not only suit practical criteria but also adhere to sustainability standards. Understanding the many forms and properties of biomaterials lays the groundwork for investigating their role in sustainable fabrication techniques, which will be covered in depth in the following sections.

The quest of sustainability has become a top priority in many industries, including materials science. As the world faces environmental concerns, climate change, and resource depletion, the creation of sustainable materials has emerged as a crucial research priority. This article gives an overview of sustainability in materials science, focusing on biomaterials and bionanocomposites. Materials science has a significant impact on our daily lives, from the devices we use to the structures we occupy. However, the manufacturing, processing, and disposal of materials have serious environmental consequences. Raw material extraction, energy usage, and waste generation all contribute to greenhouse gas emissions and pollution. Biomaterials, which are created from renewable resources including plants, animals, and microbes, provide a more sustainable alternative to traditional materials. These materials can be biodegradable, non-toxic, and compostable, which reduces environmental impact. Biomaterials have several categories, including metallic biomaterials, made from metals like titanium, stainless steel, and cobalt-chromium alloys, are employed in medical implants and equipment. Polymeric biomaterials, made from natural polymers including cellulose, starch, and proteins, are utilized for packaging, textiles, and biomedical applications. Ceramic biomaterials, made from natural sources including bones, teeth, and silicates, are employed in medical implants, tissue engineering, and medication delivery. Hence, composite biomaterials combine polymers, ceramics, and metals to provide enhanced properties.

2.7 Historical Development and Significance of Bionanocomposites

The area of bionanocomposites has experienced tremendous growth and evolution over the past few decades. The concept of bionanocomposites spans several decades, beginning with the effort of combining natural polymers with inorganic fillers leading to the advancement of nanotechnology. The journey of progression of bionanocomposites as a distinct field of research demonstrates the power of scientific innovation and advancements. The emergence of the bionanocomposites plays a crucial role in the development of sustainable and eco-friendly biomaterials. These hybrid materials are composed of biopolymers or biocompatible polymers integrated with nanoscale inorganic or organic fillers for creation of innovative materials that exhibit enhanced thermal, mechanical and barrier properties (Shchipunov 2012).

Bionanocomposites have a rich and fascinating history, beginning in the early twentieth century, when researchers started to explore the possibility of combining natural polymers with inorganic fillers. The earliest recorded attempts to develop bionanocomposites happened in the 1950s and 1960s, focusing on the investigation of natural polymers such as cellulose, starch, and proteins. During this time, the scientists discovered that the incorporation of inorganic fillers like silica, clay, calcium carbonate, etc., may significantly enhance the mechanical properties of these polymers. One of the notable early evolutions of a bionanocomposite was the synthesis of clay-filled cellulose composites in the 1960s. These early-developed composites were made by mixing cellulose fibers with clay particles and then extruded into various shapes. The resulting materials showed improved mechanical strength, thermal stability, and barrier properties compared to unfilled cellulose. In the 1980s and 1990s, nanotechnology became very eminent in research area, during which, the scientists began to explore the unique properties of materials in the nanoscale, and the concept of bionanocomposites came to existence (Lee et al. 2014). One of the important developments of this time was the recognition of the potential of layered silicates such as montmorillonite and kaolinites as a nanoscale filler for biopolymers (Cui et al. 2019).

The 2000s saw a remarkable growth in research focused on bionanocomposites specially their fabrications, characterizations and applications. These studies were centered on the growing demand for sustainable and eco-friendly materials. Bionanocomposites, with their unique combination of biodegradable biopolymers and nanoscale fillers, emerged as a promising solution to these challenges which led to the exploration of the potential of bionanocomposites possessing several advantages as compared to conventional materials, including properties such as impact resistance, fatigue resistance, stiffness, thermal conductivity, corrosion resistance, biodegradability, high specific strength, low relative density, environmentally friendly. In recent years, we have witnessed the discovery of several new high-performing composite materials, as well as advances in the fields of aviation, automobile, adhesion, packaging, biological applications, and energy storage and electronic engineering that justifies the widespread usage of bionanocomposites in applications far beyond those for which they were originally designed. Bionanocomposites have become a rapidly growing field with significant potential in a variety of sectors. As researchers continue to investigate the potential of these materials, we can expect substantial improvements in the production of sustainable and ecologically friendly materials (Bharathi and Jayas 2024).

2.8 Bionanocomposites: Integrating Biomaterials with Nanotechnology

The integration of biomaterials with nanotechnology has led to the development of bionanocomposites. These materials combine the benefits of biomaterials with the unique properties of nanoparticles, such as improved mechanical strength, thermal stability, and biological functionality. These bionanocomposites have applications in various fields, including biomedical applications that involve tissue engineering, drug delivery systems, and medical implants (Sharma et al. 2023). Several environmental applications such as water purification, air filtration, and soil remediation and industrial applications like packaging, textiles, and composite materials along with biocompatibility and biodegradability. Figure 2.3 depicts the evolution of bionanocomposites from nanocomposites.

The development of sustainable biomaterials and bionanocomposites requires addressing of several factors, including the use of renewable resources such as biomass to reduce reliance on fossil fuels. Biodegradability is another primary factor in the sustainable development of bionanocomposites that involves creation of materials that possess the ability to naturally decompose, thereby mitigating waste and environmental degradation. One of the major factors of concern is to ensure non-toxicity in the advancement of materials so that human health or the environment has no risk. Furthermore, the process of integration of bionanocomposites must be energy efficient. The fabrication of bionanocomposites may be an energy-intensive

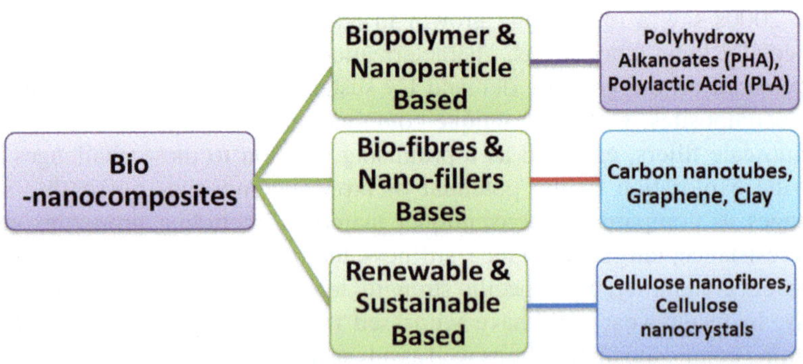

Fig. 2.3 Classification of bionanocomposites

process, contributing to further to the environmental concerns, and increased production costs too. To counter this, energy-efficient strategies such as using renewable energy sources, energy-efficient equipment, process optimization, recycling, and reuse can be employed. Moreover, techniques like in situ synthesis and mechanical grinding can greatly reduce energy consumption and waste generation. Adopting these sustainable practices in the fabrication of bionanocomposites can make the process more eco-friendly, cost-effective, and produce high-quality products. Moreover, the integration with emerging technologies by understanding the potential of integration of biomaterials and bionanocomposites through cutting-edge technologies, such as 3D printing and artificial intelligence point toward the advancement of sustainable biomaterials and bionanocomposites is essential for minimizing environmental damage and fostering eco-friendly technologies. By thoroughly understanding the properties, applications, and sustainability factors of these materials, researchers and industry professionals can collaborate to create innovative solutions that contribute to a more sustainable future.

2.9 Fabrication Techniques for Bionanocomposites

Chemical synthesis methods enable the controlled generation of nanomaterials by tuning chemical reactions and conditions, in order to regulate their size, shape, composition, and structure. Evidently, various techniques are utilized, such as the sol–gel method, which takes precursor solutions and converts them into gels to create nanomaterials, and the Precipitation Method, whose soluble precursors interact with precipitating agents to generate insoluble nanomaterials. Hydrothermal or solvothermal method is the method used for the synthesis of nanomaterials at high temperature and extreme pressure. The biological method for the synthesis of nanomaterials utilizes biological systems, such as microorganisms, plants, or enzymes, to produce nanoparticles. This method has the advantages of being environmentally friendly, cost-effective and allows the properties of nanoparticles to be modified. Microorganisms such as bacteria, fungi, and yeasts have been extensively studied as they have the ability to produce variety of nanoparticles through extraction process.

As the most appealing materials, nanomaterials made from biodegradable and sustainable polymers are gaining popularity. The most widely used nanofillers are hydroxyapatite, organic or inorganic metal nanoparticles clays, etc. One particular class of materials known as polymeric nanobiocomposites (PNBCs) is created by combining nanosized fillers with polymer materials. Because biopolymer-based hydrogel materials have high permeability and rapid electron mobilization, they have been the subject of numerous recent studies aimed at developing analyte sensors and electrode modifiers. These biopolymer hydrogels produce portable flexible and highly effective bioelectronic sensors by employing sophisticated printing techniques in electrode prototyping. Since polyaniline reinforced biocomposites have special qualities that make them a promising material for electronics applications some research has also looked into using them to fabricate electrochemical sensors and actuators. In order to improve their functional properties bionanocomposites are hybrid materials that combine biopolymers with nanostructured fillers. Because of their non-toxicity and biodegradability, these materials are widely used in sectors like water purification packaging and biomedical engineering. Their properties functionality and possible uses are all influenced by the fabrication processes. Superior mechanical, thermal, and biocompatible properties make bionanocomposites highly prized for their applications in sustainable materials biomedicine and environmental remediation.

A thorough review of the fabrication processes used to create bionanocomposites is provided in this chapter along with information on their advantages limitations and potential uses. Numerous global issues including materials science water treatment and energy conversion have important solutions thanks to nanotechnology. To improve their mechanical and thermophysical qualities, nanoparticles are mixed with a matrix material such as metals ceramics or polymers to create composite materials. The goal of all the different methods used to create polymer nanocomposites is to achieve uniform dispersion free of aggregations. The most widely used methods for creating polymer nanocomposite include electrospinning, selective laser sintering, melting-mixing, and mixing in situ polymerization. Nanoparticle aggregation is eliminated for melt mixing techniques by using water atomic layer deposition and plasma-assisted mechanochemistry. Also, high-frequency sonication has a similar function to mixing methods. Thermodynamically stable nanocomposites can be created through in situ polymerization. For creating porous structures, electrospinning is a useful and efficient technique. Additionally there are clear advantages to using selective laser sintering to create nanocomposites in order to get around the aggregation issue. The working principles of each technique, including the advantages and disadvantages, are discussed in Table 2.1.

The field of bionanocomposites has witnessed significant growth in recent years, driven by the increasing demand for sustainable and eco-friendly materials. Bionanocomposites are hybrid materials that combine biopolymers with nanomaterials, offering enhanced mechanical, thermal, and biological properties. The fabrication techniques used to synthesize these materials play a crucial role in determining their final properties and applications. In this article, we will delve into the various fabrication techniques used to synthesize bionanocomposites.

2.9.1 Solution Mixing Method

The solution mixing method which involves dispersing nanofillers in a polymer matrix with solvents mixing and solvent evaporation is a widely used technique for producing nanocomposites. The simplicity adaptability and ease of scalability of this approach are among its many advantages. After the polymer matrix has been dissolved in an appropriate solvent nanofillers like nanoparticles nanotubes or nanoclays are typically added. After that the mixture is well combined using techniques like high-shear mixing mechanical stirring or ultrasonication to guarantee

Table 2.1 Comparative table of fabrication techniques for biomaterials

S. No.	Fabrication method	Description	Strengths	Weaknesses
1.	Sol–gel processing	A chemical method involving hydrolysis and condensation reactions to create nanostructured biomaterials	Produces highly pure and homogeneous materials Enables precise control over composition and porosity	Often requires post-processing to remove solvents and byproducts Time-consuming due to slow reaction rates
2.	Electrospinning	Uses an electric field to produce ultrafine fibers from polymer solutions for scaffolds and coatings	Produces nanofibers that mimic natural ECM High aspect ratio enhances cell attachment and growth	Limited control over fiber alignment and mechanical properties Solvent toxicity can affect biocompatibility
3.	3D printing (bioprinting)	Layer-by-layer deposition to fabricate complex and customized biomaterial structures	Allows multi-material printing for functional gradients High precision and repeatability	High material costs and slow processing speed Requires biocompatible inks with controlled rheology
4.	Freeze-drying (lyophilization)	A process of freezing and sublimation to create porous structures for tissue engineering	Retains bioactivity of incorporated molecules Produces highly porous and interconnected structures	Fragile structures with low mechanical strength Long processing times and high-energy consumption
5.	Melt mixing	Mixing polymers and nanoparticles at high temperatures to form a composite	Simple and scalable process Good dispersion of nanoparticles	Limited control over microstructure High temperatures may degrade sensitive biomaterials
6.	Solution mixing method	Dispersion of nanoparticles into a polymer solution, followed by solvent evaporation or precipitation	Simple and cost-effective Ensures good dispersion of nanoparticles	Solvent residues may affect biocompatibility Requires optimization to prevent nanoparticle agglomeration

(continued)

Table 2.1 (continued)

S. No.	Fabrication method	Description	Strengths	Weaknesses
7.	In situ polymerization	Polymerization of monomers in the presence of reinforcing nanoparticles to form a nanocomposite	Strong interfacial bonding between polymer and nanoparticles Enhances mechanical and thermal properties	Requires careful control of reaction conditions Possible residual monomers may impact biocompatibility
8.	Layer-by-layer (LBL) assembly	Sequential deposition of oppositely charged materials to create multilayered nanocomposites	Precise control over thickness and composition Suitable for surface modification of implants	Time-consuming due to multiple deposition steps Requires optimization for mechanical stability
9.	Green synthesis approaches	Environmentally friendly methods using plant extracts, bacteria, or fungi to synthesize nanoparticles and biomaterials	Sustainable and eco-friendly Avoids toxic chemicals and harsh reaction conditions	Limited scalability for large-scale production Requires optimization for reproducibility and material properties

that the nanofillers are dispersed equally. In order to produce a solid nanocomposite the solvent is finally evaporated using techniques like heat treatment vacuum drying or spray drying. Since the solution mixing method enables effective dispersion and distribution of the nanofillers within the polymer matrix, it is particularly helpful for creating nanocomposites with high concentrations of nanofillers. Furthermore, this technique can be applied to the fabrication of nanocomposites with a variety of polymer matrices such as elastomers thermosets and thermoplastics. On the other hand, the solution mixing technique may also cause nanofillers to aggregate which would lower the materials mechanical strength thermal stability and electrical conductivity. Researchers have used a number of tactics to lessen this problem such as optimizing mixing conditions surface modification of nanofillers and the use of surfactants.

2.9.2 Melt Mixing Method

An approach that is frequently used to create nanocomposites is melt mixing. It involves mixing nanoparticles with a molten biopolymer under carefully regulated shear and temperature conditions. This method offers several benefits such as easy scaling high production rates and uniform dispersion of nanomaterials. Typically, the process starts with heating the biopolymer to its melting point after which nanomaterials like nanoparticles nanotubes or nanoclays are added. To ensure that the nanomaterials are dispersed evenly the mixture is then thoroughly mixed using methods like internal mixing twin-screw extrusion or single-screw extrusion. Since the melt mixing method facilitates effective dispersion and distribution of nanomaterials within the biopolymer matrix, it is highly effective for producing nanocomposites with high nanomaterial loading levels. Moreover, a variety of biopolymers such as poly(lactic acid) (PLA), poly(hydroxyalkanoates) (PHAs), and starch-based bioplastics can be used to create nanocomposites using this technique. Controlling the melt mixing method can be difficult though because the properties of the final nanocomposite and the dispersion of nanomaterials can be impacted by processing parameters like temperature shear rate and residence time. The performance and stability of the nanocomposite may be jeopardized if the melt mixing process is not optimized by material degradation such as thermal mechanical or chemical degradation.

2.9.3 In Situ Polymerization

By introducing nanoparticles into a polymer matrix while the polymerization process is underway in situ polymerization is a flexible and effective technique for producing nanocomposites. Strong interfacial contacts between nanoparticles and the polymer matrix uniform dispersion of nanomaterials and improved mechanical thermal and electrical properties in the final nanocomposite are just a few benefits of this method. A monomer or combination of monomers is typically dissolved in a solvent to start the process and then nanomaterials like nanoparticles nanotubes or nanoclays are added. The polymerization reaction is then initiated by subjecting the mixture to polymerization conditions such as heat light or catalysts. A number of polymerization techniques including reversible addition-fragmentation chain transfer (RAFT) atom transfer radical polymerization (ATRP) and free radical polymerization can be used to accomplish in situ polymerization (Ortenzi et al. 2015). The ability to effectively disperse and distribute the nanomaterials within the polymer matrix makes this technique especially helpful for creating nanocomposites with high loading levels of nanomaterials. Furthermore, a variety of polymer matrices such as thermoplastics thermosets and elastomers can be used to create nanocomposites through in situ polymerization. The resultant nanocomposites have demonstrated significant potential for a range of uses such as aerospace engineering biomedical devices and energy storage.

2.9.4 Electrospinning

For the synthesis of bionanocomposites, electrospinning is a very versatile and common method. One method is to use an electric field to turn a polymer solution containing nanomaterials into ultrafine fibers. A high surface area-to-volume ratio, the ability to create fibers with nanometer-scale widths and adjustable mechanical thermal and electrical properties are just a few benefits of this method. Before being combined with nanomaterials like nanoparticles, nanotubes, or nanoclays, the biopolymer (collagen, PLA, PHA) is generally dissolved into a solvent. To create fibers with a particular diameter and shape, the mixture is subsequently electrospun under controlled circumstances such as a high-voltage electric field. The synthesis can be achieved by electrospinning a variety of biopolymer-nanomaterial combinations. Bionanocomposites with properties tailored to particular applications. Electrospun fibers that contain silver nanoparticles for instance are suitable for wound treatment applications due to their antibacterial properties. The superior mechanical and electrical properties of carbon nanotube-containing electrospun fibers make them perfect for tissue engineering and biomedical device applications. Changes to the voltage flow rate and distance between the needle and the collector can be made to control the electrospinning process and produce fibers with the desired properties. All things considered, electrospinning is a successful technique for producing bionanocomposites with unique properties and uses.

2.9.5 Layer-by-Layer (LBL) Assembly

The layer-by-layer (LBL) assembly technique is a flexible and accurate way to create multilayered nanostructures by sequentially adsorbing oppositely charged nanoparticles and biopolymers. Because it guarantees exact control over the thickness composition and structure of the film this technique is appealing for creating thin films with specific qualities. Proteins and polysaccharides that are positively and negatively charged particles, are deposited alternately onto a substrate during the LBL assembly process. The assembly process which creates a multilayered nanostructure is driven by the electrostatic interactions between the oppositely charged species. The ability to control film thickness at the nanoscale the ability to incorporate a variety of materials and the potential to create complex nanostructures with customized properties are just a few benefits of the LBL assembly technique. Moreover, nanostructures with particular uses like drug delivery biosensing and tissue engineering can be created using the LBL assembly technique. However because it necessitates multiple deposition steps and exact control over the assembly conditions the LBL assembly technique can be labor-intensive and time-consuming. The LBL assembly method has become a potent tool for creating multilayered nanostructures in spite of these drawbacks.

2.9.6 Sol–Gel Method

The sol–gel method is a proficient and widely used technique to synthesize bionanocomposites, that involves the transition of a system from a liquid "sol" (a colloidal suspension of particles) into a solid "gel" phase. This method provides several advantages, including the ability to produce Bionanocomposites with high purity, homogeneity, and controlled nanostructure. Typically, this process involves the hydrolysis and consecutive condensation of metal alkoxides or other precursors, followed by the incorporation of biopolymers, such as polysaccharides, proteins, or nucleic acids. The obtained emulsion is then subjected to controlled conditions, such as temperature, pH, and solvent composition, to induce gelation and formation of the bionanocomposite.

The sol–gel technique allows for the synthesis of bionanocomposites with tailored properties including enhanced mechanical strength, optical properties, thermal stability, fire resistance, thermal properties, extended shelf life, biodegradability, and reducing environmental impact. This technique is particularly effective for fabricating bionanocomposites for biomedical applications such as tissue engineering, drug delivery, and biosensing. For example, sol–gel derived bionanocomposites containing silica nanoparticles and biopolymers have shown enhanced mechanical properties and biocompatibility, making them suitable for bone tissue engineering applications. Additionally, the sol–gel method can be used to synthesize bionanocomposites with unique optical, electrical, and magnetic properties, expanding their potential applications in fields such as energy storage, electronics, catalysis, marine, pollutants removal, aerospace, and biotechnology.

2.9.7 Freeze-Drying Technique

The lyophilization process also known as the freeze-drying technique is a gentle and efficient way to create bionanocomposites. This procedure involves taking the solvents out of a frozen suspension or solution that contains nanomaterials and biopolymers. Preserving the nanostructure and bioactivity of biopolymers minimizing protein denaturation and creating bionanocomposites with high surface area and porosity are just a few of the methods many benefits. The conventional steps in the freeze-drying process are as follows: (1) creating a solution or suspension with biopolymers and nanomaterials, (2) freezing the solution or suspension and (3) eliminating solvents at lower temperatures and pressures. When creating bionanocomposites for use in biomedical applications such as tissue engineering drug delivery and wound healing, this freeze-drying method is especially beneficial. For example, collagen chitosan and nanoparticle-based freeze-dried bionanocomposites have shown improved mechanical biocompatible and bioactive qualities making them appropriate for use in skin tissue engineering. Additionally the synthesis of bionanocomposites with unique characteristics like electrical conductivity thermal stability and magnetic responsiveness can be facilitated by the freeze-drying technique expanding their potential uses in a variety of industries such as biotechnology energy storage and catalysis. In conclusion, the synthesis of bionanocomposites with specific properties and uses is made possible by the freeze-drying technique.

2.9.8 Green Synthesis Approaches

Green synthesis approaches which use sustainable and environmentally friendly methods to synthesize bionanocomposites are a rapidly developing field. This method is founded on the ideas of green chemistry which seek to reduce the negative effects of chemical synthesis on the environment by utilizing renewable resources cutting waste and staying away from toxic substances. In order to create and assemble nanomaterials and biopolymers green synthesis techniques for bionanocomposites generally employ biological templates such as enzymes or microorganisms and natural solvents such as water or plant extracts. Bionanocomposites with specific qualities like sustainability biodegradability and biocompatibility can be produced using this method. Among the many benefits of using green synthesis techniques for bionanocomposites are decreased production costs, enhanced product safety and a smaller environmental impact. For instance, it has been demonstrated that using plant extracts such as curcumin or tea polyphenols can make it easier to create metal nanoparticles and bionanocomposites with antioxidant and antimicrobial qualities. Similarly bionanocomposites with specific characteristics like mechanical strength and biodegradability have been created by employing enzymes like cellulase or lipase. By offering sustainable and environmentally friendly methods for the synthesis of cutting-edge materials, green synthesis approaches for bionanocomposites have the potential to completely transform a number of industries including biomedical engineering energy storage and environmental remediation.

Green synthesis techniques are rapidly emerging discipline that encompasses the utilization of environmentally friendly and sustainable methodologies for the synthesis of bionanocomposites. These approaches are based in the principles of green chemistry that are specifically designed to minimize the environmental impact associated with chemical synthesis by employing renewable resources, reducing waste, and eschewing the use of toxic and hazardous substances. The development of bionanocomposites via green synthesis methods offers a promising route for creating advanced materials with tailored properties. This approach leverages natural solvents, such as water or plant extracts, and biological templates, including enzymes or microorganisms, to synthesize and assemble nanomaterials and biopolymers.

Green synthesis methods for bionanocomposites have several advantages such as decreased production costs less of an adverse effect on the environment and improved product safety. Tea polyphenols and curcumin for example are plant extracts that have been demonstrated to stimulate the production of metal nanoparticles and bionanocomposites with antioxidant and antimicrobial qualities. In order to create bionanocomposites with particular qualities like mechanical strength and biodegradability enzymes like lipase and cellulase have also been employed. The use of green synthesis techniques for bionanocomposites has broad ramifications for a number of industries such as energy storage environmental remediation and biomedical engineering. Researchers can develop creative environmentally friendly solutions for a range of industries by utilizing the potential of green synthesis techniques paving the way for a more sustainable future. Green synthesis methods have many benefits for bionanocomposites with a particular emphasis on decreased production costs improved product safety and a smaller environmental impact. For example, the use of plant extracts such as flavonoids or tea polyphenols has shown promise in encouraging the production of metal nanoparticles and bionanocomposites with antioxidant and antimicrobial qualities. Additionally enzymes like cellulose or lipase have been used to create bionanocomposites with particular qualities like mechanical strength and biodegradability. By providing environmentally friendly and sustainable methods for producing cutting-edge materials, green synthesis techniques for bionanocomposites have the potential to revolutionize fields such as energy storage biomedical engineering and environmental remediation.

2.9.9 3D Printing

3D printing is also referred to as bioprinting that has revolutionized the field of fabrication of bionanomaterials especially in the field biomedical engineering, regenerative medicine, and drug delivery. Bionanomaterials such as nanocomposites (e.g., PLA-hydroxyapatite), polymer-ceramic blends, nanomaterials based on carbon (e.g., gold for targeted drug delivery and silver for antimicrobial coatings), and biopolymers (e.g., carbon nanotubes for bone regeneration and graphene for conductive implants) can be fabricated precisely and cost-effectively fabricated using a variety of 3D printing techniques including Fused Deposition Modeling (FDM), Stereolithography (SLA), Selective Laser Sintering (SLS), Inkjet Bioprinting, and Electrospinning-Assisted 3D Printing. The combination of 3D printing and nanotechnology has the potential to revolutionize healthcare by enabling customized implants advanced drug delivery methods and better patient care.

Depending on how the material is deposited and processed, these 3D printing methods operate differently. A thermoplastic filament containing nanomaterials is melted and then extruded layer by layer to create structures using FDM. In order to create complex biomedical devices SLA uses a photopolymer resin with embedded nanoparticles that is selectively cured with a laser, whereas to fabricate porous and bioactive scaffolds that are appropriate for implants SLS uses a powerful laser to sinter powdered biomaterials. In order to create tissue-like structures inkjet bioprinting uses bio-inks made of living cells and nanomaterials that are precisely dispensed. Through the use of additive manufacturing and electrospinning techniques, electrospinning-assisted 3D printing creates scaffolds reinforced with nanofibers that resemble the extracellular matrix. It is anticipated that future developments in biofabrication multi-material printing and AI-assisted design will solve these issues and improve biomedical applications. Through customized implants advanced drug delivery systems and better patient care the combination of 3D printing and nanotechnology has enormous potential to revolutionize medicine possibly lowering healthcare costs and increasing access to cutting-edge treatments.

2.10 Sustainability in Bionanocomposite Design

Bionanocomposites (BNCs) are advanced materials composed of bio-based polymers (biopolymers) combined with nanoscale fillers (e.g., nanoparticles, nanofibers). These composites are increasingly being recognized for their potential in sustainable applications due to their biodegradable nature, low environmental impact, and unique mechanical properties. However, the design of sustainable BNCs is multifaceted, requiring attention to raw material sourcing, production processes, and end-of-life scenarios. Below, we delve deeper into the principles of sustainable materials design, the role of eco-friendly nanomaterials, the use of Lifecycle Assessment (LCA), and the challenges associated with achieving sustainability in BNC development (Sharma et al. 2016).

Sustainable design in bionanocomposites involves several key principles aimed at reducing the environmental impact throughout the material's life cycle from raw material extraction to disposal. The first step in sustainable bionanocomposite design is the use of renewable raw materials, such as biopolymers, that can be sourced from nature rather than petroleum-based materials. This reduces reliance on fossil resources and mitigates some of the environmental challenges associated with synthetic polymers, such as petroleum depletion and plastic waste. Biopolymers like polylactic acid (PLA), cellulose, chitosan, and starch are derived from renewable resources and are biodegradable under certain environmental conditions. For instance, PLA is derived from renewable plant-based resources like corn or sugarcane, and it can biodegrade into lactic acid in industrial composting conditions (Auras et al. 2010). PLA is widely used in food packaging, biomedical devices, and agricultural films. When compared to conventional petroleum-based plastics, PLA has a significantly lower carbon footprint, making it an attractive alternative for sustainable applications (Friedrich et al. 2020). Cellulose (the most abundant natural polymer on Earth) is another sustainable material that can serve as both a matrix or a filler in BNCs. Nanocellulose, derived from wood or agricultural residues, offers mechanical reinforcement, transparency, and biodegradability, making it ideal for use in packaging, coatings, and other applications (Bakar et al. 2020). Cellulose-based composites not only replace petroleum-based materials but also provide higher strength and lower weight.

Bionanocomposites' sustainability encompasses not only the sources of their raw materials but also the manufacturing processes. Conventional composite materials frequently use hazardous chemicals and high-energy processes. Avoiding the use of hazardous materials, minimizing waste and lowering energy consumption are the goals of sustainable bionanocomposite design. One sustainable strategy is to employ low-energy or solvent-free processes. For instance, a popular technique that does away with the need for hazardous solvents is the melt blending of biopolymers and nanofillers. This lessens chemical waste in addition to the carbon footprint (Siqueira et al. 2010). Cold pressing and melt extrusion are also viable low-energy techniques. These methods do not require high temperatures and energy-intensive steps, making them more sustainable compared to traditional processes like injection molding (Lu et al. 2014). To make bionanocomposite films and fibers for textile and packaging applications for example extrusion is utilized. High temperatures and dangerous chemicals are frequently used in traditional nanoparticle synthesis techniques. Nonetheless, methods for green chemistry are being developed to create nanoparticles in a way that is more ecologically friendly. Using bacteria, fungi, or plant extracts for biological synthesis is one such technique. Compared to traditional chemical synthesis these techniques generate fewer byproducts and employ renewable resources (Baccile et al. 2016). For example, plant-based routes to produce nanocellulose or silver nanoparticles are gaining traction for their low toxicity and sustainability.

In order to minimize environmental harm sustainable bionanocomposite design takes into account the materials end-of-life and makes sure it can be recycled or biodegraded. To make sure that bionanocomposites don't contribute to long-term pollution, bio-based polymers like PLA and polyhydroxyalkanoates (PHA) must be biodegradable. Non-toxic byproducts are produced when these materials decompose (e.g., carbon dioxide or lactic acid) that the environment absorbs. For disposable goods like packaging agricultural films or medical equipment, bionanocomposites are therefore a desirable alternative. Another important factor to take into account is the bionanocomposites' recyclability in addition to their biodegradability. Materials that are recycled and reused in closed-loop systems support the circular economy. For instance bio-based composite materials used in the automotive and construction sectors can be made to be disassembled and their materials recovered extending their lifespan and cutting down on waste (Müller et al. 2020).

The mechanical thermal and barrier qualities of bio-based polymers can be greatly improved by nanoparticles and nanofibers. However, sourcing and processing these nanomaterials in an environmentally conscious manner are crucial to preserving the sustainability of bionanocomposites. Plants algae and microbes are common biological sources of sustainable nanomaterials. Nanocellulose for example is one of the most popular sustainable nanomaterials and can be made from waste materials plant fibers or agricultural residues (Dufresne 2013a). Excellent biodegradability transparency and reinforcement make nanocellulose ideal for use in a variety of bionanocomposites such as those used in electronics packaging and biomedicine. Additionally chitosan nanoparticles—which are made from chitin a biopolymer present in crustacean shells—act as sustainable nanofillers in BNCs. Because of their antimicrobial qualities and biocompatibility, chitosan-based nanocomposites are being investigated extensively which makes them perfect for use in food packaging and biomedical applications (Venkatesan et al. 2015).

Green chemistry techniques such as biological or ecologically friendly chemical synthesis pathways that eschew dangerous reagents and solvents are pivotal in mitigating the environmental impact of microparticle production. Utilizing plant extracts fungi bacteria or algae biological methods create nanoparticles in a sustainable and safe way. Plant-based extracts for instance can be used to create silver nanoparticles (AgNPs) providing a non-toxic substitute for traditional chemical processes (Hernandez-Sierra et al. 2009). Bacterial cellulose which is produced by specific bacteria can also be readily converted into nanocellulose for use in BNCs. Another example of green synthesis which eschews harsh chemicals or solvents is the production of nanocellulose through enzymatic processes or the acidic treatment of plant residues. When compared to the use of traditional chemical treatments like sulfuric acid hydrolysis this method guarantees a minimal environmental impact.

Lifecycle Assessment (LCA) is a comprehensive method used to evaluate the environmental impact of bionanocomposites throughout their entire life cycle. This tool compares the environmental impact of materials processes and products from the extraction of raw materials to disposal assisting in the assessment of their sustainability. Bionanocomposites' environmental effects including energy use greenhouse gas emissions water consumption and waste production are measured as part of Lifecycle Assessment (LCA). Additionally it offers comprehensive quantitative information on the life cycle effects of materials which helps choose the most environmentally friendly choices. Research indicates that compared to conventional petroleum-based composites bio-based nanocomposites like those derived from PLA and nanocellulose typically have a lower carbon footprint (Pereira et al. 2019).

Additionally bio-based composites can be compared using LCA (e.g., G. PLA/nanocellulose) with traditional petroleum-based composites like carbon fiber-reinforced polymers (CFRP) or glass fiber-reinforced plastics (GFRP). Auras et al.'s comparative LCA study demonstrated how PLA-based bionanocomposites used less energy and produced fewer greenhouse gas emissions than PET-based composites (Auras et al. 2010). Similarly, bio-based composites tend to use less water and generate less solid waste than traditional materials (Müller et al. 2020). Due to their simpler processing requirements and lower melting points, bio-based composites frequently require less energy to manufacture. Furthermore compared to synthetic fibers natural fibers like flax hemp or jute have less of an impact on the environment during production and processing.

Even though bionanocomposites might have a lot to offer to the environment in attaining the desired results such as improved strength, barrier qualities and thermal stability. Their impact on the environment is still unresolved. High-performance nanomaterials like carbon nanotubes or inorganic nanoparticles may improve the qualities of BNCs, but because of their resource- and energy-intensive manufacturing processes, they may also have a greater overall environmental impact. In BNC design, a major challenge is striking a balance between these performance requirements and sustainability objectives. Sustainable nanomaterials like nanocellulose and bio-based polymers are frequently more costly to produce than their petroleum-based equivalents (Dufresne 2008). One of the biggest challenges is increasing the production of these materials to an industrial scale while keeping costs down. Although they might be more environmentally friendly, bionanocomposites are still more expensive than conventional materials in industries like construction, automotive and packaging. Research is being done to create more economical methods for mass-producing these materials.

One promising way to lessen the environmental impact of material systems is through the design of sustainable bionanocomposites. With an emphasis on green chemistry, lifecycle considerations and renewable sourcing bionanocomposites can provide improved properties with little environmental impact. However, issues like ensuring large-scale implementation high production costs and striking a balance between sustainability and performance must be resolved. Bionanocomposites have the potential to be a key component in the advancement of sustainable material science with more study and development.

2.11 Challenges in the Development and Commercialization of Bionanocomposites

Bionanocomposites (BNCs) come with a number of technical, financial, legal, and market-based obstacles in their development and commercialization. Although they have promising applications in a variety of industries including packaging biomedical devices environmental remediation and energy storage these obstacles may prevent their widespread use and slow down their large-scale adoption. A thorough examination of the difficulties encountered in the creation and marketing of bionanocomposites can be found below.

The dispersion of nanoparticles within the polymer matrix is one of the biggest obstacles in the development of bionanocomposite (BNC) materials. Nanoparticles like nanocellulose graphene oxide or clays have a tendency to clump together due to attractive forces because of their small size and large surface area (e.g., Van der Waals electrostatic interactions or hydrogen bonds). These nanoparticles become less effective due to agglomeration which degrades the BNCs general qualities. Inadequate dispersion can impair the composites mechanical thermal and barrier qualities by causing weak spots and an uneven distribution of stress. For instance clustering of the nanocellulose particles rather than uniform dispersion could drastically lower the PLA-nanocellulose composites tensile strength and flexibility. Electrical conductivity biodegradability and barrier performance are among the functional characteristics that can be impacted by irregular nanoparticle dispersion. Uneven dispersion of nanoclays in packaging applications may weaken the oxygen or moisture barrier making the composite less useful for food preservation. Functionalizing the surfaces of nanoparticles with surfactants polymers or organic molecules that increase compatibility with the polymer matrix is a popular technique to improve dispersion. For example, nanocelluloses dispersion in hydrophobic matrices like PLA can be increased by surface modification with polyvinyl alcohol (PVA) or cationic surfactants (Zhao et al. 2014).

High-shear mixing, sonication, and twin-screw extrusion are frequently used techniques to achieve better dispersion by applying mechanical forces. These processes can help break up nanoparticle agglomerates and distribute them evenly throughout the polymer matrix (Rhim et al. 2013b).

The biodegradability of bio-based polymers such as PLA, PHA or starch is the main benefit of bionanocomposites. This benefit does have a drawback though as bio-based polymers frequently have worse mechanical qualities than traditional synthetic polymers including low tensile strength high brittleness and poor thermal stability. It is still very difficult to improve mechanical qualities without compromising biodegradability (State-of-the-art advances in nanocomposite … 2023).

Nanomaterials like graphene oxide and nanocellulose can greatly improve the matrix mechanical qualities such as its flexibility impact resistance and tensile strength. The overall environmental advantages of the composite may be jeopardized though because certain nanomaterials—particularly carbon nanotubes and graphene oxide—are not biodegradable.

The properties of the material and the amount of nanoparticles are delicately balanced. Increased nanoparticle content can enhance mechanical characteristics but too much of them can cause aggregation or poor matrix compatibility which can impair biodegradability and mechanical performance. As an illustration a high concentration of nanocellulose may improve tensile strength but if the polymer matrix is overloaded may decrease the composites overall biodegradability (Friedrich et al. 2020).

To enhance mechanical properties bio-based polymers can be blended with other natural or biodegradable polymers. For instance adding PHA to PLA can improve its flexibility and impact resistance increasing its suitability for a range of uses without sacrificing its biodegradability. Achieving the ideal balance between mechanical reinforcement and biodegradability requires careful control of nanoparticle loading. In order to maximize the materials qualities while maintaining its environmental friendliness experimental designs frequently incorporate different nanoparticle contents. The commercialization of bionanocomposites is hindered by their inability to scale from laboratory-scale to industrial production. To make these materials feasible for broad use a number of scalability and economic concerns need to be resolved.

The production cost of bionanocomposites can be increased by the energy-intensive procedures used in their fabrication such as melt blending solution casting and extrusion. Furthermore, the requirement for specific tools like twin-screw extruders or high-shear mixers makes it challenging and expensive to convert laboratory-scale production to industrial production. The cost of raw materials might be lowered by investigating alternate less expensive sources of nanoparticles such as using agricultural waste to make nanocellulose.

Using plant-based waste from sectors such as agriculture for instance, using corn stover or rice husks to make nanocellulose could significantly reduce production costs. Green chemistry techniques for the synthesis of nanoparticles can reduce production costs and environmental impact. The economic impact of producing nanoparticles has been shown to be reduced by techniques like bio-based synthesis which uses mild reaction conditions and renewable resources to produce the particles.

The addition of nanoparticles to bionanocomposites as they approach commercialization raises questions about safety and legal requirements. Since nanomaterials frequently behave differently from bulk materials, it is crucial to assess the possible risks to the environment and human health. Because of their small size, nanoparticles have special characteristics that raise questions about their potential toxicity. It's capacity to break through biological barriers (skin cell membranes) prompts inquiries regarding their possible buildup within the environment or within the body. The antimicrobial qualities of silver nanoparticles for example make them popular but they can also contaminate food and the environment endangering aquatic ecosystems and human health. When exposed to specific environmental conditions like heat moisture or UV radiation nanoparticles have the potential to release potentially harmful substances over time. Long-term ecological hazards could result from the buildup of hazardous metals or other substances in the environment. The safety of nanoparticles in consumer goods must be assessed through thorough toxicological research. This involves evaluating their propensity for cellular uptake bioaccumulation and long-term environmental effects. Furthermore establishing precise safety guidelines for nanoparticles akin to current chemical safety laws may aid in reducing possible hazards (Maynard et al. 2011).

Many industries remain wary of bio-based products due to concerns that they might not function as well as conventional materials especially in demanding applications like construction or automotive parts. BNCs must for instance fulfill strict performance requirements in terms of strength resilience to environmental influences and durability to become more widely accepted in markets where synthetic polymers have historically held a dominant position. Adoption may be challenging due to the higher initial cost of bio-based and nano-enhanced materials when compared to their petroleum-based counterparts despite the potential environmental benefits. Traditional materials may be preferred by cost-conscious industries unless bionanocomposites show appreciable cost or performance gains. From technical problems with nanoparticle dispersion and material performance to financial limitations and legal obstacles, the development and commercialization of bionanocomposites present numerous and intricate challenges. Though tremendous progress is being made in each of these fields, bionanocomposites will be able to reach their full potential as sustainable substitutes for traditional materials once these obstacles are removed. With further study, creative production techniques and well-defined legal frameworks' bionanocomposites may emerge as a key component of sustainable material solutions in future.

2.12 Future Directions and Emerging Trends

Bionanocomposites represent a rapidly evolving field that integrates renewable bio-based materials with nanotechnology to create multifunctional materials with enhanced properties. As industries move toward sustainability and greener alternatives, the potential for BNCs to replace petroleum-based materials is increasingly recognized. However, the continued development of bionanocomposites requires addressing challenges related to performance, scalability, and regulatory approval. The future of BNCs will likely be shaped by several key trends and emerging technologies, which can drive innovation and expand their applications.

Nanocellulose (NC), including nanocrystalline cellulose (NCC) and nanofibrillated cellulose (NFC), is one of the most promising bio-based nanomaterials used in BNCs due to its renewable nature, biodegradability, and mechanical properties. Researchers are exploring various methods to enhance the functionality of NC, such as modifying its surface chemistry to improve its interaction with other polymers.

The surface of nanocellulose can be functionalized with hydrophobic, antibacterial, or antioxidant groups to improve its performance in specific applications, such as food packaging, medical devices, and environmental remediation (Jiang et al. 2021). NFC-based composites are particularly promising for their high surface area and mechanical strength, which make them ideal candidates for a wide range of applications from biodegradable packaging to automotive parts.

There is growing interest in combining nanocellulose with other nanomaterials such as graphene, carbon nanotubes (CNTs), and metal nanoparticles to create multifunctional BNCs. These combinations can provide improved mechanical strength, conductivity, thermal stability, and antibacterial properties. For instance, the combination of silver nanoparticles with nanocellulose in packaging materials could provide antimicrobial properties to the packaging while maintaining biodegradability.

The use of biogenic nanoparticles derived from agricultural and industrial waste is an emerging trend in BNC development. These nanoparticles are derived from renewable resources, reducing the need for fossil-derived materials. Researchers are exploring the extraction of nanomaterials from agricultural byproducts such as rice husks, corn stalks, and sugarcane bagasse. These waste materials can be processed into nano-silica, cellulose nanocrystals, and lignocellulosic fibers, which can then be integrated into bionanocomposites to improve their mechanical and barrier properties (Abdelkader et al. 2020).

One of the major drivers for the development of bionanocomposites is the increasing demand for sustainable manufacturing processes that minimize energy consumption and chemical waste. Traditional nanoparticle synthesis methods often involve the use of hazardous chemicals or energy-intensive processes. As a response, green chemistry approaches are being adopted to synthesize nanoparticles with minimal environmental impact. The use of biological agents such as bacteria, fungi, and plants to synthesize nanoparticles is gaining traction. These biological processes are often more environmentally friendly than chemical synthesis methods and can be scaled up more easily. The integration of additive manufacturing or 3D printing technologies with bionanocomposites opens up new avenues for producing complex, customized components with minimal waste. For instance, biodegradable filaments made from PLA, PHA, or starch reinforced with cellulose nanoparticles are being developed for 3D printing of bio-based components in industries such as automotive, aerospace, and medical devices (Dufresne 2013b).

One of the most promising and widely researched applications for BNCs is biodegradable food packaging. As global concerns about plastic waste grow, bio-based alternatives are being sought for food packaging materials. Nanocellulose and other nanoparticles can significantly improve the barrier properties of bio-based films, making them resistant to oxygen, moisture, and UV radiation. This is crucial for extending the shelf life of food products and reducing food waste (Rhim et al. 2013a).

Bionanocomposites made from materials like chitosan and nanocellulose are being developed for use in wound dressings. These materials can provide a moist environment for wound healing and have inherent antimicrobial properties to prevent infections (Ahmed et al. 2021). Bionanocomposites are emerging as innovative materials for environmental applications, particularly for water purification and soil decontamination. Nanocellulose-based composites are being used to adsorb heavy metals (e.g., lead, mercury) from contaminated water. The high surface area and functional groups on the surface of nanocellulose can selectively bind to metal ions, making them effective in removing pollutants from wastewater (Thostenson et al. 2005).

In future, Lifecycle Assessment (LCA) will play a crucial role in evaluating the environmental footprint of bionanocomposites. Understanding the energy consumption, waste generation, and carbon footprint throughout the production process will be essential for determining the true sustainability of BNCs compared to conventional materials. As more data becomes available, LCA tools can help optimize the synthesis of BNCs, ensuring that they not only offer superior material properties but also provide environmental benefits in terms of reduced carbon emissions and resource conservation.

The future of bionanocomposites is promising, driven by the growing demand for sustainable, high-performance materials across various industries. Advanced nanomaterials, green synthesis methods, and innovative applications will continue to fuel the development of BNCs. As research progresses, the integration of BNCs into biodegradable packaging, biomedical applications, environmental remediation, and additive manufacturing will expand. Addressing challenges related to scalability, cost, and regulatory approval will be crucial to unlocking their full potential and driving large-scale commercialization.

2.13 Conclusions

Because of their distinct qualities and wide range of uses, bionanocomposites have become a promising class of materials. The final characteristics and functions of these materials are largely determined by the fabrication processes used in their synthesis. In situ polymerization, solution mixing, melt mixing, electrospinning, layer-by-layer assembly, the sol–gel method, freeze-drying techniques and green synthesis approaches are some of the fabrication methods used in the synthesis of bionanocomposites that have been clarified by this thorough overview. The choice of a particular method depends on the specific needs of the material as each fabrication technique has unique benefits and drawbacks. These fabrication methods' ability to modify the structure, composition, and characteristics of bionanocomposites has opened up new avenues for their use in a variety of industries such as energy storage environmental remediation biomedical engineering and sustainable development. It is expected that new fabrication techniques and improvements to current methods will be made as the field of bionanocomposites develops further resulting in materials with improved performance and properties. Furthermore, the combination of bionanocomposites with other emerging technologies like biotechnology, nanotechnology, and artificial intelligence is probably going to encourage the development of novel applications. Ultimately, this thorough review gives a thorough grasp of the manufacturing processes used to create bionanocomposites highlighting their benefits, drawbacks, and uses. It is projected that bionanocomposites will have a greater impact on materials science and engineering in future as this field unfolds further. Advanced bionanocomposites which are made of biomaterials are an important field of study with enormous promise for long-term technological breakthroughs. By combining nanotechnology with biomaterials, which are made from natural renewable resources—bionanocomposites with improved mechanical, thermal, and biological qualities can be created. Numerous fields such as materials science, environmental protection, and medicine have promising uses for these materials. By utilizing biomaterials in a sustainable manner, these bionanocomposites are less dependent on synthetic polymers and other non-renewable resources and are guaranteed to be both environmentally benign and functional. By reducing energy consumption, and low chemical waste fabrication techniques like green synthesis methods and scalable processing strategies may further improve the sustainability aspect. Furthermore, the ongoing investigations of novel biodegradable and bio-based nanofillers including protein-based nanostructures, nanocellulose, and nanoclays show promises for creating materials that can be easily incorporated into biological systems and ecosystems without posing long-term environmental hazards. Because they are compatible with human tissues and have the potential for controlled release mechanisms, these cutting-edge bionanocomposites are opening the door for novel drug delivery systems scaffolds for tissue engineering and materials for wound healing in the medical field. These developments could significantly improve patient outcomes. Future studies are still needed to address issues like reproducibility understanding long-term material interactions with biological systems and scaling up fabrication techniques. All things considered, the creation of bionanocomposites based on biomaterials offers a very promising route to a future that is more efficient, clean, and sustainable for both industrial and medical applications.

References

Abdelkader F et al (2020) Agricultural waste nanomaterials for sustainable bionanocomposites. Environ Sci Technol 54(10):6543–6553

Adekoya GJ, Ezika AC, Adekoya OC, Sadiku ER, Hamam Y, Ray SS (2023) Recent advancements in biomedical application of polylactic acid/graphene nanocomposites: an overview. BMEMat. https://doi.org/10.1002/bmm2.12042

Ahmed S et al (2021) Chitosan-based bionanocomposites for wound healing. J Biomed Mater Res 109(5):712–725

Alessandrini A, Facci P (2011) Unraveling lipid/protein interaction in model lipid bilayers by atomic force microscopy. J Mol Recognit 24:387–396

Auras R et al (2010) Sustainability of polylactic acid (PLA) and its applications. Polym Eng Sci 50(7):1315–1324

Baccile N et al (2016) Green synthesis of cellulose nanomaterials: opportunities and challenges. Environ Sci Technol 50(4):1681–1689

Bakar SA et al (2020) Recent advances in cellulose nanocomposites for sustainable applications. Front Mater 7:51

Bharathi VSK, Jayas DS (2024) Evolution of bionanocomposites: innovations and applications in food packaging. Foods 13(23):3787

Cui ZK, Kim S, Baljon JJ, Wu BM, Aghaloo T, Lee M (2019) Microporous methacrylated glycol chitosan montmorillonite nanocomposite hydrogel for bone tissue engineering. Nat Commun 10:3523

Deng S et al (2022) Fabrication of biodegradable and biocompatible functional polymers for anti-infection and augmenting wound repair. Polymers 15(1):120. https://doi.org/10.3390/polym15010120

Dufresne A (2008) Polysaccharide nano crystal reinforced nanocomposites. Can J Chem 86:484–494

Dufresne A (2013a) Nanocellulose: a new ageless biorenewable material. Compos Sci Technol 72(2):136–145

Dufresne A (2013b) Nanocellulose: from fundamentals to advanced applications. Green Chem 15(4):798–810

El Mir L, Amlouk A, Barthou C, Alaya S (2007) Synthesis and luminescence properties of $ZnO/Zn_2SiO_4/SiO_2$ composite based on nanosized zinc oxide-confined silica aerogels. Phys B Condens Matter 388:412. https://doi.org/10.1016/j.physb.2006.06.151

Friedrich H et al (2020) Sustainable bioplastics: materials, properties, and applications. Nat Sustain 3(3):160–171

Furukawa H, Cordova KE, O'Keeffe M, Yaghi OM (2013) The chemistry of metal-organic frameworks for CO_2 capture, regeneration and conversion. Science 341(6149):1230444

Geim AK, Novoselov KS (2007) The rise of graphene. Nat Mater 6(3):183–191

Giner-Grau S, Lazaro-Hdez C, Pascual J, Fenollar O, Boronat T (2024) Enhancing polylactic acid properties with graphene nanoplatelets and carbon black nanoparticles: a study of the electrical and mechanical characterization of 3D-printed and injection-molded sample. Polymers 16(17):2449. https://doi.org/10.3390/polym16172449

Hedenqvist MS, Backman A et al (2006) Compos Sci Technol 66(13):2350

Hernandez-Sierra JF et al (2009) Biosynthesis and antimicrobial activity of silver nanoparticles. J Nanobiotechnol 7:1–7

Idrees H, Zaidi SZJ, Sabir A, Khan RU, Zhang X, Hassan S-U (1970) A review of biodegradable natural polymer-based nanoparticles for drug delivery applications. Nanomaterials 2020:10

Jiang W et al (2021) Functionalization of nanocellulose for high-performance bionanocomposites. Carbohydr Polym 246:116561

Kaou MH, Furkó M, Balázsi K, Balázsi C (2023) Advanced bioactive glasses: the newest achievements and breakthroughs in the area. Nanomaterials 13(16):2287. https://doi.org/10.3390/nano13162287

Katas H, Raja AB, Kumar S (2013) Chitosan nanoparticles for drug delivery. J Pharm Pharmacol 65(2):155–165
Kerman K, Saito M, Yamamura S, Takamura Y, Tamiya E (2008) Nanomaterial-based electrochemical biosensors for medical applications. Trends Anal Chem 27:585. https://doi.org/10.1016/j.trac.2008.05.004
Kim J, Lee J, Kim B (2019) Dispersion of carbon nanotubes in polymer matrix: a review. J Ind Eng Chem 71:1–13
Kohane D, Langer R (2008) Polymeric biomaterials in tissue engineering. Pediatr Res 63:487–491. https://doi.org/10.1203/01.pdr.0000305937.26105.e7
Kumar A, Kumar V, Kumar S (2017) Biodegradable and biocompatible nanomaterials for biomedical applications. J Biomed Mater Res B Appl Biomater 105(4):753–764
Lee JH, Song NB, Jo WS, Song KB (2014) Effects of nano-clay type and content on the physical properties of sesame seed meal protein composite films. Int J Food Sci Technol 49:1869–1875. https://doi.org/10.1111/ijfs.12496
Lu C et al (2014) Processing and properties of sustainable biocomposites from renewable resources. Biomacromolecules 15(3):815–827
Macha IJ, Ben-Nissan B, Müller W (2017) Kinetics and the theoretical aspects of drug release from PLA/HAp thin films. Key Eng Mater 758:113–119
Maynard AD et al (2011) Nanotechnology and nanomaterials: a primer for regulators. Nat Nanotechnol 6(6):350–358
Mohan S, Oluwafemi OS, Kalarikkal N, Thomas S, Songca SP (2016) Biopolymers—application in nanoscience and nanotechnology. In: Recent advances in biopolymers, vol 1. InTech Open, Rijeka, Croatia, pp 47–66
Moon RJ, Martini A, Nairn J, Simonsen J, Youngblood J (2011) Cellulose nanomaterials review: structure, properties and nanocomposites. Chem Soc Rev 40(7):3941–3954
Mousa MH, Dong Y, Davies IJ (2016) Recent advances in bionanocomposites: preparation, properties, and applications. Int J Polym Mater Polym Biomater 65(5):225–254. https://doi.org/10.1080/00914037.2015.1103240
Müller CM et al (2020) Life cycle assessment of bio-based composite materials. J Clean Prod 276:124019
Niinomi M (2002) Recent metallic materials for biomedical applications. Metall Mater Trans A 33(3):477–486. https://doi.org/10.1007/s11661-002-0109-2
Ortenzi MA et al (2015) Evaluation of crystallinity and gas barrier properties of films obtained from PLA nanocomposites synthesized via "in-situ" polymerization of l-lactide with silane-modified nanosilica and montmorillonite. Eur Polym J 66:478–491
Pereira SR et al (2019) Environmental impacts of bionanocomposites: a review. J Polym Environ 27(2):249–265
Rhim JW et al (2013a) Bio-based nanocomposites for food packaging applications. Food Sci Technol 19(2):134–145
Rhim JW et al (2013b) Food Sci Technol Res 19(2):135–145
Sharma S, Dua A, Malik A (2016) Bionanocomposites: a review. J Nanosci Nanotechnol 16(1):1–14
Sharma J, Gole VL, Siddiqui MA (2023) Bionanocomposites biological aspects and biomedical applications. Polym Nanocompos Fabr Appl
Shchipunov Y (2012) Bionanocomposites: green sustainable materials for the near future. Pure Appl Chem 84(12):2579–2607. https://doi.org/10.1351/PAC-CON-12-05-04
Siqueira G et al (2010) Nanocellulose in bionanocomposites. Macromolecules 43(21):8857–8868
State-of-the-art advances in nanocomposite and bio-nanocomposite polymeric materials: a comprehensive review (2023) Adv Colloid Interface Sci 318:102955
Thostenson ET, Li C, Chou T-W (2005) Nanocomposites in context. Compos Sci Technol 65:491–516
Venkatesan J et al (2015) Chitosan-based biocomposites for biomedical applications. Biomed Mater 10(3):034101

Zhao X et al (2014) Cellulose nanocrystal reinforcement in PLA-based composites. Carbohydr Polym 114:324–331

Dr. Vernica Verma is working as a research assistant, Department of Physics, University of Lucknow, and is a former assistant professor at the Department of Physics, BBDNIIT Lucknow. Dr. Vernica completed her Ph.D. in physics from the Department of Physics, University of Lucknow. She did her B.Sc. and M.Sc. in physics from the University of Lucknow in 2005 and 2007, respectively. Her specialization in research is fabrication, characterization like XRD, SEM, AFM, UV–VIS, FT-IR, Raman, gamma irradiation, etc., and humidity and gas sensors.

Dr. Priya Gupta has been working as a lecturer at Manyawar Kanshiram Government Polytechnic, Tirwa, Kannauj, since January 2022. Ms. Priya completed Ph.D. from the Integral University, Lucknow, in collaboration with Lucknow University, Lucknow, in the year 2021. She did her M.Tech. from Madan Mohan Malviya Engineering College, Gorakhpur, in 2014, and her B.Tech. from Integral University, Lucknow.

Dr. Peramjeet Singh has completed his Ph.D. from the Department of Physics, University of Lucknow, Uttar Pradesh, India, in the field of materials science. Mr. Peramjeet did his M.Sc. from Barkatullah Vishwavidyalaya, Bhopal, and his B.Sc. from GDC, Udhampur, University of Jammu, India, in 2017 and 2015, respectively. His area of research is nanotechnology, gas sensors, and the characterization of materials.

Prof. Narendra Kumar Pandey received his degree (hons.) in physics from St. Xavier's College, Ranchi, India, in 1985, and the M.Sc. degree in Physics and the Ph.D. degree from the Indian Institute of Technology Delhi, Delhi, India, in 1987 and 1995, respectively. He is currently a professor of physics in the Department of Physics, University of Lucknow, Lucknow, India. His areas of interest include solid-state sensors, especially humidity and gas sensing, nanotechnology, optical fiber, sensors, and fiber Bragg grating sensors.

Chapter 3
Biopregs and Hybrid Fibre-Reinforced Bionanocomposites

George Kwame Fobiri, Benjamin Eghan, Raphael Kanyire Seidu, and Benjamin Tawiah

Abstract With the upsurge in new technologies to obtain enhanced materials for specific applications, bionanocomposite materials have gained popularity as they combine the desirable properties of bio-based and nanomaterials. Amongst the numerous approaches to biocomposite material preparation is integrating biopolymers such as chitosan, starch, and cellulose with nanomaterials including carbon nanotube (CNT), graphene, and nanocellulose. To further improve the unique features of bionanocomposites, they are given special treatment through the concept of reinforcement where new materials are introduced into their structures to tailor their applications in engineering. Hybrid fibres and biopregs are prospects in bionanocomposite materials manufacturing. Biopregs, for instance, exhibit environmentally friendly characteristics due to their biodegradable nature, making them sustainable products worth integrating into hybrid fibre such as those obtained from natural or synthetic origins. Studies have identified numerous properties of composite materials, precisely reinforced bionanocomposites, and their prospects in high-performance products. Their properties such as high strength, lightweight, biodegradability, good thermal stability, high surface area, and excellent electrical conductivity and their successful application in various industrial sectors have been explored in past studies. Various hybrid fibre hybrid bionanocomposites such as

G. K. Fobiri (✉)
Department of Fashion Design and Textiles Studies, Kumasi Technical University, Kumasi, Ghana
e-mail: george.kfobiri@kstu.edu.gh

G. K. Fobiri · B. Tawiah
Department of Industrial Art (Textiles), Kwame Nkrumah University of Science and Technology, Kumasi, Ghana

B. Eghan
Faculty of Art, Design and Humanities, De Montfort University, Leicester, UK

R. K. Seidu
School of Fashion and Textiles, The Hong Kong Polytechnic University, Hung Hom, Kowloon, Hong Kong SAR, China

S. C. George and B. Tawiah (eds.), *Advanced Bionanocomposite Materials*, Advanced Structured Materials 241, https://doi.org/10.1007/978-981-96-8225-6_3

cellulose-based, CNT-infused, and chitosan-reinforced have been prepared previously for applications such as structural reinforcement, smart textiles, and packaging purposes due to their improved strength, biodegradability, and flexibility properties. This chapter adds to the discourse by highlighting the advancements, innovative methods, applications, challenges, and prospects of bionanocomposites reinforced with biopregs and hybrid fibres in industrial applications. Potential safety issues concerning environmental impact, chemical, and nanoparticle exposure, with the application of biopregs and hybrid fibre-reinforced bionanocomposites, are also discussed. Lastly, the chapter concludes with some insights into promising areas for biopregs and hybrid fibre-reinforced bionanocomposites. Appropriate recommendations are also given for policy directions.

Keywords Biopregs · Composite materials · Hybrid fibres · Nanomaterials · Biocomposites

3.1 Introduction

Using biopregs and hybrid fibres is an emerging technology in textile material design for sustainable textile manufacturing. The assurance that bio-based materials are environmentally friendly has shifted manufacturer's attention to reinforcing textile materials with hybrid fibres and bio-based materials such as chitosan, starch, and cellulose. For an improved-performance textile, nanomaterial composites comprising biopregs are significant. It is established that biopregs composite materials benefit from a weight reduction of about 20–30% because of higher strength-to-weight and modulus-to-weight ratios (Carvalho et al. 2024). Biopregs form a significant part of composite materials used in several areas, including automotive, sports, fashion, and aerospace. Coating or pre-impregnating textiles with biomaterials renders them more useful in composite textile production. Hybrid fibres on the other hand significantly combine the properties of fibres of different origins to enhance their strength, robustness, lightweight, biodegradability, and other desired properties for various tasks (Ismail et al. 2022; Swolfs et al. 2014). Textile fibre hybridisation helps to overcome the disadvantage of certain types of fibres with the advantage of other fibres (Ismail et al. 2022). Combining natural fibres with synthetic fibres helps to utilise the biodegradability properties of natural fibres and the excellent mechanical performance of synthetic fibres. It is said that plant fibres for instance have drawbacks in hybridisation such as their inability to combine with hydrophobic polymers due to their hydrophilic nature, which necessitates surface treatments such as pre-impregnating with chemicals for compatibility improvement (Elfaleh et al. 2023). Bionanocomposites are birthed through the increased application of nanotechnology in material design using biodegradable products guided by sustainability. Sustainable material design, sometimes called 'design for the future,' involves using green materials and procedures that cause no harm to the environment and human health (Peng et al. 2021). In bionanocomposite material manufacturing, the constituent materials are applied in their tiny state in which their chemical, physical, and biological properties are improved. These properties are hardly felt when the materials are used in their macrostate or larger state. The increased application of materials at the nanoscale level has yielded the concept of nanotechnology which has gained popularity in several sectors including the textile industry.

In material science and engineering, nanoparticles are used to enhance textile materials through nanotechnology. Nanoparticles from functional materials such as carbon, graphene, MXene, and nanocellulose are integrated with biopolymers to produce bionanomaterials. This is done through techniques such as coating, electrospinning, in situ polymerisation, sol–gel, 3D printing, and layer by layer (Azimi et al. 2020; He et al. 2018). Composites containing carbon materials for instance promise good thermal conductivity, high-temperature stability, and high strength (Agarwal 2024; Baig et al. 2021). Due to the sustainable nature of biodegradable composites, gaining support from hybrid fibres widens their applicability in many sectors. Sustainable materials with extraordinary properties have recently been sought in construction, textile and apparel, automotive, agriculture, healthcare and biomedical, aerospace, and packaging industries (Badshah et al. 2022; Joshi et al. 2020; Lambuk et al. 2023; Mazlan et al. 2024; Rahaman et al. 2024; Salleh et al. 2023; Wellbrock et al. 2020). The best approach to strengthening composite materials is through reinforcement with different fibres (Tcherdyntsev 2021; Zia et al. 2021). Composite reinforcement, especially with fibres, is achieved in various ways to obtain a new material with enhanced properties for diverse applications. This process involves the combination of a matrix, often derived from plastics, metals, or ceramics with fibre or fibre-based materials (Greb et al. 2018). The matrix performs a significant function of holding the fibres together for a successful unification (Khandelwal and Rhee 2020; Tóth et al. 2024; Yuhazri et al. 2020). Whilst the type of fibres for reinforcement is important, matrix type is also very crucial in manufacturing sustainable composites. It is observed that many polymer materials used in composite material manufacturing are expensive, involve tedious manufacturing processes, and are not eco-friendly (Sharma and Joshi 2023). Using biomaterials for the matrix in hybrid fibre-reinforced composites promises sustainable and improved composite materials.

Biomaterials are significant in sustainable materials design. However, biomaterials like biopolymers have average mechanical properties that limit their robustness and require reinforcement to increase their strength (Singh and Pandey 2023). Scholars have explored the concept of reinforcement from various perspectives. Risteska et al. (2024) studied the mechanical behaviour of polymer composites prepared with epoxy resin as a matrix and woven, nonwoven, and unidirectional glass fibre-based textiles as reinforced agents pre-impregnated, and found that nonwoven prepreg-reinforced composite had the least resistance to bending stress of 150 MPa, followed by the woven textile sample (450 MPa), and the unidirectional textile with the highest flexural strength (900 MPa). Al Mahmud et al. (2024) also investigated reinforcement of epoxy resin for a composite material through a sustainable approach by using upcycled fabrics and powdered coconut shells to obtain a result on force, bending, and impact tests of 1236.705 N, 86.76 N, and 3.33 J, respectively. In hybrid fibre reinforcement, Islam et al. (2024) realised common challenges of natural fibre-reinforced composites, including low thermal stability and high moisture absorption rate. To combat these challenges, several surface modification techniques are suggested including benzoylation treatment, mercerisation, silane treatment, acetylation process, peroxide treatment, acrylation and acrylonitrile grafting, permanganate, and stearic acid treatment (Elfaleh et al. 2023; Islam et al. 2024; Khan et al. 2024). Notwithstanding the surface modification techniques for natural fibres, it is established that the combination of natural fibres and synthetic fibres produces more enhanced fibre-reinforced composite materials (Swolfs et al. 2014; Gangil et al. 2020). It helps to reduce the cost of manufacturing composite materials (Zuo et al. 2021). Hybrid fibre-reinforced composites comprising natural and synthetic fibres and nanomaterials need recognition for sustainable and cost-effective composite material design, especially bionanocomposites.

Bionanocomposites are of several phases with at least a biological-based component such as chitosan, cellulose, and protein and a nanoparticle phase including metal oxides, nanoclays, carbon nanotubes, etc. organised at one dimension on a nanoscale, where the biopolymers cater for sustainability and eco-friendly practice whilst the inorganic components add good mechanical and thermal stability properties amongst several other desired features (Darder et al. 2007; Kumari et al. 2024; Puiggalí and Katsarava 2017; Shchipunov 2012). In bionanocomposites, an increased mechanical property is attributed to the bond between the nanoparticles and the organic polymers (Perera et al. 2022). The advancement of bionanocomposite design has called for innovative approaches to obtaining new eco-friendly hybrid materials. Past studies have focused on the prospects of bionanocomposites application in food packaging, the biomedical field, and ecological restoration (Perera et al. 2022; Nandhini et al. 2024; Phadtare et al. 2025). Also, using poly(lactic acid) and acetylated microfibrillated cellulose to produce a bionanocomposite material with tunable features (Tingaut et al. 2010), a bionanocomposite material designed by combining cellulose nanofibrils with amyloid fibrils obtained from the curli-specific gene 'A' protein to form a hybrid material with improved mechanical properties (Khatri et al. 2023). Moreover, the use of bionanocomposites in cosmetics, textile industries, automotive industries, agriculture, etc. (dos Santos et al. 2023; Joseph et al. 2020; Kodolova-Chukhontseva et al. 2022; Ravelo-Nieto et al. 2023; Seydibeyoğlu et al. 2023) is discussed in previous studies and reported to be significant materials for sustainable development. To add to the discourse, this chapter highlights the advancements made in bioprepregs and hybrid fibre-reinforced bionanocomposites to reveal the significant improvements in material design employing bionanomaterials and hybrid fibres as chief materials. Innovative methods applied in bionanocomposite and hybrid fibre composite are key to understanding the strategic formation of new materials. The chapter discusses such innovative approaches as resin transfer moulding and vacuum-assisted resin infusion, and their real-world application in bionanocomposite material design. The applications of bionanocomposites reinforced with bioprepregs and hybrid fibres, the challenges involved in the handling and utilisation, especially amid unfriendly weather conditions, regulatory issues, and cost of production are discussed in the chapter. Insights are also given into promising areas for bioprepregs and hybrid fibre-reinforced bionanocomposites such as aerospace, sports, automotive, and food packaging.

3.2 Bioprepregs Classification

Bio-based prepregs are materials that are pre-impregnated with reinforcing fibres combined with bio-based resins. The latter is widely obtained from environmentally friendly renewable sources, hence reducing the heavy reliance on petroleum-based sources. Subsequently, they are developed with good performance properties and a biodegradable nature. The fabrication of these materials consists of a polymeric matrix and the reinforcing phase (Jabbar and Nasreen 2021; Mouritz 2012) which are pre-impregnated with natural fibre sources (Carvalho et al. 2024). These composite materials exhibit good chemical stability, fire resistance, creep and chemical resistance, dampening characteristics, and reduced weight (Carvalho et al. 2024; Somarathna et al. 2024). Based on their inherent properties, they have gained increased attention from manufacturers to advance their sustainability and low-carbon emission goals.

Bio-based prepregs or bioprepregs are classified based on the reinforcing fibre types, resin or polymeric matrix, performance, and applications (Carvalho et al. 2024). Reinforcing fibres are sourced from natural fibres such as cellulose, animal and mineral fibres. For example, the performance of the composite is influenced by the mechanical properties of the natural fibres. Alternatively, properties such as chemical durability, heat resistance, and tensile strength are associated with mineral fibres to improve the eventual composites. The resin or polymeric matrix comes in two material types such as thermoplastic and thermoset. The latter in the curing process are not volatile agents, cure easily, and have good adhesion. The chemical structure of the thermoset is eliminated when the crosslink density is varied (Fronza et al. 2024). For thermoplastic, their thermal and mechanical properties can be declined via continuous reprocessing. Based on this, thermoplastics are more easily degradable than thermosets (McGregor et al. 2017). Subsequently, for performance, the composites' mechanical properties are influenced by the methods and processing conditions (Kirschnick et al. 2024). Furthermore, strength improvements are influenced when the natural reinforcement phase is increased. Due to the fibre types and performance characteristics, prepregs are further classified based on their applications, for example, in sports equipment, construction, automotive, and aerospace.

Alternatively, hybrid fibres, which combine the qualities and properties of two fibre types, are used to prepare bio-based prepregs. The integration of hybrid fibres improves performance whilst ensuring the sustainable nature of the materials. For example, extensive information on composites from hybrid carbon/flax has been reported by Luz et al. (2020), which exhibit good damping properties and mechanical performance. These materials have advanced their application for lightweight industries. Other studies reported good thermal stability and mechanical abilities from composite materials with hybrid bio-fibre/bioceramic (Islam et al. 2022), hybrid biocomposite from bioplastic, wood fibre and talc (Reddivari et al. 2024), and increased strength for hybrid bioPET (Sarker et al. 2018).

3.3 Advancements in Materials for Bioprepregs and Hybrid Fibre-Reinforced Bionanocomposites

Significant progress has been observed in reinforcing composite materials with hybrid fibres and nanomaterials technology. For the past few years, using biodegradable sources for composite materials has been the priority. This is to promote sustainability and ensure safe ecological and harmful-free manufacturing processes. As described by Jabbar and Nasreen (2021), prepregs are pre-impregnated reinforcement with a catalysed or partially cured resin/matrix to serve as foundations for forming high-performance composite materials. Prepregs tapes for instance are useful materials for composite structures in aircraft manufacturing (Mouritz 2012). Bioprepregs are, therefore, the matrix obtained from biological sources to be applied on reinforcement agents such as woven and nonwoven textiles and tows for special applications. The nature of materials used for prepregs is very significant in obtaining bioprepregs that are effective in sustainable composite material manufacturing. Several biodegradable materials methodologies have been employed to manufacture composite materials through prepreg technology. However, it is posited that the concept of prepregs production is protected by trade secrets, hence limiting information on prepregs manufacturing (Somarathna et al. 2024). This notwithstanding, a recent study discusses the preparation of a bionanocomposite film with starch from cassava root (periderm and cortex), galactomannan from locust bean, and cellulose nanofibre from cassava wastes, showing a remarkable biodegradability property of decomposing within five days (Fronza et al. 2024). Also, McGregor et al. (2017) present a methodology involving pre-impregnating natural fibres and thermoplastic composites to produce a hybrid material with improved tensile strength and surface finish. From the research, the composite tape was derived from a prepreg comprising a biomaterial (flax) and a thermoplastic poly(amide) material, resulting in a hybrid sustainable material. Moreover, advanced methodologies such as resin transfer moulding and vacuum-assisted resin infusion are adopted to manufacture a prepreg material (flax fibre impregnated with partially bio-based epoxy matrix) to increase the potential use of bio-based composites in industries (Kirschnick et al. 2024). After a careful study of the performance of each approach, it was revealed that regarding tensile strength and flexural properties, the resin transfer moulding approach was adjudged the best compared with the vacuum-assisted resin infusion. Nonetheless, the vacuum-assisted resin infusion method revealed a strong adhesion between the fibre and the matrix to produce a uniform impregnation. Figure 3.1 shows the resin transfer moulding and vacuum-assisted resin infusion approaches used for the bio-based composite material reinforced with fibres through the prepreg phenomenon.

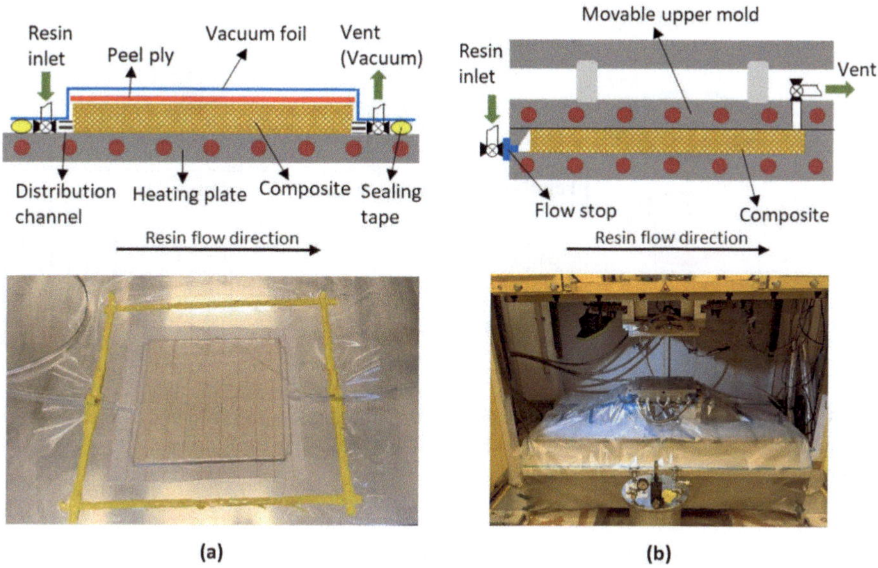

Fig. 3.1 Schematic illustration of the resin transfer moulding and vacuum-assisted resin infusion approaches to bioprepregs and hybrid fibre-reinforced biocomposite manufacturing (Kirschnick et al. 2024) (This is an open-access article distributed under the terms of the *Creative Commons CC-BY* licence)

Hybrid fibre-reinforced biocomposites are known for their biodegradable and good mechanical strength for several applications. Hybrid fibres from an amalgamation of natural and synthetic materials reinforced with nanomaterials such as nanocellulose, graphene, carbon nanotubes, and bio-based nanoparticles have been explored in past studies (Luz et al. 2020; Islam et al. 2022; Reddivari et al. 2024; Sarker et al. 2018; Signorini et al. 2024). Innovative materials and technologies are employed in this situation. In a previous study, a synthesised polyester material and cotton hybrids were used to form fully bio-based composites with improved properties (Dai et al. 2018). In this exploration, polyester was formed from itaconic acid, oxalic acid, and ethylene glycol through a polycondensation approach. The resin was then prepared by dissolving the prepolymer in dimethyl itaconate. The hybrid/composite material was formed by reinforcing the cotton fabric with the polyester resin to obtain the bio-based composite material with improved properties such as tensile strength of ~ 34 MPa, glass transition temperature of ~ 108 °C, and thermal decomposition temperature of 224–276 °C. Figure 3.2 shows a schematic illustration of the fully bio-based composite/hybrid material produced.

A recent study also explored the sustainable features of flax and hemp in hybrid composite preparation with polypropylene for reinforcement purposes (Baysal et al. 2023). In the study, unidirectional prepregs containing flax and polypropylene fibres and nonwoven mats made of hemp/polypropylene fibres are obtained for engineering

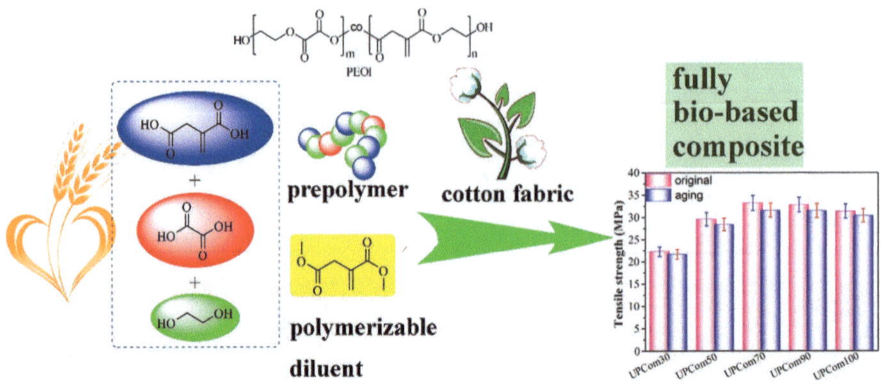

Fig. 3.2 A schematic illustration of fully bio-based cotton fabric-reinforced unsaturated polyester derived from itaconic acid. Reprinted with permission from Dai et al. (2018). Copyright 2018, American Chemical Society

purposes. It is said that the composite materials exhibit diverse properties with desirable sustainable features worth replacing traditional synthetic materials. Similar research was conducted to ascertain the characterisation of nonwoven flax fibres reinforced with polylactic acid, where bioprepregs were obtained from the fibre materials through thermal consolidation (Alimuzzaman et al. 2014). The bioprepregs acted as a sustainable approach to producing reinforced materials with eco-friendly properties. Moreover, Wang et al. (2023) developed a bioprepreg or bio-based resin as a sustainable approach to reinforcing cellulose-based materials for decorative purposes. The bioprepreg, termed aldehyde-free impregnated resin, is prepared with constituents including aqueous acrylate emulsion, water, triethylamine, ethylenediamine, polyvinyl alcohol, polyethylene glycol, titanium dioxide, and aluminium oxide. From the findings, the ability of the bioprepreg to cover pores in cellulose-based structures and help resist water makes it a good potential prepreg for reinforcing biocomposites. Butenegro et al. (2023) discuss a sustainable method of producing bio-based thermoplastic composite materials through recycled aerospace prepreg waste. It was found that the prepreg produced through recycled aerospace materials is a good reinforcement material for sustainable composites. In addition to the advancement in bioprepregs materials presented by previous studies, recent research utilises wood–poly(furfuryl alcohol) as a significant material to produce bioprepregs as a step to embracing eco-friendly composite material manufacturing (Acosta et al. 2023). The resulting laminate obtained from the wood veneers and poly(furfuryl alcohol) composites exhibited desirable characteristics deemed appropriate as sustainable prepregs. It is worth saying that bio-based composite materials have been obtained from several sources and synthesised through innovative approaches for sustainable material design.

3.4 Manufacturing Techniques for Biopregs and Hybrid Fibre-Reinforced Bionanocomposites

To set the bioproduction stage, Carvalho et al. (2024) present a general basic step for prepreg production (see Fig. 3.3). It is an essential requirement for prepreg structures to incorporate material reinforcement and a polymeric matrix. Mouritz (2012) presents it as a two-part sheet material that consists of the integration of a partially cured resin (e.g. epoxy) that is evenly spread on fibres (e.g. carbon), ensuring accurate control of fibre volume content to achieve desired quality. It is worth noting that, a hybrid composite replaces the traditional material that permits a significant weight reduction, contributing to carbon emissions reduction, especially in the automotive industry (Carvalho et al. 2024; Rahman et al. 2023). Similarly, Belzile et al. (2024) assert that composite materials are mostly derived from non-renewable sources, which offer high performance but have severe environmental implications. This underscores the need for bio-based prepregs and hybrid composite production. Bio-based prepregs are inherently sustainable, presenting an eco-friendly alternative (Odiyi et al. 2022), and their adoption necessitates the development of green matrices capable of maintaining thermochemical stability, adequate storage life, and optimal reinforcement integration (Belzile et al. 2024).

Premise to this, Belzile et al. introduced a novel unidirectional (UD) flax fibre-reinforced bio-polyurethane (BioPUR) prepreg from soy-based polyurethane. BioPUR matrix was formed from cross-linked polymeric methylene diphenyl diisocyanate (pMDI) catalysed with dibutyltin dilaurate (DBTDL) and cross-linked with

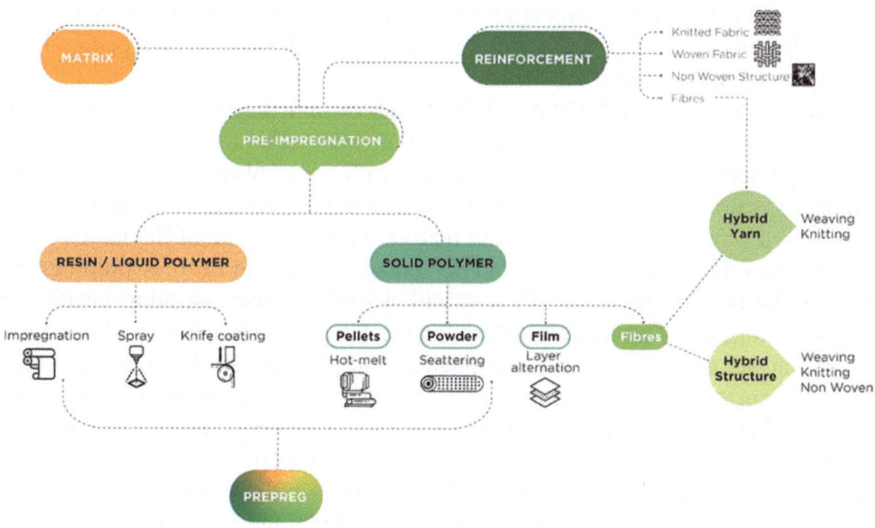

Fig. 3.3 Manufacturing prepregs process (Carvalho et al. 2024) (This is an open-access article distributed under the terms of the *Creative Commons CC-BY* licence)

glycerine. The impregnated fibres are pre-cured in an oven at 60 °C for storage and handling stability at − 18 °C, to halt curing and preserve material properties until ready for use (see Fig. 3.4). Whilst flax fibres have eco-friendliness as well as mechanical reinforcement properties, inherent hydrophilicity proved to be a challenge for consolidation of layers (Khan et al. 2018), particularly in terms of high viscosity of BioPUR in press moulding (Belzile et al. 2024). The three-roll mill treatment and oven pre-curing were applied to combat this to achieve a beta-stage material, followed by vacuum-assisted moulding in the manufacturing of laminates with a fibre volume fraction as high as 41%. Microstructural analysis using optical and SEM confirmed uniform distribution of fibres in the polyurethane matrix, negligible porosity of 0.45% and negligible interlaminar defects. The preliminary mechanical tests showed promising values of mean tensile strength 216.96 ± 16.44 MPa, tensile modulus 17.07 ± 3.07 GPa, and strain to failure of 1.36 ± 0.10%, indicating strong and stiff composite. Optical and scanning electron microscopy (SEM) confirmed uniform distribution of fibres in polyurethane matrix, zero porosity of 0.45%, and negligible interlaminar flaws. Mechanical testing showed a new composite with a mean tensile strength of 216.96 ± 16.44 MPa, tensile modulus of 17.07 ± 3.07 GPa, and strain to failure of 1.36 ± 0.10%, indicating high strength and stiffness. Furthermore, flexural tests showed a flexural strength of 231 ± 5 MPa, flexural modulus of 17.8 ± 0.8 GPa, and strain to failure of 2.32 ± 0.05%, indicating higher load-carrying capacity. Short-beam shear tests also recorded an interlaminar shear strength (ILSS) of 25.3 ± 0.4 MPa, with most samples showing elasto-plastic deformation and localised interlaminar shear failure, indicating favourable interfacial adhesion. On the other hand, some fibre–matrix debonding was observed which can be attributed to the processing conditions. Despite these imperfect features, the composite had remarkable structural integrity and mechanical properties and was a real candidate for lightweight structural applications with strength-to-weight ratio of primary concern (Belzile et al. 2024).

Recently, Margoto and Milani (2025) used vacuum bagging to create a fibre flax (FF) bioprepreg with polyfurfuryl alcohol (PFA). The method targeted better fibre alignment and mechanical properties with high fibre (60%) composites. Although ceramic polymer infiltrated by using high-temperature infiltration resulted in improved physical, mechanical, and thermal properties, one of the major disadvantages was the extremely high void percentage—24.96 in FF/PFA prepregs and 21.35 in the hybrid composite (PFA/45%FF-15%CF)—demonstrating insufficient interlayer adhesion. These flaws undermine mechanical performance and durability (Resch-Fauster et al. 2018), as highlighted by Odiyi et al. (2023), attributing them to insufficient rapid cure cycle and ineffective processing optimisation. Starvation of the resin also enhanced surface roughness and uneven polymer-matrix adhesion, limiting the composite's overall integrity. However, mechanical tests showed that the mechanical properties of the carbon fibre-reinforced PFA composite were significantly improved, with a tensile strength of 708 MPa and a Young's modulus of 61 GPa, which is 130% that of pure flax fibre composites. Flax fibre composites absorbed more energy as evidence from impact tests, probably due to their porous and flexible structure.

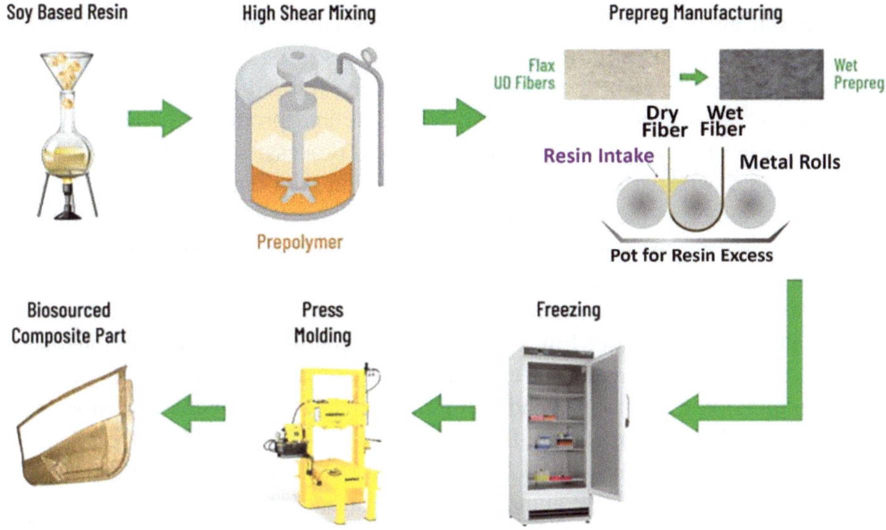

Fig. 3.4 Manufacturing prepregs (Belzile et al. 2024). (This is an open-access article distributed under the terms of the *Creative Commons CC-BY* licence)

Moreover, damping analysis at 1 Hz frequency revealed that carbon fibre composites were surpassed by flax fibre composites, with bioepoxy/35% flax fibre possessing the highest damping factor. Microstructural analysis by scanning electron microscopy (SEM) indicated fibre–matrix interfaces as mixed, whilst the bioepoxy composite had improved adhesion. A thermogravimetric analysis (TGA) indicated the carbon fibre PFA composite as the most thermally stable and has shown an overall weight loss of 15% (Margoto and Milani 2025). However, the repeated void content within the FF/PFA composite is a cause of concern for long-term reliability.

Cheng et al. (2024) discuss a novel way of fabricating thermosetting polylactic acid (PLA) composites by the use of a chemically processed PLA matrix mixed with flax fibres (FFs). The fabrication of PLA involved multiple chemical reactions of DOPO, toluene, and itaconic acid, then blending with curing agent tert-butyl peroxybenzoate (TBPB). To control the viscosity and improve impregnation, the resin was vacuum oven-dried and hot-press cured to obtain a composite with 60% of the fibre thickness. This thermal insulation is satisfactory in its overall structural integrity for the material but does have a main issue data, thus preventing premature curing in hot-melt impregnation. The viscosity of the PLA resin requires a low temperature storage to maintain its characteristics. Furthermore, the layer-by-layer stacking procedure followed by curing at 65 °C for 15 min at 1 MPa pressure was a matter of scalability, particularly process efficiency and uniform fibre-matrix distribution. For characterisation analysis, mechanical testing using an electronic universal testing machine measured a tensile strength of 128 MPa and flexural strength of 98.8 MPa, supporting average mechanical performance relative to other bio-based composites.

The material also met the UL-94 V-1 flame retardant rating (Cheng et al. 2024), suggesting potential uses for greater fire resistance.

The fabrication of hybrid composite is further explored by Song et al. (2024), investigating the use of cotton yarns to formulate and produce a continuous filament from PLA prepreg. This was a pre-treatment process, which involved soaking the cotton in 5 wt% of sodium hydroxide solution at room temperature for 2 h. This process aimed to increase the fibre barrier surface roughness and thus improve compatibility between the matrix and polymers. A made-to-measure prepreg, which utilised a rotating screw, was used to melt the resulting PLA powder in a barrel as well as homogenise it further. The treated cotton was then fed into a die for impregnation with molten PLA (see Fig. 3.5). The studies focused on increasing the quality of filaments with adjustments of crucial processing parameters such as dragging speed, immersion temperature, and cotton diameter. The mechanical characterisation validated that the tensile and flexural strengths of the PLA/TCCT composites were increased by 1.2 times and 3.1 times, respectively, at a layer width of 0.5 mm compared to pure PLA. The authors attributed these improvements to optimised processing conditions, particularly changes in nozzle temperature and printing speed (Song et al. 2024). The findings suggest that PLA/TCCT composites have promise for weight-saving applications in the automotive, consumer product, and aerospace industries where weight reduction is a significant performance criterion.

In an earlier study, Amran et al. (2015) synthesised a bio-phenolic resin (BPR)—derived from oil palm empty fruit bunch (EFB), via a liquefaction process, treated with phenol and sulphuric acid that is further subjected to formalin reaction. To prepare bioprepregs, Amran et al. explored natural rubber of different grades (Standard Malaysian Rubber Light (SMR-L-0), epoxidised natural rubber (ENR-25 and ENR-50), coupled with BPR. They further consolidated the polymer (woven glass fibres)/matrices utilising the manual hand lay-up technique, semi-cured at 60 °C and then hot-pressed at 105 °C to make the composite structure. This process enhanced the optimal compatibility of ENR-50% and BPR, showing improved impact strength and reduced water uptake. The mechanical property tests showed an impressive increase of 77.7% in Izod impact strength when using 35 wt% ENR-50. Additionally, the flexural analysis revealed that the highest flexural strength reached 29.59 MPa, with a modulus of 1.56 GPa at 15% ENR content. Scanning electron microscopy (SEM) provided further evidence of enhanced bonding between the polymer and matrix, along with a notable 47% decrease in water absorption after adding ENR-50. The results demonstrate the great promise of bio-based phenolic resin strengthened with natural rubber as a future material of choice for impact-strength and moisture-strength composites. However, the increased complexity of processing and longer curing times could need to be remediated in later optimisation for larger applications (Amran et al. 2015). Table 3.1 provides a comparative analysis of the composites discussed above, comparing their structural behaviour and mechanical properties.

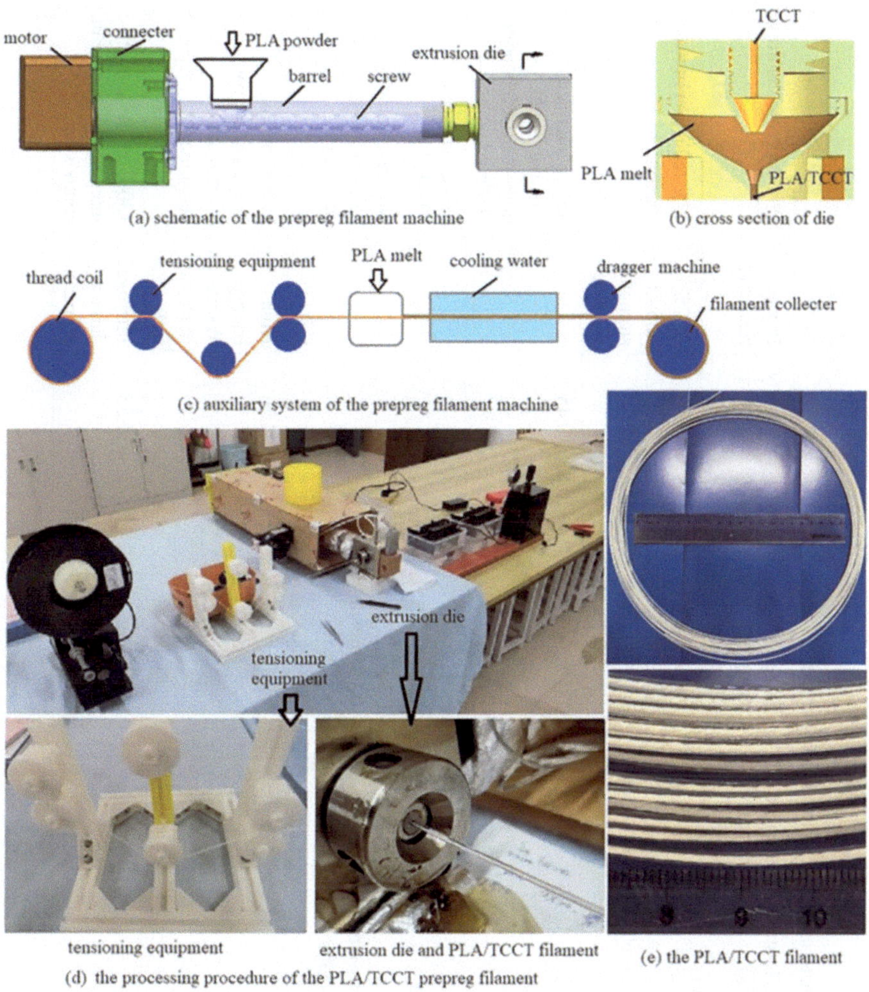

Fig. 3.5 Device and processing of the PLA/TCCT and PLA/CCT filaments (Song et al. 2024) (This is an open-access article distributed under the terms of the *Creative Commons CC-BY* licence)

3.5 Functional Properties of Biopregs and Hybrid Fibre-Reinforced Bionanocomposites

With comparable mechanical, thermal, and functional qualities, biopregs and hybrid fibre-reinforced bionanocomposites provide a sustainable substitute for conventional counterparts (Phadtare et al. 2025b). Although biopregs have been understudied, their incorporation with bionanocomposites may offer potential capacity to enhance mechanical strength, thermal stability, and barrier qualities at low filler concentrations (Andrew and Dhakal 2022).

Table 3.1 Comparative analysis of composite materials based on mechanical properties and performance

Authors	Material	Fibre content (%)	Tensile strength (MPa)	Tensile modulus (GPa)	Flexural strength (MPa)	Flexural modulus (GPa)	Porosity (%)	Impact resistance	Other notable features
Amran et al. (2015)	Bio-phenolic resin (BPR) with natural rubber				29.59	1.56		77.7% impact strength increase	47% reduction in water absorption
Belzile et al. (2024)	Flax fibre-reinforced bio-PUR	41	216.96 ± 16.44	17.07 ± 3.07	231 ± 5	17.8 ± 0.8	0.45		Strong interfacial bonding; slight fibre–matrix debonding observed
Cheng et al. (2024)	Thermosetting PLA with FF	60	128		98.8				Flame retardant UL-94 V-1
Song et al. (2024)	PLA/Cotton thread composite		3.1 × improvement over pure PLA		1.2 × improvement over pure PLA				Optimised processing parameters improved strength
Margoto and Milani (2025)	Flax fibre/PFA prepreg	60					24.96	High energy absorption in impact test	Layer bonding irregularities; void content affects performance
	Hybrid (PFA/ 45% FF-15% CF)		708	61			21.35	Improved damping properties	130% tensile strength increase versus pure flax composite

An example is Amran et al. (2015) study mentioned above that showed improved water absorption reduction of BPE composites at different ENR-50 compositions. The composites showed micro-voids in phenolic resin that trap water in composite materials that eventually compromise material performance. To mitigate this issue, ENR-50 prepregs are introduced to the composite, eliminating many micro-voids. This resulted in a water absorption rate from 11.25 to 5.95% uptake. This implies that the ENR-50 prepregs make the composite structure compact compared to the reference sample (see Fig. 3.6). This reduction is imperative to ensure longevity as well as maintain the integrity of the composite in different environments. Furthermore, the introduction of ENR-50 (35 wt%) influenced a great improvement in impact strength from 72.40 kJ/m^2 to 128.65 kJ/m^2, owing to the ability of rubber to absorb impact forces and distribute stress within the composite structure more effectively. This aligns with Kadea et al. (2023) finding, concluding that the fusion of ENR has impact strength capabilities to biocomposites, acting as an impact modifier.

Efforts from Mousa et al. (2016) express that bionanocomposites primarily comprise of matrix—the binding component; reinforcement fibres; nanofillers (e.g. cellulose nanofibres (CNFs), carbon nanotubes (CNTs), and halloysite nanotubes (HNTs))—that are property specific enhancement; compatibilisers to improve interfacial bonding between components (e.g. nanofillers and polymer matrices); plasticizers (e.g. glycerol). Bionanocomposites inform mechanical properties like tensile strength and impact resistance (Wang et al. 2022). This is evidence of the effects of the adoption of nanofillers. CNFs and CNTs significantly induce tensile modulus and strength, resulting in stiff and stronger composite(s). A review report by Mousa et al. (2016) showed that multi-walled CNT (MWCNT)-starch composites conclude

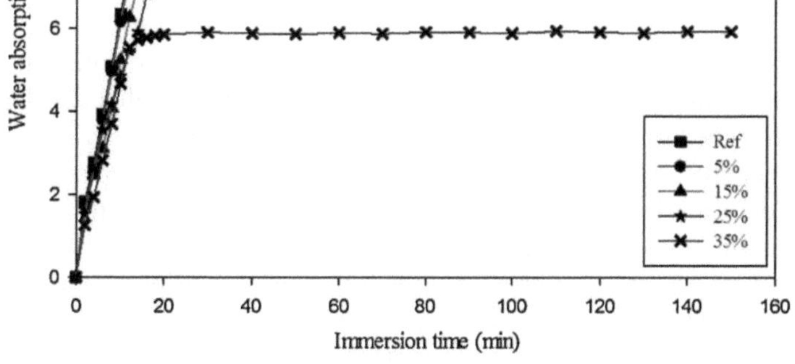

Fig. 3.6 Water uptake of composites with different compositions of ENR-50. Reprinted with permission from Amran et al. (2015). Copyright 2015 Elsevier BV

tensile strength enhancement, increasing from 2.85 MPa to 4.73 MPa. Further, HNTs ensure bending resistance that improves stress transfer at the polymer–filler interface. Nanofillers also influence the host matrix's microstructure by filling voids and reducing pore size, which improves density, reduces permeability, and enhances barrier properties against gases, liquids, and environmental degradation (Joshi et al. 2018; Yadav et al. 2024). For instance, the integration of biopolymer matrices [e.g. polyvinyl alcohol (PVA)] and nanofillers (e.g. organically modified layered silicates (OMLS), nanoclays, and nanocellulose) improves impermeability properties in composites, limiting molecular transport (Jamróz et al. 2019). Similarly, the presence of HNTs ensures these barriers by enhancing the compact structure of the respective composite (Mousa et al. 2016).

The incorporation of nanofillers into bionanocomposites improves thermal stability and heat resistance, allowing for broader application in high-temperature environments. Mousa et al. (2016) further assert that nanoclays, particularly montmorillonite, increase the decomposition temperature of biopolymers by creating a char layer that delays thermal degradation. CNTs also provide superior thermal conductivity, enhancing heat dissipation and reducing thermal expansion in composites like polylactic acid/ multi-walled (PLA/MWCNT) systems (Feng et al. 2024). Similarly, the high heat resistance and structural integrity of HNTs contribute to improved thermal stability in composites with polycaprolactone (PCL) or PLA (Mousa et al. 2016). Abotaleb et al. (2024) further demonstrated the thermal stability of HNTs, experimentally subjecting it to temperature, ranging from 25 to 1100 °C. This was characterised, revealing two major transitions: first between 400 and 500 °C, corresponding to the release of intercalated water and partial distortion of the HNT structure, and second occurring between 900 and 10 °C, marked by the collapse of the tubular structure and the exposure of alumina on the surface. These findings provide significant insights into the thermal stability of HNTs, informing future applications, especially in high-temperature environments (Abotaleb et al. 2024).

3.6 Safety Concerns, Challenges, and Regulatory Issues

Biopregs and hybrid fibre-reinforced bionanocomposites are potential composite materials with unique properties that have advanced their applications in different areas for enhanced safety, durability, and less environmental impact. However, aside from their benefits, there are some safety concerns, challenges, and regulatory issues that are widely associated with them, which could affect their full functionality in different applications. There are safety concerns with regard to the handling and utilisation of biopregs for products. Environmental conditions like humidity and temperature can have a significant impact on the safety and performance of the stored materials. This is due to the sensitivity of the forming materials to exogenous or processing conditions. For example, natural fibres used to pre-impregnate the reinforced materials undergo changes and degradation when exposed to higher temperatures and humidity (Musthaq et al. 2023), which could ultimately affect the performance of the bioprepregs. Alternatively, operators must observe standard handling precautions by wearing appropriate personal protective clothing to limit any contamination of the composing materials that could affect the eventual properties and functions. Aside from this, the processing process for bioprepregs is quite challenging and complex. This is because material composition must be tailored effectively. For example, the production or tailorability of the 'prepregs,' is challenging due to the molecular adhesion and optimal impregnation required to ensure the appropriate materials with the right properties are produced (Wiedemann et al. 1986). The choice of the right technology during processing is very important to ensure the composites are appropriately produced. This is because the polymer matrix and the reinforcement materials are critical to determining the technology needed (Carvalho et al. 2024).

Additionally, other challenges such as the cost, varying fibre quality, and limited supply of these natural fibres have affected their utilisation in the production of bioprepregs. The issue of regulation and compliance is also a critical concern in the production of bioprepregs. Compliance of composite materials with international standards is vital for effective certification and market approval. For example, the EURO 6 regulations by the European Union place serious taxes on automobiles producing 95 g/km of CO_2 emissions (Carvalho et al. 2024). This implies that composite materials must adhere to this compliance to ensure their products are lightweight which leads to reduced CO_2 emission. For hybrid fibre-reinforced bionanocomposites, there are several safety concerns, challenges, and regulatory issues that hinder the full production, and functionality of the composite materials. For example, the use of natural fibre to reinforce the bionanocomposites is affected by environmental conditions and has poor microorganism resistance, flammable, and high moisture content. These properties can cause degradation and vary the properties of the fibres, hence altering the eventual performance of the composite materials for application (Ogah et al. 2024). Furthermore, the fragility of the eventual composite materials is increased based on fibre increase loading. Due to the composition during fabrication, some composite materials could have lower thermal stability which could affect their utilisation in high-temperature applications. Additionally, bonding, interface, and compatibility of various combination materials for the composite materials are critical to ensure the desired mechanical and functional properties are achieved. The application of fibre-reinforced bionanocomposites for application, e.g. food packaging possesses serious safety concerns for the environment, animals, and human health. This is important to ensure minimal release of nanomaterials to food products that are not harmful. Thus, all fabricated composite materials must meet the required tests and standards. Migration tests are required for up to 30 days using different temperatures as stipulated in the EU regulations (Perera et al. 2022; Garcia et al. 2018). This provides core information on the associated risks for consumers and policymakers.

3.7 Applications and Future Prospects of Biopregregs and Hybrid Fibre-Reinforced Bionanocomposites

The push for eco-friendly materials due to environmental impacts and regulatory policies has advanced the production of biocomposites that utilise natural fibres in their fabrication process for reduced carbon emissions and footprint globally (Carvalho et al. 2024). This has advanced the rise in the production of biopregregs and hybrid fibre-reinforced bionanocomposites with unique benefits. The fabrication of these materials with the combined benefits of natural fibres with other component matrices for enhanced properties has advanced their application in various fields like automotive, aerospace, construction, marine, sport, and medicine amongst others. The following sections provide key examples in some application areas. The production of biopregregs requires both the polymer matrix and the reinforced materials which are pre-impregnated with natural fibre sources (Carvalho et al. 2024). These composite materials exhibit good creep and chemical resistance, dampening characteristics, and reduced weight (Carvalho et al. 2024; Knight et al. 2003), advancing their application in the automotive industry. These properties effectively influence the selection of appropriate materials for reducing CO_2 emissions and the demand for environmentally friendly designs by consumers and regulators (Kim 2012). As such manufacturers in the automotive industry are pushed to advance the adoption of eco-friendly materials with unique properties that reduce the cost of production but still maintain the quality. Subsequently, due to their dampening ability, creep resistance coupled with the high strength-to-weight ratio, these materials have the unique performance of reducing the cost of fuel, repairs, and maintenance for extended life use as compared to the conventional metallic and alternative material products (Knight et al. 2003). On the premise of this, manufacturers have advanced their design process to produce lightweight materials whilst still ensuring their safety or crashworthiness for users. This implies that safety is always paramount ahead of the cost of the materials for automotive products (Chauhan et al. 2022). Several international brands have advanced the application of biopregregs for different car parts. For example, polypreg flax/TP thermoplastic composite was fabricated with flax roving by Porcher for car interiors like dashboards for enhanced aesthetics (Nehls 2022). These materials exhibited high acoustic absorption, impact resistance, low cost, and weight with good recyclability for reduced environmental impacts. Another company, BPREGS, developed their EcoRein prepregs fabricated from thermoplastic materials reinforced with hemp or flax natural fibres, ideal for vehicle interior trims. Faurecia also developed the Flaxpreg™, a fibre-reinforced composite sandwich which is lightweight and has vibro-acoustic damping properties (Automotive World 2015). Herein, flax ribbons were used to develop a flax tape where an acrylic resin was impregnated directly to form a prepreg which was finally sandwiched with a paper honeycomb structure. Due to its good mechanical properties, it finds applications for seat backrests, door trims, package trays, and as a multi-position truck load floor for passenger vehicles. In a different approach, natural fibre-based thermoplastic prepregs were fabricated for use as interior panels for the automotive industry due its performance (Mohammadabadi et al. 2022).

The use of bioprepregs has been advanced for application in the aerospace industry, due to its lightweight materials that contribute to reducing the environmental issues, carbon footprint, and fuel consumption. A recent study by Belzile et al. (2024) fabricated bio-based prepregs using flax soy-based polyurethane composites (Fig. 3.7). Herein, flax fibres in unidirectional backed with short flax fibres mat were used for the reinforcement material which was then applied to the polyurethane matrix. Results showed that fabricated materials have good mechanical properties for application in the aerospace industry. Alternatively, bio-based intermediate prepreg materials (BIOpreg PFA) were developed by the Mitsubishi Chemical Group for aircraft structural panels (Pangarkar 2025). These materials replace the conventional phenolic panels used in commercial aircraft. Another bio-based prepreg developed from sugar cane for the aerospace industry was reported (AG FACC 2020). This innovative material poses unique properties like chemical and temperature resistance, and fire retardant, suitable for the interiors of aircraft. In the area of sports, HexPly ® nature bioprepregs were developed by Hexcel for winter sports (HEXCEL 2024). All these bioprepregs innovations in the various application areas advanced the production of lightweight materials with unique performance which further foster the frontiers for sustainable products with less environmental impacts.

Still advancing sustainable materials with unique performance for various applications, hybrid fibre-reinforced bionanocomposites present another promising composite material. These are fabricated by combining biopolymers with fibres at the nanoscale for improved functionality and mechanical properties for varying application areas (Huang et al. 2024). For automotive industries, the development of

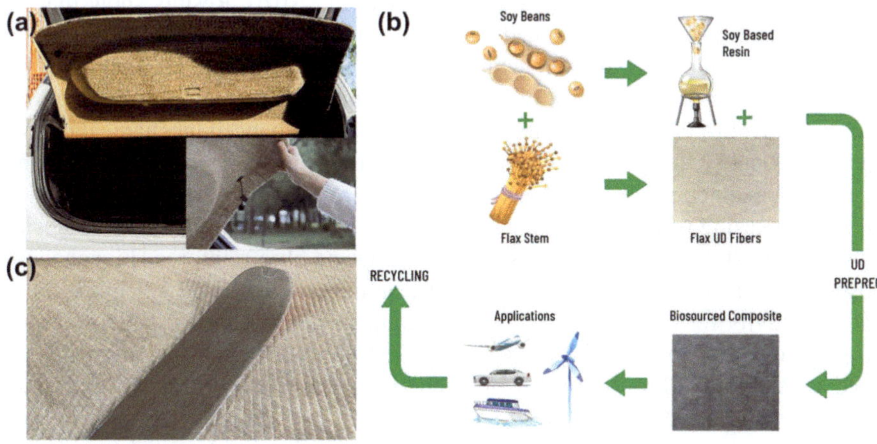

Fig. 3.7 **a** EcoRein prepregs developed BPREGS (2021), **b** production process of bio-based prepregs using flax soy-based polyurethane composites (Belzile et al. 2024) (This is an open-access article distributed under the terms of the *Creative Commons CC-BY* licence), **c** HexPly® nature bioprepregs product for winter sports (HEXCEL 2024)

fibre-reinforced bionanocomposites is gaining potential due to their unique properties for the fabrication of various automotive parts in the industry. For example, nanocellulose-reinforced bionanocomposites have higher elastic modulus, lighter weight, higher tensile strength, and velocity (Ali et al. 2022a). For example, hemp nanocellulose-reinforced bionanocomposites fabricated with a 2 wt% of nanocellulose exhibited higher impact strength, flexural strength, and tensile strength for applications in the automotive industry (Rana and Gupta 2021). Fabricated composite materials must have enhanced impact properties for automotive applications to ensure pedestrian safety and crashworthiness (Islam et al. 2022). As such, natural fibres reinforced with polymer composites are lightweight, non-toxic, and impact resistant for exterior and interior car use (Gholampour and Ozbakkaloglu 2020). For example, an impact strength of 50–80 kJ/m was recorded for cotton fibre-reinforced agro-based resin composites (Müssig 2008).

For the aerospace industry, due to their fire resistance, thermal stability, environmental benefits, lightweight and exceptional mechanical properties, fibre-reinforced bionanocomposites are gaining traction. The fabrication of bamboo composite reinforced with SiO_2 nanoparticles exhibited fracture toughness, impact resistance, flexural strength, and tensile strength (Gunasekar et al. 2024) for aircraft applications. Another study suggested the application in aircraft radome when natural fibres are used as reinforcement in the biocomposites (Haris et al. 2011). Due to the environmental impacts and limited functionality of conventional plastics for packaging, fibre-reinforced bionanocomposites are emerging as alternative materials because of their high performance for packaging. This is because efforts are advanced to improve the quality and shelf life of food products for consumers (Aprilia et al. 2024) since fabricated bionanocomposites act as protection barrier for external conditions. For example, a chitosan/carboxymethyl cellulose/ZnO bionanocomposites were fabricated for a food packaging material (Youssef et al. 2016). Results show that the novel bionanocomposites film had a significant influence on pathogenic strains and bacterial counts on the soft white cheese, hence extending its shelf life. For medical applications, the application of hybrid fibre-reinforced bionanocomposites plays a critical role because of their biocompatibility, enhanced mechanical quality, and functional performance, to advance tissue engineering, and drug delivery in a controlled manner, amongst others. The success of these materials are dependent on the processing techniques which influence the bioactivity of the nanocomposites (Joseph et al. 2020). As such the design of these hybrid fibre-reinforced bionanocomposites is critical to their functionality. For example, an increased biocompatibility and bioactivity for chitosan-based bionanocomposites can be maintained when the proportions of chitosan are high (Patra et al. 2025). Wound healing scaffolds developed with hybrid fibre-reinforced bionanocomposites aid in effective prevention and healing of infections. As such an antimicrobial suture thread from novel silk bionanocomposite embedded with silver nanoparticles was fabricated (Saraí et al. 2018). Herein, the silver nanoparticles were embedded into the cationic guar gum alkylamine biopolymeric matrix. Results show that the migration of keratinocytes enhanced the process of healing. Furthermore, a cellulose–chitosan bionanocomposite was fabricated with unique antimicrobial properties (Lin et al. 2013). Results show an excellent antibacterial activity against *S. aureus* and *E. coli* resulting in high regeneration and epithelialisation rates for healing. Alternatively, polycaprolactone/cellulose nanofibres bionanocomposites were fabricated for 3D medical plotting applications (Vidakis et al. 2023). Its unique properties enhance its application as scaffold due to its increased mechanical properties, which are critical to withstand any load. For effective drug delivery, Ali et al. (2022b) fabricated a patch from cellulose nanocrystals/chitosan/methylcellulose bionanocomposites for effective delivery of ketorolac tromethamine via the transdermal route.

As sustainability and improved properties are at the core of composite materials, the advancement of smart composites and hybridisation are critical for future research. Smart composites are gaining attention since they involve the combination of two or more materials with unique properties that improve their sensitivity to stimuli such as light, temperature, and pH amongst others (Raza et al. 2019; Zhang et al. 2019; Seyfoori et al. 2019). Hence, by advancing the frontiers for biopregs and fibre-reinforced bionanocomposites to be more smart or intelligent, these functions could expand their application areas. Alternatively, by incorporating two or more fillers or reinforcing materials in the polymer matrix, hybrid composites are formed with enhanced properties (Ismail et al. 2022; Swolfs et al. 2019). These hybrid composites comprise moisture absorption and thermal and mechanical properties that overcome that from single fibre reinforcements for diverse applications (Ziąbka et al. 2020), hence hybridisation of the composites for future studies could advance.

3.8 Conclusions

Hybrid fibres and biopregs are good substances for improving the performance of products/items in material engineering. Combining fibres from natural and artificial sources helps form hybrid material with desirable properties sourced from both materials. The chapter has highlighted innovative materials used for biopregs and hybrid fibre composites, the manufacturing techniques employed in past studies that are still effective for sustainable operations, the functional properties of biopregs and hybrid fibre-reinforced bionanocomposites, the safety concerns, and the application and prospects of biopregs and hybrid fibre-reinforced bionanocomposites. From the study, innovative techniques and materials for biopregs and hybrid-reinforced biocomposites have been identified. Materials such as cassava root (periderm and cortex), galactomannan from locust beans, and cellulose nanofibre from cassava wastes are potential biocomposite materials. Resin transfer moulding and vacuum-assisted resin infusion are methodologies adopted to manufacture a prepreg material (flax fibre impregnated with partially bio-based epoxy matrix). Composite tape is derived from a prepreg comprising a biomaterial (flax) and a thermoplastic poly(amide) material, resulting in a hybrid sustainable material. Another innovative approach is a synthesised polyester material and cotton hybrids used to form fully bio-based composites with improved properties.

It was also observed that with comparable mechanical, thermal, and functional qualities, biopregs and hybrid fibre-reinforced bionanocomposites provide a sustainable substitute for conventional counterparts. They inform mechanical properties like tensile strength and impact resistance. Nanomaterials like CNTs provide superior thermal conductivity, enhancing heat dissipation and reducing thermal expansion in composites like polylactic acid/multi-walled systems. Regarding safety concerns, environmental conditions like humidity and temperature can pose a significant impact on the safety and performance of the stored materials for bioprepreg manufacturing. Also, other challenges such as the cost, varying fibre quality, and limited supply of natural fibres have affected their utilisation in the production of bioprepregs. The application of fibre-reinforced bionanocomposites for application, e.g. food packaging possesses serious safety concerns for the environment, animals, and human health. However, due to their fire resistance, thermal stability, environmental benefits, and lightweight and exceptional mechanical properties, fibre-reinforced bionanocomposites are gaining traction in several industries. Since sustainability and improved properties are at the core of composite materials, it is recommended that the advancement of smart composites and hybridisation should be set as a priority for future research.

Declaration on the Use of Generative AI and AI-Assistive Technologies The Alternative Text Assistant AI was used to create Alt Text.

References

Abotaleb A et al (2024) Impact of thermal treatment on halloysite nanotubes: a combined experimental-computational approach. Heliyon 10(21):e39952

Acosta AP et al (2023) Wood-poly(furfuryl alcohol) prepreg: a novel, ecofriendly laminate composite. Materials 16(18):6237

AG FACC (2020) Bio-based prepreg: sustainable product design in the aerospace industry. https://www.facc.com/en/BEyond-Blog/Bio-based-prepreg-Sustainable-Product-Design-in-the-Aerospace-Industry. Accessed 29 Jan 2025

Agarwal N et al (2024) An overview of carbon-carbon composite materials and their applications. Front Mater 11

Al Mahmud MZ et al (2024) Epoxy resin composites reinforced with upcycled fabrics: mechanical, thermal, and morphological analysis. SPE Polymers 5(4):624–636

Ali N, Hoque ME (2022a) Bionanocomposites in the automotive and aerospace applications. In: Muthukumar C et al (eds) Polymer based bio-nanocomposites: properties, durability and applications. Springer Singapore, Singapore, pp 237–253

Ali MS et al (2022b) Transdermal therapeutic system: study of cellulose nanocrystals influenced methylcellulose-chitosan bionanocomposites. Int J Biol Macromol 218:556–567

Alimuzzaman S, Gong RH, Akonda M (2014) Biodegradability of nonwoven flax fiber reinforced polylactic acid biocomposites. Polym Compos 35(11):2094–2102

Amran UA et al (2015) Mechanical properties and water absorption of glass fibre reinforced bio-phenolic elastomer (BPE) composite. Ind Crops Prod 72:54–59

Andrew JJ, Dhakal HN (2022) Sustainable biobased composites for advanced applications: recent trends and future opportunities—a critical review. Compos Part C Open Access 7:100220

Aprilia S, Fitriani F, Muchtar S (2024) Bio-nanocomposites for food packaging: synthesis and surface properties. In: Sharma K et al (eds) Novel bio-nanocomposites for biomedical applications. Springer International Publishing, Cham, pp 61–82

Automotive World (2015) Faurecia Flaxpreg™ wins JEC Europe 2015 Innovation Award in the semi-products category. https://www.automotiveworld.com/news-releases/faurecia-flaxpregtm-wins-jec-europe-2015-innovation-award-semi-products-category/. Accessed 29 Jan 2025

Azimi B et al (2020) Electrospinning piezoelectric fibers for biocompatible devices. Adv Healthcare Mater 9(1):1901287

Badshah MA et al (2022) Scalable manufacturing of sustainable packaging materials with tunable thermoregulability. Nat Sustain 5(5):434–443

Baig N, Kammakakam I, Falath W (2021) Nanomaterials: a review of synthesis methods, properties, recent progress, and challenges. Mater Adv 2(6):1821–1871

Baysal A et al (2023) Mechanical characterization of hybrid biocomposites reinforced with nonwoven hemp and unidirectional flax fibers. Polym Compos 44(6):3555–3566

Belzile A et al (2024) Development of a novel flax soy-based polyurethane prepreg composite. Compos A Appl Sci Manuf 181:108136

BPREG (2021) Parcel-shelf for passenger car. https://bpreg.com/portfolio/parcel-shelf-for-passenger-car/. Accessed 28 Jan 2025

Butenegro JA et al (2023) Novel sustainable composites incorporating a biobased thermoplastic matrix and recycled aerospace prepreg waste: development and characterization. Polymers 15(16):3447

Carvalho D et al (2024) Advancing sustainability in the automotive industry: biopregs and fully bio-based composites. Compos Part C Open Access 14:100459

Chauhan V, Kärki T, Varis J (2022) Review of natural fiber-reinforced engineering plastic composites, their applications in the transportation sector and processing techniques. J Thermoplast Compos Mater 35(8):1169–1209

Cheng Z et al (2024) A novel biobased prepreg from flax and thermosetting poly(lactic acid): synthesis, curing and processing. Ind Crops Prod 222:119778

da Luz FS et al (2020) Graphene-incorporated natural fiber polymer composites: a first overview. Polymers 12(7):1601

Dai Z et al (2018) Fully biobased composites of an itaconic acid derived unsaturated polyester reinforced with cotton fabrics. ACS Sustain Chem Eng 6(11):15056–15063

Darder M, Aranda P, Ruiz-Hitzky E (2007) Bionanocomposites: a new concept of ecological, bioinspired, and functional hybrid materials. Adv Mater 19(10):1309–1319

dos Santos MS, Antunes Filho S, Backx BP (2023) Bionanotechnology in agriculture: a one health approach. Life 13(2):509

Elfaleh I et al (2023) A comprehensive review of natural fibers and their composites: an eco-friendly alternative to conventional materials. Results Eng 19:101271

Feng D et al (2024) Multifunctional performance of carbon nanotubes in thermal energy storage materials. Mater Today 75:285–308

Fronza P et al (2024) Bionanocomposite based on cassava waste starch, locust bean galactomannan, and cassava waste cellulose nanofibers. Foods 13(2):202

Gangil B et al (2020) Natural and synthetic fibers for hybrid composites. In: Hybrid fiber composites, pp 1–15

Garcia CV, Shin GH, Kim JT (2018) Metal oxide-based nanocomposites in food packaging: applications, migration, and regulations. Trends Food Sci Technol 82:21–31

Gholampour A, Ozbakkaloglu T (2020) A review of natural fiber composites: properties, modification and processing techniques, characterization, applications. J Mater Sci 55(3):829–892

Greb C et al (2018) 18—Fabrics for reinforcement of engineering composites. In: Miao M, Xin JH (eds) Engineering of high-performance textiles. Woodhead Publishing, New Delhi, pp 489–512

Gunasekar P, Anderson A, Praveenkumar TR (2024) Experimental evaluation of bamboo natural fiber composite reinforced with SiO_2 nanoparticles for aircraft applications. Aircr Eng Aerosp Technol 96(8):1108–1115

Haris MY et al (2011) Preliminary review of biocomposites materials for aircraft radome application. Key Eng Mater 471–472:563–567
He W et al (2018) Highly reactive metastable intermixed composites (MICs): preparation and characterization. Adv Mater 30(41):1706293
HEXCEL (2024) Hexcel transitions winter sports prepreg production to sustainable HexPly® nature range. https://hexcel.com/News/News-Releases/5675/hexcel-transitions-winter-sports-prepreg-production-to-sustainable-hexply-nature. Accessed 27 Jan 2025
Huang Y et al (2024) Hybrid fiber-reinforced biocomposites for marine applications: a review. J Compos Sci 8(10):430
Islam MZ et al (2022) Green composites from natural fibers and biopolymers: a review on processing, properties, and applications. J Reinf Plast Compos 41(13–14):526–557
Islam T et al (2024) Advancements and challenges in natural fiber-reinforced hybrid composites: a comprehensive review. SPE Polym 5(4):481–506
Ismail SO, Akpan E, Dhakal HN (2022) Review on natural plant fibres and their hybrid composites for structural applications: recent trends and future perspectives. Compos Part C Open Access 9:100322
Jabbar M, Nasreen A (2021) 7—Composite fabrication and joining. In: Nawab Y, Sapuan SM, Shaker K (eds) Composite solutions for ballistics. Woodhead Publishing, pp 177–197
Jamróz E, Kulawik P, Kopel P (2019) The effect of nanofillers on the functional properties of biopolymer-based films: a review. Polymers 11(4):675
Joseph B et al (2020) Bionanocomposites as industrial materials, current and future perspectives: a review. Emerg Mater 3(5):711–725
Joshi M, Adak B, Butola BS (2018) Polyurethane nanocomposite based gas barrier films, membranes and coatings: a review on synthesis, characterization and potential applications. Prog Mater Sci 97:230–282
Joshi AM, Basutkar SM, Jagadish KS (2020) Stabilized mud concrete for sustainable construction. In: Delgado JMPQ (ed) Sustainable materials in building construction. Springer International Publishing, Cham, pp 135–147
Kadea S et al (2023) Influences of epoxidized natural rubber and fiber modification on injection molded-pulp/poly(lactic acid) biocomposites: analysis of mechanical-thermal and weathering stability. Ind Crops Prod 201:116909
Khan T, Hameed Sultan MTB, Ariffin AH (2018) The challenges of natural fiber in manufacturing, material selection, and technology application: a review. J Reinf Plast Compos 37(11):770–779
Khan F et al (2024) Advances of natural fiber composites in diverse engineering applications—a review. Appl Eng Sci 18:100184
Khandelwal S, Rhee KY (2020) Recent advances in basalt-fiber-reinforced composites: tailoring the fiber-matrix interface. Compos B Eng 192:108011
Khatri V et al (2023) Bionanocomposites with enhanced physical properties from curli amyloid assemblies and cellulose nanofibrils. Biomacromol 24(11):5290–5302
Kim YK (2012) 8—Natural fibre composites (NFCs) for construction and automotive industries. In: Kozłowski RM (ed) Handbook of natural fibres. Woodhead Publishing, pp 254–279
Kirschnick U et al (2024) Manufacturing bio-based fiber-reinforced polymer composites: process performance in RTM and VARI processes. Adv Manuf Polym Compos Sci 10(1):2379205
Knight M, Curliss D (2003) Composite materials. In: Meyers RA (ed) Encyclopedia of physical science and technology (Third Edition). Academic Press, New York, NY, pp 455–468
Kodolova-Chukhontseva VV et al (2022) Application of the composite fibers based on chitosan and chitin nanofibrils in cosmetology. J Funct Biomater 13(4):198
Kumari M, Chadha M, Berry S (2024) Processing and characterization of bionanocomposites. Nanomanufacturing techniques in sustainable healthcare applications. CRC Press, pp 25–38
Lambuk L et al (2023) Progress in biomedical applications using sustainable nanoparticles. In: Wan Kamarul Zaman WS, Abdullah N (ed) Sustainable material for biomedical engineering application. Springer Nature Singapore, Singapore, pp 207–238

Lin W-C et al (2013) Bacterial cellulose and bacterial cellulose–chitosan membranes for wound dressing applications. Carbohydr Polym 94(1):603–611
Margoto OH, Milani AS (2025) Comparing flax fibre/biopolymer woven composites with carbon fibre-enhanced, partially green alternatives: mechanical performance versus sustainability. Compos Part C Open Access 16:100547
Mazlan N et al (2024) Biocomposites as aircraft materials. In: Karakoc TH, Colpan CO, Dalkiran A (eds) Sustainable materials and manufacturing techniques in aviation. Springer Nature Switzerland, Cham, pp 27–53
McGregor OPL et al (2017) Pre-impregnated natural fibre-thermoplastic composite tape manufacture using a novel process. Compos A Appl Sci Manuf 101:59–71
Mohammadabadi M, Yadama V, Smith L (2022) Wood-based prepreg for composite laminates: wood-based prepreg. Wood Fiber Sci 54(2):122–133
Mouritz AP (2012) 14—Manufacturing of fibre–polymer composite materials. In:Mouritz AP (ed) Introduction to aerospace materials. Woodhead Publishing. pp 303–337
Mousa MH, Dong Y, Davies IJ (2016) Recent advances in bionanocomposites: preparation, properties, and applications. Int J Polym Mater Polym Biomater 65(5):225–254
Müssig J (2008) Cotton fibre-reinforced thermosets versus ramie composites: a comparative study using petrochemical- and agro-based resins. J Polym Environ 16(2):94–102
Musthaq MA et al (2023) The effect of various environmental conditions on the impact damage behaviour of natural-fibre-reinforced composites (NFRCs)—a critical review. Polymers 15(5):1229
Nandhini J, Karthikeyan E, Rajeshkumar S (2024) Eco-friendly bio-nanocomposites: pioneering sustainable biomedical advancements in engineering. Discov Nano 19(1):86
Nehls G (2022) Porcher industries launches new range of flax fiber-based thermoplastic composites for automotive. https://www.compositesworld.com/products/porcher-industries-launches-new-range-of-flax-fiber-based-thermoplastic-composites-for-automotive. Accessed 29 Jan 2025
Odiyi DC et al (2022) Cure mechanism and kinetic prediction of biobased glass/polyfurfuryl alcohol prepreg by model-free kinetics. Thermochim Acta 708:179133
Odiyi DC et al (2023) A review of advancements in synthesis, manufacturing and properties of environment friendly biobased polyfurfuryl alcohol resin and its composites. Compos B Eng 267:111034
Ogah OA et al (2024) Investigating natural fiber reinforced polymer composites, biocomposites, bionanocomposites thermo-mechanical attributes emerging implementations. J Fibers Polym Compos 3(2):119–157
Pangarkar T (2025) Aerospace plastics statistics 2025. By Special Materials Technology, Aerospace and Defense Technology. https://scoop.market.us/aerospace-plastics-statistics/. Accessed 29 Jan 2025
Patra N, Gopi VS, Brahman PK (2025) Cellulose and chitosan-based bionanocomposites in biomedical applications. In: Kesari KK et al (eds) Functionalized cellulose materials: sustainable manufacturing and applications. Springer Nature Switzerland, Cham, pp 191–212
Peng C, Feng D, Guo S (2021) Material selection in green design: a method combining DEA and TOPSIS. Sustainability 13(10):5497
Perera KY, Jaiswal S, Jaiswal AK (2022) A review on nanomaterials and nanohybrids based bio-nanocomposites for food packaging. Food Chem 376:131912
Phadtare P et al (2025) Hybrid bionanocomposites as the advancements in biomedical utility. Hybrid Adv 8:100365
Puiggalí J, Katsarava R (2017) Chapter 7—Bionanocomposites. In: Jlassi K, Chehimi MM, Thomas S (eds) Clay-polymer nanocomposites. Elsevier, Amsterdam, pp 239–272
Rahaman MT et al (2024) Green production and consumption of textiles and apparel: importance, fabrication, challenges and future prospects. J Open Innov Technol Mark Complex 10(2):100280
Rahman MZ et al (2023) Advanced biopolymers for automobile and aviation engineering applications. J Polym Res 30(3):106

Rana S, Gupta M (2021) Fabrication of bionanocomposites reinforced with hemp nanocellulose and evaluation of their mechanical, thermal and dynamic mechanical properties. Proc Inst Mech Eng Part L J Mater Des Appl 235(11):2470–2481

Ravelo-Nieto E et al (2023) Textile wastewater depuration using a green cellulose based Fe_3O_4 bionanocomposite. J Environ Chem Eng 11(2):109516

Raza A et al (2019) "Smart" materials-based near-infrared light-responsive drug delivery systems for cancer treatment: a review. J Market Res 8(1):1497–1509

Reddivari BR et al (2024) Fabrication and mechanical properties of hybrid fibre-reinforced polymer hybrid composite with graphene nanoplatelets and multiwalled carbon nanotubes. Cogent Eng 11(1):2343586

Resch-Fauster K et al (2018) High performance composites produced from polyfurfuryl alcohol: enhancing the process ability at short cycle times by applying hydrophilic reinforcement. In: Proceedings of ECCM18–18th european conference on composite materials, Athens; Greece

Risteska S et al (2024) The effect of textile structure reinforcement on polymer composite material mechanical behavior. Polymers 16(24):3478

Salleh A et al (2023) The green approach-based biomaterials for tissue engineering application. In: Wan Kamarul Zaman WS, Abdullah N (eds) Sustainable material for biomedical engineering application. Springer Nature Singapore, Singapore, pp 27–47

Saraí CG-R et al (2018) Antimicrobial effect of silk and catgut suture threads coated with biogenic silver nanoparticles. In: Khan M (ed) Silver nanoparticles (Ch. 13). IntechOpen, Rijeka

Sarker F et al (2018) High-performance graphene-based natural fiber composites. ACS Appl Mater Interf 10(40):34502–34512

Seydibeyoğlu MÖ et al (2023) Review on hybrid reinforced polymer matrix composites with nanocellulose, nanomaterials, and other fibers. Polymers 15(4):984

Seyfoori A, Sarfarazijami S, Seyyed Ebrahimi SA (2019) pH-responsive carbon nanotube-based hybrid nanogels as the smart anticancer drug carrier. Artif Cells Nanomed Biotechnol 47(1): 1437–1443

Sharma A, Joshi SC (2023) Fiber-reinforced polyurethane matrix composites for engineering applications. Polyurethanes: preparation, properties, and applications. Volume 1: fundamentals. American Chemical Society, pp 101–118

Shchipunov Y (2012) Bionanocomposites: green sustainable materials for the near future. Pure Appl Chem 84(12):2579–2607

Signorini C et al (2024) Hybrid fibre-reinforced cementitious composites with short polyethylene and continue carbon fibres: influence of roving impregnation on tensile and cracking behaviour. Mater Des 248:113465

Singh PK, Pandey P (2023) Fiber reinforced biopolymer composites: a review. Macromol Symp 407(1):2100501

Somarathna Y et al (2024) Formulation of epoxy prepregs, synthesization parameters, and resin impregnation approaches—a comprehensive review. Polymers 16(23):3326

Song X et al (2024) 3D printing of continuous cotton thread reinforced poly (lactic acid). Sci Rep 14(1):30858

Swolfs Y, Gorbatikh L, Verpoest I (2014) Fibre hybridisation in polymer composites: a review. Compos A Appl Sci Manuf 67:181–200

Swolfs Y, Verpoest I, Gorbatikh L (2019) Recent advances in fibre-hybrid composites: materials selection, opportunities and applications. Int Mater Rev 64(4):181–215

Tcherdyntsev VV (2021) Reinforced polymer composites. Polymers 13(4):564

Tingaut P, Zimmermann T, Lopez-Suevos F (2010) Synthesis and characterization of bionanocomposites with tunable properties from poly(lactic acid) and acetylated microfibrillated cellulose. Biomacromol 11(2):454–464

Tóth C, Lukács NL, Kovács NK (2024) The role of the fiber–matrix interface in the tensile properties of short fiber–reinforced 3D-printed polylactic acid composites. Polym Compos 45(15):13589–13602

Vidakis N et al (2023) Optimized PCL/CNF bio-nanocomposites for medical bio-plotted applications: rheological, structural, and thermomechanical aspects. Bioprinting 36:e00311
Wang J, Liu J, Long L (2023) Bio-based impregnated resin preparation for aldehyde-free decorative paper production. Coatings 13(7):1168
Wang H et al (2022) Enhancement of strength and toughness of bio-nanocomposites with good transparency and heat resistance by reactive processing. iScience 25(7):104560
Wellbrock W et al (2020) Sustainability in the automotive industry, importance of and impact on automobile interior—insights from an empirical survey. Int J Corp Soc Responsib 5(1):10
Wiedemann G, Rothe H (1986) Review of prepreg technology. In: Pritchard G (ed) developments in reinforced plastics—5: Processing and fabrication. Springer Netherlands, Dordrecht, pp 83–119
Yadav V, Pal D, Poonia AK (2024) Nanofillers as a potential key for shaping the future of the industries. Hybrid Adv 7:100340
Youssef AM et al (2016) Enhancement of Egyptian soft white cheese shelf life using a novel chitosan/carboxymethyl cellulose/zinc oxide bionanocomposite film. Carbohyd Polym 151:9–19
Yuhazri MY, Zulfikar AJ, Ginting A (2020) Fiber reinforced polymer composite as a strengthening of concrete structures: a review. IOP Conf Ser Mater Sci Eng 1003(1):012135
Zhang Y et al (2019) High-strength, self-healable, temperature-sensitive, MXene-containing composite hydrogel as a smart compression sensor. ACS Appl Mater Interfaces 11(50):47350–47357
Zia KM et al (2021) 5—Processing of bio-based polymers. In: Zia KM et al (eds) Processing technology for bio-based polymers. Elsevier, Amsterdam, pp 151–189
Ziąbka M et al (2020) Antibacterial composite hybrid coatings of veterinary medical implants. Mater Sci Eng C 112:110968
Zuo P, Srinivasan DV, Vassilopoulos AP (2021) Review of hybrid composites fatigue. Compos Struct 274:114358

George Kwame Fobiri is an Assistant Lecturer at the Department of Fashion Design and Textiles Studies, at Kumasi Technical University. He holds a bachelor's degree in Industrial Art (Textiles) and an MFA in Textile Design from the Department of Industrial Art, Kwame Nkrumah University of Science and Technology (KNUST). He is currently pursuing a PhD in Textile Design Technology at KNUST. His research interests cover textiles, fashion, computer-aided design (CAD), e-textiles, wearable textiles, textile printing, textile dyeing, etc.

Benjamin Eghan is a Ph.D. student in Fashion and Textiles at De Montfort University, Leicester, United Kingdom. Prior to this, he earned both his MFA degree in Textile Design and his BA degree in Industrial Art (Textile option) from Kwame Nkrumah University of Science and Technology (KNUST), Kumasi, Ghana. His research interests include sustainability, consumer behavior and buying, textiles, and fashion. His Ph.D. research focuses on creating new products from fashion and textile waste, employing innovative techniques and technologies to give them a new life. His work also aims to make a social impact, particularly for rough sleepers.

Raphael Kanyire Seidu is a Ph.D. candidate in Fashion and Textile Design at the School of Fashion and Textiles at The Hong Kong Polytechnic University, Hong Kong. Before this, he gained both his MFA degree in Textile Design and BA. Degree in Industrial Art (Textile option) from Kwame Nkrumah University of Science and Technology, Ghana. He is also a Visiting Research Student at Nottingham Trent University (NTU) in the UK. He has several publications in leading journals in textile, design, technology, and material science. His recent paper was adjudged the "Second-Best Paper" presented at the 12th Sustainable Education and Development Research Conference in Somanya, Ghana. His research interest focuses on retro-reflective materials, smart textiles, cloth culture, weaving, and print technologies. He has several papers and has filed for a patent from his Ph.D. project which aims at developing smart interactive retro-reflective clothing for pedestrians' safety at night-time.

Benjamin Tawiah is the International and Industrial liaison officer for the Textiles and Fashion Section of the Department of Industrial Art, KNUST. He received his B.A. and Master of Engineering degrees from the Kwame Nkrumah University of Science and Technology, Ghana, and the Jiangnan University, China, respectively. Prior to earning his master's degree, he obtained a postgraduate Diploma in Management Information Systems from the Ghana Institute of Management and Public Administration (GIMPA) and worked with a number of Textile printing companies. After several years of industrial experience, he studied for his Ph D in 2019 from the Hong Kong Polytechnic University. He then worked as a postdoctoral Fellow at the Hong Kong Polytechnic University and the City University of Hong Kong. He is a lecturer at the Kwame Nkrumah University of Science and Technology and a postdoctoral Fellow at the Hong Kong Polytechnic University. He has published extensively in highly reputable peer-reviewed academic journals on a wide range of topics, including inkjet inks formulation, pigment dispersion, transfer printing, nanofillers, nanocomposites, flame retardant polymer composites and second-generation batteries for energy storage in advanced grid systems.

Chapter 4
Smart and Adaptive Bionanocomposite Materials Processing and Fabrication

Satyam Rawat, Peeyush Phogat, Shreya Sharma, Ranjana Jha, and Sukhvir Singh

Abstract Smart and adaptive bionanocomposite materials represent a crucial innovation in sustainable technology, addressing the need for responsive materials that minimize environmental impact. This chapter delves into the processing and fabrication techniques for creating bionanocomposites tailored for biomedical, environmental, and energy applications. By merging biocompatible matrices with nanoscale reinforcements, these composites gain advanced functionalities, including self-healing, stimuli-responsive behaviour, and adaptive morphology, allowing them to respond to factors like pH, temperature, and stress—ideal for drug delivery, biosensing, environmental monitoring, and adaptive infrastructure. Beginning with the fundamentals of bionanocomposites, the chapter discusses the integration of nanomaterials such as graphene and metal nanoparticles, which enhance durability, electrical, and thermal properties. Advanced fabrication techniques, including electrospinning, sol–gel processing, and 3D printing, allow precise control over these materials' architecture and properties. Challenges in processing, such as the need for strong matrix-nanomaterial interactions, are addressed through surface modification and functionalization methods. Further, the chapter explores adaptive fabrication control through artificial intelligence (AI) and machine learning (ML), which ensures real-time quality and reproducibility. These methods support efficient, sustainable production while allowing tailored design and functionality in composite materials. Concluding with perspectives on scalability, environmental implications, and future innovations, this chapter highlights the interdisciplinary approach necessary to harness bionanocomposites in sustainable development. By emphasizing advanced

S. Rawat · P. Phogat (✉) · S. Sharma · R. Jha
Research Lab for Energy Systems, Department of Physics, Netaji Subhas University of Technology, Dwarka, New Delhi, India
e-mail: peeyush.phogat@gmail.com

P. Phogat · S. Sharma
Council of Scientific and Industrial Research–National Institute of Science Communication and Policy Research (CSIR–NIScPR), Pusa, New Delhi, India

S. Singh
Delhi Skill and Entrepreneurship University, Dwarka, New Delhi, India

S. C. George and B. Tawiah (eds.), *Advanced Bionanocomposite Materials*, Advanced Structured Materials 241, https://doi.org/10.1007/978-981-96-8225-6_4

processing and adaptive capabilities, these materials hold transformative potential across diverse applications.

Keywords Smart bionanocomposites · Adaptive materials · Sustainable fabrication · Nanomaterial processing · Stimuli-responsive systems · Self-healing materials · AI-driven material processing

4.1 Introduction

A new class of multiphase nanostructured materials known as bionanocomposites must include particles with at least one dimension on the nanometer scale and at least one phase of biological origin. Composite materials comprising biologically derived components and particles with at least one dimension between 1 and 100 nm are called bionanocomposites. They are characterized by their ability to integrate nanomaterials with natural or synthetic biopolymer matrices to form materials with enhanced mechanical, thermal, electrical, and responsive properties. Unlike conventional composites or simple nanocomposites, bionanocomposites are uniquely defined by their biocompatibility and stimuli-responsive capabilities, making them a distinct category with diverse applications. What sets these materials apart is their ability to exhibit smart functionalities, enabling them to respond dynamically to external stimuli such as temperature, pH, light, and mechanical stress. This responsiveness, combined with their high performance and sustainability, positions smart bionanocomposites at the forefront of advanced material science. Figure 4.1 categorizes composites into conventional, nanocomposites, and bionanocomposites, highlighting their matrix materials (polymers, metals, biopolymers) and reinforcements (fibers, CNTs, nano clays). These materials hold immense promise in addressing the growing demands for sustainable, high-performance materials in diverse fields, including biomedical engineering, environmental monitoring, and energy storage. By integrating the unique properties of nanomaterials with biocompatible and often biodegradable polymers, bionanocomposites offer a pathway toward developing materials that are not only functionally superior but also aligned with global sustainability goals. The smart functionality of bionanocomposites stems from the nanoscale interactions between the matrix and the reinforcement, creating materials that can adapt their properties in real time to changing environments. This adaptivity enables their use in critical applications such as controlled drug delivery, biosensing, and environmental remediation, where conventional materials fall short. A defining characteristic of smart bionanocomposites is their advanced stimuli-responsiveness. For instance, thermal-responsive composites can alter their mechanical or structural properties with changes in temperature, a feature crucial for biomedical applications such as wound healing and drug release. pH-sensitive composites are particularly effective in targeted drug delivery systems, where they release therapeutic agents in response to the acidic microenvironment of tumours. The capacity to harness light as a stimulus has also enabled applications in photocatalysis and photo-controlled therapies. Materials such as graphene, carbon nanotubes (CNTs), and metal oxide nanoparticles exhibit exceptional electrical, thermal, and mechanical properties at the nanoscale, further enhancing the adaptivity of these composites (Shanmuganathan et al. 2010). The "smart" attribute of these composites also encompasses their ability to self-heal and repair damage autonomously. By incorporating functional nanostructures that release healing agents when a crack or defect occurs, these materials prolong their service life while maintaining their structural and functional integrity. The versatility of smart bionanocomposites lies in their ability to be tailored for specific applications. This is achieved by selecting appropriate combinations of nanomaterials and biopolymers to achieve desired functionalities. For example, in drug delivery systems, stimuli-responsive bionanocomposites can be engineered to release therapeutic agents at precise locations within the body in response to changes in pH or temperature. Such materials reduce side effects and improve treatment efficacy, showcasing the transformative potential of smart bionanocomposites in healthcare. From a sustainability perspective, bionanocomposites offer significant advantages over traditional materials. Conventional materials often rely on non-renewable resources and involve energy-intensive production processes that generate significant waste. In contrast, bionanocomposites are frequently designed with sustainability in mind, from the selection of renewable raw materials to considerations for end-of-life disposal. Biopolymers, derived from renewable resources such as plant-based cellulose or marine-sourced chitosan, serve as the matrix for these composites. These polymers are not only environmentally friendly but also capable of natural decomposition, thereby minimizing their ecological footprint. Additionally, the high functional efficiency of nanomaterials allows for the reduction of material quantities required for specific applications, further contributing to resource conservation and waste reduction. The processing and fabrication of bionanocomposites are critical to achieving their desired properties and performance. Advanced fabrication techniques such as electrospinning, sol–gel processing, and 3D printing play a pivotal role in this domain. Electrospinning, for example, enables the production of nanofiber structures that replicate the architecture of biological tissues, making it an invaluable tool in tissue engineering. Sol–gel processing allows the integration of nanoparticles into matrices, enhancing thermal stability and mechanical robustness. These methods provide precise control over the nanoscale architecture of the composites, which is essential for realizing their adaptive and responsive capabilities.

The incorporation of artificial intelligence (AI) and machine learning (ML) further augments the smart functionality of bionanocomposites. AI-driven optimization of fabrication processes ensures that these composites exhibit the required responsiveness, mechanical strength, and thermal stability. Real-time monitoring enabled by AI can dynamically adjust parameters during fabrication, ensuring that the materials produced consistently meet performance standards. For instance, AI-based tools can optimize the integration of temperature- and pH-responsive nanoparticles into biopolymer matrices, enhancing their adaptivity and ensuring reproducibility. Despite their potential, the development of smart bionanocomposites faces several challenges, particularly in achieving uniform dispersion of nanomaterials within the matrix. Nanoparticles have a natural tendency to aggregate due to their high surface

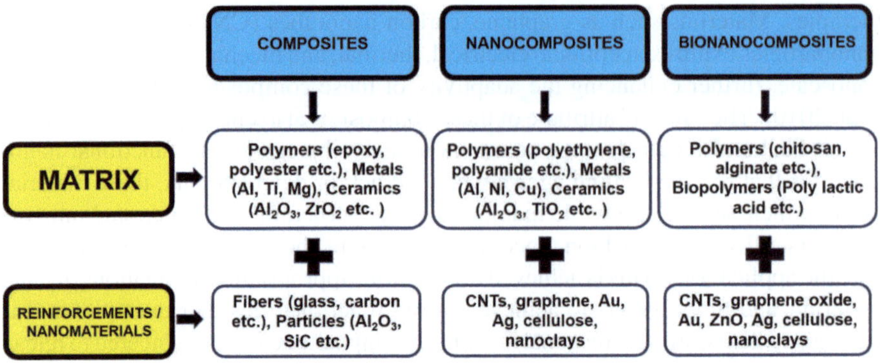

Fig. 4.1 Comparative classification of composites, nanocomposites, and bionanocomposites based on matrices and reinforcements

energy, leading to inconsistent properties in the resulting composites. To address this, surface modification and functionalization techniques are employed to improve the compatibility between nanomaterials and the matrix. Functionalization not only enhances dispersion but also allows the introduction of additional properties, such as increased biocompatibility or heightened sensitivity to environmental stimuli. These advancements are crucial for unlocking the full potential of bionanocomposites and ensuring their effectiveness across various applications (Pascual Aguilar et al. 1067).

4.2 Fundamentals of Bionanocomposite Materials

In this section, we explore the fundamental principles underlying bionanocomposite materials, which combine natural biomaterials and nanostructured reinforcements to create multifunctional materials suitable for various applications. Bionanocomposites leverage the unique properties of nanomaterials (such as high strength, conductivity, and surface area) and the biocompatibility of biomaterials to achieve adaptive, stimuli-responsive, and environmentally sustainable functionalities. This section covers the classification, properties, and characteristics of bionanocomposites, as well as the role of nanomaterials in enhancing biocompatibility, sustainability, and mechanical performance (Ariadurai).

4.2.1 Classification of Bionanocomposites

Bionanocomposites can be classified as illustrated in Fig. 4.2 based on multiple criteria, including the type of matrix material, the nature of the nanomaterials used, and the specific functionality of the composite.

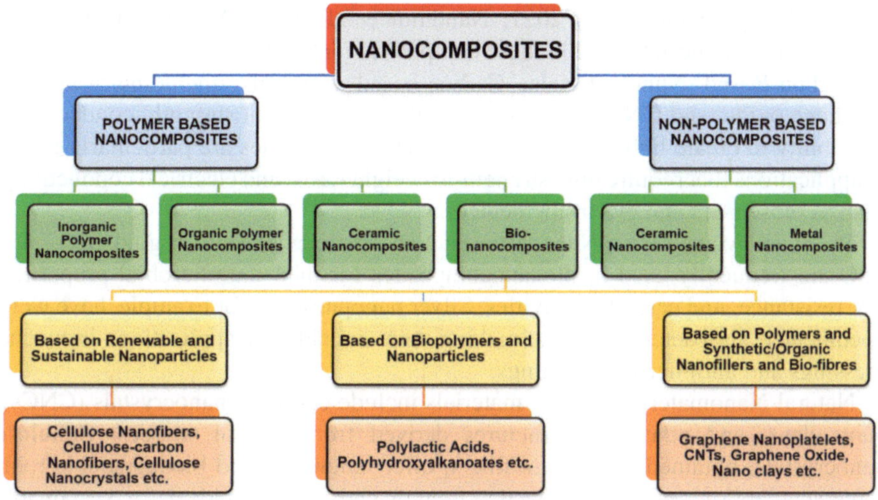

Fig. 4.2 Different types of Nanocomposites and Bionanocomposites

Matrix Type

The matrix, or primary component of the composite, determines many of the material's characteristics, such as flexibility, biocompatibility, and degradation rate (Nanocomposites with Biodegradable Polymers—Google Books 2024). There are two major types of matrices used in bionanocomposites:

Natural Biopolymer Matrix: Biopolymers derived from natural sources, such as chitosan, alginate, gelatine, and cellulose, are widely used in biomedical and environmental applications. These materials are biocompatible and biodegradable, making them suitable for applications where environmental sustainability and biological compatibility are essential. For instance, chitosan, a polysaccharide derived from chitin, is highly biocompatible, antimicrobial, and has excellent film-forming properties.

Synthetic Polymer Matrix: Synthetic polymers such as poly lactic acid (PLA), polycaprolactone (PCL), and polyethylene glycol (PEG) are often used for their tunable properties. While some synthetic polymers are biodegradable, others are selected for their mechanical strength and durability, even if they are not inherently biocompatible. However, combining synthetic matrices with biocompatible nanomaterials can improve their biological properties.

Nanomaterial Type

Nanomaterials in bionanocomposites bring distinctive properties due to their high surface area-to-volume ratio, which enables enhanced interaction with the polymer matrix (Thippeswamy et al. 2021; Nandhini et al. 2024; Aravind et al. 2022). Common nanomaterials used in these composites include:

Carbon-Based Nanomaterials: Carbon-based materials, including graphene, graphene oxide, and CNTs, are known for their mechanical strength, conductivity, and ability to create lightweight structures. These properties are particularly useful in applications that require high strength-to-weight ratios and electrical conductivity, such as biosensors and electronic devices.

Metal and Metal Oxide Nanoparticles: Nanoparticles like silver, gold, TiO_2, and ZnO are frequently used in bionanocomposites for their antimicrobial properties, UV resistance, and catalytic activities. Silver nanoparticles, for example, have well-documented antibacterial effects, making them suitable for applications in wound dressings and antimicrobial coatings.

Natural Nanomaterials: These materials include cellulose nanocrystals (CNCs), nano clays, and other nanostructures derived from natural sources. Cellulose nanocrystals, obtained from plant fibers, offer renewable and biodegradable reinforcement options with high tensile strength and stiffness, ideal for sustainable material development.

Classification by Functionality

Bionanocomposites are also classified based on their functionality, which is often determined by the intended application (Haina et al. 2020). Some common functionalities include:

Biodegradable Composites: These composites are designed to break down naturally in the environment, often through enzymatic or microbial degradation, which makes them particularly relevant for biomedical implants and drug delivery systems. Materials like PLA and PCL are frequently used due to their controlled degradation rates.

Stimuli-Responsive Composites: Stimuli-responsive, or "smart" composites, are materials that react to environmental changes such as temperature, pH, light, or mechanical stress. For instance, composites with thermal-responsive properties undergo structural or phase changes when heated, which is useful in applications such as drug delivery where controlled release is essential.

Conductive Composites: Conductive composites, often containing graphene or CNTs, are applied in electronic devices, biosensors, and tissue engineering scaffolds. The electrical conductivity of these materials allows them to interface with biological systems, such as neural tissues, and to transmit electrical signals, which is essential for biosensors and bioelectronic applications.

4.2.2 Properties and Characteristics of Bionanocomposites

Bionanocomposites combine biopolymers with nanoscale reinforcements, bringing enhanced properties that are critical to their performance in various applications, including biomedical devices, environmental sensing, and energy storage. Understanding their properties—such as mechanical strength, biocompatibility, degradability, electrical conductivity, and thermal stability—enables their precise tuning for specific applications, as compiled in Fig. 4.3. This section provides a detailed examination of these properties, using relevant equations and formulations to elucidate how each characteristic influences the material's performance.

Mechanical Strength and Elasticity

Mechanical strength in bionanocomposites results from the reinforcement of the polymer matrix with nanoscale materials, such as CNTs and graphene. The rule of mixtures is a fundamental principle in materials science that predicts composite strength as a function of both the matrix and nanomaterial properties. This relationship is described by:

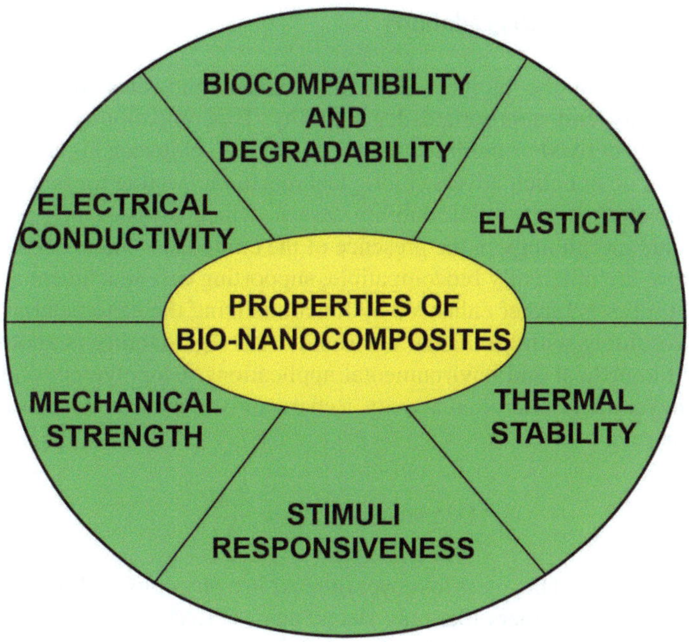

Fig. 4.3 Various properties of bionanocomposites that are critical for their performance in various applications

$$\sigma_{\text{composite}} = V_{\text{m}} \cdot \sigma_{\text{m}} + V_{\text{n}} \cdot \sigma_{\text{n}} \tag{4.1}$$

where V_{m} and V_{n} are the volume fractions of the matrix and nanomaterial, respectively, σ_{m} and σ_{n} represent the tensile strengths of the matrix and nanomaterial. As nanomaterial content increases, the composite's tensile strength also rises due to effective load transfer across the material interface. However, the compatibility and interaction between the matrix and reinforcement are crucial. The inclusion of surface-functionalized nanomaterials, such as oxidized graphene, enhances interfacial adhesion, which significantly improves strength by facilitating stress transfer.

Young's Modulus, or stiffness, of the composite is similarly enhanced by nanoscale reinforcements. Young's modulus ($E_{\text{composite}}$) can be calculated using the rule of mixtures:

$$E_{\text{composite}} = V_{\text{m}} \cdot E_{\text{m}} + V_{\text{n}} \cdot E_{\text{n}} \tag{4.2}$$

where E_{m} and E_{n} are the Young's moduli of the matrix and nanomaterial. Higher Young's modulus indicates a stiffer material, which is beneficial in load-bearing applications like structural supports or orthopaedic implants (Giacomo et al. 2013).

Biocompatibility and Degradability

For applications in the biomedical field, bionanocomposites must exhibit high biocompatibility and appropriate degradability. Biocompatibility is achieved by using naturally derived or modified biopolymers (e.g., chitosan, alginate) and nanomaterials that do not elicit adverse biological reactions. Testing for biocompatibility typically involves in vitro cell viability assays, which measure cell proliferation, adhesion, and morphology in the presence of the composite. Materials like chitosan and cellulose are inherently biocompatible, supporting cell attachment and growth. Biocompatibility is further enhanced by functionalizing the nanomaterial surface to interact favourably with biological environments. Degradability is another critical factor for biomedical and environmental applications. Biopolymers often degrade through enzymatic or microbial activity, a characteristic that can be measured by the degradation rate:

$$\text{Degradation Rate} = \frac{\Delta M}{\Delta t} \tag{4.3}$$

where ΔM represents the mass loss over time Δt in a physiological or environmental setting. Synthetic polymers like poly (lactic acid) (PLA) have controlled degradation rates, making them suitable for drug delivery systems and temporary implants. Biodegradability ensures that after fulfilling its function, the composite will decompose without leaving harmful residues, aligning with environmental sustainability goals (Kuperkar et al. 2024).

Electrical Conductivity

The electrical conductivity of bionanocomposites is essential for applications in electronics, biosensors, and tissue engineering. Conductive nanomaterials such as CNTs, graphene, and metal nanoparticles can be dispersed within the biopolymer matrix to increase conductivity. According to percolation theory, the composite exhibits conductivity only after the concentration of conductive filler surpasses a certain critical threshold, forming a continuous conductive network. The overall conductivity ($\sigma_{\text{composite}}$) is given by:

$$\sigma_{\text{composite}} = \sigma_{\text{m}} \cdot (1 - V_{\text{c}}) + V_{\text{n}} \cdot \sigma_{\text{n}} \tag{4.4}$$

where σ_{m} and σ_{n} are the conductivities of the matrix and nanomaterial, V_{c} is the critical volume fraction required for percolation. In conductive composites, surface resistivity (R_s) decreases with increasing nanomaterial content. This characteristic is particularly valuable in biosensors, where electrical signals are generated in response to biological changes. The conductive pathway formed by nanomaterials also enables the use of bionanocomposites in energy storage devices, where high conductivity allows for efficient charge and discharge cycles (Lecoublet et al. 2024).

Thermal Stability and Conductivity

Thermal properties in bionanocomposites are critical for applications in high-temperature environments, such as electronic devices and implantable biomedical devices. Adding nanomaterials like graphene or alumina increases the thermal conductivity of the composite, improving its heat dissipation capability. The effective thermal conductivity ($k_{\text{composite}}$) of the composite is modelled using the rule of mixtures:

$$k_{\text{composite}} = V_{\text{m}} \cdot k_{\text{m}} + V_{\text{n}} \cdot k_{\text{n}} \tag{4.5}$$

where k_{m} and k_{n} are the thermal conductivities of the matrix and nanomaterial, respectively. Thermally stable nanomaterials provide resistance to high temperatures, allowing bionanocomposites to maintain structural integrity in applications like electronics, which require effective heat management. By adding thermally conductive fillers, composites also benefit from improved thermal stability, reducing the risk of material deformation or breakdown at elevated temperatures (Anitha Selvasofia et al. 2024).

Stimuli Responsiveness

One of the most innovative properties of bionanocomposites is stimuli responsiveness—the ability to undergo changes in response to external stimuli, such as pH, temperature, light, or mechanical stress. These materials are designed for advanced applications, including drug delivery, tissue engineering, and self-healing materials.

pH Responsiveness: Bionanocomposites containing chitosan exhibit pH-sensitive swelling behaviour due to the protonation of amine groups in acidic environments, which facilitates the controlled release of drugs at specific sites in the body.

Thermal Responsiveness: Composites embedded with poly (N-isopropylac rylamide) (PNIPAAm) undergo a phase transition at a lower critical solution temperature (LCST). Below the LCST, PNIPAAm is hydrophilic and swells in water; above it, it becomes hydrophobic and contracts. This temperature-responsive behaviour is useful for applications that require material changes with temperature, such as smart drug carriers.

Light and Stress Responsiveness: Some composites can respond to light (photo responsive) or mechanical force (mechanoresponsive) by changing their structure or properties. This property is essential in creating materials for self-healing and adaptable systems (Sharma et al. 2023).

4.2.3 Case Study: Challenges in Large-Scale Production of Bionanocomposites for Environmental Applications

Bionanocomposites, which combine biopolymer matrices with nanostructured reinforcements, have emerged as transformative materials for addressing environmental challenges such as pollution remediation, water purification, and renewable energy storage. Their ability to integrate nanoscale functionality with environmental sustainability positions them as a viable alternative to traditional materials. However, transitioning from laboratory-scale research to large-scale industrial production introduces significant technical, economic, and operational challenges, which form the core focus of this study. The primary purpose of this case study is to analyze and address the challenges associated with the large-scale production of bionanocomposites. By identifying bottlenecks in raw material sourcing, fabrication techniques, cost-efficiency, and quality control, the study aims to propose actionable solutions for scaling up production while maintaining material functionality and sustainability. The primary purpose of this case study is to analyze and address the challenges associated with the large-scale production of bionanocomposites. By identifying bottlenecks in raw material sourcing, fabrication techniques, cost-efficiency, and quality control, the study aims to propose actionable solutions for scaling up production while maintaining material functionality and sustainability (Saraswat et al. 2023).

The methodology for this study involved a comprehensive approach combining experimental data, material synthesis, fabrication techniques, and quality assessment. Data was gathered from laboratory-scale experiments, industrial-scale trials, and existing literature to identify production challenges and opportunities. The materials used included biopolymers such as chitosan, cellulose, and alginate, chosen for their renewable origins and biodegradability. These matrices were reinforced with nanomaterials like graphene, carbon nanotubes (CNTs), and metal oxide nanoparticles (e.g., TiO_2, ZnO) due to their high surface area, conductivity, and environmental reactivity. Additional chemicals, including solvents (ethanol, acetone), surfactants, and catalysts, facilitated the synthesis and functionalization processes. The fabrication processes employed advanced equipment such as electrospinning setups for nanofiber production, sol–gel reactors for integrating nanoparticles into matrices, and 3D printing machines for creating composite membranes and devices. Analytical tools like Scanning Electron Microscopy (SEM), X-Ray Diffraction (XRD), and Fourier Transform Infrared Spectroscopy (FTIR) were used to characterize the materials' structure, morphology, and functionality. Automated quality control systems, including AI-based monitoring tools, ensured precision during production. The procedure began with the preparation of raw materials, where biopolymers were dissolved in appropriate solvents, and nanomaterials were synthesized and functionalized to enhance compatibility with the matrix. Fabrication processes such as electrospinning, sol–gel synthesis, and 3D printing were optimized to achieve the desired composite properties. Finally, rigorous quality assessments were conducted to evaluate mechanical strength, thermal stability, and environmental responsiveness, ensuring the materials met the requirements for large-scale environmental applications.

The study identified several critical findings and challenges associated with the large-scale production of bionanocomposites for environmental applications. Material sourcing emerged as a significant issue, with the availability of biopolymers such as chitosan, cellulose, and alginate being inconsistent due to seasonal and geographic factors. Nanomaterials like graphene and CNTs require energy-intensive synthesis processes, often involving hazardous chemicals, which raises concerns about sustainability. In terms of fabrication, scaling techniques like electrospinning resulted in uneven nanofiber formation, while sol–gel processing faced challenges with nanoparticle aggregation. Similarly, maintaining uniform dispersion of nanomaterials during 3D printing proved difficult, impacting the structural and functional properties of the composites. Economic barriers were also prominent, with high costs of nanomaterials and specialized equipment making production less cost-effective. Energy-intensive fabrication methods and the requirement for skilled labour further contributed to the financial constraints, particularly in applications where cost-effectiveness is critical. Quality control presented additional hurdles, as the properties of bionanocomposites—such as mechanical strength, thermal stability, and stimuli-responsiveness—depended heavily on the consistency of raw materials and precise fabrication processes. Variability in these factors often led to defects and reduced performance, necessitating the implementation of rigorous testing protocols. Despite these challenges, the potential applications of bionanocomposites in environmental fields are vast. These materials demonstrated high efficiency in water filtration by removing heavy metals, organic pollutants, and pathogens, while also being effective in pollution monitoring through sensors with enhanced sensitivity and stability. Furthermore, their integration into energy storage devices such as supercapacitors and batteries significantly improved energy density and charge–discharge efficiency. However, maintaining these advanced properties during large-scale production remains a challenge, emphasizing the need for precise fabrication techniques and real-time quality control. The findings highlight the need for interdisciplinary collaboration and innovation to overcome these challenges and unlock the full potential of bionanocomposites in environmental applications (Turcu et al. 2014).

The large-scale production of bionanocomposites is constrained by several key limitations, primarily related to sustainability, cost-effectiveness, and scalability. One major challenge lies in the energy-intensive synthesis of nanomaterials, which often involves hazardous reagents that undermine the environmental benefits of these composites. Additionally, the aggregation of nanomaterials during fabrication impacts the uniformity and performance of the final product, requiring advanced functionalization techniques to ensure compatibility with the biopolymer matrix. High production costs, stemming from expensive raw materials, specialized equipment, and skilled labor, further limit accessibility and scalability. Moreover, the lack of standardized protocols for quality control and material testing complicates efforts to maintain consistency and reproducibility in large-scale applications. To address these limitations, several improvements are proposed. Adopting green chemistry approaches for nanomaterial synthesis, such as enzyme-mediated or bio-inspired methods, can significantly reduce the environmental impact of production. The integration of ML and AI offers promising solutions for optimizing fabrication parameters, improving dispersion, and ensuring real-time quality monitoring, thereby minimizing waste and defects. Standardized protocols for material characterization and quality assessment should be established to ensure uniformity across different production scales. Collaborative efforts between academia, industry, and policymakers are crucial to developing cost-effective manufacturing methods and streamlining regulatory frameworks. Furthermore, incorporating renewable and low-cost raw materials while exploring hybrid fabrication techniques can enhance scalability and reduce overall production costs. By addressing these limitations and implementing these suggestions, the large-scale production of bionanocomposites can achieve greater efficiency, sustainability, and applicability in addressing global environmental challenges (Singh 2023). Figure 4.4 illustrates an industrial facility focused on the large-scale production of bionanocomposites. Automated machines process biopolymers and nanomaterials into membranes and sensors, while quality control systems monitor parameters. The green backdrop emphasizes the eco-friendly intent of these composites in pollution control and energy solutions.

Fig. 4.4 Industrial-scale setup for producing bionanocomposites, featuring automated machinery, quality control systems, and eco-friendly surroundings reflecting sustainability in environmental applications. This is an AI-generated image taken from ChatGPT.com

4.3 Smart Functionalities in Bionanocomposites

Smart functionalities in bionanocomposites enable these materials to respond dynamically to external stimuli, positioning them at the forefront of advanced applications in biomedicine, environmental monitoring, and adaptive systems. By integrating responsive nanomaterials with biocompatible polymers, these composites can be engineered to react to pH, temperature, mechanical stress, light, and electric or magnetic fields. Such responsiveness is often achieved by functionalizing nanomaterial surfaces or selecting polymers with inherent stimuli-sensitive properties. For example, temperature-sensitive polymers like PNIPAAm can undergo phase transitions that alter the material's structure, facilitating controlled drug release or self-healing capabilities. Additionally, composites with conductive nanomaterials such as graphene exhibit electro-sensitivity, essential in biosensors and electro-responsive drug delivery. The smart functionalities of bionanocomposites open doors to designing adaptive materials that not only enhance device performance but also advance sustainable development through precision, efficiency, and multifunctionality.

4.3.1 Temperature-Responsive Bionanocomposites

Temperature-responsive bionanocomposites are engineered materials that undergo reversible or irreversible physical and chemical changes in response to temperature fluctuations. These materials are essential for applications in biomedical devices, drug delivery systems, tissue engineering, and smart textiles, where temperature variations can serve as triggers for material responses. The temperature responsiveness of these composites is often achieved by incorporating thermally responsive polymers and nanomaterials, which interact to modulate properties like solubility, elasticity, or volume (Xu et al. 2021).

Mechanisms of Temperature Responsiveness: Temperature-responsive bionanocomposites rely on polymers with a Lower Critical Solution Temperature (LCST) or Upper Critical Solution Temperature (UCST) to exhibit changes in solubility or phase. Below the LCST, polymers are typically soluble and hydrophilic, but as temperature rises beyond the LCST, they become hydrophobic and precipitate out of solution. This behaviour is particularly beneficial for controlled drug release, as the material can be designed to release its contents only at certain temperatures. The LCST behaviour of a composite can be expressed by the Flory–Huggins theory, which describes the polymer–solvent interactions:

$$\chi(T) = \frac{\Delta H - T\Delta S}{k_{\mathrm{B}}T} \tag{4.6}$$

where $\chi(T)$ is the Flory–Huggins interaction parameter, ΔH and ΔS are the enthalpy and entropy of mixing, respectively, k_{B} is the Boltzmann constant, T is the absolute temperature. This equation illustrates that as temperature increases, the interaction parameter $\chi(T)$ changes, leading to phase separation at a critical temperature characteristic of LCST polymers. For temperature-sensitive applications, polymers like PNIPAAm are frequently used due to their LCST near body temperature (32–37 °C). Figure 4.5 illustrates the preparation of a thermo-responsive hydrogel through chitosan dissolution, CNC and PNIPAAm addition, UV photopolymerization, lyophilization, and injection into damaged tissue for regenerative healing.

Nanomaterials for Enhanced Responsiveness: Incorporating nanomaterials, such as graphene, CNTs, and metal oxide nanoparticles, into temperature-responsive polymers enhances the composite's functionality by improving heat transfer, mechanical strength, and stability. Nanomaterials with high thermal conductivity, like graphene, facilitate rapid temperature changes throughout the composite, triggering an immediate response in the polymer matrix.

For instance, Graphene and CNTs: These carbon-based materials have high thermal conductivity and contribute to faster heat distribution, resulting in quick responses to temperature changes. They also increase the composite's mechanical strength and flexibility. Metal Oxides (e.g., Fe_3O_4): Magnetic nanoparticles like Fe_3O_4 can generate heat when exposed to an alternating magnetic field, enabling remote control of temperature-triggered responses in the composite. The improved thermal conductivity k of the composite can be described as:

$$k_{\mathrm{composite}} = V_{\mathrm{m}}\Delta k_{\mathrm{m}} + V_{\mathrm{n}}\Delta k_{\mathrm{n}} \tag{4.7}$$

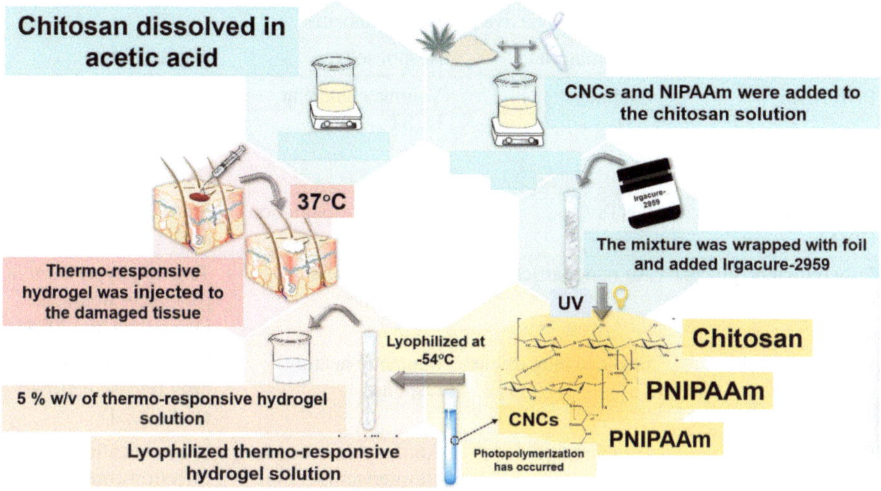

Fig. 4.5 Thermo-responsive hydrogel synthesis and application for tissue repair using chitosan, cellulose nanocrystals (CNCs), and PNIPAAm. Reprinted with permission from ref Promdontree et al. (2024). © 2024 by the authors. Licensee MDPI, Basel, Switzerland

where V_m and V_n are the volume fractions of the matrix and nanomaterial, k_m and k_n are the thermal conductivities of the matrix and nanomaterial, respectively.

Types of Temperature-Responsive Composites: The temperature-responsive behaviour in bionanocomposites can vary depending on the choice of polymer and nanomaterial. Table 4.1 summarizes several types of temperature-responsive bionanocomposites and their specific applications.

4.3.2 Self-healing Mechanisms and Material Regeneration

One of the most remarkable features of smart bionanocomposites is their ability to self-heal and regenerate after sustaining damage. Self-healing mechanisms mimic biological processes, such as wound healing in living organisms, and are designed to prolong the lifespan and functionality of materials. This capability is especially valuable in applications where materials are subjected to frequent wear and tear, such as biomedical implants, structural components, and electronic devices (Wang et al. 2024).

The self-healing process in bionanocomposites typically involves encapsulated healing agents or dynamic chemical bonds. Encapsulated healing agents, such as microcapsules filled with polymer precursors, are embedded within the composite matrix. When a crack forms, it ruptures the microcapsules, releasing the healing agents into the damaged area. These agents then polymerize or solidify, repairing the damage and restoring the material's integrity. Dynamic chemical bonds, on the other

Table 4.1 Types of temperature-responsive bionanocomposites

Polymer	Nanomaterial	Response type	Application
PNIPAAm	Graphene	Volume change at LCST	Controlled drug delivery Prinith Kaveramma et al. (2025)
Poly (vinyl alcohol) (PVA)	CNTs	Elasticity modulation with temperature	Smart textiles Dachani et al. (2024)
Poly (ethylene glycol) (PEG)	Au nanoparticles	Localized heat generation	Targeted hyperthermia for cancer treatment Debnath et al. (2024)
Polyacrylic acid (PAA)	Fe_3O_4 nanoparticles	Magnetic heating response	Remote-controlled drug release Petrovic et al. (2024)
Chitosan	Ag nanoparticles	Antibacterial release at elevated temperature	Wound healing and infection control Bhattacharjee et al. (2025)

hand, involve reversible covalent or non-covalent interactions that can reform after breaking. For instance, hydrogen bonds or disulfide bonds can facilitate self-repair at a molecular level, allowing the composite to heal multiple times. Nanomaterials, such as graphene oxide, CNTs, and metal nanoparticles, play a crucial role in enhancing self-healing efficiency. Their high surface area and reactivity improve the interaction between the healing agents and the damaged matrix. Additionally, nanomaterials can reinforce the repaired regions, ensuring that the material maintains its mechanical strength after healing. In biomedical applications, self-healing bionanocomposites can improve the durability of implants and scaffolds, reducing the need for frequent replacements and invasive procedures (Zhu et al. 2024). Figure 4.6 demonstrates a self-healing coating system comprising nano-capsules embedded in an epoxy matrix on a metal substrate. Upon mechanical damage, the nano-capsules rupture, releasing the healing agent into the affected area, facilitating autonomous repair and restoring structural integrity. This innovation enhances durability and extends the lifespan of coated materials in various applications.

4.3.3 *Stimuli-Responsive Behaviour*

Stimuli-responsive behaviour is a hallmark of smart bionanocomposites, enabling them to dynamically react to external triggers such as temperature, pH, light, or mechanical stress (Sanyal et al. 2024; Mittal et al. 2012; Bagal-Kestwal et al. 2019; Rashid and Afzal 2023). Figure 4.7 illustrates the role of stimuli-responsive nanoparticles (NPs) in responding to endogenous (e.g., pH, enzymes) and exogenous (e.g., light, magnetic fields) triggers. These nanoparticles, including gold, iron oxide, and

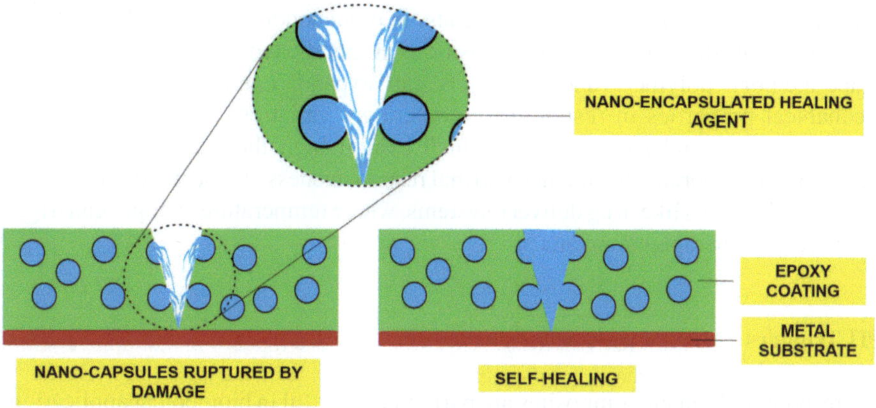

Fig. 4.6 Self-healing epoxy coating with nano-encapsulated healing agents for damage repair. Reprinted with permission from ref Pulikkalparambil et al. (2018). © 2018 Published by Elsevier B.V

quantum dots, are used in bio-imaging, targeted therapy, and biomedicine, offering precise and controlled functionality for healthcare advancements. This adaptability is achieved by integrating responsive nanomaterials into the composite matrix, which imparts sensitivity to specific environmental changes.

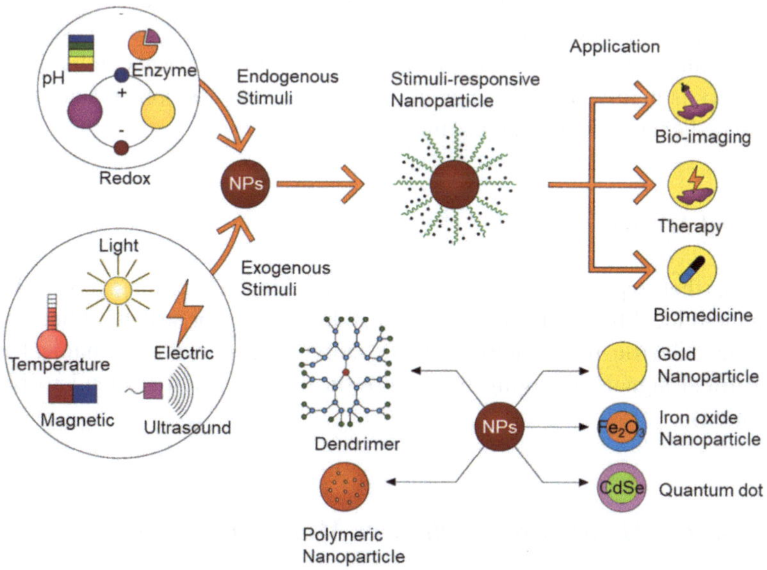

Fig. 4.7 Stimuli-responsive nanoparticles for advanced biomedical applications. Reprinted with permission from ref Pham et al. (2020). © 2020 by the authors. Licensee MDPI, Basel, Switzerland

Thermal Triggers: Thermal responsiveness in bionanocomposites is often achieved using materials that undergo phase transitions at specific temperatures. For instance, shape-memory polymers and hydrogels can expand, contract, or change their mechanical properties in response to temperature variations. Nanomaterials such as graphene or gold nanoparticles, which exhibit high thermal conductivity, are frequently incorporated to enhance thermal responsiveness. These materials are valuable in applications like drug delivery systems, where temperature changes can trigger the release of therapeutic agents.

PH Triggers

pH-responsive bionanocomposites are particularly useful in biomedical applications. For example, the acidic microenvironment of a tumour can be exploited to release drugs encapsulated within the composite. Materials like chitosan, a biopolymer that swells or contracts based on pH changes, can be combined with nanomaterials to create composites that respond to specific pH levels. This property ensures targeted delivery and minimizes side effects in treatments.

Light Triggers

Light-responsive bionanocomposites utilize nanomaterials like quantum dots, photochromic compounds, or plasmonic nanoparticles that react to specific wavelengths of light. Upon exposure to light, these materials can undergo structural or chemical changes, such as altering their shape, color, or electrical conductivity. This functionality is highly advantageous in applications such as optical sensors, where light-induced changes can generate measurable signals.

Mechanical Triggers

Composites that respond to mechanical stress are designed to adapt to physical forces, making them ideal for applications like wearable devices or structural health monitoring. Piezoelectric nanomaterials, for example, can convert mechanical energy into electrical signals, enabling the composite to sense and respond to physical deformations.

4.3.4 Shape Memory and Adaptive Morphology Changes

Shape-memory behaviour is another intriguing functionality exhibited by smart bionanocomposites. These materials can remember their original shape and return to it after being deformed, often in response to external stimuli such as heat or light. Shape memory is typically achieved by incorporating responsive polymers and nanomaterials into the composite matrix. The shape-memory effect is driven by a combination of molecular alignment and network structure. In most cases, the composite is deformed and fixed into a temporary shape under specific conditions, such as heating or cooling. When exposed to a triggering stimulus, the molecular chains in the material realign, allowing it to revert to its original configuration. Shape-memory bionanocomposites are extensively used in minimally invasive medical devices (Aston et al. 2013). For example, stents that can expand to a predetermined shape inside blood vessels are fabricated using such materials. These materials are also crucial in soft robotics, where their ability to change shape enables the creation of actuators and grippers with adaptive capabilities. In energy harvesting, shape-memory materials can improve the efficiency of devices by optimizing their geometry in response to environmental changes.

4.3.5 Sensing and Actuation Capabilities in Bionanocomposites

The integration of sensing and actuation capabilities into bionanocomposites transforms them into active components capable of detecting and responding to environmental changes (Mujawar et al. 2020).

Sensing Capabilities

Bionanocomposites equipped with nanoscale sensors can detect physical, chemical, or biological signals with high sensitivity. Nanomaterials like graphene, quantum dots, and metal oxide nanoparticles are commonly used for this purpose. These materials exhibit unique properties, such as high electrical conductivity and fluorescence, which make them ideal for sensing applications. For instance, in environmental monitoring, composites can detect pollutants or changes in pH and temperature. In healthcare, biosensors made from bionanocomposites can monitor glucose levels, detect pathogens, or measure vital signs.

Actuation Capabilities

Actuation involves the ability of a material to produce movement or mechanical work in response to a stimulus. Actuation in bionanocomposites is typically achieved through the incorporation of piezoelectric, magnetostrictive, or electroactive nanomaterials. These materials convert energy from one form to another, enabling the composite to perform mechanical actions. Piezoelectric Actuators, materials like zinc oxide nanowires, can convert mechanical stress into electrical energy, which

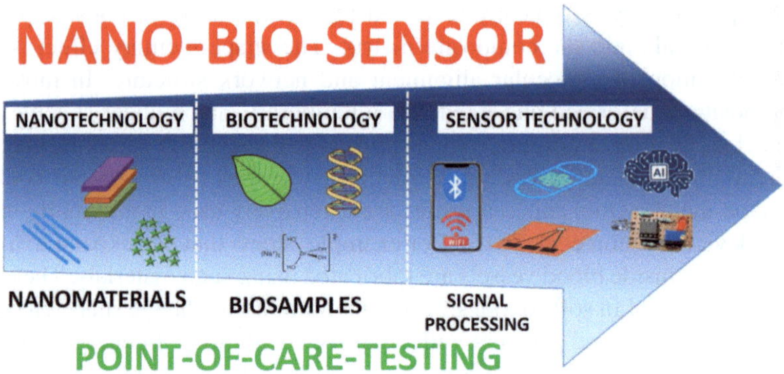

Fig. 4.8 Advanced nano-bio-sensors: bridging nanotechnology, biology, and AI for diagnostics. Reprinted with permission from ref Kulkarni et al. (2022). © 2022 by the authors. Licensee MDPI, Basel, Switzerland

is useful in energy-harvesting devices. Electroactive Polymers, polymers expand or contract when subjected to an electric field, enabling precise movements in robotic systems.

Figure 4.8 demonstrates the integration of nanotechnology, biotechnology, and sensor technology to develop nano-bio-sensors. These sensors utilize nanomaterials, bio-samples, and advanced signal processing (e.g., AI, Bluetooth) for rapid and efficient point-of-care diagnostic testing.

4.3.6 Case Study: Application of Smart Bionanocomposites in Wearable Biomedical Devices

Smart bionanocomposites represent a cutting-edge material class that integrates biological matrices with nanoscale reinforcements to achieve superior functionality, adaptability, and biocompatibility. These composites have become instrumental in the development of wearable biomedical devices, offering unique solutions for real-time health monitoring, therapeutic interventions, and personalized healthcare. The nanoscale components enhance the mechanical, electrical, and functional properties of the composites, while the biological matrices ensure compatibility with human tissues. This case study explores the design, functionality, and potential of smart bionanocomposites in wearable biomedical devices, emphasizing their ability to respond dynamically to environmental stimuli. Figure 4.9 depicts a smart wearable biomedical device using bionanocomposites, featuring flexible biopolymer substrates with nanomaterials for real-time health monitoring and wireless diagnostics. The purpose of this case study is to investigate the role of smart bionanocomposites in advancing wearable biomedical devices. It focuses on the material design, integration of nanomaterials, and the functionality achieved through these composites, particularly in real-time health diagnostics, drug delivery, and disease monitoring. Wearable biomedical devices are transforming healthcare by providing continuous monitoring and reducing reliance on invasive procedures. Smart bionanocomposites play a pivotal role in these advancements due to their adaptability, allowing devices to interface seamlessly with biological systems. Their ability to respond to physiological changes, such as variations in pH, temperature, or mechanical stress, makes them indispensable in applications like wound care, glucose monitoring, and biosensing. Highlighting the potential of these composites can guide future innovations in personalized medicine and environmentally sustainable healthcare technologies (Li et al. 2021).

The methodology for this study involved the systematic selection and integration of biopolymers, nanomaterials, and fabrication techniques to develop smart bionanocomposites for wearable biomedical devices. Biopolymers such as chitosan, alginate, and polylactic acid (PLA) were chosen for their biocompatibility, flexibility, and biodegradability, forming the primary matrix. Nanomaterials like graphene

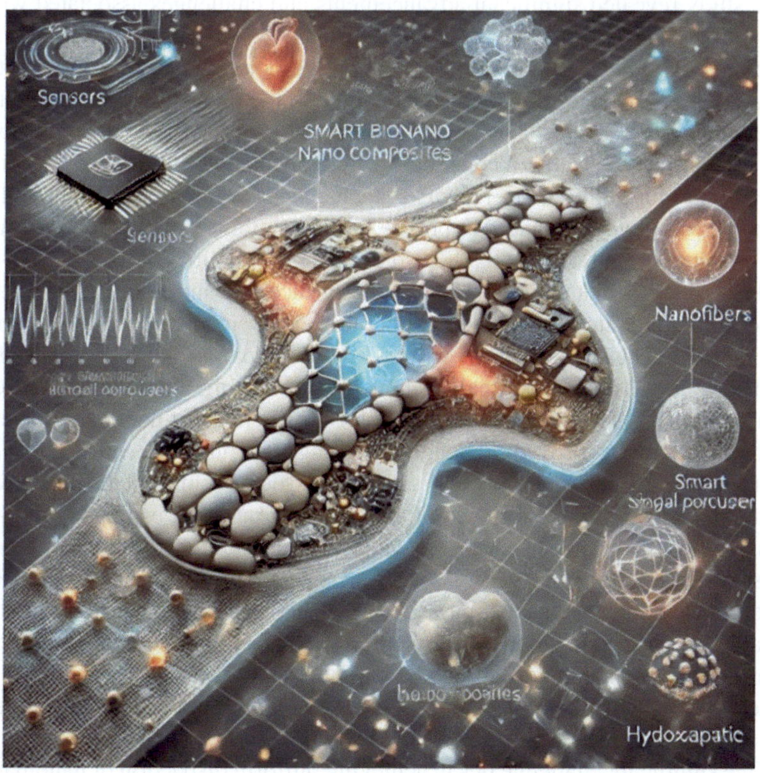

Fig. 4.9 Illustration of a smart wearable biomedical device using bionanocomposites. This is an AI-generated image taken from ChatGPT.com

and CNTs were incorporated to enhance electrical conductivity and mechanical strength, while hydroxyapatite improved biocompatibility, and quantum dots provided fluorescence-based imaging capabilities. Solvents, surfactants, and functionalization agents were used to improve the dispersion and compatibility of nanomaterials within the biopolymer matrix. The fabrication process employed advanced tools, including an electrospinning setup for creating nanofiber matrices that replicate the extracellular environment of biological tissues. These matrices were optimized using ML algorithms to ensure uniform fiber formation and consistent nanomaterial integration. Wireless communication modules and flexible electronics were integrated into the composites to enable real-time health monitoring and data transmission. Analytical techniques such as SEM and spectroscopic tools were utilized to characterize the morphology and composition of the composites. Testing involved evaluating the devices' responsiveness to environmental stimuli, such as pH and temperature, and their ability to provide precise physiological data. This comprehensive approach ensured the development of functional and adaptable wearable biomedical devices tailored for advanced healthcare applications.

The study revealed significant advancements and challenges associated with the development of smart bionanocomposites for wearable biomedical devices. These composites exhibited exceptional adaptability, responding dynamically to stimuli such as temperature and pH, which is critical for applications like wound monitoring and targeted drug delivery. The integration of nanomaterials, including graphene and CNTs, enhanced the electrical conductivity and mechanical strength of the composites, ensuring durability under repeated mechanical deformation. Additionally, the incorporation of hydroxyapatite improved biocompatibility, making these materials suitable for prolonged skin contact. The biosensing capabilities of these devices were notably enhanced by nanomaterials with high surface area and conductivity, enabling precise measurements of physiological parameters such as glucose levels, temperature, and pH. The integration of wireless communication modules allowed real-time data transmission to external systems, demonstrating the potential for remote health monitoring. However, achieving uniform dispersion of nanomaterials within the biopolymer matrix proved challenging, leading to inconsistencies in performance. ML-driven optimization partially addressed this issue by refining fabrication parameters and improving the consistency of nanofiber formation. Despite these advancements, challenges such as ensuring scalability, maintaining cost-effectiveness, and addressing sustainability concerns persist. These findings underscore the transformative potential of smart bionanocomposites while highlighting the need for continued innovation in material synthesis, fabrication processes, and device integration for further improvement and widespread adoption in healthcare.

The development of smart bionanocomposites for wearable biomedical devices faces several limitations that must be addressed to ensure their widespread adoption. One significant challenge lies in achieving uniform dispersion of nanomaterials within the biopolymer matrix. Nanomaterials, such as graphene and CNTs, tend to aggregate due to their high surface energy, which can result in inconsistent material properties and reduced device performance. Additionally, the fabrication processes, including electrospinning and integration of flexible electronics, require precise control over parameters, making them complex and time-intensive, particularly when scaling up for industrial production. The reliance on energy-intensive nanomaterial synthesis methods further raises concerns about the environmental impact and cost-effectiveness of large-scale production. Moreover, while biopolymers provide biodegradability and biocompatibility, some nanomaterials used in these composites may persist in the environment, raising sustainability and disposal concerns. Powering these devices efficiently remains another limitation, as current designs often depend on external energy sources, which can compromise their portability and long-term usability. Addressing these challenges will be crucial to unlocking the full potential of smart bionanocomposites in wearable biomedical devices (Karimzadeh et al. 2022).

To overcome the limitations associated with smart bionanocomposites in wearable biomedical devices, several improvements are recommended. Achieving uniform dispersion of nanomaterials within the biopolymer matrix can be addressed through advanced functionalization techniques or hybrid synthesis methods that enhance compatibility and reduce aggregation. Incorporating ML and AI into the fabrication process can further optimize parameters, ensuring consistent quality and reducing defects during production. Additionally, adopting green synthesis methods for nanomaterials, such as bio-inspired or enzymatic approaches, can mitigate environmental concerns and lower energy consumption. The integration of energy-harvesting technologies, including piezoelectric or thermoelectric nanomaterials, could enable self-powered devices, enhancing portability and long-term functionality. Efforts should also focus on developing fully biodegradable nanomaterials to align with environmental sustainability goals, reducing the ecological footprint of these devices. Establishing standardized protocols for quality control and testing would ensure reliability and scalability in industrial production. Collaborative research between academia, industry, and policymakers can facilitate regulatory frameworks and funding initiatives, accelerating advancements, and promoting the widespread adoption of smart bionanocomposites in wearable biomedical technologies.

4.4 Processing and Fabrication Techniques for Bionanocomposites

The processing and fabrication of bionanocomposites play a pivotal role in defining their properties and performance. These materials combine biological matrices and nanoscale reinforcements, requiring techniques that accommodate their complexity and maintain their multifunctionality. This section explores the conventional and advanced fabrication techniques for bionanocomposites, highlighting their scientific principles, capabilities, and limitations.

4.4.1 Conventional Fabrication Techniques

Casting is one of the oldest and simplest techniques for fabricating composite materials. It involves pouring a liquid mixture of the matrix and nanomaterials into a mold, followed by curing or solidifying the mixture into the desired shape. The process is straightforward, cost-effective, and suitable for various biopolymer matrices like chitosan, alginate, or gelatine. For bionanocomposites, the uniform dispersion of nanomaterials such as graphene oxide or metal nanoparticles in the liquid matrix is crucial to achieving consistent properties. Techniques like ultrasonication or mechanical stirring are employed to ensure homogeneous distribution (Zainazor et al. 2020). Despite its simplicity, casting often faces limitations in controlling the nanoscale architecture and achieving complex geometries, which restricts its use in applications requiring high precision or advanced functionalities. Molding techniques, such as injection molding and compression molding, are widely used for fabricating bionanocomposites with defined shapes. Injection molding involves injecting the molten composite material into a mold under high pressure, while compression molding compresses the material into a mold at high temperatures. These methods are efficient for mass production and allow for the fabrication of complex shapes. However, maintaining the dispersion of nanomaterials and preventing their aggregation during processing can be challenging. Additionally, the high temperatures and pressures involved in molding may degrade sensitive biological matrices, limiting its application to heat-stable systems. Extrusion is another conventional technique where the composite material is forced through a die to create continuous shapes, such as fibers or films (Kasaai 2022). This method is particularly useful for producing composites for biomedical applications, such as scaffolds or drug delivery films. While extrusion provides good control over the material's macroscopic dimensions, its ability to manipulate nanoscale features is limited. The process may also lead to alignment of nanomaterials along the extrusion direction, which can be advantageous or detrimental depending on the application.

Figure 4.10 depicts the process of fabricating bionanocomposites by combining nanoparticles with polylactic acid (PLA), a biodegradable polymer. Different fabrication techniques are highlighted, including solid-state mixing, solvent casting, in-situ polymerization, melt blending, and melt extrusion. These methods enable uniform dispersion of nanoparticles within the PLA matrix, improving the composite's mechanical, thermal, and functional properties. Bionanocomposites created through these processes are widely used in biomedical applications, packaging, and environmental sustainability due to their enhanced strength, biodegradability, and functionality (Kim et al. 2024). This approach bridges nanotechnology and polymer science, offering a sustainable solution for advanced material development with superior performance characteristics.

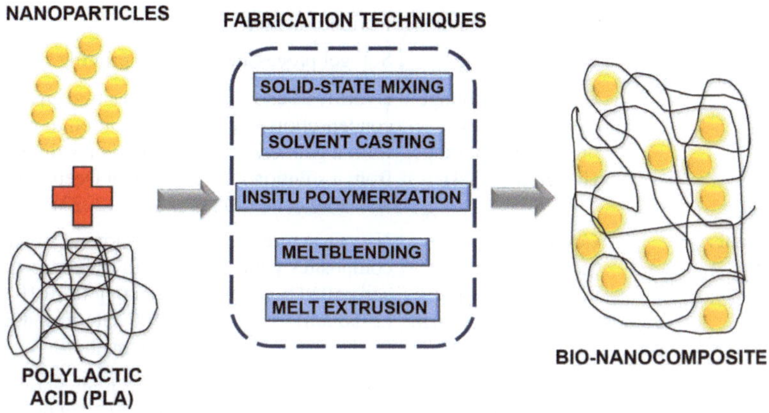

Fig. 4.10 Fabrication of bionanocomposites using PLA and nanoparticles

4.4.2 Advanced Fabrication Methods

To address the limitations of conventional techniques, advanced fabrication methods have been developed to achieve precise control over the nanoscale architecture and multifunctional properties of bionanocomposites. Table 4.2 shows a scientific comparison summarizing the features of Electrospinning, Sol–Gel Processing, and 3D Printing as fabrication methods for bionanocomposites.

4.4.3 Hybrid Fabrication Techniques for Multifunctional Composites

Hybrid fabrication techniques combine multiple processing methods to achieve bionanocomposites with enhanced properties and multifunctionality. These techniques leverage the strengths of individual methods while mitigating their limitations. For instance, a hybrid approach might involve electrospinning nanofiber mats followed by embedding them into a 3D-printed scaffold to create a composite with hierarchical structures. This combination enhances the material's mechanical strength, porosity, and bioactivity, making it suitable for tissue engineering applications. Another example is combining sol–gel processing with 3D printing to fabricate composites with nanostructured surfaces and complex geometries. The sol–gel method provides precise control over the nanoscale architecture, while 3D printing enables the creation of customized macroscopic shapes. Hybrid techniques are particularly valuable for creating multifunctional bionanocomposites that can respond to multiple stimuli or perform various roles in a single application (Yousefpour et al. 2022). However, integrating different methods requires careful optimization of

Table 4.2 Distinguishing different types of fabrication processes

Feature	Electrospinning	Sol–gel processing	3D printing
Principle	Uses a high-voltage electric field to produce nanofibers Han et al. (2024)	Hydrolysis and condensation reactions form a gel network from a solution Kalaaoui et al. (2024)	Layer-by-layer deposition of materials based on a digital model Deepak Kumar et al. (2022)
Output material structure	Continuous nanofibers with diameters ranging from 50 nm to 1 μm	Nanostructured composites with controlled pore sizes (2–50 nm)	3D-printed structures with feature resolution ranging from 10 to 100 μm (depending on the method)
Surface area	Very high (50–200 m^2/g)	Moderate to high (10–150 m^2/g depending on synthesis)	Low to moderate (1–50 m^2/g depending on the material used)
Porosity	High porosity (> 80%), ideal for tissue scaffolds and filtration	Moderate porosity (30–70%), suitable for catalysis and drug loading	Dependent on design but typically low to moderate (< 50%)
Material compatibility	Polymers, biopolymers, and nanomaterials (e.g., CNTs, silica, hydroxyapatite)	Biopolymers, ceramics, silica, and metal oxides	Wide range, including polymers, ceramics, and composite inks containing nanomaterials
Applications	Tissue engineering scaffolds, wound dressings, filtration, drug delivery Nazemiyeh et al. (2024)	Biosensors, catalysts, optical devices, and coatings Tamilselvan and Shanmugan (2024)	Personalized implants, drug delivery devices, and lab-on-a-chip systems John et al. (2023)
Control over nanoscale architecture	Excellent; nanomaterials can be aligned, and fiber size can be tuned ($\pm$ 10 nm precision)	Good; molecular-level mixing ensures uniform nanomaterial distribution but limited control over geometry	Moderate; precise geometric control but challenges in ensuring nanoscale homogeneity
Throughput	Low; limited to ~ 10m^2/h for standard electrospinning setups	Moderate; dependent on batch size and drying/curing time (e.g., 0.5–2 h for sol–gel synthesis)	High; industrial-grade 3D printers can achieve print speeds of ~ 20–100 cm^3/h
Energy consumption	~ 50–100 W for lab-scale systems (mainly from high-voltage power supply)	~ 1–5 kWh per batch, primarily for curing at elevated temperatures (100–800 °C depending on materials)	~ 1–10 kWh depending on printer type and material used (SLS is more energy-intensive than FDM)

(continued)

Table 4.2 (continued)

Feature	Electrospinning	Sol–gel processing	3D printing
Limitations	Difficulty in scaling up	Time-consuming	Uniform dispersion of nanomaterials and printability issues
Strengths	High surface area and porosity	Molecular-level uniformity and compatibility with various matrices	Unparalleled design flexibility and customization potential
Cost per unit fabrication	Low for lab-scale setups (~\$50–\$100 per gram of polymer/ nanomaterial solution)	Moderate (~\$100–\$200 per batch for sol–gel synthesis)	High initial costs for equipment (\$5–\$50 per print)

processing conditions to ensure compatibility and maintain the desired properties of the composite.

4.4.4 Case Study: Functionalization of Nanoparticles for Enhanced Biocompatibility in Tissue Engineering Applications

Nanoparticle functionalization has revolutionized tissue engineering by enabling precise control over material properties to support cellular functions, tissue regeneration, and integration with biological systems. Functionalization involves the chemical or physical modification of nanoparticles to enhance their surface properties, biocompatibility, and functionality. This process minimizes immunogenicity and cytotoxicity while promoting cellular adhesion and proliferation, making nanoparticles indispensable in scaffolds, drug delivery systems, and bioactive coatings. This case study explores the critical role of functionalized nanoparticles in tissue engineering, highlighting their techniques, challenges, and applications. The purpose of this study is to examine the impact of nanoparticle functionalization on tissue engineering. By exploring functionalization techniques, this study aims to provide insights into how nanoparticles can be tailored to enhance their compatibility with biological systems, improve scaffold performance, and address challenges in tissue regeneration. Functionalized nanoparticles offer transformative potential in tissue engineering by bridging the gap between materials science and biology. Their ability to mimic the extracellular matrix (ECM) and provide controlled release of bioactive agents enhances tissue repair and regeneration. Highlighting these applications underscores the importance of functionalization in advancing next-generation bioactive scaffolds and personalized medicine (Elkhoury et al. 2019). Figure 4.11 illustrates the process of functionalizing nanoparticles with bioactive molecules and polymers, enhancing their biocompatibility for tissue engineering applications. It shows functionalized nanoparticles interacting with cells in a 3D scaffold, highlighting their role in supporting cell adhesion and proliferation for tissue regeneration.

The methodology for this case study involved selecting and applying functionalization techniques to enhance the biocompatibility and functionality of nanoparticles for tissue engineering applications. A range of nanoparticles, including hydroxyapatite, graphene oxide, gold, and silver nanoparticles, was chosen for their intrinsic properties such as mechanical strength, conductivity, and bioactivity. Functionalizing agents like collagen, fibronectin, RGD peptides, and polyethylene glycol (PEG) were employed to modify the nanoparticle surfaces, improving their interaction with biological systems and reducing immunogenicity. Chemical coupling methods, such as salinization and carbodiimide coupling, were used to covalently

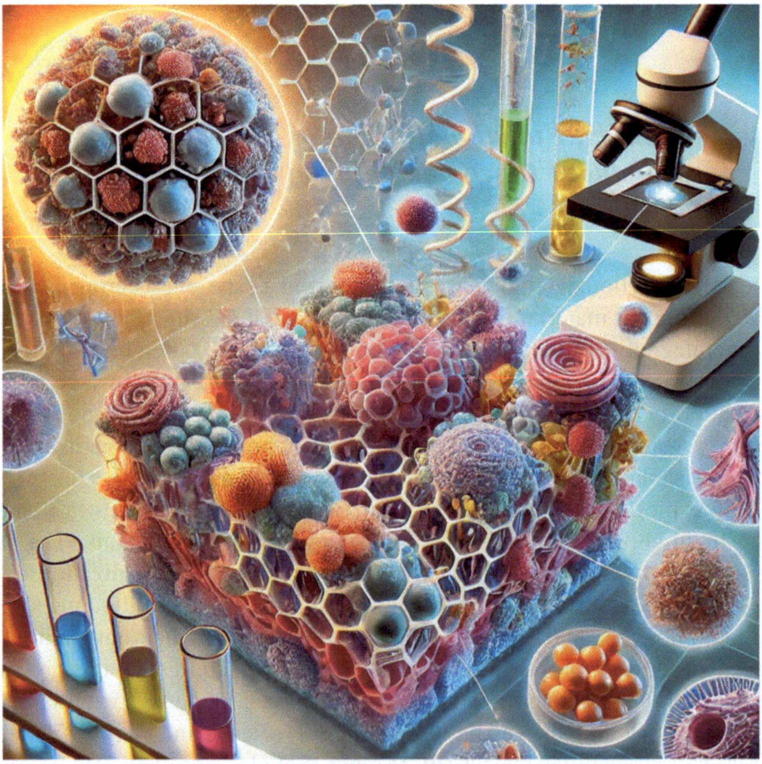

Fig. 4.11 Functionalized nanoparticles in tissue engineering: The process of enhancing biocompatibility through chemical modifications and their role in cellular interactions within scaffolds. This is an AI-generated image taken from ChatGPT.com

attach bioactive molecules to the nanoparticle surfaces, enhancing their osteoconductivity and integration into scaffolds. Physical adsorption techniques facilitated the coating of nanoparticles with biopolymers like chitosan and polyelectrolytes, which improved their biocompatibility and antimicrobial properties. Bioconjugation processes involve attaching growth factors or DNA strands to nanoparticles to promote targeted functionalities like angiogenesis and neural regeneration. Plasma treatment was also employed to create reactive surface groups on polymeric nanoparticles, enabling subsequent attachment of bioactive molecules. The functionalized nanoparticles were incorporated into 3D scaffolds, which were fabricated using advanced techniques such as electrospinning and 3D printing. Analytical tools, including SEM, TEM, and spectroscopic methods, were used to characterize the morphology, surface properties, and bioactivity of the functionalized nanoparticles. In vitro testing of the scaffolds evaluated their ability to support cell adhesion, proliferation, and differentiation, providing insights into their performance in tissue engineering applications. This comprehensive approach ensured a detailed understanding of the role and effectiveness of functionalized nanoparticles in advancing regenerative medicine.

The study demonstrated the significant impact of nanoparticle functionalization in advancing tissue engineering applications. Functionalized nanoparticles exhibited enhanced biocompatibility, enabling better interaction with biological tissues while reducing immunogenicity and cytotoxicity. For instance, hydroxyapatite nanoparticles functionalized with RGD peptides improved osteoblast adhesion and proliferation, making them ideal for bone tissue engineering. Similarly, collagen- and fibronectin-functionalized nanoparticles mimicked the ECM, creating a conducive environment for cell adhesion, differentiation, and regeneration. The controlled drug release capability of polymer-coated nanoparticles, such as those functionalized with PEG, highlighted their potential in delivering growth factors or therapeutic agents over extended periods. Functionalized nanoparticles also improved the mechanical and bioactive properties of scaffolds, ensuring structural support and promoting cellular functions. For example, silver nanoparticles functionalized with chitosan demonstrated antibacterial properties while promoting fibroblast activity, accelerating wound healing. The integration of functionalized nanoparticles into advanced fabrication methods, such as 3D printing and bioprinting, facilitated the creation of customized scaffolds with enhanced bioactivity and precise geometries. However, challenges such as nanoparticle aggregation during functionalization and the scalability of these techniques were noted. These issues often impacted the uniformity of bioactivity and material performance. Despite these challenges, the findings emphasized the transformative potential of functionalized nanoparticles in tissue engineering, paving the way for innovative solutions in regenerative medicine and personalized healthcare (Hasan et al. 2018).

Despite their transformative potential, the functionalization of nanoparticles for tissue engineering faces several limitations that need to be addressed. A primary challenge lies in the aggregation of nanoparticles during functionalization, which can significantly affect their uniform dispersion within scaffolds and reduce their overall bioactivity. The functionalization processes themselves are often complex, time-consuming, and expensive, making scalability a significant hurdle for large-scale applications. Additionally, achieving reproducibility in functionalized nanoparticle properties is challenging, as slight variations in processing conditions can lead to inconsistent results, impacting their reliability in clinical applications. Furthermore, the stringent regulatory standards for biocompatibility and safety require extensive testing and validation of functionalized nanoparticles, increasing both the time and cost of development. Environmental concerns also arise from the use of toxic chemicals in some functionalization methods, which may compromise the sustainability of the process. Addressing these limitations is essential to ensure that functionalized nanoparticles can meet the demands of advanced tissue engineering applications effectively and sustainably.

To address the challenges associated with nanoparticle functionalization in tissue engineering, several strategies can be implemented to improve efficiency, scalability, and sustainability. Advanced functionalization techniques, such as biorthogonal chemistry, can enable highly specific and stable modifications, reducing the likelihood of aggregation and enhancing reproducibility. Incorporating ML and AI into the functionalization process can optimize reaction conditions, streamline workflows, and minimize material wastage. Green chemistry approaches, such as enzyme-mediated functionalization or the use of environmentally friendly solvents, can mitigate the environmental impact of the process and align with sustainability goals. Integrating functionalized nanoparticles with emerging fabrication technologies like bioprinting and 3D printing offers opportunities to create highly customized, bioactive scaffolds with improved precision and functionality. Developing standardized protocols for functionalization and quality control would also ensure consistency and reliability across large-scale production. Collaboration between academia, industry, and regulatory bodies is crucial to streamline approval processes and reduce the time and cost associated with meeting stringent biocompatibility standards. By adopting these improvements, the functionalization of nanoparticles can become more efficient, scalable, and accessible, advancing their potential in regenerative medicine and tissue engineering (Rana et al. 2016).

4.5 Adaptive Control and Optimization in Fabrication

The fabrication of bionanocomposites is an inherently complex process that requires precise control over multiple parameters to ensure the desired material properties. Achieving this level of precision necessitates advanced methodologies for adaptive control and optimization. In recent years, the integration of ML, AI, real-time monitoring systems, and process optimization strategies has transformed the field, offering unparalleled opportunities for efficiency, precision, and sustainability in material processing. These advancements enable the production of high-quality, functional bionanocomposites tailored to specific applications.

4.5.1 ML and AI in Smart Material Processing

ML and AI have emerged as powerful tools for understanding and optimizing the fabrication of bionanocomposites. The traditional trial-and-error approach to materials processing is time-consuming, labour-intensive, and often limited in scope. AI and ML algorithms, by contrast, excel at analysing large datasets, identifying patterns, and making predictions that can guide decision-making in fabrication processes (Daghigh et al. 2020). This data-driven approach significantly accelerates the development of materials with tailored properties, for instance, ML algorithms can predict the optimal parameters for electrospinning, such as voltage, flow rate, and polymer concentration, to achieve uniform nanofiber formation (Boulogeorgos et al. 2021). Similarly, in sol–gel processing, AI models can analyze the effects of reaction conditions, such as pH and precursor concentration, on the material's final properties. By simulating different combinations of variables, these models can identify the most favourable conditions without requiring extensive experimental trials. Moreover, AI enables the exploration of parameter spaces that might otherwise remain inaccessible due to their complexity. Neural networks, for example, can model nonlinear relationships between processing conditions and material properties, providing insights into how slight adjustments can enhance performance. Reinforcement learning, a branch of AI, has also been applied to optimize fabrication processes dynamically, allowing for adaptive adjustments during production (Srinivas et al. 2025). This is particularly valuable in 3D printing, where real-time modifications to printing speed, layer thickness, or material composition can improve the quality of the final product.

Nanofiber Formation in Electrospinning

AI has been applied to optimize the electrospinning of nanofibers from bionanocomposites, such as chitosan-graphene oxide (GO) composites. In these systems, Neural networks predict the effects of parameters like polymer concentration, voltage, and flow rate on fiber diameter and alignment. For example, AI models trained on SEM images of electro-spun chitosan-GO fibers dynamically adjust spinning conditions to produce defect-free fibers. ML algorithms evaluate the effects of GO concentration on mechanical properties and cytocompatibility, ensuring the fibers are suitable for tissue engineering applications (Radacsi et al. 2018).

Stimuli-Responsive Materials

Bionanocomposites that respond to stimuli, such as temperature or pH, are optimized using AI tools. AI algorithms identify the optimal ratio of poly(N-isopropylac rylamide) (PNIPAAm) to hydroxyapatite for composites used in drug delivery. These models predict the thermal response of the composite, ensuring precise control of drug release kinetics. ML models analyze the swelling behaviour of chitosan-silica composites under various pH conditions, predicting their performance in wound healing or controlled release applications (Rezaee et al. 2024).

Nanoparticle Dispersion and Compatibility

The uniform dispersion of nanoparticles within biopolymer matrices is critical for achieving consistent material properties. AI optimizes the dispersion of CNTs in polylactic acid (PLA), ensuring uniform electrical conductivity for biosensors. ML-driven simulations evaluate the compatibility of functionalized graphene oxide with chitosan, predicting enhanced mechanical strength and antibacterial properties for biomedical implants (Mishra et al. 2023).

Real-Time Quality Control

In manufacturing processes, AI provides real-time monitoring and feedback to ensure consistent production quality. Image recognition algorithms analyze defects in bionanocomposites during fabrication. For instance, automated systems detect aggregation in cellulose nanocrystals (CNC)-based materials and adjust mixing conditions. AI-integrated sensors in 3D printing systems adjust nozzle temperature and extrusion rates to produce uniform scaffolds, such as alginate-hydroxyapatite composites, for bone tissue engineering (Gomez-Flores et al. 2024).

Predictive Material Design

ML models predict the performance of bionanocomposites under specific conditions, expediting material discovery. Gradient boosting algorithms forecast the tensile strength of collagen-silica composites based on precursor ratios and processing parameters, reducing experimental iterations. AI simulations evaluate the thermal conductivity of biopolymer-ZnO composites, optimizing their use in energy storage devices (Boulogeorgos et al. 2021).

4.5.2 Real-Time Monitoring and Quality Control Systems

Real-time monitoring is a critical component of adaptive fabrication processes, ensuring consistent quality and enabling rapid response to deviations from desired parameters. Advanced sensors and imaging systems are now commonly integrated into fabrication setups, providing continuous feedback on material properties and process conditions. These systems generate vast amounts of data, which can be processed using AI algorithms to detect anomalies, predict outcomes, and recommend corrective actions. For example, during electrospinning, optical and electron microscopy techniques can monitor fiber morphology in real time, identifying irregularities such as bead formation or non-uniform diameters. AI-powered image analysis tools can process these observations to adjust process parameters instantaneously. Similarly, in sol–gel processing, spectroscopic methods such as Fourier transform infrared (FTIR) spectroscopy or Raman spectroscopy can track chemical reactions and nanomaterial dispersion, ensuring uniformity and preventing aggregation (Sanchez-Cano et al. 2021). In 3D printing, real-time monitoring systems are often equipped with cameras and thermal sensors that track layer deposition and curing. Any discrepancies, such as misaligned layers or insufficient curing, are immediately flagged by the system, and corrective measures are implemented. ML algorithms play a pivotal role in this context, as they can predict potential defects based on historical data and real-time inputs, ensuring a higher success rate in fabrication. Quality control systems also benefit from the integration of AI-driven technologies (Menossi et al. 2022). Automated inspection processes, powered by machine vision and AI, can assess material properties such as tensile strength, thermal stability, and porosity without destructive testing. This not only reduces waste but also ensures that every batch of bionanocomposites meets stringent performance criteria.

Real-time monitoring and quality control systems are integral to ensuring precision and consistency in the manufacturing of bionanocomposites. These systems leverage advanced technologies such as sensors, imaging tools, and AI-driven algorithms to optimize fabrication processes and detect defects during production. Electrospinning, a widely used technique for fabricating bionanocomposites, provides an excellent example of how real-time monitoring enhances precision. In one study, cellulose acetate-based nanofibers containing silver nanoparticles were produced via electrospinning for antimicrobial applications. High-resolution imaging systems, combined with AI algorithms, monitored fiber formation in real-time to detect anomalies such as bead formation or diameter inconsistencies. The feedback loop enabled immediate adjustments to parameters like flow rate and voltage, ensuring the production of uniform and defect-free nanofibers. Similarly, another study focused on the development of chitosan-graphene oxide (GO) nanofibers for tissue engineering scaffolds. Real-time SEM imaging and AI tools were employed to evaluate the alignment, porosity, and morphology of the nanofibers during fabrication. These systems not only improved the uniformity of the nanofibers but also reduced material wastage by dynamically optimizing the electrospinning conditions. In the broader context of bionanocomposite manufacturing, these technologies have enabled the production of composites with high precision for a variety of real-world applications. For example, in wound healing, electro-spun bionanocomposites embedded with zinc oxide nanoparticles have been manufactured using real-time monitoring to ensure consistent antibacterial properties and structural integrity. Similarly, for environmental sensing, alginate-carbon nanotube composites have been fabricated with integrated quality control systems to ensure reproducibility and high sensitivity in pollutant detection. In energy storage, bionanocomposite electrodes made from

PLA and titanium dioxide (TiO_2) have been optimized using real-time monitoring to enhance conductivity and energy density.

4.5.3 Process Optimization for Sustainable and Efficient Fabrication

Sustainability is a major consideration in the fabrication of bionanocomposites, given the global emphasis on reducing environmental impacts. Process optimization plays a vital role in achieving this goal by minimizing energy consumption, material waste, and production time. Advanced computational models and AI algorithms facilitate this optimization by identifying the most resource-efficient pathways for material processing. For example, in electrospinning, process optimization might involve reducing the solvent-to-polymer ratio to minimize solvent waste while maintaining fiber quality. AI algorithms can model the interactions between solvents, polymers, and nanomaterials to identify compositions that achieve optimal performance with minimal environmental impact (Orooji et al. 2022). Similarly, in sol–gel processing, optimization strategies can focus on reducing reaction times and temperatures, thereby lowering energy requirements. 3D printing presents unique opportunities for sustainable fabrication due to its additive nature, which inherently minimizes material waste compared to subtractive manufacturing methods. However, the optimization of print parameters is crucial to fully realize these benefits. For instance, optimizing layer thickness and deposition rates can reduce the total printing time and energy consumption without compromising structural integrity. AI and ML tools are instrumental in these efforts, as they can predict the trade-offs between speed, precision, and material usage based on historical data and real-time monitoring. Another key aspect of sustainability in bionanocomposite fabrication is the recycling and reuse of materials. Advanced sorting and processing techniques, guided by AI, enable the recovery of nanomaterials from composite waste, reducing the demand for virgin materials. For instance, AI algorithms can identify and separate reusable nanomaterials based on their size, composition, or functional properties, making the recycling process more efficient and cost-effective. Additionally, adaptive control systems ensure that energy-intensive processes such as curing and sintering are only employed when necessary. Sensors and AI-driven models monitor the progress of these processes in real time, allowing for dynamic adjustments that minimize energy use (Kumar et al. 2024). For example, in sol–gel processing, precise control of drying and curing conditions can prevent overprocessing, thereby conserving energy and reducing production costs.

4.5.4 Case Study: AI in Optimizing the Electrospinning Process for Nanostructured Composites

Electrospinning has become an essential technique for producing nanostructured composites, offering a versatile platform for creating nanofibers with controlled properties suitable for applications in healthcare, environmental monitoring, and energy storage. The process involves applying a high-voltage electric field to a polymer solution or melt, resulting in the formation of ultrafine fibers deposited onto a collector. While electrospinning provides remarkable control over fiber morphology, achieving consistency and optimizing properties such as fiber diameter, mechanical strength, and surface texture remain significant challenges. The incorporation of AI into the electrospinning workflow has transformed the field, enabling real-time optimization, enhanced precision, and consistent production of high-performance nanofibers. This study explores the integration of AI into the electrospinning process to overcome its inherent challenges. It aims to demonstrate how AI-driven approaches streamline parameter optimization, improve fiber quality, and enable the development of advanced functional materials for various applications, such as tissue engineering, environmental sensing, and energy storage. The significance of this case study lies in the transformative impact of AI on nanostructured composite fabrication. Traditional trial-and-error methods for optimizing electrospinning parameters are inefficient, time-consuming, and often yield suboptimal results. AI algorithms, by analyzing historical data and real-time inputs, provide precise control over the process, minimizing material waste and experimental cycles. This synergy between AI and electrospinning promises to revolutionize multiple industries, enhancing material performance while reducing production costs and environmental impact. Figure 4.12 illustrates a process involving a polymer and CDP (cyclodextrin polymer) solution subjected to nozzle-free electrospinning. Hot air aids fiber formation, depositing nanofibers decorated with nanoparticles on a collector. This method enables efficient production of functionalized nanofibers for advanced applications (Subeshan et al. 2024).

The study utilized various polymer solutions, such as polyvinyl alcohol (PVA) and CDP, as matrices for electrospinning. Nanoparticles, including graphene, CNTs, and metal oxides, were incorporated as reinforcements to enhance the electrical, mechanical, and thermal properties of the nanofibers. Chemicals such as solvents (e.g., water, ethanol) and surfactants were used to optimize solution properties for electrospinning. A nozzle-free electrospinning system was employed, integrated with sensors for real-time monitoring of voltage, flow rate, and environmental conditions. Machine learning models, including regression algorithms and neural networks, were trained to predict fiber properties based on process parameters. SEM and spectroscopic techniques were used to analyze fiber morphology, diameter, and uniformity. The polymer and nanoparticle solutions were prepared and optimized for electrospinning. Real-time sensors and imaging systems monitored critical parameters, including voltage, solution flow rate, and collector distance. AI algorithms analyzed this data to predict fiber characteristics and dynamically adjusted parameters to

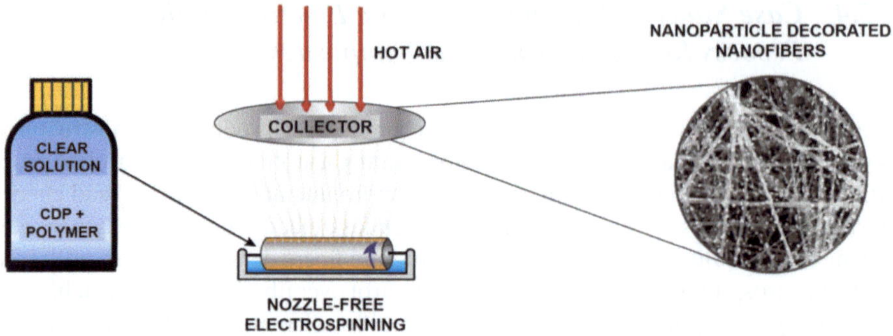

Fig. 4.12 Schematic of nozzle-free electrospinning for nanoparticle-decorated nanofibers. Reprinted with permission from Radacsi et al. (2018). © 2024 Springer Nature Limited

achieve optimal results. For example, image recognition algorithms processed SEM images to detect defects such as bead formation, allowing immediate corrections to the process. The nanofibers were functionalized by embedding nanoparticles during spinning to enhance their application-specific properties (Toscano et al. 2020).

The integration of AI significantly improved the electrospinning process by enabling real-time optimization of fiber morphology and functional performance. The AI-driven approach ensured consistent production of nanofibers with uniform diameters, enhanced mechanical strength, and tailored surface textures. In healthcare applications, nanofiber scaffolds fabricated using AI-optimized electrospinning mimicked the extracellular matrix, promoting cell adhesion, growth, and differentiation. For example, fibers with controlled porosity and alignment were developed for tissue engineering, demonstrating superior biocompatibility and regenerative potential. In environmental sensing, AI-driven electrospinning enabled the production of nanofibers embedded with conductive polymers and metal nanoparticles for detecting pollutants like heavy metals and organic contaminants. The process ensured high sensitivity and stability of sensors, even in large-scale production. Similarly, in energy storage devices, AI-guided fabrication of nanofiber electrodes with high surface area and conductivity improved charge–discharge efficiency and durability, contributing to advancements in supercapacitors and batteries. However, the study also highlighted challenges, including the need for large, high-quality datasets to train AI models and the difficulty of translating laboratory-scale AI-optimized processes into industrial production. These limitations underscore the importance of developing hybrid AI models that combine data-driven approaches with physics-based simulations for more accurate and scalable solutions.

Despite its potential, the integration of AI into electrospinning faces challenges such as the reliance on extensive datasets, which may not always be available for specific materials or processes. Computational requirements for real-time analysis and optimization can also be resource-intensive, limiting accessibility for smaller-scale operations. Additionally, the transition from laboratory to industrial-scale production requires rigorous validation and standardization of AI-optimized results, which remains a significant barrier. Future research should focus on hybrid AI models that combine data-driven methods with physics-based simulations, enabling more robust and generalizable predictions. Expanding datasets through collaborative efforts and open-access repositories can help train more effective AI models. Incorporating sustainable practices, such as green chemistry in nanoparticle synthesis, will further enhance the environmental benefits of AI-driven electrospinning. Additionally, developing cost-effective and scalable AI systems tailored for industrial applications can facilitate broader adoption across industries (Simchi and Kianfar 2024).

4.6 Applications of Smart and Adaptive Bionanocomposites

The advent of smart and adaptive bionanocomposites has opened a myriad of opportunities across various fields. By merging the unique properties of nanomaterials with the biocompatibility and adaptability of biological matrices, these materials exhibit remarkable functionalities. Their ability to respond dynamically to external stimuli such as temperature, pH, light, and mechanical stress enables their integration into diverse applications. Below, we explore their utility in four major domains: biomedical applications, environmental sensing and pollution remediation, energy storage and harvesting, and structural health monitoring and adaptive infrastructure depicted by Fig. 4.13.

4.6.1 Biomedical Applications

The biomedical sector has been one of the most significant beneficiaries of smart bionanocomposites, leveraging their biocompatibility, adaptability, and advanced functionalities for a wide range of applications. Table 4.3 highlights specific studies that exemplify the use of smart adaptive bionanocomposites in biomedical applications. It focuses on their adaptive functionalities, such as stimuli responsiveness (pH or temperature), mechanical adaptability, and signal amplification for sensing.

Drug Delivery Systems

Smart bionanocomposites are pivotal in creating controlled drug delivery systems. These composites can encapsulate therapeutic agents and release them at targeted sites in response to specific stimuli. For instance, nanomaterials such as graphene oxide and mesoporous silica nanoparticles, when embedded in biopolymer matrices like chitosan or alginate, enable pH-sensitive drug release. This is particularly advantageous in cancer therapy, where the acidic microenvironment of tumours can trigger localized drug delivery, minimizing systemic side effects. Furthermore,

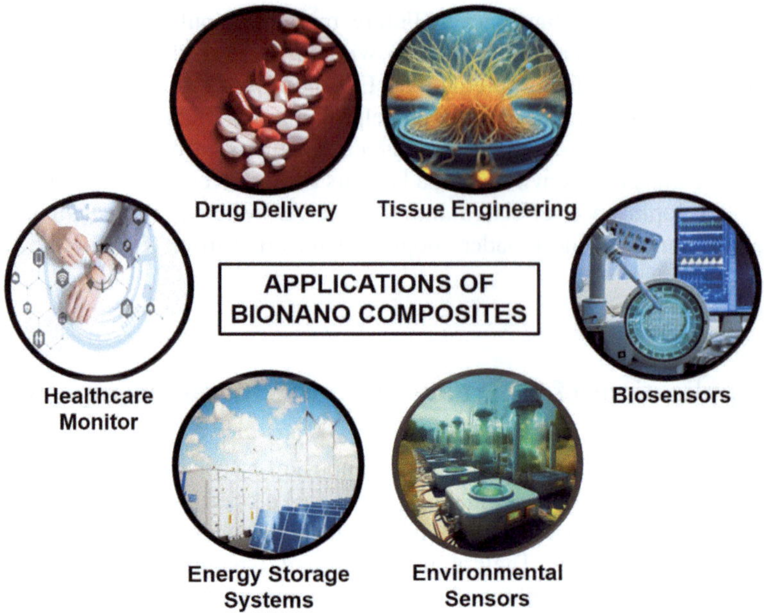

Fig. 4.13 Versatile applications of bionanocomposites in advanced fields

temperature-responsive bionanocomposites can release drugs upon encountering elevated temperatures, such as those induced by inflammation or hyperthermia.

Tissue Engineering

In tissue engineering, bionanocomposites offer scaffolds that mimic the extracellular matrix, promoting cell attachment, proliferation, and differentiation. Electro-spun nanofibers reinforced with hydroxyapatite or CNTs provide the mechanical strength and bioactivity necessary for bone and cartilage regeneration. Similarly, composites incorporating growth factors or bioactive peptides can stimulate tissue repair and regeneration. The adaptivity of these materials ensures they can conform to irregular tissue geometries, enhancing their integration with natural tissues.

Biosensing

Bionanocomposites have found extensive use in biosensors, where their high sensitivity and specificity enable the detection of biological analytes such as glucose, pathogens, and biomarkers. Nanostructures like quantum dots and gold nanoparticles embedded in bionanocomposites amplify signal transduction, facilitating real-time monitoring of physiological conditions. For example, glucose sensors employing

Table 4.3 Smart adaptive bionanocomposites in biomedical applications

Application	Bionanocomposite material	Smart functionality	Findings
Drug delivery	(1) Chitosan-graphene oxide	(1) pH-sensitive drug release (2) Stimuli-responsive (pH and temperature)	(1) Chitosan-GO composite exhibited controlled drug release in acidic tumour environments, minimizing systemic toxicity Barua et al. (2019), Asadipour et al. (2023) 2) Alginate-mesoporous silica composites released drugs in response to acidic pH and elevated temperatures, suitable for cancer therapy Barua et al. (2019), Asadipour et al. (2023)
Tissue engineering	(1) Electro-spun hydroxyapatite-polycaprolactone (PCL) nanofibers (2) Chitosan-CNT composite	(1) Adaptive scaffold for bone regeneration (2) Mechanical strength and bioactivity	(1) Hydroxyapatite-PCL nanofibers promoted osteoblast attachment and proliferation, mimicking bone extracellular matrix Christy et al. (2022), Alam et al. (2023) (2) Chitosan-CNT scaffolds supported cartilage repair, providing enhanced mechanical properties and bioactivity Christy et al. (2022), Alam et al. (2023)

(continued)

Table 4.3 (continued)

Application	Bionanocomposite material	Smart functionality	Findings
Biosensing	(1) Gold Nanoparticle-polyaniline Composite (2) Quantum dot-chitosan nanocomposite	(1) Signal amplification for glucose detection (2) Fluorescent biosensing	(1) Gold nanoparticle-polyaniline composites functionalized with glucose oxidase enabled rapid, accurate glucose sensing Palanisamy et al. (2020), George Kerry et al. (2021) (2) Quantum dot-chitosan composites detected biomarkers with high sensitivity and specificity for disease diagnostics Palanisamy et al. (2020), George Kerry et al. (2021)

bionanocomposites functionalized with glucose oxidase provide accurate, rapid measurements critical for diabetes management.

4.6.2 Environmental Sensing and Pollution Remediation

The adaptive capabilities of bionanocomposites make them invaluable for addressing environmental challenges, including monitoring pollutants and remediating contaminated ecosystems. Smart bionanocomposites equipped with responsive nanostructures serve as advanced sensors for detecting environmental pollutants (Singh 2023). For instance, composites containing metal oxide nanoparticles or graphene-based materials exhibit high sensitivity to gaseous pollutants like CO_2, NO_x, and volatile organic compounds (VOCs). These sensors change their electrical or optical properties in response to pollutant interactions, enabling precise quantification of air quality. Additionally, pH-responsive bionanocomposites can detect acidification in water bodies, providing insights into aquatic ecosystem health. Bionanocomposites also excel in pollution remediation through mechanisms like adsorption, photocatalysis, and biodegradation. Materials such as titanium dioxide and zinc oxide nanoparticles embedded in biopolymers can degrade organic pollutants under UV or visible light via photocatalysis. Similarly, magnetic bionanocomposites allow for the efficient removal of heavy metals from wastewater, as the magnetic particles can be retrieved easily after adsorbing contaminants. Biodegradable matrices further enhance environmental compatibility, ensuring the composites do not contribute

Table 4.4 Smart adaptive bionanocomposites in environmental sensing and pollution remediation

Application	Bionanocomposite material	Smart functionality	Findings
Air quality sensing	Graphene-titanium dioxide composite	Detection of gaseous pollutants (CO_2, NO_x)	Graphene-TiO_2 composites showed high sensitivity to NOx gases by altering electrical conductivity upon interaction Mohamed and Awad (2022)
Water quality sensing	pH-responsive chitosan-zinc oxide composite	Detection of water acidification	Chitosan-ZnO composites exhibited a pH-dependent optical response, suitable for monitoring aquatic ecosystem health Garg et al. (2024)
Photocatalysis	Biopolymer-titanium dioxide composite	Degradation of organic pollutants via UV light	Biopolymer-TiO_2 composites successfully degraded dyes and pesticides under UV light, preventing harmful buildup in water Afshan and Bari (2023)
Heavy metal removal	Magnetic alginate-iron oxide composite	Adsorption and magnetic retrieval of metals	Magnetic alginate-iron oxide composites efficiently adsorbed lead and arsenic from wastewater and were easily recovered Prajapati et al. (2024)
VOC detection	Polyaniline-graphene oxide composite	Responsive to VOCs)	Polyaniline-GO composites displayed changes in electrical resistance, enabling precise VOC detection in air Bilal (2024)
Oil spill cleanup	Superhydrophobic cellulose nanofiber-zinc oxide composite	Oil adsorption with reusable properties	Cellulose-ZnO composites absorbed oil from water surfaces and could be reused after simple washing Andrada Suarez et al. (2025)

to secondary pollution. Table 4.4 emphasizes studies focused on smart adaptive bionanocomposites designed to address environmental challenges. Each entry highlights the composite material, its adaptive functionality (e.g., pollutant detection, photocatalysis, adsorption), and specific findings.

4.6.3 Energy Storage and Harvesting Applications

The energy sector has increasingly adopted bionanocomposites to meet the demand for sustainable, efficient energy storage and harvesting technologies. Bionanocomposites have demonstrated remarkable potential in the fabrication of batteries and supercapacitors. Nanostructures such as graphene, CNTs, and metal sulphides embedded in biopolymers like cellulose or lignin enhance electrical conductivity and charge storage capacity. These composites are particularly relevant for developing flexible energy storage devices, which are crucial for wearable electronics. Additionally, the biodegradability of the biopolymer matrix aligns with the growing emphasis on green energy solutions. In energy harvesting, bionanocomposites facilitate the conversion of ambient energy into usable electrical power. For instance, piezoelectric bionanocomposites comprising barium titanate nanoparticles in biopolymer matrices generate electricity in response to mechanical stress. Such materials are used in wearable devices to harvest kinetic energy from human motion. Similarly, thermoelectric bionanocomposites, which convert temperature gradients into electrical energy, offer promising applications in powering remote sensors or wearable devices (Sonu et al. 2023). Table 4.5 highlights specific studies of smart adaptive bionanocomposites in the energy sector, emphasizing their functionalities and application-specific performance. Each entry focuses on the composite material, its adaptive behaviour (e.g., piezoelectricity, thermoelectricity, or enhanced conductivity), and key findings. The summarized studies demonstrate the significant potential of these materials in energy storage and harvesting, aligning with the focus on smart bionanocomposites for sustainable energy solutions.

4.6.4 Structural Health Monitoring and Adaptive Infrastructure

The construction and civil engineering sectors have begun integrating bionanocomposites into smart infrastructure systems to enhance durability, safety, and performance. Bionanocomposites embedded with sensing nanostructures enable real-time monitoring of structural integrity. For example, carbon nanotube-based composites can detect strain, cracks, or temperature variations in concrete structures. These materials exhibit changes in electrical conductivity or piezo-resistivity under mechanical stress, providing early warning signals for potential failures. Such smart materials are crucial for ensuring the safety of bridges, buildings, and other critical infrastructure. The adaptivity of bionanocomposites is particularly advantageous for creating infrastructure that responds dynamically to environmental changes. Thermally responsive composites, for instance, can regulate heat transfer in building materials, enhancing energy efficiency. Similarly, moisture-responsive composites can self-heal cracks in concrete by swelling and sealing the damaged area. These capabilities reduce maintenance costs and extend the lifespan of infrastructure, contributing to sustainable urban development. Table 4.6 summarizes studies highlighting smart adaptive bionanocomposites designed for structural health monitoring and adaptive infrastructure. Each entry includes the composite material, its adaptive functionality (e.g., piezo-resistivity, thermal regulation, or self-healing), and specific findings related to its performance in structural applications. These studies emphasize the potential

Table 4.5 Smart adaptive bionanocomposites in energy storage and harvesting applications

Application	Bionanocomposite material	Smart functionality	Findings
Supercapacitors	Graphene-cellulose nanofiber composite	Enhanced charge storage capacity	Graphene-cellulose nanofiber composites exhibited high specific capacitance and cycling stability, ideal for flexible supercapacitors Rajalakshmi et al. (2024)
Battery electrodes	CNT-lignin composite	High electrical conductivity	CNT-lignin composites demonstrated improved electron transfer and energy density in lithium-ion batteries Kannan et al. (2009)
Piezoelectric harvesting	Barium titanate-chitosan composite	Energy generation via mechanical stress	Barium titanate-chitosan composites generated electricity from human motion, enabling wearable energy harvesters Zhao et al. (2024)
Thermoelectric devices	Zinc oxide-biopolymer composite	Conversion of thermal gradients to electricity	ZnO-biopolymer composites achieved a high thermoelectric power factor, powering remote sensors effectively
Flexible energy devices	Polyaniline-graphene oxide composite	Conductive and flexible materials	Polyaniline-GO composites provided flexible, high-performance electrodes for wearable electronics Sabo et al. (2016)
Bio-supercapacitors	Lignin-carbon nanotube composite	Biodegradable energy storage material	Lignin-CNT composites offered eco-friendly supercapacitors with excellent charge–discharge efficiency

of smart bionanocomposites in enhancing safety, reducing maintenance costs, and supporting sustainable infrastructure development.

Table 4.6 Smart adaptive bionanocomposites in structural health monitoring and adaptive infrastructure

Application	Bionanocomposite material	Smart functionality	Findings
Strain sensing	Carbon nanotube-cement composite	Piezo-resistive behaviour for strain detection	CNT-cement composites exhibited electrical resistance changes under mechanical stress, enabling real-time structural monitoring in concrete
Crack detection	Graphene-polymer composite	Conductivity changes for crack identification	Graphene-polymer composites showed high sensitivity to micro-crack formation in building materials, providing early warning signals (Liu et al. 2024)
Thermal adaptivity	Phase-change material-embedded biopolymer composite	Regulation of heat transfer in buildings	Phase-change material (PCM)-biopolymer composites reduced energy consumption by dynamically controlling thermal conductivity in walls (Koncar 2019)
Self-healing concrete	Chitosan-silica nanoparticle composite	Moisture-induced self-healing capability	Chitosan-silica composites autonomously sealed cracks in concrete upon exposure to water, increasing structural longevity
Vibration monitoring	Carbon nanofiber-polyurethane composite	Sensing of mechanical vibrations	Carbon nanofiber-polyurethane composites detected vibrations and mechanical disturbances in bridge structures
Corrosion protection	Zinc oxide-biopolymer composite	Anti-corrosion coating	ZnO-biopolymer composites provided corrosion resistance in steel reinforcements, extending infrastructure durability

The image highlights the multifaceted uses of bionanocomposites, including drug delivery, tissue engineering, biosensors, healthcare monitoring, energy storage systems, and environmental sensors (Zafar et al. 2021). These materials integrate biological and nanostructured components, offering tailored properties for diverse technological and biomedical applications.

4.7 Challenges and Future Perspectives

Scaling the production of smart and adaptive bionanocomposites from laboratory settings to industrial levels presents several specific challenges related to their unique functionalities. A primary issue lies in achieving uniform dispersion of nanomaterials within biopolymer matrices while retaining their stimuli-responsive behaviour. Nanoparticles, such as graphene or CNTs, have a natural tendency to agglomerate due to their high surface energy, resulting in inconsistent adaptive properties and suboptimal performance in large-scale products. For example, composites designed for temperature-responsive drug delivery or pH-sensitive environmental sensing may lose their precision if nanoparticle distribution is uneven. Maintaining the intricate nanoscale structures necessary for adaptivity requires stringent control of processing parameters, such as temperature, flow rate, and curing conditions. Techniques like electrospinning and 3D printing, which are critical for creating these structures, often face scalability limitations. Reproducing stimuli-responsive capabilities—such as self-healing, thermal conductivity, or piezoelectricity—at industrial levels requires multi-step fabrication processes that are complex, time-intensive, and expensive. For instance, embedding self-healing agents in a composite used for adaptive infrastructure demands precise control of microcapsule integration, which can be difficult to achieve consistently on a large scale. Another significant challenge involves ensuring reproducibility and reliability while adhering to strict regulatory standards, particularly in applications like healthcare and environmental sensing. Variability in nanomaterial synthesis methods and batch-to-batch differences in raw material quality can lead to performance inconsistencies. Furthermore, the environmental footprint of producing these materials poses concerns. While biopolymers are eco-friendly, the production of nanomaterials like graphene oxide or CNTs is often energy-intensive and involves hazardous chemicals, offsetting the environmental advantages. Addressing these sustainability concerns is essential to align with global sustainability goals. The economic challenges are equally significant. The high cost of nanomaterials, advanced fabrication equipment, and skilled labour restricts the accessibility of these composites, particularly in resource-constrained settings. For example, the cost-intensive nature of hybrid techniques combining sol–gel processing with 3D printing may hinder the widespread adoption of adaptive composites for structural health monitoring or energy storage (Perera et al. 2022).

To overcome these challenges, the focus must be on developing scalable, efficient, and cost-effective methods specifically tailored for smart and adaptive bionanocomposites. The integration of ML and AI presents a transformative opportunity to address these issues. AI-driven fabrication systems can monitor and optimize parameters in real time, ensuring consistent nanoparticle dispersion and adaptive functionality. For example, AI-based predictive models can identify optimal conditions for achieving uniform piezoelectric or thermoelectric behaviour in composites used for energy harvesting.

Emerging hybrid fabrication processes, such as combining 3D printing with sol–gel processing or electrospinning, provide promising avenues for maintaining nanoscale precision while increasing scalability. These methods can integrate multiple functionalities—such as self-healing, electrical conductivity, and stimuli-responsiveness—into a single composite, broadening their application potential. For instance, hybrid processes could enable the fabrication of adaptive materials for bridges that autonomously heal cracks and simultaneously monitor strain in real time.

Sustainability is another critical area for improvement. Advancing green synthesis methods for nanomaterials, such as using bio-inspired or enzyme-mediated processes, can significantly reduce energy consumption and the environmental impact of production. Recycling strategies for bionanocomposites must be developed, including methods for recovering nanomaterials and biopolymers after use. Designing composites with biodegradable matrices that maintain their adaptive properties throughout their lifecycle could also enhance sustainability (Singh et al. 2024).

Future research should also focus on expanding the application of smart bionanocomposites in high-impact areas. In healthcare, adaptive composites can enable more precise drug delivery systems that respond to multiple stimuli simultaneously, such as temperature and pH. Affordable biosensors made from adaptive materials could revolutionize personalized medicine in low-resource settings. Similarly, in the energy sector, adaptive nanostructures could improve the performance and lifespan of supercapacitors and batteries, advancing renewable energy initiatives. In structural health monitoring, the integration of adaptive bionanocomposites could transform infrastructure by enhancing the durability and safety of critical systems. For instance, moisture-responsive composites could not only self-heal cracks in concrete but also monitor environmental changes in real time, reducing maintenance costs and extending infrastructure lifespans.

Finally, collaboration between scientists, industries, and policymakers is crucial to ensure the practical implementation of these recommendations. Developing standardized protocols for quality control, establishing regulations that balance innovation with safety, and promoting interdisciplinary partnerships can accelerate the adoption of smart and adaptive bionanocomposites for a sustainable future. By addressing these challenges and capitalizing on the potential of these materials, significant strides can be made toward achieving global sustainability and innovation goals (Bio-Based Composites: Future Trends and Environmental Aspects 2014).

4.8 Conclusion

The chapter provides a comprehensive exploration of smart and adaptive bionanocomposite materials, focusing on their processing, fabrication, and applications in sustainable technology. These materials combine biocompatible matrices with nanoscale reinforcements, designed for biomedical, environmental, and energy applications. They exhibit advanced functionalities such as stimuli responsiveness, self-healing, and adaptive morphology, achieved by merging nanomaterials like graphene or metal nanoparticles with biopolymers such as chitosan or cellulose. The chapter begins by detailing the fundamental principles of bionanocomposites, emphasizing their ability to respond dynamically to external stimuli like pH, temperature, or mechanical stress. Nanomaterials such as CNTs and metal oxides impart superior electrical, thermal, and mechanical properties to these composites, distinguishing them from conventional materials. For instance, graphene improves electrical conductivity, while biopolymers like chitosan contribute biodegradability and biocompatibility, making these materials particularly suitable for drug delivery, biosensors, and environmental monitoring.

The processing and fabrication techniques discussed are critical for tailoring the properties of bionanocomposites. Advanced methods such as electrospinning, sol–gel processing, and 3D printing allow precise control over the materials' nanoscale architecture and multifunctionality. Challenges in achieving uniform dispersion of nanomaterials are addressed through surface modification and functionalization techniques, enhancing compatibility, and enabling additional features like stimuli sensitivity. The integration of AI and ML further optimizes fabrication, providing real-time monitoring and adaptive control to improve reproducibility and efficiency. The chapter categorizes bionanocomposites based on their matrices (natural biopolymers like alginate and synthetic polymers like PLA), nanomaterials (carbon-based, metal oxides, or natural nanostructures), and their functionalities (biodegradable, stimuli-responsive, or conductive). Each category has applications tailored to its properties. For instance, biodegradable composites are used in temporary biomedical implants, while conductive composites play a pivotal role in biosensors and electronic devices. Smart functionalities, including stimuli responsiveness and self-healing, are highlighted as transformative features of bionanocomposites. These properties enable advanced applications such as pH-sensitive drug delivery, thermal-responsive scaffolds for tissue engineering, and light-responsive optical sensors. Self-healing mechanisms, achieved through encapsulated agents or dynamic chemical bonds, extend the lifespan of materials in biomedical and structural applications. These materials are integral in drug delivery, offering precise, localized therapeutic release. In tissue engineering, they provide scaffolds that mimic biological structures, enhancing cell proliferation and differentiation. Biosensors made from these composites ensure real-time health monitoring with high sensitivity and specificity. Bionanocomposites detect pollutants and aid in their removal. For instance, magnetic composites extract heavy metals from water, while photocatalytic composites degrade organic pollutants. They enhance the performance of batteries and supercapacitors, with nanostructures improving conductivity and energy density. Piezoelectric and thermoelectric composites enable energy harvesting from mechanical stress or temperature gradients. Smart bionanocomposites integrated into infrastructure materials detect strain or damage, offering early warnings and reducing maintenance costs.

Despite their potential, scaling production presents challenges, including maintaining uniform nanoscale structures, achieving economic viability, and adhering to regulatory standards. The chapter advocates for interdisciplinary collaboration and the integration of sustainable practices to overcome these hurdles. Advanced AI-driven optimization and hybrid fabrication techniques are proposed as solutions to enhance scalability and functionality. The chapter emphasizes the transformative potential of bionanocomposites in achieving sustainability goals across healthcare, environmental protection, and energy efficiency. Future developments may focus on integrating circular economy principles, recycling nanomaterials, and expanding biodegradable designs, ensuring these materials contribute meaningfully to global sustainability efforts.

Acknowledgements I would like to express my sincere gratitude to my esteemed professors and PhD Scholars for their invaluable guidance, insightful discussions, and unwavering support throughout the preparation of this book chapter. Their mentorship has been instrumental in shaping my understanding of the subject and enhancing the quality of this work. I am profoundly thankful to Prof. Anand Srivastava, the Honorable Vice Chancellor of Netaji Subhas University of Technology, for fostering a vibrant research environment and providing the resources necessary to carry out this work.

Conflict of Interest The authors declare that there is no conflict of interest regarding the publication of this chapter.

References

Afshan N, Bari A (2023) Advanced bionanomaterials for environmental clean recovery processes. In: Sustainable nanomaterials for biosystems engineering, pp 213–240. https://doi.org/10.1201/9781003333517-11

Alam MR, Shahid MA, Alimuzzaman S, Hasan MM, Hoque ME (2023) Electrospun bio-nano hybrid scaffold from collagen, Nigella sativa, and chitosan for skin tissue engineering application 38(3):234–251. https://doi.org/10.1177/08839115231162365

Andrada Suarez EE, Roca Jalil ME, Fernandez Baldo MA, Cuozzo SA (2025) Nanobiotechnology approaches for the remediation of persistent and emerging organic pollutants: strategies, interactions, and effectiveness. Environ Sci Nano. https://doi.org/10.1039/D4EN00424H

Anitha Selvasofia SD, Deepa Shri S, Manoj Kumar P, KB P (2024) Characterization of bio-nano doped phase change material (bio-nano/PCM) for building energy management. Energy Sources Part A Recover Util Environ Eff 46(1):3760–3778. https://doi.org/10.1080/15567036.2024.2322690

Aravind D et al (2022) A compendious review on fatigue properties of the bio-nanocomposites. Polym Compos 43(10):6803–6816. https://doi.org/10.1002/PC.26859

Ariadurai SA. Bio-composites: current status and future trends

Asadipour E, Asgari M, Mousavi P, Piri-Gharaghie T, Ghajari G, Mirzaie A (2023) Nano-biotechnology and challenges of drug delivery system in cancer treatment pathway: review article. Chem Biodivers 20(6):e202201072. https://doi.org/10.1002/CBDV.202201072

Aston DE, Bow JR, Gangadean DN (2013) Mechanical properties of selected nanostructured materials and complex bio-nano, hybrid and hierarchical systems. Int Mater Rev 58(3):167–202. https://doi.org/10.1179/1743280412Y.0000000012

Bagal-Kestwal DR, Pan MH, Chiang BH (2019) Processing methods for bionanocomposites. In: Bio monomers for green polymeric composite materials, pp 25–55. https://doi.org/10.1002/9781119301714.CH2

Barua S, Chen B, Baruah S (2019) Bio-nano interfacial interactions for drug delivery systems. In: Dynamics of advanced sustainable nanomaterials and their related nanocomposites at the bio-nano interface, pp 53–73. https://doi.org/10.1016/B978-0-12-819142-2.00003-3

Bhattacharjee J, Bhaskar S, Roy S (2025) Green nanomaterials for antimicrobial and anticancer applications. In: Nano-engineering at functional interfaces for multi-disciplinary applications, pp 425–444. https://doi.org/10.1016/B978-0-443-21691-6.00018-4
Bilal M (2024) Nanotechnology to monitor, remedy, and prevent pollution
Bio-Based Composites: Future Trends and Environmental Aspects (2014) Bio-based composites for high-performance materials, pp 316–327. https://doi.org/10.1201/B17601-21
Boulogeorgos AAA, Trevlakis SE, Tegos SA, Papanikolaou VK, Karagiannidis GK (2021) Machine learning in nano-scale biomedical engineering. IEEE Trans Mol Biol Multiscale Commun 7(1):10–39. https://doi.org/10.1109/TMBMC.2020.3035383
Christy PN, Basha SK, Kumari VS (2022) Nano zinc oxide and nano bioactive glass reinforced chitosan/poly(vinyl alcohol) scaffolds for bone tissue engineering application. Mater Today Commun 31:103429. https://doi.org/10.1016/J.MTCOMM.2022.103429
Dachani SR et al (2024) Innovative polymers in pharmaceutical chemistry: revolutionizing drug delivery systems. Polym-Plast Technol Mater. https://doi.org/10.1080/25740881.2024.2440531
Daghigh V et al (2020) Machine learning predictions on fracture toughness of multiscale bio-nano-composites. 39(15–16):587–598. https://doi.org/10.1177/0731684420915984
Debnath M, Khan A, Keshari R, Banerjee A, Srivastava R (2024) Impact of stimuli-responsive biomaterials in gynecological cancer therapy. Biomed Mater Devices 2(2):655–668. https://doi.org/10.1007/S44174-023-00138-0/METRICS
Di Giacomo R et al (2013) Bio-nano-composite materials constructed with single cells and carbon nanotubes: mechanical, electrical, and optical properties. IEEE Trans Nanotechnol 12(6):1026–1030. https://doi.org/10.1109/TNANO.2013.2285438
El Kalaaoui K et al (2024) Hybrid polyvinyl alcohol-silica antibacterial nanofiber fabricated by combined sol-gel and electrospinning techniques. J Solgel Sci Technol 112(1):69–83. https://doi.org/10.1007/S10971-024-06510-W/FIGURES/10
Elkhoury K et al (2019) Soft-nanoparticle functionalization of natural hydrogels for tissue engineering applications. Adv Healthc Mater 8(18):1900506. https://doi.org/10.1002/ADHM.201900506
Garg K, Sharma D, Sehgal R, Gupta R (2024) Bionanomaterials in wastewater treatment. In: Bionanomaterials for industrial applications, pp 386–412. https://doi.org/10.1201/9781003432791-26/BIONANOMATERIALS-WASTEWATER-TREATMENT-KIRTI-GARG-DIVYA-SHARMA-RUTIKA-SEHGAL-REENA-GUPTA
George Kerry R et al (2021) A comprehensive review on the applications of nano-biosensor-based approaches for non-communicable and communicable disease detection. Biomater Sci 9(10):3576–3602. https://doi.org/10.1039/D0BM02164D
Gomez-Flores A, Cho H, Hong G, Nam H, Kim H, Chung Y (2024) A critical review on machine learning applications in fiber composites and nanocomposites: towards a control loop in the chain of processes in industries. Mater des 245:113247. https://doi.org/10.1016/J.MATDES.2024.113247
Gul S, Awais M, Jabeen S, Farooq M (2020) Recent trends in preparation and applications of biodegradable polymer composites. J Renew Mater 8(10):1305–1326. https://doi.org/10.32604/JRM.2020.010037
Han L et al (2024) Electrospinning nanomaterials: a powerful strategy for wastewater treatment applications. Rev Environ Sci Bio/Technol 23(2):471–502. https://doi.org/10.1007/S11157-024-09686-3
Hasan A, Morshed M, Memic A, Hassan S, Webster TJ, Marei HES (2018) Nanoparticles in tissue engineering: applications, challenges and prospects. Int J Nanomedicine 13:5637–5655. https://doi.org/10.2147/IJN.S153758
John P, Selvam AK, Uppal M, Mohammed Adhil S (2023) Futuristic biomaterials for 3D printed healthcare devices. In: Digital design and manufacturing of medical devices and systems, pp 33–62. https://doi.org/10.1007/978-981-99-7100-8_2

Kannan AM, Renugopalakrishnan V, Filipek S, Li P, Audette GF, Munukutla L (2009) Bio-batteries and bio-fuel cells: leveraging on electronic charge transfer proteins. J Nanosci Nanotechnol 9(3):1665–1678. https://doi.org/10.1166/JNN.2009.SI03

Karimzadeh Z, Mahmoudpour M, Rahimpour E, Jouyban A (2022) Nanomaterial based PVA nanocomposite hydrogels for biomedical sensing: Advances toward designing the ideal flexible/wearable nanoprobes. Adv Colloid Interface Sci 305:102705. https://doi.org/10.1016/J.CIS.2022.102705

Kasaai MR (2022) Bio-nano-composites containing at least two components, chitosan and zein, for food packaging applications: a review of the nano-composites in comparison with the conventional counterparts. Carbohydr Polym 280:119027. https://doi.org/10.1016/J.CARBPOL.2021.119027

Kim JH, Jeong JJ, Lee JS (2024) Eco-friendly isolated nanocellulose from seaweed biomass via modified-acid and electron beam process for biodegradable polymer composites. J Compos Sci 8(7):253. https://doi.org/10.3390/JCS8070253

Koncar V (2019) Smart textiles for in situ monitoring of composites

Kulkarni MB, Ayachit NH, Aminabhavi TM (2022) Recent advancements in nanobiosensors: current trends, challenges, applications, and future scope. Biosensors 12(10):892. https://doi.org/10.3390/BIOS12100892

Kumar SD, Manohar GA, Rao PR, Mandal A (2022) SMART nano-sensors via 3D printing technology. In: Smart 3D nanoprinting, pp 111–131. https://doi.org/10.1201/9781003189404-6

Kumar A, Gautam YK, Singh N (2024) State-of-the-art developments in surface functionalized carbon-based bio/nanocomposites for theranostic antibacterials, advanced bioimaging, and molecular bioelectronics inspired biosensing platforms. J Ind Eng Chem 140:103–145. https://doi.org/10.1016/J.JIEC.2024.05.065

Kuperkar K, Atanase LI, Bahadur A, Crivei IC, Bahadur P (2024) Degradable polymeric bio(nano)materials and their biomedical applications: a comprehensive overview and recent updates. Polymers 16(2):206. https://doi.org/10.3390/POLYM16020206

Lecoublet M, Ragoubi M, Leblanc N, Koubaa A (2024) Sustainable 3D-printed cellulose-based biocomposites and bio-nano-composites: analysis of dielectric performances. Ind Crops Prod 221:119332. https://doi.org/10.1016/J.INDCROP.2024.119332

Li J, Xin M, Ma Z, Shi Y, Pan L (2021) Nanomaterials and their applications on bio-inspired wearable electronics. Nanotechnology 32(47):472002. https://doi.org/10.1088/1361-6528/ABE6C7

Liu B et al (2024) An overview of sustainable biopolymer composites in sensor manufacturing and smart cities. Adv Compos Hybrid Mater 7(5):1–20. https://doi.org/10.1007/S42114-024-00938-Y/FIGURES/8

Menossi M, Casalongué C, Alvarez VA (2022) Bio-nanocomposites for modern agricultural applications. In: Handbook of consumer nanoproducts, pp 1201–1237. https://doi.org/10.1007/978-981-16-8698-6_68

Mishra S, Yılmaz-Serçinoğlu Z, Moradi H, Bhatt D, Kuru Cİ, Ulucan-Karnak F (2023) Recent advances in bioinspired sustainable sensing technologies. Nano-Struct Nano-Objects 34:100974. https://doi.org/10.1016/J.NANOSO.2023.100974

Mittal V, Halley PJ, Avérous L (2011) Nanocomposites with biodegradable polymers abstract and keywords abstract and keywords, pp 1–38

Mohamed EF, Awad G (2022) Development of nano-sensor and biosensor as an air pollution detection technique for the foreseeable future. Compr Anal Chem 99:163–188. https://doi.org/10.1016/BS.COAC.2021.11.003

Mujawar MA, Gohel H, Bhardwaj SK, Srinivasan S, Hickman N, Kaushik A (2020) Nano-enabled biosensing systems for intelligent healthcare: towards COVID-19 management. Mater Today Chem 17:100306. https://doi.org/10.1016/J.MTCHEM.2020.100306

Nandhini J, Bellarmin M, Siva Prakash S, Sowmya Sri D, Karthikeyan E (2024) Advancements in metal oxide bio-nanocomposites for sustainable food packaging: fabrication, applications, and future prospectives. Food Bioeng. https://doi.org/10.1002/FBE2.12106

Nanocomposites with Biodegradable Polymers - Google Books (Online). Available: https://www.google.co.in/books/edition/Nanocomposites_with_Biodegradable_Polyme/kSUUDAAAQBAJ?hl=en&gbpv=1&dq=Bio-nano+composites+%2B+classification&pg=PA1&printsec=frontcover. Accessed: 24 Dec 2024

Nazemiyeh A et al (2024) Electrospinning of nanocellulose–synthetic polymer composites: a multifaceted approach to tissue engineering breakthroughs. In: Nanocellulose-based hybrid systems for tissue engineering, pp 151–169. https://doi.org/10.1039/9781837673094-00151

Orooji Y et al (2022) Valorisation of nuts biowaste: prospects in sustainable bio(nano)catalysts and environmental applications. J Clean Prod 347:131220. https://doi.org/10.1016/J.JCLEPRO.2022.131220

Palanisamy P, Chavali M, Kumar EM, Etika KC (2020) Hybrid nanocomposites and their potential applications in the field of nanosensors/gas and biosensors. In: Nanofabrication for smart nanosensor applications, pp 253–280. https://doi.org/10.1016/B978-0-12-820702-4.00011-8

Pascual Aguilar Z, Al-Ogaidi I, Suri S et al (2021) Nano and bio-composites and their applications: a review. IOP Conf Ser Mater Sci Eng 1067(1):012093. https://doi.org/10.1088/1757-899X/1067/1/012093

Perera KY, Jaiswal S, Jaiswal AK (2022) A review on nanomaterials and nanohybrids based bionanocomposites for food packaging. Food Chem 376:131912. https://doi.org/10.1016/J.FOODCHEM.2021.131912

Petrovic S, Bita B, Barbinta-Patrascu ME (2024) Nanoformulations in pharmaceutical and biomedical applications: green perspectives. Int J Mol Sci 25(11):5842. https://doi.org/10.3390/IJMS25115842

Pham SH, Choi Y, Choi J (2020) Stimuli-responsive nanomaterials for application in antitumor therapy and drug delivery. Pharmaceutics 12(7):630. https://doi.org/10.3390/PHARMACEUTICS12070630

Prajapati D et al (2024) Bionanotechnology: a paradigm for advancing environmental sustainability. Indian J Microbiol 2024:1–27. https://doi.org/10.1007/S12088-024-01389-1

Prinith Kaveramma UP, Hema Brindha M, Ajitha Nayac R, Reshma Ruth Pauline AR, Ashwin Kumar N (2025) Poly(N-isopropyl acrylamide): a thermo-responsive polymer for biotherapeutic applications. In: Synthetic polymers in drug and biotherapeutics delivery, pp 235–267. https://doi.org/10.1016/B978-0-323-95233-0.00007-8

Promdontree P, Ounkaew A, Yao Y, Zeng H, Narain R, Ummartyotin S (2024) Temperature-responsive injectable composite hydrogels based on poly(N-isopropylacrylamide), chitosan, and hemp-derived cellulose nanocrystals. Polymers 16(21):2984. https://doi.org/10.3390/POLYM16212984

Pulikkalparambil H, Siengchin S, Parameswaranpillai J (2018) Corrosion protective self-healing epoxy resin coatings based on inhibitor and polymeric healing agents encapsulated in organic and inorganic micro and nanocontainers. Nano-Struct Nano-Objects 16:381–395. https://doi.org/10.1016/J.NANOSO.2018.09.010

Radacsi N, Campos FD, Chisholm CRI, Giapis KP (2018) Spontaneous formation of nanoparticles on electrospun nanofibers. Nat Commun 9(1):1–8. https://doi.org/10.1038/s41467-018-07243-5

Rajalakshmi B et al (2024) Bio-inspired nanomaterial's for energy harvesting and storage: a green approach. In: E3S web of conferences, vol 552, p 01122. https://doi.org/10.1051/E3SCONF/202455201122

Rana D et al (2016) Surface functionalization of nanobiomaterials for application in stem cell culture, tissue engineering, and regenerative medicine. Biotechnol Prog 32(3):554–567. https://doi.org/10.1002/BTPR.2262

Rashid IA, Afzal A (2023) Smart polymers and composites. Eng Mater 634:77–99. https://doi.org/10.1007/978-981-99-0787-8_5

Rezaee K, Ansari M, Khosravi M (2024) Smart materials in medical applications: an overview of machine learning-based futures, pp 371–389. https://doi.org/10.1007/978-981-99-8398-8_25

Sabo R, Yermakov A, Law CT, Elhajjar R (2016) Nanocellulose-enabled electronics, energy harvesting devices, smart materials and sensors: a review. J Renew Mater 4(5):297–312. https://doi.org/10.7569/JRM.2016.634114

Sanchez-Cano C et al (2021) X-ray-based techniques to study the nano-bio interface. ACS Nano 15(3):3754–3807. https://doi.org/10.1021/ACSNANO.0C09563/ASSET/IMAGES/LARGE/NN0C09563_0021.JPEG

Sanyal S et al (2024) Emerging trends in smart self-healing coatings: a focus on micro/nanocontainer technologies for enhanced corrosion protection. Coatings 14(3):324. https://doi.org/10.3390/COATINGS14030324

Saraswat P, Singh S, Prasad M, Misra R, Rajput VD, Ranjan R (2023) Applications of bio-based nanomaterials in environment and agriculture: a review on recent progresses. Hybrid Adv 4:100097. https://doi.org/10.1016/J.HYBADV.2023.100097

Shanmuganathan K, Capadona JR, Rowan SJ, Weder C (2010) Biomimetic mechanically adaptive nanocomposites. Prog Polym Sci 35(1–2):212–222. https://doi.org/10.1016/J.PROGPOLYMSCI.2009.10.005

Sharma J, Gole VL, Siddiqui MA (2023) Bionanocomposites biological aspects and biomedical applications. In: Polymer nanocomposites: fabrication to applications, pp 149–171. https://doi.org/10.1201/9781003343912-11/BIONANOCOMPOSITES-JYOTI-SHARMA-VITTHAL-GOLE-MOHAMMAD-ALI-SIDDIQUI

Simchi A, Kianfar P (2024) Knowledge-based design and fabrication of advanced composites and hybrid materials for nanomedicine. In: Integrated systems: data driven engineering, pp 249–263. https://doi.org/10.1007/978-3-031-53652-6_15

Singh NB (2023) Applications of emerging nanomaterials and nanotechnology

Singh G, Preethi L, Thakur N (2024) Nano–bio dynamics: mechanisms and challenges in shaping nanomaterials for biomedical use. In: Nanoparticles in cancer theranostics: current progress in cancer management, pp 53–68. https://doi.org/10.1201/9781003463191-4/NANO

Sonu et al (2023) Agro-waste to sustainable energy: a green strategy of converting agricultural waste to nano-enabled energy applications. Sci Total Environ 875:162667. https://doi.org/10.1016/J.SCITOTENV.2023.162667

Srinivas A, Chandran KR, Nehra S, Kumar D (2025) Future aspects in the field of bio-nanomaterials toward environmental assessment. In: Bio-nanomaterials in environmental remediation, pp 295–317. https://doi.org/10.1002/9783527848546.CH12

Subeshan B, Atayo A, Asmatulu E (2024) Machine learning applications for electrospun nanofibers: a review. J Mater Sci 59(31):14095–14140. https://doi.org/10.1007/S10853-024-09994-7

Tamilselvan SN, Shanmugan S (2024) Towards sustainable solar cells: unveiling the latest developments in bio-nano materials for enhanced DSSC efficiency. Clean Energy 8(3):238–257. https://doi.org/10.1093/CE/ZKAE031

Thippeswamy BH, Maligi AS, Hegde G (2021) Roadmap of effects of biowaste-synthesized carbon nanomaterials on carbon nano-reinforced composites. Catalysts 11(12):1485. https://doi.org/10.3390/CATAL11121485

Toscano JD, Li Z, Segura LJ, Sun H (2020) A machine learning approach to model the electrospinning process of biocompatible materials. In: ASME 2020 15th international manufacturing science and engineering conference, MSEC 2020, vol 2. https://doi.org/10.1115/MSEC2020-8362

Turcu R, Craciunescu I, Nan A (2014) Upscaling of bio-nano-processes, pp 57–76. https://doi.org/10.1007/978-3-662-43899-2

Wang H, Wang W, Tao J, Liu S, Hou X, He C (2024) Anchoring mycelium on CNTs to make strong and smart self-regenerative composite materials. Compos B Eng 287:111861. https://doi.org/10.1016/J.COMPOSITESB.2024.111861

Xu L et al (2021) Temperature-responsive multilayer films based on block copolymer-coated silica nanoparticles for long-term release of favipiravir. ACS Appl Nano Mater 4(12):14014–14025. https://doi.org/10.1021/ACSANM.1C03334/SUPPL_FILE/AN1C03334_SI_001.PDF

Yousefpour F, Jamaati R, Aval HJ (2022) Synergistic effects of hybrid (HA+Ag) particles and friction stir processing in the design of a high-strength magnesium matrix bio-nano composite with an appropriate texture for biomedical applications. J Mech Behav Biomed Mater 125:104983. https://doi.org/10.1016/J.JMBBM.2021.104983

Zafar S, Nazir M, Sabah A, Jurcut AD (2021) Securing bio-cyber interface for the internet of bio-nano things using particle swarm optimization and artificial neural networks based parameter profiling. Comput Biol Med 136:104707. https://doi.org/10.1016/J.COMPBIOMED.2021.104707

Zainazor T, Fisal A, Goh EG, Sulaiman C, Sarbon NM (2020) Emerging of bio-nano composite gelatine-based film as bio-degradable food packaging: a review 1. Journal Homepage 4(4):944–956. https://doi.org/10.26656/fr.2017.4(4).365

Zhao J, Liu Q, Tong X, Wang Y, Cai K, Ji W (2024) Rational design of bio-inspired peptide electronic materials toward bionanotechnology: strategies and applications. Adv Funct Mater 34(29):2401466. https://doi.org/10.1002/ADFM.202401466

Zhu Y et al (2024) Multifunctional polyvinyl alcohol/poly-dopamine hydrogels loaded with bio-nano composites promote wound healing by repairing endogenous electric fields. Int J Biol Macromol 272:132763. https://doi.org/10.1016/J.IJBIOMAC.2024.132763

Satyam Rawat is currently pursuing an M.Sc. in Physics, specializing in the synthesis and characterization of 2D materials like reduced graphene oxide. His research focuses on developing nanocomposites of these materials for advanced applications, including supercapacitors and photodetectors. His work aims to enhance the performance and versatility of energy storage and optoelectronic devices through innovative material design and optimization.

Peeyush Phogat completed his Ph.D. in the Department of Physics at Netaji Subhas University of Technology. His current area of research is synthesis and characterization of materials, focusing particularly on applications in solar energy and capacitors. Holding a master's degree with specialization in condensed matter physics. He is driven by the pursuit of uncovering novel materials and their distinct properties, meticulously tailored for effective solar energy capture and the design of efficient capacitors. His scholarly endeavours have resulted in significant contributions towards the development of new and advanced materials for solar applications, which are evidenced by his 35 publications in esteemed SCI/SCIE journals and authorship of 14 book chapters within the renowned book series. His expertise extends comprehensively into the domain of capacitors, where his research showcases a remarkable ability to engineer materials that optimize energy storage and discharge capabilities.

Shreya Sharma is a physicist who earned her Ph.D. in Physics from Netaji Subhas University of Technology, where her research was centered on the exploration of exceptional properties exhibited by nanomaterials and their potential contributions to advancing renewable energy technologies. Her primary focus lies in investigating the unique properties of 2D materials and their electrochemical study, specifically transition metal dichalcogenides, with the aim of applying these findings to enhance solar cells, supercapacitors, photodetectors, hydrogen production, and electrochemical sensing. Her dedication to leveraging advanced materials for sustainable energy solutions reflects her vision of contributing to societal progress through innovative technology. Currently, she is working to identify sustainable energy resources, with a specific focus on blue energy foresight. She is also the author and co-editor of multiple scientific books and papers in esteemed journals. Driven by a deep commitment to environmental sustainability, Dr. Sharma's work embodies her passion for scientific advancement in service to future energy needs.

Ranjana Jha is currently serving as Vice Chancellor, IGDTUW, Delhi, India. Previously, she worked as a professor and head of Department of Physics, Netaji Subhas University of Technology. She has been actively involved in research and teaching in the field of Solar Energy Materials and Applied Physics. Received Lifetime Achievement Award in Science and several International and National awards for academics and research in Solar Energy Utilization, Materials and Device fabrication. She is the founder and in-charge of Research Lab for Energy Systems at Department of Physics. Her research interests are Solar Energy Materials and Solar Energy Utilization, Single Crystal Formation, Development of Nano-structured thin film Solar cells, and Characterization of Energy Materials for Device Applications. Published 67 manuscripts in reputed International/ National Journals. Attended/ presented papers as an author, co-author, and corresponding author in 123 International and National conferences.

Sukhvir Singh is presently working as Professor Emeritus at Delhi Skill and Entrepreneurship University, Dwarka, New Delhi, India. Previously, he worked as Ex. Emeritus Scientist at CSIR-National Physical Laboratory, New Delhi India, Ex Sr Principal Scientist, and Head Electron and Ion Microscopy Group, CSIR National Physical Laboratory, New Delhi India. Dr. Sukhvir Singh is Ph.D. in physics from Jamia Millia Islamia, New Delhi, India. Synthesis and characterization of bulk and nanostructured thin films of compound semiconductors, metal oxides, nanocomposites, and investigation of their structural, optical, electronics, and electrical properties to ascertain their application for various sensors and device applications are the main areas of his research. The development of standard reference materials (SRM) for the calibration of SEM and TEM is another area of his specialized research under him. He has also established a Centre for Nano Scale Science at CSIR-National Physical Laboratory, New Delhi where highly sophisticated and state-of-the-art advance equipment, such as XRD/HRXRD, TEM/HRTEM, SEM/EDS, AFM, TOF–SIMS, PL spectrometer and highly sophisticated specimen preparation tools for TEM and SEM are installed and serving the nation in various areas of the current research. He has published more than 150 research papers in SCI journals of high repute. He has three no. of US, South Africa and Asia patents granted in his name.

[illegible]

Chapter 5
Advanced Characterisation and Properties of Bionanocomposite Materials

Peramjeet Singh, Priya Gupta, Vernica Verma, Neetu Yadav, and Narendra Kumar Pandey

Abstract Bionanocomposites are a novel class of hybrid materials that combine nanoscale fillers with natural bio-macromolecules to provide unique features based on their complementary molecular interactions. Advanced characterisation techniques, including X-ray diffraction, scanning electron microscopy, atomic force microscopy, and transmission electron microscopy, are essential for revealing the structural and morphological characteristics of these composites. Additionally, spectroscopic methods such as Fourier-transform infrared, X-ray photoelectron spectroscopy, and Raman spectroscopy offer critical insights into their chemical interactions. Researcher could utilise these techniques to calculate the distinct properties and behaviours of bio-nano composites that might not be seen using more straightforward analysis techniques. Determining the possible uses of bio-macromolecular matrices is well-suited in various fields, especially industrial, environmental, and biomedical applications, because the addition of nanofillers enhance their mechanical, strength, thermal stability, and biocompatibility. The electrical and optical properties of bionanocomposites can be adjusted, which offers new opportunities for their application in sensors, energy storage devices, biomedical, electronics, aerospace, automotive, and other advanced technologies. Researchers can better understand the underlying mechanism influencing these material's behaviour by carefully characterising them, which will enable functional improvements and targeted enhancements. By utilising the distinct properties of both nanofillers and bio-macromolecules, these composites have the potential to spur advancements in several scientific and technical domains. The area of bionanocomposite materials research keeps changing, with particular emphasis on designing new natural matrices and functional nanomaterials, improving the connections among natural and nanomaterials feature for greater efficiency. The researcher further works on engineering techniques like 3D printing.

P. Singh (✉) · P. Gupta · V. Verma · N. Yadav · N. K. Pandey
Department of Physics, University of Lucknow, Lucknow, Uttar Pradesh, India
e-mail: ps751994lko.uni@gmail.com

P. Gupta
Manyawar Kanshiram Government Polytechnic, Tirwa, Kannauj, Uttar Pradesh, India

S. C. George and B. Tawiah (eds.), *Advanced Bionanocomposite Materials*, Advanced Structured Materials 241, https://doi.org/10.1007/978-981-96-8225-6_5

Keywords Bionano · Characterisation · Composites · Nanomaterials · Fabrication · Microscopy

5.1 Introduction

Bionanocomposites are composites comprised of bio-based materials like lipids, proteins, or polymers with nanomaterials such as fibres, tubes, or particles. The resultant hybrids combine the special properties of both materials, often yielding materials that have higher thermal, mechanical, and biological abilities as compared with their respective portions separately (Meyers et al. 2008). These composites are highly desirable in a range of areas, especially biology, mechanical engineering, and materials research (Yildizhan et al. 2018). Bionanocomposites, which interface nanotechnologies, material science, and biology, have become a focal point for development in areas like medicine, environmental tracking, farming, and electronics (Sen et al. 2024). The fabrication and identification of these nanomaterials need knowledge of both the mineral's nanoscale interactions and biological interactions (Gilbert and Banfield 2005). Researchers are particularly interested in how these small particles regulate living things, both at the biochemical and cellular levels, to improve performance while minimising possible toxicity or immunological reactions (Pandey and Prajapati 2018). Bionano nanoparticles are an exciting area in minerals research, with possibilities that include medical and ecological preservation to improved electronics and bioengineering (Karki et al. 2021). The flexibility arises from their capacity to change the ingredients, and functional qualities on a nanoscale (Lin et al. 2012). Bionanocomposite materials (BNCMs) can be formed by mixing natural and synthetic nanomaterials with biological components such as genetic material, proteins, lipids, and carbohydrates. According to recent research, those substances may be produced employing either bottom-up or top-down methods, with the objective of acquiring certain features such as biocompatibility, decomposition, and regulation of release mechanisms (Ramadhan and Ali 2025). For example, Panja et al. (2022) published research on the combination of bioactive materials with nanoparticles of metal, most notably silver and gold nanoparticles, to improve antibacterial capabilities while retaining biological compatibility for medical purposes (Panja et al. 2022). Bionanocomposite nanoparticles are formed using an array of processes, including co-precipitation technique, sol–gel synthesis, and hydrothermal techniques. Chaudhary et al. (2022) investigated the co-precipitation approach for combining biopolymers like chitosan with silver nanomaterials (Chaudhary et al. 2022). This synthesis yielded stable composites with antibacterial features which makes them suitable for wound healing applications. Despite the potential uses, integrating living organisms with nanomaterials presents stability, scalability, and toxicity issues. Zhou et al. (2021) investigated the possible dangers of bio-nano composites, with a special emphasis on the cytotoxicity of some metal-based nanoparticles (Zhou et al. 2021). To avoid detrimental effects on live organisms, they stressed the importance of meticulous engineering and management of nanoscale size, exterior functionalisation, and dosage. Patel et al. (2024) highlighted the significance of establishing greater durability and repeatable fabrication procedures to enable large-scale manufacturing of biological composite particles, specifically for pharmaceutical delivery systems. They also emphasised the future need for multifaceted composites capable of performing medical, surgical, and imaging functions at the same time (Patel et al. 2024). Technologies like sensor technology and shielding against electromagnetic interference are also interested in the spectrum dependency of the dielectric characteristics. Present transport in bionanocomposites is frequently accomplished by a variety of techniques, including diffuse transmission, tunnelling, and electrical shift. Several metallic nanoparticles may rise to form a percolating barrier, which would allow the composite to exhibit observable conductivity. Solitary charge transmission at lower concentrations of nanoparticles can lead to limited conductivity. But when the permeating barrier is broken, the conductivity increases significantly. These nanocomposites are intended to be lighter than conventional metals or ceramics while preserving strength (Ajayan 2003). These properties subjected their candidacy for aeroplanes, automobiles, and other industries where reduced weight is essential for functionality and energy economy. These nanocomposites can be resistant to environmental deterioration (Karki et al. 2021). For instance, biological parts may have natural resistance to corrosion, whereas nanomaterials might add to heat and moisture obstruction, making such materials useful in challenging situations (Eliaz 2019). Nanomaterials, including clay, carbon nanotubes, and graphite, can improve the composite's thermal stability (Kumar et al. 2009). The major benefit of these materials is that they have been synthesised from biodegradable and sustainable resources (Mitra 2014). Mechanically, characteristics may be tuned to reconcile the demand for toughness with ecological responsibility, rendering them desirable in fields like environmentally friendly building and recyclable packaging (Hussain et al. 2024). Selecting nanomaterials properly can prevent build-up or negative consequences. The physical characteristics (such as hardness and adaptability) of the nanocomposite must correspond to the tissues that it is intended to maintain or substitute (Aston et al. 2013). The composite's surface should have characteristics including imperfections, hydrophilic properties, and biochemical functionalisation to promote tissue absorption and cell adhesion (Thakur et al. 2022). The composite needs to be non-thrombogenic—that is, it shouldn't cause haemolysis or blood clotting—for applications involving blood contact (Ghosh et al. 2023). The composite could offer antibacterial qualities to stop infections, especially in surgery or healing wounds (Simchi et al. 2011). In order to facilitate healing and functional recovery, the substance should support tissue regeneration and blend in well with the biological tissue around it (Bramhill et al. 2017). These features are crucial for bio-nano composites to be successfully used in industries including implantation, pharmaceutical delivery, and tissue fabrication (Ahmad et al. 2017). Numerous materials, such as metallic oxides, sulphides, and even substances based on carbon, which are fabricated using this method, make it adaptable to a variety of uses. Comparing this technique with conventional organic synthesis processes that may require harmful or toxic solvents, it is comparatively ecologically benign because water serves as the basic solvent (Cseri et al. 2018). Despite its effectiveness on a small scale, the hydrothermal process can be difficult to scale up for industrial purposes since it requires high-pressure equipment and requires complicated variables to be controlled. Bio-nano composites, which mix nanoparticles with biological components, are a novel class of materials with high promise for environmental remediation. These composites solve difficult ecological problems by combining the special characteristics of nanoparticles with biological agents (Chatterjee and Dhar 2025). The application of bio-nano composites for ecological remediation has various advantages, including enhanced productivity, long-term viability, and cost performance (Srinivas et al. 2025). Bio-nano composite materials have shown potential for removing heavy metals and other contaminants from water (Samejo et al. 2025). These composites' nanoparticles have a wide surface area for collecting contaminants, and biological components (e.g. bacteria or fungus) can specifically target particular chemicals, which include heavy metals such as lead (Pb), cadmium (Cd), and mercury (Hg) (Asad et al. 2025). For example, biopolymer-based nanocomposites integrating biosorbents such as algae or bacteria can be used to collect heavy metal ions from polluted water and successfully remove them to levels below hazardous. Kumar et al. (2021) confirmed the efficacy of chitosan-based nanocomposites in eliminating chromium ions from water. When zinc oxide nanoparticles and chitosan, a naturally occurring biopolymer, were combined, the chromium ion removal efficiency and adsorption capacity were both increased (Kumar et al. 2021). The procedure can range from an hour to several days according to the response conditions and materials used, which could limit large-scale manufacturing. The procedure of extrusion or melting compound is used to combine the polymer and nanofiller in a melted condition. This may be accomplished with an interior mixer, a twin-screw nozzle, or a single-screw extruder. During the procedure of blending, the operating temperature ought to be tightly controlled to guarantee that the polymers melt without degrading and to encourage the nanofillers to disperse well. Bio-nano composites fabricated using melting-based synthesis techniques offer a wide range of usage, including packaging elements, automobile biocomposites, and applications in the environment (Thakur et al. 2024). Because of their extensive surface area and an inclination to aggregate, nanofillers can be difficult to disperse uniformly in the structure of polymers. To improve the mechanical qualities, the contact between the nanofillers and the polymer matrix is essential. It may be necessary to modify the outer layer or apply coupling substances for this. The qualities of the finished composite can be greatly impacted by the selection of the residence duration, screw speed, and processing temperature. The process of interconnecting contributes to the composite material's enhanced durability and robustness. In order to confirm the complete evaporated solvent, the films are usually then given a further opportunity to cure. Following fabrication, physical, thermal, morphological, and biodegradable characteristics are evaluated for the bionanocomposite films. Procedures such as XRD, Fe-SEM, TGA, AFM, Uv–Vis, and TEM are frequently employed to assess the structure and characteristics of the bionanocomposites. The method involves low-cost equipment and may be carried out in small quantities or on a big scale. A diverse selection of monomers and tiny fillers may be employed to adjust material characteristics to specific applications. When accomplished correctly, solution castings enable the consistent distribution of nanofillers, resulting in improved mechanical and functional qualities. Bio-nanocomposites fabricated using the solution casting technique offer a wide range of potential uses, including packaging, devices for health care, the food and beverages sector, and ecological uses. Nanoparticles such as graphite, carbon nanotubes, or various metal oxides can greatly enhance the combined material's conduction. When metallic nanostructures are integrated with a biopolymer matrix that has insulating properties, conducting pathways are created that promote the movement of electrons or charges. It is possible to fabricate bio-nanocomposites to possess particular dielectric characteristics which are essential for material insulators, batteries, and capacitors. The sort and extent of nanomaterials may be varied to modify the dielectric coefficient. For example, the material's total dielectric constant might increase when high-permeability nanomaterials are incorporated. The biological components aid in improving sensitivity or biodegradable properties, while the inorganic nanoparticles, like metal oxides, can possess photocatalytic or adsorptive qualities. It is possible to fabricate nanocomposites for catalytic purposes via hydrothermal synthesis. Metal nanoparticles such as Au or Ag can catalyse certain processes in bio-nano composites, e.g. while biological molecules stabilise or improve the catalytic process (Nagime et al. 2025). The combination of biological components and inorganic nanoparticles can result in bio-nanocomposites that are advantageous in the manipulation of tissues (Bagheri et al. 2025). These composites can imitate the extracellular matrix, so boosting cell adhesion, proliferation, and differentiation. Since impurity is minimised by the enclosed surroundings, the hydrothermal technique frequently results in high-purity nanoparticles. It is possible to accurately control the variables that govern the reaction to fabricate nanostructures with certain characteristics, including size, shape, and crystallinity.

5.2 Synthesis Techniques of Bionanocomposites

The synthesis of bionanocomposites is a rapidly developing subject that merges nanoparticles with natural materials. Researchers can develop composite materials with improved thermal, electrical, and functional characteristics for a variety of uses ranging from healthcare to ecological solutions, by using the right fabrication process.

5.2.1 Sol–Gel Method

A popular approach for creating bionanocomposites—materials created by combining inorganic nanoparticles with biological components—is the sol–gel process. A liquid "sol" is transformed into a solid gel using the sol–gel technique,

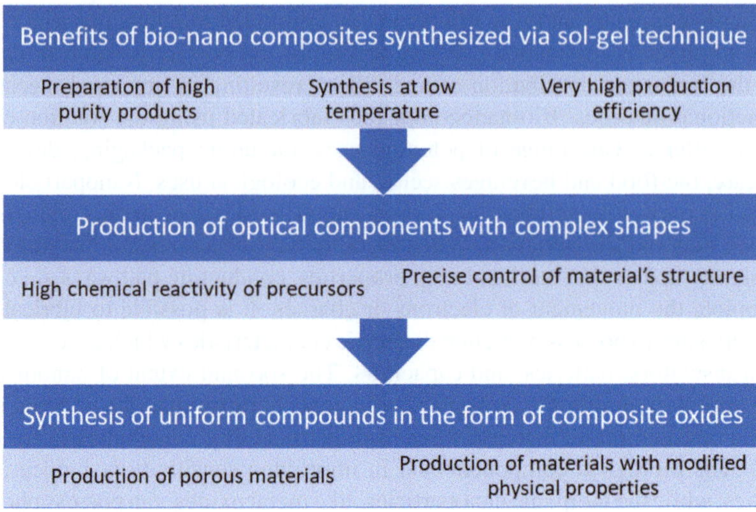

Fig. 5.1 Benefits of bio-nano composites synthesised via sol–gel technique (Bokov et al. 2021)

which has the benefit of enabling the creation of composites with precise compositions and architectures at the nanoscale. To create a sol (a colloidal suspension of nanoparticles), this precursor is subjected to hydrolysis and condensation processes with a solvent, often water or alcohol. Figure 5.1 shows different benefits of bio-nano composite materials synthesised through sol–gel methods.

5.2.2 *Melt Blending*

Melt blending is a typical technique for fabricating bionanocomposites. The technique includes softening the matrix of polymers and propagating tiny particles inside it in order to combine organic material with nanoscale additives. Natural materials such as polylactic acid, polyhydroxyalkanoates, starch-based polymers, and cellulose derivatives are the source of these biodegradable and renewable polymers (Maitra and Bhardwaj 2025). To improve the material's thermoplastic, mechanical, or hydrophobic characteristics, fillers that are nanoscale are applied. These can involve nanocellulose fibres, nanotubes of carbon, graphite, tiny clay particles, and silicon dioxide nanoparticles. Functionalisation or coatings may be required to enhance the distribution of particles inside the matrix of polymers. This might entail surfactant application, a plasma procedure, or chemical alteration to improve the tiny fillers' compatibility with the polymer matrix's properties. It is necessary to select the polymer according to how well it works with the selected nanofiller. The circumstances at which the melt is processed should be appropriate for the polymers and

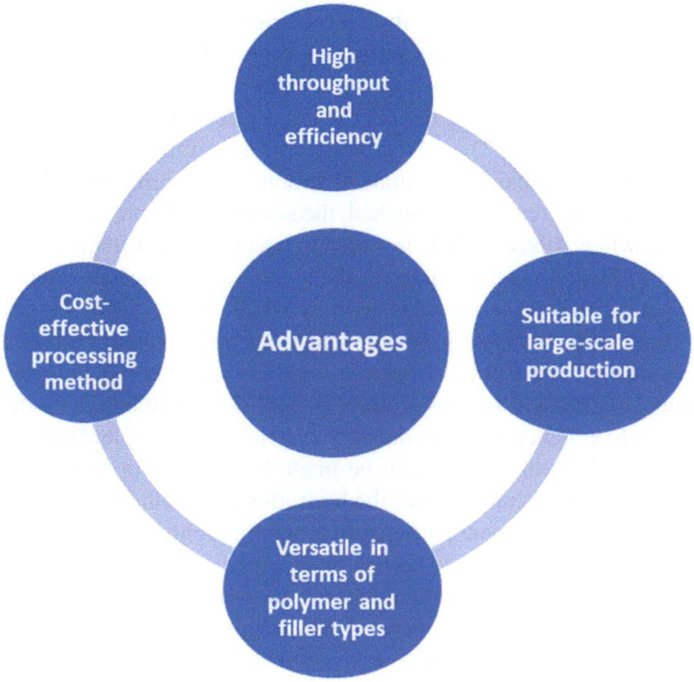

Fig. 5.2 Advantages of melt blending technique

the particles. Figure 5.2 shows various advantages of bio-nanocomposite materials via melt blending methods.

5.2.3 Hydrothermal Methods

The hydrothermal technique of fabricating bionanocomposites is a flexible and widely employed approach. In order to fabricate tiny particles, which are frequently mixed with biological components to develop complex structures, water is used as a solvent under high-temperatures and high-pressure conditions (Wang et al. 2025). Because both inorganic and biological components are integrated, these composites have special qualities. Precursor solutions are created by combining metallic oxides like molybdenum trioxide, tungsten oxide, iron oxide, and bismuth oxide, as well as other inorganic substances, in water. Furthermore, biological substances like polysaccharides, proteins, or peptides might be added to the solution. Nanoparticle size, shape, and dispersion can also be controlled by adding surfactants or stabilising agents. After preparation, the mixture is positioned in an extreme-pressure or controlled autoclave that is capable of handling the severe pressures and degrees. It is common practice to raise the temperature of the autoclave ranging from 100 to 300 °C, contingent upon the substance being produced. Because of the high temperature and high pressure, the water functions as a solvent, which encourages the creation of nanoparticles, such as semiconductors or metal oxides. Bio-nano composites can be fabricated by the interaction of the nanoparticles with biological substances in the solution. The development and integration of nanoparticles may be guided by biomolecular interactions or biomolecular templates (Krishnankutty et al. 2025). When the reaction is finished, the synthesised material is removed from the autoclave when it has cooled. In order to ascertain the dimensions, form, and composition of the composite materials, the synthesised materials are usually characterised using a variety of methods such as XRD, AFM, SPM, FE-SEM, FTIR, and XPS. Nanoparticles that can carry medications or medicinal products to certain bodily locations can be formed from bio-nano composites (Kishore et al. 2025). The biological component such as proteins or peptides might interact with particular cells or tissues to enhance the drug delivery system's targeting and effectiveness (Li et al. 2025). Sensitive biosensors can be made by combining biological materials with nanoparticles synthesised using the hydrothermal method (Amiri et al. 2025). Examples of metal oxide nanoparticles that are frequently employed in combination with biomolecules to identify certain biochemical indicators or diseases include zinc oxide, titanium oxide, bismuth oxide, and molybdenum oxide. Hydrothermally produced bionanocomposites can be used for environmental remediation, including the removal of organic contaminants or heavy metals from water.

5.2.4 In Situ Polymerisation

In situ polymerisation is an approach that forms a bionanocomposite by polymerising monomer effectively inside a framework, which is often a matrix of polymer that is composed of strengthening elements such as nanomaterials or tiny fillers (Pandey et al. 2025). By using this technique, the nanoparticles may be distributed uniformly, improving the composite's qualities. Typically, when it comes to bionanocomposites, the emphasis is on integrating biopolymers with tiny particles like ceramics, graphite, nanocellulose, other renewable resource particles. The monomers are organics compounds that will go through polymerisation. Biopolymers such as carbohydrates, polylactic acid, or polyhydroxyalkanoates are frequently used for bio-nano composites. Nanofillers that are might include natural substance such as fibre small crystals, chitosan particles, or even biologically generated materials or oxides of graphene. The nanofillers are mixed into a solution by dispersing them in an appropriate solvent, usually water or a synthetic solvent. This is important to achieve adequate dispersion and prevent nanoparticles aggregation, which might impact the qualities of the nanocomposites. The monomers undergo a polymerisation process after being added to the nanofillers particles dispersion. It comprises ring-opening polymerisation, ionic polymerisation, and free radical polymerisation. A nanocomposite forms as a result of the polymer growing around or interacting with the nanofillers. Because bionanocomposites are biodegradable and have better

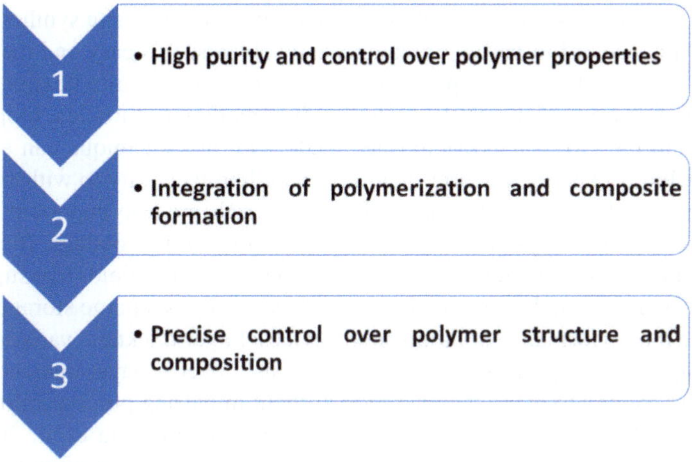

Fig. 5.3 Advantages of in situ polymerisation technique

rigidity, they are frequently employed in both food and healthcare packaging. These composites are being investigated for use in surgical instruments, healing dressings, and delivery systems for drugs, among other medical uses. Additionally, bio-based composites can be employed in the fields of renewable energy, environmental sensing, and water treatment. Figure 5.3 describes different advantages of bionanocomposite materials via in situ polymerisation technique.

5.2.5 Layer-By-Layer (LbL) Assembly

This technique is a flexible and popular method for producing bionanocomposites. It builds multilayer framework with nanoaccuracy by carefully switching a layer of complementary components, usually polyelectrolytes and nanomaterials. This technology is particularly appealing for the production of bionanocomposites because it provides for exact oversight over the dimension, content, and structure of every layer, as well as the integration of a wide range of bio-based components. To make separate solutions or suspensions, the polymeric materials and nanomaterials are dissolved in suitable solvent. Stable charge interactions are important for effective layer formation in the dispersions. The initial substance is placed into the base, which is usually a smooth surface like glass, a wafer of silicon, or materials with flexibility. After that, the substrate is taken out and allowed to dry, which enables the substrate to adhere to the surface. After that, the substrate is dipped into the oppositely charged material's solution. The second layer is formed when the two materials cling to one another via electrostatic forces. After the required number of layers has been built, further treatments like drying, heating, or linking may be used to reinforce the layers, increase stability, or optimise the mechanical characteristics of the materials. It is a sustainable and ecologically friendly technique for nanocomposite synthesis since it allows for the use of bio-based ingredients and solvents. It is easy and inexpensive, making it ideal for large-scale production. The technique is quite flexible and may incorporate a variety of materials, including inorganic nanoparticles, biopolymers, and bio-nanofillers. Bio-nanocomposites made with this technique can strengthen biodegradable packaging's barrier qualities, including its ability to withstand moisture, oxygen, and ultraviolet radiation. These materials are also important for use in the treatment of water systems and as bio-based anti-fouling coatings. It assembled bionanocomposites are being studied for use in energy-related fields like supercapacitors, battery power, and solar panels, where they can boost charge storage capacity or improve the performance of bioinspired devices. Layer thickness variations might have an impact on the performance of the materials. Layer-to-layer interactions need to be robust enough to avoid delamination without impairing processability or flexibility, particularly for applications that demand it (Plamadiala et al. 2025). The final product may need further stabilisation since the biopolymers employed in LbL assembly may be susceptible to environmental conditions like humidity, heat, or UV exposure.

5.2.6 Solution Casting

A common method to fabricate bio-based nanocomposites, especially in the fabrication of environmentally benign and recyclable materials, is solution casting. The initial step is to choose a bio-based polymer. Polymers made from organic material include cellulose, gelatin, collagen, carbohydrate, and poly lactic acid. A homogeneous solution is created by dissolving them in an appropriate solvent. The ideal thickness of the film and mechanical characteristics are attained by optimising the polymer's proportion to the solvent. The mixture of polymers contains nanofillers that are metal nanoparticles, nanotubes of carbon, graphite, nanocellulose, and nanoclays. Surface modification, mechanical stirring, and ultrasonication are method for dispersing these fillers. Once a uniform dispersion has been achieved, the polymer mixture is transferred onto a level surface, like a silicon mould or a plate of glass. After that, the solvent is let to evaporate, and the resultant mixture is evenly distributed. The volatility of the solvent will demonstrate whether this stage is carried out at normal temperature or in a controlled environment like an enclosed vacuum or a lower temperature oven. A thin layer or coating is formed when the nanofillers are incorporated into the polymer matrix during the evaporation process. With regard to the required final characteristics, post-treatment techniques like bridging, thermally annealing, or plasticiser inclusion may be used.

5.2.7 Freeze Drying (Lyophilisation)

Another common method utilised in the fabrication of bionanocomposites is freezing and drying (lyophilisation), particularly when the objective is to retain the framework and characteristics of synthetic polymers while integrating nanofillers. This approach preserves the stability of the nanocomposite structures by eliminating solvents that are from a material by condensing it and then sublimating it under low pressure. As with solutions castings, the initial phase is to choose a renewable resource polymer, like the polylactic acid, cellulose, alginate, or chitosan. A solution is comprised of dissolving a polymeric material in a solvent, such as water or organic solvents. Nanofillers must be dispersed in order to maintain homogeneity and avoid agglomeration. To establish a homogeneous mixture, methods like rapid connectivity stirring or ultrasound treatment are frequently employed. To enhance the interaction of nanofillers with a matrix of polymers, modifications to the surface may be used in some situations. Following a through mixing process, the synthesised solution is put in a container and quickly frozen. Usually, freezing is carried out at temperatures between − 40 and − 80 °C to guarantee that ice crystals form. The freezing procedure is essential since the ice particles will serve as a template for the final bio-nanocomposite's porosity structure. The material is frozen and then put in a vacuum chamber with a drastically lower pressure. The procedure usually takes place in two steps: any remaining moisture is eliminated to make sure the bio-nanocomposite is adequately dry, and the temperature and pressure are changed to allow the ice or solvent to sublimate. Depending on how complex the composite is and how big the material is, this process typically takes 24–48 h. Following lyophilisation, the bionanocomposite can undergo subsequent processing procedures, including annealing, linking, or chemical treatments, to improve its structural durability, rigidity, or practicality. Because the water is removed via sublimation, or freeze drying, it is appropriate for delicate materials that may deteriorate in the presence of gaseous solvents. By increasing the rate of biodegradation, the porous structure created by freeze drying can make the bio-nanocomposite more eco-friendly. The following fields benefit greatly from freeze-dried bio-nanocomposites: reconstruction of bone materials, tissue scaffolding, delivery systems for drugs, wound rehabilitation, consumable coatings to preserve food, and water purification or filtration. This method has the following difficulties, such as being time-consuming, costly, and having nanofiller aggregation. Figure 5.4 shows various properties of bionanocomposite materials synthesised via freeze drying (Lyophilisation) method.

5.2.8 Electrospinning

In order to develop bionanocomposites in the shape of tiny fibres, electrospinning is a frequently used technology that offers a large surface area, adaptability, and morphological customisation. Thin fibres are produced electrostatically from a mix of

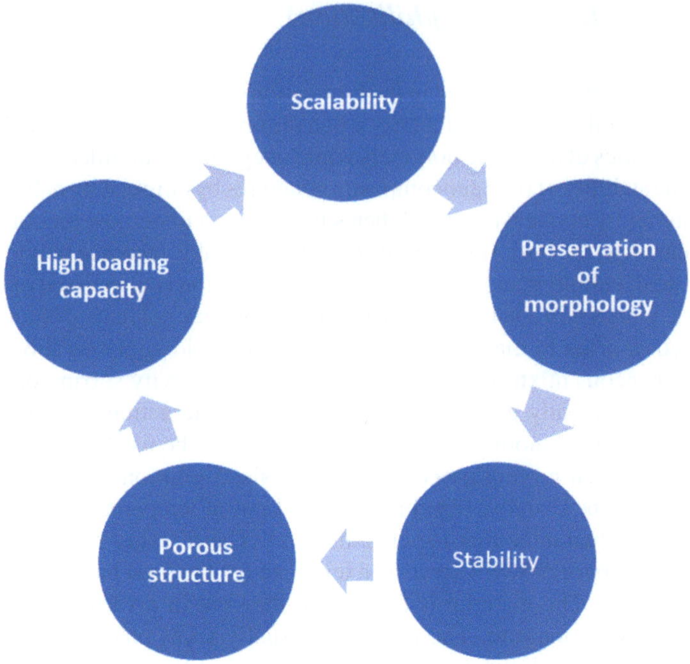

Fig. 5.4 Properties of bionanocomposites synthesised using freeze drying (Lyophilisation) techniques

polymers using this method, which has several uses, including environmental filters, tissue scaffolds, medicinal devices, wound rehabilitation, and systems to deliver drugs. The initial stage involves creating a polymer solution contains nanomaterials and biopolymers. The usual technique for doing this is to add nanofillers after dissolving the selected polymer in a solvent. The polymer solution-filled syringe (or needle) and an immobile collector are connected by a high voltage. The polymeric fibres have been enhanced with therapeutic nanostructures during the electrospinning process. By modifying process variable such as voltage, the spacing among the needle and collector, and the solution's flow velocity, the positioning and size of the nanofibers may be managed. The fibres may go through other processes after electrospinning, such as drying to get rid of leftover solvents or cross-linking to increase mechanical stability and bioactivity. The solvent selections, spacing between the needle and collector, voltages and electric field, and solution viscosity all have an impact on the electrospinning process. Nanofibers comprised of biologic polymers, like gelatin or cotton fibrin, can imitate the matrix of cells, enabling cell attachment and tissue development. It is possible to construct material fibres to release medicinal products in a regulated way while the nanoparticles do other tasks like enhancing biological activities. Electrospun fibres can serve as scaffolds to promote wound healing by delivering regulated pharmaceutical release and structural support. Electrospinning provides for precise control of fibre diameter and structure; Fibres may

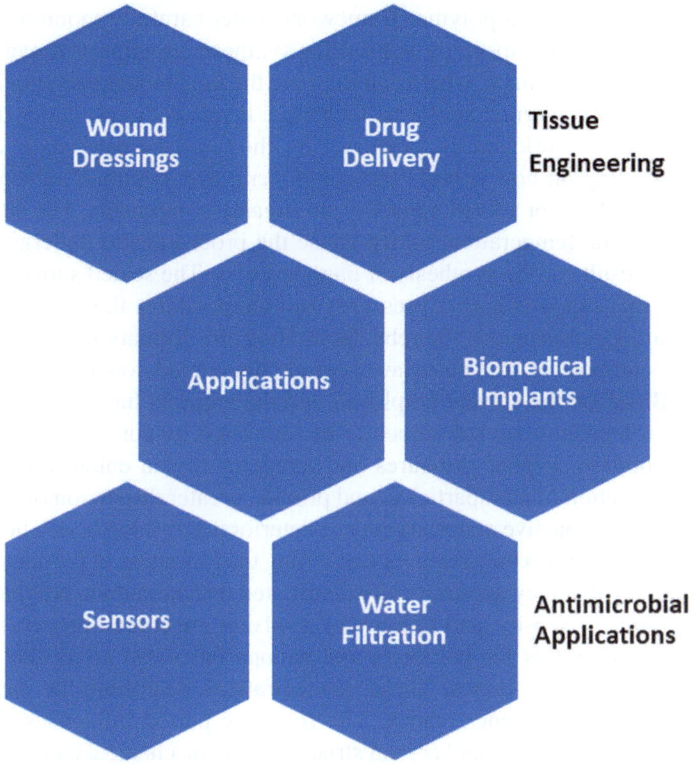

Fig. 5.5 Applications of bionanocomposite synthesised via electrospinning technique

readily include natural nanomaterials or nanoparticles to increase their biological activity and utility (Hamzehlouy et al. 2025). Due to their high surface-to-volume ratio, electrospun nanofibers are advantageous for uses such as medication administration and sensing. While the method is effective in tiny quantities, scaling up this technique for industrial use can be difficult. To eliminate remaining solvents from fibres, careful additional processing is essential, particular whenever collaborating with biopolymers. It might be challenging at times to achieve constant fibre diameter and homogeneous nanofiller dispersion. Figure 5.5 shows different applications of bionanocomposite materials fabricated via electrospinning technique.

5.2.9 Solvothermal Technique

Another well-liked methodology for developing bionanocomposites is the solvothermal method, which is particularly useful for integrating nanoparticles or bioactive substances into a polymer framework under carefully monitored circumstances. This method is especially helpful for synthesising improved nanoparticles with regulated size and shape, which can improve the final bionanocomposite's characteristics. The solvothermal synthesis technique involves dissolving or dispersing the polymeric framework and the precursor for the bio-nanomaterials in a suitable solvent. Depending on how soluble the components are, common solvents can be water-based, alcohols, or solvents made from organic compounds. The autoclave is heated to a specific temperature, which cause the procedures to undergo chemical reactions that results to the synthesis of nanoparticles. The sealed surroundings are crucial since this supports higher pressures and temperatures than can be achieved under standard circumstances, thereby facilitating the formation of nanoparticles and bio-nanocomposites. The autoclave is gradually cooled down once the reaction is finished. Based on the solvent employed and the required final product qualities, the bio-nanocomposites can subsequently be recovered by purification, centrifugation, or evaporation. Higher pressures and temperature can enhance crystallinity, hasten the fabrication of nanoparticles, and provide greater dispersion of nanofillers. However, delicate bioactive materials may be deteriorated by high heat. The solvent is crucial for breaking down the precursors, affecting the formation of the nanoparticles, and guaranteeing that they are uniformly distributed throughout the polymer matrix. The boiling point, viscosity, and polarity of a solvent are crucial process variables. Precursor material ratios for polymers and nanoparticles can affect the final bio-nanocomposite's characteristics. These concentrations determine the nanoparticle dispersion inside the polymer matrix. The response period influences the nanostructures' dimensions, form, and crystal structure. The mechanical characteristics of the resultant nanocomposite may be affected by longer response durations, which often result in bigger or more crystalline particles. The material synthesised using the solvothermal approach finds use in a number of industries, including the food sector, packaging, bactericidal and antifungal materials, pharmaceutical delivery, tissue fabrication, and sensors and diagnostics. Among the many advantages of nanocomposites are their excellent dispersion of nanoparticles, their ability to adjust the size and form of the particles, and their environmental friendliness. Although this approach has many benefits, it has drawbacks, such as high energy consumption, restricted solubility of some biopolymers, and scaling issues.

5.3 Advanced Characterisation Methods for Bionanocomposites

Bionanocomposites are materials composed of biological and nanomaterials, exhibiting unique characteristics due to the interaction of their components. The structural characteristics of these composites has significance for transforming their characteristics to certain particular applications, such as biomedical devices, healthcare equipment, and tissue engineering. Bionanocomposites are characterised using a variety of methods to comprehend their characteristics and functionality. Here are some common methods:

5.3.1 X-Ray Diffraction (XRD) of Bionanocomposites

XRD is a robust and safe technique for determining and describing the crystal structure of the synthesised nanomaterials (Ali et al. 2022). It is essential for interpreting the atom's distribution inside the composite material, which offers information on its phase composition and structure in the field of bionanocomposites. It is an essential analytical method for describing the crystalline structure of bionanocomposites. It operates by shining X-rays on a sample of bionanocomposites.

According to Bragg's Law, these X-rays are diffracted at particular angles when they interact with the material's crystal lattice:

$$n\lambda = 2d\ \sin\theta \tag{5.1}$$

where n is an integer, λ is the wavelength of the X-rays, d is the interplanar spacing, and θ is the angle of incidence. To enhance consistency and boost analytical accuracy, the bionanocomposite is generally ground to a fine powder. The sample is mounted in the XRD devices after being set on a holder. The intensity of the diffracted X-rays is measured as a function of angle, resulting in XRD patterns. The resultant diffraction pattern, sometimes known as the material's "fingerprint", provides information on its phase composition and crystal structure. The crystalline structures in the bionanocomposite were determined by XRD through the comparison of its diffraction pattern with reference data. The Scherrer equation, which uses the enlargement of diffraction peaks, is used to determine the size of crystallites. The crystal system and lattice parameters are ascertained by XRD, which provide information about the structural characteristics of the materials. The diffraction pattern analysis may be used to quantify the relative quantities of various phases in the composites. This method's high sensitivity to crystal structure makes it perfect for comprehensive structural characterisation; it is non-destructive, allowing multiple measurements to be performed on the same sample; and it works with a variety of materials, including bulk samples, films, and powders. When it comes to describing bionanocomposites, it is an essential instrument that offers important insights into their phase composition and crystalline structure. Comprehending these characteristics aids scientists in maximising the effectiveness of bionanocomposites for a range of uses, such as tissue engineering, medication delivery, and biosensing.

5.3.2 Field Emission Scanning Electron Microscopy (Fe-SEM)

It is an effective method for investigating bionanocomposites' morphology. This approach is very beneficial for bionanocomposites, as this enables researchers to see the fine details of the surface at the nanoscale, which is critical for knowing how the structure affects the composite's qualities and performance. It is a robust and versatile method for examining the outermost morphology and microstructural characteristics of bionanocomposites at the nanoscale. A highly concentrated electron beam produced by a field emission source is directed onto the sample surface by the FE-SEM. In order to develop high-resolution pictures of the surface morphology, secondary electrons are formed when the electron beam interacts with the sample. In order to enhance quality of images and avoid charging effects, bionanocomposites samples are frequently coated with a thin metallic layer made of carbon or gold. After being mounted on a holder, the sample is put in the FE-SEM chamber with a high vacuum. The concentrated electron beam creates an image by scanning the sample surface point by point and detecting the secondary electrons that are released. It produces incredibly high-resolution images that make details as little as a few nanometres visible. It is employed to investigate the size, shape, and distribution of nanoparticles in the composite. It makes clear the interface and interaction between the biological and nanomaterial components, which is essential to components, which is essential to comprehending the characteristics and functionality of the composite. The method works well for examining the structure, morphology, and any structural flaws of the surface. Because of its high resolution and magnification capabilities, FE-SEM is perfect for in-depth morphological investigations. The same sample can be analysed more than once due to its non-destructive nature. It is a vital technique for characterising bionanocomposites since it offers comprehensive information on their microstructural characteristics and surface shape. It is essential to comprehend these characteristics in order to maximise the performance of bionanocomposites in a variety of applications.

5.3.3 Transmission Electron Microscopy (TEM) of Bionanocomposites

TEM is a critical tool to investigate the internal structure and morphological features of bionanocomposites at the atomic or nanoscale level. It works by channelling an electron beam through a bionanocomposite material that is very thin. By interacting with the material during their passage through, the electrons create an image through diffraction and scattering pattern. It samples are collected by cutting them into extremely thin sections (less than 100 nm) with procedures such as ultramicrotomy or focused ion beam milling. Following that, the thin pieces are arranged for examination on a grid. After passing through the sample, the electron beam is dispersed by the material's atoms. To form high-resolution image, the scattered and transmitted electrons are measured. It may produce magnification levels of up to several million times, allowing for comprehensive analysis of the nanostructure. It is used to investigate the size, shape, and dispersion of nanoparticles in bionanocomposites. In order to comprehend the characteristics and functionality of the composite, it offers comprehensive information on the interfaces between biological and nanomaterial components. Additionally, it may show atom arrangement, flaws, and structural information. It is perfect for in-depth structural and morphological investigations because of its high resolution and magnification capabilities. Regarding the nanostructure of bio-nanomaterials, it can offer both qualities and quantitative data. It is an essential technique for characterising bionanocomposites because it provides unmatched information on their shape and internal structure. Understanding these properties enables researchers to improve the performance of bio-nanocomposites for a variety of uses.

5.3.4 Fourier Transform Infrared Spectroscopy (FTIR) of Bionanocomposites

An analytical method that is frequently used to determine and describe the molecular structure and chemical makeup of bionanocomposites is FTIR. This technique works by transmitting infrared light through a material. The sample's molecules vibrate the chemical bonds when they absorb particular infrared light frequencies. Specific functional groups and chemical bonds exhibit these absorptions. There are several ways to produce bionanocomposite samples, including thin films, pellets, and powders. Because it requires little sample preparation, it is a rapid and easy approach. Recordings of the sample's infrared spectrum reveal the transmittance or absorbance as a function of wavelength (or wavenumber). The resultant spectrum has peaks that match the sample's chemical bond's vibrational frequencies. It can detect the functional groups in bionanocomposites, offering information on their chemical composition. The approach which analyses variations in absorption peaks and aids in comprehension of the interplay across biological and nanomaterial components. It may be used to find the quantity of certain components in a composite. It helps track chemical changes throughout processes including synthesis, ageing, and degradation. A variety of sample types (solids, liquids, and gases) may be analysed using this non-destructive method that needs no sample preparation and offers quick and accurate detection of chemical bonds and functional groups. The chemical content and structural arrangement of bio-nanocomposites may be accurately determined using FTIR. It gives significant information to researchers about the interactions between different components, which aids in the design and optimisation of bionanocomposites for a variety of applications. Because of its capacity to increase sustainability, food safety, and shelf life, bionanocomposites are gaining traction in the food and packaging industry. The following are the primary uses of bionanocomposites in food and packaging. These are commonly used to create biodegradable and environmentally friendly packaging materials. Biopolymers can be mixed with nanomaterials like cellulose nanocrystals, nano clays, or graphene oxide to form bionanocomposites. Compared to typical biodegradable films, these composites have higher structural durability, adaptability, waterproofing, and barrier qualities, making them appropriate for food packaging applications (Sirbu et al. 2019). Bio-nanocomposites are utilised as carriers for food preservatives and additives that need to be released slowly or consistently. The composite's nanostructures enable the encapsulation of compounds such as antioxidants, antimicrobials, and vitamins, which are slowly released into food to maintain quality and increase shelf life. These are great for encapsulating and gently releasing preservatives like essential oils, resulting in a long-lasting impact that does not change the flavour or fragrance of the food (Muthaiah et al. 2021). Nanoparticles can also contain ingredients to improve the quality of the food while allowing for regulated release over time. These can be designed to improve the shelf life of food goods by increasing moisture barrier qualities, providing UV protection, and avoiding oxidation by mixing nanomaterials such as nano clays, cellulose nanocrystals, and chitosan with biopolymers. These can also be utilised to develop smart or intelligent food packaging solutions. These packaging methods can detect and adapt to changes in the interior or exterior environment of the food. Bionanocomposites can detect gas emissions (e.g. CO_2 or ethylene) and pH changes, suggesting rotting or ripening, and offer visual indications of food freshness (Pandey et al. 2020). The application of antimicrobial compounds in bionanocomposites can improve food safety by limiting the growth of bacteria, fungus, and other diseases (Castillo et al. 2020). Natural antimicrobial compounds, such as aromatic extracts of plants, or biopolymers like chitosan, can be included in nanocomposites to improve food shelf life while ensuring food safety.

5.3.5 Uv–Vis Spectroscopy of Bionanocomposites

Uv–Vis spectroscopy is a flexible technique used to investigate the spectral and electrical characteristics of bionanocomposites. It assesses the ability of the samples to absorb ultraviolet and visible light. When light travels through a sample, some wavelengths are absorbed by electronic transitions in molecules or nanoparticles, leading to the absorption spectrum. Bionanocomposite samples can be accomplished in a variety of ways, including formulations, suspensions, and thin films. For analysis, the material is either put directly onto a substrate or within a cuvette. The Uv–Vis spectrometer scans the sample at several wavelengths (usually 200–800 nm) and measures the absorbance or transmittance at each one. Peaks in the resultant spectrum represent the electronic transition of the composite's constituent parts. Uv–Vis spectroscopy may use distinctive absorption peaks to detect and quantify particular components in bio-nanocomposites. The approach reveals information on the optical characteristics of the nanoparticles such as dimension, form, and surface plasmon resonance. Uv–Vis can detect interactions between biological and nanomaterial components by monitoring changes in the absorption spectrum. It is used to investigate the kinetics of chemical reactions and processes inside composites by detecting variations in absorbance over time. It is non-destructive and requires little sample preparation. It is a quick and sensitive study of a wide variety of samples that yields quantitative and qualitative data. Uv–Vis spectroscopy is a valuable method for determining the optical characteristics and electrical structure of bio-nano composites. It enables researchers to better understand the relationships between different components and enhance the performance of bionanocomposites for a variety of applications. Bionanocomposites are quickly developing as a prominent class of materials with enormous potential in biomedical fields. To capitalise on their unique properties, these materials are often made up of biopolymers combined with nanoparticles. Nanomaterials have unique surface qualities and a large surface area, which improves medication loading and release profiles, allowing for tailored drug administration. Bionanocomposites, particularly those composed of chitosan and gold nanoparticles, are employed in cancer and antibacterial therapy (Vijayakumar et al. 2020). These are employed in applications related to wound healing where biopolymers such as alginate or chitosan are combined with antimicrobial nanoparticles such as silver or zinc oxide to promote healing while avoiding infections (Duarte et al. 2020). When collagen like material mixed with hydroxyapatite or CNTs can act as scaffolds for tissue regeneration (Thakur et al. 2018). These scaffolds replicate the extracellular matrix, which promotes cell proliferation and differentiation, notably in bone and cartilage repair. Bionanocomposites containing nanoparticles are employed in imaging modalities including MRI, CT, and ultrasound to increase contrast and sensitivity in illness detection, including cancer. Bionanocomposites incorporated into biosensors enable the identification of biomolecules linked to illnesses such as cancer or diabetes. Nanomaterials have a large surface area, resulting in great sensitivity and selectivity (Shao et al. 2019). Silver nanoparticles incorporated bionanocomposites have high antibacterial activity and are utilised to make coatings for surgical instruments, wound dressings, and implants, lowering the risk of infection. Some bionanocomposites exhibit antiviral activity, making them interesting candidate for antiviral drug administration and medical devices that inhibit the spread of viral diseases. Bone tissue engineering uses bio-nanocomposites that combines biopolymers such as collagen and hydroxyapatite (Wong et al. 2020). These materials are mechanically strong and bioactive, allowing for bone repair and implant integration into the surrounding bone tissue. Bio-nanocomposites are utilised to deliver chemotherapy or immune-modulating medicines to cancer cells. Gold and silica nanoparticles linked with biopolymers enable regulated release and targeting of cancer cells, boosting treatment efficacy while minimising negative effects (Sahoo et al. 2019). Bio-nanocomposite coatings, such as chitosan and hydroxyapatite nanoparticles, are utilised to improve the biocompatibility and functioning of implants. These coatings improve cell adherence and lower the likelihood of implant rejection. Bio-nanocomposites can promote stem cell proliferation and differentiation into particular cell types. These composite nanostructures can help stem cells adhere and differentiate more effectively.

5.3.6 Brunauer–Emmett–Teller (BET) Analysis of Bionanocomposites

The BET analysis is a key approach for calculating the surface area and porosity of materials, including bionanocomposites. Gas molecules physically adhering to the surface of the material is the foundation of BET analysis. The BET theory may compute the specific surface area and pore size distribution of the material by measuring the quantity of gas adsorbed at various relative pressures. Bionanocomposite samples are usually degassed under a vacuum or in an inert environment to eliminate any adsorbed pollutants or moisture. A gas, often nitrogen, is then introduced to the prepared sample at a regulated temperature. The quantity of gas adsorbed on the sample surface is evaluated at different relative pressures. The BET equation is used to assess adsorption data, yielding information on surface area and pore structure. It gives precise measurements of bionanocomposite's specific area, which is critical for recognising their reactivity and interactions with other chemicals. The approach gives information on pore size distribution and total pore volume, which is useful for adsorption, catalysis, and drug delivery applications. It aids in the design and development of bionanocomposites for a variety of applications by characterising their surface characteristics. It gives extensive information on surface area and porosity. It is suitable for a broad variety of materials, including powders, films, and porous structures. It is a non-destructive, fast analysis. BET analysis is an important method for determining the surface area and porosity of bionanocomposites. Understanding these features enables researchers to enhance the performance of bionanocomposites in applications. Nanotechnology and renewable resources have been used to provide revolutionary solutions for preserving energy, transformation, and electronics. Bionanocomposites are utilised to create sophisticated energy storage devices like batteries and super capacitors. The addition of nanomaterials such as CNTs, graphene, and nanocellulose to biopolymers can enhance their electrical conductivity, mechanical strength, and energy storage capacity (Zhang et al. 2020a). Bionanomaterials are being studied for energy harvesting applications, notably in triboelectric nanogenerators, which transform mechanical into electrical energy. Natural polymers mixed with nanomaterials such as nanocellulose, graphene oxide, and CNTs can offer the necessary mechanical flexibility, charge density, and environmental stability (He et al. 2020). Bionanocomposites are being studied for usage in photovoltaic devices, namely organic solar cells and perovskite solar cells (Liang et al. 2020). These materials have the potential to make solar technology more flexible, lightweight, and ecologically sound. The biopolymers in these composites can act as electron acceptors, while the nanomaterials enable effective charge transfer. Bionanocomposites are increasingly employed in flexible and wearable electronics, which require lightweight, elastic, and ecologically friendly materials. The addition of nanocellulose, graphene oxide, or CNTs to biopolymer matrices such as polylactic acid (PLA) or polylactic-co-glycollic acid (PLGA) improves the mechanical characteristics, conductivity, and durability of the materials (Xu et al. 2020). Bionanocomposites are gaining popularity as electromagnetic interference (EMI) shielding materials in electronic equipment. Conductive nanomaterials are incorporated into biopolymers to create the composites. These materials have high conductivity and mechanical flexibility, making them ideal for lightweight, eco-friendly EMI shielding. Bionanocomposites are also being studied in the field of bioelectronics, where materials must be biocompatible and able to interact with biological systems. Nanomaterials combined with polymers such as chitosan or gelatin can be used to fabricate devices for biological sensing devices, bioelectrodes, and tissue formation (Wang et al. 2021).

5.3.7 Thermal Analysis (TGA) of Bionanocomposites

Thermal analysis refers to the methodologies used to determine the thermal characteristics and behaviour of materials as a function of temperature. TGA examines the change in mass of a substance as it is heated, cooled, or kept at a constant temperature. The examination reveals details about the thermal stability and composition of the materials. A nanomaterial of the bio-nanocomposite is put in a TGA crucible. The sample is submitted to a temperature-controlled programme in either an inert or reactive environment. The mass of the sample is continually measured as the temperature changes. The resultant TGA curve displays mass loss versus temperature or time, revealing thermal processes including breakdown, oxidation, or volatilisation. TGA determines the breakdown temperatures of bio-nano composites, hence assessing thermal stability. The approach aids in composition determination by monitoring the mass loss of various components. It measures moisture and volatile content by measuring mass loss across certain temperature ranges. Fadeel et al. studied the properties of graphene and base materials on the human body and their environment and reported their synthesis, characterisation, and applications in various areas of society (Fadeel et al. 2018).

5.3.8 Differential Scanning Calorimetry (DSC) Analysis of Bionanocomposites

DSC monitors heat movement into or out of a substance while it is heated, cooled, or kept at a constant temperature. The research reveals information on phase transitions such as melting, crystallisation, and glass transitions. A small amount of bionanocomposite sample is put in the DSC pan. The sample is treated to a temperature protocol, usually in an inert environment. The DSC device measures the amount of heat flow necessary to keep the sample at the same temperature as the reference. The resultant DSC curve shows heat flow vs. temperature or time, demonstrating endothermic and exothermic activities. DSC detects phase transitions, such as melting and crystallisation, and calculates the accompanying temperatures. The approach determines the glass transition temperature, which is critical for understanding the material's mechanical characteristics. DSC gives thermal stability insights by evaluating changes in heat capacity and enthalpy. Thermal analysis methods such as TGA and DSC are critical for determining the thermal characteristics and behaviour of bio-nano composites. These approaches give useful information regarding thermal stability, composition, and phase transitions, which can help with the design and optimisation of bio-nano composites for a variety of applications. Bio-nano composites have recently gained popularity in architectural and aeronautical sciences due to their distinctive mix of high durability, lightweight design, biological compatibility, and durability. Such characteristics are extremely useful in sectors that need great performance and weight reduction, such as structural engineering and aeronautical engineering. In aircraft engineering, there is a growing requirement for lightweight, high-strength materials. Bio-nano composites such as those produced from natural fibres (e.g. flax, jute) and nanoparticles (e.g. CNTs or graphene) are being studied for application in load-bearing components of aeroplanes and spacecraft. These composites also offer improved resistance to cracking and environmental degradation (Zhou et al. 2021). Effective thermal insulation is required in the aerospace sector, particularly for spacecraft and satellites, to defend against high-temperature variations. Thermal insulation may be achieved by using biological composites, including aerosols, nanotubes, and polymers from natural sources. The inclusion of nanoparticles improves thermal endurance and insulating capabilities while keeping weight low; which is critical for aeronautical applications. Bio-nano composites can be used in aeronautical and structural engineering applications that need absorbance of energy and vibration dampening. These materials have applications in aerospace systems, including fuselages, landing gear, and vibration-damping panels. When nanoparticles or natural fibres such as jute, hemp, or flax are mixed into matrices, they can provide excellent damping characteristics, eliminating the requirements for extra vibration-reducing components (Khan et al. 2021). Fire safety is a critical issue in structural and aeronautical engineering. Bio-nano composites may be engineered to have improved flame-retardant qualities, making them appropriate for applications requiring fire resistance. Fireproof nanomaterials are frequently included in renewable polymer matrices, resulting in minimal combustibility (Wang et al. 2019). These systems may be integrated into aviation or government buildings that track strain, heat, and other crucial parameters in real time, assuring structural integrity and avoiding breakdowns (Pan et al. 2021).

5.3.9 Raman Spectroscopy of Bionanocomposites

An effective analytical method for examining the rotational, vibrational, and other low-frequency modes of molecules in bionanocomposites is Raman spectroscopy. Raman spectroscopy works by lighting the sample with a single monochromatic light source, often a laser. The majority of the light that interacts with the molecules in the sample is elastically dispersed. However, a tiny percentage of the light is inelastically dispersed, causing a wavelength shift. This shift in wavelength reveals information about the vibrational modes of the molecules, which may be utilised to know the chemical compositions and molecular structures. Bionanocomposites samples can be tested in various forms, including powders, films, and liquids. It is a somewhat rapid and simple method since it requires little sample preparations. The scattered light is collected and dispersed by the Raman spectrometer's diffraction grating. The resultant spectrum reveals peaks that correspond to the vibrational frequencies of the chemical bonds in the sample. Raman spectroscopy can pinpoint particular functional groups and molecular structures in bionanocomposites. The method offers comprehensive details on how biological and nanomaterial components interact. Raman spectroscopy is used to regulate quality and monitor changes in bio-nano composite manufacturing, processing, and storage. Raman mapping and imaging methods may show the distribution of distinct components inside a composite, giving spatially resolved chemical data. It is non-destructive and requires little sample preparation. It can analyse both organic and inorganic components in a wide variety of samples quickly and sensitively. It is suitable for both on-site and real-time monitoring. Raman spectroscopy is a key method for determining the chemical content and molecular structure of bio-nano composites. It gives important insights into the interactions between different components, assisting in the design and optimisation of bionanocomposites for a variety of applications. Nguyen et al. (2019) demonstrated the application of nanoparticles with magnetic properties modified with microbes for the successful removal of arsenic from polluted water (Nguyen et al. 2019). Oil spills are a huge environmental hazard since oil remains in both marine and terrestrial settings. To solve this issue, bio-nano composites were created by mixing nanomaterials, including carbon nanotubes, silica nanoparticles, and activated carbon, with hydrophobic biological components. These materials efficiently absorb and decompose oil, lowering the environmental impact of spills. Zhao et al. (2020) studied the application of hydrophobic cellulose nanocrystals coupled with microorganisms to remove crude oil (Zhao et al. 2020). The composite displayed good oil absorption capability, with microbes helping in oil biodegradation, resulting in a long-term and practical solution for oil spill clean-up. Bio-nano composites are also used to degrade organic contaminants present in industrial effluent, including pesticides, dyes, and medicines. Nanomaterials can be used to transport enzymes or microorganisms that degrade these contaminants. For example, composites having enzymes that break down aromatic compounds, such as laccases, have been utilised to remove dangerous organic contaminants from wastewater. Zhang et al. (2022) investigated the usage of a composite material made of titanium dioxide nanoparticles and laccase enzymes for the breakdown of the insecticide malathion (Zhang et al. 2022). The combination shows increased catalytic activity when compared to free enzymes, indicating a viable method for removing pesticide residues from water. Bio-nano composites can additionally be used to remediate polluted soil. These composites may be made up of nanoparticles that function as hazardous chemical adsorbents, as well as plant or microbe substances that help in bioremediation. For example, biochar-based nanocomposites including CNT or graphene have been used to remove organic and inorganic contaminants from soils. The nanocomposite revealed an improved capacity to absorb metallic substances, while the biochar component promoted plant growth, able to further detoxify the soil. Chen et al. (2020) investigated the use of biochar-based nanocomposites to remediate soil polluted with heavy metals such as copper and zinc (Chen et al. 2020).

5.3.10 Mechanical Characterisation of Bionanocomposites

Mechanical characterisation of bionanocomposites is crucial to understand their mechanical properties and performance. Various techniques are employed to assess the mechanical behaviour of these advanced materials. Here is an overview:

(a) **Tensile Testing**

  (i) **Purpose**: Determines the tensile strength, Young's modulus, and elongation at break.
  (ii) **How It Works**: A sample is subjected to a controlled tensile load until it fractures. The stress–strain curve is recorded to determine mechanical properties.
  (iii) **Applications**: Evaluate the composite's ability to withstand stretching forces and its ductility.

(b) **Compression Testing**

  (i) **Purpose**: Calculate the compressive strength and modulus.
  (ii) **How It Works**: A sample is compressed until deformation or failure occurs. The stress–strain relationship is analysed.
  (iii) **Applications**: Assesses the composite's behaviour under compressive loads, crucial for structural applications.

Ic. **Flexural Testing**

  (i) **Purpose**: Determines the flexural strength and modulus.
  (ii) **How It Works**: A sample is subjected to a bending load, and the force required to bend the sample is recorded.
  (iii) **Applications**: Evaluates the composite's resistance to bending and its stiffness.

(d) **Dynamic Mechanical Analysis (DMA)**

  (i) **Purpose**: Finds the viscoelastic properties as a function of temperature, frequency, and time.
  (ii) **How It Works**: A sample is subjected to oscillatory forces, and the storage modulus, loss modulus, and damping factor are measured.

(iii) **Applications**: Provides insights into the composite's mechanical behaviour over a range of conditions, useful for predicting performance in real-world applications.

I. **Nanoindentation**

(i) **Purpose**: Calculate hardness, elastic modulus, and other mechanical properties at the nanoscale.
(ii) **How It Works**: A sharp indenter is pressed into the sample surface, and the load–displacement response is recorded.
(iii) **Applications**: Evaluates the mechanical properties of nanoscale features within the bio-nano composite, such as individual nanoparticles or fibres.

(f) **Fatigue Testing**

(i) **Purpose**: Assesses the composite's resistance to cyclic loading and its fatigue life.
(ii) **How It Works**: A sample is subjected to repeated loading and unloading cycles until failure occurs.
(iii) **Applications**: Important for predicting the long-term durability of the composite under repeated stress.

The mechanical characterisation of bio-nano composites reveals significant facts regarding their resilience, rigidity, durability, and stability. Understanding these qualities becomes essential for enhancing the functioning of bio-nano composites in an array of applications.

5.4 Properties of Bionanocomposites

Figure 5.6 describes various properties of bionanocomposite materials.

5.4.1 Optical and Electrical Properties

The electric and optical characteristics of bio-nano composites are highly tailored due to the combined effect of biocomponents and nanoparticles. Due to various characteristics, these materials are used in various fields like sensors, electrical devices, environmental sensors, the medical industry, and sustainable energy. The potential for tailoring these characteristics to fulfil particular requirements renders bionanocomposites an interesting area for upcoming research and innovation. Several aspects, like the kind of nanomaterials employed, the biological component's content, and the synthesis methods, affect the optical characteristics of bionanocomposites. Both the reflection and absorption spectra of these materials can be tailored by changing the amount of concentration, quantity, and structure. Giacomo et al. reported a technique

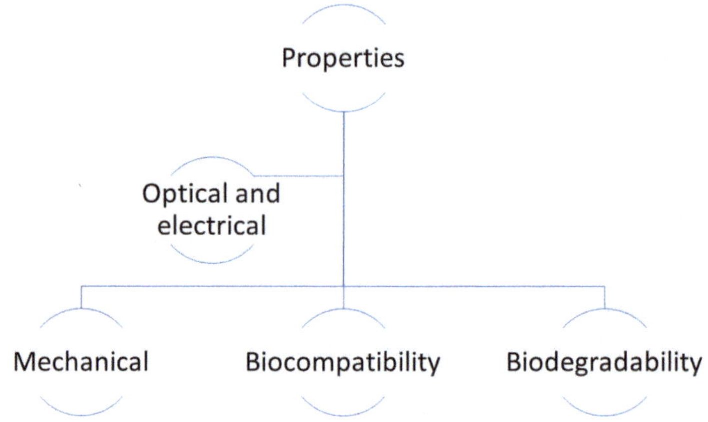

Fig. 5.6 Various properties of bionanocomposites

to fabricate a new synthetic sample that uses pure tobacco plants or fungus combined with varying amounts of carbon nanomaterials. They fabricated nanomaterials with linear electrical features, extreme temperatures sustainability up to 180 °C, thermal conductivity that increased linear nature with the environment, and, in a specific case, ocular transparency (Giacomo et al. 2013). Since cellulose is an inherently pure biopolymer, most bio-nano composites retain great optical clarity. Varying the kind and quantity of nanomaterials added to the composites will change their optical transparency. Additionally, the optical characteristics can be modified for certain uses, like paints or smart windows. Several composites have complex optical characteristics because nanomaterials have optical limiting or second-harmonic fabrication. They are consequently helpful for computing data, optical switching, and shielding. Images, medicine shipping, and medical equipment all utilise optical features such as fluorescence and luminescence. Zhang et al. studied the effect of exposure to nanoparticles on human cells and joints and reported that it majorly influences human organs like lungs, joints, and immune cells (Zhang et al. 2019). Bionanocomposites are employed in the detection of chemical substances or dangerous compounds because of their improved absorption and scattering. Applications using renewable sources of energy can benefit from these composites improved light absorption qualities. In optoelectronics components, smart coatings with transparency in light and adjustable refractive index can be employed. Particle's form and amount within the matrix of biopolymers impact the conductive properties of bio-nanocomposites. Figure 5.7 describes different factor affecting the optical and electrical properties of bio-nano composite materials.

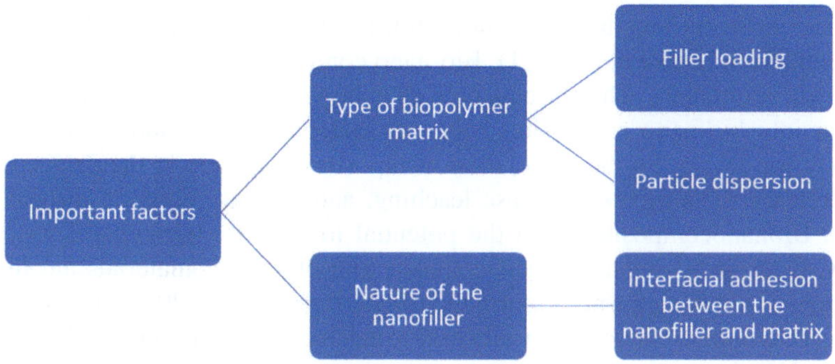

Fig. 5.7 Factor affecting the optical and electrical properties of bionanocomposites

5.4.2 Mechanical Properties

Bionano materials have an intriguing range of mechanical features because of the combined effects of their inorganic as well as organic. These composites are often intended to have exceptional qualities such as greater rigidity, firmness, hardness, and robustness, while being simultaneously more portable and biodegradable than conventional resources. These composite materials frequently have higher tensile force than pure nanoparticles alone. The addition of organic compounds, such as plant-based compounds or peptides, to nanoparticles, including CNTs, graphite, or nanomaterials, increase load-carrying ability. Using nanofillers in order can significantly improve composite rigidity. This is especially beneficial in applications requiring stiffness, such as structural materials. These materials tend to be less susceptible to breakage and cracking than their various parts because of the strengthening impacts produces by the integration of the organic and nanotube phase. Nanofillers can assist in distributing stress more uniformly throughout the material, which results in boosting its capacity for absorbing energy and preventing fracture development. Some materials are exceptionally flexible, renders their use in flexible electrical equipment, healthcare products, and bioinspired polymers. It happens precisely that the biological part may offer the composite additional ductility, which complements the stiff nanotechnology. These can exhibit enhanced durability against fatigue, or the capacity to tolerate multiple loads without breakdown. The interaction of nanomaterials with biological components helps to minimise fatigue fracture initiation and propagation, hence increasing the lifetime of the materials. These have high impact strength, which is important for absorbing energy during abrupt contraction. This is particularly useful in lighter barrier materials such as bio-based packaging, sport equipment, and armour. Nanomaterials used in composite materials, such as the nanocellulose or chitosan nanoparticles, increase the durability, water resistance, and barrier quality of the packaging material (Shah et al. 2020). Bio-nano composites can be employed as carriers for chemicals or biological control substances, providing controlled release and increasing pest control efficacy. Nanocellulose or clay-based compounds can encapsulate natural pesticides, allowing them to be released over time, increasing the length of protection while lowering the demand for frequent reapplication (Sienkiewicz et al. 2021). Bio-nano composites can function as fertilisers or growth stimulants. The introduction of nanoparticles such as nano zeolites or biochar into bio-nano composites enhances water retention, nutrient release, and soil aeration levels. Furthermore, these composites may help in the delayed monitored release of nutrients, decrease leaching, and increase nutrient absorption by plants. Bionanocomposites have the potential to increase agricultural water efficiency. Hydrogels containing nanocellulose and other nanomaterials aid in water retention, minimising the need for frequent watering (Rai et al. 2019). These composites can be employed in irrigation networks to enhance the retention of soil moisture in dry and drought-prone areas. Bionanocomposites are utilised to create smart sensors that can monitor a variety of environmental parameters such as pH, moistures, temperature, and nutrient content. These sensors can supply farmers with real-time data, enabling precision agriculture (Ghosh et al. 2020). Nanomaterials have a large surface area and conductivity, which allows for higher sensing performance. Bionanocomposites can also help to clean-up agricultural soil that has been polluted with heavy metals, pesticides, and other contaminants. For example, nanocellulose composites can absorb heavy metals, organic pollutants, and other toxins, cleansing the soil and making it suitable for plant development. The advancement of bio-nano composites-based nutrients allows for more effective nutrient delivery to plants. These nanofertilisers may release nutrients in a regulated manner, ensuring that crops have a consistent supply of critical components such as phosphate, nitrogen, and potassium. This leads to higher agricultural yields and a lower environmental effect owing to fertiliser runoff (Zhang et al. 2020b).

5.4.3 Biocompatibility

Biocompatibility is an important factor that must be considered when selecting a substance for bio or health-related uses. It refers to the capacity of the materials to coexist with organisms without triggering unfavourable response, like inflammatory disorders, poisoning, or immunological rejections. The ability of bio-nano composites is influenced by a number of aspects, particularly the nature of the original material that is biological, the kind of particles, what happens involving both steps, and how the overall structure responds in the context of life. The biocompatibility of the composite is largely influenced by the harmful effects of the nanomaterials used. Numerous nano samples like CNTs, graphite, and tiny particles, can be poisonous if not functionalised or if they leak dangerous compounds. These nanoparticles, like quarts nanomaterials or hydroxyapatite, are frequently chosen for healthcare applications. These are frequently modified with biologically relevant molecules that involve protein fragments, or plastics to increase biocompatibility and minimise cellular toxicity. This increases the compatibility of the materials with proteins and cells. The human immune system and nanoparticles can interact in various ways. Numerous nanoparticles could trigger autoimmune reactions, resulting in long-term consequences such as persistent discomfort (Xuan et al. 2023). The amount of immunological stimulation can be influenced by the substance used and the surface qualities. Biodegradable materials are often intended to reduce such reactions. The combination must promote cell proliferation, adherence, and persistence without endangering or killing cells. The materials ought to feature interfaces that promote cell adhesion, migration, and growth. Interface alterations can improve tissue interactions. The MTT test and Live/Dead staining are two examples of cell culture methods that are used to evaluate lethality and growth. The combination cannot release poisonous compounds or cause the body to experience toxic consequences. There should not be any hazardous leachates or breakdown products in the substance that might damage tissues or cells. Qamar et al. studied the toxic effect of nanomaterials on the human body and reported that properties like shape, size, and composition impact the toxicity of the nanomaterials (Qamar et al. 2024).

5.4.4 Biodegradability

The tendency of bionanocomposites to degrade gradually into lesser, harmless elements that cells may properly ingest, process, or eliminate is known as a biodegradable feature. This is crucial for clinical uses since components could have to break down in interaction with biological reactions, such as tissue design, medication delivery, and devices that are embedded. The rate of degradation of these composites ought to correspond with their biological use. The combination of materials should break down at a rate that permits the formation of new tissue and its integration with the material, e.g. in the engineering of tissues. Continuous implantation may benefit from slow breakdown, whereas partial frameworks or healing wounds may need quick deterioration. During the breakdown of these composite materials, their residuals are required to be harmless and are able to be adequately processed or eliminated by the organs. Many materials are produced using recyclable polymers or raw substances that decompose into innocuous chemicals such as CO_2 and H_2O. Water causes some composites to disintegrate. When polymers accumulate water, they break away, causing material deterioration. Organic polymers may deteriorate by enzyme metabolism, in which organisms in the body split down the substance into fragments that are smaller. External influences such as pH, temperature, and oxidation can sometimes accelerate deterioration. To forecast whether a substance will react in the body, in vitro studies of degradation are carried out in a controlled environment. In such testing, it entails inserting the material into living organisms to monitor its deterioration and the body's reaction to the by-products. It is difficult to make sure that breakdown products are not poisonous and do not cause inflammatory reactions. The differing degradation processes of nanoparticles and biological substrate might make it difficult to achieve a constant deterioration level, particularly when mixing both. Bio-nano composites are rapidly becoming essential in the cosmetics sector because of their biological origins, long-term viability, and improved functional qualities. These qualities are extremely beneficial for cosmetic formulation such as skin care, anti-ageing, conditioners, and sunscreens. Bionanocomposites can be employed in anti-ageing products to increase collagen synthesis, enhance skin suppleness, and reduce wrinkles. The incorporation of microscopic cellulose with bioactive chemicals can improve the transport of these compounds to the skin, resulting in increased effectiveness (Krishnan et al. 2020). These compounds are biologically active, encouraging cell regeneration and speeding up the rehabilitation process while also providing antibacterial protection against contamination (Xiao et al. 2020). Sunscreen products often contain nanoparticles like ZnO and TiO_2, which absorb UV light. These composites can enhance the stability and dispersion of the nanomaterials while offering further skin advantages when mixed with biopolymers like cellulose or chitosan (Hossain et al. 2019). These are frequently used in conditioners and hydration products to promote moisture retention and barrier function in the skin. Hydrogels that are contaminating cellulose particles, chitosan, and alginate have been developed to better transfer moisture to the skin. These substances can also enhance the entry of active compounds into the skin and stabilise them (Santos et al. 2020). These eco-friendly materials can reduce the environmental effect of cosmetic packaging while retaining strength and durability. Bio-nano composites can be implemented in hair care products such as conditioners, shampoos, and are made to strengthen and grow hair. Natural polymers, including keratin, chitosan, and masks strengthen and grow hair. Natural polymers, including keratin, chitosan, and cellulose, when coupled with nanoparticles, can assist in strengthening hair fibres, prevent breakage, and improve active ingredient penetration (Rossi et al. 2020). Bio-nanocomposites have emerged as a potential waste management option, enabling practical and ecologically acceptable approaches for dealing with a wide range of waste kinds, comprising plastic, heavy metals, organic contaminants, and even electronic trash. These composites have improved qualities such as large surface area, reactivity, and the capacity to interact with a range of contaminants by mixing natural, biodegradable materials and complex nanomaterials. Conventional plastics are a huge environmental problem because they end up in landfills and seas. Bio-nano composites can be developed as biodegradable polymers to replace petroleum-based products. These composites often contain renewable polymers (such as starch, chitosan, or cellulose) in addition to bioparticles, making them simpler to disintegrate in the environment (Ghaedi et al. 2020). Bio-nano composites, particularly those including natural materials such as chitosan, cellulose, or activated carbon, as well as metal oxide nanoparticles (e.g. iron oxide, zinc oxide), are highly effective at adsorbing heavy metals (such as lead, cadmium, mercury, and arsenic) from wastewater (Cai et al. 2020). Bio-nano composites are capable of removing g oil from bodies of water. These materials are frequently made up of natural fibres mixed with nanoparticles that improve their ability to absorb oil. The resultant materials are lightweight, very porous, and can absorb enormous volumes of oil in relation to their weight. Bio-nano composites were developed to help synthetic polymers biodegrade more effectively. Certain composites, for examples, contain fungi or bacterial species integrated into biodegradable nanomaterials, which speed the degradation of polymer into simpler, non-toxic molecules (Rajendran et al. 2020). Bio-nano composites are commonly utilised to remove organic contaminants from wastewater, including dyes, insecticides, and medicines. Activated carbon, chitosan, and biochar, when coupled with nanostructures, have improved adsorption properties, increasing the efficiency with which organic pollutants are removed. Bio-nano composites are also being investigated for their ability to recover precious metals such as gold, copper, and silver from e-waste. These materials can contain biopolymers and nanoparticles that attach to valuable metals in electronic trash, making them easier to collect and recycle (Yu et al. 2020). Bio-nano composites have been used to transform soil, namely to remove heavy metals, organic contaminants (like pesticide), and petroleum hydrocarbons. Natural adsorbents such as charcoal, clay, and cellulose, mixed with nanoparticles, aid in the effective removal and immobilising of contaminants from polluted soils. It is possible to employ bio-nano composites to absorb and store carbon dioxide (CO_2) from emissions from factories. Combining polymers made from biological material (e.g. chitosan, cellulose) with nanomaterials such as carbon particles, graphene, or carbon nanotubes can enhance CO_2 adsorption efficiency and reduce greenhouse gas emissions (Yu et al. 2020).

5.5 Challenges and Future Scope

Bionanocomposites are on track to be a key component of the next innovations in technology, providing imaginative solutions in the fields of material sciences, health care, and ecological fields. The potential uses in novel materials, ecological responsibility, and pharmaceuticals offer a promising future, despite ongoing difficulties with scaling, ingredient reliability, and regulatory barriers. Advances in nanotechnology and more studies into overcoming these obstacles could result in the development of more intelligent and sustainable bionanocomposite materials, which have the potential to completely transform industries in the years to come. One limiting element may be the price and accessibility of natural biopolymers and nanomaterials. Although a large number of biopolymers come from renewable resources, it is still difficult to guarantee a steady, affordable supply of these materials. Exceptional enhanced nanostructures fabricated from biocompatible components may also be costly and challenging to scale. Optimising the characteristics of the composite requires that the nanomaterials be evenly distributed and dispersed throughout the polymer matrix. However, some biological polymers are not compatible with specific kinds of nanofillers, which might result in weak interfacial bonding, aggregation, or phase separation, eventually impairing the structural and natural performance of the material. To increase compatibility, nanomaterials may need to be surface modified or functionalised, which might complicate the production procedure. Several approaches to fabricating bio-nanocomposites, such as electrospinning, solvothermal methods, and solution casting, work effectively in small- to medium-sized laboratory experiments but fail when scaled for usage in commercial settings. Industrial-scale bionanocomposites must be economically feasible, but scaling up while preserving constant quality, efficacy, and cost-effectiveness is still a significant challenge. It is important to carefully evaluate the environmental effect of bionanocomposites, despite the fact that they are frequently marketed as being more sustainable than conventional synthetic composites. This involves taking into account the composites' biodegradability, the raw materials' life cycle, and waste management. Furthermore, it is essential to guarantee the safety of nanoparticles, particularly when they are employed in biomedical applications. While bionanocomposites can have ideal structural and thermal characteristics, some biopolymers are fundamentally brittle, limiting their practicality to sectors that need excellent durability or impact resistance. The effectiveness of bio-nano composites must also be improved by enabling ongoing stability and strength, especially in the presence of extreme environmental factors, including humidity, UV rays, and high degrees. Regulations have not kept up with the quick development of bio-nano composites, particularly in food and medicinal applications. The application of bio-nano composites in delicate fields, including environmental restoration, and the packaging of food, and medicine, requires regulated testing procedures, quality control methods, and explicit regulatory standards. The long-term consequences of nanoparticles inside biological systems have not been well studied, despite the fact that many bionanocomposites are regarded as safe. A comprehensive understanding of the toxic effects, antibodies, and other possible negative consequences of nanoparticles integrated into bio-nanocomposites is necessary before broad clinical usage in medical applications. Bio-nanocomposites are becoming increasingly popular in medication transportation. Adding nanoparticles to biopolymers such as hydroxyapatite and metallic nanoparticles can allow for regulated, selected, and prolonged drug release, providing novel approaches to tissue regeneration, wound healing, and cancer treatment. Bionanocomposites having antiviral characteristics, generated by natural polymers such as chitosan and coupled with nanomaterials, show considerable potential for enhanced wound care, enabling quicker healing while avoiding infections. Improvements in the functionalisation and tailoring of 3D-printed bionanocomposites, including nanoparticles, will facilitate tissue healing and organ regeneration. The need for biodegradable and ecological packaging materials is expanding, particularly in the food industry. Bionanocomposites, which are fabricated from renewable materials such as cellulose, starch, or proteins and particles such as ceramic or cellulose nanocrystals, have good mechanical characteristics and are biodegradable, replacing conventional plastics. Heavy metals, organic pollutants, and toxic pollutants may all be eliminated from water by creating bionanocomposites. Nanoparticles incorporated in biodegradable polymers, such as clay, magnetic compounds, or activated carbon, may provide effective and long-lasting water filtration solutions. Beyond only providing reinforcement, bionanocomposites may be designed to serve a variety of purposes. For instance, sensing devices, gadgets, and wearable health surveillance frameworks require materials that are both biodegradable and electrically conductive. The incorporation of bio-nanocomposites into devices that store energy (such as power sources or capacitors) and solar panels is a fascinating discipline of future research. Nanotechnologies embedded in bio-based matrices, such as carbon-based nanoparticles or metal oxide nanoparticles, might improve the effectiveness and environmental responsibility of alternative energy sources. Materials with healing abilities are of significant importance for uses where long-term performance and reliability are critical. By modifying to the patient's unique biological circumstances, they might increase treatment effectiveness and reduce adverse effects. It will be crucial to develop standardised protocols for production, investigation, and monitoring as the area of bionanocomposites expands. Industrial accreditation and approval mechanisms for bio-nano composites, particularly for health care and food usage, can speed up their commercialisation and boost customer trust.

5.6 Conclusion

This chapter majorly focussed on various properties such as optical and electrical which makes bionanocomposites a suitable candidate for various applications like biomedical, textiles, food industry, energy storage, and agriculture sectors. These materials can be synthesised using various techniques like hydrothermal, sol–gel, melt bending, in situ polymerisation, layer-by-layer assembly, solution casting, freeze drying, electrospinning and solvothermal. These different synthesis techniques increase the applicability of bio-nano composites in various sectors of society. This chapter primarily described advanced characterisation methods like XRD, Fe-SEM, FTIR, Uv–Vis, Raman, TEM, BET, TGA, DSC and other mechanical characterisations. Future developments in bionanocomposites are expected to lead to ever more efficient, eco-friendly, and seamlessly integrated devices. The objective of the current section is to briefly investigate advanced approaches that can significantly improve our understanding of material science. Recent research in nanocomposite technology has revealed great opportunities for developing novel materials and products that are beneficial to mankind. This chapter provides an overview of bionanocomposite materials, including features, fabrication methodologies, and methods of characterisation. The methodical application of technology based on various bionanocomposite materials can significantly improve the standards of living for people.

Acknowledgements The authors acknowledge their gratitude to the Physics department, at the University of Lucknow, for their qualitative support.

Author Contribution The initial and original content of this chapter was prepared and written by Dr Peramjeet Singh, Dr. Priya Gupta, and Ms. Neetu Yadav. Prof. Narendra Kumar Pandey and Dr. Vernica Verma help with editing, reviewing, and rewriting.

Funding No specific grants from governmental, commercial, or non-profit funding organisations were given for this research.

Conflict of Interest The authors declare that they have no known financial conflicts of interest or significant personal connections that could have influenced the findings of the study discussed in this chapter.

References

Ahmad M, Manzoor K, Singh S, Ikram S (2017) Chitosan centered bio nanocomposites for medical specialty and curative applications: a review. Int J Pharm 529(1–2):200–217

Ajayan PM (2003) Bulk metal and ceramics nanocomposites. In: Nanocomposite science and technology, pp1–75

Ali A, Chiang YW, Santos RM (2022) X-ray diffraction techniques for mineral characterization: a review for engineers of the fundamentals, applications, and research directions. Minerals 12(2):205

Amiri M, Hashemi Z, Chekin F (2025) Zinc oxide nanoparticles decorated nitrogen doped porous reduced graphene oxide-based hybrid to sensitive detection of hydroxychloroquine in plasma and urine. J Mater Sci Mater Med 36(1):4. https://doi.org/10.1007/s10856-024-06847-2

Asad RU, Masood M, Maryam L, Waqas H, Raza E, Ahmad I (2025) Application of microbial technology for treatment of heavy metals contaminated wastewater. In: Smart waste and wastewater management by biotechnological approaches. Springer, Singapore pp 343–377. https://doi.org/10.1007/978-981-97-8673-2_17

Aston DE, Bow JR, Gangadean DN (2013) Mechanical properties of selected nanostructured materials and complex bio-nano, hybrid and hierarchical systems. Int Mater Rev 58(3):167–202

Bagheri L, Jafari-Gharabaghlou D, Dashti MR, Zarghami N (2025) An update on implication of POSS-based nanocomposites in bone tissue engineering: a review. J Biomater Sci Polym Ed 1–24. https://doi.org/10.1080/09205063.2025.2455234

Bokov D et al (2021) Nanomaterial by sol-gel method: synthesis and application. Adv Mater Sci Eng 1:5102014

Bramhill J, Ross S, Ross G (2017) Bioactive nanocomposites for tissue repair and regeneration: a review. Int J Environ Res Public Health 14(1):66

Cai L et al (2020) Bio nanocomposites for plastic waste biodegradation: challenges and opportunities. Environ Pollut 266:115172. https://doi.org/10.1016/j.envpol.2020.115172

Castillo PR et al (2020) Bio nanocomposites with antimicrobial activity for food packaging applications. Food Hydrocolloids 102:105622. https://doi.org/10.1016/j.foodhyd.2019

Chatterjee N, Dhar P (2025) Bio-nanocomposites: curse or miracle for the society? In: Characterization techniques in bio nanocomposites. Woodhead Publishing, pp 3–39. https://doi.org/10.1016/B978-0-443-22067-8.00001-0

Chaudhary D et al (2022) Chitosan-based bionano composites for wound healing. Int J Nanomedicine

Chen X et al (2020) Biochar-based nanocomposites for soil remediation: a review of their applications and mechanisms. Sci Total Environ 727:138424

Cseri L, Razali M, Pogany P, Szekely G (2018) Organic solvents in sustainable synthesis and engineering. In: Green chemistry. Elsevier, pp 513–553

Duarte ARC et al (2020) Bio nanocomposites for wound healing applications. Mater Sci Eng, C 109:110522. https://doi.org/10.1016/j.msec.2020.110522

Eliaz N (2019) Corrosion of metallic biomaterials: a review. Materials 12(3):407

Fadeel B, Bussy C, Merino S, Vázquez E, Flahaut E, Mouchet F, Evariste L, Gauthier L, Koivisto AJ, Vogel U, Martín C (2018) Safety assessment of graphene-based materials: focus on human health and the environment. ACS Nano 12(11):10582–10620. https://doi.org/10.1021/acsnano.8b04758

Ghaedi M et al (2020) Chitosan-based bio nanocomposites for heavy metal ions removal from water: a review. J Environ Manage 262:110288. https://doi.org/10.1016/j.jenvman.2020.110288

Ghosh A et al (2020) Nano-biosensors for precision agriculture. Agric Syst 183:102855. https://doi.org/10.1016/j.agsy.2020.102855

Ghosh A et al (2023) Recent development in polyurethanes for biomedical applications. In: Polyurethanes: preparation, properties, and applications volume 3: emerging applications. American Chemical Society, pp 163–189

Giacomo RD, Maresca B, Angelillo M, Landi G, Leone A, Vaccaro MC, Biot C, Porta A, Neitzert HC (2013) Bio-nano-composite materials constructed with single cells and carbon nanotubes: mechanical, electrical, and optical properties. IEEE Trans Nanotechnol 12(6):1026–1030. https://doi.org/10.1109/tnano.2013.2285438

Gilbert B, Banfield JF (2005) Molecular-scale processes involving nanoparticulate minerals in biogeochemical systems. Rev Mineral Geochem 59:109

Hamzehlouy A, Zarei S, Salah Othman R, Shahi F, Afshar H, Afshar Taromi A, Khonakdar HA (2025) Recent advances in biomedical applications of MXene-integrated electrospun fibers: a review. Polym Adv Technol 36(2):e70112. https://doi.org/10.1002/pat.70112

He J et al (2020) Bio nanocomposite-based triboelectric nanogenerators for sustainable energy harvesting. Nano Energy 68:104297. https://doi.org/10.1016/j.nanoen.2019.104297

Hossain MK et al (2019) Bio nanocomposites for moisturizing and skin hydration: a review of natural biopolymers in cosmetic applications. Cosmetics 6(4):57. https://doi.org/10.3390/cosmetics6040057

Hussain S, Akhter R, Maktedar SS (2024) Advancements in sustainable food packaging: from eco-friendly materials to innovative technologies. Sustain Food Technol 2(5):1297–1364

Karki S et al (2021) Nanocomposite and bio-nanocomposite polymeric materials/membranes development in energy and medical sector: a review. Int J Biol Macromol 193:2121–2139

Khan A et al (2021) Vibration damping and impact resistance of bio nanocomposites for structural applications. Mater Today Commun 26:101697. https://doi.org/10.1016/j.mtcomm.2020.101697

Kishore A, Singh C, Kaur G (2025) Bio-nanomaterials: an introduction. In: Bio-nanomaterials in environmental remediation: industrial applications, pp1–46. https://doi.org/10.1002/9783527848546.ch1

Krishnan S et al (2020) Chitosan and cellulose-based bionanocomposites in cosmetic formulations: applications in anti-aging and skin care. J Cosmet Dermatol 19(4):884–893. https://doi.org/10.1111/jocd.13267

Krishnankutty RE, Aswani R, Visakh PM (2025) Nanoformulations for agricultural applications. Wiley

Kumar P et al (2009) Nanoscale particles for polymer degradation and stabilization—trends and future perspectives. Prog Polym Sci 34(6):479–515

Kumar P et al (2021) Chitosan-based nanocomposites for heavy metal ion removal: synthesis and application. J Environ Manage 278:111413

Li X, Xu J, Yao S, Zhang N, Zhang BT, Zhang ZK (2025) Targeting drug delivery system to skeletal muscles: a comprehensive review of different approaches. J Cachexia Sarcopenia Muscle 16(1):e13691. https://doi.org/10.1002/jcsm.13691

Liang Y et al (2020) Bio nanocomposites for organic solar cells. J Mater Chem A 8(29):14707–14730. https://doi.org/10.1039/D0TA04242B

Lin N, Huang J, Dufresne A (2012) Preparation, properties and applications of polysaccharide nanocrystals in advanced functional nanomaterials: a review. Nanoscale 4(11):3274–3294

Maitra J, Bhardwaj N (2025) Development of bio-based polymeric blends–a comprehensive review. J Biomater Sci Polym Ed 36(1):102–136. https://doi.org/10.1080/09205063.2024.2394300

Meyers MA et al (2008) Materials: Structure and mechanical properties. Prog Mater Sci 53(1):1–206

Mitra BC (2014) Environment friendly composite materials: bio composites and green composites. Def Sci J 64(3)

Muthaiah SJ et al (2021) Bio nanocomposites for controlled release of active agents in food packaging. Food Control 119:107445. https://doi.org/10.1016/j.foodcont.2020.107445

Nagime PV, Shaikh NM, Singh S, Chandak VS, Chidrawar VR, Nweye EP (2025) Metallic nanostructures: an updated review on synthesis, stability, safety, and applications with tremendous multifunctional opportunities. Pharm Nanotechnol. https://doi.org/10.2174/0122117385358312250108180301

Nguyen TA et al (2019) Magnetic nanoparticles functionalized with microorganisms for arsenic removal from contaminated water. Environ Sci Pollut Res 26(5):4604–4613

Pan Y et al (2021) Bio nanocomposites for structural health monitoring and sensing in aerospace and civil engineering. Sens Actuators, A 324:112684. https://doi.org/10.1016/j.sna.2021.112684

Pandey A et al (2020) Smart and intelligent food packaging materials based on bio nanocomposites. Food Packag Shelf Life 24:100488. https://doi.org/10.1016/j.fpsl.2020.100488

Pandey RK, Prajapati VK (2018) Molecular and immunological toxic effects of nanoparticles. Int J Biol Macromol 107:1278–1293

Pandey M, Nazar R, Elella MH, Praharaj S, Makhado E, Rout D, Gilani EH (2025) A comprehensive review of recent developments in biomedical materials based on graphene-modified bio-nanocomposites. BioNanoScience 15(1):125. https://doi.org/10.1007/s12668-024-01757-7

Panja S et al (2022) Development of bioactive bionano composites for antimicrobial applications. J Nanomater

Patel P et al (2024) Future prospects of bionano composites in drug delivery and diagnostics. Adv Drug Deliv Rev

Plamadiala I, Croitoru C, Pop MA, Roata IC (2025) Enhancing polylactic acid (PLA) performance: a review of additives in fused deposition modelling (FDM) filaments. Polymers 17(2):191. https://doi.org/10.3390/polym17020191

Qamar W, Gulia S, Athar M, Ahmad R, Imam MT, Chandra P, Singh BP, Haque R, Hassan MI, Rahman S (2024) An insight into impact of nanomaterials toxicity on human health. PeerJ 12:e17807. https://doi.org/10.7717/peerj.17807

Rai PK et al (2019) Nanocomposites for improving water retention and moisture management in agriculture. Sci Total Environ 669:903–912. https://doi.org/10.1016/j.scitotenv.2019.03.163

Rajendran R et al (2020) Nanocomposites in electronic waste management: opportunities for sustainable recycling. Environ Chem Lett 18(1):17–36. https://doi.org/10.1007/s10311-019-00947-w

Ramadhan SA, Ali DS (2025) Innovations in core–shell nanoparticles: advancing drug delivery solutions and precision medicine. OMICS J Integr Biol. https://doi.org/10.1089/omi.2024.0182

Rossi M et al (2020) Bio nanocomposites in sustainable packaging materials: from design to environmental impact. Environ Sci Pollut Res 27:4406–4416. https://doi.org/10.1007/s11356-019-07084-4

Sahoo SK et al (2019) Nanoparticle-based bio nanocomposites for cancer therapy and immunotherapy. Biotechnol Adv 37(4):107334. https://doi.org/10.1016/j.biotechadv.2019.03.003

Samejo S, Baig JA, Algethami FK, Kazi TG, Akhtar K (2025) Green synthesized magnesium oxide nanohybrid as an effective adsorbent for removing toxic metals from contaminated water. Int J Environ Sci Technol 1–14. https://doi.org/10.1007/s13762-025-06335-2

Santos A et al (2020) Bio nanocomposite-based hair care products: recent advancements in enhancing hair strength and growth. J Cosmet Sci 71(4):347–358. https://doi.org/10.1111/jocs.12583

Sen A et al (2024) Bio-nanocomposites: innovative solutions for addressing issues in health, agriculture, energy and environmental domains. Nano-Structures Nano-Objects 39:101270

Shah DU et al (2020) Sustainable biopolymer-based nanocomposites for food packaging: a review. Food Res Int 132:109084. https://doi.org/10.1016/j.foodres.2020.109084

Shao Y et al (2019) Bio nanocomposites-based biosensors for disease detection: from design to application. Sensors 19(6):1435. https://doi.org/10.3390/s19061435

Sienkiewicz M et al (2021) Application of biopolymer-nanocomposites for controlled delivery of pesticides. Ind Crops Prod 159:112989. https://doi.org/10.1016/j.indcrop.2020.112989

Simchi A, Tamjid E, Pishbin F, Boccaccini AR (2011) Recent progress in inorganic and composite coatings with bactericidal capability for orthopaedic applications. Nanomed Nanotechnol Biol Med 7(1):22–39

Sirbu D et al (2019) Bio nanocomposites in food packaging. Sustain Agric Rev 39:221–243. https://doi.org/10.1007/978-3-030-16443-2_8

Srinivas A, Chandran KR, Nehra S, Kumar D (2025) Future aspects in the field of bio-nanomaterials toward environmental assessment. Bio-Nanomaterials Environ Remediat Ind Appl 295–317. https://doi.org/10.1002/9783527848546.ch12

Thakur VK et al (2018) Bio nanocomposites for bone tissue engineering: preparation, characterization, and biological evaluation. J Nanosci Nanotechnol 18(3):1984–1997. https://doi.org/10.1166/jnn.2018.15691

Thakur A et al (2022) Recent advancements in surface modification, characterization and functionalization for enhancing the biocompatibility and corrosion resistance of biomedical implants. Coatings 12(10):1459

Thakur D, Bareen MA, Gupta A, Saha S, Sahu JK (2024) Frontiers in 3D printing for biobased food packaging. Food Sci Biotechnol 1–21. https://doi.org/10.1007/s10068-024-01770-2

Vijayakumar V et al (2020) Bio nanocomposites in drug delivery and biomedical applications. J Drug Deliv Sci Technol 55:101449. https://doi.org/10.1016/j.jddst.2019.101449

Wang W et al (2019) Flame-retardant bio nanocomposites for aerospace and structural applications. Compos A Appl Sci Manuf 122:274–281. https://doi.org/10.1016/j.compositesa.2019.04.018

Wang J et al (2021) Bio nanocomposites for bioelectronics: Trends and applications. Biosens Bioelectron 182:113134. https://doi.org/10.1016/j.b

Wang S, Wang X, Luo Y, Liang Y (2025) A comprehensive review of conventional and stimuli-responsive delivery systems for bioactive peptides: from food to biomedical applications. Adv Compos Hybrid Mater 8(1):12. https://doi.org/10.1007/s42114-024-01053-8

Wong XY, Sena-Torralba A, Alvarez-Diduk R, Muthoosamy K, Merkoçi A (2020) Nanomaterials for nanotheranostics: tuning their properties according to disease needs. ACS Nano 14(3):2585–2627. https://doi.org/10.1021/acsnano.9b08133

Xiao F et al (2020) Bio nanocomposites for sunscreens and UV protective cosmetics: advances and challenges. Int J Cosmet Sci 42(3):205–212. https://doi.org/10.1111/ics.12624

Xu C et al (2020) Bio nanocomposite materials for flexible electronics and wearable devices. Sci Adv 6(18):eaaz2632. https://doi.org/10.1126/sciadv.aaz2632

Xuan L, Ju Z, Skonieczna M, Zhou PK, Huang R (2023) Nanoparticles-induced potential toxicity on human health: applications, toxicity mechanisms, and evaluation models. MedComm 4(4):e327. https://doi.org/10.1002/mco2.327

Yildizhan S et al (2018) Bio-composite materials: a short review of recent trends, mechanical and chemical properties, and applications. Eur Mech Sci 2(3):83–91

Yu Y et al (2020) Carbon dioxide capture by bio nanocomposites: a review of materials and mechanisms. Environ Sci Technol 54(6):3557–3574. https://doi.org/10.1021/acs.est.9b07694

Zhang Y et al (2020a) Bio nanocomposite electrodes for supercapacitors and lithium-ion batteries. Nano Today 33:100867. https://doi.org/10.1016/j.nantod.2020.100867

Zhang Y et al (2020b) Bio nanocomposite fertilizers for enhanced nutrient release and crop growth. Environ Sci Technol 54(14):8859–8870. https://doi.org/10.1021/acs.est.0c01757

Zhang H et al (2022) Titanium dioxide and laccase-based composites for the degradation of malathion pesticide. Sci Total Environ 810:152103

Zhang S, Ren Q, Qi H, Liu S, Liu Y (2019) Adverse effects of fine-particle exposure on joints and their surrounding cells and microenvironment. ACS Nano 13(3):2729–2748. https://doi.org/10.1021/acsnano.8b08517

Zhao Y et al (2020) Hydrophobic cellulose nanocrystals and bacteria composites for oil spill clean-up. Environ Sci Technol 54(12):7882–7890

Zhou X et al (2021) Biodegradable and sustainable bio nanocomposites in concrete reinforcement for structural applications. Constr Build Mater 300:124317. https://doi.org/10.1016/j.conbuildmat.2021.124317

Zhou X et al (2021) Toxicity and safety assessment of bionano composites. Nanotoxicology

Peramjeet Singh has completed his Ph.D. from the Department of Physics, University of Lucknow, Uttar Pradesh, India, in the field of material science. He did his M.Sc. from Barkatullah Vishwavidyalaya, Bhopal, and his B.Sc. from GDC, Udhampur, University of Jammu, India, in 2017 and 2015, respectively. His area of research is nanotechnology, gas sensors, and the characterization of materials.

Priya Gupta has been working as a lecturer at Manyawar Kanshiram Government Polytechnic, Tirwa, Kannauj, since January 2022. She has completed a Ph.D. from Integral University, Lucknow, in collaboration with Lucknow University, Lucknow, in the year 2021. She did her M. Tech. from Madan Mohan Malviya Engineering College, Gorakhpur, in 2014, and her B.Tech. from Integral University, Lucknow.

Vernica Verma was a former assistant professor at the Department of Physics, BBD, Lucknow, and completed her Ph.D. in Physics from the Department of Physics, University of Lucknow. She did her B.Sc. and M.Sc. in Physics from the University of Lucknow in 2005 and 2007, respectively. Her specialisation in research is fabrication, characterization like XRD, SEM, AFM, UV–VIS, FT-IR, Raman, gamma irradiation, etc., and humidity and gas sensors.

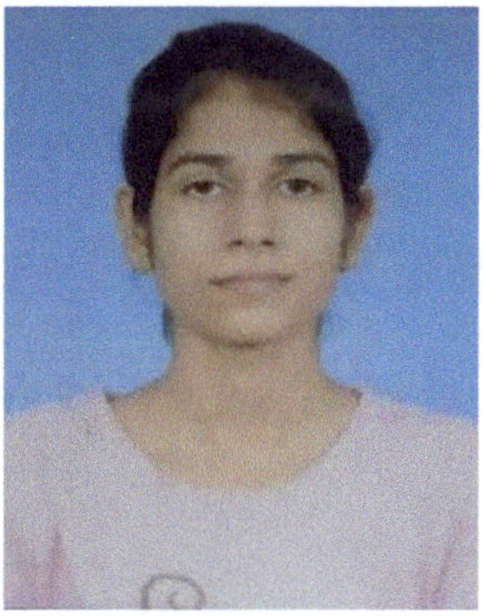

Neetu Yadav is currently pursuing Ph.D. in Material Science, Department of Physics, University of Lucknow. She did her M.Sc. in Physics from University of Lucknow in 2019 and B.Sc. from Kumaun University in 2017. Her specialisation in research is Fabrication, Characterization like XRD, SEM, AFM, UV–VIS, FT-IR, Raman, Gamma irradiation etc., and humidity and Gas Sensors.

Narendra Kumar Pandey received his degree (hons.) in Physics from St. Xavier's College, Ranchi, India, in 1985, and the M.Sc. degree in Physics and the Ph.D. degree from the Indian Institute of Technology Delhi, Delhi, India, in 1987 and 1995, respectively. He is currently a professor and head of department with the Department of Physics, University of Lucknow, Lucknow, India. His areas of interest include humidity and gas sensing, nanotechnology, and optical fibre, sensors, and fibre Bragg grating sensors.

Chapter 6
Wound Healing and Controlled Drug Release Using the Synergistic Role of Bionanocomposites

Shana Shaji, Jerin Jose, Jeny Jose, Juwel Jose, and Alessandro Distante

Abstract Bionanocomposites, hybrid materials that combine biological molecules with nanosized materials, have developed as a radical solution in the domain of therapeutics and drug delivery, by addressing the critical challenges in modern public health. These pioneering composites connect the exclusive properties of nanomaterials, viz. tunable physical and chemical characteristics, high surface area with enhanced bioactivity together with the specificity and biocompatibility of biological mechanisms. This collaboration has led to transformative improvements in the design and delivery of therapeutic agents. One of the crucial contributions of bionanocomposites falls in their capability to achieve targeted drug delivery, enabling the exact transportation of medicines to exact cells or tissues by curtailing the side effects. This is particularly important during the treatment of intricate diseases, viz. cancer, where conventional therapies sometimes affect other healthy tissues. Besides, these composites simplify controls drug release, safeguarding sustained therapeutic efficacy over time and refine patient compliance. In vaccine development, bionanocomposites performance as potent adjuvants and antigen delivery systems, improving immune responses and aiding the creation of next-generation vaccines coupled with improved safety and profiles. They also play a vital role in immunotherapy, offering

S. Shaji · A. Distante
Istituto Scientifico Biomedico Euro Mediterraneo (ISBEM), Brindisi, Italy
e-mail: distante@isbem.it

J. Jose
University of Catania, Sicily, Italy

J. Jose (✉)
Amal Jyothi College of Engineering, Kanjirappally, Kerala, India
e-mail: jerinjose100@gmail.com

J. Jose
Bar-Ilan University, Ramat Gan, Israel

J. Jose
Caritas Hospital, Kottayam, Kerala, India

S. C. George and B. Tawiah (eds.), *Advanced Bionanocomposite Materials*, Advanced Structured Materials 241, https://doi.org/10.1007/978-981-96-8225-6_6

stages for the delivery of immunomodulatory agents to treat diseases like infectious diseases and autoimmune disorders. Additionally, the antimicrobial properties of certain bionanocomposites have been leveraged in producing wound dressings and coatings for medical devices, reducing infection risks and encouraging healing. Despite these challenges, advancements, viz. scalability, safety, and regulatory hurdles persist blocks to their extensive acceptance. This chapter reviews the groundbreaking role of bionanocomposites in therapeutics and drug delivery, underlining their transformative role in public health. By mentioning the existing developments, limitations, and future research in this field, the chapter grip on its capacity to transform healthcare delivery and refine global health results. The present works also mention some full and partial green manufacturing practices to communicate the sustainability opportunities and aspects to the researchers.

Keywords Wound healing · Antimicrobial properties · Green manufacturing techniques · Nanocomposites

6.1 Introduction

In the ever-evolving domain of public health, the incorporation of biomaterials with nanotechnology has opened doors for innovation. Bionanocomposites, a hybrid class of materials merging biopolymers with nanostructures (1–100 nm), have developed as a groundbreaking solution for modern public health challenges. These nanostructures can be in the form of nanofiber, nanotubes or nanoparticles (Mousa et al. 2016). These advanced materials show amazing properties, viz. improved biocompatibility, mechanical strength, and antimicrobial effectiveness, making them ideal for environmental and medical applications. These advanced properties are due to the combination of properties contributed by both matrix and reinforcing material (Nandhini et al. 2024). With the rise of antibiotic resistance and infectious diseases, reduced toxicity, biodegradability, and chronic health conditions, there is an urgent necessity for adaptable materials that can back disease prevention, treatment and diagnosis. Thus, they are a demanding candidate for medical biomaterial applications (Kalsoom Khan et al. 2017).

Bionanocomposites have shown great promise in various healthcare domains, from wound healing and specific drug delivery to biosensors and antimicrobial coatings. Their ability to get engineered from the molecular level allows for precise control over their functionality, opening the way for personalized medicine and targeted therapies. Moreover, these materials are often derived from sustainable and biodegradable sources, aligning with global efforts to reduce medical waste and environmental pollution. By using the capability of nanotechnology, researchers have developed bionanocomposites capable of responding to physiological stimuli, ensuring controlled and efficient therapeutic effects.

One of the most exciting developments in this field is the use of nanocellulose, chitosan, and graphene-based composites for antimicrobial and antiviral applications. These materials have demonstrated exceptional efficacy in combating drug-resistant bacteria and emerging viral threats, contributing a capable substitute to traditional disinfection methods. Additionally, bionanocomposites are revolutionizing the domain of regenerative medicine, providing scaffolds which mimic the extracellular matrix and promote tissue repair. The incorporation of nanoparticles such as silver, gold, copper, and zinc oxide further enhances their antibacterial properties, making them invaluable in wound care and surgical settings. They help to arrest contaminations and thereby speed up the therapeutic process (Thakkar et al. 2010; Ali et al. 2020).

Beyond direct medical applications, bionanocomposites are also transforming public health by improving water purification, food safety, and air filtration systems. Their multifunctionality extends to biosensing technologies, enabling rapid and accurate disease detection, which is crucial for outbreak control and epidemiological surveillance. As research in this domain continues to expand, interdisciplinary association and teamwork between materials scientists, biomedical engineers, and healthcare professionals remain vital to unlocking the full potential of bionanocomposites.

Despite their immense promise, challenges such as large-scale production, regulatory approval, and potential toxicity must be carefully addressed. A deeper understanding of their long-term connections with the biological systems is necessary to ensure their safe and effective implementation in public health initiatives. Nevertheless, the future of bionanocomposites remains incredibly bright, with ongoing research driving the development of next-generation materials tailored to the dynamic needs of modern medicine. As we stand on the brink of a new era in biomedical technology, bionanocomposites offer a revolutionary path towards safer, more efficient, and sustainable healthcare solutions.

This chapter explores on the manufacturing techniques, wound healing, targeted drug delivery applications and challenges of bionanocomposites, shedding light on their transformative impact on public health. Few fully and partially green manufacturing techniques are also included in the chapter to impart the sustainability thoughts in the researchers. By delving into recent advancements and future prospects, we aim to provide a comprehensive understanding of how these materials are shaping the future of healthcare.

6.2 Fundamentals and Composition

Bionanocomposites combine the biopolymers (e.g. alginate, chitosan, PLA) with nanoscale reinforcements (1–100 nm), such as metallic or polymeric nanoparticles, to create hybrid structures optimized for biomedical use (Nandhini et al. 2024; Pande and Sanklecha 2017). Natural biopolymers provide biocompatibility and low toxicity, while the nanoparticles enhance thermal stability, mechanical strength, and functional properties like antimicrobial activity and controlled drug release (Nandhini et al. 2024; Bharathi and Jayas 2024). For example, alginate-based nanocomposites permit sustained drug delivery, while collagen composites mimic the extracellular matrix for tissue regeneration (Nandhini et al. 2024; Râpă et al. 2020). Synthetic biopolymers, viz. PLA are also widely used due to their compatibility and biodegradability with nanofillers such as silver nanoparticles (AgNPs), which improve antimicrobial efficacy in the prosthetic coatings (Râpă et al. 2020; Dawwam et al. 2022).

6.3 Desired Properties for Health Applications

There are some desired properties for a bionanocomposite to be selected for the healthcare applications. Some of them are mentioned below. Non-toxicity and biocompatibility are paramount for bionanocomposites, ensuring minimal immune response and safe integration with biological systems, as seen in PLA-collagen-AgNPs composites tested for fibroblast viability (Râpă et al. 2020; Dawwam et al. 2022). The bionanocomposites must possess the antimicrobial property as it should resists bacterial infections, particularly in the wound healing and implants domain. Sustainability and biodegradability allows materials to break down into non-toxic by-products, dropping the environmental impact and removing the need for surgical procedures (Nandhini et al. 2024; Wani and Mundada 2021). Mechanical durability is very critical for prosthetics, where materials like cellulose or chitosan nanocomposites deliver high flexibility and tensile strength (Nandhini et al. 2024; Bharathi and Jayas 2024). This property is very essential for the tissue scaffolds and body implants. Functional properties, such as the ability to release drugs (e.g. doxorubicin-loaded chitosan-magnetic nanocomposites) or scavenge pathogens, further distinguish these materials in targeted therapies (Nandhini et al. 2024; Pande and Sanklecha 2017). Responses to temperature, moisture, pH and magnetic fields also important in this context. For diagnostic and bioimaging purposes, the optical capabilities are also demanding (El-Aziz et al. 2024). Within the intended physiological conditions, chemical and thermal stability of these bionanocomposites must be taken care of (Krzykowska et al. 2024; Parvinzadeh Gashti 2025). It should maintain the function and structure of composites and therefore, must resist the enzymatic degradations.

6.4 Manufacturing Techniques

The effective fabrication of bionanocomposites for healthcare applications depends on the advanced manufacturing techniques that guarantee the proper incorporation of bio-based materials with nanoscale components. These methods have a vital role in attaining the anticipated biocompatibility, mechanical strength, and exact functionality essential for applications viz. drug delivery, wound healing, and the biomedical implants. Several manufacturing techniques, extending from solution-based processes, viz. electrospinning and solvent casting to super-precision techniques, viz. self-assembly and 3D printing, permits on the proper personalization of material properties to tackle the exact medical necessities. The choice for an appropriate manufacturing technique mainly depends on aspects, viz. intended application, material composition, processing conditions, and scalability. Green synthesis methods are also mentioned in this part. Commonly adopted and some selective techniques along with their merits and demerits are mentioned below.

6.4.1 Ionic Gelation

This is a mild gelation technique using the electrostatic interactions between polyelectrolytes (e.g. chitosan) and oppositely charged cross-linkers (e.g. phosphotungstate anions). It aids in precise control over the proton-conductive spots in materials intended for the biomedical membranes or fuel cells. Recent researches include the supports of porous alumina to establish a flat chitosan assembly with a uniform proton conductivity (Szliszka et al. 2009). Its advantages include room-temperature processing and biocompatibility. The slower diffusion of cross-linkers limits its scalability (Janjua 2019; Singh et al. 2024). Ionic gelation is broadly used in bioencapsulation, drug delivery, and the wound healing fields. Alginate-Ca^{2+} has been successfully carried out using this method.

6.4.2 Solution Casting

It is a solvent-based method where polymers (e.g. chitosan, PLA) are first dissolved and mixed with nano-reinforcements, viz. cellulose nanocrystals and ZnO, and then casted onto the substrates. It is dried then to remove the solvent. Solvent casting harvests semi-crystalline films with superior tensile strength (e.g. PLA/chitosan films exhibited a higher secant modulus in the range of 80–90% than the normal melt-blended variants (Cheerarot and Saikrasun 2023). Even though it is ideal for lab-scale biomedical film fabrication, challenges such as inefficiency in solvent removal, thermal degradation, and the environmental concerns from chloroform/dichloromethane are some disadvantages. (Cheerarot and Saikrasun 2023; Mathew and Oksman 2014). It has major application in biodegradable scaffolds. Its cost-effective and simple nature are the advantages that attracts the researchers. A slow drying time is the disadvantage of the process.

6.4.3 Melt Blending

In this process, thermoplastic polymers (PLA, PCL) are reinforced with nanoparticles (CNCs, Ag) above their melting points with the help of extruders. This solvent-free technique improves the industrial scalability and eliminates the thermal degradation problems understood in the solvent casting. For example, PLA/chitosan composites processed through melt blending showed enhanced smoother surfaces and healthier chitosan dispersion but with lower crystallinity (Cheerarot and Saikrasun 2023; Agbakoba et al. 2023). Its drawbacks are polymer degradation at high temperatures and sensitivity to moisture (Cheerarot and Saikrasun Dec. 2023).

6.4.4 Sono-co-precipitation (Ultrasonification)

It is a green synthesis method combining ultrasound with chemical precipitation to integrate nanoparticles viz. Co_3O_4, ZnO into bio templates such as cellulose nanocrystals. It uses sound waves of high frequency in the band of 20–100 kHz for disperseing the nanoparticles to the polymer material. For instance, carboxylated cellulose nanocrystals enables ZnO nanoparticle synthesis in the temperature range 35–80 °C, achieving antimicrobial bionanocomposites with precise particle sizes (50 nm–1 μm). Ultrasound improves the reaction homogeneity and condenses the aggregation, even though the calcination phases (e.g. 500 °C for Co_3O_4) rises the energy costs (Janjua 2019; Singh et al. 2024). This process guarantees a uniform distribution of nanoparticles like silver, gold, graphene, carbon nanotubes (CNTs) or hydroxyapatite in biopolymers. Its applications include tissue engineering scaffolds and antimicrobial coatings.

6.4.5 Additive Manufacturing (3D Printing)

A layer-by-layer extrusion-based deposition of nanoparticle and biopolymer composites to generate custom shapes are carried out in this procedure. 3D printing of the polylactic acid-based composites incorporated with cellulose nanocrystals is an example to mention (0.1–1 wt%, CNC). Enhanced formulations presented an improved crystallization of 5–15% with 20 °C advanced thermal stability, enabling customizable prosthetics and antimicrobial coatings. Melt-extruded filaments generally retains flexibility through the bio-based plasticizers (PEG, PVOH) (Agbakoba et al. 2023). Challenges including the nozzle clogging due to CNC agglomeration and limited CNC loading thresholds are also reported (Singh et al. 2024; Agbakoba et al. 2023). Patient-specific inserts, organ printing, and bone scaffolds are the major applications cited in the literatures. However, there are limitations too for this advanced and high precision manufacturing process. Demand of expensive equipments and the costly nature for the completion of the procedures are few to mention.

6.4.6 Green Co-precipitation

Green co-precipitation uses the plant extracts (e.g. peels of pomegranate) as bio-reducing components for the metal oxide synthesis. For example, $Zn(OH)_2$/ZnO bionanocomposites prepared at a temperature of 60 °C with 38 nm crystallites showed the antimicrobial and photocatalytic properties. This technique removes toxic solvents but necessitates a precise control on bio-reductant reaction times and concentrations (Dawwam et al. 2022). Wound healing and antimicrobial materials have applications in green co-precipitation. The control over the shape and size of nanoparticles is unfortunately a limitation to this technique.

6.4.7 Lyophilization (Freeze Drying)

Freeze drying or Lyophilization is a procedure in which a solution comprising nanoparticles and biopolymers are frozen and then sublimated under vacuum to create a porous structure. In some cases, a secondary stage of desorption is also carried out to eliminate any water remaining confirming long-term stability. Degradation of specific sensitive biomolecules are not happening here. Without the need of chemical additives and high temperatures, bioactive compounds are preserved here. This makes freeze drying an attractive choice for the applications, viz. highly porous drug carriers and scaffolds for tissue engineering. This dehydration technique improves stability in the fabricated bionanocomposites. Due to this, the shelf life is improved by maintaining bioactivity. Due to the nature of this process, it is considered as a green technique of fabrication.

6.4.8 Hot-Press Compounding

Melt-blended pellets (e.g. PLA/chitosan) are being compressed under a heat of 180 °C at a pressure of 5 tons to films. This ends up in mechanically robust, amorphous materials with excellent smooth surfaces, appropriate for flexible medical packaging. But it reduces the water adsorption capability compared to other solvent-cast films (Cheerarot and Saikrasun 2023). Energy-intensive heating and restricted film thickness controls are the notable drawbacks for hot press compounding technique (Cheerarot and Saikrasun 2023).

Techniques like ionic gelation, solution casting, and melt blending are common, with innovations like sono-co-precipitation enabling efficient integration of ZnO

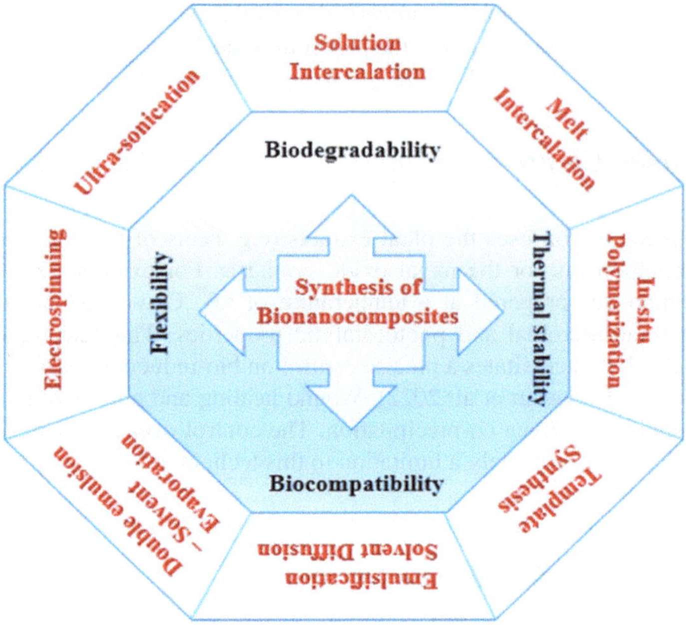

Fig. 6.1 Fabrication routes of bionanocomposites (Nandhini et al. 2024)

nanoparticles into cellulose templates for the antimicrobial applications (Dawwam et al. 2022). Solvent-free processes (e.g. melt extrusion) and UV-A exposure reduce environmental footprint along with enhanced material stability (Bharathi and Jayas 2024; Dawwam et al. 2022). For prosthetics, approaches, viz. solvent casting or 3D printing optimizes the structural precision and biocompatibility, as understood in PLA-based composites reinforced with AgNPs for antimicrobial coatings (Bharathi and Jayas 2024; Dawwam et al. 2022).

Other manufacturing techniques for the bionanocomposites are depicted in Fig. 6.1.

'Solvent casting' and '3D printing' can be considered as partially green techniques. If water-based solvents are considered in solvent casting it can be called partially green. If non-toxic or biodegradable bio-inks are used in 3D printing, it can be honoured under green synthesis method.

6.4.9 Applications in Prosthetics

Bionanocomposites like PLA-collagen-AgNPs blends balance mechanical robustness with antibacterial properties, ideal for prosthetic limbs or implants (Râpă et al. 2020). Their tunable porosity and surface roughness (e.g. 46.5 nm in optimized PLA composites) enhance cell adhesion and tissue integration (Râpă et al.

2020). Green-synthesized cellulose-ZnO nanocomposites offer dual functionality, preventing bacterial colonization while maintaining biocompatibility (Dawwam et al. 2022). These materials also enable personalized designs through additive manufacturing, catering to patient-specific anatomical needs (Bharathi and Jayas 2024; Dawwam et al. 2022). By merging sustainability with performance, bionanocomposites are redefining prosthetic innovation, reducing infection risks, and improving long-term usability (Râpă et al. 2020; Dawwam et al. 2022). The integration of natural polymers, advanced nanoparticles, and eco-conscious synthesis methods positions bionanocomposites as pivotal tools in modern medicine, bridging sustainability with high-performance healthcare solutions.

6.5 Bionanocomposites for Antimicrobial and Wound Healings

The process of wound healing in human body is an intricate phenomenon, including a tightly orchestrated sequence of cellular and biochemical events. This complex cascade typically unfolds a sequence of some overlapping phases, including haemostasis and inflammation, proliferation, maturation, remodelling and at last the wound contraction. These stages are pictorially represented in Fig. 6.2.

These phases often intersect or occur simultaneously, making it challenging to delineate clear boundaries between them concurrently (Steenfos 1994). Immediately following the injury, the haemostasis phase is triggered, resulting in the cessation of bleeding. This process involves the release of various platelet factors from degranulating thrombocytes, as well as the coagulation cascade, where prothrombin is converted into thrombin, ultimately forming a stable fibrin network that seals the wound and prevents excessive bleeding.

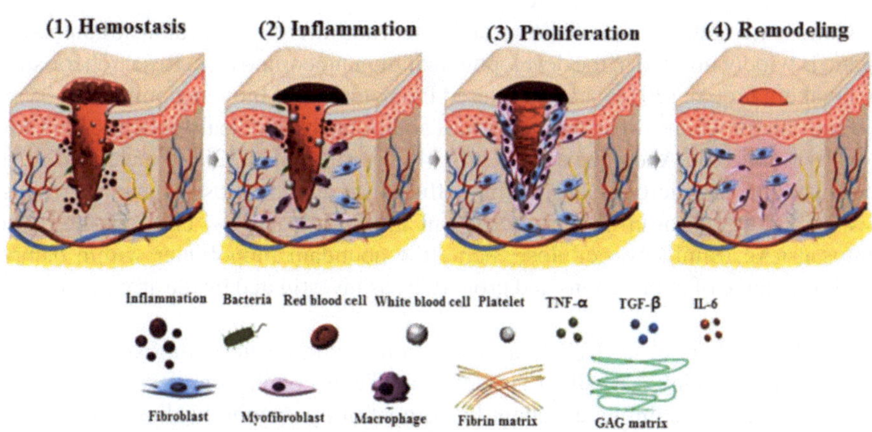

Fig. 6.2 Stages of wound healing (Trinh et al. 2022)

As haemostasis concludes, the inflammation phase commences, marked by the issue of certain inflammatory factors, viz. thromboxanes, histamines, and prostaglandins from mast cells. These factors induce vasoconstriction, reducing blood loss, while polymorphonuclear neutrophils (PMNs) arrives the wound site, attracted by the growth factors and fibronectins. PMNs issue free radicals, phagocytose debris, and thereby kills the bacteria through respiratory burst. As inflammation subsides, macrophages replace PMNs, producing factors that enhance angiogenesis, re-epithelialization, and tissue regeneration (Sharma et al. 2012).

Following the inflammation phase, the proliferation and migration phase begins, lasting a maximum of 21 days post-injury. During this phase, the wound starts to rebuild with healthy granulation tissue, characterized by angiogenesis, the development of extracellular matrix (ECM), and collagen deposition. Injured mesenchymal cells are converted into some fibroblast cells, enabling the cell migration and regeneration of tissues. The wound continues to strengthen during this stage, along with the restructuring of the fibroblast cells and creation of new tissues (Sharma et al. 2012).

The concluding segment of the wound healing is remodelling phase, where collagen formation and organization occur, facilitated by collagenases and matrix metalloproteinases. Over time, the cellular activity declines, and the blood vessels in the affected area are reduced. Contraction of wound takes place, and the type III collagen is remodelled into collagen of type I, enhancing the wound's tensile strength, ultimately leading to complete wound closure.

Nature has endowed its living beings with a diverse array of biomaterials that hold essential characteristics, including biocompatibility, non-toxicity, biodegradability, and the ability to mimic the host's extracellular situation. Furthermore, many of these biomaterials are non-immunogenic, making them ideal for wound healing dressing materials. An effective wound dressing should encompass multiple factors beyond mere exudate absorption. It should maintain a particular moist interface among the wound and dressing, preventing the infection, and create an ideal atmosphere that helps the healing with minimal scarring.

Chitosan, a natural, cost-effective, and potent antimicrobial and anti-inflammatory agent, is widely utilized in the development of bionanocomposites. Researchers, such as Jayakumar et al. (2011), have explored the potential of bionanocomposites based on chitosan for wound dressing and various additional biomedical applications. Additionally, studies by Abdelgawad et al. (2014) and Tariq et al. (2015) have confirmed the enhanced antimicrobial and healing efficacy of metal-chitosan bionanocomposites, including chitosan-nZnO, chitosan-nCu, chitosan-nTiO_2, chitosan-nAg, and chitosan-nAu. Similarly, cellulose, being the abundant biopolymer within nature, exhibits a range of applications and properties, as investigated by the aforementioned researchers.

Notably, Sandri et al. (2014) a novel bionanocomposite comprising chitosan, montmorillonite clay, and silver sulfadiazine. Their conclusions revealed that this bionanocomposite exhibits excellent biocompatibility, wound healing, and antimicrobial properties, making it an excellent candidate in the handling of various chronic skin-related diseases.

Developments in nanotechnology have directed to the evolution of novel nanocomposites that exhibit remarkable properties, making them ideal for wound dressing applications.

One such example is hyaluronic acid-based nanocomposites, which have demonstrated exceptional porosity, swelling, biocompatibility, and biodegradation properties (Anisha et al. 2013). Moreover, these nanocomposites have shown haemostatic potential and antibacterial properties, making them very effective against a variety of microbes, including *Pseudomonas aeruginosa*, *Staphylococcus aureus*, *Escherichia coli*, and *Klebsiella pneumoniae.* These findings suggest that hyaluronic acid-based nanocomposites are promising candidates for wound dressing applications, particularly for chronic wounds infected with bacteria.

Another notable example is guar gum-loaded carboxymethyl nanocomposites, which have been shown to enhance re-epithelial growth while minimizing inflammatory cell presence (Orsu and Matta 2020). This suggests that these nanocomposites can effectively reduce microbial infections, making them a suitable choice for wound dressing materials.

Nanocomposites based on silk have also been explored for the wound healing applications, owing to their exceptional permeability to oxygen and water vapours, tensile strength, thermal resistance, and antimicrobial characteristics (Chouhan and Mandal 2020). These characteristics make silk-based nanocomposites an attractive option for wound dressing applications.

Along with the modern cancer therapy, bionanocomposites have been explored for ecofriendliness. Eco-friendly bio-nanocomposites have been manufactured with an intention of antimicrobial functionalities, helping as wound dressings and drug delivery systems. These eco-friendly novel materials deliver not only the therapeutic agents but also offer better structural support for tissue regeneration applications (Nandhini et al. 2024).

Bionanocomposites have developed as a promising class of materials in the realm of antimicrobial and wound healing treatments. These composites synergistically combine biopolymers with nanoparticles, resulting in materials that displays an enhanced mechanical property, biocompatibility, and antimicrobial efficacy. A summary of the biomaterials research mentioning the nanoparticles used, their corresponding biological role in wound healings are provided in the Table 6.1 for easy reference.

The integration of nanoparticles, viz. zinc oxide, silver, and titanium dioxide into biopolymer matrices has been extensively studied for their antimicrobial properties. For instance, silver nanoparticles are renowned for their exceptional antimicrobial activity and successful targeting of a wide range of pathogens. Similarly, zinc oxide nanoparticles have demonstrated efficacy against various bacterial strains, including multi-drug-resistant species, making them invaluable in combating infections (Pino et al. 2023).

In the context of wound healing, bionanocomposites offer multifunctional benefits. They not only stop microbial colonization but also promote tissue renewal and repair. The incorporation of nanoparticles into biopolymers can enhance cell production and differentiation, a crucial step in wound curing process. For example,

Table 6.1 Comparison of different biomaterials and their biological role in would healing (Chhabra and Bhati 2022)

Biomaterial	Characteristics	Nano-biomaterial	Monomer units and linkage	Biological role in wound healing	References
Chitosan	It is non-toxic, bio-degradable anti-fungal, biocompatible and anti-microbial in nature	Chitosan can be structured into nanotubes, nanorods, nanofibrils and nanoparticles	*N*-acetyl glucosamine connected by β-1, 4 glycosidic links	It offers non-protein matrix for 3D progress of tissue It stimulates neutrophils infiltration and movement It plays as a haemostat in initial blood coagulation cascade It boosts fibroblast, PMN leucocytes and macrophages for cell growth and phagocytosis	Wong et al. (2012), Shukla et al. (2013)
Cellulose	It is biocompatible, non-allergic and non-toxic. crystal-like behaviours, sustainability and biodegradability	Nanocrystals, nanofibers, nano whiskers, nanorods, nanofibrils, and nanocellulose	β-d-Glucose connected through β-1,4-glycosidic linkage	It helps in the captivation of exudations developing in the consumption of cell debris (for example necrotic tissue and fibrinous coating) and the porous cellulose constitution copycats ECM (extra cellular matrix) of skin therefore assisting in tissue regeneration	Pei et al. (2015), Portela et al. (2019), Moon et al. (2016)
Silk fibroin	It is biocompatible, bio-degradable and many physical properties of it can be modified. Bioresorbable and non-immunogenic substance	Silk can transform in to silk nanoparticles, nanofilms, nanocomposites	Disulphide bonds	Silk provokes cell migration and multiplication It activate collagen formation and boost epithelization by enhancing keratinocyte and fibroblast proliferation	Patil et al. (2020), Shen et al. (2015)

(continued)

Table 6.1 (continued)

Biomaterial	Characteristics	Nano-biomaterial	Monomer units and linkage	Biological role in wound healing	References
Collagen	It is non-toxic, biodegradable and anti-inflammatory in nature	Nanofibers, nanoparticles	Amino acids connected through amide linkage	Collagen fascinates fibroblast and increase collagen deposition It stimulates tissue growth, autolytic debridement, reepithelization and angiogenesis. deactivates matrix metalloproteinases (MMPs)	Barroso et al. (2020), Deng et al. (2018)
Gelatin	It is biodegradable, extremely absorbent, non-immunogenic and biocompatible in nature	Nanocomposites, nanofilms	Sugar and amino acids	It accelerates signal transduction and cell union in well-synchronized evolution of wound healing course	Galdopórpora et al. (2019)
Pectin	Because of the existence of great number sof esters and galacturonic acid it holds anti-inflammatory properties Pectin is biocompatible, biodegradable and Antithrombogenic in nature	Nanocomposites, Nanofilms	D-galacturonic acid connected by α-(1 → 4)-glycosidic bonds	It is hydrophilic in nature. Pectin permits elimination of wound exudates from wound bed. Similarly it allows conservation of acid setting which is impermeable to microbes	Munarin et al. (2012), Kodoth et al. (2019)
Hyaluronic acid	Bacteriostatic, non-immunogenic, decomposable, Highly biocompatible and possess high water retention capacity	It can be architecture in to nanoparticles and nanotubes	β-d-glucuronic acid and *N*-acetyl-d-glucosamine connected by β-1,4 and β-1,3 glycosidic links	Hyaluronic acid accommodates crucial phases of wound healing like inflammation, cellular migration, and angiogenesis, It also fuels TGFβ-1, b-FGF, PDGF, and EGF growth factors production and improve keratinocytes production in wound bed	Snetkov et al. (2019)

chitosan-based nanogels have been developed to encapsulate interleukin-2, stimulating the immune scheme and advancing the wound healing process. Hydrogel-based bionanocomposites have gained attention due to their superior water content and capacity to preserve a moist atmosphere, which is conducive to wound healing. These hydrogels can be engineered to respond on to changes within the wound's micro-environment, releasing therapeutic agents as needed. For instance, hydrogels incorporating silver sulfadiazine-loaded halloysite nanotubes have shown potential as antimicrobial and biodegradable wound dressings (da Silva et al. 2021). The development of bionanocomposites also emphasizes sustainability and environmental friendliness. Green synthesis methods, such as the use of plant extracts for nanoparticle production, have been explored to reduce the ecological footprint of these materials. For example, green silver nanoparticles have been incorporated into collagen scaffolds, resulting in materials with fungicidal activity suitable for wound healing applications (Trotta et al. 2024).

Mechanical properties of bionanocomposites are important for their role in wound healing tasks. The addition of nanoparticles can enhance the tensile strength and elasticity of the biopolymer matrix, providing better support for the regenerating tissue. Studies have shown that incorporating nano-zinc oxide into biocomposites improves their mechanical properties, making them more durable and effective as wound dressings (Pino et al. 2023). Moreover, the controlled release of antimicrobial agents from bionanocomposites ensures sustained action against pathogens, dropping the frequency of dressing changes and refining the patient compliance. This controlled release can be achieved by altering the connections within the biopolymer matrix and nanoparticles, allowing for a gradual release of the therapeutic agents. For instance, quaternized starch/carboxymethyl cellulose bionanocomposites incorporating the cinnamon essential oil nanoemulsions have been developed for their antibacterial properties and potential in wound healing (Moustafa et al. 2024).

The biocompatibility of bionanocomposites is paramount, as materials used in wound healing must not elicit adverse reactions. Biopolymers, viz. collagen alginate, and chitosan are commonly used due to their inherent biodegradability and biocompatibility. When combined with nanoparticles, these biopolymers can enhance the overall performance of the composite without compromising safety. For example, chitosan-based hydrogels have been utilized for their antimicrobial properties and ability to promote tissue regeneration. Infection control remains a significant challenge in wound management, and bionanocomposites offer a proactive approach by resisting bacterial colonization and biofilm formation. The antimicrobial properties of the incorporated nanoparticles interrupt the microbial cell skins and affect with their metabolic processes, successfully reducing infection rates. For instance, hydrogels based on antimicrobial and antifouling zwitterionic nanocomposite have been developed to prevent bacterial adhesion and promote the wound curing deprived of triggering secondary damage (Dai et al. 2021).

The flexibility of bionanocomposites permits for the integration of various healing agents, including antibiotics, anti-inflammatory compounds and growth factors, tailored to specific wound types and healing stages. This adaptability makes them appropriate for an extensive range of applications, from acute injuries to chronic wounds. For example, nanogels have been researched to encapsulate antimicrobial drugs, anti-inflammatory agents, and various growth factors to enable the growth of fresh tissues and blood vessels.

In summary, bionanocomposites epitomise a multifaceted solution in antimicrobial and wound healing treatments. Their exclusive combination of biopolymers and nanoparticles offers improved antimicrobial efficacy, encourages tissue regeneration, and delivers a mechanical support to the healing tissues. The ongoing researches in this domain continue to expand the potential suitability of bionanocomposites, targeting sharply to improve patient outcomes and advanced wound care management.

6.6 Targeted Drug Delivery and Controlled Release of Bionanocomposites

Controlled release mechanisms are pivotal in the modern drug delivery systems, aiming to administer therapeutic agents at predetermined rates, durations, and targeted locations to enhance efficacy and patient compliance. These systems are designed to maintain optimal medicine concentrations inside the therapeutic window, minimizing side effects and reducing dosing frequency. Different proteins and their role in the targeted and precise drug delivery systems are shown in Fig. 6.3.

One primary mechanism of meticulous release of drugs is diffusion-controlled system is where the drug circulates through a membrane or polymer matrix. In reservoir systems, the drug core is encased in a polymer coating, and release occurs as the drug diffuses through this barrier. The rate of this release is inclined to the polymer's thickness and properties. In contrast, matrix systems disperse the drug

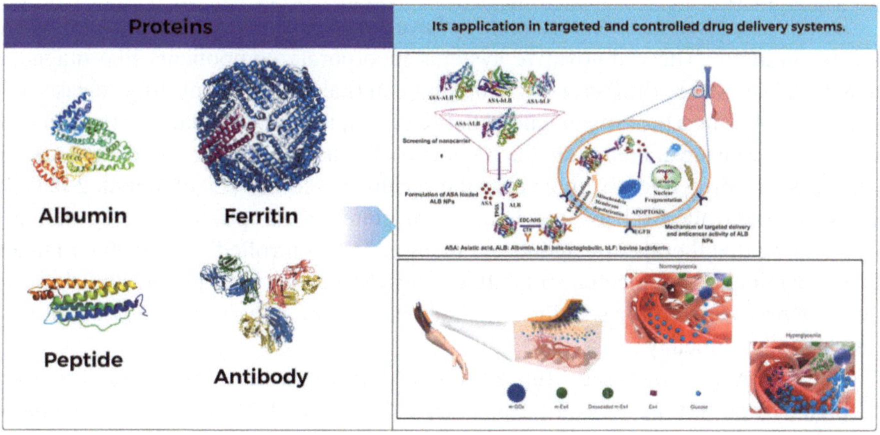

Fig. 6.3 Different proteins and their role in drug delivery system (Tahir et al. 2025)

uniformly inside a polymer matrix, and release happens as the drug diffuses out of the matrix material. The porosity of polymer and the drug's solubility significantly affect the release kinetics (Siegel and Rathbone 2012). Lyophilization-based manufacturing of the intended bionanocomposite will give better results. Another mechanism is where the drug or its carrier dissolves over time. In encapsulation dissolution systems, the drug is coated with a slow-dissolving material, and release occurs as the coating dissolves in bodily fluids. Alternatively, in matrix dissolution systems, the drug is embedded in a matrix that gradually dissolves, releasing the drug. The dissolution rate of the carrier material dictates the drug release profile.

Osmotic-controlled release uses the osmotic pressure to perform the drug release. These systems fundamentally consist of an osmotic agent and a drug reservoir encircled within a semi-permeable membrane. Water influx through the membrane creates a pressure, pushes the medicine through an orifice at a specific & controlled rate. This mechanism generally offers consistent release profiles, mainly independent of external conditions (Siegel and Rathbone 2012). Swelling-controlled release involves polymers that will expand upon absorbing bodily fluids. The swelling enhances the mesh size of the polymeric network, facilitating the medicine diffusion. The rate of this release is ruled by the polymer's swelling kinetics and the medicine's diffusion rate through the used swollen matrix (Raval et al. 2010).

Erosion-controlled release relies on the slow degradation of the matrix polymer. As the matrix continues to erode, the medicine is released. This mechanism is particularly useful for delivering drugs over an extended period, with the erosion rate of the polymer determining the specific release kinetics. Stimuli-responsive release systems are engineered to deliver the drugs in response to specific physiological triggers, viz. temperature variations, pH, or due to the presence of certain enzymes. For example, polymeric sensitive to pH can release drugs when they encounter the acidic environment of the stomach or the neutral pH of the intestines. This targeted method guarantees that the medicine is released at the anticipated site of action (Siegel and Rathbone 2012).

Developments in the latest technology have led to the growth of remote-controlled release systems. These innovative systems incorporate components like magnetic nanoparticles within drug carriers, enabling external activation of drug release. For instance, applying a flashing magnetic field can tempt localized heating, triggering the release of the encapsulated drug. Such systems offer accurate control over the location and timing of drug release, enhancing therapeutic outcomes (Ziegler et al. 2024). In summary, controlled release mechanisms are integral to optimizing drug delivery by ensuring that therapeutic agents are released in a controlled, predictable manner. Understanding these mechanisms allows for the design of sophisticated delivery systems personalized for specific medical needs, ultimately improving patient care and treatment efficacy.

The advances of focused drug delivery mechanisms have transformed modern medicine by permitting controlled, precise, and competent therapeutic involvements. Bionanocomposites have appeared as capable candidate for transporting drugs with superior specificity by minimizing the possible side effects. These materials have the advantages of nanotechnology, ensuring improved drug stability, prolonged circulation, and site-specific release. In the facet of focused drug delivery, bionanocomposites can be custom engineered to answer numerous biotic activates such as temperature, pH, or even the external stimuli. These external stimuli can be ultrasound based or magnetic field based. Their adaptability makes them an effective for dealing with the cancer, chronic ailments, and wound healing where the conventional routes of drug delivery fail. As mentioned before, one significant application of bionanocomposites is in the cancer treatment. Biodegradable polymer reinforced with nano particles carrying the anticancer agents has confirmed the ability for targeted and controlled drug delivery. These can specifically rise the close attention of drugs in tumour tissues, resulting in an enhanced treatment. For instance, natural-based polymeric nanobiocomposites have been attempted for their possibility in cancer treatment, merging diagnostic and therapeutic proficiencies (Mondal et al. 2023).

The adaptability of bionanocomposites permits for the reinforcement of therapeutic agents, chemotherapeutics, and proteins. For the improvement of the delivery of hydrophobic cancer drugs, Pullulan-based bioconjugates have been exploited. By creating micelles which are self-assembled, pullulan derivatives can carry various drugs, viz. paclitaxel and doxorubicin. This improves the overall stability. Additionally, the modifications such as folate conjugations enable these arrangements to aim directly the cancer cells thereby avoiding the overexpression of these receptors on the tumour cells.

Bionanocomposites with special stimuli-responses is manufactured to get an on-demand type drug release. Composite membranes which are triggered magnetically reinforces magnetite nanoparticles inside a polymer matrix. This polymer matrix should be thermosensitive. When an oscillating magnetic field is experienced, these membranes will control the rate of drug releases thus providing a scheduled dosing at very accurate controls. These arrangements have proved sustained functionality and better biocompatibility. Yusefi et al. (2021) reported the easily synthesized 5-Fluorouracil reinforced magnetic cellulose bionanocomposites intended for the colorectal cancer treatment. In the present work, Fe_3O_4-nanoparticles were reinforced on rice straw cellulose for carrying 5-fluorouracil. Multifunctional properties of these MC/5-FU bionanocomposites were confirmed. TEM and SEM investigations proved the presence of Fe_3O_4 nanofillers enclosing the cellulose matrix. The maximum drug release was identified at a pH 7.4 and a temperature of 44.2 °C.

Poursadegh et al. (2024) studied on the preparation of a chitosan-magnetic graphene quantum dot bionanocomposite which are pH-sensitive. It was intended for the drug delivery applications. Here, a blend of chitosan-based hydrogel with various amounts of magnetic graphene quantum dots were prepared. They were then considered for the pH-sensitive controlled release of Methotrexate. The outcomes from the drug release and swelling behaviour revealed that these hydrogel beads had experienced a pH-based swelling and a release of the Methotrexate. The results from these also suggests that the fabricated bionanocomposite and magnetic hydrogel beads have decent capability for an implantable pH-sensitive delivery system intended for the treatment of tumours. Few routes for the drug delivery in humans are shown in Fig. 6.4.

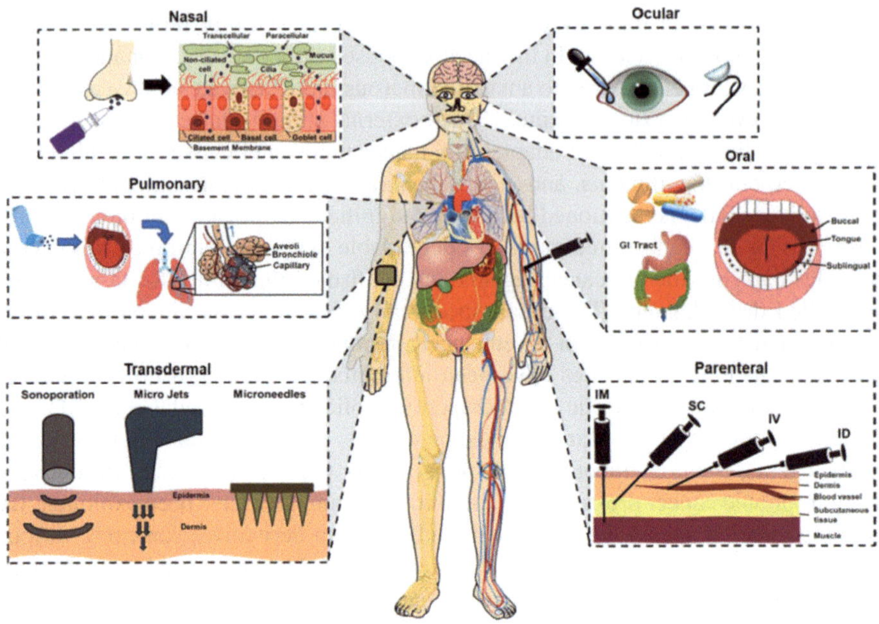

Fig. 6.4 Few routes for drug delivery (Benoit et al. 2020)

The incorporation of the therapeutic and diagnostic functions into a single composite platform has opened doors to the advancements of theragnostic systems. MnO_2-based nanoplatforms have been attempted for contrast-enhanced magnetic resonance imaging (MRI) and targeted drug delivery. This releases necessary chemotherapeutic negotiators in response to the specific stimuli detected within the micro-environment of tumour. It simultaneously enhances the imaging contrast, thus assisting in both diagnosis and treatments (Shi et al. 2018). To manufacture the drug delivery scaffolds, natural polymers, viz. gelatin are mixed with the PLA, being a synthetic polymer. This combination is used to give the antibodies like tetracycline. The attempt yielded a better antibacterial activity and controlled drug release (Moya-Lopez et al. 2022). Tahir et al. (2025) reported on the role of protein-based bionanocomposites in contributing the growth of drug delivery at specific locations. By combining the proteins with the therapeutic agents, these composites can deliver superior controlled release to specific targets and profiles. Various disease models has explored this system. In cancer treatments, nanocarriers having proteins are excellent for the selective release of chemotherapeutic agents.

Bionanocomposites signify a promising option for targeted drug delivery. They are now important for the proper health treatment. Its capability to combine controlled release, specific delivery, and biocompatibility, coupled with multifunctionality makes it highly demanded for next-generation treatment strategies. Ongoing developments in this domain will definitely open new potentials for improvements in treatments in a range of medical circumstances.

6.7 Current Challenges of Bionanocomposites in Healthcare

Even though the applications of bionanocomposites are very promising, there are numerous concerns about its production. Nanomaterials can penetrate the blood–brain barrier and enter the organs of the human body. A major concern is with its compatibility and its toxicity, which can trigger immune responses and tissue degeneration. The long-term stability of bio-nanocomposites in biological environments is still facing a huge problem. As many of these materials degrade over time, they can lead to unexpected biological activity. The large-scale production of these composites is still in the laboratory phase. The problems due to the high costs associated with operating at low pressures and temperature needs to be dealt with. The interaction of these composites with everyday materials and components is still under study. The integration of bio-nanocomposites with existing waste management systems is another puzzle to solve. For tissue engineering, there is still, a lack of appropriate biomaterials.

6.8 Recent Developments in Bionanocomposites for Healthcare

Integration of antimicrobial-based nanoparticles into the system of bionanocomposites can reduce the risk of infections and deaths by the medical equipment. Researchers have positively combined biopolymers like cellulose, chitosan, and alginate for the surge in bioactivity of nanocomposites. The practice of 3D printing of bionanocomposites has demonstrated hopeful results for making patient-focused implants. Novel smart drug delivery arrangements are being explored to ensure sustained and controlled drug release. Research has revealed the great capability of the antimicrobials in bionanocomposites. This qualifies their capacity for wound healing applications. The addition of metallic nanoparticles viz. gold and silver with the biopolymers is another productive and exciting research domain on which researchers are concentrating. Graphene-based bionanocomposites have captured attractions in regenerative medicine. The latest research has confirmed the aids of using bionanocomposite coatings on medical devices with the intention of reducing infections and biofilm formations.

6.9 Future Prospects of Bionanocomposites in Healthcare

Cautious calculation of a nanoparticle's potential toxicity is mandatory to guarantee that it will not harm the patient. Getting advanced medical materials through regulatory approval is often difficult and lengthy. The reproducibility of the clinically developed bionanocomposites is very important for their widespread use. With the potential growth in the practice of Artificial Intelligence (AI) and machine learning with bionanocomposites, it is intended to progress treatment monitoring and diagnostics. Bionano interfaces are in the course of deployment to allow direct interaction between bionanocomposites and biological systems. Hybrid bionanocomposites with inorganic and organic materials are trying to offer various multifunctional biomedical solutions. Research concepts for injectable bionanocomposites for nominal invasive treatments are also under development. Scientists are exploring the possibility of sustainable bionanocomposite materials to diminish toxicity concerns. Multi-responsive bionanocomposites that are proficient in adjusting to environmental stimuli will widely expand the therapeutic applications. The gene delivery applications using bionanocomposites are getting attention concerning genetic disorders.

6.10 Conclusions

Bionanocomposites represent a revolutionary advancement in modern medicine, particularly in public health, by offering innovative solutions for drug delivery, antimicrobial treatments, wound healing, and controlled release mechanisms. These composite materials, which integrate biopolymers with nanoparticles, have demonstrated significant potential in improving therapeutic outcomes through enhanced biocompatibility, biodegradability, and targeted delivery.

One of the key applications of bionanocomposites is in antimicrobial and wound-healing treatments. Their incorporation into wound dressings helps prevent bacterial infections while promoting tissue regeneration. Materials, viz. zinc oxide, silver, and titanium dioxide nanoparticles are widely used for their strong antimicrobial properties, dipping the risk of infections and supporting faster wound closure. The use of hydrogels and biopolymer-based scaffolds has further enhanced wound care, ensuring sustained drug release while maintaining a favourable healing environment.

Another essential feature of bionanocomposites is their role in controlled release mechanisms, which enable precise, sustained, and precise drug delivery. Various release mechanisms, such as swelling-controlled, diffusion-controlled, erosion-controlled, and stimuli-responsive systems, allow for optimal drug bioavailability and improved patient compliance. Smart drug carriers, such as magnetic and pH-sensitive nanocomposites, ensure that therapeutic agents are delivered at the right place and time, reducing systemic side effects and enhancing treatment efficiency.

A major advancement in nanomedicine is the targeted drug delivery capabilities of bionanocomposites. These materials facilitate the selective transport of drugs to diseased cells, improving therapeutic efficiency and reducing toxicity to healthy tissues. In cancer therapy, for example, functionalized polymeric nanocomposites enable the controlled release of chemotherapy drugs at tumour sites, minimizing damage to non-cancerous cells. Additionally, theranostic platforms combining imaging and treatment capabilities offer new possibilities in precision medicine, improving both diagnosis and therapy in a single system.

The growing body of research on bionanocomposites underscores their vast potential to transform healthcare. Their applications in infection control, chronic disease management, regenerative medicine, and even environmental sustainability highlight their multidisciplinary importance. As nanotechnology continues to evolve, the advances of eco-friendly, scalable and cost-effective bionanocomposite formulations will further enhance their clinical applicability.

In conclusion, bionanocomposites stand at the forefront of next-generation biomedical innovations. Their ability to provide effective antimicrobial treatment, accelerate wound healing, enable controlled drug release, and facilitate targeted therapy makes them invaluable tools in modern healthcare. With continuous advancements in nanotechnology and material science, these smart biomaterials will perform a gradually crucial role in answering public health challenges and improving patient outcomes worldwide.

Declaration During the preparation of this book chapter, the author(s) used Chatgpt, Perplexity in order to coordinate the information in the right way. After using these tools, the author(s) tried to review and edit the content as needed and take(s) full responsibility for the content of the publication.

References

Abdelgawad AM, Hudson SM, Rojas OJ (2014) Antimicrobial wound dressing nanofiber mats from multicomponent (chitosan/silver-NPs/polyvinyl alcohol) systems. Carbohydr Polym 100:166–178. https://doi.org/10.1016/j.carbpol.2012.12.043

Agbakoba VC, Hlangothi P, Andrew J, John MJ (2023) Preparation of cellulose nanocrystal (CNCs) reinforced polylactic acid (PLA) bionanocomposites filaments using biobased additives for 3D printing applications. Nanoscale Adv 5(17):4447–4463. https://doi.org/10.1039/d3na00281k

Ali MdA et al (2020) Advancements in plant and microbe-based synthesis of metallic nanoparticles and their antimicrobial activity against plant pathogens. Nanomaterials 10(6):1146. https://doi.org/10.3390/nano10061146

Anisha BS, Biswas R, Chennazhi KP, Jayakumar R (2013) Chitosan–hyaluronic acid/nano silver composite sponges for drug resistant bacteria infected diabetic wounds. Int J Biol Macromol 62:310–320. https://doi.org/10.1016/j.ijbiomac.2013.09.011

Barroso A, Mestre H, Ascenso A, Simões S, Reis C (2020) Nanomaterials in wound healing: from material sciences to wound healing applications. Nano Select 1(5):443–460. https://doi.org/10.1002/nano.202000055

Benoit DSW, Overby CT, Sims Jr KR, Ackun-Farmmer MA (2020) Drug delivery systems. In: Biomaterials science. Elsevier, pp 1237–1266. https://doi.org/10.1016/B978-0-12-816137-1.00078-7

Bharathi VSK, Jayas DS (2024) Evolution of bionanocomposites: innovations and applications in food packaging. Multidisciplinary Digital Publishing Institute (MDPI). https://doi.org/10.3390/foods13233787

Cheerarot O, Saikrasun S (2023) Effects of different preparation methods (solvent casting and melt blending) on properties of chitosan-filled polylactic acid biocomposite films. J Elastomers Plast 55(8):1173–1198. https://doi.org/10.1177/00952443231198468

Chhabra P, Bhati K (2022) Bionanomaterials: advancements in wound healing and tissue regeneration. In: Recent advances in wound healing. IntechOpen. https://doi.org/10.5772/intechopen.97298

Chouhan D, Mandal BB (2020) Silk biomaterials in wound healing and skin regeneration therapeutics: from bench to bedside. Acta Biomater 103:24–51. https://doi.org/10.1016/j.actbio.2019.11.050

da Silva GLP, de A. Morais LC, Olivato JB, Marini J, Ferrari PC (2021) Antimicrobial dressing of silver sulfadiazine-loaded halloysite/cassava starch-based (bio)nanocomposites. J Biomater Appl 35(9):1096–1108. https://doi.org/10.1177/0885328221995920

Dai Z et al (2021) An antifouling and antimicrobial zwitterionic nanocomposite hydrogel dressing for enhanced wound healing. ACS Biomater Sci Eng 7(4):1621–1630. https://doi.org/10.1021/acsbiomaterials.1c00039

Dawwam GE, Al-Shemy MT, El-Demerdash AS (2022) Green synthesis of cellulose nanocrystal/ZnO bio-nanocomposites exerting antibacterial activity and downregulating virulence toxigenic genes of food-poisoning bacteria. Sci Rep 12(1). https://doi.org/10.1038/s41598-022-21087-6

Deng A, Yang Y, Du S, Yang S (2018) Electrospinning of *in situ* crosslinked recombinant human collagen peptide/chitosan nanofibers for wound healing. Biomater Sci 6(8):2197–2208. https://doi.org/10.1039/C8BM00492G

El-Aziz MEA, Tohamy H-AS, Youssef AM, El Desouky FG (2024) Preparation and optimization of grafted hydroxyethyl cellulose, polypyrrole, and nitrogen-doped carbon quantum dots bionanocomposites for electrical, optical, and photoluminescence multicoloring applications. Int J Biol Macromol 278:134965. https://doi.org/10.1016/j.ijbiomac.2024.134965

Galdopórpora JM, Morcillo MF, Ibar A, Perez CJ, Tuttolomondo MV, Desimone MF (2019) Development of silver nanoparticles/gelatin thermoresponsive nanocomposites: characterization and antimicrobial activity. Curr Pharm des 25(38):4121–4129. https://doi.org/10.2174/1381612825666191007163152

Janjua MRSA (2019) Synthesis of Co_3O_4 Nano aggregates by co-precipitation method and its catalytic and fuel additive applications. Open Chem 17(1):865–873. https://doi.org/10.1515/chem-2019-0100

Jayakumar R, Prabaharan M, Sudheesh Kumar PT, Nair SV, Tamura H (2011) Biomaterials based on chitin and chitosan in wound dressing applications. Biotechnol Adv 29(3):322–337. https://doi.org/10.1016/j.biotechadv.2011.01.005

Kalsoom Khan A et al (2017) Carrageenan based bionanocomposites as drug delivery tool with special emphasis on the influence of ferromagnetic nanoparticles. Oxid Med Cell Longev 2017(1). https://doi.org/10.1155/2017/8158315

Kodoth AK, Ghate VM, Lewis SA, Prakash B, Badalamoole V (2019) Pectin-based silver nanocomposite film for transdermal delivery of Donepezil. Int J Biol Macromol 134:269–279. https://doi.org/10.1016/j.ijbiomac.2019.04.191

Krzykowska B et al (2024) Polymer bionanocomposites based on a P3BH/polyurethane matrix with organomodified montmorillonite—mechanical and thermal properties, biodegradability, and cytotoxicity. Polymers (Basel) 16(18):2681. https://doi.org/10.3390/polym16182681

Mathew AP, Oksman K (2014) Processing of bionanocomposites: solution casting, pp 35–52. https://doi.org/10.1142/9789814566469_0018

Mondal A, Nayak A, Chakraborty P, Banerjee S, Nandy B (2023) Natural polymeric nanobiocomposites for anti-cancer drug delivery therapeutics: a recent update. Pharmaceutics 15(8):2064. https://doi.org/10.3390/pharmaceutics15082064

Moon RJ, Schueneman GT, Simonsen J (2016) Overview of cellulose nanomaterials, their capabilities and applications. JOM 68(9):2383–2394. https://doi.org/10.1007/s11837-016-2018-7

Mousa MH, Dong Y, Davies IJ (2016) Recent advances in bionanocomposites: preparation, properties, and applications. Int J Polym Mater Polym Biomater 65(5):225–254. https://doi.org/10.1080/00914037.2015.1103240

Moustafa H, Nasr HE, Youssef AM (2024) Development of antibacterial carboxymethyl cellulose/quaternized starch bionanocomposites based on cinnamon essential oil nanoemulsion for wound healing applications. Biomass Convers Biorefin 14(21):27477–27489. https://doi.org/10.1007/s13399-022-03403-2

Moya-Lopez C et al (2022) Multifunctional PLA/gelatin bionanocomposites for tailored drug delivery systems. Pharmaceutics 14(6):1138. https://doi.org/10.3390/pharmaceutics14061138

Munarin F, Tanzi MC, Petrini P (2012) Advances in biomedical applications of pectin gels. Int J Biol Macromol 51(4):681–689. https://doi.org/10.1016/j.ijbiomac.2012.07.002

Nandhini J, Karthikeyan E, Rajeshkumar S (2024) Eco-friendly bio-nanocomposites: pioneering sustainable biomedical advancements in engineering. Discover Nano 19(1):86. https://doi.org/10.1186/s11671-024-04007-7

Nandhini J, Karthikeyan E, Rajeshkumar S (2024) Eco-friendly bio-nanocomposites: pioneering sustainable biomedical advancements in engineering. Springer. https://doi.org/10.1186/s11671-024-04007-7

Orsu P, Matta S (2020) Fabrication and characterization of carboxymethyl guar gum nanocomposite for application of wound healing. Int J Biol Macromol 164:2267–2276. https://doi.org/10.1016/j.ijbiomac.2020.07.322

Pande VV, Sanklecha VM (2017) Bionanocomposite: a review

Parvinzadeh Gashti M et al (2025) Plasma-assisted hydroxyapatite/chitosan bionanocomposite films with improved thermal stability, biomineralization and optical absorption properties. ChemNanoMat 11(3). https://doi.org/10.1002/cnma.202400577

Patil PP, Reagan MR, Bohara RA (2020) Silk fibroin and silk-based biomaterial derivatives for ideal wound dressings. Int J Biol Macromol 164:4613–4627. https://doi.org/10.1016/j.ijbiomac.2020.08.041

Pei Y et al (2015) Effectively promoting wound healing with cellulose/gelatin sponges constructed directly from a cellulose solution. J Mater Chem B 3(38):7518–7528. https://doi.org/10.1039/C5TB00477B

Pino P, Bosco F, Mollea C, Onida B (2023) Antimicrobial nano-zinc oxide biocomposites for wound healing applications: a review. Pharmaceutics 15(3):970. https://doi.org/10.3390/pharmaceutics15030970

Pino P, Bosco F, Mollea C, Onida B (2023) antimicrobial nano-zinc oxide biocomposites for wound healing applications: a review. Pharmaceutics 15(3). https://doi.org/10.3390/pharmaceutics15030970

Portela R, Leal CR, Almeida PL, Sobral RG (2019) Bacterial cellulose: a versatile biopolymer for wound dressing applications. Microb Biotechnol 12(4):586–610. https://doi.org/10.1111/1751-7915.13392

Poursadegh H, Barzegarzadeh M, Amini-Fazl MS (2024) Preparation of pH-sensitive chitosan-magnetic graphene quantum dot bionanocomposite hydrogel beads for drug delivery application: emphasis on effects nanoparticles. Polyhedron 247:116705. https://doi.org/10.1016/j.poly.2023.116705

Râpă M et al (2020) Development of bionanocomposites based on PLA, collagen and AgNPs and characterization of their stability and in vitro biocompatibility. Appl Sci (Switzerland) 10(7). https://doi.org/10.3390/app10072265

Raval A, Parikh J, Engineer C (2010) Mechanism of controlled release kinetics from medical devices. Braz J Chem Eng 27(2):211–225. https://doi.org/10.1590/S0104-66322010000200001

Sandri G et al (2014) Montmorillonite–chitosan–silver sulfadiazine nanocomposites for topical treatment of chronic skin lesions: in vitro biocompatibility, antibacterial efficacy and gap closure cell motility properties. Carbohydr Polym 102:970–977. https://doi.org/10.1016/j.carbpol.2013.10.029

Sharma R, Ahuja M, Kaur H (2012) Thiolated pectin nanoparticles: preparation, characterization and ex vivo corneal permeation study. Carbohydr Polym 87(2):1606–1610. https://doi.org/10.1016/j.carbpol.2011.09.065

Shen G, Hu X, Guan G, Wang L (2015) Surface modification and characterisation of silk fibroin fabric produced by the layer-by-layer self-assembly of multilayer alginate/regenerated silk fibroin. PLoS ONE 10(4):e0124811. https://doi.org/10.1371/journal.pone.0124811

Shi Y, Guenneau F, Wang X, Hélary C, Coradin T (2018) MnO_2-gated nanoplatforms with targeted controlled drug release and contrast-enhanced MRI properties: from 2D cell culture to 3D biomimetic hydrogels. Nanotheranostics 2(4):403–416. https://doi.org/10.7150/ntno.28046

Shukla SK, Mishra AK, Arotiba OA, Mamba BB (2013) Chitosan-based nanomaterials: a state-of-the-art review. Int J Biol Macromol 59:46–58. https://doi.org/10.1016/j.ijbiomac.2013.04.043

Siegel RA, Rathbone MJ (2012) Overview of controlled release mechanisms. In: Fundamentals and applications of controlled release drug delivery. Springer US, Boston, MA, pp 19–43. https://doi.org/10.1007/978-1-4614-0881-9_2

Singh S, Bala N, Kaur B (2024) Green sustainable bionanocomposites materials: short review. J Manage Eng Sci 1(3):98–106. https://doi.org/10.61552/jmes.2024.03.003

Snetkov P, Morozkina S, Uspenskaya M, Olekhnovich R (2019) Hyaluronan-based nanofibers: fabrication, characterization and application. Polymers (Basel) 11(12):2036. https://doi.org/10.3390/polym11122036

Steenfos HH (1994) Growth Factors and Wound Healing. Scand J Plast Reconstr Surg Hand Surg 28(2):95–105. https://doi.org/10.3109/02844319409071186

Szliszka E, Czuba ZP, Domino M, Mazur B, Zydowicz G, Krol W (2009) Ethanolic extract of propolis (EEP) enhances the apoptosis—inducing potential of TRAIL in cancer cells. Molecules 14(2):738–754. https://doi.org/10.3390/molecules

Tahir D, Syarifuddin S, Noor EEM, Heryanto H, Mohamed MA (2025) Advancements in protein-based bionanocomposites for targeted and controlled drug delivery systems: a comprehensive review. J Drug Deliv Sci Technol 106:106698. https://doi.org/10.1016/j.jddst.2025.106698

Tahir D, Syarifuddin S, Noor EEM, Heryanto H, Mohamed MA (2025) Advancements in protein-based bionanocomposites for targeted and controlled drug delivery systems: a comprehensive review. J Drug Deliv Sci Technol [Online]. Available: https://api.semanticscholar.org/CorpusID:276389478

Tariq M, Alam MdA, Singh AT, Iqbal Z, Panda AK, Talegaonkar S (2015) Biodegradable polymeric nanoparticles for oral delivery of epirubicin: in vitro, ex vivo, and in vivo investigations. Colloids Surf B Biointerfaces 128:448–456. https://doi.org/10.1016/j.colsurfb.2015.02.043

Thakkar KN, Mhatre SS, Parikh RY (2010) Biological synthesis of metallic nanoparticles. Nanomedicine 6(2):257–262. https://doi.org/10.1016/j.nano.2009.07.002

Trinh X-T et al (2022) A comprehensive review of natural compounds for wound healing: targeting bioactivity perspective. Int J Mol Sci 23(17):9573. https://doi.org/10.3390/ijms23179573

Trotta F et al (2024) In vitro characterization of bionanocomposites with green silver nanoparticles: a step towards sustainable wound healing materials. Nano Select 5(2). https://doi.org/10.1002/nano.202300087

Wani SD, Mundada AS (2021) A review: emerging trends in bionanocomposites. Int J Pharm Res Technol 11(1). https://doi.org/10.31838/ijprt/11.01.01

Wong K, Lee, Zhang, Liu, Sun, Che (2012) Enhancement of anticancer efficacy using modified lipophilic nanoparticle drug encapsulation. Int J Nanomedicine, 731. https://doi.org/10.2147/IJN.S28783

Yusefi M et al (2021) 5-Fluorouracil loaded magnetic cellulose bionanocomposites for potential colorectal cancer treatment. Carbohydr Polym 273:118523. https://doi.org/10.1016/j.carbpol.2021.118523

Ziegler R, Ilyas S, Mathur S, Goya GF, Fuentes-García JA (2024) Remote-controlled activation of the release through drug-loaded magnetic electrospun fibers. Fibers 12(6):48. https://doi.org/10.3390/fib12060048

Shana Shaji is a medical doctor graduated from Government Ayurveda Medical College, Tripunithura, India. With a strong foundation in traditional Indian medicine, she has expanded her expertise to include modern research methodologies and cutting-edge technologies like Machine Learning and Artificial Intelligence in the domain of public health. Her research interests focus on environment and public health, maternal nutrition, and the use of image recognition using Machine learning & deep learning to study brain tumours. She has published three international conference papers in field of public health. Currently, she is associated with ISBEM—Istituto Scientifico Biomedico Euro Mediterraneo, Bindisi, Italy as a research associate.

Jerin Jose, currently working as a postdoctoral researcher in the Department of Civil Engineering and Architecture, University of Catania, Italy. The area of research is in the 'Recyclability of Polymer materials used in Additive manufacturing'. He has also completed another Postdoctoral fellowship from University of Salento, Italy. Jerin has completed Ph.D., Masters and Bachelors from Karunya Institute of Technology and Sciences, Tamil Nadu, India. His research interests include Biocomposites, Machine learning for aerospace structural health monitoring, Metal matrix composites, syntactic foams, agro waste based reinforcements. He is also an Assistant Professor in Amal Jyothi College of Engineering (Autonomous), Kerala, India from 2020 onwards.

Jeny Jose is currently pursuing a Ph.D. in an interdisciplinary project at Bar-Ilan University in Israel and the University of California, Davis. Her research focuses on the development of new materials through intercalation methods for various optoelectronic applications. Her Master's thesis, titled *Zirconium and Cissus quadrangularis-Doped Hydroxyapatite*, was presented as a poster at the Nano-Florida Virtual Conference. As a part of her Ph.D. work titled *Unlocking Optoelectronic Potential: Metal Intercalation Strategies for Tuning Tungsten Diselenide*, Jeny presented her findings at the prestigious Gordon Research Conference in Massachusetts, United States.

Juwel Jose is a graduate in Bachelor of science for Mathematics from Madras Christian College. He also completed Post Graduate Diploma in Data Science and currently working at Caritas Hospital, Kottayam as 'Medical Research Trainee'. He is an enthusiast in medical research and its subclasses. Juwel has carried out various machine learning/deep learning based prediction research works viz. White blood cells classification, Oral Squamous Cell Carcinoma (OSCC) Classification, Predictions of Kidney, Autism & Parkinson's diseases, Brain Tumor Segmentation, Early Detection and Prediction of Brain Stroke Severity, Spine X-Ray classification, Ovarian cancer subtype classification and Breast cancer classification.

Dr. Alessandro Distante is a renowned Italian cardiologist and biomedical researcher. As Professor at the University in Pisa and Section Director of CNR IFC in Lecce, he made significant contributions to the field of cardiology, namely on cardiovascular disease prevention, diagnosis, and treatment, focusing as well on clinical and experimental use of ultrasound in Cardiology. On 1997, he set up the EUROECHO meeting, now the largest of the world, within the outstanding initiatives of the European Society of Cardiology. His research has been published in numerous international journals, and he has collaborated with various institutions worldwide. His expertise is recognised thanks to his involvement in prestigious research projects and academic affiliations. Now retired, Dr. Distante's legacy has continued through his 25 years of scientific direction of ISBEM, a non-profit social enterprise located in Mesagne (Apulia), which inspires future generations of researchers, firstly increasing their number and then promoting their scientific biomedical mission through innovative projects. He promotes the GIFT Paradigm to benefit all people of the world by producing knowledge, transferring results and exchanging best practices. Throughout his career, as of March 2025, his H-index and total citation counts as 55 and 13,221respectively.

Chapter 7
Bionanocomposites in Energy Applications: The Future of Clean Energy

Amirhossein Raouffard, Alireza Fatahi, Abbas Mohammadi, and Atiyeh Derakhshandehpour

Abstract The increasing global energy demand, combined with the urgent environmental concerns associated with conventional energy sources, presents significant obstacles to the sustainable advancement of modern society. Addressing these challenges has driven research efforts toward developing affordable and biodegradable alternatives that can replace toxic and expensive materials in energy technologies. Bionanocomposites have emerged as a promising solution in this arena, garnering substantial interest due to their unique electrochemical characteristics, renewable origins, and environmental compatibility. This book chapter delves into the advancements in bio-nanocomposites and their potential to revolutionize energy storage and conversion devices, such as batteries, supercapacitors, solar cells, and fuel cells. The focus is on the integration of biopolymers with nanostructured inorganic materials to enhance energy storage and conversion efficiency. Specifically, improvements have been observed in the development of key components like electrodes, separators, current collectors, and electrolytes. These improvements are primarily attributed to the ability of biopolymer-based matrices to facilitate better loading of active materials and the creation of uniform, porous structures that optimize electrochemical performance.

Keywords Bionanocomposites · Energy storage · Energy conversion · Clean energy · Supercapacitors · Batteries · Solar cells · Fuel cells · Water splitting

7.1 Introduction

The increasing global demand for energy, particularly in developed and developing countries, has necessitated a shift toward more sustainable energy technologies as viable alternatives to conventional fossil fuel-based electricity generation. Reliance

---

A. Raouffard · A. Fatahi · A. Mohammadi (✉) · A. Derakhshandehpour
Department of Chemistry, University of Isfahan, Isfahan, Iran
e-mail: abbaspolymer66@gmail.com

S. C. George and B. Tawiah (eds.), *Advanced Bionanocomposite Materials*, Advanced Structured Materials 241, https://doi.org/10.1007/978-981-96-8225-6_7

on fossil fuels has led to significant environmental concerns, including climate change and global warming, primarily due to greenhouse gas emissions from electricity generation, which have increased exponentially in recent decades (Ang et al. 2022).

To mitigate the adverse environmental impacts associated with fossil fuel combustion, stringent regulations have been implemented to limit their usage, prompting the search for cleaner and renewable energy sources. Energy derived from inexhaustible sources is classified as renewable. These sources are naturally replenished within a relatively short timeframe. Unlike finite fossil fuels, renewable energy sources such as solar, wind, biomass, hydropower, geothermal, and tidal energy offer a sustainable pathway for energy generation. These energy sources can either be utilized directly or stored in energy storage systems, including batteries, for later use (Deshmukh et al. 2023).

In recent years, several innovative approaches have been developed to optimize energy harvesting while minimizing environmental impact. The necessity to reduce fossil fuel dependency has accelerated the emergence of next-generation energy solutions, such as electric vehicles (EVs) powered by lithium-ion batteries, thermoelectric materials, fuel cells, and photovoltaic (PV) systems, among others (Bensebaa 2013). One promising strategy to enhance the efficiency and sustainability of these technologies is the utilization of recycled nanomaterials derived from industrial and agricultural waste (Abd Elkodous et al. 2022).

With the increasing production of industrial and agricultural waste, environmental challenges associated with their disposal have become a significant global concern. Proper waste management not only prevents environmental pollution but also creates opportunities for recycling and reusing these materials in various industries. A novel and efficient approach in this regard is the conversion of waste into advanced nanomaterials for energy storage and conversion applications. Recycled nanomaterials such as carbon nanotubes (CNTs), carbon quantum dots (CQDs), graphene, and metal nanoparticles possess unique properties, including high surface area, chemical stability, electrical conductivity, and catalytic activity, making them highly valuable in lithium-ion batteries, fuel cells, supercapacitors, and solar cells. These cutting-edge technologies not only reduce production costs and energy consumption but also play a crucial role in mitigating greenhouse gas emissions and advancing renewable energy solutions (Abd Elkodous et al. 2022). Figure 7.1 provides an overview of the interconnections between energy storage, energy conversion, and the diverse value-added nanomaterials utilized in energy-related applications.

Alongside recycled nanomaterials, bionanocomposites have emerged as an important research area due to their biocompatibility, biodegradability, and easy accessibility of raw materials. These nanocomposites are typically composed of natural polymers such as chitosan, cellulose, and proteins combined with carbon-based or inorganic nanomaterials, finding applications in biomedical engineering, energy, smart packaging, and biosensors (Bunekar and Green 2020).

Bionanocomposites exhibit unique physical, chemical, and biological phenomena and properties at the nanoscale, compared to atoms and molecules or bulk materials. In this regard, they have diverse applications in various fields of engineering and technology (Bunekar and Green 2020; Dwivedi and Dwivedi 2023; Talukdar 2019).

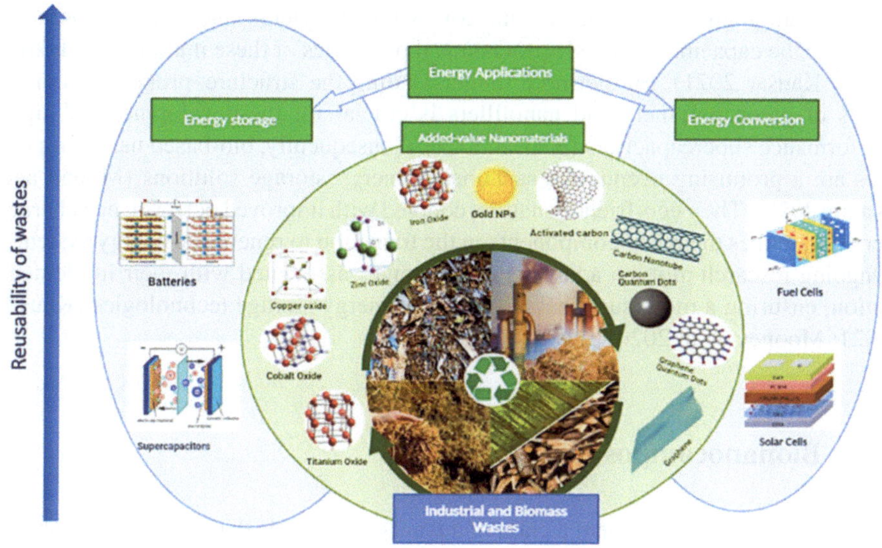

Fig. 7.1 Diagram illustrating the process of recycling nanomaterials (NMs) derived from biomass and industrial byproducts, along with their potential uses in energy applications. Adapted with permission from Degruyter (Abd Elkodous et al. 2022)

The energy industry is one of the fields in which nanoscale science and engineering continue to be successfully applied. Bionanocomposites have high surface-to-volume ratios due to their small dimensions and are suitable for catalytic and energy storage applications. At the nanoscale, bionanocomposites may exhibit quantum behavior, which is useful in optical and electronic applications. Many bio-nanocomposites are composed of biodegradable materials, which makes them suitable for environmental applications (Bunekar and Green 2020; Dwivedi and Dwivedi 2023; Talukdar 2019).

In the energy industry, bionanocomposites are used in solar cells to enhance light absorption and improve charge transfer. Materials such as bio-inspired nanocrystals are used due to their excellent photovoltaic properties. In addition, bio-based nanocomposites are used in batteries and supercapacitors to increase energy density and improve performance. Carbon nanotubes and graphene, which are derived from biological sources, are used in electrodes due to their high electrical conductivity and charge storage capacity. Bio-based nanocomposites are also used as catalysts in energy processes such as hydrogen production and the conversion of CO_2 into useful fuels. Platinum nanoparticles on carbon-based bio-based nanocomposites improve catalytic efficiency in fuel cells (Wang et al. 2020; Chen et al. 2020).

Furthermore, bio-based nanocomposites are used in green energy systems such as nanofluids and phase change materials to improve efficiency and stability. Despite their promising properties, the full implementation of green polymer nanocomposites in advanced energy applications remains challenging. Further research is necessary to optimize their performance and scalability (Bunekar and Green 2020; Kausar 2021).

Incorporating green nanofillers with conductive polymers and nanocarbons can enhance the capacitance and electrochemical properties of these materials (Talukdar 2019; Kausar 2021). Furthermore, understanding the structure–property relationships of green polymers and nanofillers is crucial for the development of high-performance supercapacitors (Kausar 2021). Consequently, bio-based nanocomposites are a promising avenue for sustainable energy storage solutions (Mohan and Kanny 2020). Their eco-friendly nature, coupled with improved performance characteristics, makes them key components in the transition to renewable energy systems. Ongoing research aims to address the challenges associated with their implementation, ensuring a more sustainable future for energy storage technologies (Kausar 2021; Mooney et al. 2020; Siwal et al. 2020).

7.2 Bionanocomposites

Bionanocomposites represent a class of advanced materials that integrate biological polymers with nanoscale fillers, resulting in multifunctional composites with superior properties and environmentally sustainable characteristics. These materials exhibit a multiphase architecture, typically consisting of a biopolymer matrix reinforced with inorganic or organic nanofillers that enhance mechanical strength, thermal stability, and chemical resistance.

The biopolymers used in bionanocomposites can be classified based on their origin into natural biopolymers and synthetic biopolymers. Natural biopolymers, which are obtained from renewable biological sources such as plants, animals, and microorganisms, offer significant advantages, including biodegradability, biocompatibility, and wide availability, making them highly suitable for the development of sustainable materials (Jeevanandam et al. 2022; Karki et al. 2021; Anwer et al. 2023). Figure 7.2 illustrated various biopolymer structures and sources (Table 7.1).

In addition to natural biopolymers, synthetic biopolymers offer suitable solutions by mimicking the properties of their natural counterparts while exhibiting greater flexibility in terms of processing and performance. A notable example is polylactic acid (PLA), a biodegradable thermoplastic synthesized from renewable raw materials such as corn starch and sugarcane. Due to its excellent mechanical properties and biocompatibility, PLA is widely used in biomedical implants, food packaging, and agricultural films (Karki et al. 2021).

The incorporation of nanofillers into biopolymer matrices significantly enhances the structural and functional properties of bionanocomposites. These nanofillers, which typically measure between 1 and 100 nm, include nanoclays (Anwer et al. 2023; Sen et al. 2024), carbon nanotubes (Huang 2020; Goswami et al. 2021; Ates et al. 2020), and metal nanoparticles (Ag, Au, and Fe_3O_2) (Ates et al. 2020; Burlec et al. 2023), metal oxides (TiO_2, M_2O_2, SiO_2, and Zn). Hybrid nanomaterials are essential components in bio-nanocomposites and energy storage systems due to their

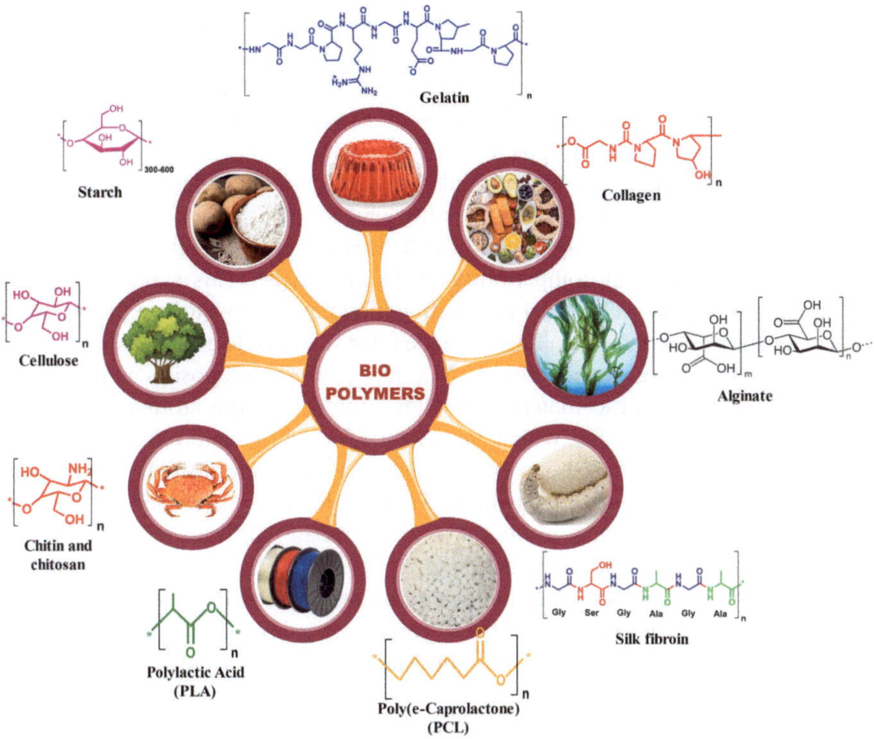

Fig. 7.2 Structures and sources of biopolymers that can be used for developing bionanocomposites. Adapted with permission from Elsevier (Anwer et al. 2023)

Table 7.1 Properties and applications of biopolymers (Karki et al. 2021; Anwer et al. 2023; Sen et al. 2024)

Property	Natural biopolymers	Synthetic biopolymers
Source	Renewable and sustainable	Renewable or artificially synthesized
Biodegradability	High	Varies, often high
Customizability	Limited	High
Cost	Generally lower	May be higher
Applications	Packaging, energy, agriculture, and medicine	Medicine, energy, and specialty packaging

exceptional mechanical strength, electrical conductivity, and surface area. Their addition improves key properties, including mechanical reinforcement, thermal conductivity, and gas barrier performance. Furthermore, the functionalization of nanofillers enables specific interactions with the polymer matrix, enabling advanced capabilities such as antimicrobial properties and controlled drug release (Ates et al. 2020; Mallick et al. 2023).

Integrating MXene with biopolymers such as cellulose has led to the development of advanced bionanocomposites with promising implications for clean energy technologies. Cellulose is a natural and renewable polymer that is highly valued for its mechanical strength, biodegradability, and tunable surface chemistry. When combined with MXene, it significantly enhances the mechanical durability, flexibility, and environmental compatibility of the composite. This synergy results in materials with superior electrical conductivity, higher energy storage capacity, and high mechanical flexibility, making them ideal for sustainable energy solutions. One of the most promising applications of MXene-cellulose composites is in supercapacitor technology. These materials provide efficient charge storage and exhibit fast charge–discharge cycles, which are essential for next-generation energy storage systems. Figure 7.3 shows the synthesis and characterization of MXene/cellulose composites for flexible supercapacitors. Their high surface area and conductivity enable them to outperform conventional electrode materials. Furthermore, MXene-based nanocomposites exhibit remarkable electrocatalytic properties, especially in hydrogen evolution and oxygen evolution reactions, thereby facilitating the production of sustainable hydrogen fuel. These composites play a key role in optimizing catalytic efficiency and stability by utilizing a large surface area and tunable electronic structure. Beyond supercapacitors and electrocatalysis, MXene-cellulose materials are being investigated for applications in fuel cells, where they contribute to efficient ion transport and long-term durability. Their versatility extends to wearable energy storage devices and flexible energy harvesting systems, furthering their potential impact in renewable energy technologies.

Looking ahead, research on MXene-based bionanocomposites continues to advance, focusing on improving material stability, optimizing large-scale manufacturing techniques, and increasing integration into commercial energy storage devices. By refining surface interactions and enhancing mechanical properties, scientists aim to develop materials that not only exhibit superior performance but also align with principles of sustainability and energy efficiency. Continued innovation in this field underscores the potential of MXene and biopolymer composites to advance the future of clean energy technologies (Shaheen et al. 2024; Zhao et al. 2025; Al-Fakih et al. 2024).

7.2.1 Synthesis and Fabrication Techniques

Bionanocomposites are synthesized using various methods, each tailored to optimize the interaction between biopolymers and nanofillers to achieve desired material properties. These methods leverage different processing techniques to integrate nanofillers like nanoparticles, nanotubes, or nanoclays with biopolymer matrices such as cellulose, chitosan, and polylactic acid (PLA). The most widely used methods are solution

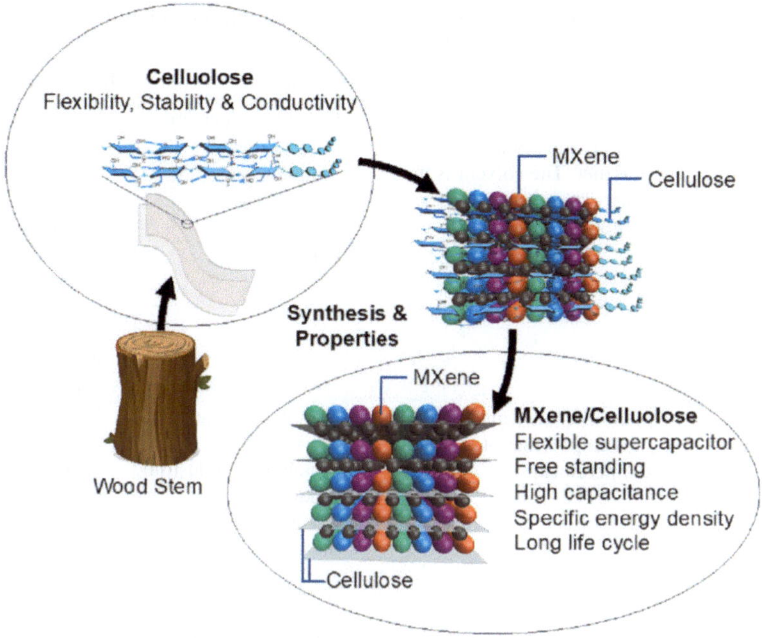

Fig. 7.3 Synthesis and characteristics of MXene/cellulose composites for flexible supercapacitors. Adapted with permission from Elsevier (Shaheen et al. 2024)

intercalation, melt intercalation, in situ polymerization, sol–gel processes, extrusion, template synthesis, electrospinning, and Layer-by-Layer (LbL) techniques are summarized in Table 7.2 and Fig. 7.4 (Anwer et al. 2023; Ates et al. 2020; Sen et al. 2024).

7.3 Energy Applications

Energy plays a crucial role in the development trajectories of both developed and developing nations. The increasing global population and rising standards of living have led to a substantial increase in energy demand. Concurrently, conventional energy resources, primarily based on fossil fuels, are finite. Furthermore, the utilization of these fuels is associated with environmental pollution and global warming (Chen et al. 2023; Zeng et al. 2024).

To address the global energy crisis, extensive scientific efforts have been directed toward developing novel energy solutions and optimizing existing technologies. These challenges can be mitigated through two primary approaches. The first involves the efficient conversion of inexhaustible energy sources, such as solar radiation, water,

Table 7.2 Methods of synthesis for bio-nanocomposites: advantages and disadvantages (Anwer et al. 2023; Ates et al. 2020; Sen et al. 2024)

Method	Description	Advantages	Disadvantages
Solution intercalation	Nanofillers are dispersed in a solvent and combined with a polymer. The solvent is then evaporated, leaving the composite material	– Uniform dispersion of nanofillers – Simple to execute – Suitable for thermosetting polymers	– Solvent removal can be challenging – Potential environmental concerns due to solvent use – Limited scalability for industrial applications
Melt intercalation	Nanofillers are mixed with a molten polymer at elevated temperatures	– No solvent required, eco-friendly – Suitable for industrial production – Compatible with thermoplastics	– High processing temperatures can degrade some biopolymers – Inefficient dispersion of nanofillers in some cases
In situ polymerization	Nanofillers are dispersed in a monomer, and polymerization occurs directly within the composite matrix	– Strong interfacial bonding – Customizable polymer properties – Effective for integrating reactive nanofillers	– Requires precise control of reaction conditions – Can involve toxic initiators or catalysts – Often expensive
Sol–gel process	Metal alkoxides are hydrolyzed and condensed to form a networked gel, which incorporates the polymer and nanofiller	– High control over nanostructure morphology – Excellent for ceramic and metal oxide composites – Produces thermally stable materials	– Expensive precursors – Sensitive to moisture during synthesis – Limited scalability for industrial use
Electrospinning	Uses high-voltage electric fields to produce nanofibers from a polymer solution, often integrating nanoparticles during the process	– Produces nanofibrous structures with high surface area – Ideal for biomedical and filtration applications – Facilitates controlled morphology of fibers	– High setup cost – Requires fine control over environmental and processing conditions – Limited scalability for large-scale production

(continued)

Table 7.2 (continued)

Method	Description	Advantages	Disadvantages
Template synthesis	Biomolecules or templates (e.g., microorganisms, cells) are used to guide the formation of nanostructures, which are then integrated with the polymer	– Precise control of nanostructure size and shape – Uses renewable resources as templates – Environmentally friendly	– Limited to specific biomaterial availability – Can be time-intensive – Template removal may introduce additional processing steps
Extrusion	Polymer and nanofiller are melted and mixed in a screw extruder, often followed by shaping through dies	– Cost-effective and scalable – Suitable for continuous production – Compatible with various polymers and nanofillers	– Restricted to thermoplastic materials – High shear forces may cause filler aggregation – Limited control over final nanostructure dispersion
Layer-by-layer (LbL)	Thin layers of polymers and nanofillers are sequentially deposited onto a substrate using electrostatic, covalent, or other interactions to build up the composite layer-by-layer	– High precision in film thickness and composition – Allows multi-functionalization – Excellent control over material properties	– Time-intensive for large-scale applications – Requires precise control of deposition conditions – Limited thickness achievable per cycle

and oxygen reduction reactions, into usable forms like electricity and fuel. Technologies such as fuel cells, solar cells, and water-splitting catalysts play a pivotal role in this domain. The second approach focuses on the development of cost-effective, environmentally sustainable, and high-performance energy storage systems. Among energy storage technologies, lithium-based batteries, including lithium-ion, lithium-air, and lithium-sulfur batteries, are particularly promising due to their high energy density and efficiency. Supercapacitors are yet another extremely effective energy storage option; they can store and release energy quickly within seconds. A sustainable and low-carbon future depends on these developments in renewable energy technology (Ates et al. 2020; Xie et al. 2022).

High mechanical strength, thermal stability, and superior conductivity are just a few of the special attribute combinations that make bionanocomposites extremely promising for energy applications. Because of these qualities, they may be integrated

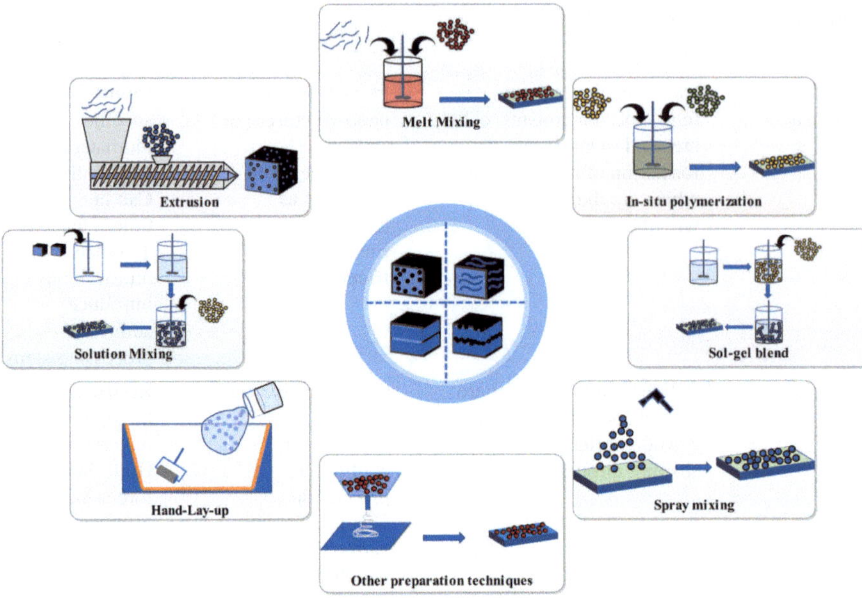

Fig. 7.4 Different processing techniques for the fabrication of biopolymer nanocomposites derived from renewable resources. Adapted with permission from Elsevier (Anwer et al. 2023)

with renewable energy technologies, including energy storage devices, solar cells, and wind turbine parts. By adding nanofillers to biopolymer matrices, electrical and thermal conductivity are increased, improving device energy efficiency. Additionally, its flexibility and light weight make it easier to create strong, small parts that are essential for portable and renewable energy systems (Ang et al. 2022; Sezali et al. 2024).

Bio-based nanocomposites can help increase photovoltaic efficiency in solar energy applications by improving energy conversion efficiency through optimized light absorption and charge transport properties. Due to the high strength-to-weight ratio of these materials, wind turbine blades last longer and perform better, reducing wear and operating costs. Similarly, bio-based nanocomposites provide ion transport and charge retention in energy storage devices such as batteries and supercapacitors, resulting in increased energy density and longer lifetime. Table 7.3 shows recent studies on the use of nanocomposites in the field of clean energy.

By addressing sustainability issues and reducing carbon footprints, the renewable and biodegradable properties of bio-based nanocomposites complement the environmental goals of the renewable energy industry. Through ongoing advancements in materials science, bionanocomposites offer the potential to transform the energy landscape, making renewable energy technologies more efficient, sustainable, and economically viable. Figure 7.5 illustrates various energy-rich nanocomposite and bionanocomposite materials and their applications in the energy sector, highlighting their role in enhancing energy conversion, storage, and sustainability.

Table 7.3 Studies on the use of nanocomposites in the field of clean energy

Number	Composite structure	Application	Result	References
1	$CuAuTiO_2$/chitosan hydrogel composite as a photocatalyst	Hydrogen production	The $CuAuTiO_2$/ChTPP composite achieved a hydrogen production rate of 1790 μmol g^{-1} h^{-1}, 12 times higher than TiO_2/ChTPP, due to band gap reduction (3.09 eV), better charge separation ($R_{ct} = 2175\ \Omega$), and enhanced visible light absorption (590 μmol g^{-1} h^{-1})	Ramírez et al. (2024)
2	Sodium alginate as a polymer binder in phosphorus/carbon anodes for sodium-ion batteries	Binder sodium-ion batteries	Using sodium alginate as a binder, the electrode showed a high reversible capacity of 1064 mAh g^{-1} and 90.1% capacity retention after 200 cycles at 800 mA g^{-1}	Xiao et al. (2020)
3	Chitosan-reduced graphene oxide-$CuInS_2$ composite (Chi@RGO-CIS)	Solar cells	Chitosan-reduced graphene oxide-$CuInS_2$ composite (Chi@RGO-CIS) under gamma irradiation outperforms platinum as a counter electrode in dye-sensitized solar cells, achieving an efficiency of 12.21% compared to 9.96%	Areerob et al. (2022)
4	Manganese dioxide–chitosan nanocomposite	Supercapacitors, rechargeable batteries, electrochemical sensors, and anti-corrosion coatings	The manganese dioxide–chitosan nanocomposite exhibits a high specific capacitance (424 F/g) and excellent cycling stability, with enhanced ionic and electronic conductivity, forming porous and crack-free films	Hassan et al. (2014)

(continued)

Table 7.3 (continued)

Number	Composite structure	Application	Result	References
5	Sulfonated polysaccharides (alginate and carboxymethylcellulose) as improved binders for $Li_4Ti_5O_{12}$ electrodes in lithium-ion batteries	Binder batteries	Sulfonation of alginate (Alg) and carboxymethyl cellulose (CMC) enhances ionic conductivity and improves $Li_4Ti_5O_{12}$ electrode performance in lithium-ion batteries. The CMC-S electrode retains 153.2 mAh g^{-1} at the 100th cycle, compared to 117.6 mAh g^{-1} for the non-sulfonated version	He et al. (2020)
6	Fluorinated corn starch-based binder for silicon anodes in lithium-ion batteries	Binder batteries	This study presents a fluorinated corn starch-based binder for silicon anodes in lithium-ion batteries, offering greater stability and capacity retention compared to commercial binders. Electrodes using this binder retained 88% of their initial capacity after 100 cycles	Jin et al. (2021a)
7	N-doped rGO/C@Si composite anode using chitosan, reduced graphene oxide (rGO), and silicon for lithium-ion batteries	Batteries	This study developed an N-doped rGO/C@Si composite anode using chitosan, reduced graphene oxide (rGO), and silicon for lithium-ion batteries. The composite enhances electrical conductivity and mitigates silicon's volume expansion, retaining a capacity of 1115.8 mAh g^{-1} after 150 cycles	Yu et al. (2020)

(continued)

Table 7.3 (continued)

Number	Composite structure	Application	Result	References
8	Silver sulfide/nickel titanate/chitosan nanocomposite for photocatalysis, water splitting, and antibacterial applications	Water splitting	This study developed a silver sulfide/nickel titanate/chitosan nanocomposite for photocatalysis, water splitting, and antibacterial applications. The nanocomposite achieved complete degradation of metronidazole within 40 min under visible light, exhibited a hydrogen production rate of 62.35 μmol/g, and demonstrated strong antibacterial activity against *Streptococcus pneumoniae* and *Escherichia coli*	Wang et al. (2021)
9	ZnO/ZnS nanocomposite-based dye-sensitized solar cell (DSSC) using a chitosan-based polymer electrolyte	Solar cell (DSSC)	This study developed a ZnO/ZnS nanocomposite-based DSSC using a chitosan polymer electrolyte, which enhanced ionic conductivity and stability. The protective ZnS layer on ZnO reduced electron recombination, resulting in a 1.6% power conversion efficiency (PCE)	Praveen et al. (2019)
10	Chitosan-based nanocomposite with sulfonated titanium dioxide for fuel cells	Fuel cells	This paper presents a chitosan-based nanocomposite with sulfonated titanium dioxide for fuel cells. The membranes show enhanced proton conductivity (0.035 S/cm), comparable to Nafion 117 (0.033 S/cm), along with improved thermal and mechanical stability	Ahmed et al. (2021)

(continued)

Table 7.3 (continued)

Number	Composite structure	Application	Result	References
11	Chitosan–Prussian blue (KCoFe-CS) nanocomposite on nickel foam	Water splitting	The chitosan–Prussian blue (KCoFe-CS) nanocomposite on nickel foam enhances OER/HER efficiency and stability, achieving 98 and 96% Faradaic efficiencies and enabling stable water splitting at 1.55 V and 20 mA cm^{-2}	Dhanasekaran et al. (2023)
12	The boron- and nitrogen-doped porous carbon chitosan-based nanocomposite with iron-cobalt (BNPC-FeCo-900)	Water splitting	The boron- and nitrogen-doped porous carbon chitosan-based nanocomposite with iron-cobalt (BNPC-FeCo-900) serve as a bimetallic catalyst for both the hydrogen evolution reaction (HER) and oxygen evolution reaction (OER), exhibiting outstanding performance with low overpotentials (186 and 320 mV), small Tafel slopes (82 and 55 mV dec^{-1}), and low charge transfer resistance	Sam and Cao (2024)
13	Nickel oxide/ chitosan-coated polypyrrole nanotube (NiO/CHI@PPy-NT) nanocomposite	Microbial fuel cell (wastewater treatment systems and sustainable bioenergy generation from organic sources)	The NiO/CHI@PPy-NT nanocomposite, used as an anode in a single-chamber microbial fuel cell (SMFC), achieved a power density of 755 mW m^{-2}, which is 4.4 times higher than that of the bare stainless steel anode (170 mW m^{-2}). Additionally, its charge transfer resistance was 165.3 Ω, which is 4.9 times lower than that of the bare stainless steel anode (817 Ω)	Geetanjali et al. (2022)

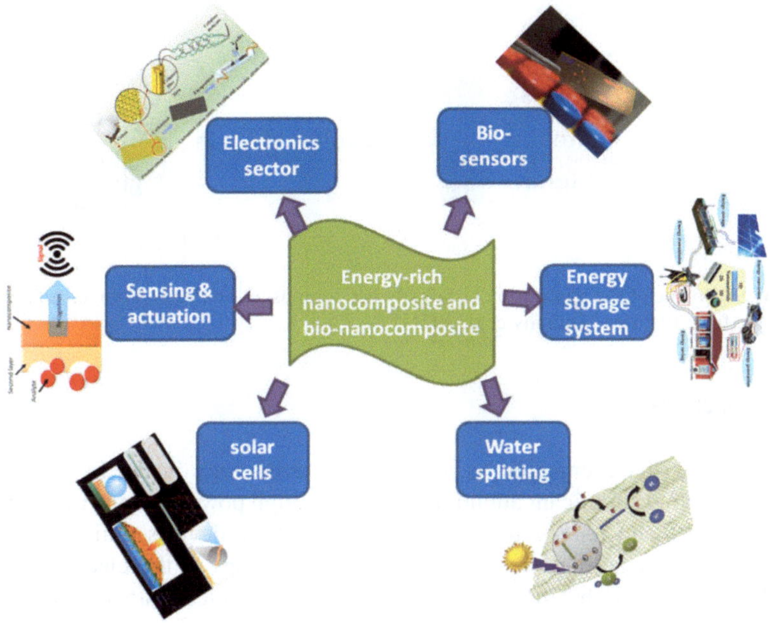

Fig. 7.5 Energy-dense nanocomposite and bio-nanocomposite materials and their applications in the energy sector. Adapted with permission from Elsevier (Karki et al. 2021)

7.3.1 Energy Storage Systems

Energy storage systems based on bionanocomposites have attracted significant attention in the energy industry due to their unique characteristics, such as biodegradability, biocompatibility, low cost, lightweight, and flexibility. These materials are widely used in applications such as batteries, solar cells, and supercapacitors. Bionanocomposites significantly enhance the performance of energy storage devices due to their nanoscale structure. These improvements include increased energy density, efficiency, durability, and cost-effectiveness. Moreover, the use of these environmentally friendly materials represents an important step toward the development of sustainable technologies and reducing dependence on non-renewable resources. As shown in Fig. 7.6, in this category, the most recent developments in devices for energy storage that make use of biopolymers, specifically, in batteries, and supercapacitors are discussed (Patra et al. 2024; Mahajan and Sharma 2024).

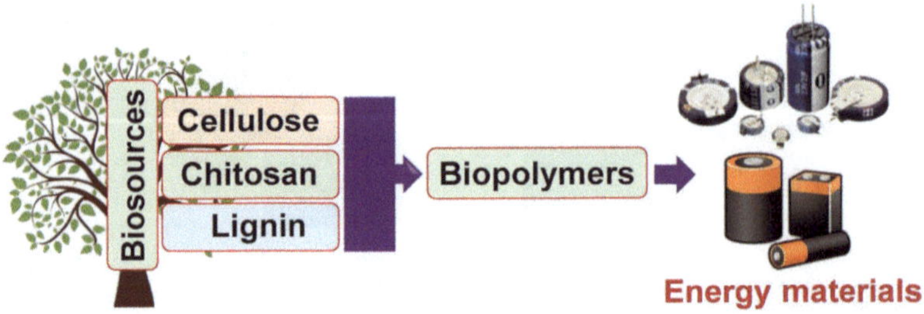

Fig. 7.6 Schematic of the application of biopolymer composite for energy storage. Adapted with permission from Elsevier (Anthony et al. 2021)

Supercapacitors

Supercapacitors have emerged as a promising technology in the field of energy storage, attracting significant attention due to their high power density, rapid charge–discharge rates, and long-term cycling stability. These devices utilize two mechanisms for energy storage: reversible Faradaic reactions at the electrode surface (Pseudocapacitance) and the mechanism based on the electrostatic adsorption of ions at the electrode–electrolyte interface (EDLC).

Despite their advantages, supercapacitors face limitations such as low specific capacitance, high self-discharge rates, and suboptimal discharge efficiency. The performance of SCs is heavily influenced by the choice of electrode materials, prompting extensive research into various candidates, including metal oxides (e.g., Co_3O_4, MnO_2), conductive polymers (e.g., polyaniline, polypyrrole), and carbon-based materials. Carbonaceous materials, such as activated carbon, carbon nanotubes, graphene, and templated carbons, are particularly attractive due to their high surface area, tunable pore structures, and excellent conductivity.

A significant trend in electrode material development is the utilization of biomass-derived carbon. Renewable sources such as cellulose, chitin, chitosan, and lignin offer sustainable alternatives, often containing heteroatoms (O, N, S) that enhance pseudocapacitive behavior and electrochemical performance. Structural optimization and heteroatom doping strategies further improve capacitance and stability, making biomass-derived carbons a viable alternative to traditional carbon materials.

Electrolyte selection plays a crucial role in determining the overall performance of SCs. Conventional liquid electrolytes, though widely used, pose safety and environmental concerns due to toxicity and leakage risks. Polymer electrolytes, which integrate ionic conductivity with mechanical stability, have gained interest as safer alternatives. However, their production often involves resource-intensive processes. Biomass-derived polymer electrolytes offer an eco-friendly solution, leveraging materials such as cellulose, starch, and chitosan. These bio-based electrolytes exhibit desirable properties, including renewability, low cost, and superior mechanical strength, while also addressing sustainability concerns in SC fabrication.

Recent studies highlight the effectiveness of biomass-derived materials in SC applications. For instance, cellulose-based porous carbon structures exhibit high surface area and conductivity, achieving specific capacitances up to 358 F/g with a retention of 97% over 10,000 cycles.

Chitosan-based gel polymer electrolytes and composite electrodes, particularly when combined with graphene oxide or polyaniline, have demonstrated capacitances exceeding 600 F/g with excellent cyclic stability. These materials play a crucial role in supercapacitor technology. Figure 7.7a illustrates the schematic representation of a supercapacitor in charged and discharged states, while Fig. 7.7b highlights chitosan- and chitin-based polymers used in the fabrication of electrode materials for supercapacitors.

Lignin-derived porous carbon electrodes, benefiting from phenolic functional groups, contribute to enhanced redox activity, achieving capacitances up to 728 F/g with excellent durability. Additionally, nitrogen-doped porous carbons synthesized from proteins and chitin show promising electrochemical properties, with specific capacitances ranging from 300 to 400 F/g and retention rates above 90% over extended cycling (Ahmed et al. 2023; Lin et al. 2021; Dar et al. 2024). Table 7.4 shows some studies on the application of bionanocomposites in supercapacitors and the results.

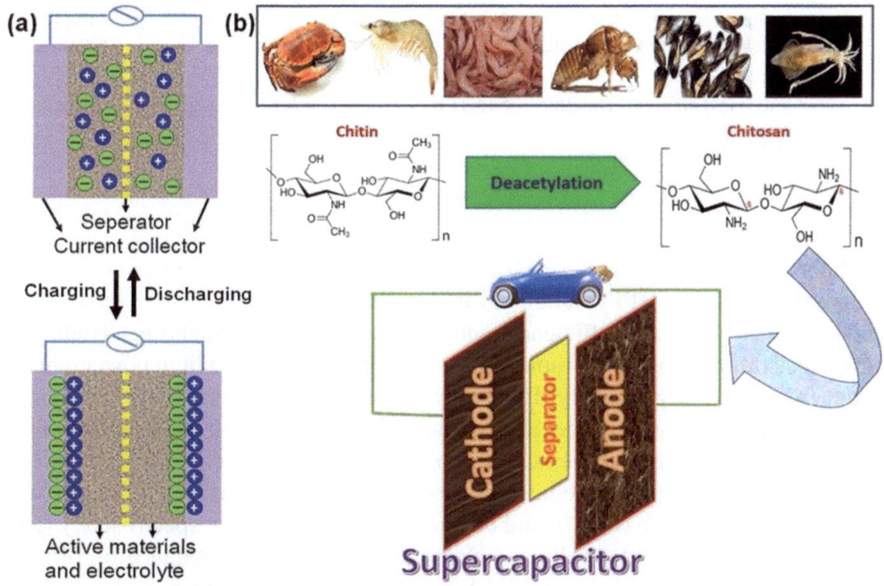

Fig. 7.7 **a** Schematic diagram of a charged and discharged supercapacitor. Adapted with permission from Elsevier (Chen and Dai 2013). **b** Polymers derived from chitosan and chitin used in the fabrication of electrode materials for supercapacitors. Adapted with permission from Elsevier (Vinodh et al. 2021)

Table 7.4 Studies on the application of bionanocomposites in supercapacitors and their results

Number	Composite structure	Specific capacitance	Cycling performance	Density	References
1	Chitosan-graphene-polyaniline	1519 F g^{-1}	1000 cycles (88%)	0.1 A g^{-1}	Torvi et al. (2018)
2	Chitosan/graphene oxide—multi-walled nanotubes (MWCNT)/ polyaniline (CS/GM/PANI)	609.2 F g^{-1}	500 cycles (96%)	5 A g^{-1}	Ghasem Hosseini and Shahryari (2017)
3	Core–shell nanofibrillated cellulose/polypyrrole/tubular graphitic carbon nitride	158 F g^{-1}	2000 (~ 91%)	5 mA cm^{-2}	Li et al. (2019)
4	Chitosan-ZnO/polyaniline	587.15 F g^{-1}	1000 (80%)	175 mA cm^{-2}	Pandiselvi and Thambidurai (2014)
5	Nitrogen-doped biomass activated carbon using the agricultural waste corn silks as raw materials and $ZnCl_2$	358.0 F g^{-1}	1000	0.5 A g^{-1}	Sun et al. (2018)
6	Polypyrrole@cobalt oxyhydroxide/cellulose fiber composite	571.3 F g^{-1}	1000 (93.02%)	0.2 A g^{-1}	Yang et al. (2020a)
7	Sodium carboxymethyl cellulose/collagen composite aerogel	185.3 F g^{-1}	5000 (90.2%)	0.5 A g^{-1}	Yu et al. (2018)
8	Poly(vinyl alcohol)-tetraethyl orthosilicate-manganese dioxide-poly(o-anisidine) (PVA-TEOS-MnO_2-POA)	548.23 F g^{-1}	1000 cycles (95%)	0.1 A g^{-1}	Mulla et al. (2024)

Tammela et al. (2015) developed an asymmetric supercapacitor utilizing polypyrrole-cellulose (PPy-cellulose) and carbon nanofibres (C-nf) electrodes to enhance energy storage efficiency and environmental sustainability. Traditional supercapacitors, despite their rapid charge–discharge cycles and high power density, suffer from low energy density. This study addresses this limitation by employing a PPy-cellulose composite as the positive electrode and its heat-treated derivative (C-nf) as the negative electrode, extending the operational voltage to 1.6 V in aqueous electrolytes. Electrochemical characterization via cyclic voltammetry (CV), galvanostatic charge–discharge (GCD), and electrochemical impedance spectroscopy (EIS) demonstrated that the PPy-cellulose electrode exhibited a capacitance of 146 F/g, while the C-nf electrode showed 59 F/g. The asymmetric device achieved 25 F/g at 1.6 V, significantly outperforming symmetric counterparts (45 F/g at 0.8 V), yielding an energy density of 33 J/g, more than double that of conventional PPy-cellulose supercapacitors (~ 14 J/g). The device retained 77% of its capacitance after 1000 cycles, confirming its stability.

Salleh et al. (2021) explored chitosan as a biopolymer binder in graphene-based supercapacitor electrodes, optimizing its concentration for enhanced electrochemical performance. At 10 wt% chitosan, the electrodes exhibited a specific capacitance of 135.65 F/g, with 87.2% retention after 1000 cycles. An asymmetric supercapacitor, utilizing MnO_2–graphene–chitosan as the positive and graphene–chitosan as the negative electrode, achieved an energy density of 80.40 Wh/kg at 2894.61 W/kg, retaining 74.18% capacitance after 1000 cycles. Cyclic voltammetry confirmed efficient charge storage behavior, supporting chitosan's viability as a sustainable electrode binder.

Shang et al. (2021) developed nitrogen-doped hierarchically porous carbon (N-HPC) electrodes from chitin nanofibers (ChNF) and zeolitic imidazolate frameworks (ZIF-8) as bio-templates. ChNF was extracted from crab shells, integrated with ZIF-8, and carbonized at 800–1000 °C under nitrogen to yield N-HPC electrodes with enhanced electrochemical properties. The electrodes exhibited a specific capacitance of 182.5 F/g in a three-electrode system and 128.5 F/g in a symmetric two-electrode setup, with 90% retention over 5000 cycles. The supercapacitor delivered an energy density of 4.46 Wh/kg at 50 W/kg. This study demonstrates a scalable, eco-friendly approach for high-performance bio-derived electrode materials in next-generation energy storage devices.

Li et al. (2021) developed high-performance, flexible all-solid-state supercapacitors (ASCs) by integrating engineered textile electrodes with bacterial cellulose (BC)-reinforced hydrogel electrolytes for wearable electronics. Polyester fiber textiles were modified via electrostatic self-assembly with reduced graphene oxide (RGO) and polyaniline (PANI), yielding electrodes with an areal capacitance of 1221 mF/cm^2, 94.4% retention after 10,000 charge–discharge cycles, and 97.9% retention after 2000 bending cycles. A BC-polyacrylamide (PAM) hydrogel electrolyte was synthesized, exhibiting high ionic conductivity (125 mS/cm), tensile strength (330 kPa), and stretchability (1300%). The integrated ASC achieved an areal capacitance of 564 mF/cm^2, an energy density of 50.1 μWh/cm^2, and a power density of 20 mW/cm^2, maintaining stability after 3000 bending cycles. This study presents a robust strategy for next-generation flexible supercapacitors with enhanced electrochemical and mechanical properties.

Chen et al. (2021) synthesized nitrogen- and oxygen-co-doped hierarchical porous carbon (HPC) electrodes from chitosan using a one-step carbonization-activation process with potassium nitrate (KNO_3) and magnesium nitrate ($Mg(NO_3)_2 \cdot 6H_2O$). The HPC exhibited a high surface area (922 m^2/g) and hierarchical porosity, enhancing charge storage. The HPCMgK-600 variant demonstrated a capacitance of 225.3 F/g at 1 A/g in a three-electrode system, retaining 88% capacitance over 10,000 cycles. In a symmetric configuration, it achieved an energy density of 10.46 Wh/kg at 500.08 W/kg, while in an all-solid-state supercapacitor, it maintained 70.8 F/g at 1 A/g. This research highlights a scalable and sustainable approach for bio-derived carbon materials in high-performance supercapacitors.

The integration of biomass-derived carbon electrodes and polymer electrolytes into SCs represents a significant advancement in sustainable energy storage. These materials offer high electrochemical performance while mitigating the environmental impact associated with conventional SC components. Future research should focus on optimizing synthesis techniques, improving scalability, and further enhancing electrochemical properties to facilitate the commercialization of eco-friendly SCs. By leveraging renewable resources, these advancements contribute to the development of sustainable, high-performance energy storage solutions, aligning with global efforts toward greener technologies.

Cui et al. (2023) developed a flexible, self-supporting MXene/biomass/chitosan (MBC) carbon aerogel electrode to enhance supercapacitor performance. Using hydrothermal synthesis, electrostatic self-assembly, and concentration difference permeation (CDP), they integrated positively charged radish flakes with negatively charged MXene nanosheets, ensuring homogeneous distribution. Among different MBC compositions, MBC-6 (6% MXene doping) exhibited optimal electrochemical properties with a BET surface area of 225.96 m^2/g, enabling efficient ion transport. In a three-electrode system with 3 M H_2SO_4, MBC-6 achieved a volumetric capacitance of 1801.4 mF/cm^3 at 2 mV/s, an energy density of 33.4 Wh/L, and 82% capacitance retention over 50,000 cycles, attributed to MXene's conductivity and low charge transfer resistance.

This study by Wei et al. (2022) addresses the limitations of biomass-derived porous carbon microspheres (PCMs) in supercapacitor applications due to insufficient reactive functional groups. To enhance electrochemical performance, MXene was integrated with chitosan-derived porous carbon microspheres (CPCM) via electrostatic interaction, forming a sandwich-like CPCM/MXene composite.

The CPCM framework preserved the MXene nanosheet dispersion while preventing reaggregation, thereby improving ion transport kinetics and electrolyte accessibility. Structural characterization revealed a high specific surface area (2387 m^2/g) and pore volume (1.39 cm^3/g) for CPCM, which decreased upon MXene integration (1649 m^2/g and 0.75 cm^3/g, respectively). Electrochemical impedance spectroscopy (EIS) demonstrated lower charge transfer resistance (0.16 Ω) and Warburg resistance (2.55 Ω) in CPCM/MXene, compared to CPCM alone.

In a three-electrode system with 1 M H_2SO_4, CPCM/MXene exhibited a specific capacitance of 362 F/g at 0.5 A/g and retained 93.87% of its capacitance after 10,000 cycles at 10 A/g. Cyclic voltammetry (CV) revealed a quasi-rectangular profile with prominent redox peaks (0.2–0.6 V), indicating electric double-layer capacitance (EDLC) and Faradaic pseudocapacitance contributions. In a two-electrode system with 1 M Na_2SO_4, CPCM/MXene exhibited stable electrochemical performance up to 2.0 V, with a capacitance of 194.5 F/g at 1 A/g and an energy density of 27.8 Wh/kg at 500 W/kg, retaining 22.8 Wh/kg even at 20,062.1 W/kg. This work advances biomass-derived porous carbon-based electrodes and provides an effective strategy for enhancing supercapacitor performance via MXene integration.

The study by Pu et al. (2022) investigates the synthesis and electrochemical performance of nitrogen-doped MXene derived from chitosan for high-efficiency supercapacitors. Chitosan, a nitrogen-rich biopolymer, serves as a doping precursor, enhancing MXene's capacitance, stability, and charge storage capability. Electrochemical tests, including cyclic voltammetry, galvanostatic charge–discharge, and electrochemical impedance spectroscopy, confirm significant performance improvements. The optimized material achieves a capacitance of 286.28 F/g at 3 A/g, an energy density of 21.47 Wh/kg, and a power density of 2343.75 W/kg, while retaining 98% capacitance after 10,000 cycles. These enhancements result from improved electron transport, increased surface reactivity, and inhibited MXene layer restacking, facilitating efficient ion diffusion and pseudocapacitive charge storage.

The integration of biomass-derived carbon electrodes and polymer electrolytes into SCs represents a significant advancement in sustainable energy storage. These materials offer high electrochemical performance while mitigating the environmental impact associated with conventional SC components. Future research should focus on optimizing synthesis techniques, improving scalability, and further enhancing electrochemical properties to facilitate the commercialization of eco-friendly SCs. By leveraging renewable resources, these advancements contribute to the development of sustainable, high-performance energy storage solutions, aligning with global efforts toward greener technologies.

Batteries

Batteries, functioning as electrochemical energy storage devices, are indispensable in modern energy supply systems, powering a wide range of applications from portable electronics to large-scale energy grids. The energy storage capacity of a battery is directly correlated with the size of its individual cells, whereas the nominal voltage is governed by the number of cells connected in series.

Conventional battery technologies, including nickel–cadmium and lithium-ion systems, rely on metal-based and inorganic active materials, which have demonstrated commercial viability and widespread adoption. However, concerns persist regarding their environmental impact, safety risks, and the complexities associated with recycling, particularly due to the potential leakage of toxic substances. To mitigate these limitations, recent advancements have explored the integration of biomaterials, particularly biopolymer nanocomposites, into battery architectures.

The incorporation of nanofillers further enhances the composite's thermal stability, mechanical resilience, and ionic conductivity. Additionally, these nanostructures contribute to the suppression of lithium dendrite formation, a major safety concern in lithium-based batteries, by modulating ion transport pathways and reinforcing electrode integrity. In electrolyte applications, biopolymer nanocomposites can serve as solid or gel electrolytes, with porous architectures that integrate ionic liquids or retain liquid electrolytes, improving ionic conductivity while maintaining environmental sustainability. Figure 7.8a schematically depicts a lithium-ion battery and its operation, while Fig. 7.8b shows how biopolymers have been used in battery development.

Furthermore, nanocomposite-based separators provide a mechanically robust yet highly ion-permeable barrier, improving thermal stability and reducing the risk of

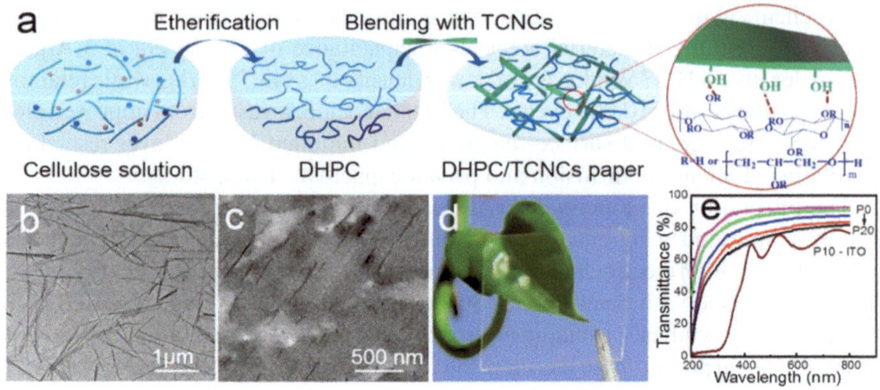

Fig. 7.8 Schematic diagram of the fabrication of cellulose-based nanocomposite papers (**a**), TEM image of TCNCs (**b**), TEM image (**c**) and photograph (**d**) of P10, and optical transmittance under UV–vis light of pure DHPC and nanocomposite papers (**e**). Adapted with permission from ACS (Cheng et al. 2018)

short circuits. These innovations highlight the potential of biopolymer nanocomposites in advancing battery technologies toward safer, more sustainable, and higher-performance energy storage solutions (Kausar 2021; Patra et al. 2024; Ikram et al. 2021).

The study by Jin et al. (2021b) proposes an advanced method for improving silicon-based anodes in lithium-ion batteries (LIBs) by modifying silicon nanoparticles with an alkaline chitosan (CTS) solution. The proposed technique involves dispersing ~ 150 nm silicon nanoparticles in an alkaline CTS solution containing LiOH and urea, leading to controlled oxidation and the formation of a Si/SiO_2/CTS precursor. Subsequent coagulation, regeneration, and pyrolysis at 700 °C in an inert atmosphere yield a nitrogen-doped carbon-coated silicon composite with an outer SiO_2 layer acting as a buffer. The resulting material has a BET surface area of 22.46 $m^2\ g^{-1}$, with mesopores enhancing ion transport. Electrochemical testing revealed that the $Si@SiO_2@N$–C anode exhibited a high initial discharge capacity of 2310.5 mAh g^{-1} (coulombic efficiency: 60.31%) and retained 1026 mAh g^{-1} after 200 cycles at 420 mA g^{-1}, outperforming commercial silicon anodes (597.9 mAh g^{-1}). It also demonstrated superior rate capability, maintaining 1864.5 mAh g^{-1} even at 4200 mA g^{-1}. Nyquist analysis indicated a reduced charge transfer resistance (280.8 Ω), attributed to nitrogen doping, which enhances conductivity. This study contributes to LIB advancements by introducing a novel CTS etching method that integrates SiO_2 and N-doped carbon for enhanced structural stability and conductivity.

Gou et al. (2020) developed a biodegradable gel polymer electrolyte (GPE) to enhance lithium-ion battery (LIB) performance while addressing safety and sustainability concerns. Traditional liquid electrolytes pose fire hazards, and solid polymer electrolytes suffer from low ionic conductivity. GPEs provide a balance between these extremes, but most commercial options are synthetic and non-biodegradable. This study introduces a cellulose-based GPE with strong electrochemical properties and environmental benefits.

The process involves TEMPO oxidation of eucalyptus pulp, converting hydroxyl (–CH_2OH) groups into carboxyl (–COOH) groups to obtain microcrystalline cellulose fiber (MCCF). Further homogenization yields nanocrystalline cellulose fiber (NCCF). Varying ratios of NCCF (0–6%) were used to create membranes, which were dried and later soaked in a liquid electrolyte (1 M $LiPF_6$ in EC/DMC/EMC 1:1:1). NCCF addition enhanced mechanical strength (from 26.9 MPa at 0% NCCF to 59.6 MPa at 6% NCCF) but reduced electrolyte absorption (344.3–237.1%) due to decreased pore space. The membranes exhibited high thermal stability (220–310 °C), surpassing commercial polyethylene and polypropylene separators.

Electrochemical analysis revealed high ionic conductivity, with CCG0 (0% NCCF) at 2.04 mS cm^{-1} and CCG1 (1% NCCF) at 1.84 mS cm^{-1}. Higher NCCF levels reduced conductivity due to compact pore structures. The GPEs maintained electrochemical stability up to 4.5 V, outperforming commercial Celgard separators (~ 4.2 V). However, increasing NCCF content raised interfacial resistance (R_{ct}), from 225 Ω (CCG0) to 1300 Ω (CCG6), showing a trade-off between mechanical robustness and ion transport.

Yang et al. (2020b) investigated the enhancement of battery separators using nanofiber-reinforced membranes. A composite separator, CCG1 (1% NCCF), exhibited superior lithium-ion battery performance, retaining 91% capacity after 50 cycles at 0.2 C and delivering 96.25 mAh g^{-1} at 2 C—outperforming commercial separators. This improvement is due to its porous structure, which facilitates ion transport, and the hydrophilic nature of cellulose, which enhances electrolyte absorption.

Bacterial cellulose (BC) membranes reinforced with aramid nanofibers (ANFs) (2–8% concentration) showed significant mechanical enhancements. The 4% ANF/BC membrane achieved a 2.5-fold increase in tensile strength (104.7 MPa vs. 41.3 MPa for pure BC), but excessive ANF content caused mechanical deterioration due to fiber agglomeration. Electrochemically, adding 2% ANFs increased ionic conductivity from 2.25 mS/cm (pure BC) to 12.54 mS/cm, while reducing interfacial resistance from 391 to 172 Ω, improving charge transport. However, higher ANF concentrations led to conductivity reduction due to membrane densification.

The 2% ANF/BC membrane exhibited a discharge capacity of 157 mAh/g with 93% retention after 100 cycles, surpassing the commercial Celgard 2340 separator (141 mAh/g). Despite a slight reduction in electrochemical stability due to the polarity of aramid groups, the optimized balance of mechanical strength, ionic conductivity, and interfacial compatibility significantly enhanced battery performance.

Tan and colleagues (2021) developed a high-performance anode material for lithium-ion batteries (LIBs) using a composite of nickel oxide (NiO) and carbon derived from natural cotton. By utilizing cotton as a sustainable carbon source, the researchers synthesized NiO/C composites through a hydrothermal process followed by calcination. This method produced a hierarchical porous structure where NiO nanosheets were embedded in a carbon matrix, significantly improving electrochemical properties. The synthesized and fabricated materials exhibited a high specific surface area of 314.4 m^2/g and a total pore volume of 0.539 cm^3/g, as determined by the Brunauer–Emmett–Teller (BET) model. Electrochemical analysis showed that the composite achieved a reversible specific capacity of 727 mAh/g after 150 cycles at a current density of 100 mA/g. Even at a higher current density of 2 A/g, it maintained a capacity of 476 mAh/g, indicating its excellent rate capability and stability. Furthermore, after 150 cycles, the material retained 83.9% of its initial capacity, outperforming conventional graphite anodes. This improved electrochemical performance was attributed to the synergistic interaction between the NiO nanostructures and the carbon matrix, which effectively reduced the volumetric expansion and mechanical stress during charge–discharge cycles and enhanced the structural integrity. These features position this composite as a promising candidate for the next generation of lithium-ion battery anodes.

Yang and colleagues (2020) created an innovative method to improve the electrochemical characteristics of $Li_2ZnTi_3O_8$ (LZTO), a stable and reasonably priced anode material with low ionic and electronic conductivity, in order to improve the performance of lithium-ion batteries (LIBs). In order to overcome these constraints, the scientists created a composite material by combining LZTO and TiO_2 and covering it with chitosan-derived nitrogen-doped carbon (N–C). The N–C@LZTO/TiO_2 composite was made using a hydrothermal method and calcination after LZTO was formed by a solid-state reaction. Battery performance significantly improved, according to the electrochemical tests. After 100 cycles, the composite's specific discharge capacity was 266.3 mAh/g at 200 mA/g, which is 80% more than that of bare LZTO. Additionally, the composite maintained a discharge capacity of 100.7 mAh/g at a high current density of 2000 mA/g, while bare LZTO only managed to reach 38.3 mAh/g. The composite's charge transfer resistance (R_{ct}) significantly decreased from 51.0 Ω (bare LZTO) to 13.6 Ω (EIS), indicating improved electronic conductivity and enhanced diffusion of lithium ions.

Tan and colleagues (2021) explored the development of carbon nanocomposites generated from cotton and nickel oxide (NiO) as a sustainable solution for LIB anodes. Through a hydrothermal reaction and calcination, the researchers created NiO/C composites using natural cotton as the carbon source. This produced a hierarchical porous structure with NiO nanosheets embedded inside a carbon matrix. By improving ion transport and buffering structural stress during cycling, this structure significantly increased electrochemical performance. The BET analysis revealed that the NiO/C composite had a total pore volume of 0.539 cm^3/g and a high specific surface area of 314.4 m^2/g. Compared to conventional graphite anodes, electrochemical measurements showed an outstanding specific capacity of 727 mAh/g with 83.9% retention after 150 cycles at 100 mA/g. With a capacity of 476 mAh/g, the composite demonstrated outstanding rate performance and stability even at a high current density of 2 A/g. The improved cycling stability and capacity retention were ascribed to the carbon matrix and NiO nanosheets' synergistic interaction, which reduced mechanical stress and volume expansion during charge–discharge cycles.

7.3.2 *Energy Conversion Devices*

Bionanocomposites, a unique class of materials combining biopolymers and nanoscale reinforcements, have gained significant attention in energy conversion devices due to their tunable physicochemical properties.

The polymer matrix, typically derived from renewable biopolymers such as chitosan, cellulose, polylactic acid (PLA), or polyhydroxyalkanoates (PHA), serves as a lightweight, flexible scaffold that can accommodate a high loading of nanofillers. These nanofillers, such as carbon nanotubes, graphene, or metal-oxide nanoparticles, impart enhanced thermal conductivity, electrical conductivity, and mechanical strength. The chemistry of these materials involves the formation of strong interfacial interactions—such as hydrogen bonding, van der Waals forces, or covalent grafting—between the polymer matrix and nanofillers. Tailored polymer-nanofiller interfaces are critical in achieving superior energy conversion efficiencies in applications such as photovoltaic cells, fuel cells, hydrogen generation and storage, and thermoelectric devices (Powell et al. 2021).

Photovoltaics (Solar Cells)

Photovoltaic cells are responsible for converting light energy, primarily from sunlight, into electrical energy. The solar cells that enable this conversion are composed of semiconductors. The optical material, which absorbs light, is organized in layers and modules to create a complete system for transforming light energy into electricity. Since the development of the first inorganic solar cell in 1954, silicon has been the most commonly used material for photovoltaic fabrication. However, due to several limitations, including the fact that silicon is a rigid and costly material, its price is directly tied to the cost of photovoltaic solar cells. In contrast, polymer nanocomposites hold promise for photovoltaic applications due to their lower manufacturing costs, material availability, and high conversion efficiency (Powell et al. 2021).

Bionanocomposites have emerged as promising materials in the field of photovoltaics due to their unique combination of organic and inorganic components, which can synergistically enhance the efficiency and sustainability of solar cells. These materials typically integrate biopolymers with nanoscale additives, including metal oxides (e.g., TiO_2, ZnO), quantum dots, or carbon-based nanostructures like graphene. The polymer matrix serves as a flexible, processable medium, while the nanostructures enhance charge carrier mobility, light absorption, and stability. Incorporating biopolymers into hybrid perovskite or organic photovoltaic systems is made possible by functionalizing them with conductive groups or cross-linkers, which further optimizes their structural and electrical characteristics. For example, it has been demonstrated that cellulose nanocrystals (CNCs) enhance the performance of the photoactive layer by improving the dispersion of quantum dots in polymer matrices (Strange et al. 2008).

Praveen et al. (2020) investigated the possibility of using chitosan-based polymer electrolytes (CPEs) in dye-sensitized solar cells (DSSCs) as a solid-state substitute for traditional liquid electrolytes. In order to improve ion mobility and film flexibility, chitosan, a biodegradable biopolymer, was dissolved in formic acid and combined with a KI/I_2 redox with PEG-400 plasticizer to optimize it as a solid polymer electrolyte (SPE). Because this formulation stopped solvent loss and degradation, DSSC stability was increased. The power conversion efficiency (PCE) of DSSCs employing CPEs was 5.58%, which is 2.4 times more than that of undoped ZnO-based cells, when combined with Li-doped ZnO (5% Li) nanostructures as photoanodes. A decrease in charge transfer resistance (R_{ct}) to 261 Ω cm^2 was demonstrated by electrochemical impedance spectroscopy (EIS) study, enhancing charge mobility and reducing recombination losses. Notably, Lorentz force effects that improved charge carrier diffusion and extraction caused PCE to rise to 5.83% when an external magnetic field of 200 Oe was applied.

Rahman et al. (2021) investigated chitosan-based solid polymer electrolytes (SPEs) as an alternative to conventional liquid electrolytes (LEs) in dye-sensitized solar cells (DSSCs). Chitosan was selected as the host material due to its functional groups that facilitate ion transport. Sodium iodide (NaI) was incorporated at varying concentrations to enhance ionic conductivity, with the 30 wt% NaI formulation achieving the highest conductivity of 1.11×10^{-4} S cm^{-1} at room temperature. DSSCs using chitosan-NaI SPEs exhibited a short-circuit current density (J_{sc}) of 0.32 mA cm^{-2}, an open-circuit voltage (V_{oc}) of 0.7 V, a fill factor (FF) of 0.30, and an overall power conversion efficiency (η) of 0.06%. While lower than LE-based DSSCs (typically exceeding 0.3%), the solid-state electrolyte provided superior long-term stability, retaining 100% efficiency after one month, compared to an 85% performance loss in LE-based cells due to solvent evaporation. Electrochemical stability analysis showed that the chitosan-NaI electrolyte remains stable up to 1.85 V, suitable for DSSC operation below 1 V. Ionic transference analysis confirmed that charge transport is predominantly facilitated by iodide (I^-) anions, minimizing electron recombination and enhancing stability.

Briscoe et al. (2015) investigated biomass-derived carbon quantum dots (CQDs) as sensitizers in ZnO nanorod-based solar cells, focusing on chitosan-derived CQDs. Chitosan was hydrothermally carbonized to produce CQDs functionalized with amine groups, which facilitated strong electrostatic interactions with ZnO nanorods, leading to a uniform coating and enhanced visible light absorption. The chitosan-CQD-sensitized solar cells, incorporating CuSCN as the hole conductor and gold contacts, achieved a power conversion efficiency (PCE) of 0.061%, the highest among single-CQD devices. This improvement was attributed to effective charge transport and reduced recombination losses at the ZnO/CQD interface. The devices recorded a short-circuit current density (J_{sc}) of 0.500 mA/cm^2, an open-circuit voltage (V_{oc}) of 275 mV, and a fill factor (FF) of 0.44. Compared to CQDs from glucose and chitin, chitosan-derived CQDs provided an optimal balance of light absorption, charge transport, and interfacial stability due to their amine-rich surface chemistry.

Cheng et al. (2018) developed transparent and flexible nanocomposite papers for solar cells using cellulose. Traditional cellulose papers suffer from high roughness and poor mechanical properties, limiting their optoelectronic applications. Figure 7.8a shows the schematic illustration of the fabrication of cellulose-based nanocomposite papers, Fig. 7.8b presents the TEM image of TCNCs, Fig. 7.8c shows the TEM image of P10, and Fig. 7.8d displays the photograph of P10. Additionally, Fig. 7.8e illustrates the optical transmittance under UV–vis light of neat DHPC and nanocomposite papers. To overcome these issues, the researchers modified cellulose into O-(2,3-Dihydroxypropyl) cellulose (DHPC), which offers high transparency, adhesiveness, and flexibility. To enhance mechanical performance, tunicate cellulose nanocrystals (TCNCs) were incorporated as reinforcing agents. These marine-derived nanocrystals, with a high aspect ratio (L/W $\approx$ 72), formed a percolating network within the DHPC matrix, significantly strengthening the substrate. The optimized nanocomposite (10% TCNC) maintained 85% transparency at 550 nm and exhibited an 8.5-fold increase in toughness compared to pure DHPC. Dynamic mechanical analysis (DMA) showed a substantial increase in tensile storage modulus, from 0.55 GPa (pure DHPC) to 2.21 GPa (15 wt% TCNC). The nanocomposite paper was used as a substrate for inverted polymer solar cells, featuring a conductive ITO layer. These flexible solar cells achieved a power conversion efficiency (PCE) of 4.98%, with a short-circuit current density (J_{sc}) of 13.65 mA/cm^2, an open-circuit voltage (V_{oc}) of 0.71 V, and a fill factor (FF) of 51.37%. The conductive ITO layer demonstrated stable electrical performance under repeated bending, confirming the substrate's durability and flexibility.

Li et al. (2024) developed a cellulose-based film by combining cellulose nanofibers (CNF) with zinc oxide (ZnO)-modified MXene (ZNM-MXene) to enhance solar evaporation and thermal energy conversion. This innovative material efficiently absorbs sunlight, retains heat, and facilitates water transport, making it highly effective for seawater desalination and power generation. The study demonstrated a high evaporation rate of 1.27 kg/m^2 per hour and an impressive energy conversion efficiency of 82.15%. When integrated with a thermoelectric (TE) module, the system generated 15.32 times more power than a conventional TE module. Experimental results confirmed that ZnO particles increased surface roughness, reduced light reflection, and enhanced photothermal properties, enabling the material to absorb up to 93.46% of incident sunlight. The film's thermal conductivity decreased from 0.41 W/m K (CNF@MXene) to 0.27 W/m K, improving heat retention. Additionally, its water absorption properties were enhanced, as indicated by a reduction in water contact angle from 54.22 to 41.63°, demonstrating increased hydrophilicity.

Fuel Cells

Fuel cells based on bio-composite materials represent a significant leap toward sustainable energy solutions that harness the inherent benefits of natural and renewable resources. Using materials such as biomass-derived carbon, clays, and other elements inspired by nature, these bio-composites are a cost-effective and environmentally friendly alternative to conventional fuel cell components. The unique properties of bio-composites, such as high specific surface area, porosity, and thermal stability, make them ideal candidates for fabricating electrodes, electrolytes, and catalysts. For example, biomass-derived carbons offer excellent biocompatibility and electrochemical performance, while natural clays such as montmorillonite and kaolinite improve ionic conductivity and mechanical stability when used in polymer matrices.

Bio-composite materials have gained attention for their role in advancing microbial fuel cells (MFCs), offering a sustainable and environmentally friendly approach to energy generation. These materials, incorporating green polymers and nanocomposites, are critical for enhancing the efficiency and ecological compatibility of MFC components. Recent research emphasizes the utilization of bio-derived polymers, including cellulose, chitosan, and polyaniline, which offer large surface areas and superior electrical conductivity when combined with metal oxides and graphene oxide nanoparticles (Kausar 2024).

Although biocomposite-based fuel cells have many benefits, their broad use is hindered by issues, including scalability, material property consistency, and the requirement for additional optimization to improve lifetime and performance. However, current studies are investigating novel synthesis techniques and hybrid composite architectures to overcome these obstacles. Bio-composite fuel cells, which combine affordability and excellent performance, have the potential to completely transform the energy industry by providing a route to cleaner, more sustainable energy production for a range of uses, including stationary power generation and mobility (Jaleh et al. 2023).

The study by ul Haque et al. (2020), a composite biowand based on polyaniline (PAni), reduced graphene oxide (rGO), and chitosan (CHI) is investigated for enzymatic glucose biofuel cells (EFCs). The composite improves electron transport, enzyme stability, and electrochemical performance when immobilized with ferritin (Frt) and graphene oxidase (GOx) on a glassy carbon (GC) electrode. Strong enzymatic activity was demonstrated with a constant biowand current density of 0.02 ± 3.5 mA cm^{-2} in 20 mM glucose solution. The Hammer technique, hydrazine reduction, and aniline polymerization were used to create the rGO-PAni composite, which was subsequently ultrasonically mixed with chitosan. Structural stability was ensured by hydrogen bonding and electrostatic interactions. Electron transfer was enhanced by the inclusion of ferritin and optimal enzyme loading (3.80×10^{-8} mol cm^{-2}), which reduced resistance and improved performance. At 4 °C, the bioanode showed resilience by retaining 95% of its initial current after one week.

Yousefi et al. (2020) investigated self-assembled chitosan/montmorillonite (CHI/MMT) nanocomposites as cost-effective membranes for microbial fuel cells (MFCs), enhancing electricity generation and wastewater treatment while overcoming limitations of ceramic separators. The optimized nanocomposite (CMMT-1/2) significantly reduced oxygen diffusion (oxygen mass transfer coefficient: 9.1×10^{-6} cm/s), creating a favorable anaerobic environment that promoted exoelectrogenic bacterial growth. As a result, power density increased to 229.12 ± 18.5 mW/m^2, over three times that of unmodified ceramic, while current density reached 1422.22 ± 41.2 mA/m^2, and coulombic efficiency improved to 86.97 ± 13.2%. The nanocomposite also lowered internal resistance, with ohmic resistance decreasing by 73.2%, and charge transfer resistance at the anode and cathode dropping by 96.44% and 66.14%, respectively. Proton conductivity reached 222.73 ± 22.7 mS/cm, surpassing Nafion membranes.

7.3.3 Water Splitting Catalysts

Water splitting has become a pivotal process in advancing sustainable hydrogen production, addressing the global shift from fossil fuels to cleaner energy sources. Advances in bio-nanocomposites derived from natural biopolymers, including chitosan, chitin, and cellulose, have highlighted their significant potential in catalyzing photocatalytic, electrocatalytic, and photoelectrocatalytic hydrogen generation. These biopolymers possess unique structural and chemical attributes that enhance catalyst performance and reaction efficiency. Figure 7.9 illustrates the mechanism of electrochemical water splitting.

Water splitting entails the decomposition of water into hydrogen and oxygen through processes such as photocatalysis, electrocatalysis, or their combination in photoelectrocatalysis. Photocatalysis relies on light-driven electron excitation

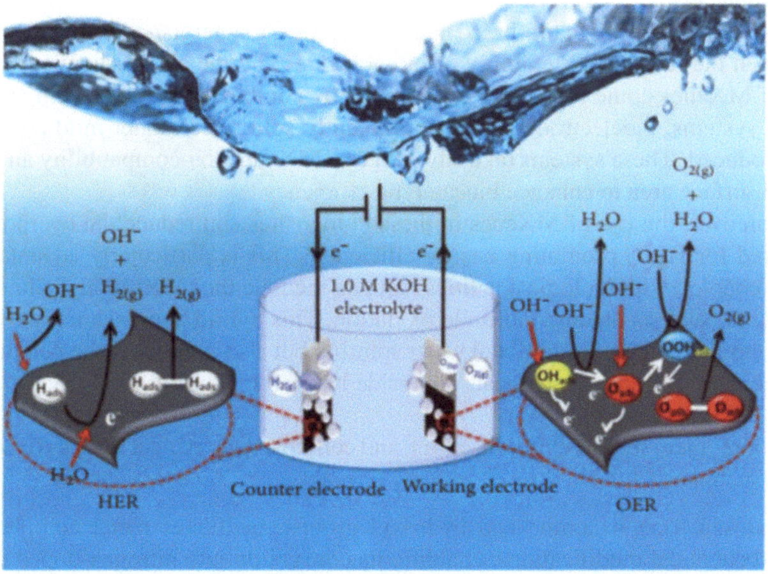

Fig. 7.9 The mechanism of electrochemical water splitting. Adapted with permission from Wiley Online Library (Keshipour et al. 2023)

in semiconductors, creating electron–hole pairs that drive hydrogen evolution at the conduction band and oxygen evolution at the valence band. Incorporating bio-nanocomposites enhances this process by improving charge separation and reducing recombination, significantly boosting photocatalytic efficiency.

Electrocatalysis uses catalysts to lower the overpotential of the hydrogen evolution reaction (HER) and oxygen evolution reaction (OER), where biopolymer-based materials provide high surface area and abundant active sites for improved electron transfer and catalytic performance. Photoelectrocatalysis merges the benefits of light-driven photocatalysis with electrochemical processes, leveraging biopolymer supports to improve the stability and activity of semiconductor electrodes (Keshipour et al. 2023).

Because of its structural and electrical adaptability, MXenes are useful in water splitting. They can fine-tune the interlayer spacing thanks to their layered structure, which is essential for maximizing electron mobility and ion transport. By effectively adsorbing and activating water molecules, MXenes' surface functional groups enhance their catalytic activity. Furthermore, the strong bio-nanocomposite structure is fostered by the high conductivity and mechanical stability of MXenes, which improve their integration with biological molecules.

Recent research shows that heterostructured MXenes can be designed through surface modification or composites with other nanomaterials, which allows for expanded active sites and enhances charge transfer—both essential in HER and OER processes. It has been shown that introducing heteroatoms such as nitrogen or sulfur into the MXene network modulates the electronic properties and enhances catalytic performance.

By adding MXenes to bio-based nanocomposites, a synergistic platform is created in which the functional flexibility of biomolecules and the electrochemical properties of MXenes come together. For example, by combining MXenes with enzyme-based systems, bioelectrodes that can efficiently split water under mild conditions are produced. These systems take advantage of MXenes' biocompatibility and large active surface area to enhance biocatalytic efficiency.

Moreover, the use of MXenes in these composites can reduce the overpotential required for HER, promoting energy efficiency. This is particularly advantageous when coupled with biological entities that can mediate the OER through alternative pathways, such as organic molecule oxidation. This hybrid approach not only mitigates the challenges of traditional water splitting but also introduces the potential for producing value-added chemicals alongside hydrogen (Zhao et al. 2025; Sundarraj et al. 2025).

Biopolymers such as chitosan, chitin, and cellulose play critical roles as functional catalyst supports.

- Chitosan, rich in amine and hydroxyl groups, facilitates metal complexation and chemical modifications, enabling its conversion into nitrogen-doped carbon materials. These materials enhance electron conductivity and reduce charge recombination rates, exemplified by chitosan-based composites like Pd/CS and

Cu@CS, which have achieved hydrogen production rates exceeding 2.5×10^5 μmol H_2 g^{-1} h^{-1} under optimal conditions.
- Chitin, derived from the deacetylation of chitin, possesses a higher nitrogen content, further boosting its catalytic activity. As a sustainable precursor for porous carbon materials, chitin supports catalysts such as Ru/Chitin composites, which demonstrate excellent stability and reusability, with minimal degradation over successive cycles.
- Cellulose, the most abundant natural polymer, offers high versatility through its nanocrystalline and nanofibrillated forms, which provide extensive surface areas and active sites for catalyst deposition. Cellulose-derived carbon materials effectively support semiconductors like TiO_2 and transition metals, enhancing HER and OER processes. For instance, cellulose-based nitrogen-doped carbon supports have achieved impressively low overpotentials of 55 mV for HER (Keshipour et al. 2023).

Thangarasu et al. (2023) investigated water splitting for sustainable hydrogen production, focusing on the hydrogen evolution reaction (HER) at the cathode and oxygen evolution reaction (OER) at the anode. The study introduced a Fe_2O_3/Ni nanocomposite supported on microcrystalline cellulose (MCC) as a low cost, environmentally friendly electrocatalyst, addressing the limitations of expensive platinum-group metal (PGM)-based catalysts.

The Fe_2O_3/Ni-MCC composite exhibited high catalytic activity, achieving a HER overpotential of 163 mV and an OER overpotential of 360 mV at 10 mA/cm^2, significantly enhancing reaction efficiency. The synergistic interaction between Fe_2O_3 and Ni facilitated charge transfer and intermediate formation (OH and OOH species**), improving reaction kinetics. MCC acted as a natural binder and dispersing agent, eliminating synthetic polymers, and enhancing catalyst stability, conductivity, and hydrophilicity, thereby optimizing electrode–electrolyte interactions. This study positions the Fe_2O_3/Ni-MCC composite as a scalable, cost-effective electrocatalyst for hydrogen production via water splitting.

Keshipour and Faezeh Eyvari-Ashnak (Keshipour and Eyvari-Ashnak 2023) explored photocatalytic water splitting for hydrogen (H_2) and oxygen (O_2) generation using sunlight. They developed a novel Au@CoPc@NC photocatalyst, comprising gold nanoparticles (Au), cobalt(II)-phthalocyanine (CoPc), and nitrogen-doped carbon (NC) derived from chitosan, designed to optimize light absorption, charge separation, and catalytic efficiency. With a band gap of 2.01 eV, the photocatalyst efficiently utilized visible light for energy conversion. Under optimized conditions, it achieved a hydrogen production rate of 792 mol H_2 per mol Au per hour, using a 200 W xenon lamp ($\lambda > 420$ nm) at neutral pH. The Z-scheme heterojunction enhanced charge transfer while minimizing electron–hole recombination, improving photocatalytic performance. The catalyst exhibited strong hydrogen evolution activity, particularly in acidic conditions, and maintained long-term stability with minimal nanoparticle aggregation.

Medany et al. (2024) developed a nickel cobaltite-chitosan (NC@Chit) composite with dual functionality in energy storage and water splitting. For electrochemical capacitors, NC@Chit exhibited a specific capacitance of 345 F/g at 1 A/g, surpassing pure nickel cobaltite (234 F/g) and retaining 93% of its capacitance after 3000 cycles, demonstrating excellent stability.

In water splitting, NC@Chit showed high efficiency for both the oxygen evolution reaction (OER) and the hydrogen evolution reaction (HER). It achieved OER overpotential of 310 mV at 10 mA/cm^2 with a Tafel slope of 67 mV/dec, and HER overpotential of 240 mV with a Tafel slope of 71 mV/dec, marking it as a competitive non-noble metal-based electrocatalyst. Electrochemical impedance spectroscopy (EIS) confirmed enhanced charge transfer efficiency, with NC@Chit reducing charge transfer resistance (11 Ω cm^2 vs. 37 Ω cm^2 for unmodified GC/NC).

Wang et al. (2021) investigated the $Ag_2S/NiTiO_3$/Chitosan (AS/NTO/CS) nanocomposite for visible-light-driven photocatalytic water splitting, focusing on the hydrogen evolution reaction (HER). The heterojunction structure of the nanocomposite enhanced light absorption, charge separation, and catalytic efficiency, significantly improving hydrogen production.

Under visible light irradiation, AS/NTO/CS achieved a hydrogen evolution rate of 62.35 μmol/g, surpassing AS/NTO (38.40 μmol/g) and vastly outperforming pure $NiTiO_3$, which showed negligible activity. Gas chromatography with nitrogen carrier gas validated the hydrogen yield. The synergistic effect of Ag_2S, $NiTiO_3$, and chitosan-reduced electron–hole recombination, ensuring efficient charge migration and improving photocatalytic stability.

The use of biopolymers in catalysis offers numerous advantages, including sustainability, biocompatibility, and cost-effectiveness. Their functional groups enable seamless integration with diverse metals and semiconductors, while their conversion into doped carbon materials improves electronic properties and broadens application possibilities. Despite these benefits, challenges such as limited thermal stability and solubility in acidic environments need to be addressed to maximize their catalytic potential. The use of bio-based nanocomposites in water splitting represents a significant advance toward sustainable hydrogen production. Researchers have developed highly efficient and environmentally friendly catalysts for HER and OER by exploiting the intrinsic properties of cellulose, chitosan, and chitin. Overcoming existing barriers and expanding these technologies for industrial applications will require further creativity in the modification of bio-based polymers and hybridization with advanced materials.

7.4 Challenges and Future Perspectives

Adoption of bionanocomposites in energy-related applications is beneficial but several issues need to be solved to fulfill the maximum potential, scalability, and commerciality of the same. The major demerit of the biopolymers is their low thermal stability which can be a matter of concern with regard to the durability of the product especially in high-temperature applications.

However, there is a problem of solubility in extreme pH conditions in some catalytic and energy storage systems. Another major issue is the development of large-scale synthesis of the bio-nanocomposites with good quality and performance. Although the laboratory experiments have shown their potential, their industrialization requires sophisticated processing, economical production, and optimum design of the material. Replicability and high efficiency are critical and cannot be achieved through biopolymers without their precise functionalization and combination with nanomaterials.

Furthermore, the long-term stability of the structure of the bio-nanocomposite-based electrodes and catalyst is still a problem. Prolonged contact with the natural environment, including humidity, oxygen, and UV light, is known to degrade these materials. Coating, surface treatment, and enhanced material development are therefore very important in increasing the service life of the materials.

Another restriction is the controlled allocation of nanofillers in the biopolymer matrices because this process is critical for the modulus, conductivity, and chemical resistance. To enhance the performance of energy storage devices like lithium-ion batteries and supercapacitors, it is crucial to avoid agglomeration and ensure uniform distribution of the nanoparticles.

Despite all these challenges, bio-nanocomposites are very promising for future clean energy solutions. Further work in material innovation, molecular engineering, and hybrid composite designs will be required to circumvent the present difficulties. The development of sustainable and cost-efficient production processes will be of great importance for the implementation of the above-listed measures. Also, the use of new biopolymer sources and the integration of new advanced nanomaterials, such as MXenes and metal–organic frameworks (MOFs), can further improve the functional properties of bionanocomposites.

Future work may also entail combining bionanocomposites with smart energy systems, such as self-healing materials and adaptive interfaces, to enhance efficiency and sustainability. The application of artificial intelligence (AI) and machine learning (ML) in the field of materials science can accelerate the discovery of new bionanocomposites with enhanced electrochemical behavior. Furthermore, biomimetic nanocomposites that emulate the hierarchical organization of natural materials may result in the development of new materials with improved mechanical, electrical, and chemical properties. Interdisciplinary cooperation between researchers, industries, and policy makers will be crucial to accelerate the commercialization of bionanocomposites. Thus, the establishment of standard procedures, life-cycle analyses, and the increase of knowledge about the degradation processes will be important to enhance their application in energy technologies.

7.5 Conclusion

The novel class of materials known as bio-based nanocomposites has great potential for use in renewable energy applications. Their particular combination of stability, biocompatibility, and electrochemical innovations renders them viable substitutes for energy storage and conversion systems of the future. This chapter has examined their many uses, such as in photocatalytic water splitting, batteries, supercapacitors, and fuel cells.

Bio-based nanocomposites are a more sustainable option than the current conventional energy sources but have their challenges of technical and financial barriers to their commercial use. To ensure that they can be used more frequently, their sustainability, effectiveness, and scalability must be enhanced together with the use of sophisticated material design and advanced techniques. Moreover, the research-to-industry gap in applying the knowledge generated in laboratories will be closed through the collaborative effort of engineers, chemists, and materials scientists. With increasing energy demand in the world, there will be an increasing need for energy materials that are sustainable and high-performing. Bio-based nanocomposites are a possible way to achieve sustainable energy in the future through renewable resources and the ability to tailor the materials for specific applications. This field will be mostly defined by research and investments in this area, as renewable energy technology of the future.

Furthermore, combining bio-based nanocomposites with advanced fabrication techniques, such as 3D printing and electrospinning, will enable the creation of highly advanced and efficient energy systems. Their expanding role in emerging fields, including wearable energy storage, flexible electronics, and bio-inspired energy harvesting, underscores their adaptability and importance in achieving sustainable energy. By addressing current challenges and leveraging technological innovations, bio-based nanocomposites will play a fundamental role in the future of clean energy development.

References

Abd Elkodous M, Hamad HA, Abdel Maksoud MIA, Ali GAM, El Abboubi M, Bedir AG et al (2022) Cutting-edge development in waste-recycled nanomaterials for energy storage and conversion applications. Nanotechnol Rev 11:2215–2294. https://doi.org/10.1515/ntrev-2022-0129

Ahmed S, Arshad T, Zada A, Afzal A, Khan M, Hussain A et al (2021) Preparation and characterization of a novel sulfonated titanium oxide incorporated chitosan nanocomposite membranes for fuel cell application. Membranes 11

Ahmed S, Sharma P, Bairagi S, Rumjit NP, Garg S, Ali A et al (2023) Nature-derived polymers and their composites for energy depository applications in batteries and supercapacitors: advances, prospects and sustainability. J Energy Storage. 66:107391

Al-Fakih GOA, Ilyas RA, Atiqah A, Atikah MSN, Saidur R, Dufresne A et al (2024) Advanced functional materials based on nanocellulose/Mxene: a review. Int J Biol Macromol 278:135207

Ang T-Z, Salem M, Kamarol M, Das HS, Nazari MA, Prabaharan N (2022) A comprehensive study of renewable energy sources: classifications, challenges and suggestions. Energy Strat Rev 43:100939

Anthony LS, Vasudevan M, Perumal V, Ovinis M, Raja PB, Edison TNJI (2021) Bioresource-derived polymer composites for energy storage applications: brief review. J Environ Chem Eng 9:105832

Anwer AH, Ahtesham A, Shoeb M, Mashkoor F, Ansari MZ, Zhu S et al (2023) State-of-the-art advances in nanocomposite and bio-nanocomposite polymeric materials: a comprehensive review. Adv Colloid Interface Sci 318

Areerob Y, Hamontree C, Sricharoen P, Limchoowong N, Laksee S, Oh W-C, Pattarith K (2022) Novel gamma-irradiated chitosan-doped reduced graphene-$CuInS_2$ composites as counter electrodes for dye-sensitized solar cells. RSC Adv 12:15427–15434. https://pubs.rsc.org/en/content/articlehtml/2022/ra/d2ra01749k

Ates B, Koytepe S, Ulu A, Gurses C, Thakur VK (2020) Chemistry, structures, and advanced applications of nanocomposites from biorenewable resources. Chem Rev 120:9304–9362. https://doi.org/10.1021/acs.chemrev.9b00553

Bensebaa F (2013) Clean energy. Interface Sci Technol 19:279–383

Briscoe J, Marinovic A, Sevilla M, Dunn S, Titirici M (2015) Biomass-derived carbon quantum dot sensitizers for solid-state nanostructured solar cells. Angew Chem Int Ed 54:4463–4468

Bunekar N, Tsai TY (2020) Green nanocomposites. emerging carbon-based nanocomposites for environmental applications, pp 55–70. [Internet]. https://doi.org/10.1002/9781119554882.ch3

Burlec AF, Corciova A, Boev M, Batir-Marin D, Mircea C, Cioanca O et al (2023) Current overview of metal nanoparticles' synthesis, characterization, and biomedical applications, with a focus on silver and gold nanoparticles. Pharmaceuticals 16:1410. https://www.mdpi.com/1424-8247/16/10/1410/htm

Chen T, Dai L (2013) Carbon nanomaterials for high-performance supercapacitors. Mater Today. https://www.sciencedirect.com/science/article/pii/S136970211300223X

Chen S, Skordos A, Thakur VK (2020) Functional nanocomposites for energy storage: chemistry and new horizons. Mater Today Chem 17:100304. https://www.sciencedirect.com/science/article/pii/S2468519420300641

Chen K, Weng S, Lu J, Gu J, Chen G, Hu O et al (2021) Facile synthesis of chitosan derived heteroatoms-doped hierarchical porous carbon for supercapacitors. Microporous Mesoporous Mater 320

Chen L, Hu Y, Wang R, Li X, Chen Z, Hua J et al (2023) Green building practices to integrate renewable energy in the construction sector: a review. Environ Chem Lett 22(2):751–784. https://doi.org/10.1007/s10311-023-01675-2

Cheng Q, Ye D, Yang W, Zhang S, Chen H, Chang C et al (2018) Construction of transparent cellulose-based nanocomposite papers and potential application in flexible solar cells. ACS Sustain Chem Eng 6:8040–8047

Cui Y, Shi Q, Liu Z, Lv J, Wang C, Xie X et al (2023) MXene/biomass/chitosan carbon aerogel (MBC) with shared cathode and anode for the construction of high-efficiency asymmetric supercapacitor. Chem Eng J 472:144701

Dar MA, Majid SR, Satgunam M, Siva C, Ansari S, Arusalan P et al (2024) Advancements in supercapacitor electrodes and perspectives for future energy storage technologies. Int J Hydrogen Energy 70:10–28

Deshmukh MKG, Sameeroddin M, Abdul D, Abdul SM (2023) Renewable energy in the 21st century: a review. Mater Today Proc 80:1756–1759

Dhanasekaran T, Bovas A, Radhakrishnan TP (2023) Hydrogel polymer-PBA nanocomposite thin film-based bifunctional catalytic electrode for water splitting: the unique role of the polymer matrix in enhancing the electrocatalytic efficiency. ACS Appl Mater Interfaces 15:6687–6696. https://doi.org/10.1021/acsami.2c18006

Dwivedi S, Dwivedi N (2023) Bionanotechnology towards green energy: innovative and sustainable approach, pp 1–320

Geetanjali, Dhillon SK, Kundu PP (2022) Development of polypyrrole nanotube coated with chitosan and nickel oxide as a biocompatible anode to enhance the power generation in microbial fuel cell. J Power Sources 539:231595

Ghasem Hosseini M, Shahryari E (2017) A novel high-performance supercapacitor based on chitosan/graphene oxide-MWCNT/polyaniline. J Colloid Interface Sci 496:371–381

Goswami AD, Trivedi DH, Jadhav NL, Pinjari DV (2021) Sustainable and green synthesis of carbon nanomaterials: a review. J Environ Chem Eng 9:106118

Gou J, Liu W, Tang A (2020) A renewable gel polymer electrolyte based on the different sized carboxylated cellulose with satisfactory comprehensive performance for rechargeable lithium ion battery. Polymer 208

Hassan S, Suzuki M, El-Moneim AA (2014) Synthesis of MnO_2-chitosan nanocomposite by one-step electrodeposition for electrochemical energy storage application. J Power Sources 246:68–73

He C, Gendensuren B, Kim H, Lee H, Oh ES (2020) Electrochemical performance of polysaccharides modified by the introduction of SO_3H as binder for high-powered $Li_4Ti_5O_{12}$ anodes in lithium-ion batteries. J Electroanal Chem 876

Huang B (2020) Carbon nanotubes and their polymeric composites: the applications in tissue engineering. Biomanuf Rev 5:1–26. https://doi.org/10.1007/s40898-020-00009-x

Ikram R, Mohamed Jan B, Abdul Qadir M, Sidek A, Stylianakis MM, Kenanakis G (2021) Recent advances in chitin and chitosan/graphene-based bio-nanocomposites for energetic applications. Polymers 13:3266

Jaleh B, Nasri A, Eslamipanah M, Nasrollahzadeh M, Advani JH, Fornasiero P et al (2023) Application of biowaste and nature-inspired (nano)materials in fuel cells. J Mater Chem A Mater 11:9333–9382

Jeevanandam J, Vadanasundari V, Pan S, Barhoum A, Danquah MK (2022) Bionanotechnology and bionanomaterials: emerging applications, market, and commercialization. In: Bionanotechnology: emerging applications of bionanomaterials, pp 3–44

Jin B, Wang D, Song L, Cai Y, Ali A, Hou Y et al (2021a) Biomass-derived fluorinated corn starch emulsion as binder for silicon and silicon oxide based anodes in lithium-ion batteries. Electrochim Acta 365

Jin H, Zhu M, Liu J, Gan L, Gong Z, Long M (2021b) Alkaline chitosan solution as etching phase to design Si@SiO_2@N-carbon anode for lithium-ion battery. Appl Surf Sci 541

Karki S, Gohain MB, Yadav D, Ingole PG (2021) Nanocomposite and bio-nanocomposite polymeric materials/membranes development in energy and medical sector: a review. Int J Biol Macromol 193:2121–2139

Kausar A (2021) Green nanocomposites for energy storage. J Compos Sci 5:202

Kausar A (2024) Progressive green nanocomposites for microbial fuel cells—state-of-the-art and technical advancements. Polym-Plast Technol Mater 63:2151–2169

Keshipour S, Eyvari-Ashnak F (2023) Chitosan-derived nitrogen-doped carbon as a support of cobalt(II)-phthalocyanine/gold nanoparticles for photocatalytic water splitting. ACS Omega 8:41624–41632. https://doi.org/10.1021/acsomega.3c05801

Keshipour S, Hadidi M, Gholipour O (2023) A review on hydrogen generation by photo-, electro-, and photoelectro-catalysts based on chitosan, chitin, cellulose, and carbon materials obtained from these biopolymers. Adv Polym Technol 2023. https://doi.org/10.1155/2023/8835940

Li F, Dong Y, Dai Q, Nguyen TT, Guo M (2019) Novel freestanding core-shell nanofibrillated cellulose/polypyrrole/tubular graphitic carbon nitride composite film for supercapacitors electrodes. Vacuum 161:283–290

Li X, Yuan L, Liu R, He H, Hao J, Lu Y et al (2021) Engineering textile electrode and bacterial cellulose nanofiber reinforced hydrogel electrolyte to enable high-performance flexible all-solid-state supercapacitors. Adv Energy Mater 11

Li Y, Fu C, Wang Z, Huang L, Chen L, Liao G et al (2024) Novel cellulose-based films with highly efficient photothermal performance for sustainable solar evaporation and solar-thermal power generation. J Clean Prod 458:142416

Lin S, Wang F, Shao Z (2021) Biomass applied in supercapacitor energy storage devices. J Mater Sci 56:1943–1979. https://doi.org/10.1007/s10853-020-05356-1

Mahajan P, Sharma M (2024) A comprehensive review on developments and future perspectives of biopolymer-based materials for energy storage. Energy Storage 6:e634. https://doi.org/10.1002/est2.634

Mallick P, Gadtya AS, Tripathy D, Satpathy SK, Moharana S (2023) Carbon–metal hybrid nanomaterials for high technologies. In: Nanoparticles reinforced metal nanocomposites: mechanical performance and durability, pp 273–295. https://doi.org/10.1007/978-981-19-9729-7_9

Medany SS, Nafady A, Soomro RA, Hefnawy MA (2024) Construction of chitosan-supported nickel cobaltite composite for efficient electrochemical capacitor and water-splitting applications. Sci Rep 14:2453. https://www.nature.com/articles/s41598-023-49692-z

Mohan TP, Kanny K (2020) Green nanofillers for polymeric materials. Adv Struct Mater 126:99–138

Mooney M, Nyayachavadi A, Rondeau-Gagné S (2020) Eco-friendly semiconducting polymers: from greener synthesis to greener processability. J Mater Chem C Mater 8:14645–14664. https://pubs.rsc.org/en/content/articlehtml/2020/tc/d0tc04085a

Mulla MS, Yaseen M, Torvi A, Laisa CP, Kariduraganavar MY (2024) Development of novel poly(o-anisidine) nanocomposite flexible membrane electrodes and characterization of their supercapacitor applications. Ionics 31:2149–2160. https://doi.org/10.1007/s11581-024-05983-7

Pandiselvi K, Thambidurai S (2014) Chitosan-ZnO/polyaniline ternary nanocomposite for high-performance supercapacitor. Ionics 20:551–561

Patra N, Ramesh P, Donthu V, Ahmad A (2024) Biopolymer-based composites for sustainable energy storage: recent developments and future outlook. J Mater Sci Mater Eng 19:1–15. https://doi.org/10.1186/s40712-024-00181-9

Powell MD, LaCoste JD, Fetrow CJ, Fei L, Wei S (2021) Bio-derived nanomaterials for energy storage and conversion. Nano Select 2:1682–1706. https://doi.org/10.1002/nano.202100001

Praveen E, Peter IJ, Kumar AM, Ramachandran K, Jayakumar K (2019) Performance of ZnO/ZnS nanocomposite based dye-sensitized solar cell with chitosan-polymer electrolyte. Mater Today Proc 35:27–30

Praveen E, Peter IJ, Kumar AM, Ramachandran K, Jayakumar K (2020) Boosting of power conversion efficiency of 2D ZnO nanostructures-based DSSC by the Lorentz force with chitosan polymer electrolyte. J Inorg Organomet Polym Mater 30:4927–4943

Pu L, Zhang J, Jiresse NKL, Gao Y, Zhou H, Naik N et al (2022) N-doped MXene derived from chitosan for the highly effective electrochemical properties as supercapacitor. Adv Compos Hybrid Mater 5:356–369

Rahman NA, Hanifah SA, Mobarak NN, Ahmad A, Ludin NA, Bella F, Su'ait MS (2021) Chitosan as a paradigm for biopolymer electrolytes in solid-state dye-sensitised solar cells. Polymer 230:124092. https://www.sciencedirect.com/science/article/pii/S0032386121007151

Ramírez O, Lopéz-Frances A, Baldoví HG, Saldías C, Navalón S, Leiva A et al (2024) Hydrogel composites based on chitosan and $CuAuTiO_2$ photocatalysts for hydrogen production under simulated sunlight irradiation. Int J Biol Macromol 273

Salleh NA, Kheawhom S, Mohamad AA (2021) Chitosan as biopolymer binder for graphene in supercapacitor electrode. Results Phys 25:104244

Sam DK, Cao Y (2024) Iron-cobalt nanoparticles embedded in B,N-doped chitosan-derived porous carbon aerogel for overall water splitting. ACS Appl Mater Interfaces 16:32311–32321. https://doi.org/10.1021/acsami.4c06141

Sen A, Kurian MS, Narayanan DP, Abraham A, Thomas TS, Jayalakshmi PS et al (2024) Bio-nanocomposites: innovative solutions for addressing issues in health, agriculture, energy and environmental domains. Nano-Struct Nano-Obj 39:101270

Sezali NAA, Ong HL, Villagracia AR, Hoang TD (2024) Bio-based nanomaterials for energy application: a review. Vietnam J Chem 62:1–12. https://doi.org/10.1002/vjch.202300158

Shaheen I, Akkinepally B, Hussain I, Hussain S, Rosaiah P, Qureshi A et al (2024) Fabrication of MXene/cellulose composite-based flexible supercapacitor: synthesis, properties, and future perspectives. J Energy Storage 87:111513

Shang Z, An X, Liu L, Yang J, Zhang W, Dai H et al (2021) Chitin nanofibers as versatile bio-templates of zeolitic imidazolate frameworks for N-doped hierarchically porous carbon electrodes for supercapacitor. Carbohydr Polym 251

Siwal SS, Zhang Q, Devi N, Thakur VK (2020) Carbon-based polymer nanocomposite for high-performance energy storage applications. Polymers 12:505

Strange M, Plackett D, Kaasgaard M, Krebs FC (2008) Biodegradable polymer solar cells. Sol Energy Mater Sol Cells 92:805–813

Sun K, Zhang Z, Peng H, Zhao G, Ma G, Lei Z (2018) Hybrid symmetric supercapacitor assembled by renewable corn silks based porous carbon and redox-active electrolytes. Mater Chem Phys 218:229–238

Sundarraj S, Vadivel N, Murthy AP, Theerthagiri J, Choi MY (2025) MXene electrocatalysts: transformative approaches in hydrogen production with alternative anode reactions. Small 21

Talukdar K (2019) Nanomaterials-based composites for energy applications

Tammela P, Wang Z, Frykstrand S, Zhang P, Sintorn IM, Nyholm L et al (2015) Asymmetric supercapacitors based on carbon nanofibre and polypyrrole/nanocellulose composite electrodes. RSC Adv 5:16405–16413. https://pubs.rsc.org/en/content/articlehtml/2015/ra/c4ra15894f

Tan Y, Li Q, Lu Z, Yang C, Qian W, Yu F (2021) Porous nanocomposites by cotton-derived carbon/NiO with high performance for lithium-ion storage. J Alloys Compd 874

Thangarasu S, Baby N, Bhosale M, Lee J, Jeong C, Oh TH (2023) Fe_2O_3/Ni nanocomposite electrocatalyst on cellulose for hydrogen evolution reaction and oxygen evolution reaction. Int J Mol Sci 24

Torvi A, Naik S, Kariduraganavar M (2018) Development of supercapacitor systems based on binary and ternary nanocomposites using chitosan, graphene and polyaniline. Chem Data Collect 17–18:459–471

ul Haque S, Nasar A, Inamuddin, Rahman MM (2020) Applications of chitosan (CHI)-reduced graphene oxide (rGO)-polyaniline (PAni) conducting composite electrode for energy generation in glucose biofuel cell. Sci Rep 10

Vinodh R, Sasikumar Y, Kim HJ, Atchudan R, Yi M (2021) Chitin and chitosan based biopolymer derived electrode materials for supercapacitor applications: a critical review. J Ind Eng Chem 2021. https://www.sciencedirect.com/science/article/pii/S1226086X21004676

Wang H, Yao CJ, Nie HJ, Yang L, Mei S, Zhang Q (2020) Recent progress in integrated functional electrochromic energy storage devices. J Mater Chem C. https://pubs.rsc.org/en/content/articlehtml/2020/tc/d0tc03934a

Wang J, Sun J, Huang J, Fakhri A, Gupta VK (2021) Synthesis and its characterization of silver sulfide/nickel titanate/chitosan nanocomposites for photocatalysis and water splitting under visible light, and antibacterial studies. Mater Chem Phys 272

Wei L, Deng W, Li S, Wu Z, Cai J, Luo J (2022) Sandwich-like chitosan porous carbon spheres/MXene composite with high specific capacitance and rate performance for supercapacitors. J Bioresour Bioprod 7:63–72

Xiao W, Sun Q, Banis MN, Wang B, Li W, Li M et al (2020) Understanding the critical role of binders in phosphorus/carbon anode for sodium-ion batteries through unexpected mechanism. Adv Funct Mater 30

Xie Y, Khoo KS, Chew KW, Devadas VV, Phang SJ, Lim HR et al (2022) Advancement of renewable energy technologies via artificial and microalgae photosynthesis. Bioresour Technol 363:127830

Yang S, Sun L, An X, Qian X (2020a) Construction of flexible electrodes based on ternary polypyrrole@cobalt oxyhydroxide/cellulose fiber composite for supercapacitor. Carbohydr Polym 229

Yang Y, Huang C, Gao G, Hu C, Luo L, Xu J (2020b) Aramid nanofiber/bacterial cellulose composite separators for lithium-ion batteries. Carbohydr Polym 247

Yousefi V, Mohebbi-Kalhori D, Samimi A (2020) Start-up investigation of the self-assembled chitosan/montmorillonite nanocomposite over the ceramic support as a low-cost membrane for microbial fuel cell. Int J Hydrogen Energy 45:4804–4820. https://www.sciencedirect.com/science/article/pii/S0360319919344398

Yu M, Han Y, Li J, Wang L (2018) Magnetic N-doped carbon aerogel from sodium carboxymethyl cellulose/collagen composite aerogel for dye adsorption and electrochemical supercapacitor. Int J Biol Macromol 115:185–193

Yu L, Liu J, He S, Huang C, Gong Z, Gan L et al (2020) N-doped rGO/C@Si composites using sustainable chitosan as the carbon source for lithium-ion batteries. Appl Surf Sci 501

Zeng Q, Li C, Magazzino C (2024) Impact of green energy production for sustainable economic growth and green economic recovery. Heliyon 10:e36643

Zhao Z, Murad M, Pei C, Park HS, Yu X (2025) Rational design of heterostructured MXene-based nanomaterials in electrocatalytic water splitting. ChemCatChem 17

Amirhossein Raouffard a master's student in polymer chemistry at the University of Isfahan, specializing in pressure-sensitive adhesives and polyurethanes. My research focuses on developing advanced formulations and optimizing the mechanical and thermal properties of these materials. With extensive experience in polymer chemistry, I have dedicated my studies to developing innovative products with industrial and environmental applications. I have participated in research projects related to polyurethanes and pressure-sensitive adhesives and have published my findings in this field. My goal is to enhance smart technologies in polymer chemistry through advanced and applied research, contributing to the progress of this field.

Alireza Fatahi a Ph.D. candidate in Polymer Chemistry at the University of Isfahan. My research primarily focuses on the synthesis of polyurethanes, smart coatings, and self-healing polymers. With a strong background in polymer chemistry, I am dedicated to developing innovative materials that have practical applications in both industrial and environmental fields. I have actively participated in various research projects and have published several papers in reputable scientific journals. As a passionate researcher, I aim to contribute to the advancement of polymer science by exploring new solutions and technologies.

Dr. Abbas Mohammadi is an accomplished Associated Professor of Polymer Chemistry at the University of Isfahan, where he also serves as the Head of the Polymer Research Center. His research is fundamentally focused on the synthesis, modification, and application of advanced polyurethane systems and polymer nanocomposites. A central theme of his work is developing sustainable and high-performance materials to address critical industrial and environmental challenges. His core research interests include creating flame-retardant polymers, highly efficient anticorrosion and antibacterial coatings, and advanced materials for wound healing and oil/water separation. Dr. Mohammadi has a proven track record of securing competitive grants from national bodies like the Iranian National Science Foundation (INSF) and leading impactful, industry-collaborative research projects. His scholarly output is extensive and influential, demonstrated by a robust publication record in high-impact international journals covering topics from waterborne polyurethane nanocomposites for corrosion protection to innovative antibacterial coatings and bioactive wound dressings. Dr. Mohammadi is also a recognized peer reviewer for numerous prestigious international journals and holds significant editorial roles in national polymer science publications. His contributions extend beyond the laboratory through authored books and book chapters on polyurethanes and smart coatings, active participation in international conferences, and the training of the next generation of scientists through a wide array of undergraduate and graduate courses. His work is characterized by a successful translation of fundamental polymer science into practical, real-world solutions for industry and healthcare.

Atiyeh Derakhshandehpour is a master's student in Polymer Chemistry at Isfahan University, where she also earned her bachelor's degree in Applied Chemistry. Her current research focuses on polyurethane, exploring its properties and applications in various industries. Atiye is passionate about advancing the field of polymer chemistry and aims to contribute to innovative solutions through her research.

Chapter 8
Bionanocomposites in Environmental Remediation and Water Purification

Bismark Sarkodie, Benjamin Tawiah, Benjamin K. Asinyo, and Bin Fei

Abstract Environmental sustainability has emerged as a guiding concept for development across the globe. Numerous pollutants in the air, water, and soil have gobbled up the environment, causing catastrophic damage to humans and other living species. Tackling water shortage and pollution is critical for combating global environmental challenges, necessitating the development of novel solutions for efficient water, air purification, or pollution control. Novel bionanocomposites are proof of green solutions that show prospects for remediation of the environment from pollutants, before and even after their application owing to their eco-friendliness. This chapter discusses bionanocomposites, prevalent air, and water pollutants and their negative impact, bionanocomposites fabrication methods, remediation strategies, and the application of bionanocomposites in environmental remediation. The significance and performance of bionanocomposites and hybrid bionanocomposites are discussed along with their physicochemical properties, including large surface area, thermal stability, and varied functionalization, which are crucial in dealing with the issues of environmental pollution. A case study in China on air and water purification and regulations has been provided. Furthermore, challenges faced in the fabrication and application of bionanocomposites and future outlook have been discussed.

B. Sarkodie (✉) · B. Tawiah · B. K. Asinyo
Department of Industrial Art (Textiles), Kwame Nkrumah University of Science and Technology, Kumasi, Ghana
e-mail: bismarksarkodie@yahoo.com

B. Tawiah
e-mail: benjamin.tawiah@knust.edu.gh

B. K. Asinyo
e-mail: asinyoboss@cass.edu.gh

B. Sarkodie
School of Environment, Key Laboratory of Yellow River and Huaihe River Water Environment and Pollution Control, Ministry of Education, Henan Key Laboratory for Environmental Pollution Control, Henan Normal University, Xinxiang, China

B. Tawiah · B. Fei
School of Fashion and Textiles (SFT), The Hong Kong Polytechnic University, Hung Hom, Kowloon, Hong Kong SAR, China
e-mail: bin.fei@polyu.edu.hk

S. C. George and B. Tawiah (eds.), *Advanced Bionanocomposite Materials*, Advanced Structured Materials 241, https://doi.org/10.1007/978-981-96-8225-6_8

Keywords Bionanocomposites · Pollutants · Water treatment · Air treatment · Nanomaterials

8.1 Introduction

Bionanocomposites, also known as biohybrid materials, are made up of biopolymers and organic or inorganic solids with different structures and morphologies that combine at the nanoscale scale (Zafar et al. 2016; Medhi and Kalita 2024). They comprise biological or biodegradable constituent(s) and fillers with at least one dimension ranging from 1 to 100 nm. Due to the outstanding characteristics that exist in these novel hybrids as structural or functional materials, scientists are investing heavily in this research area, with potential applications as constituents of heterogeneous catalysts and optical, magnetic, and electrochemical devices, among others. Furthermore, bionanocomposites meet national and customers' preferences for more biodegradable and sustainable goods, selecting materials with lower ecological damage (Pires et al. 2023; Puiggalí and Katsarava 2017).

The environment is clouded with pollutants in the air, water, and soil, mostly from industries and households. These air, water, and soil pollutants have detrimental effects on the livelihood of mankind and other living organisms. In 2015, the United Nations General Assembly established the 2030 Sustainable Development Goals, which stress a coherent route with many perks as a key strategy for dealing with the complex difficulties encountered in the quest for environmental sustainability (Lu et al. 2020). Composites of many different forms and sources have been employed in remedying the environment from harmful pollutants. However, the desire for more bionanocomposites stems from the world's massive annual plastic usage, which is believed to be about 300 million tons per year, with the bulk of these plastics being petro-based synthetic material such as polyacrylonitrile, polypropylene, polyester, and many inorganic materials. The latter are often recognized as non-biodegradable and become secondary pollutants after use. The inability to appropriately deal with persistent plastic pollution poses a huge threat to the development of every nation and the existence of humanity. Applied remedies such as garbage buildup in landfills are unsustainable in the long run owing to the scarcity of disposal sites caused by ongoing urban expansion. Burning plastic garbage as an alternative treatment strategy emits high levels of carbon dioxide and harmful chemicals such as dioxins, which harm the ecology.

Based on the aforementioned issues, focusing relentlessly on the fabrication and development of bionanocomposites and their application in environmental remediation and water purification is key to a green environment. The chapter covers bionanocomposite, properties, production processes, certain pollutants in the environment, and treatment procedures. The use of bionanocomposites in pollutant treatment, particularly water purification, has also been critically considered. There is also a case study of the transient period of air and water purification in China from

2016 to 2023, with an emphasis on key air and water pollutant removal (Mousa et al. 2016).

8.2 Properties of Bionanocomposites

The functionalities of bionanocomposites hinge on their principal ingredients. Bionanocomposites are largely composed of nanofillers and biopolymers (natural or synthetic). Nanofillers include metal oxides, metals, mineral particles (Montmorillonite), and carbon nanomaterials, whereas biopolymers include chitosan, chitin, starch, polylactic acid, and polyhydroxyalkanoates. Figure 8.1 depicts a variety of biodegradable polymers utilized in the production of bionanocomposites. Bionanocomposites have excellent mechanical characteristics, biocompatibility, and biodegradability. They are largely successful because of their large surface area and increased surface reactivity, which make it possible to eliminate pathogenic microbes and other pollutants in air, soil, and water more efficiently than their micro- or macro-scale counterparts. These qualities are exceedingly important in the development of ecologically friendly materials (green nanocomposites) for cleaning up the environment. Bionanocomposites based on starch, cellulose, polyhydroxybutyrate, poly- (butylene succinate), polylactic acid, and polycaprolactone have been researched extensively. Bionanocomposites, which are biodegradable materials reinforced with engineered nanostructures, provide additional optical, electrical, electromagnetic shielding, and magnetic properties, resulting in the birth of a diverse range of advanced materials for use in environmental remediation and other fields. Furthermore, the modification of nanofillers with functionalized biopolymers results in an appealing alternative for pollutant removal. Chitosan/clay bionanocomposites can combine adsorbent characteristics and ion exchange capability to recover dyes and heavy metals (Zafar et al. 2016). The usage of biopolymers results in extremely renewable and biodegradable materials. For instance, biopolymers like nanocellulose are ideal for bionanocomposites because of their high aspect ratio, low density (1.6 g/cm^3), low coefficient of thermal expansion, and high tensile strength (500 MPa). In a word, bionanocomposites have a wide surface area and strong reactivity because they contain functional groups (such as hydroxyl, amine, and carboxyl groups) and active sites.

8.3 Fabrication Strategies

Fabrication methods are a critical aspect of every production chain. They do influence the constituents and also the outcome of production. A variety of fabrication processes are employed to pretreat materials before the combination of nanofillers and polymers. This is mostly dint to the poor compatibility of numerous polymers

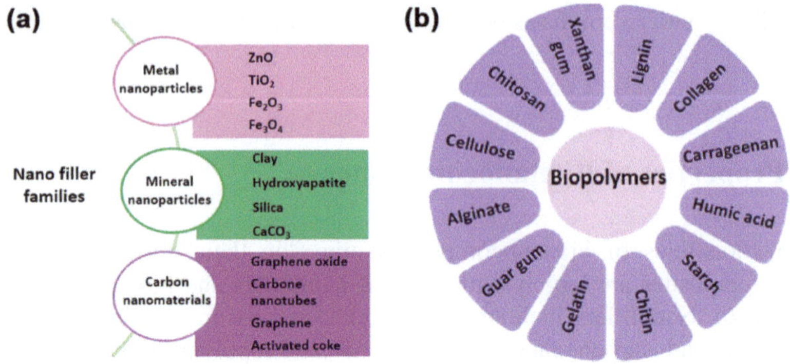

Fig. 8.1 Types of **a** nanofillers and **b** biopolymers used in bionanocomposites. Reprinted with permission from Chkirida et al. (2021)

with nanofillers meant for bionanocomposites. For example, the hydrophilic character of nanocellulose could hinder its employment as a filler in nanocomposite systems containing hydrophobic particles/polymers. This is primarily due to the tendency for poor interaction/blending of nanocellulose with hydrophobic polymers. This leads to enhanced water absorption and filler aggregation owing to hydrogen bonding. Consequently, surface modification of nanocellulose is necessary to ensure its compatibility with the hydrophobic component of the bionanocomposite. Briefly, some surface modification techniques enhance the miscibility of nanocellulose with hydrophobic polymers including silylation, grafting, acetylation, and the use of surfactants (Mousa et al. 2016). In this section, key priority is placed on the fabrication methods for bionanocomposites, but not on pretreatment methods although the latter is also significant in bionanocomposites fabrication. Bionanocomposites are typically synthesized through methods such as in-situ polymerization, solution casting, solution blending, solution intercalation, melt processing, emulsification, extrusion, drying techniques, electrohydrodynamic processes, cryogelation, and complex coacervation, which facilitate the formation of various nanostructures, that includes nanoemulsions, nanohydrogels, nanofibers, nanospheres, nanoliposomes, and nanosponges (Pires et al. 2023). Figure 8.2 illustrates several manufacturing techniques utilized in the production of bionanocomposites.

In situ polymerization method: Nanoparticles are mixed with the liquid monomer or monomer solution beforehand in this method. After that, polymerization starts by applying heat, radiation, or the proper initiators. When utilizing a high-shear mixer, the monomer's low viscosity—as opposed to its melt viscosity allows for uniform particle mixing. Moreover, the decreased viscosity and elevated diffusivity lead to an increased rate of monomer diffusion into the interlayer area. Nanocomposite morphology can also be regulated by adjusting reaction conditions and modifying the surface of the filler.

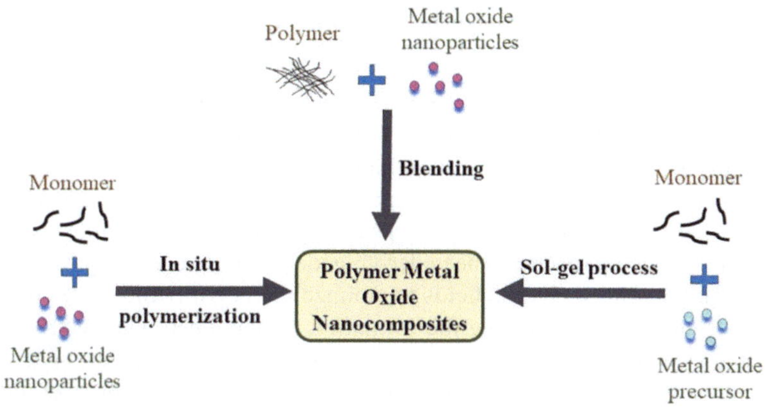

Fig. 8.2 Three fundamental approaches to synthesizing bionanocomposites. Reprinted with permission from Pandey et al. (2021) (Monomeric or polymeric materials are combined with particles, precursors of metals, metal-oxide, and metal–organic framework to obtain nanocomposite via fabrication methods such as in-situ polymerization, sol–gel method, solvent blending, etc.).

Melt processing: Nanoparticles can be directly incorporated into a molten polymer without the need for a solvent as the medium. This procedure omits solvent usage and is compatible with industrial polymer extrusion and mixing methods. It provides a cost-effective method for producing polymer nanocomposites. A diverse array of polymer/clay nanocomposites has been synthesized by this method, including nylon 6, polystyrene, and polypropylene. Melt intercalation provides a straightforward method for the synthesis of nanocomposites. Nevertheless, it is essential to meticulously "fine-tune" the surface chemistry of the filler to enhance its compatibility with the polymer matrix.

Extrusion methods: Extrusion methods provide a compelling approach in which the active material is incorporated into a biopolymer nanocarrier solution, often composed of polysaccharides, allowing the efficient encapsulation of a wide variety of hydrophilic or hydrophobic active chemicals. This procedure involves producing bionanocomposites with a predetermined cross-sectional profile by extruding material via a die of the specified shape, with or without thermal aid. In laboratory-scale procedures, the solution is introduced into a syringe and subsequently propelled via a nanoscale needle into a gelling medium. These procedures are deemed beneficial due to their gentle operational parameters regarding temperature or solvents. Furthermore, an extruder may be operated with controlled temperatures and speeds (Pires et al. 2023).

The sol–gel method: The sol–gel technique is a bottom-up method that consists of two reaction steps: the production of sol and gel. A polymerization event coupled with hydrolysis causes solid nanoparticles to first disperse in liquid monomer (sol), then agglomerate to create a three-dimensional interconnected network (gel) across the liquid phase. As a nucleating agent, the polymer helps layered crystals form and leads to the creation of nanocomposite structures because it percolates the layers as the crystals grow (Pandey et al. 2021; Bokov et al. 2021).

The electrohydrodynamic process: Electrospinning and electro-spraying are beneficial innovative methods for nanoencapsulation, facilitating the creation of environmentally benign, active electrospun bionanocomposites. This method is distinguished by its ability to operate at ambient temperatures, cheap cost, and scalability, making it particularly suitable for encapsulating delicate and volatile compounds compared to other approaches. The underlying idea of the two techniques is based on the use of high-voltage electric fields to atomize the polymer/bioactive component solution or emulsion, facilitating the formation of nanofibers via electro-spinning or particles via electro-spraying. The capacity to execute this operation at reduced temperatures guarantees the stability of the active compounds. Moreover, the large surface area and the nanoscale dimensions of the fiber/droplet promote rapid and efficient activity. Consequently, the future possibilities and aspirations for this subject are substantial, necessitating further study to enhance the understanding of size and shape, as well as to perform further tests to expand the variety of bionanocomposites (Pires et al. 2023).

8.4 Environmental Pollutants

Environmental pollutants are characterized as hazardous compounds that are released into our surroundings. They are harmful to people and other organisms. They are injurious solids, liquids, or gases generated at elevated amounts that diminish the integrity of our environment. They are also called anthropogenic pollutants since they are caused by human activities, particularly at home and in industries. Given the significant advancements in technology, society, and service supply caused by the Industrial Revolution, it has substantially increased the volume of pollutants released by the manufacturing sectors into the environment. Global environmental pollution is unequivocally seen as an international public health concern with several dimensions. Social, economic, and legislative issues, along with lifestyle habits, are interconnected with this significant problem. Environmental pollutants can be classified as air, water, soil, and occasionally solid pollutants, all of which adversely impact both mankind and the ecosystem (Priyadarshanee et al. 2022; Manisalidis et al. 2020).

Air pollutants: Air pollution is defined as the persistent presence of pollutants in the atmosphere in significant concentrations. Carbon monoxide, carbon dioxide, nitrogen monoxide, nitrogen dioxide, sulfur trioxide, volatile organic compounds, dioxins, and polycyclic aromatic hydrocarbons are a few examples of air pollutants. The six main air pollutants identified by the World Health Organization are lead, sulfur oxides, nitrogen oxides, carbon monoxide, particulate matter, and ground-level ozone. Human-induced air pollution constitutes a significant public health globally, responsible for around 9 million fatalities annually. The carbon footprint produced by autos and industry has posed a significant concern in recent years. The problem exacerbated in emerging countries due to expanding industrialization, unregulated urbanization, and overpopulation. This results in poor air quality, especially in countries where social injustices are prevalent and sustainable environmental management is not well understood. High levels of carbon monoxide can cause severe poisoning when inhaled. The human body may experience acute poisoning or chronic toxicity from heavy metals like lead, depending on the extent of exposure. Among the ailments caused by the substances are mostly respiratory problems such as COP, asthma, and bronchiolitis, along with lung cancer, heart attacks, central nervous system dysfunction, and skin diseases. Finally, climatic change induced by environmental degradation influences the geographical distribution of several infectious illnesses, as do natural catastrophes (Manisalidis et al. 2020; Sivaramanan 2015).

Water pollutants: Water pollution consists of elevated amounts of organic, inorganic, and biological contaminants that adversely impact the purity of water. A variety of contaminants can be found in water, including agricultural pollutants, thermal pollutants, radioactive pollutants, suspended solids, nutrients, pathogens, organic and inorganic pollutants, and others.

Figure 8.3 illustrates the cycle of water in countries involving the principal wastewater sources and their treatment processes. Organic and inorganic contaminants are mostly released by industrial effluents and sewage into aquatic environments. Several inorganic materials such as arsenic, antimony, boron, beryllium, barium, chloride, calcium, copper, cadmium, chromium, cobalt, lead, and zinc, among others, are identified in wastewater. These inorganic contaminants are non-biodegradable and cause hazards to aquatic ecosystems and the health of humanity. In summary, the primary water contaminants are heavy metals, synthetic colors, perfluoroalkyl compounds, and antibiotics such as tetracycline and sulfamethoxazole, originating from pharmaceutical businesses and sewage. Figure 8.4 depicts several water contaminants identified in industrial effluent.

Soil pollutants: Soil pollution arises from the introduction of chemicals or the dumping of waste materials, including heavy metals, hydrocarbons, fungicides, herbicides, and insecticides into the soil. Soil contaminants are sometimes comparable with water pollutants. The primary classifications of soil pollutants are heavy metals and organic contaminants. Recently, contamination from pharmaceuticals and plastics has emerged as a significant threat. Soil contaminants may readily infiltrate the soil, impacting the organisms residing inside it. The European Union's soil policy for 2030 emphasizes the overarching goal of zero pollution: by 2050, soil contamination must be diminished to levels deemed non-harmful to human health and the environment (Vieira 2024 #1105).

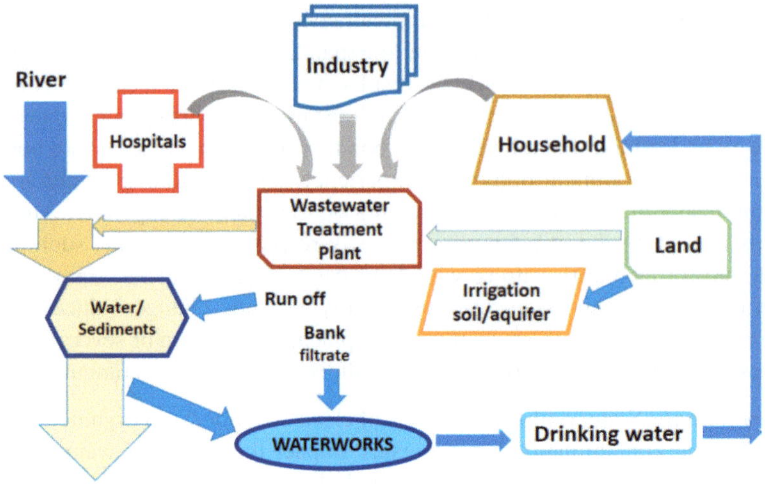

Fig. 8.3 Emerging contaminants' sources and environmental pathways. Reprinted from open access article (Vasilachi et al. 2021) by MDPI under CC BY 4.0 (Wastewater is obtained from different anthropogenic activities in hospitals, industries, and homes, among others. They are transferred into a water treatment plant to remove water pollutants using various remediation techniques such as filtration, adsorption, photocatalysis, etc.)

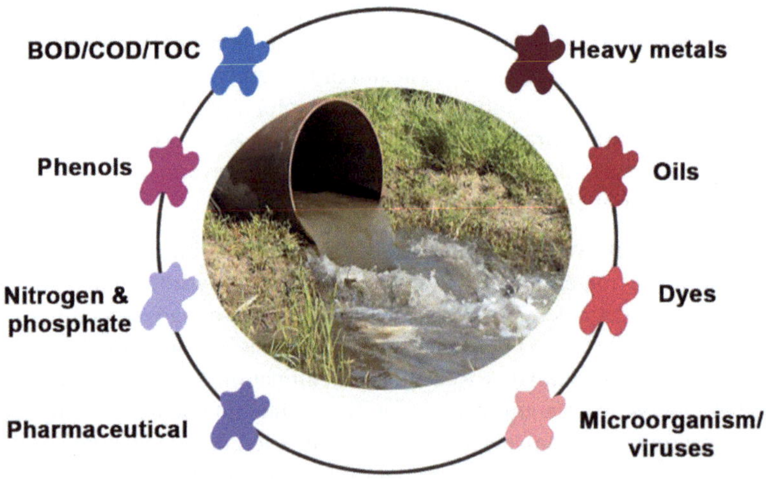

Fig. 8.4 Key pollutants in wastewater

8.5 Remediation Techniques

Diverse techniques are employed to remediate environmental contaminants (comprising organic and inorganic pollutants). The principal treatment techniques include coagulation and flocculation, distillation and boiling, adsorption, ion exchange, membrane filtration, catalytic processes, and biological approaches. Each approach possesses distinct capabilities, benefits, and downsides, rendering it unsuitable for all sorts of contaminants. Nevertheless, integrated strategies are typically more beneficial than singular strategies. When selecting the suitable technique for pollution removal, both the efficacy of pollutant elimination and cost efficiency must be considered (Mohammadi et al. 2022). There are three primary sorts of remedial procedures. They encompass physical, chemical, and biological methods. Figure 8.5 illustrates the various methods applicable in removal pollutants from the environment. Most studies have employed the use of physical methods such as adsorption in removing water and air. State-of-the-art research works encompassing various advanced oxidation processes under chemical methods that display high efficiency in pollutant removal have become a pathway to ensuring green ecosystem.

8.5.1 Physical Methods

The removal of pollutants from municipal wastewater is achieved by several methods, including adsorption, flocculation, precipitation, ion exchange, and membrane filtration, which are frequently regarded as the most ecologically friendly and cost-effective treatment alternatives. Bionanocomposites are utilized as adsorbents, aptly

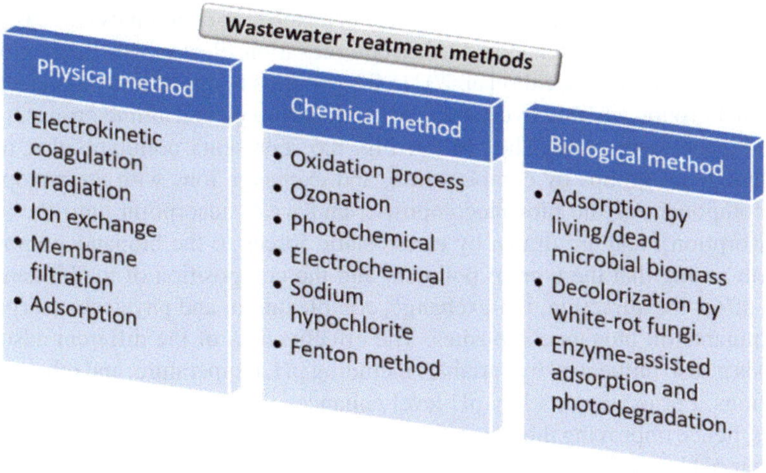

Fig. 8.5 Categories of the various methods applied in water treatment

referred to as nano sorbents. With decades of investigation, the adsorption strategy is thought to be the most thoroughly investigated technique. The process by which gas or liquid molecules are moved to a solid sorbent surface and held there by chemical and/or physical intermolecular interactions is known as adsorption. The movement of ions between two electrolyte solutions or between an electrolyte and a complex is referred to as an ion exchange. Absorption occurs throughout the material's bulk, whereas adsorption occurs only on the surface. The elimination of contaminants by nanosorbents can occur through either adsorption or absorption mechanisms. Nonetheless, several bionanocomposites documented are involved in the degradation of agrochemicals. Nanoparticles have been documented to degrade agrochemicals like DDT, lindane, endosulfan, atrazine, and aldrin. Conversely, nanoparticles integrated into biological materials improve the biosorption process, an environmentally sustainable technique for the removal of agrochemicals.

The elimination of contaminants using nano sorbents adhered to pseudo-first-order and pseudo-second-order kinetics. The pseudo-first-order model indicates that the adsorption rate is related to the number of vacant spots on the nanoparticle surface (Sebastian et al. 2020). According to the pseudo-first-order adsorption kinetic model, the adsorbate's diffusion across the adsorbent surface controls the adsorption rate. Some adsorption processes that require long time to occupy the adsorption sites are described by pseudo-second-order adsorption kinetics. The pseudo-second-order model posits that chemisorption is the rate-limiting mechanism, in contrast to the pseudo-first-order model, which attributes this role to physisorption. The adsorption process has shown success in eliminating dyes, organic contaminants, and metals from diverse industrial wastewater. It is essential to ascertain the thermodynamic characteristics, including standard free energy change (ΔG°), to evaluate the viability of the process. For example, if $\Delta G^{\circ} < 0$, the process is spontaneous and viable; conversely, if $\Delta G^{\circ} \geq 0$, it is not. Recent investigations have extensively examined adsorption using activated carbon, economical commercial products, and biosorption. Figure 8.6 illustrates the variety of adsorption techniques for heavy metal removal, including ion exchange, surface complexation, electrostatic interaction, and physical adsorption (Chkirida et al. 2021). The process is significantly affected by the functional groups, notably hydroxyl, carbonyl, carboxyl, and amine groups, present in the bionanocomposite. The "M" in Fig. 8.6 represents pollutants that interact with functional groups by complexation, and exchange ions with the nanoparticle filler component of the bionanocomposite, and weak adsorption onto the surface (physisorption), and are drawn by electrostatic forces to the bionanocomposite. It is worth noting that the type of pollutant and the composition of the bionanocomposite affect the attraction, ion-exchange, complexation, and physical adsorption of contaminants on bionanocomposites. The effectiveness of the different adsorption methods is also influenced by variables including pH, temperature, and other solution conditions. For example, a low pH level enhances the protonation of bionanocomposites, hence improving the adsorption of anionic pollutants, whereas deprotonation at higher pH levels favorably influences the adsorption of cationic pollutants. Protonation and deprotonation inhibit the adsorption of cationic and anionic contaminants by electrostatic attraction, respectively.

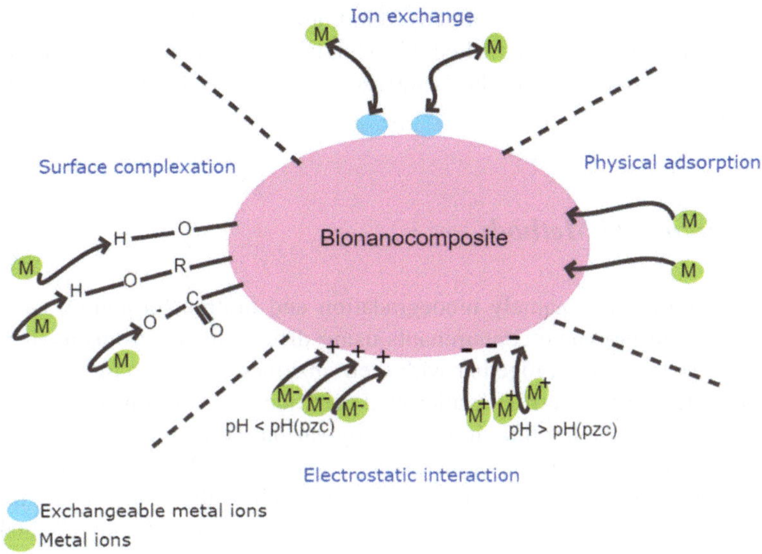

Fig. 8.6 Primary mechanisms in the adsorption of pollutants on bionanocomposite. Reprinted with permission from Chkirida et al. (2021)

Membrane technology is increasingly being used to treat municipal and industrial wastewater due to its increased energy efficiency, versatility, and ease of use. These methods can be broadly divided into four categories based on the size of the filter's pores: microfiltration, ultrafiltration, nanofiltration, and reverse osmosis. Pollutants with sizes of 100–1000 nm, 5–100 nm, and 1–5 nm are removed using reverse osmosis membranes, microfiltration, ultrafiltration, and nanofiltration, respectively. Air filtration employing bionanocomposites may be categorized into two types: physisorption and chemisorption, as elaborated in the adsorption mechanism. Negative pollutants stick to the filter surface or to atoms or molecules that are embedded in the filtering system in the first case. In chemisorption, impurities engage in either catalytic or non-catalytic chemical interactions with the filter membrane surface or embedded molecules/atoms, leading to the creation of simpler compounds. Both chemisorption and physisorption make it easier to get rid of gaseous molecules. Alongside electrostatic interactions, hydrogen bonding, van der Waals forces, hydrophobic interactions, and polar-polar interactions also occur during the physisorption of pollutants onto bionanocomposites. Filters are generally constructed in layers to enhance their effectiveness in eliminating pollutants. The preservation of a low-pressure drop (ΔP), indicative of the pressure variation before and after filtration, provides optimal performance of the face mask or filters. A low ΔP signifies optimal fluid/airflow through the filter mask/membrane, ensuring excellent breathability for face masks. In the context of porosity, ΔP, chemical functionalization techniques, filler blending, and layered filters, it is crucial to consider the nuances of how the functionalization process, as well as the blending and arrangement of fibers, influence filtration efficiency and user comfort. The accessibility of active/binding sites in the filtering system is the sole constraint in the sorption method. If no binding sites are there for the impending contaminants to adhere to the filter's surface, the residual pollutant molecules may traverse the filter.

8.5.2 Biological Methods

Biological approaches, namely biodegradation and biotransformation, can eliminate organic and inorganic contaminants using diverse microorganisms, including bacteria, algae, and fungi, together with their enzymes. Research indicates that the decontamination process utilizing microbial communities results in alterations to the microbial community structure, influenced by various contaminants in the medium and the operational parameters of the process (Dalvi et al. 2021). Biofiltration technology is an advantageous and sustainable approach to the removal of hazardous contaminants from aquatic environments. Diverse biological technologies, including bio trickling filters and biofilters, have been devised for air and water treatment, characterized by their cost-effectiveness, ease of fabrication, and simplicity of operation. A notable benefit of this procedure is the conversion of pollutants into biodegradable waste that may degrade within a typical timeframe, hence producing no secondary pollution. Figure 8.7 illustrates the entrapment and breakdown of pollutants by a microbial biofilter.

The second group comprises microbiological bioremediation techniques, including biotechnological approaches enhanced by ecotone planting or natural sorbents such as kaolin clays and various forms of zeolites. The last approach, microbial bioremediation, is among the most prevalent and promising techniques for the restoration of polluted surface waters and other environmental elements. In comparison to other technologies, bioremediation techniques are non-invasive and do not disrupt the trophic relationships within the food web of aquatic environments. These technologies are part of the environmental biotechnology sector and leverage current scientific advancements. Companies providing water treatment services provide many sorts of biopreparations in both solid and liquid forms. The choice of microorganisms and their composition are sometimes regarded as proprietary information of an organization. Biopreparations are created from the consortia of microorganisms chosen for a particular type of contamination. The positive aspect of this method is the rapid and efficient biodegradation of contaminants present in aquatic environments. For instance, the biodegradation of hydrocarbon-polluted medium with varying chain lengths, saturation levels, and configurations necessitates the interaction of a diverse assemble of microorganisms from many taxonomic groups. There exist intricate ecological and physiological interrelations among them. The most significant source of these bacteria is natural habitats with certain levels of hydrocarbon pollution. Besides biosorption, microorganisms can induce bio-oxidation of pollutants by oxidizing them in the ecosystem by enzyme

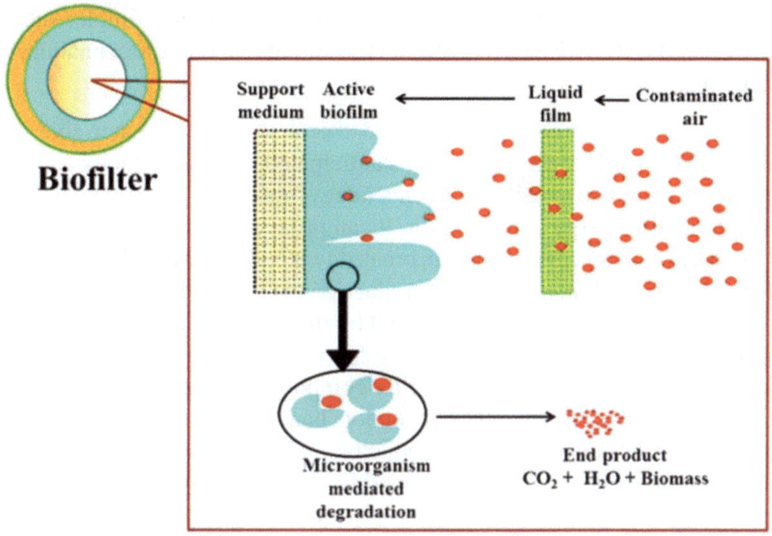

Fig. 8.7 Mechanism in air purification using biofilter. Reprinted with permission from Meena et al. (2021) (The air pollutant goes through the liquid film which does not have a substantial effect on the degradation of pollutant. In the later stage, microbes capture and release enzymes that facilitate the degradation of air pollutants into simple compounds such as CO_2 and H_2O)

secretion or altering ambient circumstances via their metabolic processes, contingent upon the pollutant type. Three major groups of bio-oxidants or suppliers of bio-oxidant enzymes are microorganisms, including bacteria, fungi, and algae. The primary source of these compounds' oxidation is typically the enzymes that these bio-oxidants produce. The enzymes can be collected from microorganisms and plants and employed indirectly, or they can be secreted directly by the microorganisms in the process environment (Mohammadi et al. 2022). Temperature, pH, metal ion concentration, and oxygenation conditions, if not maintained, affect the activity of microorganisms, making it a major challenge in bioremediation (Bala et al. 2022; Odoh et al. 2019). These microorganisms are supported by polymeric materials or bionanocomposites are used in environmental remediation.

8.5.3 Chemical Methods

Advanced oxidation processes, notably photocatalysis, the Fenton method, electrocatalysis, and ozonation, have garnered significant attention among various chemical cleanup strategies in recent years. The remediation of environments using advanced oxidation processes is a beneficial chemical strategy that employs generated radicals, highly reactive electrons or ions, and neutral species by exposing polluted medium (air, water, etc.) to high-energy electromagnetic radiation. These radicals primarily target pollutant molecules to facilitate their decomposition into simpler substances, mostly CO_2 and H_2O. Advanced oxidation techniques encompass ozonation, plasma treatment, and photocatalysis.

Photocatalysis: Solar energy is considered the most superior renewable resource available on the planet, owing to the Sun's anticipated longevity. Photocatalytic technology is considered an excellent "green strategy" due to its environmental benefits, which include low energy consumption. It can produce oxidizing agents like superoxide and hydroxyl radicals, which can break down pollutants like dyes in wastewater and aid in air purification, including the reduction of CO_2. When a bionanocomposite is used for photocatalysis, light is absorbed by the composite catalyst, which causes photoexcited electrons to move from the valence band (VB) to the conduction band (CB) of the main semiconductor nanofiller component in the composite.

Photoinduced electrons in the conduction band reduce surface-adsorbed electron acceptors and donors on semiconductors, which are appropriately polluting, while holes in the valence band are oxidizing them. The efficacy of photocatalytic bionanocomposite materials is ultimately determined by the separation of photoinduced electrons and holes. The bionanocomposite's mechanism is shown in the upper right corner of Fig. 8.8, showing how electrons are excited from the organic polymer's highest occupied molecular orbital (HOMO) to its lowest unoccupied molecular orbital (LUMO) and from the valence band (VB) to the conduction band (CB). The contact between the polymer and metal oxide enables the transfer of carriers to promote the generation of pollutant-degrading radicals.

Electrocatalysis: This technique is essential in several renewable energy conversion technologies. This discipline is comparatively nascent. Significant efforts were dedicated to the development of high-performance electrocatalysts, which depend substantially on advanced structural engineering techniques. Electrocatalysis is a catalytic process that entails oxidation and reduction processes by direct electron transfer, necessitating electrocatalysts to reduce the overpotential of these reactions. The mechanism of surface catalytic processes in photocatalysis closely resembles that of electrocatalysis. The reduction reaction sites (like H_2 evolution sites) work as cathodes, while the oxidation reaction sites (like O_2 evolution sites) act as anodes in a nanoparticle photocatalyst, which is essentially a cohesive microelectrochemical cell. Whether or not the reduction and oxidation reactions take place on different electrodes is what separates the two comparable processes.

Electrokinetic methods such as electrocoagulation, electrochemical oxidation, electroflotation, electroosmosis, and electrophoresis, facilitate the breakdown of diverse contaminants by electrochemical processes. They are particularly advantageous for the remediation of industrial wastewater, including textile effluent. Electrokinetic treatments have effectively removed pollutants such as reactive dye, phenolic compounds, and heavy metals.

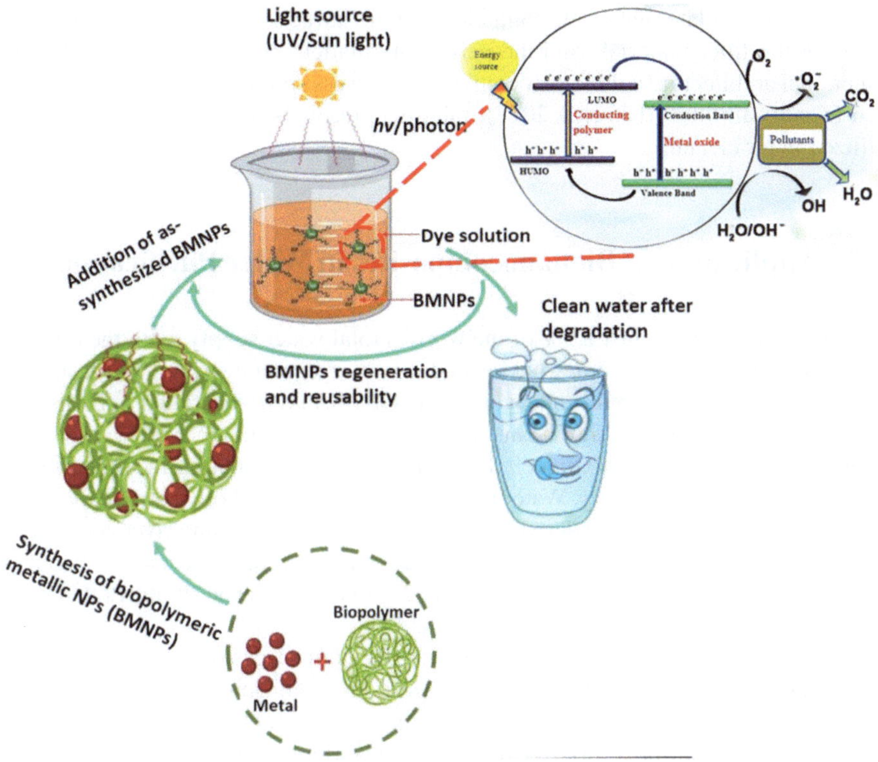

Fig. 8.8 Schematic representation of bionanocomposite fabrication via single, dual or multiple fabrication techniques (such as co-precipitations, green synthesis, in-situ precipitations, ex-situ precipitations, and/or hydrothermal techniques) and degradation of pollutant into clean water via photocatalysis. Reprinted with permission from Emmanuel et al. (2024) (Metal particles are combined with biopolymers to form bionanocomposite which is then immersed in wastewater. Light irradiation occurs to facilitate the production of radicals to degrade pollutants. In the upper-right corner is a likely process for the photocatalytic destruction of contaminants. Reprinted with permission from Pandey et al. (2021))

8.6 Application of Bionanocomposites in Air Purification

There is an urgent need to create an air-filtering material that is both environmentally benign and multifunctional to eliminate the secondary environmental contamination that is linked to non-biodegradable composites. The fabrication of multifunctional bionanocomposite membranes as potential options for filter innovation has been a focus in air purification (Shen et al. 2023; Lan et al. 2022). Bionanocomposites are utilized to investigate their potential as highly efficient air filtration materials. Bionanocomposites, due to their many functional groups and active sites, interact with contaminants. This facilitates the initial physical capture of particulate matter, while the functional groups of bionanocomposites additionally draw in pollutants, including particulate matter, through electrostatic attraction and dipole interactions between the filter materials and the particulate matter. Research indicates that the functional architectures of bionanocomposites possess significant potential for use in eco-friendly air filters (Liu et al. 2017). Table 8.1 presents several bionanocomposites utilized in air cleanup.

8.7 Application of Bionanocomposites in Water Purification

Freshwater constitutes just 2.5% of the world's total water supply, with the quantity of accessible water being somewhat lower due to its irregular distribution. Freshwater has been essential for human endeavors related to life and productivity; yet, this vital resource has encountered significant water security issues in the river basins of the twenty-first century. Approximately 80% of the global population faces significant threats to water sustainability. Water integrity, encompassing both water scarcity and water contamination, has hindered economic growth by disturbing river ecosystems and the commodities and services they provide to populations, particularly in developing nations (Tang et al. 2022; Mishra 2023; Oki and Kanae 2006; Pratap et al. 2023; Fahad et al. 2015).

Wastewater from the chemical sector mostly comprises organic and inorganic contaminants. These contaminants are hazardous, mutagenic, carcinogenic, and predominantly non-biodegradable. Comprehensive treatment of discharges produced by diverse chemical operations is important. A variety of physical, chemical, and biological techniques have been evaluated for the treatment of wastewater generated by the chemical, biological, food, pharmaceutical, pulp and paper, dye, and textile industries. The selection of wastewater treatment technologies depends on the kind, characteristics, and concentration of pollutants. The water treated must be environmentally sustainable and reusable. A multitude of research has examined the relationship of biosorbents with magnetic nanoparticles due to their magnetic characteristics. The magnetic properties of this material facilitate its recovery from aquatic environments, surpassing traditional separation methods such as centrifugation and filtering. Generally, oil spills float on the water's surface for a period of time. Considering these buoyant properties, bionanocomposite adsorbents designed for this application must be low-density, readily recoverable, and unsinkable to effectively interact with surface-suspending oils and optimize oil removal, as seen in Fig. 8.9b. Ferromagnetic bionanocomposites are the optimal candidates. Bacteria and other organic contaminants from sewage, fertilizers, agricultural runoff, forestry, food processing, tree and brush debris, industrial waste, and other sources are examples of organic water pollutants. Chemical waste from industrial byproducts, acid rain from industrial or volcanic emissions, acid mine drainage, acidification of lakes from water flowing from acidic soils, carbon dioxide emissions and runoff, inorganic salts, mineral acids, metals, trace elements, metal compounds, metal–organic complexes, sulfates, and cyanides are examples of inorganic water pollutants. One of the main sources of wastewater is the manufacturing industry (Mohammadi et al. 2022; Mishra and Maiti

Table 8.1 Bionanocomposites, synthesis, efficiency, and mechanism in removing air pollutants

Bionanocomposites	Synthesis	Performance	Novelty	Mechanism
Amine-modified biochar beads	Pyrolysis *Pollutant*: CO_2	CO_2 capture efficiency was obtained at 10.1 mmol g^{-1}, which was much greater than those of other sorbents with commercial activated carbon	The presence of nitrogen groups in the beads greatly enhanced the sorption capacity	Well-developed pore structure with a high specific surface area of 328.6 $m^2\ g^{-1}$ enhanced CO_2 adsorption (Nguyen and Lee 2016)
Mesoporous carbon stabilized MgO nanoparticles synthesized by pyrolysis of $MgCl_2$ preloaded waste biomass	Pyrolysis *Pollutant*: CO_2	Maximum capacity for CO_2 capture was observed at 5.45 mol kg^{-1}, which was 10 times greater than other MgO based CO_2 sorption	Inexhaustible $MgCl_2$ in seawater and the abundantly available biomass waste can be utilized to prepare mesoporous carbon stabilized MgO nanoparticles (mPC-MgO) for CO_2 capture	The high CO_2 capture and favorable selectivity were mainly attributed to their special structure (i.e., surface area, functional groups, and the MgO nanoparticles) (Liu et al. 2013)
Ag nanowire (AgNW)/ bacterial cellulose (BC) membrane/filter	In-situ cultivation *Pollutant*: Staphylococcus aureus strain	The filtration efficiency of the BC/AgNWs composite filter paper for $PM_{2.5}$ and PM_{10} is as high as 99.7% and 99.8%, respectively, which are on par with commonly used 3 M masks on the market	The addition of AgNW endows the BC filter with antibacterial activity and increases the porosity and air permeability	Surprisingly, AgNW incorporation achieves high-efficiency filtration through electrostatic adsorption (Wu et al. 2022)

(continued)

Table 8.1 (continued)

Bionanocomposites	Synthesis	Performance	Novelty	Mechanism
CNFs/chitosan (2:1) filter paper	Emulsion freeze-drying technique *Pollutant*: Particulate matter (PM) of different sizes and CO_2	Filtration of 92.52% $PM_{0.1}$, 96.88% $PM_{0.3}$, 97.77% $PM_{0.5}$, 99.56% $PM_{2.5}$ and 4.8 mmol/g CO_2 adsorption	Surface modification of CNFs by chitosan was able to tailor CNF surface properties without monotonous and unpractical solvent-exchange protocols to make effective sustainable air nanofilters	Higher CO_2 adsorption as a result of a decrease in pore diameter and an increase in specific surface area. Furthermore, high CNF concentrations are anticipated to increase the number of OH groups on the CNF surface, which could lead to more amino groups reacting with the surface OH groups and changing the surface, all of which would increase CO_2 adsorption (Sepahvand et al. 2021)
Polyester (PE)/cellulose/ silica bionanocomposite membrane	Ultrasonic irradiation and solution casting *Pollutants*: CO_2, N_2, O_2, CH_4	The highest selectivity for CO_2/N_2 was 19.11 cm^2/s and the highest selectivity for CO_2/CH_4 and O_2/N_2 were 17.5 and 6.09 cm^2/s, respectively	Cellulose/silica was used for enhancing strength and thermal resistance, and affinity for gas molecules of PE membranes	Cellulose/silica, having hydroxyl groups containing surfactant act as pseudo cross-linker to improve interaction with polyester and affinity for gas (Ahmadizadegan et al. 2018)
Magnesium oxide embedded nitrogen self-doped biochar composites	Ion exchange *Pollutant:* Lead	Large adsorption capacity (893 mg/g), a very short equilibrium time (< 10 min), and a large throughput (~ 4450 BV) It also quickly and efficiently removed Cd^{2+} and tetracycline	Properly engineered biochar material's surface for enhancement in pollutant-bionanocomposite interaction	Surface coordination of Pb^{2+} and functional groups (C=O or O=C–O, pyridinic, pyridonic, and pyrrolic N) enhance adsorption and ion exchange (Ling et al. 2017)

2017). Multiple investigations have documented the adsorption of heavy metals on bionanocomposites, including Hg, Cd, Pb, Cr, Ni, and As. Kumar et al. (2020) investigated the Hg absorption from simulated wastewater utilizing a bionanocomposite derived from chitosan. Due to the abundance of amino and hydroxyl functional groups of chitosan, infused with Brassica leaf extract, a maximum adsorption capacity of 50 mg/g over 45 min at 60 °C was observed. Padilla-Ortega et al. (2016) effectively examined the removal of cadmium ions utilizing a bionanocomposite in the form of macroporous foams, which consisted of chitosan intercalated inside vermiculite interlayers. The investigation indicates that the adsorption capability of the bionanocomposite is three times more than that of the individual units. For the same objective, Methionine-modified bentonite/Alginate bionanocomposite was also developed by Ahmad and Mirza (2015). The bionanocomposite efficiently adsorbed 30.86 mg g^{-1} Pb (II) and 217.39 mg g^{-1} Cd (II) at 303 K, alongside good recyclability. Mirza and Ahmad (2018) designed an efficient bionanocomposite utilizing Xanthan gum and montmorillonite for lead (Pb) absorption from industrial and simulated wastewater. It was reported that the bionanocomposite has the potential to adsorb 187 mg/g of Pb in 4 h. Some bionanocomposites, their performance, and their mechanism in water treatment from pollutants have been provided in Table 8.2.

8.8 Application of Bionanocomposites in the Removal of Other Pollutants

Another medium that falls victim to pollution in the world is soil. Bionanocomposites are also utilized for the elimination of contaminants from the soil. The application of diverse agrochemicals has led to enhanced productivity; yet, it has environmental drawbacks, including pollution, heightened resistance to pesticides and herbicides, and potential risks for consumers. Promoting an eco-friendly environment may entail employing bionanocomposites during the nursery and planting phases or utilizing them for pollution remediation. To address these challenges, nano-formulations are developed utilizing bio-based components (such as chitosan, lignin, cellulose, and fibers) that serve as an alternative to mitigate dangers linked to traditional methods. Herbal formulations of essential oils applied to bionanocomposites such as chitosan-based, poly (ethylene glycol)-based, and poly(ε-caprolactone)-based composites have been thoroughly investigated for the sustainable development of pest resistance. Moreover, Chitosan-based nanocomposites have been utilized as delivery vehicles for bioactive chemicals owing to their low toxicity, biocompatibility, and biodegradable properties. Research findings indicate that lignin-based nano- and microcarriers serve as excellent delivery mechanisms within the plant. This approach significantly aids in the elimination of soil pollutants, including plastic trash and other non-degradable components or compounds utilized in farming and harvesting. The use of bionanocomposites for the elimination of pollutants, including heavy metals, pesticides, sulfates, and hydrocarbons from the soil, parallels water remediation

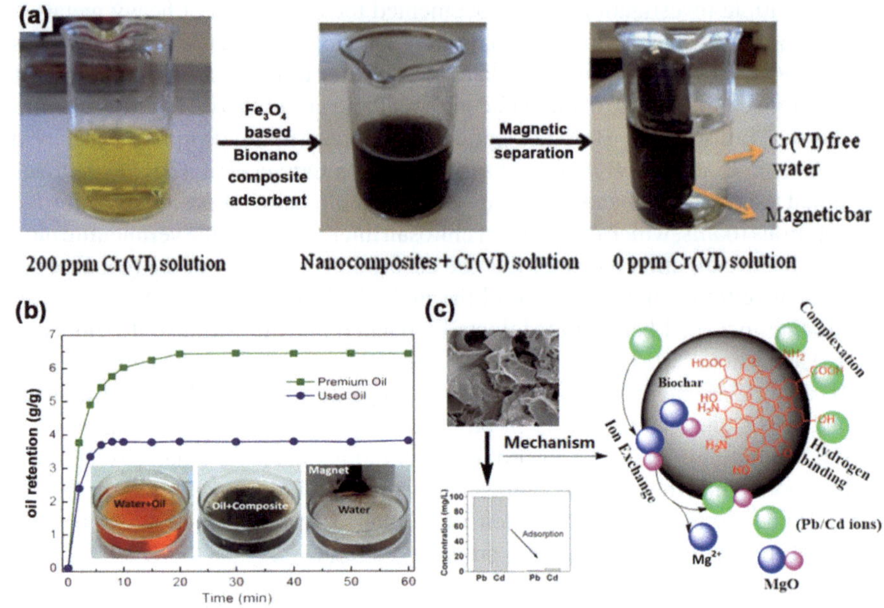

Fig. 8.9 **a** Illustration of Cr (VI) removal from aqueous solution using Polypyrrole/Fe_3O_4 nanocomposite. (Biocomposite prepared by the incorporation of Fe_3O_4 into polypyrrole efficiently removed Cr(VI) from water via ion exchange and reduction on its surface). Reprinted with permission from Bhaumik et al. (2011); **b** enhanced removal and retention of oil via adsorption using coconut shell based activated carbon–iron oxide magnetic nanocomposite prepared by in situ co-precipitation technique, with high recoverability using an external magnet. Reprinted with permission from Raj and Joy (2015); and **c** removal of lead and cadmium metals in water via ion-exchange between lead and cadmium on magnesium oxide-based self-doped biochar composites rich in nitrogen. Reprinted with permission from Ling et al. (2017)

processes, as analogous contaminants are also present in water and occasionally in air.

8.9 Case Study of Air and Water Purification and Regulations in China

China's economy has advanced significantly over the last thirty years. China's gross domestic product (GDP) grew by a factor of 43 between 1990 and 2017, averaging 10% per year. Fossil fuels, unique minerals, and precursors have been the main drivers of this incredible economic achievement in China. The simultaneous presence of water, soil, and air pollution in China has significantly impeded sustainable social and economic growth in recent years. Water and air pollution are particularly concerning due to their direct impact on human health (Tang et al. 2022).

Table 8.2 Bionanocomposites, remediation technique, fabrication method, and application in water pollutant removal

Bionanocomposite	Remediation technique	Fabrication method	Performance	Novelty	Mechanism
Chitosan/ lignosulfonate	Adsorption of lead	Free-radical polymerization	525 mg/g in 225 min	Utilization of lignosulfonate (papermaking industry waste) together with a sodium–chitosan adsorbent	Electrostatic attraction and chemical interaction (Zhang et al. 2021)
Biomimetic SiO_2@chitosan	Adsorption of mercury	Dissolution-deposition method	204.1 mg/g in 30 min	A novel chitosan-based nanocomposite adsorbent consisting of a biomimetic SiO_2 micro-/ nanostructure and chitosan nanoparticles coating on the surface	The hollow leaf-like SiO_2 scaffold within the adsorbent has a stable chemical property; while on the surface SiO_2, the chitosan nanoparticles provide a large number of active sites such as amino and hydroxyl groups for adsorbing heavy metal ions. The special SiO_2 structure also prevents the agglomeration and loss of chitosan, which enables the efficient contact between the functional groups of chitosan and heavy metal ions (Liu et al. 2019)

(continued)

Table 8.2 (continued)

Bionanocomposite	Remediation technique	Fabrication method	Performance	Novelty	Mechanism
Anchoring nanoscale zero-valent iron within bacterial cellulose particles	Adsorption of Co^{2+} and Sr^{2+}	In-situ chemical reduction method	107.10 mg/g (Co^{2+}) and 64.96 mg/g (Sr^{2+}) maximum adsorption capacity Exceptional removal efficiency of Co^{2+} (92.53%) and Sr^{2+} (58.62%) removal in natural seawater systems	Bacterial cellulose particles were for the first time used as the supporting material for nanoscale zero-valent iron for adsorption of heavy metals	High adsorption capacity is controlled by redox and hydroxyl complexation (Cheng et al. 2024)
Bi_2O_3/cellulose acetate	Photocatalytic degradation of Rhodamine B	Electrospinning	About 95% dye degradation at 90 min, which was far better than cellulose acetate and Bi_2O_3/ polyacrylonitrile nanofiber membranes	The flexible bionanocomposite membrane demonstrates easy recyclability after several runs with no significant decline in performance	Photocatalytic degradation of dye by OH radicals (Sarkodie et al. 2024)

(continued)

Table 8.2 (continued)

Bionanocomposite	Remediation technique	Fabrication method	Performance	Novelty	Mechanism
Iron and zirconium modified luffa fiber	Adsorption of As (V)	Grafting	It grants 70–80 times higher adsorption capacity than luffa fiber. The adsorption capacities of the grafted materials were 2.55–2.89 mg/g	Biodegradable Luffa plant fibers were successfully grafted with Fe and Zr and used as adsorbent	Grafting with Fe and Zr increased the zeta potential and zero point of charge of luffa fiber (from pH 3.9 to 7.4 for Fe grafting and to 7.6 for Zr grafting), due to chemical bonding of the metals, possibly with the hydroxyl and carboxylic groups to promote adsorption (Nguyen et al. 2020)
Zirconium oxide-modified biochar derived from pomelo peel	Adsorption of sulfate	Chemical precipitation method	A maximum sulfate adsorption capacity of 35.21 mg/g, which is much higher than that of pomelo peel biochar (1.02 mg/g) was achieved	Zirconyl chloride was used to modify the pomelo peel biochar to obtain more recyclable and eco-friendlier adsorbent	The adsorbent has a high specific surface area and total pore volume and contains numerous zirconium oxides and hydroxyl groups to promote adsorption (Ao et al. 2020)

(continued)

Table 8.2 (continued)

Bionanocomposite	Remediation technique	Fabrication method	Performance	Novelty	Mechanism
Bacterial cellulose/polydopamine/Bi_2MoO_6 composite nanofiber membranes	Adsorption and photocatalytic degradation of Rhodamine B	Vacuum extraction	96% dye degradation in saturated O_2 under 150 min irradiation	Recyclable photocatalyst was obtained The activity of Bacterial cellulose/polydopamine/Bi_2MoO_6 was significantly enhanced by introducing O_2 into the catalytic system The adsorption photocatalytic coupling reaction of the membrane catalyst accelerates the rate of pollutant degradation compared to single photocatalytic reaction	Degradation of dye by OH and superoxide radicals produced during photocatalysis (Tao et al. 2024)

(continued)

Table 8.2 (continued)

Bionanocomposite	Remediation technique	Fabrication method	Performance	Novelty	Mechanism
Silk-Au Bionanocomposite	Adsorption and degradation of methylene blue	Gelation	Methylene blue degradation in 15 min	The addition of Au NPs allows the coupling of the dye removal capacities of silk with its in-situ degradation by catalysis	Electrostatic-based interaction (Belda Marín et al. 2022)
ZIF-8 modified bacterial cellulose/ sodium alginate composite aerogel	Adsorption of tetracycline hydrochloride	Chemical cross-linking and in situ growth in freeze-drying	A maximum adsorption capacity of 746.27 mg/g	The use of bacterial cellulose, a biodegradable and recyclable material	High surface area promotes adsorption of antibiotic (Yang et al. 2023)
Chitosan/Ag bionanocomposite	Photocatalytic degradation of sodium fluorescein dye and adsorption of heavy metals	Blending	Good sodium fluorescein dye photodegradation rate, heavy metals removal and antimicrobial activity against gram-negative bacteria like *E. coli*, and gram-positive bacteria like *G. bacillus*	Chitosan as green reactant in removal of dye, heavy metals and microbes and improve catalyst recycling without any loss	Reduction of dye molecules in photocatalysis and electrostatic interaction with heavy metals in adsorption (Al-Sherbini et al. 2019)

Air purification and regulatory frameworks in China: In the past, the main source of carbon dioxide (CO_2) emissions and a few air pollutants was fossil fuels, particularly coal. China is now the world's largest energy consumer and carbon emitter, surpassing the United States in CO2 emissions in 2007 and surpassing the U.S. in energy consumption in 2009. Peak yearly emissions of sulfur dioxide, nitrogen oxides, and primary fine particulate matter ($PM_{2.5}$) were recorded in China in 2012, 2006, and respectively, at 2.5, 4.6, and 1.5 times the 1990 levels (Fig. 8.10). In 2017, China accounted for 23.2% of worldwide energy consumption and 27.6% of global CO_2 emissions. Over 70% of the 338 municipal cities in China fell short of complying with the National Ambient Air Quality Standard, with $PM_{2.5}$ non-compliance being the most significant air pollution issue. China confronted the dual challenges of enhancing national air quality and addressing climate change (Lu et al. 2020).

Figure 8.11 shows the air quality in China from the year 2016 to 2023. Good air has taken the stage in the various categories of air quality assessment in China over the eight years. Over the years, excellent air quality gradually increased, demonstrating the positive manifestation of a clean environment (Ministry of Ecology and Environment, the People's Republic of China 2023).

According to a 2020 report on the SOEE in China, the control of particulate matter ($PM_{2.5}$) and ozone has been promoted in a coordinated way, the 2020 VOC control plan was issued and five rounds of supervision and assistance have been carried out in the prevention and control of ozone pollution. Clean fuel heating, which uses bionanocomposites to filter or remove pollutants, has supplanted coal heating in homes in Beijing, Tianjin, Hebei, and the neighboring areas, as well as Fen Wei Plain. A total of 39,000 industrial furnaces in the plain areas have been inspected. Ultra-low emission has been achieved of about 950 million kilowatts of

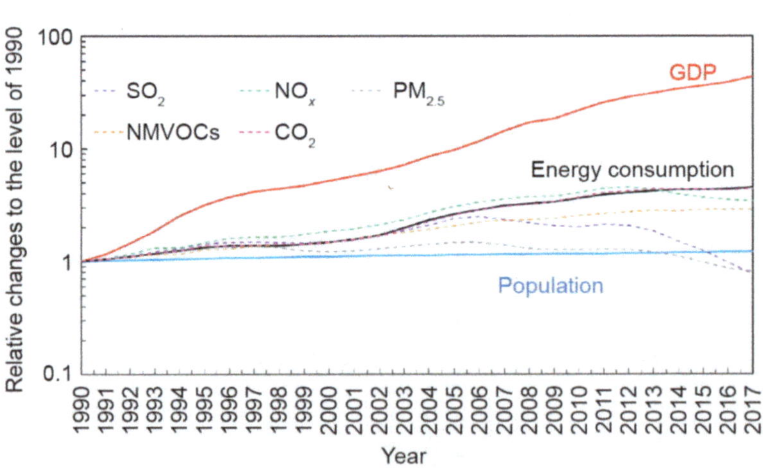

Fig. 8.10 Population, GDP, and energy consumption trends from 1990 to 2017; 1990 numbers are taken as 1; emissions of SO_2, NO_x, main $PM_{2.5}$, non-methane volatile organic compounds (NMVOCs), and CO_2 are also included. Reprinted with permission from Zheng et al. (2018), an open access article under CC BY-NC-ND 4.0 by Elsevier

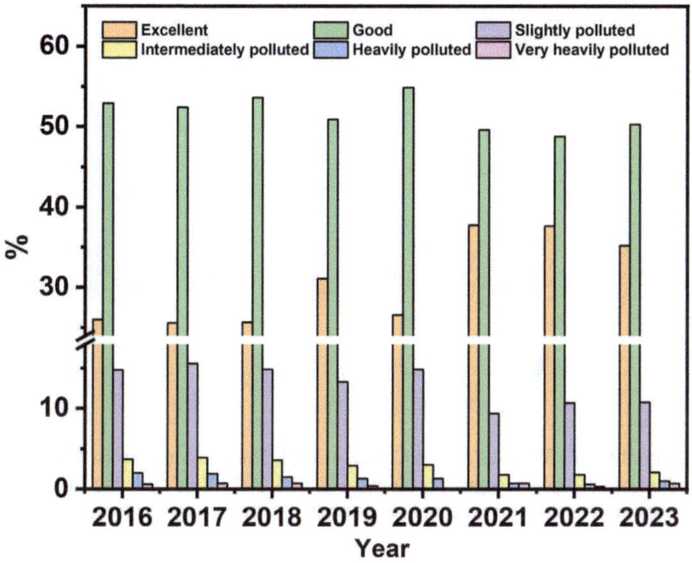

Fig. 8.11 Air quality from 2016 to 2023 in China (Ministry of Ecology and Environment, the People's Republic of China 2023). Data for this figure was obtained from the 2023 Report on the State of the Ecology and Environment (SOEE) in China, freely available at https://english.mee.gov.cn/Resources/Reports/soe/

coal-fired power generation units. The percentage of days meeting air quality standards increased to 87.0% (84.5% as the target) among cities at or above the prefecture level. Among cities failing to meet air quality standards at or above the prefecture level, the average concentration of $PM_{2.5}$ was 28.8% lower than that in 2015 (18% as the target) (Ministry of Ecology and Environment, the People's Republic of China, 2021).

According to the 2023 report, 203 of the 339 cities at or above the prefecture level (sometimes known as "APL cities") nationwide met the national air quality standard, making up 59.9% of the total; 136 cities, or 40.1% of the total, did not meet the standard (Fig. 8.12). Out of the 339 cities, 105 (or 31.0%) did not meet the fine particulate matter ($PM_{2.5}$) standard, 79 (23.3%) did not meet the ozone (O3) standard, 58 (17.1%) did not meet the inhaling particulate matter (PM) standard, and one (0.3%) did not meet the nitrogen dioxide (NO_2) standard. All cities met the carbon monoxide (CO) and sulfur dioxide standards. In a nutshell, 2020 recorded 59.9% attainment of national air quality, which was 13.3% better than the year 2019. However, no significant improvement from 2020 to 2023 has been achieved (Ministry of Ecology and Environment, the People's Republic of China 2023). Although environmentally friendly materials have been employed in reducing pollution in China, the persistency of the pollutants calls for further action. Consequently, research into more advanced technology and materials for the production of efficient bionanocomposites for air purification is key.

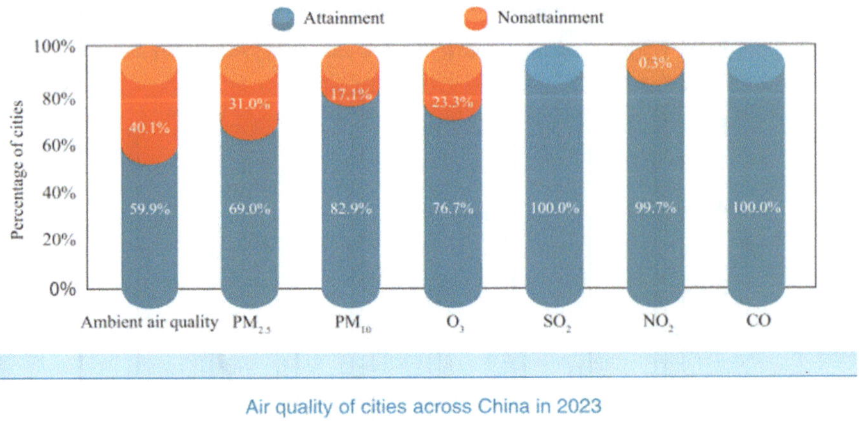

Fig. 8.12 Air quality of 339 cities across China in 2023. Data obtained from Ministry of Ecology and Environment, the People's Republic of China (2023), 2023 Report on the State of the Ecology and Environment in China, freely available at https://english.mee.gov.cn/Resources/Reports/soe/

Water purification and regulatory frameworks in China: Water purification and restrictions in China: In addition to air pollution, water shortage represents one of the most formidable challenges of our era, impacting the lives of humans and other organisms in China. China is experiencing escalating water constraints, significantly exacerbated by water pollution. Similar to air pollution, a major drawback of China's rapid economic growth has been the substantial rise in water contamination. China's water pollution presents a multifaceted issue, characterized by the presence of many toxins and sources of contamination. In the past two decades, China has undergone swift industrialization and urbanization. Consequently, by 2018, the urban population had risen by over 70% compared to 2001, while GDP had surged by over 700%.

The growth of population and urbanization will inevitably result in heightened demand for water supplies. Indeed, there has been a steady increase in China's surface water resources and availability per person in recent years. Therefore, the amount of residential sewage outflow has been increasing at the same time and will continue to do so. These include odorous, almost black rivers mainly from ammonia (NH_3–N), biochemical oxygen demand (BOD_5), chemical oxygen demand (COD), eutrophic lakes or reservoirs mainly from high concentrations of nitrogenous and phosphorous materials, and toxic water bodies mainly from heavy metals or organic compounds. The aforementioned conventional pollutants and hazardous compounds may originate from point-source pollution, non-point-source pollution, and/or internal contamination (Tang et al. 2022). Since the assertion by Chinese President in 2005 that "Clean waters and lush mountains are gold and silver," China's national development policy has emphasized environmental conservation over economic growth. Numerous municipal wastewater treatment facilities (WWTPs) have been established in the last two decades (Tang et al. 2022). The predominant technologies employed in municipal wastewater treatment plants (WWTPs) in China are anaerobic-anoxic-oxic systems, which utilize an anaerobic tank, anoxic tank, and oxic tank, as well as oxidation ditches, which collectively represent over 50% of the current WWTPs (Zhang et al. 2016). Additional technologies, including advanced oxidation processes, absorption, and ion exchange, are also utilized in industries for the elimination of pollutants from water and air.

China has made significant strides in ecological and environmental governance, and the country's overall ecological and environmental quality has been continuously improving. Throughout the nation, surface water quality kept becoming better. The percentage of sections with excellent and good water quality (Grades I to III) was 89.4% in 2023, up 1.5% points from 2022, according to Fig. 8.13a. Like 2022, the percentage of sections with water quality below Grade V was 0.7%. Furthermore, Fig. 8.13b provides a detailed examination of dynamism from 2016 to 2023. The proportion of surface water sections with water quality meeting Grade I-II standards in China rose from 67.8% in 2016 to 89.4% in 2023, which is an increase of 21.6% in water purity. The proportion of sections that are inferior to the Grade V standard fell from 8.6% to 0.7%, respectively, a decrease of 7.9%. This implies the water quality was improved from 2016 to 2023 and water inferior to Grade V reduced significantly over the years. Undoubtedly, the categorization of water quality is mostly founded on the quality of the main water basins in China. Figure 8.13c shows a graphical comparison of the main water basins in China for 2016 and 2013. The water quality of the mainstream of the Yangtze River and the Yellow River remained stable at Grade II, with a significantly low % of water inferior to Grade V. The main vehicle for water treatment in China has been WWTP. Figure 8.14 illustrates that the number of WWTP, capacity, and ratio increases every year. The significant increase in WWTPs buttresses the reason for the improvement in water quality over the years (Ministry of Ecology and Environment, the People's Republic of China 2023).

To treat water and air, China has used a variety of treatment methods including physical treatment technology, primarily membrane separation, and adsorption, due to the several benefits of membrane separation, such as ease of use, significant treatment effectiveness, and the ability to remove multiple pollutants at once. In China's Printed Circuit Board Wastewater Treatment Engineering Technical Specification (HJ 2058–2018), the process (ultrafiltration + reverse osmosis) has been noted as an effective treatment technique for Ni-containing wastewater effluent with the potential for reuse for various applications. Among the various existing treatment processes in China, membrane separation is mainly used as an advanced treatment (Li et al. 2022) for large-scale applications.

8.10 Challenges and Future Outlook

Bionanocomposites hold a considerable promise for environmental cleanup; nonetheless, they confront many challenges. Addressing the challenging issues is critical to their long-lasting success in implementation and incorporation into current

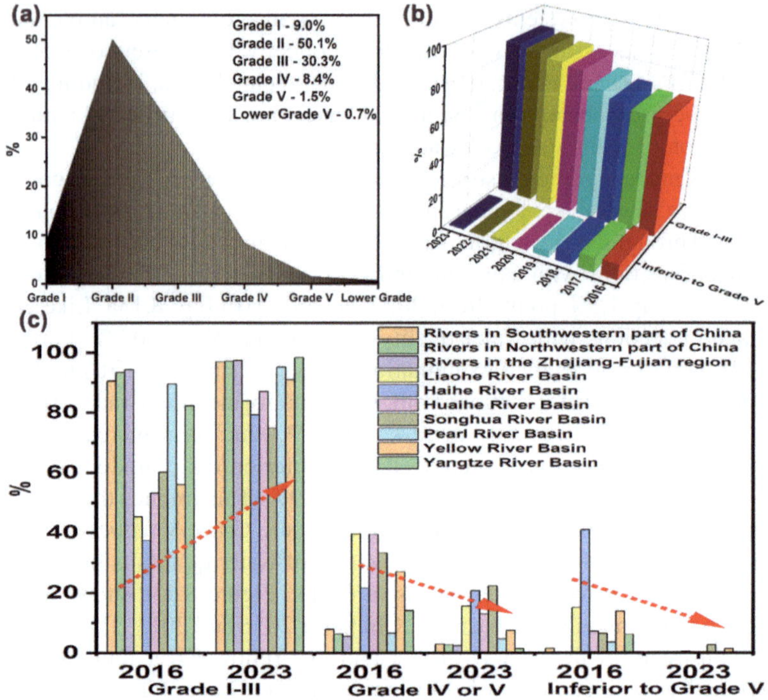

Fig. 8.13 Water quality in the years 2023 (**a**), 2016–2023 (**b**), and 7 major river basins and rivers in the Southwestern, Northwestern, and Zhejiang-Fujian regions of China (**c**). Data for this figure was obtained from the 2023 Report on the State of the Ecology and Environment in China, freely available at https://english.mee.gov.cn/Resources/Reports/soe/

remedial techniques. The following are three key issues in the manufacture and application of bionanocomposites in environmental remediation.

(1) Despite the adoption of "Near Zero Emission" (NZE) for firms with high pollutant output, recycling rates of industrial wastewater remain low compared to advanced countries. Most industrial firms lack advanced wastewater treatment equipment for particular water quality and using standard treatment technology results in poor efficiency and greater costs for the state when purifying water. Although bionanocomposites are favored over alternative options for environmental cleanup, their synthesis may be costly when considering the cost of ingredients and manufacturing processes. Therefore, they might not be suitable for large-scale applications. The development of more cost-effective yet unique bionanocomposites should be prioritized to protect the environment at all costs.
(2) Achieving proper interaction between bionanocomposite components, particularly metal/metal oxide particles, and biopolymers, is still challenging. Well-dispersed nanoparticles within the composite matrix are difficult to obtain, hence compromising overall performance. Concurrently, nanoparticle mobility impairs structural integrity and functions under a variety of conditions during

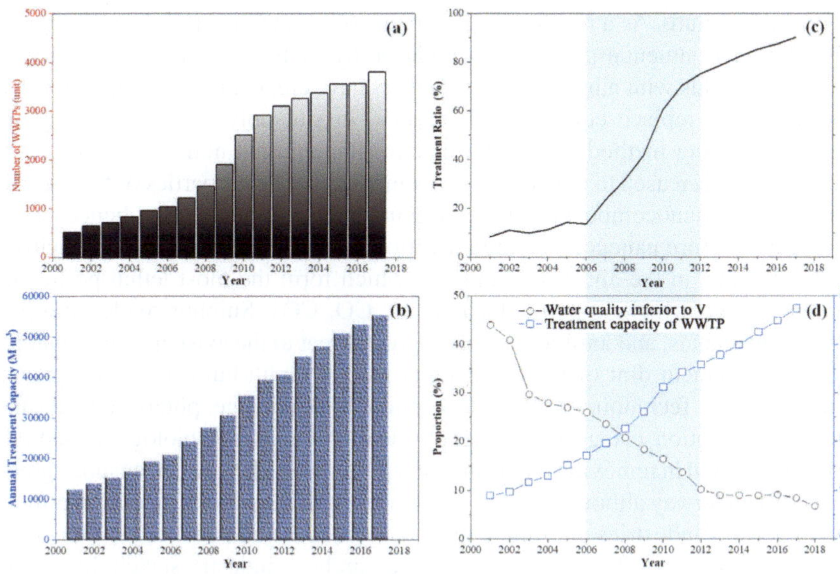

Fig. 8.14 Statistical trend of Chima's effort in combating water pollution—China's capacity of domestic sewage treatment from 2000 to 2018. **a** The quantity of WWTPs; **b** Yearly capacity for treatment; **c** treatment ratio; **d** riverine water quality changes and daily treatment capacity. Data for the plots were obtained from Tang et al. (2022) with permission

preparation and application. The above-mentioned issue stems principally from the differences in chemical makeup and condition of each ingredient. To achieve the full functioning of the produced bionanocomposite, advances in the development of innovative ways for improving particle and biopolymer interaction are required.

(3) Continuous effort in innovation is needed to successfully handle operational issues in bionanocomposites production, especially parameters and filler intercalation since they influence the outcome of fabrication. Furthermore, a large proportion of bionanocomposites utilized in environmental remediation include metal or metal oxide nanoparticles that are not biodegradable. This presents a risk after use, despite the fact that the majority of bionanocomposites are constructed with bio-based materials. Intensive research on bio-based nanomaterials is needed to limit the usage of metal/metal oxide nanoparticles.

8.11 Conclusion

Environmental remediation from pollutants released from industries and homes has been a crucial agenda to the development of countries. Ecofriendly remediation path using bionanocomposites to reduce the manifestation of secondary pollutants has become a key player in fulfilling the green target of the world dint of their biodegradable nature. As a result, using bionanocomposite materials is a big step in solving the environmental problems brought on by traditional plastics. By mixing bio-based materials with nanofillers, researchers have created novel composites that could eventually replace conventional plastics. In situ polymerization, filtration, the solution mixing method, extrusion, electrospinning, and numerous more techniques have all been used to create bionanocomposites. The properties of the various materials for bionanocomposites play a role in enhancing interaction, hence, some are pretreated before nanocomposite fabrication to improve compatibility. Environmental pollutants in the air, water, and soil which form the most lethal pollutants in the environment include particular matters, CO, CO_2, Sulphur oxide, nitrogen oxides, heave metals, and antibiotics pose a great threat to the existence of humanity and our environment dint of the detrimental reaction with human system, plants, soil, animals, etc. Techniques such as adsorption, ion exchange, photocatalysis, and microbial degradation, categorized under physical, chemical, and biological methods have been explored in removing environmental pollutants. The quest to ensure a green world is still underway although several challenges with decentralization of treatment plants in various industries and municipals have been a huge challenge, especially in developing countries. It is imperative that researchers, industry stakeholders, and policymakers collaborate to expedite the advancement and broad implementation of bionanocomposite technology. In the end, this will help create a future marked by greater sustainability and ecological awareness.

References

Ahmad R, Mirza A (2015) Sequestration of heavy metal ions by Methionine modified bentonite/alginate (Meth-bent/Alg): a bionanocomposite. Groundw Sustain Dev 1(1–2):50–58

Ahmadizadegan H et al (2018) Synthesis and characterization of polyester bionanocomposite membrane with ultrasonic irradiation process for gas permeation and antibacterial activity. Ultrason Sonochem 41:538–550

Al-Sherbini AA et al (2019) Utilization of chitosan/Ag bionanocomposites as eco-friendly photocatalytic reactor for bactericidal effect and heavy metals removal. Heliyon 5(6):e01980

Ao H et al (2020) Adsorption of sulfate ion from water by zirconium oxide-modified biochar derived from pomelo peel. Sci Total Environ 708:135092

Bala S et al (2022) Recent strategies for bioremediation of emerging pollutants: a review for a green and sustainable environment. Toxics 10(8):484

Belda Marín C et al (2022) Silk bionanocomposites for organic dye absorption and degradation. Appl Sci 12(18):9152

Bhaumik M et al (2011) Enhanced removal of Cr(VI) from aqueous solution using polypyrrole/Fe_3O_4 magnetic nanocomposite. J Hazard Mater 190(1):381–390

Bokov D et al (2021) Nanomaterial by sol-gel method: synthesis and application. Adv Mater Sci Eng 2021(1):5102014

Cheng R et al (2024) Anchoring nanoscale zero-valent iron within bacterial cellulose particles for boosting efficient adsorption of Co(II) and Sr(II) from seawater: dual system and varying adsorption mechanisms. J Environ Sci

Chkirida S, Zari N, Bouhfid R (2021) Insight into the bionanocomposite applications on wastewater decontamination. J Water Process Eng 43:102198

Dalvi V et al (2021) Removal of pollutants from wastewater via biological methods and shifts in microbial community profile during treatment process, pp 19–38
Emmanuel SS et al (2024) A pragmatic review on bio-polymerized metallic nano-architecture for photocatalytic degradation of recalcitrant dye pollutants. J Polym Environ 32(1):1–30
Fahad A et al (2015) Wastewater and its treatment techniques: an ample. Indian J Sci Technol 12:25
Kumar J, Arland S, Gour P (2020) Mercury removal from simulated waste water by chitosan nano composite embedded with leaf extract of *Brassica gongylodes*. Mater Today Proc 26:728–739
Lan X et al (2022) Biodegradable trilayered micro/nano-fibrous membranes with efficient filtration, directional moisture transport and antibacterial properties. Chem Eng J 447:137518
Li Q-G et al (2022) Heavy metal-contained wastewater in China: discharge, management and treatment. Sci Total Environ 808:152091
Ling L-L et al (2017) Magnesium oxide embedded nitrogen self-doped biochar composites: fast and high-efficiency adsorption of heavy metals in an aqueous solution. Environ Sci Technol 51(17):10081–10089
Liu W-J et al (2013) Mesoporous carbon stabilized MgO nanoparticles synthesized by pyrolysis of $MgCl_2$ preloaded waste biomass for highly efficient CO_2 capture. Environ Sci Technol 47(16):9397–9403
Liu X et al (2017) Soy protein isolate/bacterial cellulose composite membranes for high efficiency particulate air filtration. Compos Sci Technol 138:124–133
Liu J et al (2019) A biomimetic SiO_2@chitosan composite as highly-efficient adsorbent for removing heavy metal ions in drinking water. Chemosphere 214:738–742
Lu X et al (2020) Progress of air pollution control in china and its challenges and opportunities in the ecological civilization era. Engineering 6(12):1423–1431
Manisalidis I et al (2020) Environmental and health impacts of air pollution: a review. Front Public Health 8:14
Medhi U, Kalita E (2024) Lignocellulosic-based bio-nanocomposites: synthesis, properties, and biomedical applications. Novel bio-nanocomposites for biomedical applications. Springer, pp 113–145
Meena M et al (2021) Biological-based methods for the removal of volatile organic compounds (VOCs) and heavy metals. Environ Sci Pollut Res 28(3):2485–2508
Ministry of Ecology and Environment, the People's Republic of China (2021) 2020 report on the state of the ecology and environment in China, vol 30
Ministry of Ecology and Environment, the People's Republic of China (2023) 2023 report on the state of the ecology and environment in China, in Ministry of Ecology and Environment
Mirza A, Ahmad R (2018) Novel recyclable (Xanthan gum/montmorillonite) bionanocomposite for the removal of Pb (II) from synthetic and industrial wastewater. Environ Technol Innov 11:241–252
Mishra RK (2023) Fresh water availability and its global challenge. Br J Multidiscip Adv Stud 4(3):1–78
Mishra S, Maiti A (2017) The efficiency of *Eichhornia crassipes* in the removal of organic and inorganic pollutants from wastewater: a review. Environ Sci Pollut Res 24:7921–7937.
Mohammadi SA et al (2022) Biological oxidation methods for the removal of organic and inorganic contaminants from wastewater: a comprehensive review. Sci Total Environ 843:157026
Mousa MH et al (2016) Recent advances in bionanocomposites: preparation, properties, and applications. Int J Polym Mater Polym Biomater 65(5):225–254
Nguyen M-V, Lee B-K (2016) A novel removal of CO_2 using nitrogen doped biochar beads as a green adsorbent. Process Saf Environ Prot 104:490–498
Nguyen TTQ et al (2020) Iron and zirconium modified luffa fibre as an effective bioadsorbent to remove arsenic from drinking water. Chemosphere 258:127370
Odoh CK et al (2019) Status, progress and challenges of phytoremediation—an African scenario. J Environ Manage 237:365–378
Oki T, Kanae S (2006) Global hydrological cycles and world water resources. Science 313(5790):1068–1072

Padilla-Ortega E et al (2016) Ultrasound assisted preparation of chitosan–vermiculite bionanocomposite foams for cadmium uptake. Appl Clay Sci 130:40–49

Pandey B et al (2021) Photocatalytic-sorption processes for the removal of pollutants from wastewater using polymer metal oxide nanocomposites and associated environmental risks. Environ Nanotechnol Monitor Manage 16:100596

Pires JRA et al (2023) Current applications of bionanocomposites in food processing and packaging. Polymers 15(10):2336

Pratap B et al (2023) Wastewater generation and treatment by various eco-friendly technologies: possible health hazards and further reuse for environmental safety. Chemosphere 313:137547

Priyadarshanee M, Mahto U, Das S (2022) Mechanism of toxicity and adverse health effects of environmental pollutants. Microbial biodegradation and bioremediation. Elsevier, pp 33–53

Puiggalí J, Katsarava R (2017) Chapter 7—Bionanocomposites. In: Jlassi K, Chehimi MM, Thomas S (eds) Clay-polymer nanocomposites. Elsevier, pp 239–272

Raj KG, Joy PA (2015) Coconut shell based activated carbon–iron oxide magnetic nanocomposite for fast and efficient removal of oil spills. J Environ Chem Eng 3(3):2068–2075

Sarkodie B et al (2024) Boosting the recyclability of flame-made Bi_2O_3 via insitu immobilization in cellulose acetate nanofibers for efficient photocatalytic degradation of rhodamine B dye. Ceram Int 50(12):21951–21957

Sebastian A, Nangia A, Prasad MNV (2020) Advances in agrochemical remediation using nanoparticles. Agrochemicals detection, treatment and remediation. Elsevier, pp 465–485

Sepahvand S et al (2021) Preparation and characterization of air nanofilters based on cellulose nanofibers. Int J Biol Macromol 182:1392–1398

Shen R et al (2023) Biodegradable electrospun nanofiber membranes as promising candidates for the development of face masks. Int J Environ Res Public Health 20(2):1306

Sivaramanan S (2015) Air pollution sources, pollutants and mitigation measures. https://doi.org/10.13140/2.1.5106.8485

Tang W et al (2022) Twenty years of China's water pollution control: experiences and challenges. Chemosphere 295:133875

Tao Y et al (2024) Preparation of BC/PDA/Bi_2MoO_6 composite nanofiber membranes and efficient removal of organic pollutants from wastewater. Colloids Surf, A 702:134809

Vasilachi IC et al (2021) Occurrence and fate of emerging pollutants in water environment and options for their removal. Water 13(2):181

Vieira DCS et al (2024) Soil pollution in the European Union—An outlook. Environ Sci & Policy 161:103876. https://doi.org/10.1016/j.envsci.2024.103876

Wu A et al (2022) Rational design of bacterial cellulose-based air filter with antibacterial activity for highly efficient particulate matters removal. Nano Select 3(1):201–211

Yang X et al (2023) Zeolitic imidazolate framework-8 modified bacterial cellulose/sodium alginate composite aerogel for efficient removal of tetracycline hydrochloride. J Ind Text 53:15280837231204374

Zafar R et al (2016) Polysaccharide based bionanocomposites, properties and applications: a review. Int J Biol Macromol 92:1012–1024

Zhang QH et al (2016) Current status of urban wastewater treatment plants in China. Environ Int 92–93:11–22

Zhang F et al (2021) Preparation of chitosan/lignosulfonate for effectively removing Pb(II) in water. Polymer 228:123878

Zheng B et al (2018) Trends in China's anthropogenic emissions since 2010 as the consequence of clean air actions. Atmos Chem Phys 18(19):14095–14111

Bismark Sarkodie is a lecturer in the Department of Industrial Art, Textiles and Fashion section, KNUST, Ghana. He had his first and second postdoctoral fellowships at the Advanced Fiber Materials Engineering Research Center of Anhui Province, School of Textile and Clothing in Anhui Polytechnic University and School of Environment in Henan Normal University, China, respectively. He obtained his doctorate in the School of Materials Science and Engineering, East China University of Science and Technology in 2022, and his master's in the School of Textiles and Clothing, Jiangnan University in the year 2018. His scientific interests include polymer nanomaterials, metal and metal-oxide nanomaterials, and their application in the field of textiles and catalysis. His current research focus is on the synthesis of photocatalytic nanofibrous composites using spray synthesis and electrospinning strategies, and their application in the degradation of pollutants.

Benjamin Tawiah is a senior lecturer and a visitng postdoctoral fellow at the Hong Kong Polytechnic University. He got his Ph.D. in 2019 from the Hong Kong Polytechnic University where he worked as a postdoctoral Fellow and also served as a research assistant at the City University of Hong Kong. Dr. Tawiah has been engaged in basic and applied research in many areas including filament nanocomposites, flame retardant polymer composites, and second-generation batteries for energy storage in grid systems. He has written and published several research articles, book chapters, and books.

Benjamin K. Asinyo is an Associate Professor in the Department of Industrial Art, KNUST, Kumasi. He has been lecturing officially for over ten (10) years. In addition, he is a research fellow, designer, and curriculum developer. He instructs in Textiles, Curriculum Development, Research Methodology, and Computer Applications in Design.

Bin Fei is a Full Professor at the School of Fashion and Textiles (SFT), The Hong Kong Polytechnic University with over 20 years of lecturing experience. He obtained a B.Sc. Chemistry from the Department of Applied Chemistry, Chinese University of Science & Technology in 1996 and Ph.D. from Changchun Institute of Applied Chemistry, Chinese Academy of Sciences in 2003. He has since then engaged in basic and applied research of sustainable Textile Materials. He has published over 250 research articles, and several book chapters with over 10,000 citations in Google Scholar, i10 index of 157, and h-index of 57. He is the research leather of the ESG sustainable fashion hub and the MA program leader of the globally enviable fashion program at SFT, PolyU.

Chapter 9
Bionanocomposites and Their Potential Applications in Agriculture, Food Processing, and Food Packaging

Ahmed Fathy Mostafa Ghanem

Abstract Nanotechnology has found strong workability and applicability in various cutting-edge areas. In particular, the employment of nanomaterials in the field of agriculture and its applications has led to significant advances in the last decade, not only in food processing but also in food packaging. Building on this point, this chapter focuses on nanocomposites developed for agriculture and food applications, covering their types, methods, characteristics, functions, trends, challenges, and prospects. To narrow the focus, a special class of nanocomposites will be discussed in this report: bionanocomposites. Specifically, these materials combine biopolymers as matrices, such as polysaccharides, polycaprolactone, and polylactic acid, with nanostructures as fillers, like natural, carbonaceous metal, and metal oxide nanomaterials. These formulations can be fabricated using several physical and chemical approaches in order to ultimately acquire improved properties relevant to the target applications. In particular, bio-nanocomposites have shown superior performance when used as fertilizers, pesticides, herbicides, soil modifiers, food additives, food flavor, food colorant, or even as food packaging materials. There is no doubt that these compositions present promising alternatives to frequently used petrochemicals, such as polyethylene. More information will be provided about the role of nanofillers in addressing the drawbacks of some commonly used biopolymers, particularly in terms of physicochemical properties, including barrier, mechanical, and thermal behaviors, as well as compatibility and degradability. In conclusion, despite some limitations related to commercialization, such as concerns about nanotoxicity, government regulations, public awareness, and cost, the use of bionanocomposites could significantly contribute to sustainable food production, starting from eco-friendly cultivation and ending with the consumer. Certainly, this would have positive environmental, economic, and social impacts.

Keywords Biopolymers · Nanomaterials · Bionanocomposites · Agro and food applications

A. F. M. Ghanem (✉)
Packaging Materials Department, Chemical Industries Research Institute, National Research Centre, Dokki, Giza, Egypt
e-mail: af.ghanem@nrc.sci.eg; afghanem@kentech.ac.kr; a7mdghanem@gmail.com

Korea Institute of Energy Technology, KENTECH, Naju, Republic of Korea

S. C. George and B. Tawiah (eds.), *Advanced Bionanocomposite Materials*, Advanced Structured Materials 241, https://doi.org/10.1007/978-981-96-8225-6_9

9.1 Introduction

The nexus between agriculture, food processing, and food packaging areas represents the interconnected relationship in the food chain to ensure the security, safety, sustainability, and efficiency of the food products from the farm till received by the consumer (Kagan 2016; Castell-Perez et al. 2017; Knorr et al. 2020). Agriculture, the starting point of the food chain, is the primary step in producing the raw materials for food production (Yap and Al-Mutairi 2024). It involves crop cultivation (grains, fruits, vegetables), livestock farming (meat, dairy, eggs), and aquaculture (fish, seafood). Food processing, converting the raw materials into consumable products (Bellinghouse and Bellinghouse 2009), adds value to the agricultural product by improving its shelf life, safety, and nutrition. It includes primary processing (washing, drying, milling), secondary processing (baking, fermentation, preservation), and tertiary processing (ready-to-eat foods). Food packaging, ensuring safety and extending shelf life (Bandera 2005), protects processed food, ensuring hygiene, convenience, and preservation. It encompasses primary packaging (direct food contact, e.g., films, trays), secondary packaging (grouping products, e.g., boxes), and tertiary packaging (bulk transport, e.g., pallets).

Interlinkage between these three sectors happens when high-quality raw materials are subjected to processing as food, and the processed food requires appropriate packaging to retain freshness, and the packaging materials used need to be sustainable to reduce the environmental impact, benefiting agriculture. However, each sector has many challenges. The agriculture sector suffers from soil degradation, low fertility, pest resistance, overuse of pesticides, water scarcity, poor irrigation efficiency, and post-harvest losses (Alam 2014; Aktar et al. 2009; Ingrao et al. 2023; Shahbazi et al. 2025). The food processing shows some drawbacks, including food spoilage, contamination, nutrient loss during processing, excessive food waste, and high energy consumption in preservation (Augusto 2020). The food packaging segment faces undesirable shortages like plastic waste causing pollution, weak barrier properties of bioplastics, a lack of smart packaging for food safety, and the toxicity of synthetic packaging additives (Ibrahim et al. 2022; Han et al. 2018). Bionanocomposites act as a bridge across agriculture, food processing, and packaging by offering sustainable, smart, and functional solutions across agriculture, food processing, and food packaging, addressing challenges like food waste, contamination, inefficient resource use, and environmental pollution. Nano-enhanced fertilizers are for controlled nutrient release and soil improvement. Nano-encapsulated biopesticides are for targeted delivery and reduced chemical use. Bio-nanocomposite-based hydrogels can be designed for water retention. Bionanocomposite-based coatings can be utilized for delayed ripening and microbial protection. Antimicrobial bio-nanocomposite coatings inhibit bacterial growth. Nanoencapsulation is by bionanocomposites of vitamins, antioxidants, and bioactive compounds for controlled release. Bionanocomposites with oxygen scavengers and moisture absorbers tend to extend shelf life.

Biodegradable bio-nanocomposite films can be utilized as alternatives to conventional plastics. Nanofiller-reinforced biopolymers enhance gas and moisture resistance. Use of natural bio-based nanomaterials with antimicrobial and antioxidant properties eliminates poisonous packaging materials.

This chapter explores the impacts of bionanocomposites to overcome the challenges in agriculture, food processing, and food packaging. Particularly, the reader will start with an overview of composite materials and nano substances as an introduction to bionanocomposites. Then, the impact of nano fillers in the biopolymer's matrices would be illustrated intensively to recognize the properties of bio-nanocomposites and understand the basics. This will open the dialogue to identify their significance as promising materials for several applications. Classification, methods of preparation, and characterizations were also discussed in more details. Finally, the light will be shed on the evolution, innovation, and commercialization of bionanocomposites.

9.2 Overview

Before narrating about the bionanocomposites, one should understand the meaning of "composite material." By definition, the composite is a combination of two or more materials of different characteristics in order to eventually obtain a novel material that has unique physicochemical properties compared to the pristine components (Arulprasanna and Omkumar 2024) (Fig. 9.1). Generally, the components of the composite material were included in different ratios. However, the main factors that control this ratio are the acquired properties required for a specific application and the mode of interactions within this mixture. Particularly, the aim of mixing one or more materials is to improve some undesired properties of one or other constituents. It is worth differentiating between the words "Mixing" and "Composite." Despite the meaning of composite is the combination of materials, it is conditioned.

The constituent materials, as a matrix and as a filler, must keep their separate phases at the microscopic or macroscopic level and work synergistically to enhance the overall performance (Pereira and Martins 2014). This condition is not available in mixtures since when two materials are mixed and a homogenous mixture has resulted in a single phase, in such a case, it cannot be considered as a composite. That is because both materials lost their physical phases, like what happens in alloys. For instance, the crystalline structure of a salt is lost when dissolved in water. Thus, the mixture of salt and water is a non-composite material. It could be claimed that every composite is a mixture of substances, but not every mixture is a composite. Principally, the major features demanded to be improved in general are strength, capability to operate in harsh conditions, robustness, reusability, weight, toughness, conductivity, efficiency, and stability (Egbo 2021). Chawla (2012) presented a comparison between the composite material and its constituents in terms of some physical and mechanical properties (Fig. 9.2).

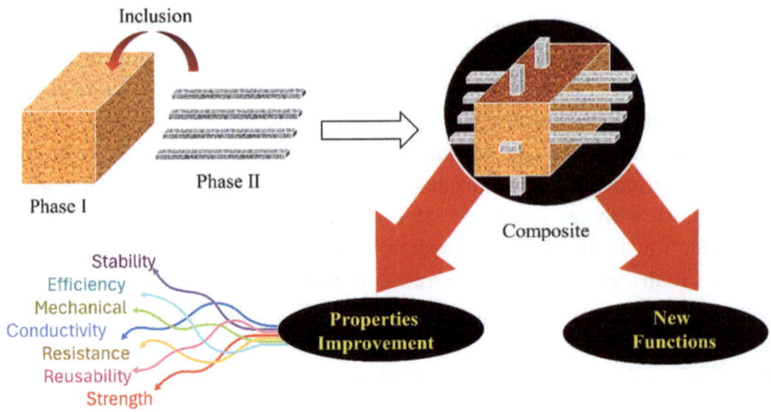

Fig. 9.1 General concept of composite formation

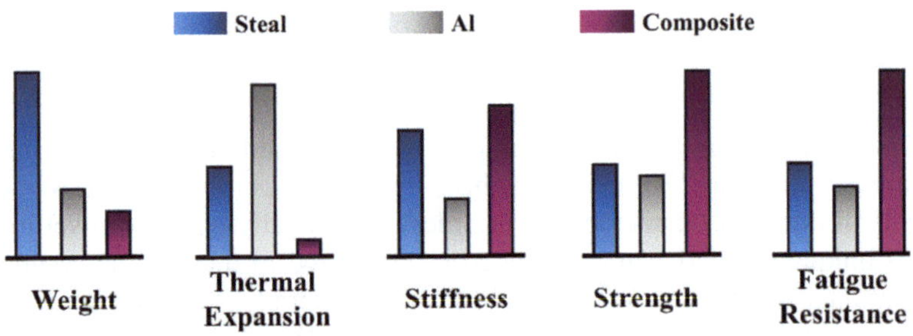

Fig. 9.2 Comparison between a metallic composite and its constituents

However, the target is not only the improvement of the aforementioned characteristics, but also new functions can be imparted to the obtained composite thanks to the presence of some ingredients carrying certain properties. Indeed, the tailoring of a composite material mainly depends on the kind of materials that are mixed to form this formulation. Moreover, designing a specific composite has a strong relation to the application. Hence, a certain manufacturing way shall be followed since the aim is not to mix some materials but to meet the requirements of an exact application. Therefore, to design a specific composite, one should first determine the demands of the application that the composite would be exploited in, the materials of the composite that should meet the application requirements, and finally select the proper method of fabrication. Some case studies will be presented in this chapter, illustrating this approach in more detail.

Particularly, composites have several classifications relying on the composed materials, the properties, the manufacturing process, the function, the structure, and the application. However, composite materials can be simply and more generally classified into inorganic, organic, and hybrid according to the kind of mixed materials; the matrix and the reinforcement substances (Arulprasanna and Omkumar 2024; Barbero

2010). Inorganic composites are mixed inorganic species that might be metals (such as Al, Cu, and Ti), or ceramics (like glass and alumina) (Sumithra et al. 2023; Mahajan and Aher 2012). These composites can be used under harsh conditions of temperature and pressure in addition to corrosion due to their strength and stability (Sumithra et al. 2023). On the other hand, polymer composites are the most common organic composites (Duraikkannu et al. 2021). Polymer combination is a process of mixing two or more different polymers to obtain a new matrix with enhanced functionalities. Particularly, this kind of composite is characterized by lightweight, strength, and stability. However, sometimes polymer composites involve the addition of a reinforcing filler for further improvement in the properties, especially the strength and stiffness. The fillers could also be organic fibers such as carbon and natural fibers (Sumithra et al. 2023). Thus, there are organic composites containing carbon fibers that have strong applicability in automotive and aerospace (Ozkan et al. 2020; Das 2001). Scientists combined the inorganic and organic constituents to produce hetero mixtures named hybrid composites (García-Martínez and Collar 2020). This type of mixture consolidates both properties of organic and inorganic phases. The produced hybrid composites enjoy several characteristics regards that improved mechanical properties, resilience, and resistance against degradation (Sathishkumar et al. 2014). Figure 9.3 provides more details about the sub-classes for each type of composite material, including some of their examples (Shahzad 2012; Kollia et al. 2020; Chandekar et al. 2020; Zhu et al. 2004; Song et al. 2013; Gil 2009; Paul and John 2018; Zubair et al. 2020; Behbahani et al. 2011; Nunes and Ramanathan 1995; Dursun and Duran 2019; Cafri et al. 2012; Sadik et al. 2014; Zhao et al. 2017; Choi and Bansal 2005; Pihtili 2009; Cui et al. 2021; Javni et al. 2002; Mazhoud et al. 2017; Najah et al. 2023; Choudhary et al. 2021). No doubt, hybrid composites are applied in a wide range of fields, including energy (Wu et al. 2022), water treatment (Kotobuki et al. 2021), medicine (Victor and Muthu 2014), electronics (Yang et al. 2020), construction (Afzal et al. 2020), sports (Gao et al. 2024), marine (Zhang et al. 2023), automotive (Velasco-Parra et al. 2020), and aerospace (Samir et al. 2024).

Almost three decades later, a new evolution in material science occurred. Most of the known materials were obtained at the nanoscale, the one billionth of a meter. Therefore, a new terminology appeared in the scientific community: nanocomposites. According to the aforementioned discussion, when at least one of the combined materials of a composite becomes nanosized, the resulting material can be called a nanocomposite (Omanović-Mikličanin et al. 2020). By definition, the nanomaterial's size should be less than 100 nm (nm) (Pokropivny et al. 2007). Sometimes, the sizes might increase partially to 300 and 500 nm in some cases but still called nano matter (Szafraniec et al. 2017). This may happen if these materials demonstrate similar properties to the less-sized nanoparticles. In-depth, fabricating materials with dimensions within 1–100 nm significantly changes the material's physicochemical properties (Ulusoy 2023). The features of a material in the nanoscale are completely different from its bulk counterpart. The mechanical, conductance, thermal, physical, optical, and chemical behavior could be altered totally by decreasing the size of the material up to 10^{-9} m (Gavrilenko 2016; Seifi et al. 2020; Asha and Narain 2020; Ghorai 2013). In brief, the materials at the nanoscale show an improvement

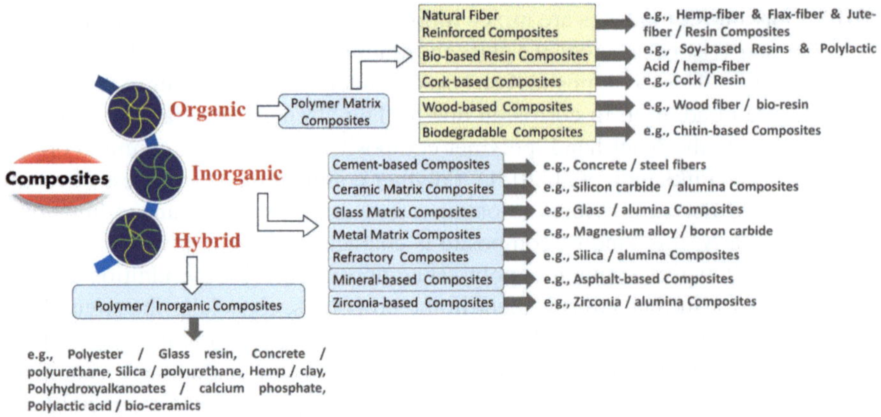

Fig. 9.3 Classification of composites according to the kind of combined materials and some examples for each type

in the atomic bonding and a decrease in the defect density, which particularly could improve the strength, hardness, and toughness of the material (Tang et al. 2021; Wu et al. 2020). Moreover, by reducing the size of the material, the surface-to-volume ratio and the surface area increase dramatically (Otero et al. 2023; Pozzi et al. 2024). This means that the number of atoms exposed to the environment becomes much higher. No doubt, this would enhance the material's reactivity with the surroundings. Nanosubstance displays surface plasmonic effects when subjected to lower dimensions (Wu et al. 2011). This means that the optical properties can be changed apparently thanks to the quantum effect that causes the material to produce much fluorescence depending on its size as described in Chhatre et al. (2012), Yang et al. (2024), Mogensen and Kneipp (2014). Also, because of the phenomenon of the quantum confinement effect, the materials at the nanometer scale exhibit exceptional conductivity (Oxandale et al. 2023). Magnetic ability can be influenced by the quantum mechanical effects as well (Katwal 2023). Other properties can also be detected for the nanoscale substances, such as surface wettability, lightweight, and biocidal activity in some cases (Zhang et al. 2024; Tasca and Antiochia 2020; Li et al. 2022). Actually, these superior characteristics of nanomaterials encourage people to investigate these matters in a wide range of applications including medicine (Nasimi and Haidari 2013), e.g., drug delivery, treatment, and diagnosis; electronics (Franklin 2015; Vidor et al. 2016), e.g., transistors and flexible devices; energy (Gungure and Jule 2025), e.g., fuel cells, solar cells, batteries, and supercapacitors; environmental protection (Ibrahim et al. 2016), e.g., water treatment and air purification; coatings (Somasundaram and Kumaravel 2019), e.g., self-cleaning and antimicrobial surfaces; flame retardancy (Bo et al. 2024); water-repelling (Bae et al. 2010); agriculture (Mehrazar et al. 2015), e.g., fertilizers, pesticides, herbicides; food processing (Curulli 2020; Jafarizadeh-Malmiri et al. 2019), e.g., additives, corruption sensing, and protection; and packaging (Adeyeye and Ashaolu 2021; Bumbudsanpharoke et al. 2025; Zhang et al. 2019), e.g., food storage, pharmaceuticals, and electronics.

Figure 9.4 shows the classification of the nanomaterials based on several perspectives, such as size, shape, composition, source, and applications (Alkaç et al. 2021; Buzea and Pacheco 2017; Khan 2017; Rahman 2011; Alshammari et al. 2023; Alezzi et al. 2024; Ghanem et al. 2014, 2019; Ghanem and Abdel Rehim 2018; Yassin et al. 2019; El-Gendi et al. 2020; Badawy et al. 2021a; Abdelhak et al. 2022; Al-Balakocy et al. 2022; Ali et al. 2024).

Particularly, in the field of composite science, the nanomaterials can be incorporated into the main matrix of the composite as a reinforcement filler or a modifier in order to improve the properties at the nanoscopic level rather than the macro and microscopic scales. In such a case, the produced formulation is called a nanocomposite. Therefore, in literature, numerous nanocomposites have been developed for a wide range of applications.

As mentioned in the previous composites' classification, Fig. 9.3, polymers, metals, or ceramics can be used as matrices to host mostly all kinds of nanomaterials. During this chapter, light will be shed on a specific class of nanocomposites, the hybrid polymer nanocomposites. The main matrix of this composite is polymers and the fillers are nanomaterials. As commonly known, polymers, according to their origin, can be categorized into synthetic, natural, and semi-synthetic (Alves et al.

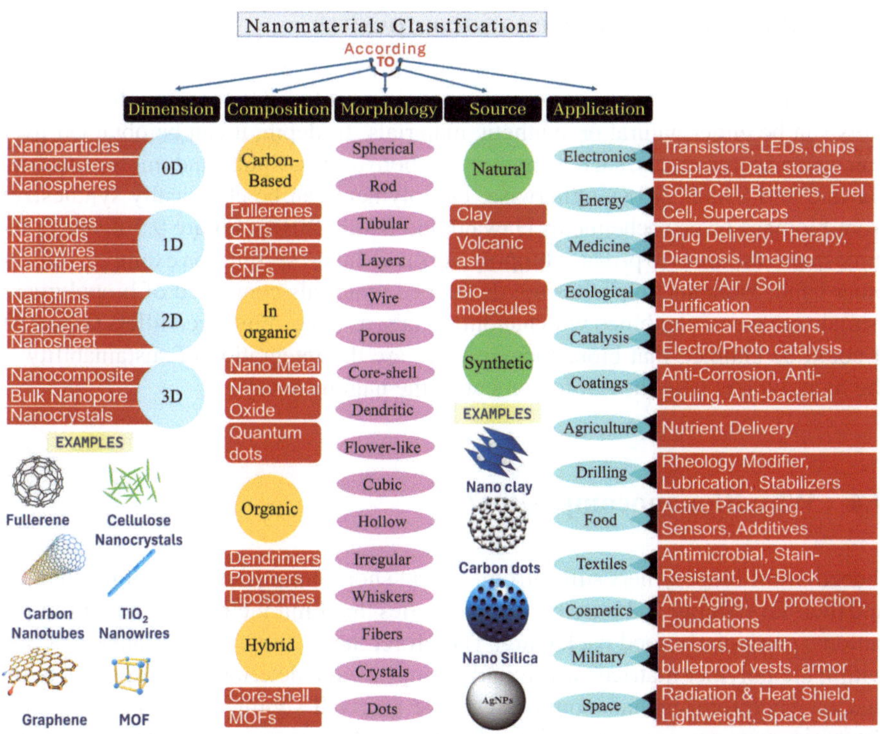

Fig. 9.4 Detailed classification of nanomaterials according to size, morphology, composition, origin, and applications with some known examples

2020). Synthetic polymers are chemically prepared usually from petroleum products, like polyethylene, polypropylene, polyvinyl chloride, and polystyrene, which mostly cause environmental burden (Parisi et al. 2015). Natural polymers are derived from natural sources such as plants, e.g., cellulose and starch; animals, e.g., collagen and keratin; and organisms, e.g., DNA and silk (Olatunji 2015). Particularly, these kinds of polymers are eco-friendly and called biopolymers. Semi-synthetic polymers are natural polymers but chemically modified to tune their properties such as cellulose acetate and esterified starch (Seremeta and Sosnik 2023; Nishitha et al. 2024; Zarski et al. 2021). Furthermore, polymers that are prepared from natural monomers are called semi-synthetic polymers or bio-based polymers such as polylactic acid and polycaprolactone (Cywar et al. 2022). No doubt, thousands of articles have been released, over the last years simultaneously with the evolution of nanoscience and applications, about the synthesis, characterization, and evaluation of polymers containing nanomaterials in several fields. However, through the few upcoming pages, the last two types of polymers and their nanocomposites will be discussed in more detail including the potential applications in the agriculture and food industry.

9.3 Biopolymers and Bionanocomposites

Biopolymers and bionanocomposite are two strongly related materials. Biopolymers can be either natural or synthetic materials. In-depth, it can be obtained from renewable bioresources, such as protein, carbohydrates, lipids, and microorganisms (Jawaid and Swain 2018). Otherwise, biopolymers can be chemically synthesized from biomass or petrochemicals (Jawaid and Swain 2018). Figure 9.5 displays the three main resources to produce biopolymers and the most common examples. On the other hand, bionanocomposites are developed materials composed of biopolymers, as main matrices, combined with nanomaterials, as nano-reinforced fillers (Fig. 9.5). It could be claimed that bionanocomposites are the integration of sustainability of the biopolymers with the reinforcement of the nanomaterials.

9.4 Why Bionanocomposites?

Indeed, most biopolymers suffer from several shortcomings in terms of mechanical and thermal stability and poor barrier properties that limit their usage in many applications, particularly food packaging. Therefore, the incorporation of nanomaterials within the polymer matrices was found to improve the deteriorated properties of the matrix. Therefore, bionanocomposites have received significant attention because the main matrices are made of eco-friendly polymers as an alternative to traditional petroleum-based plastics. Moreover, the inclusion of the nanoparticles into the bio-based polymer not only improves the mechanical strength, thermal stability,

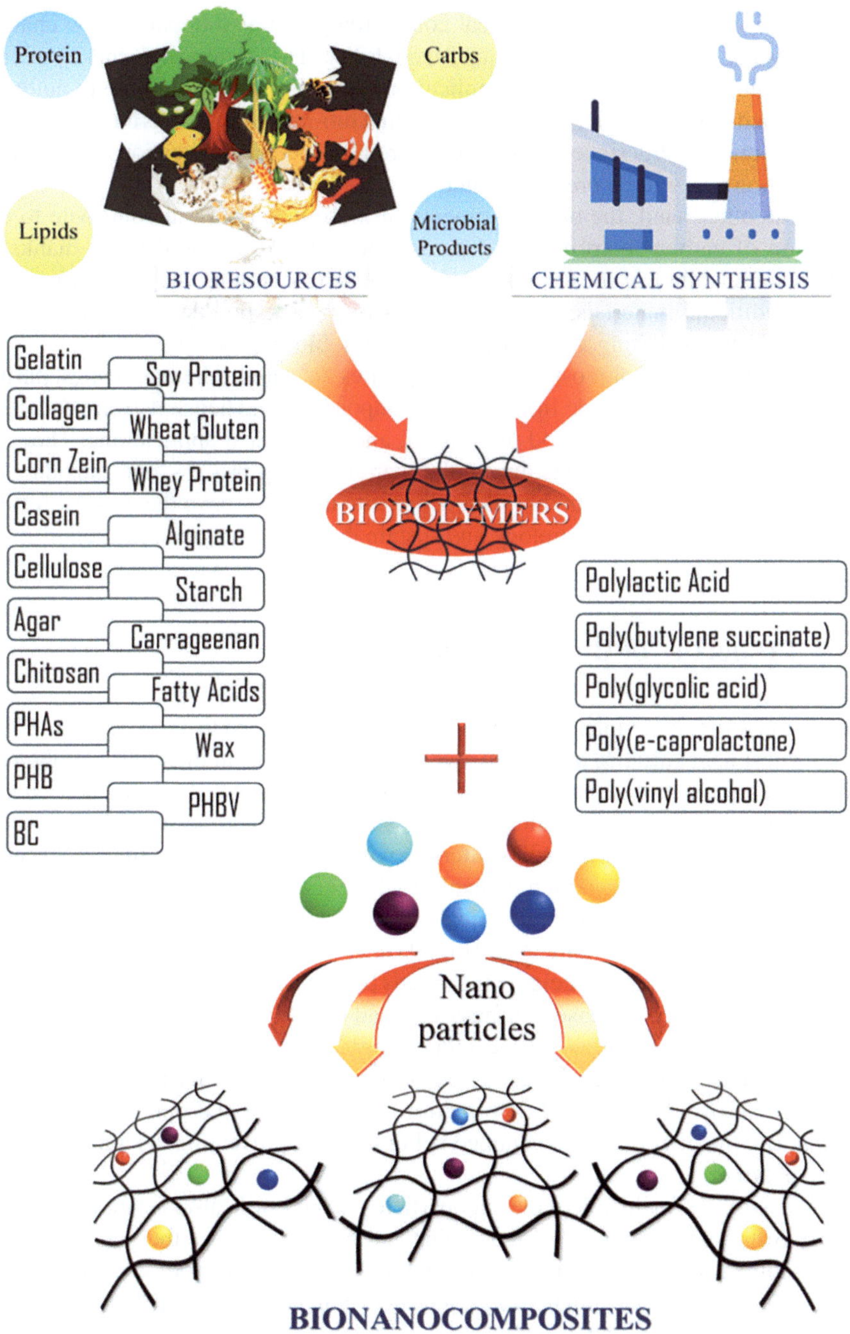

Fig. 9.5 Sources of biopolymers with some common types and the formation of bionanocomposites. PHB and PHBV: poly(3-hydroxybutyrate) (PHB) and its copolymer with 3-hydroxyvaleric acid (PHBV), PHAs: polyhydroxyalkanoate, and BC: bacterial cellulose

permeability properties, and electrical conductivity, but also impartes new characteristics such as biocidal potential and wettability. These enhancements nominate bionanocomposites for numerous applications starting from agriculture to automotive and medical devices. It could be claimed that bio-nanocomposites have versatility and multifunctionality since they can be tailored for specific applications with certain specifications, as will be presented later.

On the other hand, the bio-constituents of these materials can contribute to the reduction of carbon footprint and decrease the reliance on fossil fuels, thanks to the biodegradability of most bio-nanocomposites. Concerning the production cost, unlike traditional materials, bio-nanocomposites can be produced at lower prices, particularly by exploiting agro-waste or industrial bio-byproducts.

Research proved that most bionanocomposites are safe for human beings, particularly those are totally made from natural components, and hence they are excellent candidates for agriculture, food processing, and biomedical applications. They are also non-toxic for the laborers during the manufacturing process, and thus they do not pose any risks. No doubt, bio-nanocomposite materials pave the way for further advancement in current technology and sustainability for the green future.

9.5 Properties of Bionanocomposites

The characteristics of bionanocomposites are strongly related to the synergistic effect of the guest nanoparticles in the polymer matrix. Herein is an illustration of the role of the incorporated nanoparticles in terms of some features, such as mechanical, thermal, barrier, optical, biological, surface, and electrical properties. Mechanically, the perfect dispersion of the nanomaterials within the biopolymer matrix, because of the strong interfacial interaction thanks to their high surface area-to-volume ratio, leads to improvement in the mechanical characteristics in terms of tensile strength, stiffness, elongation, and toughness (Fig. 9.6).

The presence of nanoparticles in the spaces between the polymer chains decreases their slide and hence increases stiffness and load transfer. In addition, the nanofillers can form a rigid network with the biopolymer chain which increases the toughness and elongation at the break by absorbing energy and dissipating the stress, creating microvoids during the deformation. In the case of hydrophobic nanofillers, mechanical integrity can be maintained in humidity by preventing the quick swelling of the polymer.

Thermally, the inclusion of the nanoparticles in the low heat-resistant biopolymer matrix restricts the mobility of the chains and hence increases the glass transition temperature and melting temperature, making the polymer more thermally stable (Fig. 9.7). High thermal conductivity nanofillers enhance the overall thermal conductivity of the polymer, which is favorable in many applications such as food packaging and coatings. Moreover, nanoparticles decrease the biopolymer's coefficient of thermal expansion, allowing the polymer to afford the change in temperature and keep its structural integrity.

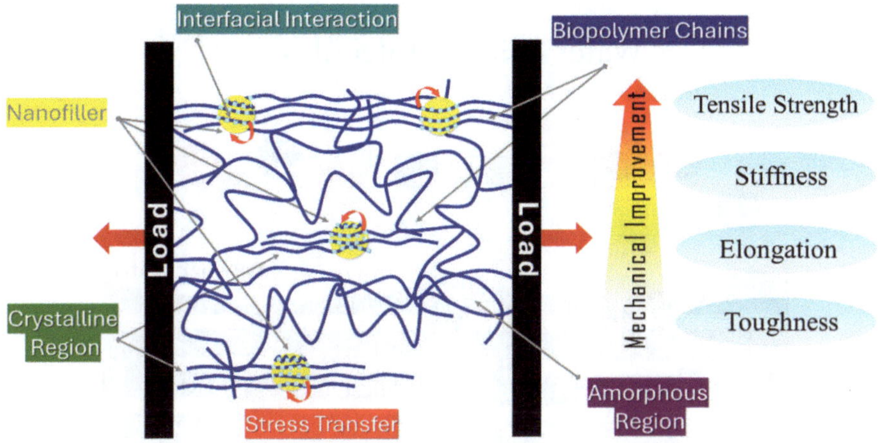

Fig. 9.6 Schematic representation illustrating the role of nanoparticles in the improvement of biopolymers' mechanical properties

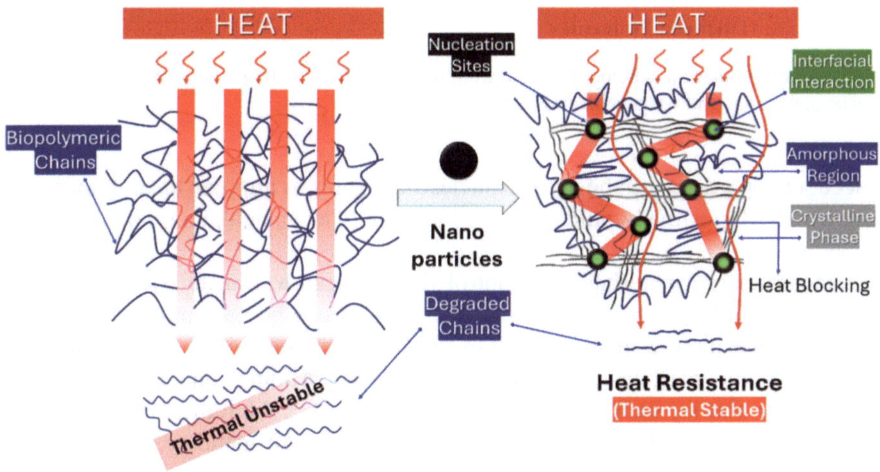

Fig. 9.7 Schematic representation illustrating the influence of nanoparticles in the increment of biopolymers' thermal stability

Among the most crucial characteristics of biopolymers, particularly in food packaging and protective coatings, are barrier properties against the permeability of gases, liquids, and light. Anchoring the nanofillers between the polymer chains leads to blocking the free spaces against the diffusion of moisture and gases, i.e., decreasing the diffusion coefficient. Thus, the diffused molecules take a longer path or more winding paths to travel through the polymer matrix, i.e., tortuosity (Fig. 9.8).

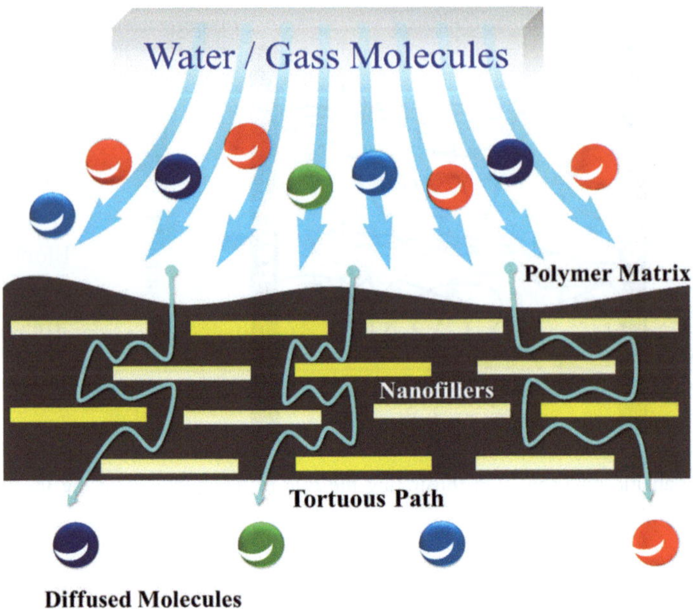

Fig. 9.8 Representation scheme demonstrating the nanofillers as barrier agents for the polymer matrix against gas and water molecules

Indeed, nanoparticles can be used as nucleating agents to improve the polymer crystallinity which positively influences the material's mechanical, thermal, and barrier properties.

Nanostructured materials can control the absorption, transmission, scattering, and reflection of light by the polymer matrix. This can be accomplished thanks to the surface plasmon resonance, refractive index switching ability, and UV-blocking effect of the nanoparticles, depending on their types. Adding some nanofiller to the bio-polymer matrix, such as silica and titania, increases the light transmission and reduces the light scattering. However, quantum dots and metal nanoparticles like gold and silver enhance light absorption. Metal oxides, such as zinc oxide nanoparticles, protect the polymer from degradation by harmful UV rays, i.e., increasing the photostability of the polymer (Fig. 9.9). Particularly, this is an important feature for food processing and packaging.

Indeed, biopolymers have poor intrinsic electrical properties that limit their conductivity applications, like bioelectronics, energy, and sensors. Interestingly, the incorporation of nanoparticles enhances the electrical properties of biopolymers. Metallic, metal oxides, and carbonaceous nanoparticles induce electrical pathways within the polymer matrix, forming percolation networks that allow efficient transportation of charge. The presence of the nanoparticles in the biopolymers increases the dielectric constant and reduces resistance. Not only the electrical, but also the ionic conductivity can be improved by the nanoparticles.

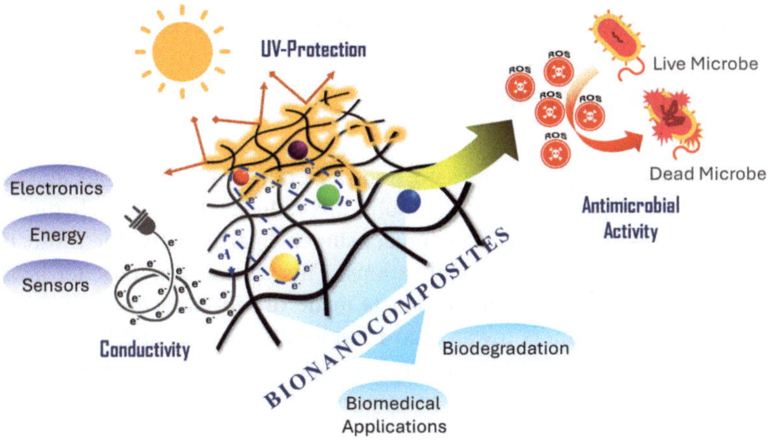

Fig. 9.9 Representation scheme demonstrating the nanofillers as UV-blocking, conducting, antimicrobial, and bioactive agents for the biopolymer matrix

Unfortunately, most biopolymers have no or limited biocidal activity. Nanoparticles impart biopolymers a biological performance against pathogens, making them favorable for biomedical and food packaging applications. However, not all types of nanoparticles have the same activity, but they are also different in their mode of action. Some particles, like metallic and metal oxide, inhibit microbial growth by generating reactive oxygen species that interact with the cell membrane causing mutation and cell death. Carbon nanomaterials like graphene can be used as a nano knife that causes rupture and cell leakage. Moreover, thanks to the high conductivity of carbon nanoparticles, it disturbs the electrical system on the membrane surface by withdrawing the charges responsible for bioprocesses required for cell life, causing cell death.

To maximize the benefits of nanoparticles' inclusion in the biopolymer matrix, crucial challenges should be considered. Aggregation should be avoided, i.e., the better dispersion of the nanomaterials in the biopolymer matrix, the higher the applicability of the bio-nanocomposites. Achieving strong interfacial interaction between the nanocomposite components and controlling the filler ratios primarily improves the distribution and, hence, acquires distinguishable properties.

9.6 Significance of Bionanocomposites

9.6.1 Social Significance

Numerous bionanocomposites are safe, non-poisonous, biodegradable, and biocompatible, as previously mentioned. No doubt, this reduces the risk of chemical exposure and achieves high protection for consumers, particularly for those used in food packaging and biomedical applications. This contributes to better health and safety standards. Moreover, using products made of bio-nanocomposites increases the awareness of consumers toward eco-friendly alternatives and hence influences their choices. They will become more conscious of selecting safer and greener products. Indeed, increasing public awareness of these products causes pressure on manufacturers to adopt sustainable practices by using bionanocomposites.

9.6.2 Industrial Significance

Using bio-nanocomposites paves the way for companies to innovate and develop new products with specific properties, such as high strength, biodegradability, and barrier performance, for a wide range of industries, including packaging, agriculture, and healthcare. This will open new markets and create business opportunities for corporations to invest in research, development, production, and commercialization of bionanocomposites. Unlike traditional plastics, the production of bionanocomposites is mostly dependent on renewable resources, i.e., low waste output and less energy consumption. This will be cost-effective, highly efficient, and eco-friendly in the long term, which contributes to sustainable manufacturing processes.

9.6.3 Economic Significance

In the beginning, the production cost of products based on bionanocomposites might be high because of the need for special manufacturing processes. However, in the long run, their social and ecological benefits can compensate for the overall expenses. Economically, biodegradable packaging products reduce landfill management costs by eliminating the disposal costs associated with non-compostable plastic products. Also, the increasing demands for bionanocomposites support the establishment of green technologies, which will create job opportunities and induce economic growth. Particularly, the research and development in bioplastics and nanomaterial sectors are rapidly growing and can be included in new manufacturing and contribute to waste management. As is commonly known, the economy of most countries relies on using fossil fuels as a source of energy and petrochemical products. Without a doubt, the use of renewable resources in the production of bionanocomposites can decrease the dependence on fossil fuels. This leads to a more stable supply chain and less economic vulnerability associated with unstable prices of oil and gas.

9.6.4 Environmental Significance

Certainly, conventional plastics pose an ecological threat. Environmental pollution can be controlled by using sustainable biomaterials. Therefore, there is an urgent need to pay more attention to bionanocomposites to replace non-degradable plastics thanks to their rapid and safe breakdown in oceans, soil, and ecosystems. No one can deny that the production of bionanocomposites reduces the release of greenhouse gases compared with the production of synthetic plastics and their composites. This helps to mitigate carbon footprint and control climate change. A further benefit can be recorded for the use of bionanocomposites by contributing to the circular economy because of their reusability, recyclability, and degradability.

9.7 Classification of Bionanocomposites

Bionanocomposites can be categorized from different perspectives, such as the matrix type, added filler type, interaction nature between components, and application. This helps in understanding the structure-properties relationship of bio-nanocomposites. Figure 9.10 illustrates the classification of bio-nanocomposites in more detail.

In-depth, the biopolymer matrix used to formulate the bionanocomposite can be biodegradable and bio-based polymers, such as polysaccharide, e.g., cellulose, starch, chitosan, and alginate; protein, e.g., gelatin, soy protein, and whey protein; lipid, e.g., natural waxes and phospholipids, or others like polylactic acid and polyhydroxyalkanoates. Also, the main polymer matrix can be composed of biodegradable natural polymers, blended, grafted, or copolymerized with synthetic biopolymers or non-compostable ones such as polylactic acid/polycaprolactone and starch/polyvinyl alcohol. However, the inclusion of nanomaterials as fillers in the biopolymer matrix exhibits different dispersion levels. As a common example, layered materials like graphene and clay, as inorganic nanomaterial, can illustrate the possible interactions between the polymer matrix and the filler (Fig. 9.11). The polymer chain can either be intercalated between the layers and increase the interlayer spacing, exfoliate the layers achieving a high dispersion level and maximum property enhancement, or phase-separated, causing material failure.

Particularly, intercalation and exfoliation, caused via the dispersion of nanofillers in the biopolymer matrix, induce many aspects of the interaction between the components of bionanocomposites. The better interfacial interaction might involve physical interaction, hydrophobic/hydrophilic, hydrophobic/hydrophobic, π–π stacking, electrostatic attraction, van der Waals forces, or chemical bonding. The stronger the interfacial interaction is, the better dispersion is maintained and accordingly improves

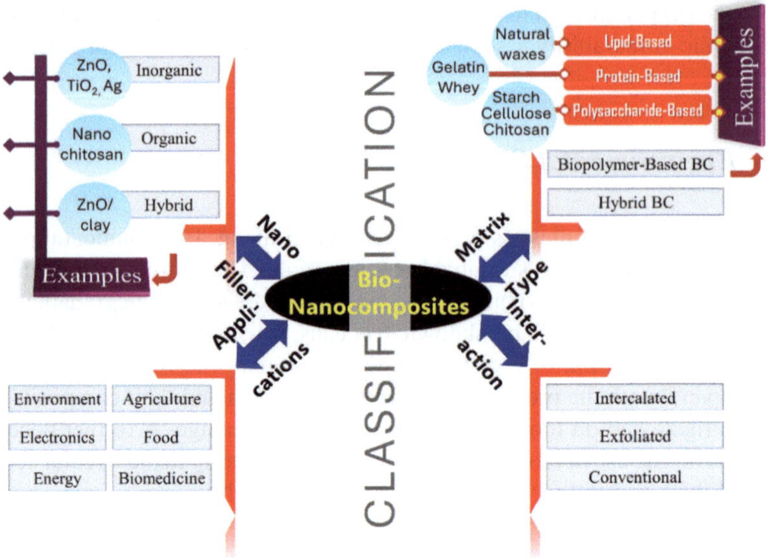

Fig. 9.10 Classification of bionanocomposites according to types of matrices and nanofiller as well as the mode of interaction and potential applications

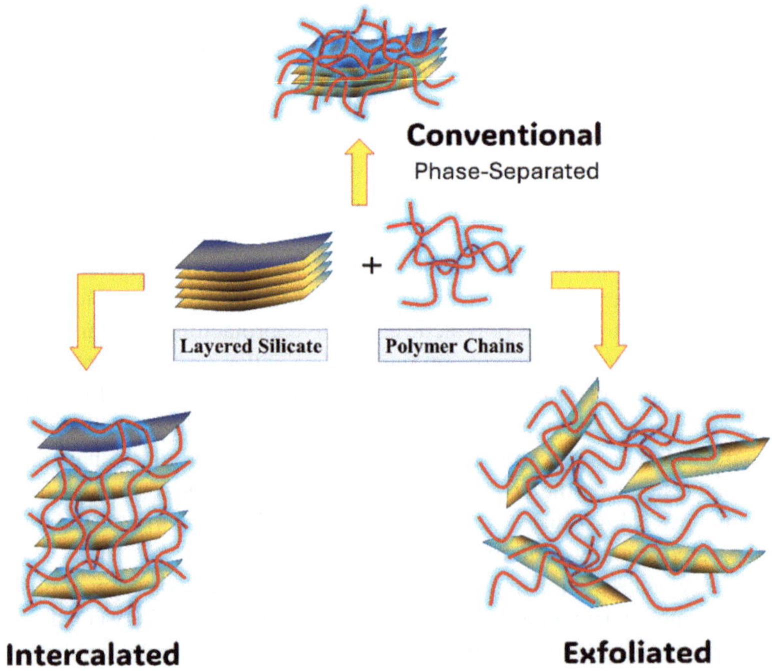

Fig. 9.11 Level of dispersion of layered nanofiller in biopolymeric matrix

the properties. Another category can be added to this classification as previously mentioned for the nanomaterials (Fig. 9.4). Bionanocomposites can be classified according to their morphology into nanoparticles, nanospheres, nanogels, nanofibers, and nanomicelles (Menossi et al. 2022). Therefore, the bulk bio-composites materials such as alginate/chitosan blend can be formulated at the nanoscale to obtain bionanocomposites, alginate/chitosan nano-capsules (Srujana and Bhagat 2022; Saberi Riseh et al. 2024).

9.8 Preparation Methods of Bio-nanocomposites

Indeed, several techniques are being used to prepare polymer bionanocomposites on the industrial scale or even in the lab. Nevertheless, each method has pros and cons, as well as the target application. The key factor is to achieve better dispersion of the nanofillers into the biopolymer matrix. Therefore, the synthesis protocol depends mainly on the type of polymer, the added nanofiller, the final properties of the product, and the target application. Table 9.1 depicts an intensive comparison between several preparation methods of bio nanocomposites.

9.9 Characterization of Bionanocomposites

Characterization of bionanocomposites is crucial to investigate their morphology, chemical structure, physicochemical properties, barrier characteristics, and biological performance. Various techniques and instruments are being used worldwide. Electron microscopes, including scanning (SEM) and transmission (TEM), as well as atomic force microscopes, are usually used to examine the surface morphology of the nanoparticles and their dispersion within the biopolymer matrix. Particularly, the atomic force microscope provides extra information about the roughness of the surface and sometimes determines the nanomechanical properties of nanolayers interacting with polymer chains. The chemical structure and the surface functional groups, along with the nature of the interaction between the bionanocomposites components, can be discussed by Fourier transform infrared spectroscopy (FTIR), Raman spectroscopy, and X-ray photoelectron spectroscopy (XPS). X-ray diffraction (XRD) indicates the crystalline structure of the produced bio-nanocomposites, which might refer to the possible interactions between the nanoparticles and the polymer matrix. For instance, in the case of exfoliation of layered nanomaterials, the XRD pattern confirmed what happened for the nanomaterials, i.e., the vertical lines disappeared, indicating the amorphous structure. Thermogravimetric analysis (TGA) and differential scanning calorimetry (DSC) evaluate the thermal stability of the biopolymer matrix before and after the addition of nanofillers by determining weight loss, as well as crystallization, melting, and glass transition temperatures. Mechanical testing

Table 9.1 Common preparation methods for bio-nanocomposites (Ahari et al. 2022; Thakur et al. 2022; Kamal et al. 2022; Mhd Haniffa et al. 2016)

Method	Description	Advantages	Disadvantages	Applications	Examples	References
In-situ polymerization	Dispersion of nanofiller in the monomer solution followed by polymerization	High dispersion level and strong interfacial interaction between polymer chains and nanoparticles	Use of solvents and uncontrollable polymerization conditions	Coatings, packaging, biomedicine, and water treatment	PLA/TiO_2 NWs	Li et al. (2011)
					PLA/ PVAc@CNC	Geng et al. (2018)
					PAM/PVA/ $CaCO_3$ NPs	Abo-ouf et al. (2023a)
Melt extrusion	Addition of nanoparticles to the polymer melt in a twin-screw extruder under heating conditions	Solvent-free, environmentally friendly, industrially applicable, suitable for thermoplastic biopolymers	Lower nanofiller dispersion in the polymer matrix, unsuitable for some nanoparticles and biopolymers due to high processing tempertature	Packaging films, and agricultural mulch films	PLA/clay	Paramitha et al. (2020)
					PCL/Graphene	Ghanem et al. (2021)
Solution blending	Dispersion of the nanofiller in a solution of polymer, followed by casting and solvent drying up	Simple, high dispersion level, economic, suitable for water-soluble and thermally unstable biopolymers	Some biopolymers are water-insoluble and require toxic solvents, and time-consuming process for achieving high dispersion and drying	Food packaging, coatings, and water treatment	PCL/RGO	Ghanem et al. (2024)
					Chitosan/ZnO	Lin et al. (2021)
					Starch/clay	Chung et al. (2010)
					CMC/Pom. NPs	Khalid et al. (2024a)
					PAM/PVA/ $CaCO_3$	Abo-Ouf et al. (2023b)
					PVA/Biochar/ TiO_2	Ghanem et al. (2025)

(continued)

Table 9.1 (continued)

Method	Description	Advantages	Disadvantages	Applications	Examples	References
Electrospinning	A well-dispersed mixture of polymer and nanofiller electro-spun under high voltage, forming a bio-nano-composite mat	High surface area, high-strength nano bio-nanocomposite fibers and can be loaded with active materials to control their release	Special equipment is required, and limited industrial applications	Biomedicine and active food packaging	PVA/Ag NPs	Augustine et al. (2018)
					PVA/PVP/ZnO	Khaleel et al. (2024)
					PVA/Cs/TA/ZnO	He et al. (2024)
Layer-by-layer assembly	Electrostatic attraction between multilayers of an oppositely charged polymer mixture containing a well-dispersed nanofiller	Ultra-thin bio-nanocomposite	Time-consuming process	Coatings and packaging	Chitosan/ cellulose nanowhiskers	de Mesquita et al. (2010)
					Chitosan/ manganese oxide nanoflake	Chen et al. (2009)
Bio-templating	Using bio-templates to anchor the additive particles via ex-situ blending	Eco-friendly and sustainable	Limited industrial scale and non-homogeneous distribution	Biomedicine, packaging, water treatment, and energy	BC/zeolite	Ibrahim et al. (2022), Abd El Wahab et al. (2023)
					CNC/ZnO NPs	Badawy et al. 2021b)
					PES/C/ZnO NPs	Darwesh et al. (2023)

measures the tensile strength, Young's modulus, elongation, elasticity, stiffness, flexibility, and viscoelastic properties of the formed bionanocomposites compared with the neat matrix. For barrier analysis, two main instruments should be used: water vapor transmission rate (WVTR) and gas transmission rate (OTR). These analyses are crucial in food packaging applications to estimate the permeability of the packaging materials against the diffusion of water vapors and gases, particularly oxygen. The results provide insight into the food shelf life. The impact of nanofiller on the biodegradability of the biopolymer matrix can be assessed by burying it in soil/compost (in vivo), or by exposing it to specific enzymes (in vitro) and following the loss in weight or the deterioration in the mechanical properties. Finally, not all these characterizations should be performed together, but mainly it depends on the target application and its requirements and specifications.

9.10 Applications

Generally, there are many benefits of nanotechnology in the sector of sustainable agriculture (Menossi et al. 2022). Nanomaterials were found to increase the productivity of the plant by reducing agrochemical losses and their costs, slowing water release, and sensing soil conditions. Also, nanosubstances can improve agrochemical availability with limited levels, deliver genes and drug molecules to specific sites, stimulate plant growth, act against diseases, minimize the loss of nutrients, and enhance food quality. Ecologically, nanoparticles could not only reduce soil biodiversity losses but also improve its quality. They provide protection and remediation of water, as well as diminish the emission of toxic gases into the air.

Since biopolymers have a wide range of applications in various industries, bionanocomposites can significantly improve such manufacturing thanks to their intriguing properties. Particularly, bio-nanocomposites contribute to advances in the sectors of agriculture, food processing, and packaging. In this section, the role of bio-nanocomposites in these applications will be highlighted.

In modern agricultural applications, active compounds can be encapsulated using bionanocomposites to control their release, i.e., reduction of agrochemical doses, diminished losses for evaporation, reduced phytotoxicity, avoiding harmful chemicals, and crop productivity (Menossi et al. 2022; Saberi Riseh et al. 2024). In this direction, specifically polysaccharides-based bio-nanocomposites are used as vehicles to encapsulate the active ingredients to elaborate controlled release for agrochemicals, such as fertilizers and pesticides (Merino et al. 2018; Mondéjar-López et al. 2024). However, other bionanocomposites including protein-based and biopolyesters-based nanocomposites can be also used for this purpose (Menossi et al. 2022; Zhao et al. 2008). Particularly, alginate/chitosan-based bionanocomposites in the form of nano-capsules can be used as carriers for versatile agrochemicals. For instance, nano-capsules of *Acetamiprid* using alginate/chitosan can be used as a pesticide in several soils of different pH due to their stability (Kumar et al. 2015). The same nano-capsules were loaded with *paraquat dichloride* and used as a herbicide with improved properties than the free form (dos Santos Silva et al. 2011). *Gibberellic acid*, a growth hormone for the plant, was also included in this capsule to control its release depending on the changes in temperature and pH (Santo Pereira et al. 2017). In an interesting study, *imazapic* and *imazapyr*, as herbicides, were loaded and optimized on alginate/chitosan nano-capsules (Maruyama et al. 2016). The results confirmed high entrapment efficiency with lower toxicity. Alginate/chitosan nano-capsules loaded with copper oxide nanoparticles as a smart delivery nano-fertilizer were fabricated by Leonardi et al. (2021). The proposed bionanocomposite displayed the benefits in the seedling growth of *Fortunella margarita* and the developing of epigean and hypogean parts. Other nano-capsules were also investigated for other herbicides, insecticides, and pesticides. Chitosan/polylactic acid/*imidacloprid* nano-formulations presented prolonged release of the pesticide depending on the particle size (Li et al. 2011). Nanogels composed of chitosan/cashew gum, as the nano-biopolymer matrix, and *Lippia sidoides,* as an essential oil filler, were optimized in terms of gum/chitosan ratios, cashew gum weight %, and oil ratio (Abreu et al. 2012). The results emphasized the independence of oil release, as the insecticide, on these ratios. Bionanocomposites based on cellulose nanofibers/starch/dimethyl phthalate were designed as herbicides (Patil et al. 2018). The authors presented that the addition of cellulose nanofibers caused significant reinforcement for the starch matrix and improved the controlled release performance of dimethyl phthalate from the starch granules, as well as decreasing the initial release rate. Zein protein nanoparticles incorporated with *geraniol* and *R-citronellal*, as pesticides, were formulated using the antisolvent precipitation method (de Oliveira et al. 2018). The study findings indicated high loading capacity and good physicochemical stability with lower toxicity levels. Indeed, several studies have focused on using chitosan-based bio-nanocomposites as entrapment agents for fertilizers (Riseh et al. 2023). To name but a few, NPK, nitrogen (N), phosphorus (P), and potassium (K), were loaded into chitosan nanoparticles to improve the leaf number, plant height, and leaf area of the coffee seedlings (Ha et al. 2019). Chitosan-silicon nano-fertilizer was prepared via encapsulation of silicon, for its slow release, in the chitosan nano-matrix to enhance maize growth and its yield (Kumaraswamy et al. 2021). Nano-fertilizers based on tripolyphosphate, chitosan, and zinc oxide were formulated to control phosphorus leaching (Dimkpa et al. 2023). The results indicated significantly decreased shoot phosphorus levels compared to the conventional fertilizer and mobilization of the resident potassium, sulfur, magnesium, and calcium from the soil into the plant, which improved the overall nutritional quality. Indole-3-acetic acid, a plant hormone, was encapsulated into phenyl succinic acid-modified chitosan nanoparticles, and the release was evaluated (Ding 2024). Another study, performed by Kassem et al., discussed the development of biodegradable bio-nanocomposite coating based on cellulose nanofibers and engineered biochar hybrids for controlled phosphate fertilizer release (Kassem et al. 2022). The incorporation of cellulose/engineered biochar as a coating material significantly enhanced fertilizer performance. It increased the crushing strength of fertilizer granules by 30.29% compared to uncoated ones and improved soil water holding capacity by 7.78%, maintaining 3.00% water retention after 25 days. Moreover, phosphorus release was effectively controlled, with a 43.90% reduction in phosphorus leaching over an 80-day soil incubation period.

Additionally, the nanoparticles composing bionanocomposites can enhance soil structure and porosity. Zewdu et al. emphasized that nanoclay could improve soil structure in terms of root growth and microbial activity by promoting particle aggregation, increasing porosity, and improving air and water circulation (Zewdu et al. 2024). Thanks to their hydrophilic nature, bionanocomposites can fill the spaces between soil particles, which increases soil moisture by absorbing much moisture from the surrounding environment. This enhances the cohesion of particles and the stability of their overall structure. This is particularly beneficial in arid and semi-arid regions. Different carboxymethyl cellulose hydrogels incorporated with nanocellulose and nanoclay were prepared as soil conditioning and nutrient carriers to improve cucumber growth (Bauli et al. 2021). The prepared formulations showed the potential to reduce fertilizer loss and improve water use in agriculture. Nanoclay and nano lignin were incorporated into the starch matrix as a soil conditioner and UV-C blocking (Rodrigues et al. 2025). The results confirmed that the incorporation of 7% nanoclay into starch notably enhanced its cation exchange capacity and swelling behavior. Further addition of nano lignin improved the nanocomposite's resistance to photodegradation and thermal degradation. This formulation also exhibited excellent water retention, controlled release of NPK nutrients, reduced leaching of NH_4^+, $H_2PO_4^-$, and K^+ ions, and showed antimicrobial activity against both gram-positive and gram-negative bacteria. These properties underline its biodegradability and suitability as an eco-friendly soil conditioner. Field trials on cherry tomatoes confirmed enhanced seedling growth, validating its effectiveness under real agricultural conditions.

Moreover, bionanocomposites can maintain the mechanical structure of the soil. Arabani et al. (2023) studied a bionanocomposite based on rice fiber and nanoclay. The results indicated a significant improvement in the shear and compressive strength of soil, besides elasticity. Nutrient deficiency, such as zinc, in soils, poses a significant global challenge as a key micronutrient limitation that adversely affects crop productivity. Nutrients are essential for various physiological processes in plants, including carbohydrate metabolism, protein synthesis, enzyme activation, and the formation of nucleic acids and lipids (Augusto 2020; Ibrahim et al. 2022). For instance, the lack of zinc can lead to stunted growth, disrupted neurological functions, impaired nitrogen metabolism, and reduced efficiency in photosynthesis (Han et al. 2018; Arulprasanna and Omkumar 2024). Recognized as the fourth most critical yield-limiting nutrient, following nitrogen, phosphorus, and potassium, zinc deficiency is not only an agricultural issue but also a public health concern. According to the World Health Organization (WHO), approximately 800,000 deaths annually are attributed to insufficient zinc intake (Kumari et al. 2024). To address this issue, bio-nanocomposites can support the growth of beneficial soil contents by supplying nutrients (e.g., Zn, Fe, Cu) required for the processes that take place in the plant (Rivera-Solís et al. 2024). Chitosan/ZnO bio-nanocomposites could improve root elongation, enhance nutrient uptake in crops, and develop the quality of the plant due to biological activity and micronutrient provision (Rani et al. 2024; Zungu et al. 2023; Li et al. 2024a; Tomke and Rathod 2020). Starch-based bio-nanocomposites have been reported to improve water retention and promote better germination rates in soils. Starch/Zn@nanoclay bionanocomposite hydrogel was investigated as a new type of seed-coating material for *Oryza sativa* plant (Kumari et al. 2024). The results indicated improvement in water retention, micronutrient retention of Zn, germination percentage, vegetative growth, and yield compared to uncoated seeds. Kumar et al. developed a slow-release formulation of zinc and copper micronutrients using bionanocomposites based on poly(vinyl alcohol) and starch, reinforced with carbon nanofibers (Kumar et al. 2018). This system demonstrated a release profile of approximately 2.8 mg/g and 5.3 mg/g/ carbon nanofibers for zinc and copper, respectively, specifically tested on chickpea plants. Another promising development involved urea–starch bionanocomposites incorporating hydroxyapatite nanoparticles, designed for the controlled release of phosphorus and nitrogen. The results indicated sustained phosphorus availability after four weeks of soil incubation and a notable reduction in ammonia volatilization (Giroto et al. 2017). Three-dimensional bionanocomposite network structures are capable of absorbing large amounts of water and soluble fertilizers without disintegrating. Once absorbed, these resources can be gradually released into arid and semi-arid soils, providing sustained moisture and nutrients essential for plant development. A water reservoir system based on cellulose hydrogels reinforced by cellulose nanofibers was developed for agricultural uses (Das et al. 2023). The prepared hydrogels displayed a remarkable swelling ratio of $\sim$1070% with high mechanical strength. On the other hand, bio-nanocomposites may have the ability to improve soil texture and reduce erosion in loamy soils, especially in regions prone to surface runoff. That is because several studies investigated the inclusion of nanoparticles and biopolymers in different systems to control the erosion of soils in different conditions (Orts et al. 2000; Fatehi et al. 2021; Kumar et al. 2022; Chang et al. 2015; Alkhaza'leh et al. 2025; Abdelhak 2022; Patra et al. 2022). Another function promoted by bio-nanocomposites is to improve the soil quality. Bio-nanocomposites can be used as a cleaning agent and pH buffering for the contaminated soil. Nanomaterials can bind heavy metals and toxins, such as Pb, As, Cd, and Cr, helping clean soils (Liao et al. 2021; Shaghaleh et al. 2025; Mohanaprakash et al. 2025; Medina-Pérez et al. 2019; Ali et al. 2023).

Although bionanocomposites are derived from biodegradable and generally safe materials, the fate and behavior of nanoparticles in soil ecosystems remain a subject of active research. Concerns regarding nanoparticle accumulation, potential toxicity to soil microbiota, and long-term soil health need to be addressed through lifecycle assessment and eco-toxicological studies. No doubt, bio-nanocomposites represent a frontier in sustainable agriculture, offering multifunctional benefits that directly address soil degradation challenges. By improving soil structure, enhancing nutrient management, and supporting microbial ecosystems, these materials contribute significantly to soil quality and hence crop productivity.

In recent years, the integration of bionanocomposites in the food processing has gained significant attention as a strategic approach to enhancing food safety, extending shelf life, and mitigating microbial contamination. In the fields of food industry and packaging, bio-nanocomposites have proven their workability and feasibility. Bionanocomposites exhibit improved mechanical, thermal, and barrier properties, along with antimicrobial and antioxidant functionalities that directly contribute to safer food systems. One of the most crucial roles of bio-nanocomposites in the food industry is the inhibition of microbial growth during and after processing (Babaei-Ghazvini et al. 2021; Paidari et al. 2023). Pathogenic microorganisms such as *Escherichia coli*, *Listeria monocytogenes*, and *Salmonella* are major concerns in food safety (Lin et al. 1996; Silva 2023; Rhoades et al. 2009). Bio-nanocomposites containing active compounds such as metal-based nanoparticles or even natural-based nanoparticles have shown strong bactericidal effects due to their ability to disrupt microbial cell walls, generate reactive oxygen species, and destroy cell DNA (Babaei-Ghazvini et al. 2021; Amiri et al. 2022).

Similar to agrochemical encapsulation for agriculture applications, food-grade bioactive ingredients can also be encapsulated to control their release in food over time. Particularly, encapsulation protects them from degradation and enhances their bioavailability. Utilizing bio-composite nanoparticles paved the way for the advancement of functional foods and nutraceuticals. Among bioactive compounds that are successfully encapsulated are vitamins, polyphenols, probiotics, fatty acids, antibiotics, and antioxidants. These compounds have several actions, such as biocidal and antioxidant activity, that protect foods from degradation via microbial attack or variation in the storage conditions. For instance, chitosan/β-carotene-based nano-emulsion was prepared via homogenization and used as a food additive with improved heat stability and controlled release (Hou et al. 2012). Clove essential oil was entrapped into chitosan nano-capsules and used as a natural preservative to extend the shelf life of blueberries (Xu et al. 2023). In this study, the obtained nano-capsules have a good antioxidant activity compared with the commercial antioxidant and a high antibacterial activity against pathogens naturally growing on blueberries. Nouri published exciting work investigating the encapsulation of some essential oils in chitosan nano-capsules and the performance, immune responses, and intestinal bacteria population in the broiler chickens were evaluated (Nouri 2019). The results emphasized that, especially, entrapment of thyme essential oil into chitosan nano-capsules improved its herbal activity in the broiler chickens. Particularly, phenolic compounds have strong antioxidant activity and have many purposes as anti-allergenic, anti-inflammatory, anti-atherogenic, anti-thrombotic, and anti-microbial agents in food products and other biological applications (Assadpour and Mahdi Jafari 2019). These phytochemicals are commonly abundant in plants (Bjelakovic et al. 2004). Phenolic compounds are frequently used as additives in food thanks to their antioxidant capabilities as a trap for free radicals, which leads to extending the shelf-life of food products without any negative effect on their sensory or nutritional qualities (Assadpour and Mahdi Jafari 2019). Indeed, different biopolymers and bio-nanocomposites of distinguishing morphology were exploited as hosting matrices for bioactive compounds used in the food industry. Therefore, Table 9.2 presents some selected studies of several bionanocomposites loaded with polyphenols, vitamins, food antimicrobial agents, essential oils, food colorants, food flavors, and minerals/salts that showed further advancements in food processing technology.

Table 9.2 Selected examples of bionanocomposites formulations and their potential applications in food processing

Bio-nanocomposite structure	Biopolymer	Filler	Potential applications	References
Nano-emulsions	Soy protein	Catechins	Antioxidant	Bhushani et al. (2016)
	Lactoferrin/alginate multilayer	Curcumin	Food additive	Pinheiro et al. (2016)
	Maltodextrin/whey protein	Folic acid	Vitamin	Assadpour et al. (2016)
	Lauric arginate/lectin	Thymol, eugenol	Antimicrobial agents	Ma et al. (2016)
	Modified starch	β-carotene	Food colorants	Liang et al. (2013)
	Triacylglycerol/Starch	Peppermint oil	Food flavor	Liang et al. (2012)
Nanoparticles	β-lactoglobulin/alginate	Curcumin	Liquid food nutraceutical	Hosseini et al. (2015)
	Chitosan	Quercetin	Antioxidant	Zhang et al. (2008)
	Chitosan/zein complex	Vitamin C	Food additive	Dong and Wang (2016)
	Zein	Thymol, carvacrol	Antioxidant, Antimicrobial	Wu et al. (2012)
	Chitosan/carrageenan	Nisin	Antimicrobial	Chopra et al. (2014)
	Chitosan	Lutein	Food colorants	Arunkumar et al. (2013)
	Starch	Menthone, menthol, limonene	Food flavor	Ades et al. (2012)
	Sodium alginate	Iron salt	Food minerals	Katuwavila et al. (2016)
	Guar gum	Selenium	Food minerals	Soumya et al. (2013)
Nanofibers	Zein	Gallic acid	Antioxidant	Neo et al. (2013)
	Polyvinyl alcohol	Ethyl vanillin	Food flavor	Lević et al. (2014)
Nanohydrogels	Chitosan	Copper, zinc, manganese, iron	Antimicrobial	Du et al. (2009)

Over the last few years, hundreds of articles have been published discussing the possible contribution of bionanocomposites to sustainable food packaging. The authors approached the application from different angles. Some articles have focused on improving the physicochemical properties, including mechanical and thermal stability along with barrier characteristics, of the packaging materials by optimizing nanofiller addition to the biopolymer matrix. Other studies investigated the incorporation of active ingredients in a free or a nano form to protect the food from spoilage over time or to detect its decay (Dena et al. 2021). Interesting research considered all features. Indeed, the inclusion of nanofiller usually improves the barrier properties against the diffusion of water vapor and oxygen gas toward the food product, decreasing the rate of corruption. However, another feature can be added if the nanofiller has a biocidal activity that retard the probable bacterial growth, i.e., shelf-life extension. Table 9.3 involves several research articles presenting the inclusion of nanofillers in the biopolymer matrix and their role in food-active packaging.

Furthermore, several articles have also been released discussing multiple-component bionanocomposites as packaging materials and the acquired properties. For instance, agar/nanocrystalline cellulose/savory essential oil bionanocomposite films were obtained (Atef et al. 2015). Incorporating savory essential oil (SEO) into agar-bio-nanocomposite films affected both their mechanical and functional properties. At low concentrations, SEO was evenly dispersed throughout the film matrix, while higher concentrations (1.5%) led to oil accumulation and cracks. The addition of SEO decreased tensile strength, Young's modulus, increased elongation at break, increased opacity, and slightly affected water vapor permeability. UV light transmission was also reduced, which can benefit food packaging applications. The films exhibited enhanced antimicrobial activity, particularly against gram-positive bacteria such as *Listeria monocytogenes*, *Staphylococcus aureus*, and *Bacillus cereus*, compared to gram-negative bacteria like *Escherichia coli.* This bio-nanocomposite can act as an active packaging material to improve food safety and extend shelf life. Another study involved ternary bi-nanocomposite packaging films based on poly(lactic acid) reinforced with cellulose nanocrystals and lignin nanoparticles at 1% and 3% by weight, respectively, were developed (Yang et al. 2016). These nanofillers were incorporated in poly(lactic acid) and glycidyl methacrylate-grafted poly(lactic acid) via melt extrusion. The results proved an improvement in transparency and UV-blocking performance. The dual nanofiller system also promoted nucleation and crystallization, leading to higher crystallinity compared to binary composites. Mechanically, the ternary films exhibited greater strength and stiffness compared with the pristine. Both nanofillers contributed to antibacterial activity, effectively reducing the growth of the plant pathogen *Pseudomonas syringae* pv. *tomato.* These films demonstrated promising potential for sustainable antimicrobial packaging in the food industry. Wen et al. produced bio-nanocomposite based on polylactic acid nanofibers loaded with cinnamon essential oil and β-cyclodextrin via the electrospinning technique (Wen et al. 2016). The obtained films displayed excellent antimicrobial activity against *E. coli* and *S. aureus,* extending the shelf life of meat, suggesting its potential applications in active food packaging. Starch/halloysite/nisin bio-nanocomposite films were developed to extend the shelf life

Table 9.3 Selected bionanocomposite packaging materials and their characteristics

Biopolymer matrix	Nano filler	Function	References
Polylactic acid	Cellulose nanocrystals	High oxygen barrier	Fortunati et al. (2012)
	Cellulose/lignin nanostructure	Mechanical and antibacterial	Yang et al. (2016)
	Magnesium oxide	Antibacterial, mechanical, UV-block, and barrier	Swaroop and Shukla (2018)
Potato starch	Starch nanocrystals/catechin	Mechanical and thermal	Sessini et al. (2016)
Polycaprolactone	Nano montmorillonite	Mechanical stability	Marras et al. (2008)
	Reduced graphene oxide	Thermal, barrier, and antimicrobial	Ghanem et al. (2021, 2024)
Pea starch	Clay/ZnO	Mechanical stability	Yu et al. (2009)
	Cellulose whiskers	Mechanical and thermal stability	Svagan et al. (2009)
Whey protein	Zein	Mechanical stability and high water vapor barrier	Oymaci and Altinkaya (2016)
	Nano montmorillonite	Mechanical stability	Wakai and Almenar (2015)
Cellulose acetate	Nano montmorillonite		Hassan-Nejad et al. (2009)
Chitosan	Bacterial cellulose nanoribbons		Velásquez-Cock et al. (2014)
Hydroxy propyl methyl cellulose	Chitosan	Mechanical stability and high barrier	Moura et al. (2009)
Starch	Chitin whiskers	Mechanical, barrier, and antimicrobial	Qin et al. (2016)
Chitosan	Silver	Mechanical stability, high barrier, and antibacterial	Qin et al. (2019)
Cellulose		Mechanical, thermal, transparency, and antibacterial	Chen et al. (2020)

(continued)

Table 9.3 (continued)

Biopolymer matrix	Nano filler	Function	References
Agar, alginate, and chitosan	Copper oxide	UV-blocking, mechanical, and antibacterial	Shankar et al. (2017)
Gelatin	Zinc oxide/selenium/cellulose nanofibers	Barrier, mechanical, antioxidant, and antibacterial	Ahmadi et al. (2021)
Chitosan	Titanium dioxide/red apple pomace extract	Antibacterial, barrier, Thermal, mechanical, and UV-blocking	Lan et al. (2021)
Zein/alginate	Titanium dioxide/betanin	Mechanical, surface hydrophobicity, antioxidant, and antibacterial	Amjadi et al. (2020a)
Carboxy methyl cellulose	Zinc oxide	Antibacterial	Li et al. (2021)
Soy protein	Zinc oxide/cinnamaldehyde	Mechanical stability, barrier, and antifungal	Wu et al. (2019)
Pectin	Cinnamaldehyde nano-emulsions	Antimicrobial	Otoni et al. (2014)
Cupuassu puree-pectin	Chitosan	Mechanical and barrier	Melo et al. (2019)
Guar gum	Silver/copper	Mechanical stability	Arfat et al. (2017)
Chitosan/gelatin	Quercetin	Mechanical, barrier, antioxidant, antibacterial	Yadav et al. (2020)
Carboxy methyl cellulose	Pomegranate	Mechanical, barrier, and antimicrobial	Khalid et al. (2024b)
Chitosan	Zinc oxide	Antimicrobial	Sasidharan et al. (2024)
Gelatin/chitosan	*L. nobilis* nano-emulsion	Thermal stability and antimicrobial	Ghaderi et al. (2024)
Chitosan/alginate	Lemongrass nano-emulsion/silver nanoparticles	Mechanical, barrier, and antibacterial	Pandey et al. (2024)

(continued)

Table 9.3 (continued)

Biopolymer matrix	Nano filler	Function	References
Chitosan	Titanium dioxide nano-particles/ thymol	Antimicrobial	Li (2025)
	Chitin whiskers	Mechanical and barrier	Rubentheren et al. (2015)
	Cellulose nanocrystal/carvacrol	Antioxidant and antibacterial	Li et al. (2024b)
Alginate/gelatin	Nanoclay	Mechanical, barrier, and antibacterial	Perera et al. (2023)

of the soft cheese by inhibition of *Listeria monocytogenes* (Meira et al. 2016). Briefly, the produced films were used as active antimicrobial packaging materials compared with pure starch film. Mechanical testing demonstrated improved properties with the addition of halloysite, but the inclusion of nisin led to lower Young's modulus and tensile strength. Thermal stability also decreased in films containing nisin by 20 °C. The antimicrobial activity of the films applied to the surface of Minas Frescal cheese inoculated with *L. monocytogenes*, the films with 6 g/100 g nisin, completely inhibited the pathogen after 4 days. These results suggest that starch/halloysite films with nisin are effective as antimicrobials for controlling food-borne contamination. Another lab has developed active bionanocomposite films made of fish gelatin and chitosan nanoparticles and studied the addition of *Origanum vulgare*L. essential oil as antibacterial packaging films (Hosseini et al. 2016). The physical analysis showed that the presence of essential oil enhanced the films' crystallinity without a significant effect on the thermal stability. Moreover, the nanocomposite films were more flexible but less resistant and had low water vapor permeability depending on the essential oil concentration. The produced bioactive films, containing 1.2% (w/v) of essential oil, exhibited a notable antimicrobial activity against *Staphylococcus aureus*, *Listeria monocytogenes*, *Salmonella enteritidis*, and *Escherichia coli*. Quinoa-protein/chitosan nanoparticles/thymol bio-nanocomposites obtained via inkjet printing (Caro et al. 2016). The prepared films displayed good antimicrobial activity against *L. innocua*, *S. aureus*, *Salmonella typhimurium*, *Enterobacter aerogenes*, and *E. coli*. These films can be used as new packaging materials to extend the shelf life of fresh fruits. Bionanocomposite films, containing various proportions of ginger essential oil, chitosan, and montmorillonite, were developed and evaluated for their antibacterial effect on chilled beef (Zhang et al. 2021). The film showed reduced tensile strength, but enhanced elongation at break and water vapor permeability. The migration of phenols from the films increased exponentially, reaching equilibrium after about 48 h. The bio-nanocomposite films effectively slowed the rise in pH, hue angle, and moisture content of chilled beef, while also reducing lipid oxidation and inhibiting surface microbial growth. These findings suggest that these bio-nanocomposite films could serve as a sustainable alternative to traditional non-degradable plastic films in food packaging. A study aimed to create gelatin/chitosan nanofiber/zinc oxide nanoparticle bionanocomposite films incorporated with betanin nanoliposome for fresh beef preservation (Amjadi et al. 2020b). The resulting film demonstrated strong mechanical properties and high surface hydrophobicity with significant antibacterial activity and antioxidant properties. The film effectively controlled bacterial growth, lipid oxidation, and changes in pH and color in the beef samples. Chitosan/polyvinyl alcohol/castor oil/titanium dioxide nanoparticles bionanocomposite food packaging films were prepared, investigated, and evaluated using white bread as a model product (Devi et al. 2024). The study found that incorporating castor oil and titanium dioxide nanoparticles into films significantly enhanced their antimicrobial properties, light barrier, thermal stability, optical transparency, and mechanical strength. Additionally, these modifications reduced water solubility, moisture content, and both water vapor and oxygen permeability. The modified film extended the bread's shelf life to 10 days, outperforming commercial polyethylene packaging (which only extended the bread's shelf life to 6 days). These films demonstrate superior packaging properties, making them a promising and sustainable alternative to traditional plastic packaging in food industry applications. Eshaghi et al. (2024) developed bionanocomposite packaging films based on multiple ingredients: carboxymethyl cellulose/myrrh gum/titanium dioxide nanoparticles/dill essential oil. The study demonstrated that incorporating essential oil and titanium dioxide nanoparticles into carboxymethyl cellulose/myrrh gum composite films reduced moisture content and water vapor permeability, improved tensile strength, and boosted antimicrobial and antioxidant properties. The components were well-integrated, with no significant aggregation or cracking observed. The optimized carboxymethyl cellulose/myrrh gum film (1.5%/0.25%) containing 0.5% titanium dioxide NPs and 2.25% essential oil showed the best overall performance and is suitable for use as an edible film. Functionally, this bioactive nanocomposite film extended the shelf life of refrigerated fish fillets by 12 days, effectively inhibiting microbial growth and reducing oxidation, compared to the control sample. Indeed, this chapter is not enough to present all studies performed using enormous bionanocomposites as packaging materials for food preservation.

In food processing chains, bionanocomposites can be empolyed as active packaging to delay spoilage, but also in intelligent packaging to detect food spoilage or contamination. Bionanocomposites active packaging materials can release antimicrobial or antioxidant agents in a controlled manner, thereby prolonging the shelf life of processed foods such as dairy, meat, and ready-to-eat meals. However, intelligent packaging represents an advanced technology in the food industry. It depends on using smart materials to monitor food safety and the surrounding environment. Critical parameters, such as temperature and pH, can be measured, enabling better quality control and enhanced food safety (Biji et al. 2015). By incorporating nanostructures as reactive components, bio-based packaging materials can detect gases, chemical contaminants, and pathogens, while also responding to environmental changes. This technology not only helps consumers identify fresh products but also has the potential to significantly reduce foodborne illnesses (Jiang et al. 2015).

Various intelligent packaging technologies have been developed for food applications, including different types of sensors, indicators, and radio-frequency identification systems (Biji et al. 2015). Sensor technologies can interact with both internal factors, e.g., food components and headspace gases, and external environmental conditions, including pH, temperature, and light, generating responses that reflect the food's current state (Kuswandi et al. 2011). Biosensors, which combine biological sensing elements with transducers, are particularly valuable for detecting pathogens, pesticides, allergens, and other critical factors affecting food quality and safety (Kuswandi et al. 2011). Nanosensors made from bionanocomposites could change their colors when they detect gases produced by food spoilage, providing consumers with clear visual warnings about product safety (Duncan 2011).

Quality indicators form another important category of intelligent packaging solutions. These devices directly show food quality through reactions with microbial metabolites produced during food spoilage. Time–temperature indicators can reveal deviations from recommended storage conditions by showing pH changes (Liu et al. 2023). Researchers have developed an innovative temperature monitoring system using chitosan/gold nanoparticle bionanocomposites in order to detect the freezing conditions and thermal history of fresh products (Wang et al. 2018). This smart material exploits the localized surface plasmon resonance properties of the gold nanoparticles, which undergo a distinct color change from pink to dark gray upon nanoparticle aggregation during freezing. The indicator showed a clear color transition after 24 h of freezing. This might be valuable for monitoring frozen seafood and meats. However, under thermal conditions, the indicator exhibited another behavior. Indeed, these types of bionanocomposite sensors offer significant advantages over conventional monitoring systems in terms of requiring no external devices for readout, being cost-effective, and being environmentally friendly. Similarly, gas indicators for oxygen and carbon dioxide typically rely on color changes resulting from chemical or enzymatic reactions. Conventional oxygen indicators often use methylene blue for its reversible color-changing properties (Dobrucka and Cierpiszewski 2014; Vu and Won 2014). Gelatin/tapioca starch/zinc oxide bio-nanocomposite films are as gas sensing and antimicrobial packaging materials (Azizun et al. 2021). The highest response for ammonia gas was obtained at pH 7 with a quick response time of 10 s. Polyvinyl alcohol/chitosan/xylan/nano-hydroxyapatite was assessed as a gas sensor packaging film (Vadivel et al. 2019). The color change response of the nanocomposite film in an ammonia environment was studied at different time intervals via UV–vis. The colorimetric behavior of the prepared formulations can be used as a tool for the real-time fish freshness method. Stimuli-responsive packaging systems can alter their properties in response to specific triggers like temperature or pH or release trapped compounds (Brockgreitens and Abbas 2016; Fuciños et al. 2015; Azizi-Lalabadi et al. 2023). In addition to their use in smart and active packaging, bionanocomposites can be directly applied as edible coatings on food products (Bratovcic 2021; Jafarzadeh et al. 2021; de Oliveira Filho et al. 2022). Such coatings can form semi-permeable barriers that control gas exchange and moisture loss, while simultaneously providing antimicrobial protection.

The technology shows promising potential for integration into active and smart packaging systems across food, pharmaceutical, and biomaterial applications, providing a reliable, visual method to ensure product quality and safety throughout the supply chain. Bio-nanocomposites represent a promising frontier in food processing, offering a combination of structural, functional, and active properties to enhance food safety. Their use in antimicrobial films, active coatings, and intelligent packaging systems can significantly mitigate microbial risks and improve shelf life. Table 9.4 depicts the functional performance of bio-nanocomposites in agriculture, food processing, and packaging, Finally, bionanocomposites have not only these applications but also, they have several other applications, such as energy (Ramesh et al. 2021), water treatment (Dar et al. 2020), medicine (Nandhini et al. 2024), and engineering (Wei et al. 2012).

Table 9.4 Summary of functional performance of bio-nanocomposites in agriculture, food processing, and food packaging

Application	Functions	Action
Agriculture	• Soil conditioning • Anti-microbial agent • Controlled release of fertilizers, pesticides, and nutrients • Soil remediation	• Soil structure and water retention improvement • Root development • Enhanced nutrient delivery • Chemical runoff control
Food processing	• Antimicrobial food coatings • Nutrient encapsulation • Real-time monitoring	• Shelf-life extension • Microbial inhibition • Food safety improvement • Nutrient bioavailability improvement • Less-used preservatives • Spoilage detection
Food packaging	• Barrier enhancement • Active packaging (gas scavenging, UV resistance, antioxidant, and antimicrobial) • Strengthening agents • Smart indicators (pH, temperature, and pathogens)	• Thermal and mechanical improvement • Decrease gas and moisture diffusion • Long product expiry date • Environmental safety • Real-time freshness monitoring • Biodegradability • Pathogen inhibition

9.11 Evolution of Bionanocomposites

By increasing the populations and the demands for enhanced sustainable materials as well as the ecological impacts of traditional synthetic polymers, bionanocomposites were found to exploit the pros of both biopolymers and nanomaterials together to acquire developed materials with superior features overcoming the shortcomings of main matrices and used as alternatives to frequently used polymers or even their nanocomposites. Despite the term bionanocomposite being relatively modern, the usage of biopolymers has been for centuries, remarkably before the twentieth century, in traditional applications such as paper making, textiles, and food. With the birth of nanoscience and the obtaining of versatile nanomaterials by the beginning of the 1980s and continuing into the 1990s, researchers started to incorporate nanomaterials into the polymer matrices and study the properties of the produced

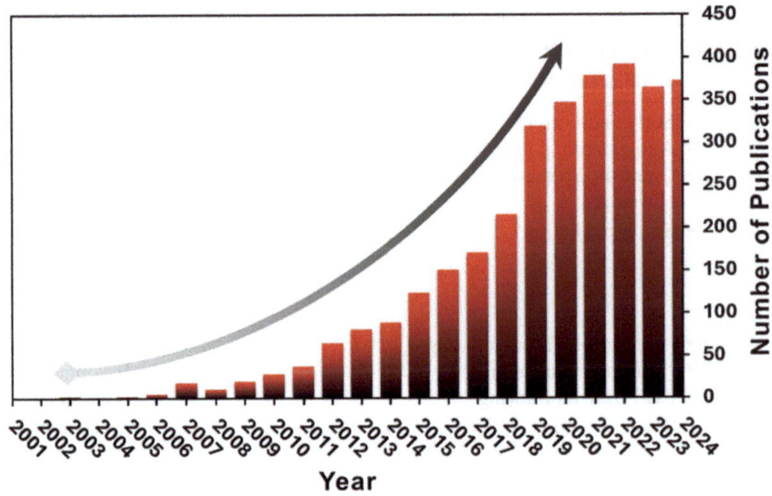

Fig. 9.12 Annual publications were collected from PubMed using the keywords for search: "Bionanocomposites"

polymer nanocomposites. For instance, Golovoy (1988) studied the flammability of polymer-layered silicate nanocomposites early. Carbon nanotubes and fullerenes were added to conducting polymers to improve their electrical and optical properties (Coleman et al. 1998; Yoshino et al. 1999). By the beginning of the 2000s, the concept of bionanocomposites gained prominence, and people worldwide explored the addition of nanomaterials to biopolymers. Nanocomposites of biodegradable polyesters and starch with nano silicates, in addition to cellulose nanofiber-reinforced matrices, were the first bio-nanocomposites obtained during the first half of the 2000s (Ray et al. 2002; Sinha Ray et al. 2002; Maiti et al. 2002; Pantoustier et al. 2002; Kalambur and Rizvi 2004; Bhatnagar and Sain 2005; Oksman and Sain 2005). Onwards till this moment, numerous published papers have become available for versatile bionanocomposites as the number of publications increased dramatically over the quarter of this century (Fig. 9.12).

9.12 Innovations and Commercialization

The intersection of nanoscience and biopolymeric science paved the way for the technology of bionanocomposites since these kinds of materials enjoy unique characteristics over traditional ones. Indeed, the evolution of the bio-nanocomposites passes through several phases. As discussed in the previous section, the research in bionanocomposites commenced at the beginning of the 2000s and increased quickly during the last decade. This was in parallel to the global production of biopolymers (Fig. 9.13). However, by increasing the population and the demand to find

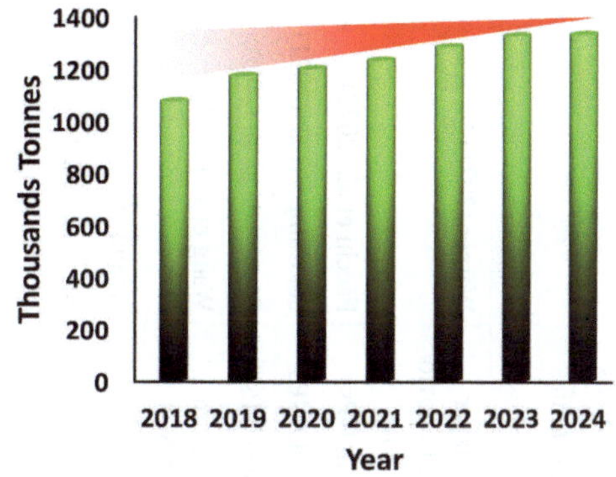

Fig. 9.13 Global production capacities of biopolymers between 2018 and 2024

safe alternatives for traditional plastics, manufacturers have paid strong attention to bionanocomposites. Therefore, early collaboration was initiated between the business and scientific sectors to eventually commercialize a wide range of bionanocomposites from the 2010s till the present. Table 9.5 depicts some companies that produce the components of some common bio-nanocomposites and their potential applications.

9.13 Regulation and Legislation

The safe and effective integration of nanotechnology into the agriculture sector necessitates a balanced approach between technological advancement and regulatory oversight. Regulatory frameworks must evolve alongside increasing public acceptance to ensure environmental and human health protection while facilitating innovation. Legislation plays a critical role not only in the safe application of nanomaterials but also in market authorization and public awareness. Unlike conventional chemicals, nanomaterials possess unique properties and interactions that require specialized protocols for assessing toxicity, functionality, and environmental behavior. Thus, regulatory evaluations should follow a "case-by-case" or "product-by-product" approach due to the distinct characteristics of each nanomaterial formulation (He et al. 2019). Mitter and Hussey propose several actions to enhance nanotechnology regulation in agriculture (Mitter and Hussey 2019) such as:

(I) *Capacity building in regulatory agencies*

There is a need for specialized tools, qualified personnel, validated analytical methods, and sufficient financial resources to ensure proper identification, characterization, and oversight of nanomaterials.

(II) *Integration of emerging technologies*

Table 9.5 Selected companies producing the components of some common bionanocomposites and their potential applications

Manufacturer	Product name	Description	Characteristics	Applications	Published studies	
					Bionanocomposites	References
BASF SE (Germany) www.basf.com	Ecovio®	Polylactic acid blended with poly butylene adipate-terephthalate and bio-based contents	Biodegradable bioplastic, high performance and strong like polyethylene plastics	• Packaging	Layer-by-layer assembled Ecovio/PEI/Nanoclay	Werner (2021)
					Ecovio/Ag NPs	Facchi et al. (2022)
					Ecovio/ZnO NPs	del Campo et al. (2021)
Natureworks LLC (USA) www.natureworksllc.com	Ingeo™	Polylactic acid bio-polymers	Faster crystallization speeds, high levels of crystallinity, high stiffness, high crystalline melting temperatures, high degrees of stress-induced crystallization, biodegradable, recyclable, moldable, UV resistant, hypoallergenic, and lightweight	• Food and beverage packaging • Agriculture • Electronics	Ingeo/SiO_2 NPs	Sepulveda et al. (2020)
					Ingeo/Calcium carbonate NPs	Wang et al. (2021)
					Ingeo/CNT/GNP	Adami et al. (2021)
					Ingeo/NC/Ag NPs	Ng et al. (2021)
					Ingeo/Nanoclays	Oliver-Ortega et al. (2021)

(continued)

Table 9.5 (continued)

Manufacturer	Product name	Description	Characteristics	Applications	Published studies	
					Bionanocomposites	References
Novamont (Italy) https://www.novamont.com/	Mater-BI®	Starch-based bioplastics with a high content of renewable raw materials	Biodegradable, compostable, properties are completely like traditional plastics, resistant to humidity and pathogens, and free of heavy metals	• Packaging • Agriculture	Materi-Bi/SiO_2 NPs	Milionis et al. 2014)
					Materi-Bi/Nanoclay	Cavallaro et al. (2018)
					Materi-Bi/Bacterial cellulose nanofibers	Nainggolan et al. (2013)
Tecnaro (Germany) https://www.tecnaro.de/en/	Arboblend®	Brand of biopolymers such as polyhydroxyalkanoates, polycaprolactone, polyester, starch, polylactic acid, lignin, and natural resins	Biodegradable, rigid, moderate to high strength and thermal resistance, food-compatible, good printable and bondable, good colorable, scratch resistant, flame-retardant, and transparent or translucent	• Packaging • Office • Supplies • Elastic • Cards • Toys • Athletics • Equipment • Plant jars • Lab • Supplies • Filaments	Arboblend V2/ Ag NPs	Mazurchevici et al. (2021), Motaş et al. (2022)

(continued)

Table 9.5 (continued)

Manufacturer	Product name	Description	Characteristics	Applications	Published studies	
					Bionanocomposites	References
Blue Goose Biorefineries Inc. (Canada) https://bluegoose biorefineries. com/	BGB Ultra™	Nanocrystalline cellulose gel	Uniform crystal size, thixotropic, form chiral nematic network, birefringent dried films, charge switchable, and free sulfate half ester moiety	• Packaging • 3D printing • Batteries • Water treatment • Drilling fluids • Sensors • Membranes • Lubrication	Gellan gum/BGB Ultra	Pereira et al. (2018)
					Polyacrylonitrile/ BGB Ultra/SiNPs/ AgNPs	Ren et al. (2017)
					BGB Ultra/GNP/ PLA	Mohd (2022)
					Pt/RGO/BGB Ultra	Dong et al. (2023)
					BGB Ultra/CuO NPs	Hisham (2020)
					PLA/modified BGB Ultra	Stepanova et al. (2020)
Stora ENS (Finland) https://www.sto raenso.com/	NBSK	Bleached softwood kraft pulp	Long and slender fibers, excellent tensile properties, form uniform sheets, biodegradable, and recyclable	• Paper, board, and tissue uses	CNF (NBSK)/ Bio-epoxy resin	Nissilä et al. (2021)

(continued)

Table 9.5 (continued)

Manufacturer	Product name	Description	Characteristics	Applications	Published studies	
					Bionanocomposites	References
Golden-shell Biochemical Co. Ltd. (China) https://www.jinke-chitin.com/	Chitosan	Natural linear poly-saccharides biopolymers	Nontoxic, highly soluble in acid, different molecular weights and deacetylation degrees, lower tensile strength, high permeability, and biodegradable	• Drug delivery • Packaging • Water treatment • Biomedical	Salicylic acid/gentamicin/chitosan nanoparticles	Ji et al. (2011)
					Chitosan/Au NPs	Silva et al. (2019)
					Chitosan/CNC	Tian et al. (2017)
Borregaard LignoTech (Norway) https://www.borregaard.com/	Ligno-sulfonates	Highly cross-linked natural biopolymer of sulfonated lignin derivatives	Water-soluble, salt-tolerant, thermally stable, dispersants, binding agents, crystal growth modifiers, emulsion stabilizers, complexing and chelating agents	• Agriculture • Industrial binders • Leather tanning • Oil well drilling • Batteries • Water treatment	Lignosulfonate/Ag NPs	Milczarek et al. (2013)
					Chitosan/ligno-sulfonates NPs	Kim et al. (2013); Al-Rashed et al. (2019)
					Lignosulfonate/Au NPs	Karbowska et al. (2017)
					Lignosulfonate/Pd NPs	Płócienniczak-Bywalska et al. (2023)
Loba chemicals (India) https://www.lobachemie.com/	Carbo-hydrate polymers	Natural biopolymers including acacia, cellulose, CMC, dextine, gum, and starch	Biodegradable, biocompatible, various physicochemical properties depending on the type and the grade	• Food • Medicine • Eco-technology	Rubber/starch NPs	Rajisha et al. (2014)
					Chitosan/PVP/starch NPs	Poonguzhali et al. (2018)

(continued)

Table 9.5 (continued)

Manufacturer	Product name	Description	Characteristics	Applications	Published studies	
					Bionanocomposites	References
Southern Clay Products, Inc. (USA) https://scprod.lookchem.com/	Sodium montmorillonite	Natural nanoclay	Naturally occurring aluminosilicate layered mineral, low cost, hydrophobic, and strong	• Food • Medicine • Water treatment • Energy • Agriculture	Chitosan/PVA/nanoclay	Koosha et al. (2015)
					Eudragit RS/nanoclay	Liu et al. (2020)
					PCL/nanoclay	Yusoh et al. (2018)
					Starch/CNC/nanoclay	Olusanya et al. (2023)
Nanografi Nanotechnology (Turkey) https://nanografi.com	CNFs	Cellulose nanofiber	Particle size: 4–10 nm wide, length:1–3 μm, powder, cotton sourced, carboxymethylated, hydrophilic, water dispersible, and carboxyl content: 2.5 mmol/g	• Body armor • Flexible batteries • Filters • Agriculture • Food	CNF/CuO NPs/Ag	Vidakis et al. (2022)
					CA/CNF	Acarer-Arat et al. 2024)
					PLA/Cu_2O/CNF	Petousis et al. (2022)
					PCL/CNF	Vidakis et al. (2023)

Regulatory bodies can improve monitoring and enforcement using advanced technologies such as real-time sensors, data analytics, and novel toxicological approaches to address safety concerns effectively.

(III) *Promotion of international cooperation*

Given the global nature of agricultural production and trade, harmonized regulatory frameworks and open communication between nations are essential to ensure consistency in risk assessments and to support the commercialization of nano-enabled agricultural products.

Currently, regulatory responses vary globally. The European Union (EU) has introduced more stringent nanomaterial regulations, including legal definitions, mandatory labeling, and specific risk assessment criteria (He et al. 2019; Amenta et al. 2015). In contrast, the United States follows a more flexible approach, emphasizing risk assessments based on scientific data and requiring safety submissions only in certain cases. This regulatory diversity highlights the urgent need for standardized policies that ensure both innovation and safety in the agricultural use of nanotechnology (Menossi et al. 2022). Particularly, U.S. regulations emphasize monitoring nanomaterials primarily during the manufacturing and processing stages to prevent food contamination. Regulatory decisions are based exclusively on scientific risk assessments. Furthermore, among all registered nanoproducts, the submission of safety data is only mandated in select cases.

In fact, a significant gap remains in scientific understanding regarding the toxicological implications and potential migration of nanoscale fillers, additives, and reinforcement agents from packaging materials into food products. A recent review of global legislative approaches to nanotechnology in the food sector highlights inconsistencies and limited information exchange between countries. This regulatory fragmentation may compromise environmental and human health protection and hinder the international trade of innovative nano-enabled food packaging solutions. The European Union (EU) has taken a leading role in defining and regulating nanomaterials within the food sector. Regulation (EU) No. 1169/2011 provides a legal definition of "engineered nanomaterials" as deliberately manufactured materials with dimensions or internal structures in the nanoscale range ($\leq$100 nm) (European Parliament Council 2011a). This regulation mandates transparent labeling to inform consumers and to guide enforcement across various sectors. In terms of food contact materials, Regulation (EU) No. 10/2011 addresses plastic materials and articles intended for food contact, emphasizing that nanoparticles may exhibit different toxicological behaviors compared to their bulk counterparts (European Parliament Council 2011b). Consequently, risk assessments must be performed on a case-by-case basis (European Commission 2009). Nanomaterials are only permitted if explicitly listed in the regulation. Currently, titanium nitride nanoparticles have been authorized. Also, the regulation states that nanoparticles should not be covered by the functional barrier concept, which allows the migration, below a given detection limit, of certain non-authorized substances present in multilayer materials. Additionally, Regulation (EC) No. 450/2009 governs active and intelligent food contact materials.

It stipulates that any intentional release of nanoparticles into food matrices necessitates a separate safety assessment and potentially a food additive authorization. In the EU market, even if packaging materials meet non-EU standards (e.g., U.S. FDA), suppliers must still provide a Declaration of Compliance to demonstrate safety under EU law (Molenveld et al. 2015).

Overall, while the EU has developed a structured regulatory approach, most non-EU countries have yet to implement similarly rigorous or explicit frameworks. Their guidelines remain largely implicit, offering only general direction to industry stakeholders without formalized legal enforcement. This disparity underscores the need for harmonized international regulation to ensure both safety and innovation in using nanotechnology in agriculture and food applications.

9.14 Conclusions

Bionanocomposites represent an innovation in materials science, particularly in agriculture, food processing, and packaging. By integrating biopolymers with nanomaterials, these composites offer a unique combination of sustainability, enhanced functionality, and superior performance. Their mechanical reinforcement, thermal stability, and barrier properties make them viable alternatives to conventional plastics, which are associated with environmental degradation.

In agriculture, bionanocomposites have revolutionized controlled-release fertilizers, pesticides, and herbicides, enhancing crop yields while reducing the environmental impact of excessive agrochemical usage. Their role in food processing is equally transformative, as they facilitate the encapsulation of bioactive compounds, improving food quality, nutrition, and shelf life. In packaging, bionanocomposites provide antimicrobial and antioxidant properties, ensuring food safety while reducing reliance on synthetic preservatives and non-biodegradable plastic.

Despite their significant potential, commercialization remains a challenge. Factors such as nanotoxicity concerns, regulatory barriers, production costs, and public perception slow their widespread adoption. However, ongoing research and technological advancements are addressing these limitations, paving the way for large-scale applications. Governments and industries worldwide are increasingly investing in bio-based and sustainable solutions, recognizing the environmental, economic, and social benefits of bionanocomposites.

Looking forward, bionanocomposites are expected to play a pivotal role in the shift toward circular economies, where materials are designed for reuse, biodegradability, and minimal environmental impact. Collaboration between scientists, policymakers, and industry leaders will be essential to overcoming commercialization barriers and fostering sustainable innovations. As research continues to refine their properties and applications, bionanocomposites will likely emerge as key contributors to greener technologies, enhancing both human well-being and environmental sustainability.

Acknowledgements The author gratefully acknowledges the financial support provided by the National Research Centre of Egypt for parts of the work cited in this chapter, under grant number 13010307. The author also thanks the Korea Institute of Energy Technology (KENTECH) for providing the academic environment that supported the finalization and submission of this book chapter.

References

Abd El Wahab RM et al (2023) Assessment of a green zeolite/bacterial cellulose nanocomposite membrane as a catalyst to produce biodiesel from waste cooking oil. Energy Sources Part A Recov Util Environ Effects 45(4):10350–10365

Abdelhak M (2022) Soil improvement in arid and semiarid regions for sustainable development. In: Natural resources conservation and advances for sustainability. Elsevier, Amsterdam, pp 73–90

Abdelhak A et al (2022) Synthesis and evaluation of poly (isobutylene-alt-maleic anhydride)-co-poly (ethylene glycol) and its copper oxide nanocomposite for enhancing the performance of water-based drilling fluids in HPHT Wells. Egypt J Chem 65(9):349–360

Abo-Ouf RM, Ghanem AF, Youssef AM, Ayad DM, Abdelaala MY (2023a) Polyacrylamide/polyvinyl alcohol hydrogel and its calcium carbonate nanocomposite as adsorbents for methylene blue. MANSOURA J Chem 63(5):6–14

Abo-Ouf RM, Ghanem AF, Youssef AM, Ayad DM, Abdelaala MY (2023b) MANSOURA J Chem

Abreu FO et al (2012) Chitosan/cashew gum nanogels for essential oil encapsulation. Carbohyd Polym 89(4):1277–1282

Acarer-Arat S et al (2024) Heavy metal rejection performance and mechanical performance of cellulose-nanofibril-reinforced cellulose acetate membranes. ACS Omega 9(41):42159–42171

Adami R et al (2021) Electromagentic properties of filaments containing nanofillers for 3D printing

Ades H et al (2012) Complexation with starch for encapsulation and controlled release of menthone and menthol. LWT-Food Sci Technol 45(2):277–288

Adeyeye SAO, Ashaolu TJ (2021) Applications of nano-materials in food packaging: a review. J Food Process Eng 44(7):e13708

Afzal A et al (2020) Role of polymeric composite in civil engineering applications: a review. Polym Plastics Technol Mater 59(10):1023–1040

Ahari H et al (2022) Bio-nanocomposites as food packaging materials; the main production techniques and analytical parameters. Adv Colloid Interface Sci 310:102806

Ahmadi A et al (2021) Functional biocompatible nanocomposite films consisting of selenium and zinc oxide nanoparticles embedded in gelatin/cellulose nanofiber matrices. Int J Biol Macromol 175:87–97

Aktar MW, Sengupta D, Chowdhury A (2009) Impact of pesticides use in agriculture: their benefits and hazards. Interdiscip Toxicol 2(1):1

Alam A (2014) Soil degradation: a challenge to sustainable agriculture. Int J Sci Res Agric Sci 1(4):50–55

Al-Balakocy NG et al (2022) Photoluminescent and photochromic smart window from recycled polyester reinforced with cellulose nanocrystals. Luminescence 37(9):1575–1584

Alezzi MM et al (2024) Fabrication of graphene oxide/boron nitride/titanium nitride nanocomposites for improving performance of water based drilling fluids in low-pressure low-temperature wells. Egypt J Chem 67(13):2197–2205

Ali Q et al (2023) Nanoremediation for heavy metal contamination: a review. Hybrid Adv 4:100091

Ali R et al (2024) Synergistic impact of biochar nanorods on the performance of polyacrylamide matrix: UV-assisted degradation of phenol and biological activity. Egypt J Chem 67(3):431–448

Alkaç İM et al (2021) Nanomaterials and their classification. In: Nanomaterials for direct alcohol fuel cells. Elsevier, Amsterdam, pp 17–33

Alkhaza'leh H et al (2025) Nanoparticles in soil reclamation: a review of their role in reducing soil compaction. Air Soil Water Res 18:11786221241311725
Al-Rashed MM et al (2019) Mechanism and factors influencing formation and stability of chitosan/lignosulfonate nanoparticles. Macromol Chem Phys 220(1):1800338
Alshammari BH et al (2023) Organic and inorganic nanomaterials: fabrication, properties and applications. RSC Adv 13(20):13735–13785
Alves TF et al (2020) Applications of natural, semi-synthetic, and synthetic polymers in cosmetic formulations. Cosmetics 7(4):75
Amenta V et al (2015) Regulatory aspects of nanotechnology in the agri/feed/food sector in EU and non-EU countries. Regul Toxicol Pharmacol 73(1):463–476
Amiri MR et al (2022) Antibacterial, antifungal, antiviral, and photocatalytic activities of TiO_2 nanoparticles, nanocomposites, and bio-nanocomposites: recent advances and challenges. J Public Health Res 11(2):22799036221104150
Amjadi S et al (2020a) Preparation and characterization of TiO_2 NPs and betanin loaded zein/sodium alginate nanofibers. Food Packag Shelf Life 24:100504
Amjadi S et al (2020b) Multifunctional betanin nanoliposomes-incorporated gelatin/chitosan nanofiber/ZnO nanoparticles nanocomposite film for fresh beef preservation. Meat Sci 167:108161
Arabani M, Shalchian MM, Rahimabadi MM (2023) The influence of rice fiber and nanoclay on mechanical properties and mechanisms of clayey soil stabilization. Constr Build Mater 407:133542
Arfat YA et al (2017) Deciphering the potential of guar gum/Ag–Cu nanocomposite films as an active food packaging material. Carbohydr Polym 157:65–71
Arulprasanna A, Omkumar M (2024) A review on composites: selection and its applications
Arunkumar R, Prashanth KVH, Baskaran V (2013) Promising interaction between nanoencapsulated lutein with low molecular weight chitosan: characterization and bioavailability of lutein in vitro and in vivo. Food Chem 141(1):327–337
Asha AB, Narain R (2020) Nanomaterials properties. In: Polymer science and nanotechnology. Elsevier, Amsterdam, pp 343–359
Assadpour E, Mahdi Jafari S (2019) A systematic review on nanoencapsulation of food bioactive ingredients and nutraceuticals by various nanocarriers. Critic Rev Food Sci Nutr 59(19):3129–3151
Assadpour E et al (2016) Optimization of folic acid nano-emulsification and encapsulation by maltodextrin-whey protein double emulsions. Int J Biol Macromol 86:197–207
Atef M, Rezaei M, Behrooz R (2015) Characterization of physical, mechanical, and antibacterial properties of agar-cellulose bionanocomposite films incorporated with savory essential oil. Food Hydrocolloids 45:150–157
Augustine R et al (2018) Electrospun polyvinyl alcohol membranes incorporated with green synthesized silver nanoparticles for wound dressing applications. J Mater Sci Mater Med 29:1–16
Augusto PE (2020) Challenges, trends and opportunities in food processing. Curr Opin Food Sci 35:72–78
Azizi-Lalabadi M et al (2023) The impact of essential oils on the qualitative properties, release profile, and stimuli-responsiveness of active food packaging nanocomposites. Crit Rev Food Sci Nutr 63(13):1822–1845
Azizun N et al (2021) Effect of pH on functional, gas sensing and antimicrobial properties of bio-nanocomposite gelatin film for food packaging application. J Food Sci Technol 58:3338–3345
Babaei-Ghazvini A, Acharya B, Korber DR (2021) Antimicrobial biodegradable food packaging based on chitosan and metal/metal-oxide bio-nanocomposites: a review. Polymers 13(16):2790
Badawy AA et al (2021a) Utilization and characterization of cellulose nanocrystals decorated with silver and zinc oxide nanoparticles for removal of lead ion from wastewater. Environ Nanotechnol Monitor Manage 16:100501

Badawy AA et al (2021b) Utilization and characterization of cellulose nanocrystals decorated with silver and zinc oxide nanoparticles for removal of lead ion from wastewater. Environ Nanotechnol Monitor Manage 16:100501

Bae GY et al (2010) Superhydrophobic PET fabrics achieved by silica nanoparticles and water-repellent agent. Fibers Polym 11:976–981

Bandera L (2025) Impact of packaging materials on food safety and shelf life

Barbero EJ (2010) Introduction to composite materials design. CRC Press, Boca Raton

Bauli CR et al (2021) Eco-friendly carboxymethyl cellulose hydrogels filled with nanocellulose or nanoclays for agriculture applications as soil conditioning and nutrient carrier and their impact on cucumber growing. Colloids Surf A 623:126771

Behbahani H, Nematollahi B, Farasatpour M (2011) Steel fiber reinforced concrete: a review

Bellinghouse VC, Bellinghouse VC (2009) Food processing: methods, techniques and trends. Nova Science Publishers

Bhatnagar A, Sain M (2005) Processing of cellulose nanofiber-reinforced composites. J Reinf Plast Compos 24(12):1259–1268

Bhushani JA, Karthik P, Anandharamakrishnan C (2016) Nanoemulsion based delivery system for improved bioaccessibility and Caco-2 cell monolayer permeability of green tea catechins. Food Hydrocolloids 56:372–382

Biji K et al (2015) Smart packaging systems for food applications: a review. J Food Sci Technol 52:6125–6135

Bjelakovic G et al (2004) Antioxidant supplements for prevention of gastrointestinal cancers: a systematic review and meta-analysis. Lancet 364(9441):1219–1228

Bo L et al (2024) Recent remediation strategies for flame retardancy via nanoparticles. Chemosphere 354:141323

Bratovcic A (2021) Physical–chemical, mechanical and antimicrobial properties of bio-nanocomposite films and edible coatings. Int J Res Appl Sci Biotechnol 8(5):151–156

Brockgreitens J, Abbas A (2016) Responsive food packaging: recent progress and technological prospects. Compr Rev Food Sci Food Saf 15(1):3–15

Bumbudsanpharoke N et al (2025) The rise of nanotechnology in pharmaceutical and healthcare packaging: applications and future prospective. Packag Technol Sci 38(1):3–31

Buzea C, Pacheco I (2016) Nanomaterials and their classification. Springer, New Delhi, pp 3–45

Cafri M et al (2012) Boron carbide/magnesium composites: processing, microstructure and properties. J Eur Ceram Soc 32(12):3477–3483

Caro N et al (2016) Novel active packaging based on films of chitosan and chitosan/quinoa protein printed with chitosan-tripolyphosphate-thymol nanoparticles via thermal ink-jet printing. Food Hydrocolloids 52:520–532

Castell-Perez E et al (2017) Food processing and waste within the nexus framework. Curr Sustain Renew Energy Rep 4:99–108

Cavallaro G et al (2018) Filling of mater-Bi with nanoclays to enhance the biofilm rigidity. J Funct Biomater 9(4):60

Chandekar H, Chaudhari V, Waigaonkar S (2020) A review of jute fiber reinforced polymer composites. Mater Today Proc 26:2079–2082

Chang I et al (2015) Soil treatment using microbial biopolymers for anti-desertification purposes. Geoderma 253:39–47

Chawla K (2012) Composite materials: science and engineering, vol 1. Springer, New York, NY

Chen X et al (2009) Biopolymer-manganese oxide nanoflake nanocomposite films fabricated by electrostatic layer-by-layer assembly. Mater Sci Eng C 29(1):284–287

Chen Q-Y et al (2020) A one-pot synthesis and characterization of antibacterial silver nanoparticle–cellulose film. Polymers 12(2):440

Chhatre A et al (2012) Color and surface plasmon effects in nanoparticle systems: case of silver nanoparticles prepared by microemulsion route. Colloids Surf A Physicochem Eng Aspects 404:83–92

Choi SR, Bansal NP (2005) Mechanical behavior of zirconia/alumina composites. Ceram Int 31(1):39–46

Chopra M et al (2014) Surfactant assisted nisin loaded chitosan-carageenan nanocapsule synthesis for controlling food pathogens. Food Control 37:158–164

Choudhary N et al (2021) Reinforcement of polylactic acid with bioceramics (alumina and YSZ composites) and their thermomechanical and physical properties for biomedical application. J Vinyl Additive Technol 27(3):612–625

Chung Y-L et al (2010) Preparation and properties of biodegradable starch–clay nanocomposites. Carbohydr Polym 79(2):391–396

Coleman JN et al (1998) Percolation-dominated conductivity in a conjugated-polymer-carbon-nanotube composite. Phys RevB 58(12):R7492

Cui C et al (2021) Fatigue performance of concrete–polyurethane composite materials under compression. J Transp Eng Part B Pavements 147(3):04021030

Curulli A (2020) Nanomaterials in electrochemical sensing area: applications and challenges in food analysis. Molecules 25(23):5759

Cywar RM et al (2022) Bio-based polymers with performance-advantaged properties. Nat Rev Mater 7(2):83–103

Dar OA et al (2020) Bionanocomposites in water treatment. In: Bionanocomposites. Elsevier, Amsterdam, pp 505–518

Darwesh OM et al (2023) Application of microalgal-ZnO-NPs for reusing polyester/cotton blended fabric wastes after modification by cellulases enzymes. Waste Dispos Sustain Energy 5(4):471–482

Das S (2001) The cost of automotive polymer composites: a review and assessment of DOE's lightweight materials composites research

Das D, Bhattacharjee S, Bhaladhare S (2023) Preparation of cellulose hydrogels and hydrogel nanocomposites reinforced by crystalline cellulose nanofibers (CNFs) as a water reservoir for agriculture use. ACS Appl Polym Mater 5(4):2895–2904

de Mesquita JP, Donnici CL, Pereira FV (2010) Biobased nanocomposites from layer-by-layer assembly of cellulose nanowhiskers with chitosan. Biomacromol 11(2):473–480

de Moura MR et al (2009) Improved barrier and mechanical properties of novel hydroxypropyl methylcellulose edible films with chitosan/tripolyphosphate nanoparticles. J Food Eng 92(4):448–453

de Oliveira JL et al (2018) Zein nanoparticles as eco-friendly carrier systems for botanical repellents aiming sustainable agriculture. J Agric Food Chem 66(6):1330–1340

de Oliveira Filho JG et al (2022) Bio-nanocomposite edible coatings based on arrowroot starch/cellulose nanocrystals/carnauba wax nanoemulsion containing essential oils to preserve quality and improve shelf life of strawberry. Int J Biol Macromol 219:812–823

del Campo A et al (2021) Accelerated disintegration of compostable Ecovio polymer by using ZnO particles as filler. Polym Degrad Stab 185:109501

Dena ASA et al (2021) User-friendly lab-on-paper optical sensor for the rapid detection of bacterial spoilage in packaged meat products. RSC Adv 11(56):35165–35173

Devi N et al (2024) Novel chitosan-based smart bio-nanocomposite films incorporating TiO_2 nanoparticles for white bread preservation. Int J Biol Macromol 267:131367

Dimkpa CO et al (2023) Chitosan and zinc oxide nanoparticle-enhanced tripolyphosphate modulate phosphorus leaching in soil. ACS Agric Sci Technol 3(6):487–498

Ding Z (2024) Chitosan-based nanofertilizers for sustainable agriculture

Dobrucka R, Cierpiszewski R (2014) Active and intelligent packaging food-Research and development—a review. Pol J Food Nutr Sci 64(1)

Dong L et al (2023) In-situ synthesis of Pt nanoparticles/reduced graphene oxide/cellulose nanohybrid for nonenzymatic glucose sensing. Carbohydr Polym 303:120463

Dong F, Wang Y (2016) Encapsulation of vitamin C by a double-layer zein/chitosan structure with improved stability and controlled release. Int Nanomed Nanosurg 2(1)

dos Santos Silva M et al (2011) Paraquat-loaded alginate/chitosan nanoparticles: preparation, characterization and soil sorption studies. J Hazard Mater 190(1–3):366–374

Du W-L et al (2009) Antibacterial activity of chitosan tripolyphosphate nanoparticles loaded with various metal ions. Carbohydr Polym 75(3):385–389

Duncan TV (2011) Applications of nanotechnology in food packaging and food safety: barrier materials, antimicrobials and sensors. J Colloid Interface Sci 363(1):1–24

Duraikkannu SL et al (2021) A review on phase-inversion technique-based polymer microsphere fabrication. Colloid Interf Sci Commun 40:100329

Dursun GM, Duran C (2019) Glass alumina composites for functional and structural applications. Ceram Int 45(9):12550–12557

Egbo MK (2021) A fundamental review on composite materials and some of their applications in biomedical engineering. J King Saud Univ Eng Sci 33(8):557–568

El-Gendi A et al (2020) Antifouling and antimicrobial polyethersulfone/hyperbranched polyester-amide/Ag composite. RSC Adv 10(41):24169–24175

Eshaghi R, Mohsenzadeh M, Ayala-Zavala JF (2024) Bio-nanocomposite active packaging films based on carboxymethyl cellulose, myrrh gum, TiO_2 nanoparticles and dill essential oil for preserving fresh-fish (*Cyprinus carpio*) meat quality. Int J Biol Macromol 263:129991

European Commission (2009) Commission Regulation (EC) No 450/2009 of 29 May 2009 on active and intelligent materials and articles intended to come into contact with food. Off J Eur Union, pp 3–11

European Parliament and Council (2011a) Regulation (EU) No 1169/2011 of the European Parliament and of the Council of 25 October 2011 on the provision of food information to consumers. Off J Eur Union L 304, p 18e63

European Parliament and Council (2011b) Regulation (EU) No 10/2011 of the European Parliament and of the Council of 14 January 2011 on plastic materials and articles intended to come into contact with food. Off J Eur Union, pp 1–89

Facchi DP et al (2022) Composite filter with antimicrobial and anti-adhesive properties based on electrospun poly (butylene adipate-co-terephthalate)/poly (acid lactic)/Tween 20 fibers associated with silver nanoparticles. J Memb Sci 650:120426

Fatehi H et al (2021) Biopolymers as green binders for soil improvement in geotechnical applications: a review. Geosciences 11(7):291

Fortunati E et al (2012) Effects of modified cellulose nanocrystals on the barrier and migration properties of PLA nano-biocomposites. Carbohyd Polym 90(2):948–956

Franklin AD (2015) Nanomaterials in transistors: from high-performance to thin-film applications. Science 349(6249):aab2750

Fuciños C et al (2015) Functional characterisation and antimicrobial efficiency assessment of smart nanohydrogels containing natamycin incorporated into polysaccharide-based films. Food Bioprocess Technol 8:1430–1441

Gao Y, Guo W, Baghaei S (2024) Ceramic based nanocomposites with alumina-carbon nanotube reinforcement for improved energy absorption in sports-related injuries: microstructural analysis and low velocity impact. Ceram Int 50(7):11129–11137

García-Martínez J-M, Collar EP (2020) Organic–inorganic hybrid materials. Polymer 13(1):86

Gavrilenko VI (2016) Optics of nanomaterials. Jenny Stanford Publishing, New Delhi

Geng S et al (2018) Well-dispersed cellulose nanocrystals in hydrophobic polymers by in situ polymerization for synthesizing highly reinforced bio-nanocomposites. Nanoscale 10(25):11797–11807

Ghaderi F et al (2024) Potential of gelatin/chitosan nanocomposite films loaded with *L. nobilis* nanoemulsions for preservation of refrigerated ostrich meat-based hamburgers. J Food Meas Charact 18(5):3474–3491

Ghanem AF, Abdel Rehim MH (2018) Assisted tip sonication approach for graphene synthesis in aqueous dispersion. Biomedicines 6(2):63

Ghanem A et al (2014) Photocatalytic activity of hyperbranched polyester/TiO_2 nanocomposites. Appl Catal A 472:191–197

Ghanem AF et al (2019) Enhancement the photocatalytic and biological activity of nano-sized ZnO using hyperbranched polyester. J Inorg Organomet Polym Mater 29:928–938
Ghanem AF et al (2021) Investigation of water sorption, gas barrier and antimicrobial properties of polycaprolactone films contain modified graphene. J Mater Sci 56:497–512
Ghanem AF et al (2024) Polycaprolactone composite films infused with hyperbranched polyester/reduced graphene oxide: influence on biodegradability, gas/water transport and antimicrobial properties for sustainable packaging. RSC Adv 14(9):5740–5753
Ghanem AF et al (2025) Polyvinyl alcohol film comprising biochar modified titanium dioxide nanocomposites as decoloring and disinfectant agents. Sci Rep 15(1):11423
Ghorai S (2013) Chemical, physical and mechanical properties of nanomaterials and its applications. The University of Iowa, Iowa
Gil LJ (2009) Cork composites: a review. Material 2(3):776–789
Giroto AS et al (2017) Role of slow-release nanocomposite fertilizers on nitrogen and phosphate availability in soil. Sci Rep 7(1):46032
Golovoy A (1988) Chemistry and technology of polymer additives
Gungure AS, Jule LT (2025) Nanomaterials for energy applications: batteries, super-capacitors, and solar cells. In: Exploring nanomaterial synthesis, characterization, and applications. IGI Global, Hershey, PA, pp 101–124
Ha NMC et al (2019) Preparation of NPK nanofertilizer based on chitosan nanoparticles and its effect on biophysical characteristics and growth of coffee in green house. Res Chem Intermed 45:51–63
Han JW et al (2018) Food packaging: a comprehensive review and future trends. Comp Rev Food Sci Food Saf 17(4):860–877
Hassan-Nejad M et al (2009) Bio-based nanocomposites of cellulose acetate and nano-clay with superior mechanical properties. In: Macromolecular symposia. Wiley Online Library, New York, NY
He X, Deng H, Hwang H-M (2019) The current application of nanotechnology in food and agriculture. J Food Drug Anal 27(1):1–21
He Y et al (2024) Development of a multifunctional active food packaging membrane based on electrospun polyvinyl alcohol/chitosan for preservation of fruits. Int J Biol Macromol 277:134636
Hisham S et al (2020) Statistical approach for prediction of thermal properties of CNC and CNC-CuO nanolubricant using Response Surface Methodology (RSM). IOP Conf Ser Mater Sci Eng 788(1):012016
Hosseini SMH et al (2015) Nanocomplexes arising from protein-polysaccharide electrostatic interaction as a promising carrier for nutraceutical compounds. Food Hydrocolloids 50:16–26
Hosseini SF et al (2016) Development of bioactive fish gelatin/chitosan nanoparticles composite films with antimicrobial properties. Food Chem 194:1266–1274
Hou Z et al (2012) Effect of chitosan molecular weight on the stability and rheological properties of β-carotene emulsions stabilized by soybean soluble polysaccharides. Food Hydrocolloids 26(1):205–211
Ibrahim SM et al (2022) Effective single and contest carcinogenic dyes adsorption onto A-zeolite/bacterial cellulose composite membrane: adsorption isotherms, kinetics, and thermodynamics. J Environ Chem Eng 10(6):108588
Ibrahim ID et al (2022) Need for sustainable packaging: an overview. Polymers 14(20):4430
Ibrahim RK et al (2016) Environmental application of nanotechnology: air, soil, and water. Environ Sci 23:13754–13788
Ingrao C et al (2023) Water scarcity in agriculture: an overview of causes, impacts and approaches for reducing the risks. Heliyon 9(8):e18507
Jafarizadeh-Malmiri H et al (2019) Nano-additives for food industries. Springer, Cham, pp 41–68
Jafarzadeh S et al (2021) Application of bio-nanocomposite films and edible coatings for extending the shelf life of fresh fruits and vegetables. Adv Colloid Interface Sci 291:102405

Javni I et al (2002) Effect of nano-and micro-silica fillers on polyurethane foam properties. J Cellular Plastics 38(3):229–239

Jawaid M, Swain SK (2018) Bionanocomposites for packaging applications. Springer, Cham

Ji J et al (2011) Preparation, characterization and in vitro release of chitosan nanoparticles loaded with gentamicin and salicylic acid. Carbohydr Polym 85(4):803–808

Jiang X et al (2015) Future perspectives towards the use of nanomaterials for smart food packaging and quality control. Part Part Syst Charact 32(4):408–416

Kagan CR (2016) At the nexus of food security and safety: opportunities for nanoscience and nanotechnology. ACS Publications, pp 2985–2986

Kalambur SB, Rizvi SSH (2004) Starch-based nanocomposites by reactive extrusion processing. Polym Int 53(10):1413–1416

Kamal A et al (2022) Fabrication techniques of polymeric nanocomposites: a comprehensive review. Proc Inst Mech Eng Part C J Mech Eng Sci 236(9):4843–4861

Karbowska B, Rębiś T, Milczarek G (2017) Mercury-modified lignosulfonate-stabilized gold nanoparticles as an alternative material for anodic stripping voltammetry of thallium. Electroanalysis 29(9):2090–2097

Kassem I et al (2022) Cellulose nanofibers/engineered biochar hybrid materials as biodegradable coating for slow-release phosphate fertilizers. ACS Sustain Chem Eng 10(46):15250–15262

Katuwavila NP et al (2016) Alginate nanoparticles protect ferrous from oxidation: potential iron delivery system. Int J Pharm 513(1–2):404–409

Katwal R (2023) A review: properties and diverse applications of smart magnetic quantum dots. Nano-Struct Nano-Object 35:101001

Khaleel MR, Hashim FS, Alkhayatt AHO (2024) Preparation, characterization, and the antimicrobial activity of PVA-PVP/ZnO nanofiber films via indigenous electrospinning setup. J Mol Struct 1310:138325

Khalid SA et al (2024a) Free-standing carboxymethyl cellulose film incorporating nanoformulated pomegranate extract for meat packaging. Carbohydr Polym 332:121915

Khalid SA et al (2024b) Free-standing carboxymethyl cellulose film incorporating nanoformulated pomegranate extract for meat packaging. Carbohydr Polym 332:121915

Khan ZH (2017) Nanomaterials and their applications, vol 84. Springer, Cham

Kim S et al (2013) Chitosan–lignosulfonates sono-chemically prepared nanoparticles: characterisation and potential applications. Colloids Surf B Biointerf 103:1–8

Knorr D, Augustin MA, Tiwari B (2020) Advancing the role of food processing for improved integration in sustainable food chains. Front Nutr 7:34

Kollia E et al (2020) Effect of water aging on the mechanical properties of flax fiber/bio-based resin composites. J Appl Polym Sci 37(13):48787

Koosha M et al (2015) Nanoclay-reinforced electrospun chitosan/PVA nanocomposite nanofibers for biomedical applications. RSC Adv 5(14):10479–10487

Kotobuki M et al (2021) Ceramic-polymer composite membranes for water and wastewater treatment: bridging the big gap between ceramics and polymers. Molecules 26(11):3331

Kumar S et al (2015) Development and evaluation of alginate–chitosan nanocapsules for controlled release of acetamiprid. Int J Biol Macromol 81:631–637

Kumar R, Ashfaq M, Verma N (2018) Synthesis of novel PVA–starch formulation-supported Cu–Zn nanoparticle carrying carbon nanofibers as a nanofertilizer: controlled release of micronutrients. J Mater Sci 53(10):7150–7164

Kumar SA et al (2022) Review on biopolymer stabilization—a natural alternative for erosion control. Advances in sustainable materials and resilient infrastructure. Springer, pp 185–200

Kumaraswamy R et al (2021) Chitosan-silicon nanofertilizer to enhance plant growth and yield in maize (*Zea mays* L.). Plant Physiol Biochem 159:53–66

Kumari P et al (2024) Environmentally benign approach to formulate nanoclay/starch hydrogel for controlled release of zinc and its application in seed coating of *Oryza sativa* plant. Int J Biol Macromol 257:128278

Kuswandi B et al (2011) Smart packaging: sensors for monitoring of food quality and safety. Sens Instrum Food Qual Saf 5:137–146

Lan W et al (2021) Development of red apple pomace extract/chitosan-based films reinforced by TiO_2 nanoparticles as a multifunctional packaging material. Int J Biol Macromol 168:105–115

Leonardi M et al (2021) Smart nanocomposites of chitosan/alginate nanoparticles loaded with copper oxide as alternative nanofertilizers. Environ Sci Nano 8(1):174–187

Lević S et al (2014) Thermal, morphological, and mechanical properties of ethyl vanillin immobilized in polyvinyl alcohol by electrospinning process. J Therm Anal Calorim 118:661–668

Li D (2025) Development and evaluation of titanium dioxide/chitosan nanocomposite coatings for enhanced food preservation and nutrient retention. Alex Eng J 121:484–491

Li Y et al (2011) Synthesis and characterization of bionanocomposites of poly (lactic acid) and TiO_2 nanowires by in situ polymerization. Polymer 52(11):2367–2375

Li M, Huang Q, Wu Y (2011) A novel chitosan-poly (lactide) copolymer and its submicron particles as imidacloprid carriers. Pest Manag Sci 67(7):831–836

Li X et al (2021) Characterization and antibacterial activity of edible films based on carboxymethyl cellulose, *Dioscorea opposita* mucilage, glycerol and ZnO nanoparticles. Food Chem 349:129208

Li T et al (2022) Lightweight and flexible graphene foam composite with improved damping properties. Nanomaterials 12(8):1260

Li Y et al (2024a) Enhancing post-harvest quality of tomato fruits with chitosan oligosaccharide-zinc oxide nanocomposites: a study on biocompatibility, quality improvement, and carotenoid enhancement. Food Chem 454:139685

Li C et al (2024b) Green and facile fabrication of multifunctional cellulose nanocrystal and carvacrol together reinforced chitosan bio-nanocomposite coatings for fruit preservation. Int J Biol Macromol 265:130651

Liang R et al (2012) Physical and antimicrobial properties of peppermint oil nanoemulsions. J Agric Food Chem 60(30):7548–7555

Liang R et al (2013) Stability and bioaccessibility of β-carotene in nanoemulsions stabilized by modified starches. J Agric Food Chem 61(6):1249–1257

Liao Q et al (2021) Simultaneous immobilization of Pb, Cd and As in soil by hybrid iron-, sulfate- and phosphate-based bio-nanocomposite: effectiveness, long-term stability and bioavailablity/bioaccessibility evaluation. Chemosphere 266:128960

Lin M-H et al (2021) Hybrid ZnO/chitosan antimicrobial coatings with enhanced mechanical and bioactive properties for titanium implants. Carbohydr Polym 257:117639

Lin C-M, Fernando SY, Wei C (1996) Occurrence of *Listeria monocytogenes*, *Salmonella* spp., *Escherichia coli* and *E. coli* O157:H7 in vegetable salads. Food Control **7**(3):135–140

Liu X et al (2020) Clay-polymer nanocomposites prepared by reactive melt extrusion for sustained drug release. Pharmaceutics 12(1):51

Liu Y et al (2023) Principle, development and application of time–temperature indicators for packaging. Packag Technol Sci 36(10):833–853

Ma Q, Davidson PM, Zhong Q (2016) Nanoemulsions of thymol and eugenol co-emulsified by lauric arginate and lecithin. Food Chem 206:167–173

Mahajan GV, Aher VS (2012) Composite material: a review over current development and automotive application. Int J Sci Res Publ 2(11):1–5

Maiti P, Batt CA, Giannelis EP (2002) Biodegradable polyester/layered silicate nanocomposites. MRS Proc 740:I5.3

Marras SI et al (2008) Biodegradable polymer nanocomposites: the role of nanoclays on the thermomechanical characteristics and the electrospun fibrous structure. Acta Biomater 4(3):756–765

Maruyama CR et al (2016) Nanoparticles based on chitosan as carriers for the combined herbicides imazapic and imazapyr. Sci Rep 6(1):19768

Mazhoud B et al (2017) Mechanical properties of hemp-clay and hemp stabilized clay composites. Construct Build Mater 155:1126–1137

Mazurchevici SN et al (2021) Nanocomposite biopolymer arboblend V2 nature AgNPs. Polymers 13(17):2932

Medina-Pérez G et al (2019) Remediating polluted soils using nanotechnologies: environmental benefits and risks. Pol J Environ Stud 28(3)

Mehrazar E, Rahaie M, Rahaie S (2015) Application of nanoparticles for pesticides, herbicides, fertilisers and animals feed management. Int J Nanoparticles 8(1):1–19

Meira SMM et al (2016) Starch-halloysite nanocomposites containing nisin: characterization and inhibition of Listeria monocytogenes in soft cheese. LWT-Food Sci Technol 68:226–234

Melo PTS et al (2019) Combining cupuassu (*Theobroma grandiflorum*) puree, pectin, and chitosan nanoparticles into novel edible films for food packaging applications. J Food Sci 84(8):2228–2233

Menossi M, Casalongué C, Alvarez VA (2022) Bio-nanocomposites for modern agricultural applications. Handbook of consumer nanoproducts. Springer, pp 1201–1237

Merino D, Casalongué C, Alvarez VA (2018) Polysaccharides as eco-nanomaterials for agricultural applications. In: Handbook of ecomaterials, pp 2709–2730

Mhd Haniffa MAC et al (2016) Review of bionanocomposite coating films and their applications. Polymers 8(7):246

Milczarek G et al (2013) One-step synthesis of lignosulfonate-stabilized silver nanoparticles. Colloids Surf B Biointerf 105:335–341

Milionis A, Ruffilli R, Bayer IS (2014) Superhydrophobic nanocomposites from biodegradable thermoplastic starch composites (Mater-Bi®), hydrophobic nano-silica and lycopodium spores. RSC Adv 4(65):34395–34404

Mitter N, Hussey K (2019) Moving policy and regulation forward for nanotechnology applications in agriculture. Nat Nanotechnol 14(6):508–510

Mogensen KB, Kneipp K (2014) Size-dependent shifts of plasmon resonance in silver nanoparticle films using controlled dissolution: monitoring the onset of surface screening effects. J Phys Chem C 118(48):28075–28083

Mohanapragash AG et al (2025) Nanotechnology in sustainable soil remediation: exploring a review on carbon-based, clay-based, dendrimer, polymer, and nanocomposite based nanomaterials for environmental cleanup. Int J Environ Anal Chem 1–26

Mohd S et al (2022) Cellulose nano crystal/graphene nano platelets hybrid nanofillers reinforced polylactic acid biocomposites: mechanical and morphological properties. IOP Conf Ser Earth Environ Sci 1102(1):012005

Molenveld K, Van den Oever M, Bos HL (2015) Biobased packaging catalogue. UR-Food & Biobased Research, Wageningen

Mondéjar-López M et al (2024) A review on the encapsulation of "eco-friendly" compounds in natural polymer-based nanoparticles as next generation nano-agrochemicals for sustainable agriculture and crop management. Int J Biol Macromol 280:136030

Motaş JG et al (2022) Influence of silver nanoparticles on mechanical characteristics of arboblend v2 nature lignin-based polymer

Nainggolan H et al (2013) Mechanical and thermal properties of bacterial-cellulose-fibre-reinforced Mater-Bi® bionanocomposite. Beilstein J Nanotechnol 4(1):325–329

Najah M et al (2023) Mechanical properties of calcium phosphate reinforced polyhydroxyalkanoate (PHA) biocomposite. J Thermoplast Compos Mater 36(8):3294–3311

Nandhini J, Karthikeyan E, Rajeshkumar S (2024) Eco-friendly bio-nanocomposites: pioneering sustainable biomedical advancements in engineering. Discov Nano 19(1):86

Nasimi P, Haidari M (2013) Medical use of nanoparticles: drug delivery and diagnosis diseases. Int J Green Nanotechnol 1:1943089213506978

Neo YP et al (2013) Encapsulation of food grade antioxidant in natural biopolymer by electrospinning technique: a physicochemical study based on zein–gallic acid system. Food Chem 136(2):1013–1021

Ng W et al (2021) Hybridization of 5-nitroisatin and silver nanoparticles for boosting the antimicrobial performance of poly (lactic acid)/nanocellulose nanocomposite films. Food Packag Shelf Life 30:100763

Nishitha M, Narayana B, Sarojini BK (2024) Studies on guar gum and cellulose acetate superabsorbent hydrogel beads with high water retention properties for rejuvenation of dryland wettability. Indian J Chem Technol (IJCT) 31(3):425–437

Nissilä T et al (2021) Ice-templated cellulose nanofiber filaments as a reinforcement material in epoxy composites. Nanomaterials 11(2):490

Nouri A (2019) Chitosan nano-encapsulation improves the effects of mint, thyme, and cinnamon essential oils in broiler chickens. Br Poult Sci 60(5):530–538

Nunes P, Ramanathan LV (1995) Corrosion behavior of alumina-aluminum and silicon carbide-aluminum metal-matrix composites. Corrosion 51(08):610–617

Oksman K, Sain M (2005) Cellulose nanocomposites: processing, characterization and properties. American Chemical Society (ACS), Washington, DC

Olatunji O (2015) Natural polymers: industry techniques and applications. Springer, Cham

Oliver-Ortega H et al (2021) Nanocomposites materials of PLA reinforced with nanoclays using a masterbatch technology: a study of the mechanical performance and its sustainability. Polymer 13(13):2133

Olusanya JO, Mohan TP, Kanny K (2023) Effect of nanoclay (NC) on mechanical and thermal properties of cellulose nanoparticles (CNPs) filled cornstarch bioplastic film. J Polym Res 30(7):262

Omanović-Mikličanin E et al (2020) Nanocomposites: a brief review. Health Technol 10(1):51–59

Orts WJ, Sojka RE, Glenn GM (2000) Biopolymer additives to reduce erosion-induced soil losses during irrigation. Ind Crops Prod 11(1):19–29

Otero J, Borsoi G, Monasterio-Guillot L (2023) The boom in nanomaterials for built heritage conservation: Why does size matter? Materials 16(8):3277

Otoni CG et al (2014) Antimicrobial and physical-mechanical properties of pectin/papaya puree/cinnamaldehyde nanoemulsion edible composite films, vol 41, p 188

Oxandale SW et al (2023) Enhanced thermoelectric performance via quantum confinement in a metal oxide semiconductor field effect transistor for thermal management. Commun Mater 4(1):85

Oymaci P, Altinkaya SA (2016) Improvement of barrier and mechanical properties of whey protein isolate based food packaging films by incorporation of zein nanoparticles as a novel bionanocomposite. Food Hydrocolloids 54:1–9

Ozkan D et al (2020) Carbon fiber reinforced polymer (CFRP) composite materials, their characteristic properties, industrial application areas and their machinability. Springer, Cham, pp 235–253

Paidari S et al (2023) Bio-nanocomposites and their potential applications in physiochemical properties of cheese: an updated review. J Food Meas Charact 17(3):2595–2606

Pandey S, Sekar H, Gundabala V (2024) Development and characterization of bilayer chitosan/alginate cling film reinforced with essential oil based nanocomposite for red meat preservation. Int J Biol Macromol 279:135524

Pantoustier N et al (2002) Biodegradable polyester layered silicate nanocomposites based on poly (ϵ-caprolactone). Polym Eng Sci 42(9):1928–1937

Paramitha T et al (2020) Properties of clays reinforced PLA nanocomposites by melt extrusion technique. Malays J Fundam Appl Sci 16(4):453–457

Parisi OI, Curcio M, Puoci F (2015) Polymer chemistry and synthetic polymers. Springer, Cham, pp 1–31

Patil MD et al (2018) Tuning controlled release behavior of starch granules using nanofibrillated cellulose derived from waste sugarcane bagasse. ACS Sustain Chem Eng 6(7):9208–9217

Patra SK et al (2022) Prospects of hydrogels in agriculture for enhancing crop and water productivity under water deficit condition. Int J Polym Sci 2022(1):4914836

Paul V, John MJ (2018) Composites from natural fibers and bio-resins. In: Biocomposites. Jenny Stanford Publishing, pp 1–25
Pereira C, Martins M (2014) Flame retardancy of fiber-reinforced polymer composites based on nanoclays and carbon nanotubes. In: Polymer green flame retardants. Elsevier, Amsterdam, pp 551–595
Pereira DR et al (2018) Nanocellulose reinforced gellan-gum hydrogels as potential biological substitutes for annulus fibrosus tissue regeneration. Nanomed Nanotechnol Biol Med 14(3):897–908
Perera KY, Jaiswal AK, Jaiswal S (2023) Extending cheese shelf-life using eco-friendly sodium alginate-gelatin films reinforced with nanoclay. Food Biosci 56:103304
Petousis M et al (2022) Multifunctional material extrusion 3D-printed antibacterial polylactic acid (PLA) with binary inclusions: the effect of cuprous oxide and cellulose nanofibers. Fibers 10(6):52
Pihtili H (2009) An experimental investigation of wear of glass fibre–epoxy resin and glass fibre–polyester resin composite materials. Eur Polym J 45(1):149–154
Pinheiro AC, Coimbra MA, Vicente AA (2016) In vitro behaviour of curcumin nanoemulsions stabilized by biopolymer emulsifiers—effect of interfacial composition. Food Hydrocolloids 52:460–467
Płócienniczak-Bywalska P et al (2023) Lignosulfonate-assisted in situ deposition of palladium nanoparticles on carbon nanotubes for the electrocatalytic sensing of hydrazine. Molecules 28(20):7076
Pokropivny V et al (2007) Introduction to nanomaterials and nanotechnology. Tartu University Press, Tartu
Poonguzhali R et al (2018) Nanostarch reinforced with chitosan/poly (vinyl pyrrolidone) blend for in vitro wound healing application. Polym-Plastics Technol Eng 57(14):1400–1410
Pozzi M et al (2024) Visualization of the high surface-to-volume ratio of nanomaterials and its consequences. J Chem Educ 101(8):3146–3155
Qin Y et al (2016) Effects of chitin nano-whiskers on the antibacterial and physicochemical properties of maize starch films. Carbohydr Polym 147:372–378
Qin Y et al (2019) Preparation and characterization of antioxidant, antimicrobial and pH-sensitive films based on chitosan, silver nanoparticles and purple corn extract. Food Hydrocolloids 96:102–111
Rahman M (2011) Nanomaterials. BoD–Books on Demand
Rajisha K et al (2014) Preparation and characterization of potato starch nanocrystal reinforced natural rubber nanocomposites. Int J Biol Macromol 67:147–153
Ramesh M, Maniraj J, Rajesh Kumar L (2021) Biocomposites for energy storage. In: Biobased composites: processing, characterization, properties, and applications, pp 123–142
Rani N, Kusum, Hooda V (2024) Chitosan/ZnO nanocomposites for improving the growth and reducing the toxicity of Zn in *Sorghum bicolor* (L.) Moench plants. Acta Physiol Plant 46(6):67
Ray SS et al (2002) New polylactide/layered silicate nanocomposite: nanoscale control over multiple properties. Macromol Rapid Commun 23(16):943–947
Ren S et al (2017) Electrospun nanofibers made of silver nanoparticles, cellulose nanocrystals, and polyacrylonitrile as substrates for surface-enhanced Raman scattering. Materials 10(1):68
Rhoades J, Duffy G, Koutsoumanis K (2009) Prevalence and concentration of verocytotoxigenic *Escherichia coli*, *Salmonella enterica* and *Listeria monocytogenes* in the beef production chain: a review. Food Microbiol 26(4):357–376
Riseh RS, Vazvani MG, Kennedy JF (2023) The application of chitosan as a carrier for fertilizer: a review. Int J Biol Macromol 252:126483
Rivera-Solís LL et al (2024) Biostimulation of plants with nanocomposites: a new perspective to improve crop production. In: Nanocomposites for environmental, energy, and agricultural applications. Elsevier, Amsterdam, pp 217–276

Rodrigues JS et al (2025) Designing sustainable soil conditioners: nanocomposite-based thermoplastic starch for enhanced soil health and crop performance. Int J Biol Macromol 297:139747

Rubentheren V et al (2015) Processing and analysis of chitosan nanocomposites reinforced with chitin whiskers and tannic acid as a crosslinker. Carbohydr Polym 115:379–387

Saberi Riseh R et al (2024) Nano/micro-structural supramolecular biopolymers: innovative networks with the boundless potential in sustainable agriculture. Nano-Micro Lett 16(1):147

Sadik C, El Amrani IE, Albizane A (2014) Recent advances in silica-alumina refractory: a review. J Asian Ceram Soc 2(2):83–96

Samir S et al (2024) Introduction to polymer composites in aerospace. Curr Nat Sci Eng 1(4):329–339

Santo Pereira AE et al (2017) Chitosan nanoparticles as carrier systems for the plant growth hormone gibberellic acid. Colloids Surf B 150:141–152

Sasidharan S et al (2024) Innovative use of chitosan/ZnO NPs bio-nanocomposites for sustainable antimicrobial food packaging of poultry meat. Food Packag Shelf Life 43:101298

Sathishkumar T et al (2014) Glass fiber-reinforced polymer composites—a review. J Reinf Plast Compos 33(13):1258–1275

Seifi H et al (2020) A review on current trends in thermal analysis and hyphenated techniques in the investigation of physical, mechanical and chemical properties of nanomaterials. J Anal Appl Pyrol 149:104840

Sepulveda J et al (2020) Effect of functionalized silica nanoparticles on the mass transfer process in active PLA nanocomposite films obtained by supercritical impregnation for sustainable food packaging. J Supercritic Fluids 161:104844

Seremeta KP, Sosnik A (2023) Natural and semi-natural polymers. Biomaterials and biopolymers. Springer, Cham, pp 55–70

Sessini V et al (2016) Processing of edible films based on nanoreinforced gelatinized starch. Polym Degrad Stab 132:157–168

Shaghaleh H et al (2025) Soil flushing coupled with aminated-nanocellulose/MOF hydrogel nanocomposites adsorbents: a novel sustainable remediation strategy for Cr (VI)-contaminated agricultural soils. Sep Purif Technol 353:128440

Shahbazi F et al (2025) Losses in agricultural produce: a review of causes and solutions, with a specific focus on grain crops. J Stored Prod Res 111:102547

Shahzad A (2012) Hemp fiber and its composites—a review. J Compos Mater 46(8):973–986

Shankar S, Wang L-F, Rhim J-W (2017) Preparation and properties of carbohydrate-based composite films incorporated with CuO nanoparticles. Carbohydr Polym 169:264–271

Silva FVM (2023) Pasteurization of food and beverages by high pressure processing (HPP) at room temperature: inactivation of *Staphylococcus aureus*, *Escherichia coli*, *Listeria monocytogenes*, *Salmonella*, and other microbial pathogens. Appl Sci 13(2):1193

Silva IO et al (2019) Multifunctional chitosan/gold nanoparticles coatings for biomedical textiles. Nanomaterials 9(8):1064

Sinha Ray S et al (2002) New polylactide/layered silicate nanocomposites. 1. Preparation, characterization, and properties. Macromolecules 35(8):3104–3110

Somasundaram S, Kumaravel V (2019) Application of nanoparticles for self-cleaning surfaces. Springer, Cham, pp 471–498

Song Y et al (2013) Mechanical properties of poly (lactic acid)/hemp fiber composites prepared with a novel method. J Polym Environ 21:1117–1127

Soumya RS et al (2013) Preparation and characterization of selenium incorporated guar gum nanoparticle and its interaction with H9c2 cells. PLoS ONE 8(9):e74411

Srujana S, Bhagat D (2022) Chemical synthesis of chitosan (CS)–sodium alginate (ALG) nanoparticles. Nanotechnol Environ Eng 7(1):289–296

Stepanova M et al (2020) Composite biomaterials based on poly (L-lactic acid) and functionalized cellulose nanocrystals. J Renew Mater 8(4):383–395

Sumithra G et al (2023) Review on composite classification, manufacturing, and applications. Mater Today Proc. https://doi.org/10.1016/j.matpr.2023.04.637

Svagan AJ, Hedenqvist MS, Berglund L (2009) Reduced water vapour sorption in cellulose nanocomposites with starch matrix. Compos Sci Technol 69(3–4):500–506

Swaroop C, Shukla M (2018) Nano-magnesium oxide reinforced polylactic acid biofilms for food packaging applications. Int J Biol Macromol 113:729–736

Szafraniec J et al (2017) Chitosan-based nanocapsules of core-shell architecture. Polymer 62(7–8):509–515

Tang H et al (2021) Effects of defect and temperature on the mechanical performance of WS2: a multiscale analysis. J Phys Cham C 125(4):2680–2690

Tasca F, Antiochia R (2020) Biocide activity of green quercetin-mediated synthesized silver nanoparticles. Nanomaterials 10(5):909

Thakur M et al (2022) Introduction to biorenewable nanocomposite materials: methods of preparation, current developments, and future perspectives. In: Biorenewable nanocomposite materials, vol 2: Desalination and wastewater remediation. ACS Publications, Washington, DC, pp 1–24

Tian X et al (2017) Cationic surface modification of nanocrystalline cellulose as reinforcements for preparation of the chitosan-based nanocomposite films. Cellulose 24(1):163–174

Tomke PD, Rathod VK (2020) Facile fabrication of silver on magnetic nanocomposite (Fe_3O_4@ Chitosan–AgNP nanocomposite) for catalytic reduction of anthropogenic pollutant and agricultural pathogens. Int J Biol Macromol 149:989–999

Ulusoy U (2023) A review of particle shape effects on material properties for various engineering applications: from macro to nanoscale. Minerals 13(1):91

Vadivel M et al (2019) Bioactive constituents and bio-waste derived chitosan/xylan based biodegradable hybrid nanocomposite for sensitive detection of fish freshness. Food Packag Shelf Life 22:100384

Velasco-Parra JA, Ramón-Valencia BA, Mora-Espinosa WJ (2020) Composite materials reinforced with ceramic waste and matrix of unsaturated polyester for applications in the automotive industry. Dyna 87(212):251–258

Velásquez-Cock J et al (2014) Influence of the acid type in the production of chitosan films reinforced with bacterial nanocellulose. Int J Biol Macromol 69:208–213

Victor SP, Muthu J (2014) Polymer ceramic composite materials for orthopedic applications—relevance and need for mechanical match and bone regeneration. J Mechatron 2(1):1–10

Vidakis N et al (2022) Multi-functional 3D-printed vat photopolymerization biomedical-grade resin reinforced with binary nano inclusions: the effect of cellulose nanofibers and antimicrobial nanoparticle agents. Polymers 14(9):1903

Vidakis N et al (2023) Optimized PCL/CNF bio-nanocomposites for medical bio-plotted applications: rheological, structural, and thermomechanical aspects. Bioprinting 36:e00311

Vidor FF, Meyers T, Hilleringmann U (2016) Inverter circuits using ZnO nanoparticle based thin-film transistors for flexible electronic applications. Nanomaterials 6(9):154

Vu CHT, Won K (2014) Leaching-resistant carrageenan-based colorimetric oxygen indicator films for intelligent food packaging. J Agric Food Chem 62(29):7263–7267

Wakai M, Almenar E (2015) Effect of the presence of montmorillonite on the solubility of whey protein isolate films in food model systems with different compositions and pH. Food Hydrocolloids 43:612–621

Wang Y-C et al (2018) Chitosan and gold nanoparticles-based thermal history indicators and frozen indicators for perishable and temperature-sensitive products. Food Control 85:186–193

Wang S et al (2021) Comparative study on water vapour resistance of poly (lactic acid) films prepared by blending. Filling Surf Deposit 11(12):915

Wei T et al (2012) Tough bio-based elastomer nanocomposites with high performance for engineering applications. Adv Eng Mater 14(1–2):112–118

Wen P et al (2016) Fabrication of electrospun polylactic acid nanofilm incorporating cinnamon essential oil/β-cyclodextrin inclusion complex for antimicrobial packaging. Food Chem 196:996–1004

Werner CFW (2021) Layer-by-layer assembly of nanoclays for improved barrier properties of biobased polymers. Norwegian University of Life Sciences, Ås
Wu J-L et al (2011) Surface plasmonic effects of metallic nanoparticles on the performance of polymer bulk heterojunction solar cells. ACS Nano 5(2):959–967
Wu Y, Luo Y, Wang Q (2012) Antioxidant and antimicrobial properties of essential oils encapsulated in zein nanoparticles prepared by liquid–liquid dispersion method. LWT-Food Sci Technol 48(2):283–290
Wu J et al (2019) Enhanced physico-mechanical, barrier and antifungal properties of soy protein isolate film by incorporating both plant-sourced cinnamaldehyde and facile synthesized zinc oxide nanosheets. Colloids Surf, B 180:31–38
Wu Q et al (2020) Mechanical properties of nanomaterials: a review. Nanotechnol Rev 9(1):259–273
Wu H et al (2022) Polymer-/ceramic-based dielectric composites for energy storage and conversion. Energy Environ Mater 5(2):486–514
Xu Y et al (2023) Clove essential oil loaded chitosan nanocapsules on quality and shelf-life of blueberries. Int J Biol Macromol 249:126091
Yadav S et al (2020) Preparation, physicochemical and biological evaluation of quercetin based chitosan-gelatin film for food packaging. Carbohydr Polym 227:115348
Yang W et al (2016) Synergic effect of cellulose and lignin nanostructures in PLA based systems for food antibacterial packaging. Eur Polym J 79:1–12
Yang Y, Deng H, Fu Q (2020) Recent progress on PEDOT: PSS based polymer blends and composites for flexible electronics and thermoelectric devices. Mater Chem Front 4(11):3130–3152
Yang Q et al (2024) Plasmonic nanomaterial-enhanced fluorescence and Raman sensors: multifunctional platforms and applications. Coord Chem Rev 507:215768
Yap CK, Al-Mutairi KA (2024) A conceptual model relationship between Industry 4.0—food-agriculture nexus and agroecosystem: a literature review and knowledge gaps. Foods 13(1):150
Yassin MA et al (2019) Green synthesis of cellulose nanofibers using immobilized cellulase. Carbohydr Polym 205:255–260
Yoshino K et al (1999) Electrical and optical properties of conducting polymer-fullerene and conducting polymer-carbon nanotube composites. Fullerene Sci Technol 7(4):695–711
Yu J et al (2009) Preparation and characterization of glycerol plasticized-pea starch/ZnO–carboxymethylcellulose sodium nanocomposites. Biores Technol 100(11):2832–2841
Yusoh K, Kumaran SV, Ismail FS (2018) Surface modification of nanoclay for the synthesis of polycaprolactone (PCL)–clay nanocomposite. In: MATEC Web of Conferences. EDP Sciences
Zarski A, Bajer K, Kapuśniak J (2021) Review of the most important methods of improving the processing properties of starch toward non-food applications. Polymers 13(5):832
Zewdu D, Krishnan M, Raj PN (2024) Nanoclay for climate change adaptation and mitigation: a critical review. In: Nanoclay-based sustainable materials, pp 151–166
Zhang Y et al (2008) Physicochemical characterization and antioxidant activity of quercetin-loaded chitosan nanoparticles. J Appl Polym Sci 107(2):891–897
Zhang S et al (2019) Recent advances in nano-materials for packaging of electronic devices. J Mater Sci Mater Electron 30(15):13855–13868
Zhang Y-P et al (2021) Preparation and characterization of bio-nanocomposites film of chitosan and montmorillonite incorporated with ginger essential oil and its application in chilled beef preservation. Antibiotics 10(7):796
Zhang Z-Q et al (2023) Progress in polymer-ceramic hybrid antifouling coatings. Chin J Polym Sci 41(7):995–1001
Zhang J et al (2024) Wettability and bactericidal properties of bioinspired ZnO nanopillar surfaces. Langmuir 40(14):7353–7363
Zhao R, Torley P, Halley PJ (2008) Emerging biodegradable materials: starch-and protein-based bio-nanocomposites. J Mater Sci 43(9):3058–3071
Zhao H et al (2017) High-efficiency heating characteristics of ferrite-filled asphalt-based composites under microwave irradiation. J Mater Civil Eng 29(6):04017007

Zhu J et al (2004) Curing and mechanical characterization of a soy-based epoxy resin system. J Appl Polym Sci 91(6):3513–3518

Zubair M et al (2020) Chitosan/chitin-based composites for food packaging applications. In: Handbook of chitin and chitosan. Elsevier, Amsterdam, pp 641–670

Zungu B et al (2023) Zn nutrients-loaded chitosan nanocomposites and their efficacy as nanopriming agents for maize (*Zea mays*) seeds. Front Chem 11:1243884

Ahmed Fathy Mostafa Ghanem is a Researcher at the Packaging Materials Department, National Research Centre (NRC), Egypt. He holds a PhD in Chemistry from Ain Shams University (2021), with research experience in France, Poland, and the USA. He is currently a postdoctoral scholar at the Korea Institute of Energy Technology (KENTECH). Dr. Ghanem has authored over 30 peer-reviewed articles, holds two patents, and has contributed to 15 research projects. His work focuses on polymer nanocomposites for applications in food packaging, energy, and water treatment. He collaborates internationally and serves as a reviewer for several leading scientific journals.

Chapter 10
Bionanocomposite Materials in Health and Pharmaceutical Sciences

Promila Sharma, Anita Thakur, Vijay Singh, Manpreet Kaur, Rituka Rani, and Ashok Kumar Malik

Abstract Biomedical nanocomposites, an emerging class of ingenious materials, encompass interdisciplinary fields including chemistry, biology, physics, material research, and manufacturing and have brought about a new era in pharmaceutical sciences and the healthcare industry. They are versatile for several biomedical applications, including antimicrobial agents, drug carriers, tissue regeneration, wound care, dentistry, bioimaging, bone filler, cancer treatment, and diagnostics, owing to their unique features, which include high surface area, biocompatibility, biodegradability, and high mechanical strength. Besides being biodegradable, bionanocomposites improve the effectiveness of the medications they contain by enabling better bioavailability, controlled release, and targeted drug delivery. As a result, it reduces the amount of medication needed, further minimizing the environmental footprint. The extensively utilized bionanocomposites and their processing methods in medical applications are thoroughly examined in this chapter. Chitosan nanocomposites, silica nanoparticles in hydrogels, graphene oxide-based composites, collagen nanocomposites, polymer nanocomposites with hydroxyapatite, starch nanocomposites, liposomes with nanoparticles, and other recently researched bionanocomposites are introduced and briefly explained in this chapter. Their significant functions are highlighted for their versatility and potential for advancement in biomedical technologies.

P. Sharma (✉) · A. K. Malik
Department of Chemistry, Punjabi University, Patiala, Punjab, India
e-mail: promilaattri07@gmail.com

P. Sharma · A. Thakur
Department of Chemistry, RIMT University, Mandigobindgarh, Punjab, India

V. Singh
Department of Botany, Mata Gujri College, Fatehgarh Sahib, Punjab, India

M. Kaur
Department of Punjabi, Govt. Ranbir College, Sangrur, Punjab, India

R. Rani
Department of English, GSSS Talwan, Talwan, Punjab, India

S. C. George and B. Tawiah (eds.), *Advanced Bionanocomposite Materials*, Advanced Structured Materials 241, https://doi.org/10.1007/978-981-96-8225-6_10

Keywords Biopolymers · Eco-friendly · Bionanocomposites · Nanoreinforcements

Highlights

1. Bionanocomposite composition explores the incorporation of naturally occurring biopolymers and nanoscale reinforcements.
2. These substances offer enhanced rigidity, temperature resistance, and biocompatibility, making them ideal for therapeutic delivery, tissue engineering, and imaging applications.
3. Considers potential challenges and the need for additional research to ensure compliance for biological ecosystems.
4. These substances may be easily tailored to fulfil specific needs, facilitating customized treatment and targeted therapy.

10.1 Introduction

Polymers have outperformed other conventional materials in terms of adaptability, resilience, and efficiency. The aforementioned types of biopolymers, referred to as green composites, biocomposites, bionanocomposites, or nanocomposites, demonstrate more effective humidity and gas barrier capabilities, stiffness, combustibility, thermal properties, and optical properties, which enhance shelf life and reduce environmental impact (Akter et al. 2022). Today, a number of industries, including materials science and health care, as well as the transportation, textile, construction, food, biomedical, and manufacturing sectors, are investigating the use of bionanocomposites. The introduction of bionanocomposites, a class of materials made a distinction by their intricate engineered structures, heralds an era of innovation in the search for revolutionary solutions for health care and pharmaceuticals (Aminnezhad et al. 2024). They have attracted a lot of captivations as a prospective candidate for application in biomedical as biomaterials because of their peculiar characteristics (Aravind et al. 2022) as seen in Fig. 10.1.

Biomedical nanocomposites, a growing category of biological components, have opened up new possibilities in the healthcare field. Adding these products tends to enhance the traits that these components add strength to things that could not endure independently. The biopolymer that carries the nanoparticles can improve the stiffness of the composite whilst also improving its biological properties and vice versa. This expands the composite's options and allows it to be used more frequently. Bionanocomposites and nanoparticles enable the medicine's eco-compatibility through biodegradability, resulting in ecological sustainability and medical treatment leading to better improvements.

Subsequent breakthroughs have been achieved in the formation of bionanocomposites, which integrate nano-sized fillers and have superior characteristics to native biopolymers. Bionanocomposites consist of sophisticated biopolymer composites

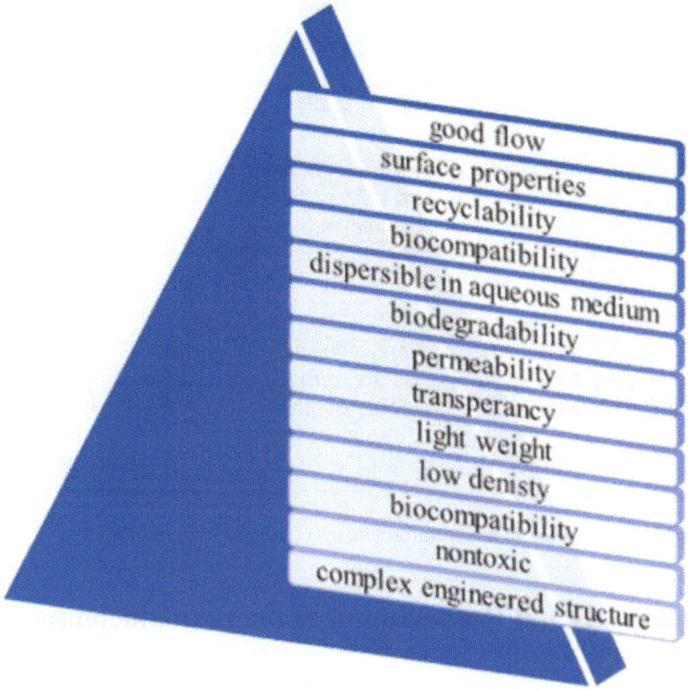

Fig. 10.1 Unique properties of bionanocomposites

that are reinforced with nano-sized variables, offering distinct benefits for environmentally friendly components in the pharmaceutical trade (Singh and Shagolsem 2021).

In bionanocomposites, the continuous phase is often a biopolymer framework, with the dispersed phase incorporating possibly inorganic or organic nanoreinforcements. Several of the most frequently implemented nanoreinforcements are listed in Fig. 10.2.

Biocomposites often combine biologically active substances (e.g. DNA, amino acids, and lipids) with nanostructures (e.g. nanoparticles, nanofibres, and nanotubes) to form composites with distinct mechanical, physiological, and chemical properties. Their potential in health care is tremendous, including drug delivery, tissue engineering, diagnostics, wound healing, burn dressing, bone fillers, antibacterial agents, fungal infection treatment, and more (Singh et al. 2021).

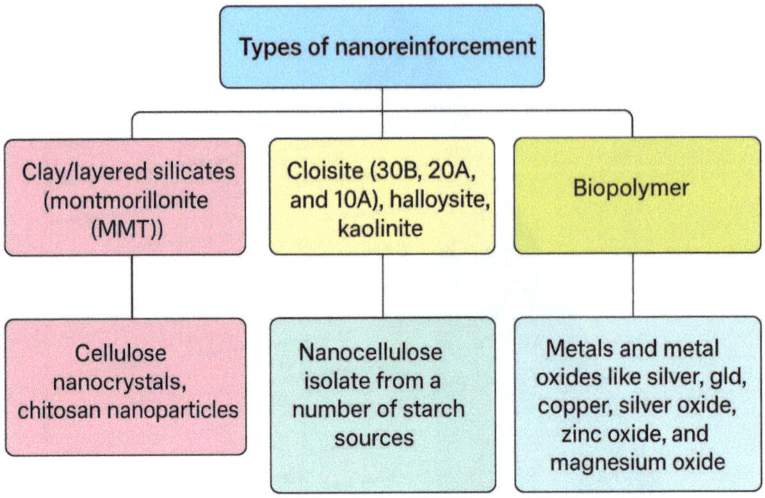

Fig. 10.2 Inorganic or organic nanoreinforcements combined with biopolymer

10.2 Synthesis and Properties of Some Bionanocomposites

Bionanocomposites are synthesized through the interaction of biodegradable or biopolymers with nanoparticles to improve their tensile, thermodynamics, and barrier capabilities. Synthesis requirements frequently incorporate the selection of biopolymers matrix (chitosan, starch, cellulose, polylactic acid (PLA), polyhydroxyalkanoates (PHA), gelatin, alginate, etc.,) and nanomaterials having high surface area, dispersion stability, and compatibility with the polymer matrix (graphene and carbon nanotubes (CNTs), silver (Ag), gold (Au), zinc oxide (ZnO), montmorillonite and halloysite nanotubes, hydroxyapatite (HAp), silica nanoparticles) (Zia et al. 2020; Idumah 2021). Some common conventional procedure approaches for designing bionanocomposites as biocatalysts are shown in Fig. 10.3.

Bionanocomposites properties are classified into various categories, as shown in Fig. 10.4.

Bionanocomposites showing unique mechanical functions have generated a great deal of interest in imitating bone structure. The incorporation of nanofillers improves surface roughness and porosity, bringing it closer to the bone itself and facilitating osteoblast spreading and bone rebuilding. Other examples of inorganic ceramic nanofillers that are native to bone are tricalcium phosphate, bioactive glass, and glass ceramic (apatite–wollastonite).

Organic bionanomaterials in the biomedical area have received a lot of attention. Silk fibroin, chitosan, and biodegradable polymers like poly(lactic–glycolic) acid (PLGA) are examples of organic nanomaterials. Organic bionanomaterials in the therapeutic area have received a lot of attention. Silk fibroin, chitosan, and biodegradable polymers like poly(lacticco-glycolic) acid (PLGA) are examples of organic

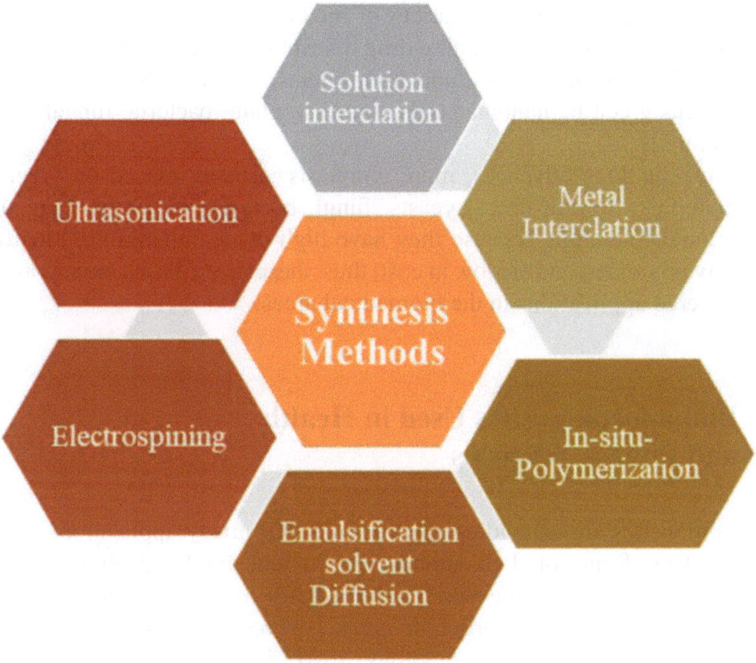

Fig. 10.3 Synthesis methods of bionanocomposites

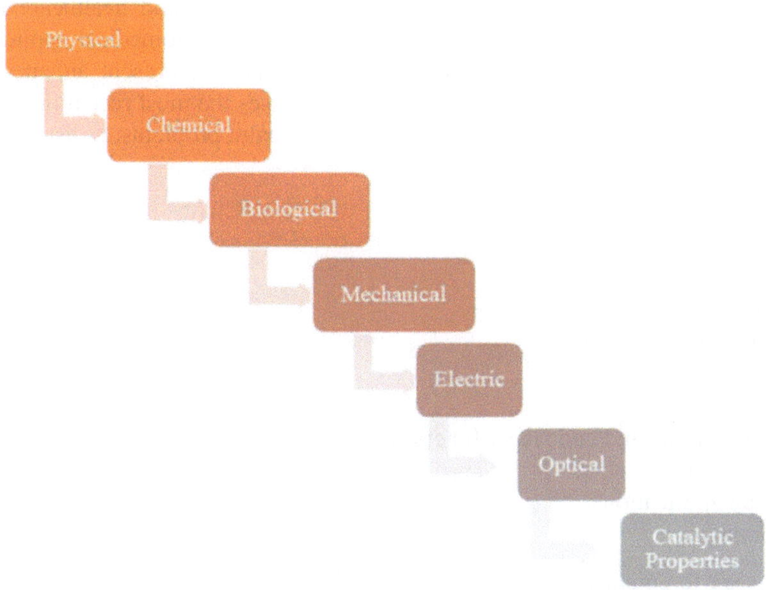

Fig. 10.4 Properties of bionanocomposites

nanomaterials. It demonstrates a variety of medicinal uses, including regenerative medications, for effectively managing drug distribution in cells. (Singh et al. 2021).

Organic bionanomaterials include lipoproteins, DNA or RNA, peptides, and so on, whereas green bionanomaterials comprise plant, bacteria, fungal, and viral bionanomaterials.

As nanotechnology advances, more emphasis is being placed on green-based bionanomaterials. Several crops, yeasts, fungi, bacteria, and viruses are used to produce bionanomaterials because they have high cytocompatibility, low toxicity, eco-friendly processes, and are low in cost; thus, these green-based bionanomaterials have discovered applications in the biomedical areas.

10.3 Bionanocomposites Used in Health and Pharmaceuticals

Bionanocomposites are expanding rapidly field with many applications in various fields nowadays. Some of their applications include biodegradability to manage toxic ecological problems, the design of specialty food packaging, and biomedical speciality applications. Particular areas covered correspond to applications such as defence, information technology, EMI shielding, insulation, and automobiles as green materials, with bionanocomposites based on biopolymer and nanocomposite selected (Harussani et al. 2022). Their main advantage, being renewable, has led to their growing use in the pharmaceutical and healthcare industries in recent years. Bionanocomposites demonstrate the biomedical applications of biopolymer-based nanocomposites. Bionanocomposites present an interesting opportunity in the medical field. This section discusses the most recent discoveries in bionanocomposites for the treatment of several health problems, as listed below:

- Drug delivery.
- Scaffolds.
- Tissue engineering.
- Wound healing.
- Bone.
- Enzyme immobilization.
- Medical implant.
- Haemodialysis.
- Antimicrobial and anticancer functionalities.
- Anti-anaemia.
- Dental applications.
- Bioimaging.

We seek to emphasise some of the most significant recent breakthroughs made by researchers in the field of bionanocomposites for medical applications. The

Table 10.1 Biomedical applications of biopolymer based on alginate

Biopolymer	Nanocomposite	Synthesis method	Biomedical application
Alginate	Zinc oxide nanoparticles	Coprecipitation, hydrothermal synthesis	UV-blocking wound dressings, antibacterial applications
	Chitosan	Ionic gelation, coacervation	Wound dressings, antibacterial applications
	Magnetic nanoparticles	Coprecipitation, emulsion methods	Magnetic resonance imaging, targeted drug delivery
	Silver nanoparticles	Ionotropic gelation, coprecipitation	Antibacterial Wound dressings, controlled drug release
	Gold nanorods	Seed-mediated growth, surface functionalisation	Photothermal therapy, drug delivery

bionanocomposites evaluated for particular uses in the biomedical field are discussed as follows:

10.3.1 Alginate-Based Bionanocomposites

In this context, the objective of this section was to focus on alginate-based films that include varying amounts of the multifunctional ZnO-rGO/sepiolite composite, chitosan, iron, silver, and gold nanoparticles for use in the medical field. Over the past years, antimicrobial and antioxidant properties have been extensively researched to develop active materials for biomedical applications (Table 10.1) (Alves et al. 2021). Nanoalginates are now the most advanced type of alginate polymer. Surprisingly, nanoalginate particles or fibres are not viable for a wide range of applications unless combined with another polymer or binary component. Recent developments in the realm of nanofibres and nanoparticle forms of alginate known to alginate-based bionanocomposites are found to show many biological treatments (Zia et al. 2020).

10.3.2 Chitosan-Based Bionanocomposites

Given the prospect of increasing biomedical expectations throughout a broad spectrum of disorders, there is an urgent need for the development of sophisticated

Table 10.2 Biomedical applications of biopolymer based on chitosan

Biopolymer	Nanocomposite	Synthesis method	Biomedical application
Chitosan	Poloxamer	Solvent casting, blending	Drug delivery, wound healing, tissue engineering
	Nanohydroxyapatite	Electrospinning, precipitation	Bone tissue engineering, dental applications
	Graphene oxide	Electrostatic assembly, covalent bonding	Neural tissue engineering, drug delivery
	Carbon nanotubes	Electrostatic assembly, covalent bonding	Antibacterial coatings, biosensors, drug delivery
	Silver nanoparticles	Ionic gelation, coprecipitation	Wound dressings, antimicrobial applications

biocompatible materials with enhanced biological functionality. Chitosan, a naturally occurring cationic polyelectrolyte copolymer material, stands out for its exceptional physiological capabilities, making it a viable base substance for the development of multifunctional biomedical materials (Table 10.2). Massive emphasis has been directed to the development of high-performance composites, which are obtained via strategic alteration of chitosan's structure or its derivative, as well as the incorporation of other biopolymers (Desai et al. 2023). Aside from its primary use in horticulture and agriculture for plant defence and crop yield enhancement, recent studies have shown a number of new applications in the fields of bioimaging sensors, targeted drug delivery, advanced biomedical devices, antimicrobial properties, stimuli-responsiveness, and water solubility. Nanoparticles embedded in chitosan matrices have recently been found to have superior biological activity, as cells proliferate/differentiate more rapidly when they connect to nanocomposites instead of bulk or micron-sized substrates/scaffolds. This section discusses some chitosan-based nanocomposites utilised in surgical wound dressings, drug delivery, and biological sensing (Ahmad et al. 2020).

10.3.3 Cellulose and Starch-Based Bionanocomposites

The outstanding biological compatibility, complete degradation without detrimental contaminants, inexpensive and widespread availability, and sustainability of cellulose and starch-based bionanocomposites make them ideal for a wide range of applications in biomedicine (Table 10.3) (Namazi and Mosadegh 2011; Orue et al. 2016).

Table 10.3 Biomedical applications of biopolymers based on cellulose and starch

Biopolymer	Nanocomposite	Synthesis method	Biomedical application
Cellulose	Nanocellulose	Acid hydrolysis, electrospin	Drug delivery, wound healing, tissue engineering
	Gold nanoparticles	In situ synthesis, chemical reduction	Photothermal therapy, biosensing
	Chitosan, silver nanoparticles	Coprecipitation, blending	Antibacterial, wound dressings
	Magnetic nanoparticles	Coprecipitation, emulsion methods	Magnetic resonance imaging, targeted drug delivery
	Silver nanoparticles	In situ reduction, impregnation	Antimicrobial wound dressings. Infection control
	Hydroxyapatite	Ultrasonic irradiation	Bone tissue
	PVA–CNC hydrogels	–	Ophthalmic biomaterial
Starch	Clay nanocomposites	Melt blending, solution casting	Controlled drug release, food packaging
	Silver nanoparticles	Microwave-assisted synthesis	Antimicrobial wound healing
	Zinc oxide nanoparticles	Hydrothermal synthesis, precipitation	UV-blocking wound dressings, antibacterial applications
	Starch, mesoporous silica nanoparticles	Sol–gel, coprecipitation	Controlled drug release, bioimaging
	Gold nanoparticles	Citrate reduction, surface modification	Photothermal therapy, drug delivery

10.3.4 Collagen and Gelatin-Based Bionanocomposites

In the pursuit of bionanocomposites that are both versatile and compatible with biological organs, there has been a lot of demand for collagen-based nanomaterials for reconstructing and replacing bodily tissues such as tendons, skin, cardiovascular transplants, cardiac valves, jaws, molars, and bones (Mahmood et al. 2022; Biswas 2022; Radhakrishnan 2020). This section reviews collagen-based composites mixed with bioceramic components such as hydroxyapatite, β-tricalcium phosphate (β-TCP), and calcium silicate to boost the regeneration and repair of tissues. Carbon-based and graphene-based biocomposites can promote brain growth in neural stem cells and antimicrobial activity devices for health care (Siddiqui et al. 2024).

Gelatin's adaptive functionalization allows straightforward and efficient binding as well as incorporation with other materials (e.g. metal nanoparticles, carbon minerals, and polymeric materials with functional properties) to synthesize hybrid

Table 10.4 Biomedical applications of biopolymer based on collagen and gelatin

Biopolymer	Nanocomposite	Synthesis method	Biomedical application
Collagen	Hydroxyapatite	Self-assembly, freeze-drying	Orthopaedic implants, bone tissue engineering
	Chitosan, silver nanoparticles	Coprecipitation, blending	Antibacterial wound dressings, infection control
	Gold nanorods	Seed-mediated growth, surface functionalisation	Photothermal therapy, biosensing
	Silver nanoparticles	In situ reduction, coprecipitation	Wound dressings, antibacterial coatings
	Magnetic nanoparticles	Coprecipitation, emulsion methods	Wound dressings, antibacterial coatings
Gelatin	Magnetic nanoparticles	Coprecipitation, emulsion methods	Hyperthermia cancer treatment, targeted drug delivery
	Graphene oxide	Solvent casting, crosslinking,	Drug delivery, tissue engineering, wound healing
	Gold nanoparticles	Citrate reduction, surface modification	Photothermal therapy, drug delivery
	Titanium dioxide nanoparticles	Coprecipitation, hydrothermal synthesis	UV-blocking wound dressings, drug delivery
	Silver nanowires	Electrospinning, co-electrodeposition	Antibacterial coatings, wound healing

materials with improved thermomechanical, physiochemical, and biological properties. Gelatin-based hydrogels have distinct functional properties that make them suitable for a variety of environmental, industrial, and medicinal applications (Table 10.4) (Parashar et al. 2021; Bera et al. 2021; Mushtaq et al. 2022).

10.3.5 Pullulan, Polyhydroxyalkanoate (PHA), and Polylactic Acid (PLA)-Based Bionanocomposites

Pullulan is an anaerobically developed, easily soluble in water glucan gum synthesised from carbohydrates that come from the fungal organism *Aureobasidium pullulans*. It offers various biological advantages, including non-toxicity, non-mutagenicity, biodegradability, non-carcinogenicity, and haemocompatibility. It offers various pharmacological positive aspects, including non-toxicity, non-mutagenicity, biodegradability, non-carcinogenicity, and haemocompatibility (Manivannan et al. 2023; Coltelli et al. 2020; Roy 2024).

Polyhydroxyalkanoate, a biodegradable polymer derived from microorganisms and plants, is widely employed in biomedical applications and devices such as sutures, heart valves, bone scaffolds, and drug administration of substances of medicinal interest. PHA bionanocomposites are physiologically stable and biodegradable and can be used in a variety of medical applications, including drug delivery, cancer detection, bone replacement, and brain surgery (Pulingam et al. 2022; Fernandez-Bunster and Pavez 2022).

PLA-based biological composites in a variety of implantation and medical devices can replace and/or regenerate deteriorated or traumatised cells or organs, hence promoting recovery and fixing defects and increasing patients' quality of life (Table 10.5) (Kuram 2025; Kiran 2021).

Table 10.5 Biomedical applications of biopolymer based on pullulan, PHA, and PLA

Biopolymer	Nanocomposite	Synthesis method	Biomedical application
Pullulan	Gold nanoparticles	Coprecipitation, electrostatic assembly	Cancer therapy, drug delivery
	Silver nanocluster	Green synthesis, reduction methods	Antibacterial coatings, drug delivery
	Carbon nanotubes	Coprecipitation, thermal treatment	Drug delivery, biosensing
	Nanoparticles	Coprecipitation, emulsion methods	Magnetic resonance imaging, targeted drug delivery
	Zinc oxide nanoparticles	Hydrothermal synthesis, precipitation	UV-blocking wound dressings, antibacterial applications
PHA	Montmorillonite nanocomposites	Melt blending, solvent casting	Wound dressings, tissue engineering, controlled drug release
	Graphene oxide	Solution blending, Freeze-drying	Drug delivery, tissue engineering, wound healing
	Chitosan–silver nanoparticles	Coprecipitation, blending	Antibacterial wound dressings, infection control
	Silver nanoparticles	In situ reduction, impregnation	Antimicrobial, medical implants
	Gold nanorods, surface functionalization	Seed-mediated growth	Photothermal therapy, bioimaging
PLA	Calcium phosphate	Solvent casting, electrospinning	Bone tissue engineering, drug delivery
	Gold nanoparticles	Citrate reduction, surface modification	Photothermal therapy, drug delivery
	Silver nanofibres	Electrospinning, electrodeposition	Antibacterial coatings, wound healing
	Magnetic nanoparticles	Coprecipitation, emulsion methods	Magnetic resonance imaging, targeted drug delivery
	Zinc oxide nanoparticles	Hydrothermal synthesis, precipitation	UV-blocking wound dressings, antibacterial applications

PHAs, PLA, and their composites have a significant prospect of use in agriculture, biomedicine, and packaging materials. Over time, the prudent use of diverse low-cost and locally plentiful biomass by-products has prepared the way for the long-term manufacturing of these biopolymers (Boey et al. 2021).

10.4 Legal and Ethical Considerations in the Application of Bionanocomposites

Polymeric nanocomposites play a transformational role in biomedical innovation. Polymeric nanocomposites exhibit different properties as a result of the integration of nanoscale fillers into polymer matrices, with applications including drug delivery systems, tissue engineering, diagnostic imaging, implantable devices, biosensors, and therapeutics. Nonetheless, problems such as biocompatibility, toxicity issues, and stringent regulatory scrutiny are highlighted as critical factors in clinical translation. Polymeric nanocomposites provide a game-changing synergy in tissue engineering by improving mechanical strength and biocompatibility for the creation of biomimetic scaffolds. Despite long-term biocompatibility and regulatory hurdles, polymeric composites are transformational tools in the field of regenerative medicine. Polymeric nanocomposites have the potential to revolutionize diagnostic imaging, resulting in improved medical diagnostics and early illness identification. The integration of biomedicine and AI and a trend towards personalised medicine have been suggested as future trends that will transform the field of health care and pharmaceuticals.

10.5 Future Prospectus

Within the months and years to come, the biomedical field will have an enormous potential dependency on bionanocomposites. Bionanocomposites have a promising future, owing to their ability to improve therapeutic outcomes, foster sustainable healthcare practices, and accelerate advances in diagnostic and treatment techniques. As we strive for greener and more effective healthcare surroundings, bionanocomposites serve as an inspiration of inventiveness, providing a road to transformative biomedical solutions that prioritize efficacy, sustainability, and patient-centred care.

Although there are now insufficient methods of manufacturing and the expenses for raw materials are prohibitively costly, it is projected that this industry will have a large scope in the future. The cost of production must be reduced. Furthermore, their biodegradability, antibacterial characteristics, and intelligent functionality make them perfect for next-generation medical solutions. To fully realise their promise, problems such as safety, biocompatibility, and regulation must be overcome.

The most significant applications for these bionanocomposites include construction, transportation, and health care. A significant increase can be expected in the near future because they are an environmentally favourable technique that could replace waste-generating corporations. Bionanocomposites are expected to play a larger role in personalised and precision medicine as research progresses.

10.6 Conclusion

An array of bionanocomposites show promise for industrial uses in a variety of sectors, including airliners, medicine, building, and communications. Bionanocomposite materials are a type of material that combines biological and nanomaterials to form innovative nanocomposites with distinctive characteristics. Bionanocomposites are transforming the medical profession by providing sophisticated, multifunctional, and extremely effective treatments for healthcare problems. Their uses in biomedicine, implants, medication delivery, and diagnostics make them critical to the future of medicine. Bionanocomposites are important in pharmaceutical and healthcare applications because of their biocompatibility, controlled drug release, antibacterial qualities, and regeneration capacities. These composites have high mechanical strength, biocompatibility, and bioactivity because they combine biopolymers (alginate, cellulose, starch, chitosan, collagen, gelatin, pullulan, PHA, and PLA) with nanoscale fillers from various categories, including silicate-based materials, nanoparticles from inorganic materials, carbon-based nanomaterials, and metal/metal oxides. These qualities make them ideal for a variety of applications, including matrix creation, medication administration, tissue engineering, bone filler, dental applications, imaging, biosensors, and wound dressing.

References

Ahmad M et al (2020) Chitin and chitosan-based bionanocomposites. Bionanocomposites. Elsevier, pp 145–156

Akter M, Uddin MH, Tania IS (2022) Biocomposites based on natural fibers and polymers: a review on properties and potential applications. J Reinf Plast Compos 41(17–18):705–742

Alves Z et al (2021) Design of alginate-based bionanocomposites with electrical conductivity for active food packaging. Int J Mol Sci 22(18):9943

Aminnezhad S et al (2024) Novel therapeutic applications of nanoparticles and bio-nanocomposites in modern biomedicine. Biocomposite nanomaterials and their applications. Springer, pp 181–193

Aravind D et al (2022) A compendious review on fatigue properties of the bio-nanocomposites. Polym Compos 43(10):6803–6816

Bera H, Hossain CM, Saha S (2021) Biopolymer-based nanomaterials in drug delivery and biomedical applications. Academic Press, New York, NY

Biswas MC et al (2022) Recent advancement of biopolymers and their potential biomedical applications. J Polym Environ 30(1):51–74

Boey JY et al (2021) A review of the applications and biodegradation of polyhydroxyalkanoates and poly(lactic acid) and its composites. Polymers (Basel) 13(10):1544
Coltelli M-B et al (2020) Pullulan for advanced sustainable body-and skin-contact applications. J Funct Biomater 11(1):20
Desai N et al (2023) Chitosan: a potential biopolymer in drug delivery and biomedical applications. Pharmaceutics 15(4):1313
Fernandez-Bunster G, Pavez P (2022) Novel production methods of polyhydroxyalkanoates and their innovative uses in biomedicine and industry. Molecules 27(23):8351
Harussani M et al (2022) Recent applications of carbon-based composites in defence industry: a review. Defence Technol 18(8):1281–1300
Idumah CI (2021) Novel trends in conductive polymeric nanocomposites and bionanocomposites. Synth Met 273:116674
Kiran I et al (2021) Graphene functionalized PLA nanocomposites and their biomedical applications. In: Graphene based biopolymer nanocomposites. Springer, Singapore, pp 83–105
Kuram E (2025) PLA-based biocomposites and bionanocomposites in the biomedical and healthcare applications. Natural fiber-reinforced PLA composites. Elsevier, pp 211–236
Mahmood A et al (2022) Recent progress in biopolymer-based hydrogel materials for biomedical applications. Int J Mol Sci 23(3):1415
Manivannan M et al (2023) Review on applications of Pullulan in bone tissue engineering: blends and composites with natural and synthetic polymers. Polym Polym Compos 31:09673911231192810
Mushtaq F et al (2022) Preparation, properties, and applications of gelatin-based hydrogels (GHs) in the environmental, technological, and biomedical sectors. Int J Biol Macromol 218:601–633
Namazi H, Mosadegh M (2011) Bio-nanocomposites based on naturally occurring common polysaccharides chitosan, cellulose and starch with their biomedical applications. In: Recent developments in bio-nanocomposites for biomedical applications, pp 379–397
Orue A et al (2016) Bionanocomposites based on thermoplastic starch and cellulose nanofibers. J Thermoplast Compos Mater 29(6):817–832
Parashar P et al (2021) Gelatin-based nanomaterials in drug delivery and biomedical applications. Biopolymer-based nanomaterials in drug delivery and biomedical applications. Elsevier, pp 407–426
Pulingam T et al (2022) Biomedical applications of polyhydroxyalkanoate in tissue engineering. Polymers 14(11):2141
Radhakrishnan S et al (2020) Collagen based biomaterials for tissue engineering applications: a review. In: Processes and phenomena on the boundary between biogenic and abiogenic nature. Springer, Cham, pp 3–22
Roy S et al (2024) Biomedical applications of Pullulan. Biopolymers for biomedical applications. Wiley, Hoboken, NJ, pp 253–285
Siddiqui VU et al (2024) Collagen-based biocomposites. Polymer composites derived from animal sources. Elsevier, pp 149–169
Singh TV, Shagolsem LS (2021) Biopolymer based nano-structured materials and their applications. In: Nanostructured materials and their applications, pp 337–366
Singh KR, Nayak V, Singh RP (2021) Introduction to bionanomaterials: an overview. In: Bionanomaterials: fundamentals and biomedical applications, pp 1-1–1-21
Singh SV et al (2021) Bionanocomposites and their multifunctional applications. In: Bionanomaterials: fundamentals and biomedical applications. IOP Publishing, Bristol, pp 4-1–4-46
Zia KM et al (2020) Bionanocomposites: green synthesis and applications. Elsevier, Amsterdam

Promila Sharma is presently working as an Assistant Professor at the Department of Chemistry, under the School of Basic, Applied, and Biosciences at RIMT University, Mandigobindgargh-Punjab. She has a total academic experience of 4 years of teaching experience. She obtained her Sc Degree with a specialization in inorganic chemistry from Punjabi University, Patiala. She qualified for NET in chemical sciences. She also has five a of research experience 5 years with ten publications in peer-reviewed journals and two papers in International conferences and has one book chapter to her credit. Her research interest is synthesizing and characterizing heterocyclic and bis-heterocyclic compounds. In medicinal chemistry, heterocyclic compounds play an essential role by demonstrating different biological activities. Further emphasis is on their application as sensors to develop new methodologies and remediation for detecting metals and explosive materials. Computational studies using DFT are also her expertise to shed light on the design and development of synthesized hybrids and to correlate the photophysical properties of synthesized probes with their experimental characteristics.

Anita Thakur is working as an Assistant Professor at the Department of Chemistry, under the School of Basic, Applied, and Biosciences at RIMT University, Punjab. She has a total academic experience of 4 years of teaching experience. She obtained her Sc Degree with a specialization in physical chemistry from Punjabi University, Patiala. After that, she joined M. Phil in 2012 and Ph. D. (2015–2019) from the same institute. She also has five a of research experience 5 years with seven publications in peer-reviewed journals and one paper at an International Conference and has one book to her credit. Dr. Anita has made significant contributions to the field of surface chemistry, nanotechnology, and water and wastewater treatment through her research publications. Her studies have not only expanded scientific understanding but also hold promise for the development of potential therapeutic interventions.

Vijay Singh is currently working as an Assistant Professor in, Department of Botany, Mata Gujri College, Fatehgarh Sahib, Punjab. He obtained his Sc Degree from Himachal Pradesh University in 2008. He then joined M. Phil in 2010 and Ph. D. (2012–2017) at Punjabi University, Patiala. He has 52 research publications, including 11 book chapters and one conference paper to his credit. His research interest encompasses plant cytogenetics, molecular biology, and phytochemistry, particularly focusing on the behavior of chromosome numbers in morphology, biochemistry, and physiology of plants. His work holds promise for advancements in future technologies and has been recognized both locally and at international scientific forums.

Manpreet Kaur is working as an assistant professor at the department of Punjabi at Govt. Ranbir college, Sangrur. She has a total academic experience of 2 years of teaching at Yadavindra Public School, Patiala, and has recently joined at Govt. Ranbir college, Sangrur. She has obtained her degree of M.Ed degree from Desh Bhagat University, Mandi Gobindgarh. After that, she acquired her degree of Master's degree in Economics and Punjabi from Punjabi University Patiala. **Her** wealth of knowledge enriches both academic and research endeavors at the college.

Rituka Rani is working as an English mistress in the govt. School of Punjab. She is a gold medalist during her post graduation and B.ed Degree. She is exploring her research with collaborations where she actively contributes to the advancement of scientific understanding. With a passion for diverse aspects, she plays a crucial role in shaping the intellectual landscape within the department, fostering a culture of excellence and innovation.

Ashok Kumar Malik is working as a professor in the chemistry department of Punjabi University, Patiala. He got the title of Humboldtian Top 2% of World-Ranking-Scientists by Stanford-Elsevier in 2023. He is working as coordinator of the GIAN Program. His research area is very wide, with 300 research articles. He guided about 30 PhD students.

Chapter 11
Conductive Bionanocomposites in Electrical and Electronic Applications: Toward Green Electronics

Sujata Deb

Abstract In the realm of sustainable and green electronic materials, conductive bionanocomposites (CBNCs) have propelled technological advancements to new heights across the globe. These composites are typically created by integrating biodegradable polymers with electrically conductive nanomaterials, for instance, graphene, carbon nanotubes (CNTs), and metal nanoparticles. This innovative approach offers exceptional properties over conventional electronic materials, making them ideal for an extensive range of electrical and electronic applications. This chapter aims to provide a broad overview of conducting bionanocomposites, composed of various categories of biopolymer matrices and conductive nanomaterials for eco-friendly electrical and electronic applications. It systematically outlines the fundamentals of CBNCs, including synthesis techniques, environmental and industrial impact, and diverse applications, emphasizing their potential to revolutionize green electronics. The evolution of CBNCs represents a significant step toward more sustainable and eco-friendly electronics.

Keyword Biodegradable polymers · Electrically conductive nanomaterials · Conductive bionanocomposites (CBNCs) · Green electronics

11.1 Introduction

Bionanocomposites (BNCs) are an enthralling class of composite materials that have components of biological (bio-based polymers) origin and nanoparticles (1–100 nm) (Darder et al. 2007; Kumawat et al. 2024). Most of the materials for daily use including packaging are nothing but made up of synthetic plastics. Over the last decade of the twentieth century, they have grown rapidly and are found to meet both the commercial and industrial market requirements (Xia et al. 2021;

S. Deb (✉)
Associate Professor Department of Physics, University of Science and Technology Meghalaya (USTM), Ri-Bhoi, Techno City, Killing Road, Baridua, Meghalaya, India
e-mail: sujatabmk@gmail.com

S. C. George and B. Tawiah (eds.), *Advanced Bionanocomposite Materials*, Advanced Structured Materials 241, https://doi.org/10.1007/978-981-96-8225-6_11

Ncube et al. 2021), however, these resources are exhaustible, and as plastics are highly non-biodegradable waste, their usage increases every year and are acting as a threat to global environment. To avoid difficulties caused by synthetic polymers, there is a renewed interest in natural polymers, in place of synthetic polymers or petroleum-based (Zhao et al. 2023). The difference between petroleum-based or synthetic polymers and natural polymers is in biocompatibility and biodegradability. The distinguished difference in the solubility properties between them is that mainly synthetic polymers are soluble in organic solvents, in contrast, natural polymers are soluble in polar solvents, mostly in water, which determines their areas of applications. The other advantages of biopolymer include its wide abundance and relatively cost-effectiveness. Interest in composites of biological origin with the addition of nanoparticles, was initiated by researchers in the early 1990s. This study has led to an innovative era of study and application of composite materials with an extensive range of environmental benefits. Furthermore, conductive bionanocomposites (CBNCs) in green electronic materials, have advanced technology to new frontiers across the globe.

This chapter aims to demonstrate an inclusive overview of the various developments and applications of electrically conductive bionanocomposites (CBNCs), emphasizing their potential to revolutionize green electronics. The chapter begins with the fundamentals of CBNCs, stating the impact of the integration of functional nanofillers into biopolymers. *Percolation Theory and Excluded Volume Theory of Percolation,* which provide a framework for scheming and optimizing both biomatrices and nanofillers with desired properties are discussed. The chapter further includes a discussion on historical development, different categories of matrices and nanofillers, synthesis techniques, and diverse applications, highlighting the possibility of using CBNCs for sustainable and environmentally friendly electronics. The regulatory and safety considerations are also suggested in this chapter which is then followed by the future outlook and implications.

11.1.1 Fundamentals of Conductive Bionanocomposites

Over the past decades, conductive bionanocomposites (CBNCs) have gained significant attention as green materials with the potential to revolutionize various interdisciplinary fields from medicine to electronics and beyond. Generally, bio-components like proteins, DNA, and carbohydrates, themselves at nanoscale dimensions are not inherently electrically conductive. Many bio-components are considered electrically insulators with some exceptions. This limitation restricts their direct application in areas demanding high electrical conductivity (Selvakumaran and Lubineau 2014; Zhang et al. 2021; Amdursky et al. 2014; Gray and Ellis Jr 1994). However, it is also possible to tailor or improve their electrical properties by physical or chemical methods (Tarhini and Tehrani-Bagha 2023). Among the various methods, incorporating conductive nanomaterials into their structure or chemically modifying their surface to create conductive polymer composites is considered the most effective and appropriate approach. Conductive materials can be categorized into intrinsic and filled types based on their structure. Intrinsic electrically conductive materials include polyacetylene (PA), polyaniline (PANI), and polypyrene (PPy). However, intrinsic conductive materials generally exhibit low electrical conductivity and achieving higher levels of electrical conductivity is technically challenging (Shakir et al. 2020; Lyu et al. 2020; Gómez et al. 2021). In contrast, filled conductive materials like carbon nanotubes (CNTs), metals, metal oxides, and metal sulfides (Ariga et al. 2019; Dubey et al. 2020), provide exceptional electrical conductivity, through mechanisms like interconnection within the polymer or by electron tunneling (Bertolini et al. 2020; Zhao et al. 2021). Thus, making them widely used to enhance biopolymers conductivity (Wang et al. 2018; Samarjeet et al. 2020; Hong et al. 2022).

Carbon nanotubes (CNTs), are one-dimensional nanomaterials, formed by folding graphene sheets. They possess enhanced electrical, chemical, and mechanical properties. CNTs can be categorized as single-walled CNTs (SWCNTs) and multi-walled CNTs (MWCNTs). CNTs-based biodegradable electronic devices are reported by many researchers (Chen et al. 2017; Hwang et al. 2015). Graphene is a single layer of carbon atoms arranged in a hexagonal lattice (two-dimensional). Because of its excellent conductivity, it is considered an effective nanofiller to enhance the electrical conductivity of biopolymers (Young et al. 2012). Ling et al. also utilized graphene to develop novel conductive nanocomposites and to fabricate sensors (Ling et al. 2018). Table 11.1 displays the electrical properties of commonly used carbon conductive fillers.

Table 11.1 Electrical properties of carbon-based reinforced polymer composites

Filler	Matrix	Electrical property			References
		Percolation threshold	Filler content	Electrical conductivity (S/m)	
Carbon fiber (CNF)	Polypropylene (PP)	0.75 vol%	2.5 vol%	10	Paleo et al. (2014)
Multi-walled Carbon nanotubes (CNTs)	Polypropylene (PP)	1.07 wt%	15 wt%	0.23	Verma et al. (2015)
Carbon black	Polypropylene (PP)	2.75 vol%	6.5 vol%	2	Gong et al. (2016)
Graphite nanoplatelet	Polypropylene (PP)	3 vol%	9.3 vol%	1.7	Jun et al. (2018)
Graphene	Polypropylene (PP)	1.2 wt%	3.5 wt%	0.9	Ghislandi et al. (2013)

11.1.2 Importance of Percolation Threshold Concentration for CBNCs

One of the primary concerns with carbon-based fillers in CBNCs is their electrical percolation behavior and the associated percolation threshold. Some theories provide a framework for designing and optimizing both biomaterials and nanocomponents with desired properties. Percolation theory for conductive bionanocomposites is used to understand how the conductivity of these composites enhances as the concentration of conductive nanoparticles increases (Xue and Sun 2016). Thus, for the material to conduct electricity, an optimal concentration of nanofiller should be added to the biopolymer. As shown on the right side of Fig. 11.1, a conductive network of particles must be formed within the polymer matrix. With low nanofiller concentrations, the particles will remain untouched by each other, and the material will display dielectric behavior as shown on the left side of Fig. 11.1. But, as the concentration of nanofiller increases to a critical value, electrical conductivity is found to rise by ~ 10 orders of magnitude (highlighted zone in Fig. 11.1).

This critical value of the concentration of filler, at which enough conductive nanoparticles form a continuous network throughout the composite is known as the percolation threshold concentration (PTC). Achieving and surpassing this threshold is crucial to enhancing the electrical conductivity of BPNCs. To maintain the desired electrical properties of the material, the PTC must be reduced as much as possible. An advanced way to decrease the percolation threshold is to blend one polymer to which conductive nanoparticles have been added with another polymer with no added nanoparticles. The mixture of two or more polymers is called a blend, and under precise processing conditions, a co-continuous morphology is obtained. This

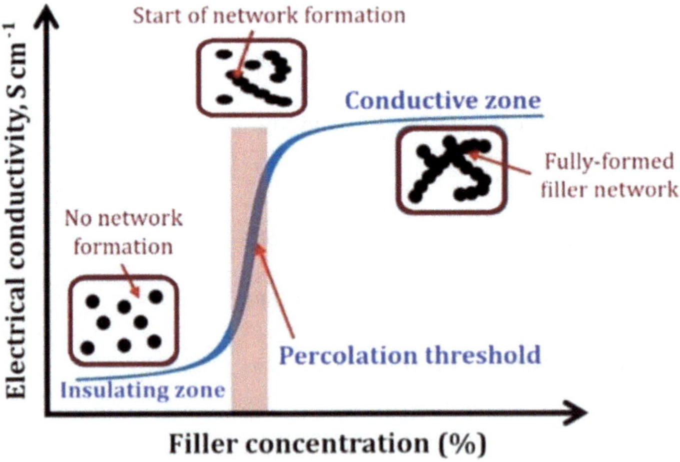

Fig. 11.1 Schematic representation of the percolation threshold principle (https://www.etsmtl.ca/en/news/electrically-conductive-polymer-composites)

will result in a reduction in PTC (https://www.etsmtl.ca/en/news/electrically-conductive-polymer-composites). Factors of the nanoparticles, including their size, shape, aspect ratio, and dispersion significantly affect the percolation threshold.

The ***Excluded Volume Theory of Percolation*** deals with the percolation threshold. This theory focuses on the concept of extended volume, which is the volume around an object that cannot be occupied by other objects. In ***percolation theory***, this concept helps to determine the critical concentration of objects needed for a continuous path to form through the system (Tamjid et al. 2024; Xu et al. 2008). This observation matches well with both experimental and theoretical reported works related to CBNCs (Dey et al. 2014). Furthermore, the orientation of the nanoparticles in various directions within the polymer matrix influences the electrical conductivity of the BPNCs (Hammami et al. 2020; Zhang et al. 2024). Thus, a comprehensive understanding of the fundamental mechanisms that govern the electrical conductivity and overall functionalities of CBNCs is very much required for future innovations in the electrical and electronic sectors. The various types of CBNCs with applications are displayed in Table 11.2.

Functional Properties of CBNCs

The integration of nano-sized particles into bio components enhances their conductive properties and other functionalities. Conductive BPNCs have strong π-conjugation with, optical, electronic, and magnetic properties like metals. As particles get smaller, their surface area increases (Supova et al. 2011). Owing to the very large surface area (S.A) to volume (V) ratio, the composite's properties are influenced by the numerous interactions and linkages among its mixed constituents. Thus, conductive bionanocomposites (CBNCs) are considered advanced materials with enhanced properties like improved electrical conductivity, mechanical strengths, and biocompatibility. Because of these unique properties, CBNCs find wide-ranging applications in medical fields, electronics and other semiconductor industries, textile industries, water purification industries, and many more (Kuznetsova et al. 2023; Joseph et al. 2020; Saied and Ward 2020; Zhang et al. 2018). Biopolymers, such as chitosan and cellulose, and, including synthetic polymers PLA, PVA, PVP, and PCL, are used as matrices and nanomaterials include metal, inorganic nanoparticles, and carbon nanostructures (Tarhini and Tehrani-Bagha 2023; Paleo et al. 2014). Since synthetic polymers are non-biodegradable, and are a major threat to environmental problems, the present study focuses on biodegradable polymer nanocomposites, which serve as a friendly alternative for conductive BPNCs (Salehi et al. 2021).

The electrical properties of CBNCs are significantly affected by several fundamental factors like the filler content, diffusion, shape, aspect ratio, orientation, size, conductivity, its interaction with the matrix, nature of the polymer matrix, surface energy, and preparation method of polymer composites (Sohi et al. 2011). Dielectric properties play a vital role in many applications such as electronics and biomedical devices. The dielectric, and electrical properties of the CBNCs are significantly affected by the dispersal of conductive nanoparticles and the interface between the

Table 11.2 Different categories of CBNCs and their subsequent applications

CBNCs	Conductive nanoparticles	Biopolymer matrix	Synthesis method	Applications	References
CS/PAA/PPy/ Ag-nanoparticles	Ag-nanoparticles	Chitosan/ polyacrylic acid/ polypyrrole	Casting method	Conducting bionanocomposites hydrogel	Youssef et al. (2019)
Poly(vinyl pyrrolidone)/tin oxide (PVP/ SnO_2) nanocomposite	Tin oxide (SnO_2)	Poly (vinyl pyrrolidone)	Solution casting method	Flexible-type advanced electronic devices	Dhatarwal et al. (2020)
Chitin/MoS_2 Nanosheet	(MoS_2) nanosheets	Chitin	At first, MoS_2 nanosheets were created by the glycerol/urea system. KOH/urea aqueous solution was then used to dissolve chitin at low temperatures with the dispersion of the MoS_2 nanosheets	Biomass-based dielectric capacitors	Chen et al. (2020)
Chitosan-Ag nanocomposites	Silver nanoparticles	Chitosan	Chemical method	UV and electromagnetic shielding materials	Patel et al. (2021)
Cellulose/Boron nitride nanosheet	Boron nitride nanosheet	Cellulose	Chemical route	Dielectric energy storage devices	Lao et al. (2018)

nanoparticles and the biopolymer matrix. To enhance and optimize these properties, the nanofiller and polymer matrix with upright compatibility and interaction are selected. Figure 11.2 shows a schematic representation of conductive filler and insulating polymer matrix in insulating composites and conductive composites. Despite their tremendous potential, CBNCs need to overcome a few challenges such as maintaining stability, ensuring even dispersion of conductive nanoparticles throughout the bio-component matrix, and scaling up production for industrial uses (Joseph et al. 2020).

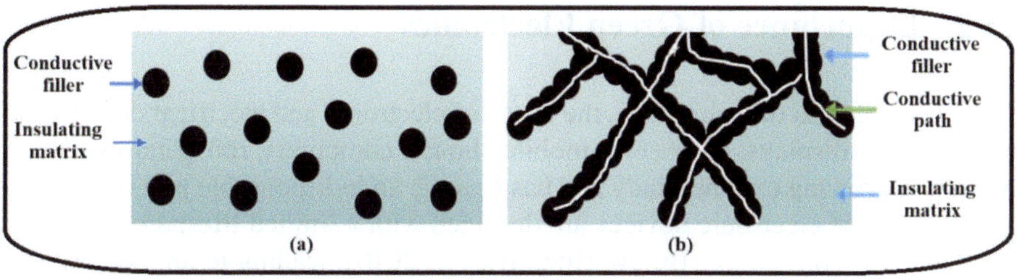

Fig. 11.2 Schematic of **a** conductive filler and insulating polymer matrix in insulating composites and **b** conductive composites

11.2 Historical Development of CBNCs

Bionanocomposites (BNCs) are quite recent innovations in materials science, integrating biological materials with nanoscale entities to develop composites with unique functional properties (Ruiz-Hitzky et al. 2008). Driven by the demand for sustainable and recyclable materials, the development of bionanocomposites (BNCs) began in the late twentieth century, however, the term was introduced first in 2004 (Shchipunov 2012). A variety of common terms like: nano biocomposites, green composites, bioplastics, and biohybrids are interchangeably used for bio nanocomposites. However, the term "bionanocomposites" greatly differs from the term—"biocomposites" which are made partially from biopolymers, but they do not consist of nano-sized additives. The period between 1990 and 2000, saw the emergence of conductive bio nanocomposites, which combined biodegradable polymers with conductive nanomaterials. Advancements in this domain of composites began around the beginning of the twenty-first century, with more focus on enhancing the biocompatibility, conductivity properties, and mechanical properties of BNCs. These categories of advanced materials found applications in sensors, medical devices, energy storage devices, wearable electronics, and many more. Of late, advanced synthesis techniques are being adopted to develop highly conductive and biocompatible bionanocomposites. Figure 11.3 displays the evolution of CBNCs.

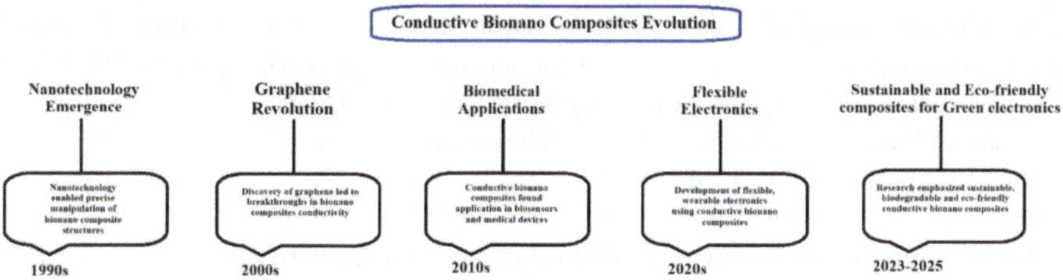

Fig. 11.3 Evolution of CBNCs

11.3 Importance of Green Electronics

With the evolution of technology, the usage of electronic and electrical devices such as LCD/LED displays, solar cells, mobile phones, computers, refrigerators, etc., by humans is growing exponentially and has become an indispensable part of our lives. However, these electronic devices are designed with a limited lifespan, brought on by planned obsolescence (Bulow 1986; Bidgoli 2010), leading to an increase in e-waste production and damage to ecology; the most perturbing characteristics of these discarded electronic products. This waste comprises a mixture of two kinds of materials, one that is valuable such as gold or copper, or recyclable constituents, and one that is toxic and non-biodegradable, posing a threat to human health and the environment. According to statistics, in 2024, it was reported that approximately 62 million tons of electronic waste were generated globally, of which only about 22.3% of this e-waste was properly collected or incinerated and recycled (https://ewastemonitor.info/the-global-e-waste-monitor-2024/) and the rest 77.7% was accumulated in the environment. This low recycling rate leads to significant waste of valuable raw materials and poses serious environmental and health risks due to the hazardous substances contained in e-waste. The countries that produce the most e-waste (in Kilo Tons) include: China, the US, and India (https://www.statista.com/chart/24291/e-waste-by-country/). Improving e-waste management also requires the ways the electronic devices are fabricated. Substituting conventional materials used in electronic systems (or devices) with biodegradables can lessen the impact of environmental pollution. "Green electronics" represents electronic systems designed with biodegradable and sustainable materials. To reduce e-waste, use of CBNCs in electronic devices can be one of the promising solutions to realize the biodegradable and sustainable future of the electronic industry. CBNCs as cited are evolved materials that combine biological entities with nanomaterials and provide enhanced or unique properties compared to their counterparts such as electrically conductive, and biocompatible, and are suitable for diverse applications from biotechnology to electronics while being eco-friendly. The world of electronic devices has been revolutionized with emerging green flexible and printed electronics using conductive bionanocomposites. Proteins derived from black soldier fly (insect) are suitable for fabricating flexible electronics. Recently, it has been demonstrated that black soldier fly can successfully bioconvert the Organic Fraction of Municipal Solid Waste (OFMSW) substrate and yield beneficial biomacromolecules. Testa, Edoardo, et al. utilized BSF pupae proteins in integration with conductive carbon black/serinol pyrrole, carboxymethyl cellulose, and glycerol as a bioconverter of the OFMSW (Testa et al. 2025). This biodegradable conductive bionanocomposite is found to be a suitable material for flexible electronic devices as shown in Fig. 11.4.

Bozo et al., demonstrated the use of blends such as polylactic acid (PLA)-polyhydroxybutyrate (PHB) as flexible substrates for conductive multi-walled carbon nanotubes (MWCNTs), for the fabrication of electronics components (Bozó et al. 2021). Kwon et al., reported the construction of green conductive fibers, with conductive poly(3,4-ethylenedioxythiophene) (PEDOT) coated with regenerated cellulose

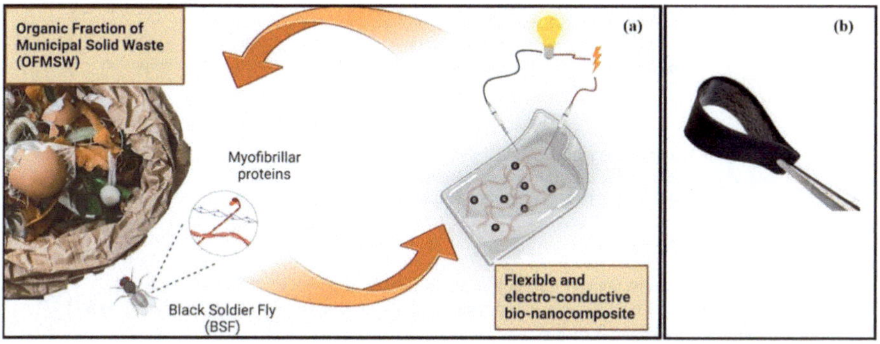

Fig. 11.4 **a** Schematic of conductive bionanocomposites from black soldier fly (BSF) proteins to green flexible electronics and **b** Prepared highly flexible bionanocomposite material

(RC), through a wet-spinning process and vapor-phase polymerization (VPP). The obtained PEDOT/RC-VPP turned on a green LED, demonstrating the flexibility and stability of green conductive fibers. This approach can be unified into several electronic textiles to produce wearable green electronics (Kwon et al. 2022). Saadi et al., developed a multi-functional bionanocomposite with an algae-derived carbohydrate biopolymer and boron nitride nanosheet for green electronic applications, such as supercapacitors and flexible interconnects (Saadi et al. 2024). CBNCs-based electronic devices further include field-effect transistors (FET), resistive memory, resistive random-access memory (RRAM), artificial neural network simulation, and many more. With a collective effort, a more sustainable and eco-friendly future can be built.

11.4 Classification of Conductive Bionanocomposites (CBNCs)

Generally, the matrix, reinforcement, and interface are the fundamental components of a composite.

Matrix: The continuous phase of a composite is known as the matrix. It not only holds the reinforcement in place, but also transfers loads between reinforcements, and protects this support from damage. The matrix includes: a polymer, metal, or ceramic (Saini et al. 2017).

Reinforcement: The dispersed phase implanted within a matrix is termed reinforcement. It is responsible for imparting strength and stiffness to the composite. For example, carbon fibers, are widely used for reinforcement in composites. These fibers can give exceptional strength and stiffness to a composite material, making them stronger than steel, yet significantly lighter (Saini et al. 2017; Pech-Canul et al. 2018).

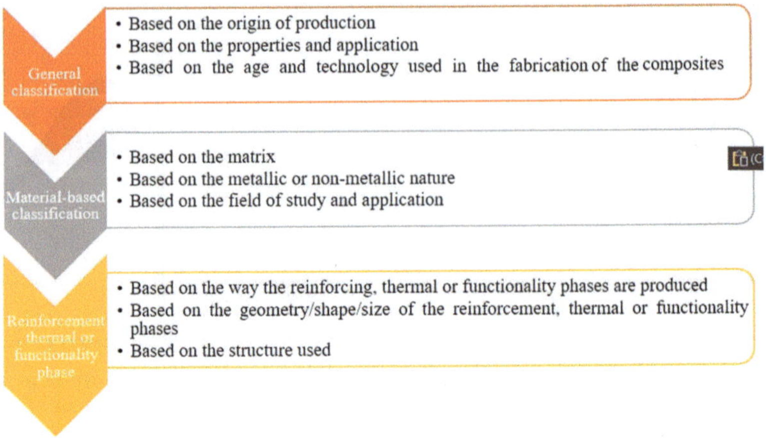

Fig. 11.5 Classification of CBNCs

Interface: The boundary region between the matrix and the reinforcement is the interface. Its core function is to enable load transfer between the matrix and reinforcement and plays a crucial role in defining the overall performance of the composite.

Updating the classification, composite materials can be divided into three main groups and subsequent subgroups (Pech-Canul et al. 2018), as shown in Fig. 11.5.

Based on the sources, composites can be classified into natural composites, synthetic composites, and hybrid composites. The constituent(s) of biological origin include a wide variety of sources of biomaterials (Lau et al. 2018; Karimah et al. 2021). They are valued for their sustainability, biodegradability, and lower environmental impact.

11.5 Classification Based on Types of Biopolymers

Polymers of biological origin are the prime constituents for bionanocomposites. They can be classified as (Shchipunov 2012; Saini et al. 2017):

- Carboxymethyl cellulose
- Chitosan
- Alginate
- Gelatin
- Guar gum
- Pectin
- Aliphatic polyester-based nanocomposites
- Polysaccharides-based nanocomposites
- Polypeptide–polymer nanocomposites.

The various structural formulae of polysaccharides are shown in Fig. 11.6.

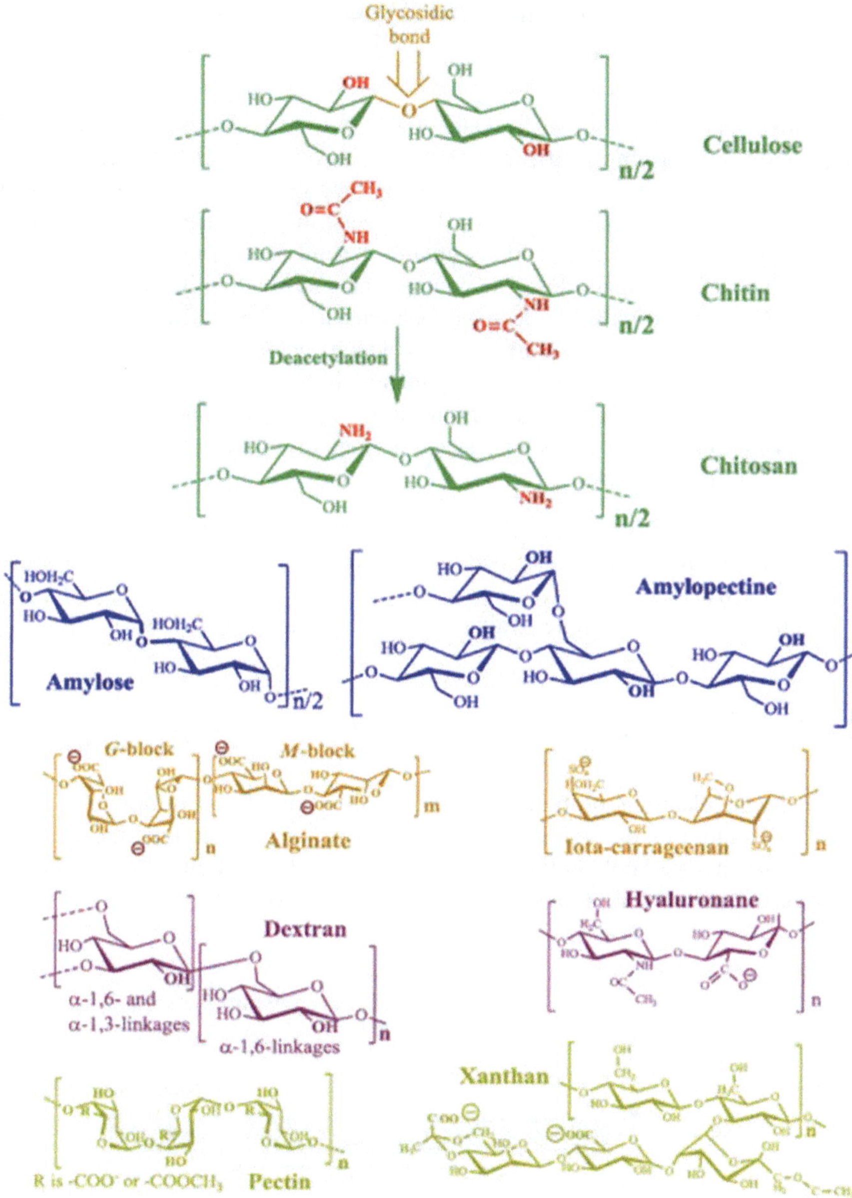

Fig. 11.6 Structural formulae of polysaccharides (Shchipunov 2012)

11.6 Classification Based on the Type of Conductive Nanomaterials

Conductive nanomaterials can be categorized into several types based on their composition, dimensions, and properties. The introduction of nanoparticles into the biopolymer matrix not only reinforces the biopolymer but also influences the other properties such as physical, optical, magnetic thermal, mechanical, and electrical properties. These modifications greatly depend on the interactions between biomolecules and nanoparticles. The nanoparticles have a highly extended surface area that is increased with the decrease of dimensions. One of the characteristics of nanoparticles is the large surface area, which increases the area of contacts and subsequently effectiveness of interactions with the biomolecules present in the bionanocomposite matrix. Another controlling factor is the shape of a nanoparticle; considering the geometry into account (Fig. 11.7), three main categories of nanoparticles are as follows:

a. 0 dimensional, for example, spherical silica nanoparticles, nanoclusters, etc.
b. 1 dimensional, for example, carbon nanotubes with diameter < 100 nm or cellulose whiskers.
c. 2 dimensional, for example, graphene sheets of 1 nm thick.

11.6.1 Various Categories of Conductive Nanocomposites

Studies on the electrical properties of CBNCs offer significant information on the conductive nature of the materials. Zhi Shen Tong et al. examined the electrical conductivity in doped polyaniline (PANI) (Tong et al. 1992). Leung et al. calculated the electrical and optical properties of polyaniline (PANI) copolymers. These novel

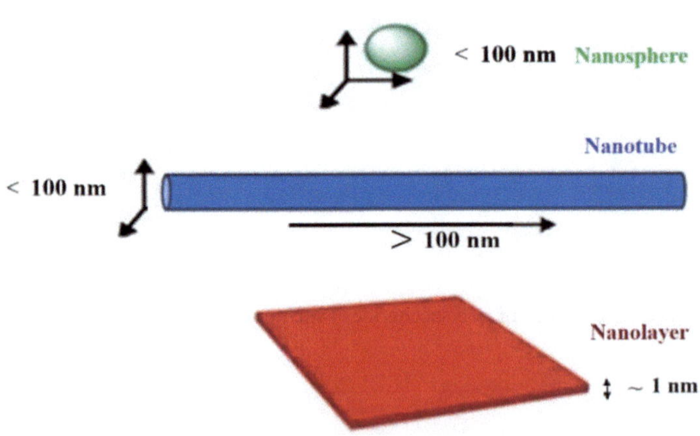

Fig. 11.7 Classification based on dimensions

copolymers find applications in nano-electronic devices, optoelectronics, sensors, molecular wires, switches, etc. (Leung et al. 2002). Tanaka et al. investigated the optical and electrical properties of doped polyaniline (PANI) and their temperature dependence (Tanaka et al. 1999). The conductivity of the doped PANI is the maximum than base polymeric material.

Carbon-Based Nanocomposites

Carbon-based fillers such as carbon black, graphite, glassy carbon, carbon nanotubes, and graphene have been effectively utilized in creating polymeric composites with notable electrical properties (Tarhini and Tehrani-Bagha 2023; Paleo et al. 2014; Sanjinés et al. 2011). The rapid advancement of telecommunications and electronic technologies has led to electromagnetic (EM) radiation pollution and electromagnetic interference (EMI), which poses a serious threat to human health. Additionally, this interference can hinder the performance of electronic devices. Consequently, there is an urgent need for highly efficient electromagnetic interference (EMI) shielding materials to ensure the reliable operation of electronic devices and protect human health from EM radiation pollution. Generally, metals and metallic composites are used as EMI shielding materials due to their excellent electrical conductivity and strong electromagnetic wave reflection capabilities. However, their high density, susceptibility to corrosion, difficult processability, and low mechanical properties limit their use in highly integrated electronic devices. Additionally, traditional metal-based EMI shielding materials can cause secondary EM radiation pollution by reflecting electromagnetic waves, affecting further the reliability and accuracy of precision electronic devices. The conductive polymer composites with highly conductive carbon fillers like carbon black, carbon nanofibers, multi-walled carbon nanotubes, graphene, carbon fibers, and their hybrids have gained attention as lightweight EMI shielding materials. The conductive polymer composites offer noteworthy advantages over the traditional metal-based EMI shielding materials including lightweight, high corrosion resistance, excellent thermal, mechanical, and electrical properties, and adjustable EMI shielding performances (Luo et al. 2018; Khan et al. 2015; Banerjee et al. 2020).

Cellulose-Based/Hybrid/Metal-Based Nanocomposites

Cellulose, the most abundant bio-based material, stands out due to its renewability, biodegradability, biocompatibility, non-toxicity, and environmental friendliness. These qualities make it a promising alternative to exhaustible fossil resources. The combination of cellulose nanocrystals and nanocarbons including fullerenes, nanotubes, and graphene-based materials has recently emerged as a new class of hybrid composites, displaying excellent physicochemical characteristics and performances (Thomas et al. 2018). These hybrids show a wide range of applications from energy storage, and optoelectronics to biomedical, food packaging, and many more

(Trache et al. 2020). Mendoza et al. investigated the potential use of well-designed conducting polymer/metal-based nanocomposites in creating high-performance electrocatalyst-based nanocomposites for applications in electrochemical energy conversion (Mendoza et al. 2024).

11.7 Synthesis Techniques for the Preparation of CBNCs

Various techniques like ex-situ, in-situ, and single-pot methods are available for fabricating CBNCs by incorporating conductive nanoparticles into the polymer matrix. These methods have different approaches. In the ex-situ method, conductive nanoparticles are prepared separately before being incorporated into the biopolymer matrix. However, this approach has certain disadvantages, it can sometimes lead to agglomeration, which may affect the composite's properties. The in-situ method involves synthesizing conductive nanoparticles directly within the biopolymer matrix. This approach has a more uniform distribution of nanoparticles and strong interfacial bonding between the biopolymer and the nanoparticles; however, it is more complex. The single-pot method integrates multiple steps into one process. Each method has its advantages (Chatterjee et al. 2016). Thermoplastic polymer nanocomposites are synthesized by solution blending, in-situ polymerization, and melt blending. These techniques are used due to their environment friendliness, easy processability, and cost-effectiveness. Recently, various other synthesis techniques, as shown in Fig. 11.8 such as green synthesis, electrospinning, and processing under supercritical conditions have also gained interest.

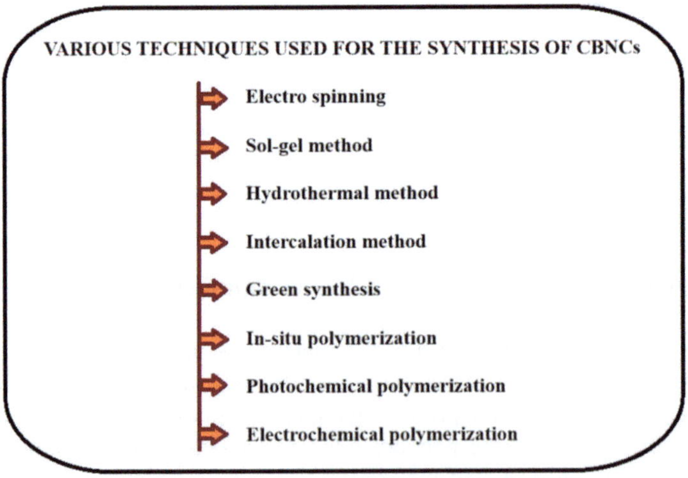

Fig. 11.8 Various synthesis techniques of conductive bionanocomposites (CBNCs)

11.8 Application of CBNCs in Electrical and Electronic Devices

The various industrial and environmental challenges have engrossed significant attention in different conducting bionanocomposites for diverse applications (Min et al. 2023; Dubey et al. 2019). The vast consumption and limited availability of petroleum-based polymers are among the industrial challenges that have led to the development of renewable and biodegradable substitutes. Furthermore, the risks of plastic residues in the environment. In this regard electrically conducting polymer bio nanocomposites (CBNCs) have evolved as promising new-generation hybrid materials for a wide range of applications (Fig. 11.9). BNCs are a novel class of composite materials and provide several essential features. Polymeric matrices include either natural or synthetic, both biodegradable and non-biodegradable. Economic and environmental concerns encourage the application of biodegradable polymers derived from sustainable resources such as biomass to reduce non-biodegradable nature.

Largely, numerous functional components, for instance, substrate, dielectric, active layer, conducting, and semiconducting layers, together constitute any electronic devices/systems. By substituting the traditional materials (used for these components) with biodegradable and biocompatible materials, the generation of e-waste can be controlled to a greater extent. Several electronic devices using biodegradable and biocompatible materials as substrates, dielectric, etc., have been reported. Huang, Tsung-Syun, et al., demonstrated polymethyl methacrylate (PMMA) as a suitable dielectric layer for pentacene-based organic thin film transistors (OTFTs), exhibiting exceptional electric characteristics. PMMA was used to replace SiO_2 to confer the transfer characteristics of OTFTs (Huang et al. 2007). Bystrenova et al., reported an organic semiconductor, Pentacene, as a suitable substrate that might open new paths for flexible organic electronics (Bystrenova et al. 2008). Because of various superior features like sustainability, biodegradability, etc., nanocellulose (NC) is the most common suitable ecological nanomaterial as substrates. Park et al., successfully fabricated highly sensitive flexible and transparent organic phototransistors (OPTs) for detecting light in the visible range (400–750 nm) using biodegradable cellulose nanofibrillated fiber substrate (Park

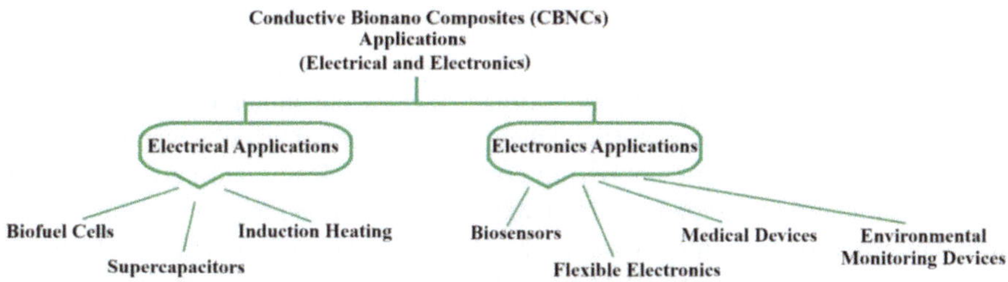

Fig. 11.9 Applications of CBNCs in electrical and electronics fields

et al. 2018). Essentially, biodegradable and biocompatible properties enable CBNCs-based electronic devices for a wide range of applications in the biomedical fields as well. These diverse applications display biodegradable and biocompatible materials as promising resources for environmental sustainability.

11.8.1 Sensors

The significant biopolymers used for sensing purposes are chitosan, cellulose, alginate, etc. (Shukla et al. 2014; Buk et al. 2017; Yao et al. 2018). Sharma et al. have developed a novel multi-component matrix consisting of chitosan (Chi)/polypyrrole-nanotubes (PPy-NTs) and gold nanoparticles (Au-NPs) nanocomposites-based glucose biosensor by immobilizing glucose oxidase onto ITO electrode. The GOx/Chi/PPy-NTs/Au-NPs electrode was found to exhibit a good linearity of 3–230 μM ($R^2 = 0.98$) toward the oxidation of glucose with a sensitivity of ca. 149 μA μM^{-1} mL and a detection limit of 3.10 μM. The large surface area, size, and quantum effect of metal nanoparticles contributed toward enhanced properties like fast response time of the biosensors while the existence of chitosan added stability and reproducibility to the biosensor (Sharma and Kumar 2016). Luo et al. investigated sophisticated porous structures and improved mechanical properties of layer-by-layer assembled bacterial cellulose (BC)/graphene oxide (GO) nanocomposites. Their results showed that BC/GO hybrid nanocomposite can be applied to flexible sensors, chemical sensors, temperature sensors, proximity sensors, energy storage devices, flexible bending actuators, battery separators, and supercapacitors (Luo et al. 2018). Kumar et al. (2012) developed volatile organic compounds (VOC) sensors using conductive biopolymer nanocomposites by dispersing multi-wall carbon nanotubes as "guest conducting filler" in biopolymer, poly(lactic acid) (PLA) as "host matrix" through solution mixing technique. A schematic representation of a vapor sensing device is shown in Fig. 11.10.

11.8.2 Energy Storage and Smart Textiles/Wearable Electronics

Traditional silicon-based electronic storage devices encounter the von Neumann bottleneck problem, which restricts their overall performance. Overcoming this bottleneck is important for improving computing efficiency and performance. As the number of electronic gadgets is increasing, non-volatile memory, which retains stored information even after power is switched off, is becoming increasingly crucial in modern daily life. Recently, RRAM has been considered an ideal device because of its non-volatile nonlinear conduction characteristics, offering the possibility to overcome the von Neumann bottleneck. Additionally, silicon-based storage devices face

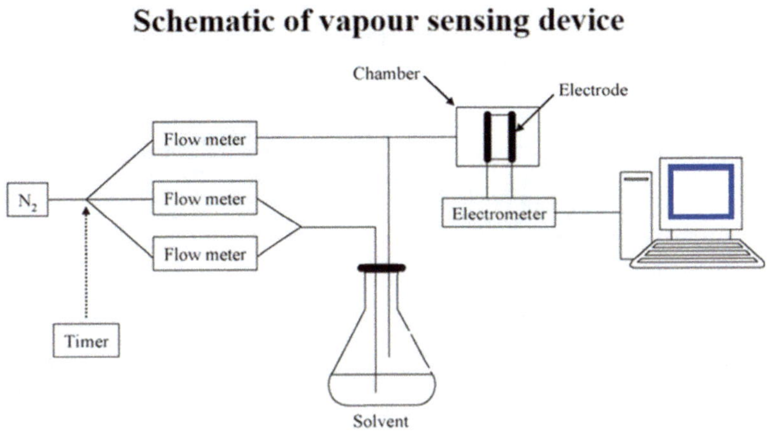

Fig. 11.10 Schematic of vapor sensing device

challenges in terms of scalability, storage density, and cost efficiency. CBNCs have gained significant attention as potential candidates for the next generation of green electronics due to their biodegradability and environment-friendly bioelectronics properties and are likely to challenge traditional silicon-based devices.

Recently, CBNCs-based electronic devices including energy storage devices have been successively reported (Fig. 11.11), which reflects their application in data storage, resistive random access memory (RRAM) and field-effect transistors (FET), displaying exceptional characteristics like flexibility, stability, transparency, and eco-friendly (Xing et al. 2020).

Su-Ting Han et al. reported a high-performance flexible resistive random access memory (RRAM) with a three-layered structure, which was fabricated by utilizing solution-processed MoS_2 nanosheets sandwiched between poly(methyl methacrylate) polymer layers as shown in Fig. 11.12. This hybrid device presented robust access speed and retention ability (Han et al. 2016).

CBNCs have also emerged as promising composites for wearable technology due to their unique tunable properties. Authors reported the use of biodegradable conductive bionanocomposites from Black Soldier Fly Proteins for flexible electronic devices (Testa et al. 2025). Schmidt et al., introduced electrically conductive liquid “Electrofluids” and demonstrated their use as stretchable and flexible materials with tunable properties, created using different solvents such as glycerol, polydimethylsiloxane (PDMS), hexadecane, and ethylene glycol, to control the percolation and microstructure of conductive particle networks. They used carbon black (CB) as a conductive filler; altering its concentration in various solvents (polar and nonpolar) electrical conductivity was studied (Schmidt et al. 2024).

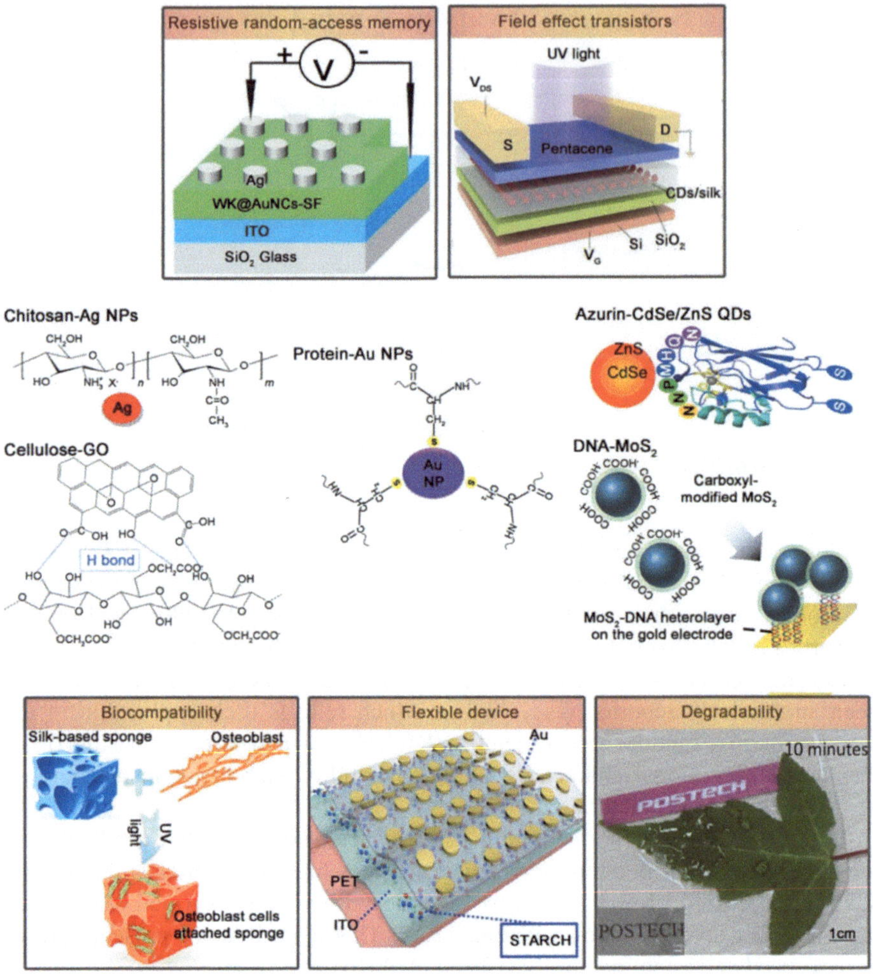

Fig. 11.11 RRAM and FET composed of different biocomposites (Xing et al. 2020)

11.9 Environmental Impact

The rapid progress of technology has come with considerable environmental costs.

Definition of Electronic-Waste by the United Nations (UN):

> Any discarded product with a battery or plug, and features toxic and hazardous substances such as mercury, that can pose a severe risk to human and environmental health.

The International Telecommunication Union (ITU) specifies that e-waste is one of the largest and most complex waste streams in the world. The mapping of e-waste

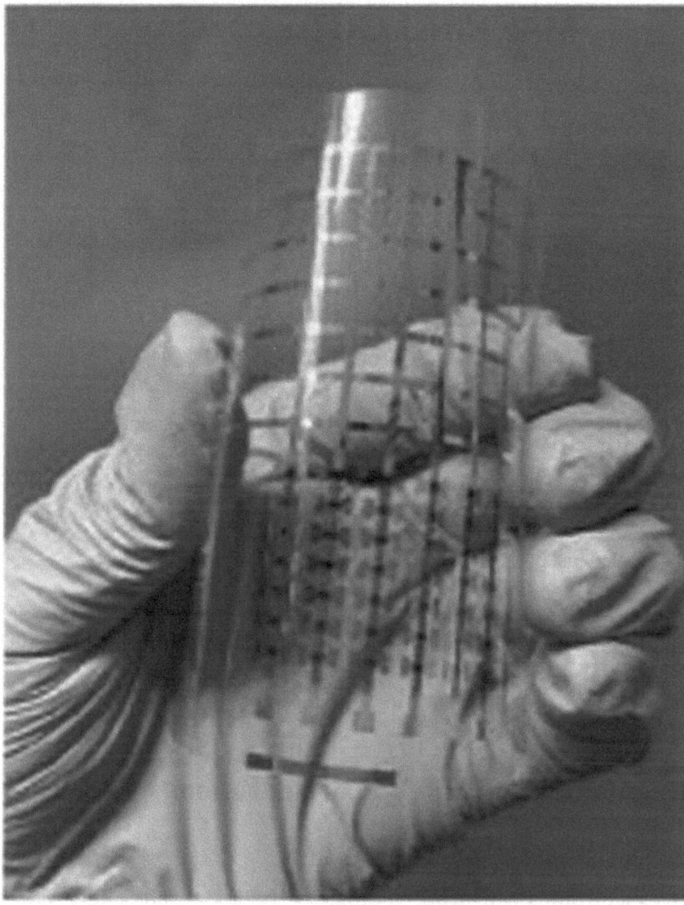

Fig. 11.12 Optical image of the fabricated memory device (Han et al. 2016)

as detailed in the report "A New Circular Vision for Electronics—Time for a Global Reboot" is shown in Fig. 11.13.

When electronic items get near the end of their useful life, they are discarded, and this leads to the generation of e-waste. This issue involves the lifecycle of the devices everyone uses; starting from manufacturers to investors, traders, consumers, and policy-makers, and thus, every user has a social responsibility toward its proper recycling and disposal approaches.

Compared to traditional non-degradable materials that are used in electrical and electronic devices, conductive bionanocomposites (CBNCs) can significantly participate as biodegradable, energy efficient, and less toxic with appropriate reprocessing and disposal methods (Tamjid et al. 2024). Because of these factors, electrical and electronic devices like supercapacitors, batteries, biosensors, medical devices, and flexible electronics incorporating CBNCs may greatly contribute to more sustainable

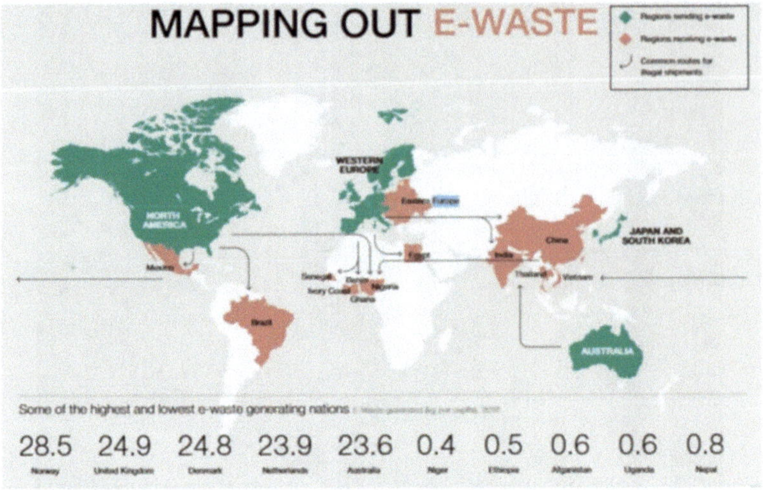

Fig. 11.13 Mapping of e-waste as in the report "A New Circular Vision for Electronics—Time for a Global Reboot" (Isibor 2024)

products, leading to better management of e-waste, and minimizing environmental damage.

11.10 Regulatory and Safety Considerations

As nanotechnology enables cutting-edge innovation across various sectors from medicine to electronics, ensuring regulatory policies and safety considerations for the protection of human health and the environment becomes critical. The regulatory area for nanomaterials is evolving alongside developments in nanoscience and nanotechnology (https://www.genevaenvironmentnetwork.org/resources/updates/the-growing-environmental-risks-of-e-waste/).

To keep pace with the rapid advancements in nanotechnology, regulatory bodies worldwide, such as the Environmental Protection Agency (EPA) and the Food and Drug Administration (FDA) in the United States and the more stringent regulations through frameworks like the Registration, Evaluation, Authorization and Restriction of Chemicals (REACH) and the European Chemicals Agency (ECHA) in the EU, are continuously updating guidelines and standards to ensure the safe use of nanomaterials, including biocompatibility studies, implementing safety, and environmental impact evaluations, thereby promoting greater transparency and accountability in the use of these materials (Chávez-Hernández et al. 2024). The cooperation between countries plays a key role in implementing global regulatory frameworks on nanomaterials. An outline of the regulatory frameworks of a few countries is shown in Table 11.3.

Table 11.3 Regulatory frameworks on nanomaterials of a few countries (Chávez-Hernández et al. 2024)

Country	Framework
United States	• Toxic substances control act • Draft guidance for the industry considering whether an FDA-regulated product involves the applications of nanotechnology • Nanotechnology regulatory science research plan • National nanotechnology initiative • Nanotechnology research and development act • Federal insecticide, fungicide, and rodenticide act • Clean air act • Comprehensive environmental recovery, compensation, and liability act • Resource conservation and recovery act • Federal food, drug, and cosmetic act • Voluntary cosmetic registration program
European countries	• Globally harmonized system • Action Plan for Europe 2005–2009 • Scientific committee on emerging and newly identified risk • European classification, labeling, and packaging of substances and mixtures • Approved code of practice • Risk assessment guideline of the control of substances, hazardous to health regulations • ACOP of the dangerous, substances and explosive atmosphere regulation • EU cosmetic regulation 1223/2001 • Directive 76/768/EEC
Canada	• Canadian Environmental Protection Act • Pest control product act fertilizer, feeds, food and drug act
Australia	• Food standards Australia-New Zealand • National Industrial Chemical Assessment Scheme • Therapeutic good administration • Australian pesticides and veterinarian medicines authority
Korea	• Act of registration and evaluation of chemicals • Development promotion act
China	• Act of registration and evaluation of chemicals • Development promotion act

11.11 Global Market Overview and Prospects of CBNCs

According to the report by market researchers, the conductive polymers in the global market was valued at \$4.4 billion in 2024 and is estimated to grow at a CAGR (Compound annual growth rate) of over 8.4% from 2025 to 2034. A brief outline of the conductive polymers market report attributes is shown in Table 11.4.

The Market Research Report provides a comprehensive analysis of the global Conductive Polymer Composites Market and analyzes the policies of top global companies with an emphasis on Conductive Polymer Composites portfolios with

Table 11.4 Conductive polymers market report attributes

Report attribute	Details
Base year:	2024
Conductive polymers market size in 2024:	USD 4.4 billion
Forecast period:	2025–2034
Forecast period 2025–2034 CAGR:	8.4%
2034 value projection:	USD 9.7 billion
Historical data for:	2021–2024
No. of pages:	220
Tables, charts and figures:	205
Segments covered:	Conduction mechanism, application
Growth drivers:	• Expanding global semiconductor and electronics industry • Increasing demand for light weight components • Growing U.S. solar power industry
Pitfalls challenges:	• Lack of stability under certain environmental conditions and higher cost of production

https://www.gminsights.com/industry-analysis/conductive-polymers-market

a wide range of industry parameters to accelerate the global market of Conductive Polymer Composites. The Conductive Polymer Composites market has been segmented into different categories:

a. **By type**
b. Conjugated Polymer-Based
c. Resin-Based.
d. **By Application**
e. Electromagnetic Interference (EMI) Shielding
f. Antistatic Solutions
g. Sensors and Actuators
h. Printed Electronics
i. Others.

The drive for renewable energy demands the use of conductive biopolymers in modern electronic devices like smartphones, wearables, and automotive electronics. Further, the growing emphasis on energy efficiency across various industries propels the implementation of conductive biopolymer composites in energy-saving appliances, contributing toward market growth. For instance, in March 2024, KEMET, a leading global electronic components supplier for aerospace, defense, avionics, and mission-critical applications, introduced the first T581 Polymer Tantalum Surface Mount capacitors, meeting the requirements for Military Performance Specification Sheets MIL-PRF-32700/2. This release marks a significant industry advancement. In March 2024, Covestro inaugurated a new production plant (the world's first solvent-free melt process) for polycarbonate copolymers, marking a major advancement

Fig. 11.14 Conductive polymers market size, by conduction mechanism (https://www.gminsights.com/industry-analysis/conductive-polymers-market)

in polymer manufacturing technology, followed by a new product line based on polycarbonate copolymers in April 2024.

Based on the conduction mechanism, the conductive polymer market is divided into two segments: conducting polymer composites and inherently conductive polymers. The conducting polymer composites sector leads the market and these composites for being lightweight and efficient materials are widely used in the automotive and electronics industries. Thus, making them chosen over traditional conductive materials. The market size of conductive polymers, done by conduction mechanism, 2021–2034 is shown in Fig. 11.14.

11.12 Future Outlook and Implications

Functional conductive bionanocomposites find wide technological applications owing to their conductive, flexible, and eco-friendly features, providing an effective solution for e-waste management and environmental safety. However, achieving excellent device performance while meeting eco-friendly requirements remains a challenge. The electrical conductivity of CBNCs is influenced by both the biopolymer matrices and nanofillers. However, creating devices that are both stable and biodegradable remains a great challenge. A substantial amount of work has been undertaken by researchers worldwide to find appropriate methodologies and functional nanocomposites. Most natural biopolymers are electrical insulators, which restricts their application. Conversely, synthesized conductive/semiconductive polymers like PANI, PPy, and PEDOT, are biocompatible but not biodegradable, limiting their use in electronic devices. To obtain efficient conductive or semiconductive biomaterials with biodegradability, integrating conductive nanofillers such as CNTs, GO, graphene, and SiO_2, into biopolymers (biodegradable) can greatly improve their conductive or dielectric properties. Developing novel biopolymers with enhanced conductive properties or adopting different approaches like merging natural and synthetic polymers for enhancement, can be well thought-out. Thus, these aspects can be carefully investigated for future high-performance and eco-friendly electrical and electronic devices.

11.13 Conclusion

Driven by increasing demand for biodegradable, sustainable, and environmentally friendly materials, the conductive bionanocomposites are gaining considerable attention as components of green electronics. Both biopolymer matrix and nanofillers play a crucial role in the formation of a novel conductive composite. Intrinsic conductive polymers have limitations while integration of functional nanofillers, such as graphene and carbon nanotubes, into a polymer matrix greatly enhances the electric conductivity property. This chapter presents a comprehensive overview of the various aspects of conductive bionanocomposites such as the fundamentals, different categories of conductive biopolymers nanomaterials, synthesis techniques used, and electrical and electronic applications. Further, environmental impact, regulatory and safety measures, and the global market overview are also discussed. By integrating these conductive nanocomponents with a biopolymer matrix, researchers can develop more sustainable, eco-friendly, and innovative polymer nanocomposite solutions for a wide range of applications in various industries, leading toward green electronics.

References

Amdursky N, Marchak D, Sepunaru L, Pecht I, Sheves M, Cahen D (2014) Electronic transport via proteins. Adv Mater 26(42):7142–7161

Ariga K, Ito M, Mori T, Watanabe S, Takeya J (2019) Atom/molecular nanoarchitectonics for devices and related applications. Nano Today 28:100762

Banerjee P, Bhattacharjee Y, Bose S (2020) Lightweight epoxy-based composites for EMI shielding applications. J Electron Mater 49:1702–1720

Bertolini MC, Ramoa S, Merlini C, Barra G, Soares BG, Pegoretti A (2020) Hybrid composites based on thermoplastic polyurethane with a mixture of carbon nanotubes and carbon black modified with polypyrrole for electromagnetic shielding. Front Mater 7:174

Bidgoli H (2010) The handbook of technology management, supply chain management, marketing and advertising, and global management, vol 2. Wiley

Bozó E, Ervasti H, Halonen N, Shokouh SH, Tolvanen J, Pitkanen O, Jarvinen T, Palvolgyi PS, Szamosvolgyi A, Sápi A, Konya Z (2021) Bioplastics and carbon-based sustainable materials, components, and devices: toward green electronics. ACS Appl Mater Interfaces 13(41):49301–49312

Buk V, Emregul E, Emregul KC (2017) Alginate copper oxide nano-biocomposite as a novel material for amperometric glucose biosensing. Mater Sci Eng C 74:307–314. https://doi.org/10.1016/j.msec.2016.12.003

Bulow J (1986) An economic theory of planned obsolescence. Q J Econ 101(4):729–749

Bystrenova E, Jelitai M, Tonazzini I, N. Lazar A, Huth M, Stoliar P, Dionigi C, Cacace MG, Nickel B, Madarasz E, Biscarini F (2008) Neural networks grown on organic semiconductors. Adv Funct Mater 18(12):1751–1756

Chatterjee B, Kulshrestha N, Gupta PN (2016) Nano composite solid polymer electrolytes based on biodegradable polymers starch and poly vinyl alcohol. Meas J Int Meas Confed. 82:490–499. https://doi.org/10.1016/j.measurement.2016.01.022

Chávez-Hernández JA, Velarde-Salcedo AJ, Navarro-Tovar G, Gonzalez C (2024) Safe nanomaterials: from their use, application, and disposal to regulations. Nanoscale Adv 6(6):1583–1610

Chen Z, Zhang A, Wang X, Zhu J, Fan Y, Yu H, Yang Z (2017) The advances of carbon nanotubes in cancer diagnostics and therapeutics. J Nanomater 2017:1–13

Chen H, Li X, Yu W, Wang J, Shi Z, Xiong C, Yang Q (2020) Chitin/MoS2 nanosheet dielectric composite films with significantly enhanced discharge energy density and efficiency. Biomacromol 21:2929–2937. https://doi.org/10.1021/acs.biomac.0c00732

Darder M, Aranda P, Hitzky E (2007) Bionanocomposites: a new concept of ecological, bioinspired, and functional hybrid materials. Adv Mater 19:1309–1319

Dey KK, Kumar P, Yadav RR, Dhar A, Srivastava AK (2014) CuO nanoellipsoids for superior physicochemical response of biodegradable PVA. RSC Adv 4(20):10123–10132

Dhatarwal P, Choudhary S, Sengwa RJ (2020) Effectively nanofiller concentration tunable dielectric properties of PVP/SnO_2 nanodielectrics. Mater Lett 273:127913

Dubey N, Kushwaha CS, Shukla SK (2019) A review on electrically conducting polymer bionanocomposites for biomedical and other applications. Int J Polym Mater Polym Biomater 69(11):709–727. https://doi.org/10.1080/00914037.2019.1605513

Dubey N, Kushwaha CS, Shukla SK (2020) A review on electrically conducting polymer bionanocomposites for biomedical and other applications. Int J Polym Mater Polym Biomater 69(11):709–727

Ghislandi M, Tkalya E, Marinho B, Koning CE (2013) Electrical conductivities of carbon powder nanofillers and their latex-based polymer composites. Compos Part A Appl Sci Manuf 53:145–151

Gómez IJ, Vázquez Sulleiro M, Mantione D, Alegret N (2021) Carbon nanomaterials embedded in conductive polymers: a state of the art. Polymers 13:745

Gong T, Peng SP, Bao RY, Yang W, Xie BH, Yang MB (2016) Low percolation threshold and balanced electrical and mechanical performances in polypropylene/carbon black composites with a continuous segregated structure. Compos Part B Eng 99:348–357

Gray HB, Ellis WR Jr (1994) Electron transfer. Bioinorg Chem: 315–363

Hammami I, Benhamou K, Hammami H, SoretoTeixeira S, Arous M, Kaddami H, Graça MPF, Costa LC (2020) Electrical, morphology and structural properties of biodegradable nanocomposite polyvinyl-acetate/cellulose nanocrystals. Mater Chem Phys 240:122182

Han ST, Zhou Y, Chen B, Wang C, Zhou L, Yan Y, Zhuang J, Sun Q, Zhang H, Roy VA (2016) Hybrid flexible resistive random access memory-gated transistor for novel nonvolatile data storage. Small 12(3):390–396

Hong G, Choi S, Yoo D, Oh T, Song Y, Yeon JH (2022) Moisture dependence of electrical resistivity in under-percolated cement-based composites with multi-walled carbon nanotubes. J Mater Res Technol 16:47–58

https://ewastemonitor.info/the-global-e-waste-monitor-2024/

https://www.etsmtl.ca/en/news/electrically-conductive-polymer-composites

https://www.genevaenvironmentnetwork.org/resources/updates/the-growing-environmental-risks-of-e-waste/

https://www.gminsights.com/industry-analysis/conductive-polymers-market

https://www.statista.com/chart/24291/e-waste-by-country/

Huang TS, Su YK, Wang PC (2007) Study of organic thin film transistor with polymethylmethacrylate as a dielectric layer. Appl Phys Lett 91(9)

Hwang JY, Kim HS, Kim JH, Shin US, Lee SH (2015) Carbon nanotube nanocomposites with highly enhanced strength and conductivity for flexible electric circuits. Langmuir 31:7844–7851

Isibor PO (2024) Regulations and policy considerations for nanoparticle safety. In: Environmental nanotoxicology: combatting the minute contaminants. Springer, Cham, pp 295–316

Joseph B, Krishnan S, Sagarika VK, Tharayil A, Kalarikkal N, Thomas S (2020) Bionanocomposites as industrial materials, current and future perspectives: a review. Emergent Mater 3:711–725

Jun YS, Um JG, Jiang G, Lui G, Yu A (2018) Ultra-large sized graphene nano-platelets (GnPs) incorporated polypropylene (PP)/GnPs composites engineered by melt compounding and its thermal, mechanical, and electrical properties. Compos Part B Eng 133:218–225

Karimah A, Ridho MR, Munawar SS, Adi DS, Damayanti R, Subiyanto B, Fatriasari W, Fudholi A (2021) A review on natural fibers for development of eco-friendly bio-composite: characteristics, and utilizations. J Mater Res Technol 13:2442–2458

Khan M, Tahir MN, Adil SF, Khan HU, Siddiqui MR, Al-warthan AA, Tremel W (2015) Graphene based metal and metal oxide nanocomposites: synthesis, properties and their applications. J Mater Chem A 3(37):18753–18808

Kumar B, Castro M, Feller J-F (2012) Poly (lactic acid)–multi-wall carbon nanotube conductive biopolymer nanocomposite vapour sensors. Sens Actuators, B Chem 161(1):621–628

Kumawat YK, Sharma K, Kumar V (2024) Bio-nanocomposites: fundamentals and recent advances. In: Sharma K, Tiwari SK, Kumar V, Kalia S (eds) Novel bio-nanocomposites for biomedical applications. Springer series on polymer and composite materials. Springer, Cham. https://doi.org/10.1007/978-3-031-69654-1_1

Kuznetsova LS, Arlyapov VA, Plekhanova YV, Tarasov SE, Kharkova AS, Saverina EA, Reshetilov AN (2023) Conductive polymers and their nanocomposites: application features in biosensors and biofuel cells. Polymers 15(18):3783

Kwon G, Lee K, Jeon Y, Jeong M, Kim J, You J (2022) Hybrid nanoarchitectonics with conductive polymer-coated regenerated cellulose fibers for green electronics. ACS Sustain Chem Eng 10(40):13444–13452

Lao J, Xie H, Shi Z, Li G, Li B, Hu GH, Yang Q, Xiong C (2018) Flexible regenerated cellulose/boron nitride nanosheet high-temperature dielectric nanocomposite films with high energy density and breakdown strength. ACS Sustain Chem Eng. 6:7151–7158. https://doi.org/10.1021/acssuschemeng.8b01219

Lau K-T, Hung P-Y, Zhu M-H, Hui D (2018) Properties of natural fibre composites for structural engineering applications. Compos B Eng 136:222–233

Leung LM, Tan KH, Lam TS, WeiDong H (2002) Electrical and optical properties of polyacetylene copolymers. React Funct Polym 50(2):173–179

Ling S, Wang Q, Zhang D, Zhang Y, Mu X, Kaplan DL, Buehler MJ (2018) Integration of sti graphene and tough silk for the design and fabrication of versatile electronic materials. Adv Funct Mater 28:1705291

Luo H, Dong J, Yao F, Yang Z, Li W, Wang J, Xu X, Hu J, Wan Y (2018) Layer-by-layer assembled bacterial cellulose/graphene oxide hydrogels with extremely enhanced mechanical properties. Nano-Micro Letters 10:1–10

Lyu LY, Kirihara K, Okigawa Y, Hasegawa M, Ding WX, Wang Y, Mukaida M, Zhou Y, Wei Q (2020) Extracting carrier mobility using a photoinduced charge transfer reaction: from conducting polymers to nanocarbon materials. Org Electron 78:105615

Mendoza AE, Andronescu C, Olean-Oliveira A (2024) Design of conducting polymer/metal-based nanocomposites as electrocatalysts for electrochemical energy conversion. Synth Metals: 117662

Min J, Jung Y, Ahn J, Lee JG, Lee J, Ko SH (2023) Recent advances in biodegradable green electronic materials and sensor applications. Adv Mater 35(52), Article 2211273. https://doi.org/10.1002/adma.202211273

Ncube LK, Ude AU, Ogunmuyiwa EN, Zulkifli R, Beas IN (2021) An overview of plastic waste generation and management in food packaging industries. Recycling 6(1):12

Paleo AJ, Sencadas V, van Hattum FWJ, Lanceros-Me´ndez S, Ares A (2014) Carbon nanofiber type and content dependence of the physical properties of carbon nanofiber reinforced polypropylene composites. Polym Eng Sci 54:117–128

Park J, Seo JH, Yeom SW, Yao C, Yang VW, Cai Z, Jhon YM, Ju BK (2018) Flexible and transparent organic phototransistors on biodegradable cellulose nanofibrillated fiber substrates." *Advanced Optical Materials* 6, no. 9 (2018): 1701140.

Patel GB, Singh NL, Singh F, Kulriya PK (2021) Effect of swift heavy ions irradiation on physicochemical and dielectric properties of chitosan and chitosan-Ag nanocomposites. Radiat Phys Chem 181:109288. https://doi.org/10.1016/j.radphyschem.2020.109288

Pech-Canul MI, Valdez S, Acosta-Enríquez MC, Hdz-García HM, Acosta-Enríquez EB, Flores-García JC (2019) A revamped classification of composite materials. In: Materials science and technology 2018, MS and T 2018. Association for Iron and Steel Technology, AISTECH, pp 1319–1325

Ruiz-Hitzky E, Darder M, Aranda P (2008) In bio-inorganic hybrid nanomaterials. In: Ruiz-Hitzky E, Ariga K, Lvov YM (eds). Wiley, pp 1–40

Saadi MA, Likhi FH, Nath MD, Jayan R, Zahin F, Thakur MS, Yuan Y, Islam MM, Panat R, Karim A, Ajayan PM (2024) Algae-derived nacre-like dielectric bionanocomposite with high loading hexagonal boron nitride for green electronics. ACS Nano 18(48):33081–33096

Saied MA, Ward AA (2020) Physical, dielectric and biodegradation studies of PVC/silica nanocomposites based on traditional and environmentally friendly plasticizers. Adv Nat Sci Nanosci Nanotechnol. https://doi.org/10.1088/2043-6254/ab9d17

Saini RK, Bajpai AK, Jain E (2017) Fundamentals of bionanocomposites. Biodegrad Biocompatible Polym Compos Process Prop Appl: 351–377

Salehi MH, Golbaten-Mofrad H, Jafari SH, Goodarzi V, Entezari M, Hashemi M, Zamanlui S (2021) Electrically conductive biocompatible composite aerogel based on nanofbrillated template of bacterial cellulose/polyaniline/nano-clay. Int J Biol Macromol 173:467–80. https://doi.org/10.1016/j.ijbiomac.2021.01.121

Samarjeet SS, Qibo Z, Nishu D, Vijay KT (2020) Carbon-based polymer nanocomposite for high-performance energy storage applications. Polymers 12:505

Sanjinés R, Abad MD, Vâju C, Smajda R, Mionić M, Magrez A (2011) Electrical properties and applications of carbon based nanocomposite materials: an overview. Surf CoatS Technol 206(4):727–733

Schmidt DS, Kraus T, González-García L (2024) Electrofluids with tailored rheoelectrical properties: liquid composites with tunable network structures as stretchable conductors. ACS Appl Mater Interfaces 16(33):43942–43950

Selvakumaran L, Lubineau G (2014) Electrical behavior of laminated composites with intralaminar degradation: a comprehensive micro-meso homogenization procedure. Compos Struct 109:178–188

Shakir MF, Rashid IA, Tariq A, Nawab Y, Afzal A, Nabeel M, Naseem A, Hamid U (2020) Emi shielding characteristics of electrically conductive polymer blends of PS/PANI in microwave and IR region. J Electron Mater 49:1660–1665

Sharma A, Kumar A (2016) Study of structural and electro-catalytic behaviour of amperometric biosensor based on chitosan/polypyrrole nanotubes-gold nanoparticles nanocomposites. Synthetic Metals 220:551–559

Shchipunov Y (2012) Bionanocomposites: green sustainable materials for the near future. Pure Appl Chem 84(12):2579–2607

Shukla SK, Parlak O, Shukla SK, Mishra S, Turner AP, Tiwari A (2014) Self-reporting micellar polymer nanostructures for optical urea biosensing. Ind Eng Chem Res 53:8509–8514. https://doi.org/10.1021/ie5012799

Sohi NJS, Bhadra S, Khastgir D (2011) The effect of different carbon fillers on the electrical conductivity of ethylene vinyl acetate copolymer-based composites and the applicability of different conductivity models. Carbon 49:1349–1361. https://doi.org/10.1016/j.carbon.2010.12.001

Supova M, Martynkova GS, Barabaszova K (2011) Effect of nanofillers dispersion in polymer matrices: a review. Sci Adv Mater. https://doi.org/10.1166/sam.2011.1136]

Tamjid E, Najafi P, Khalili MA, Shokouhnejad N, Karimi M, Sepahdoost N (2024) Review of sustainable, eco-friendly, and conductive polymer nanocomposites for electronic and thermal applications: current status and future prospects. Discover Nano 19(1):29

Tanaka J, Satake K, Miyamae T (1999) Temperature dependence of optical and electrical properties of doped polyacetylene. Synth Met 101(1–3):371–372

Tarhini A, Tehrani-Bagha AR (2023) Advances in preparation methods and conductivity properties of graphene-based polymer composites. Appl Compos Mater 30:1737–1762

Testa E, Barbera V, Fasoli E, Giese U, Belviso MR, Rossini P, Bruno D, Tettamanti G, Orlando M, Molla G, Casartelli M (2025) Electroconductive bionanocomposites from black soldier fly proteins for green flexible electronics. ACS Sustain Chem Eng

Thomas B, Raj MC, Joy J, Moores A, Drisko GL, Sanchez C (2018) Nanocellulose, a versatile green platform: from biosources to materials and their applications. Chem Rev 118:11575–11625

Tong ZS, Wu MZ, Ding JL, Hu QH, Zhu M, Chen L, Xu XL, Wan MX, Zhou HX (1992) Ion induced electrical conductivity versus chemical doping in polyaniline (PANI). Nucl Instrum Methods Phys Res Sect B Beam Interact Mater Ats 71(1):26–32

Trache D, Thakur VK, Boukherroub R (2020) Cellulose nanocrystals/graphene hybrids—a promising new class of materials for advanced applications. Nanomaterials 10(8):1523

Verma P, Saini P, Choudhary V (2015) Designing of carbon nanotube/polymer composites using melt recirculation approach: effect of aspect ratio on mechanical, electrical and EMI shielding response. Mater des 88:269–277

Wang C, He D, Cui Y (2018) Research progress in electrically conductive structural composites. J Mater Eng 46:1–13

Xia Q, Chen C, Yao Y, Li J, He S, Zhou Y, Li T, Pan X, Yao Y, Hu L (2021) A strong, biodegradable and recyclable lignocellulosic bioplastic. Nat Sustain 4(7):627–635

Xing X, Chen M, Gong Y, Lv Z, Han S-T, Ye Z (2020) Building memory devices from biocomposite electronic materials. Sci Technol Adv Mater 21(1):100–121

Xu W, Raychowdhury S, Jiang DD, Retsos H, Giannelis EP (2008) Dramatic improvements in toughness in poly (lactide-co-glycolide) nanocomposites. Small 4(5):662–669

Xue Q, Sun J (2016) Electrical conductivity and percolation behavior of polymer nanocomposites. Polym Nanocomposites: Electr Therm Prop: 51–82

Yao J, Ji P, Wang B, Wang H, Chen S (2018) Color-tunable luminescent acrofibers based on CdTe QDs-loaded bacterial cellulose nanofibers for pH and glucose sensing. Sens Actu B Chem 254:110–119. https://doi.org/10.1016/j.snb.2017.07.071

Young RJ, Kinloch IA, Gong L, Novoselov KS (2012) The mechanics of graphene nanocomposites: a review. Compos Sci Technol 72:1459–1476

Youssef AM, El-Aziz MEA, Abd El-Sayed ES, Moussa MA, Turky G, Kamel S (2019) Rational design and electrical study of conducting bionanocomposites hydrogel based on chitosan and silver nanoparticles. Int J Biol Macromol 140:886–894. https://doi.org/10.1016/j.ijbiomac.2019.08.199

Zhang H, Peng M, Cheng T, Zhao P, Qiu L, Zhou J, Lu G, Chen J (2018) Silver nanoparticles-doped collagen–alginate antimicrobial biocomposite as potential wound dressing. J Mater Sci 53(21):14944–14952

Zhang W, Wei L, Ma Z, Fan Q, Ma J (2021) Advances in waterborne polymer/carbon material composites for electromagnetic interference shielding. Carbon 177:412–426

Zhang Z, Hu L, Wang R, Zhang S, Fu L, Li M, Qi X (2024) Advances in Monte Carlo method for simulating the electrical percolation behavior of conductive polymer composites with a carbon-based filling. Polymers 16(4):545

Zhao L, Xia W, Zhang P (2021) Economical conductive graphite-filled polymer composites via adjustable segregated structures: construction, low percolation threshold, and positive temperature coefficient effect. J Appl Polym Sci 138

Zhao X, Wang Y, Chen X, Yu X, Li W, Zhang S, Meng X, Zhao ZM, Dong T, Anderson A, Aiyedun A (2023) Sustainable bioplastics derived from renewable natural resources for food packaging. Matter 6(1):97–127

Sujata Deb is an **Associate Professor** in the Department of **Physics** at University of Science and Technology Meghalaya (USTM), India. She holds a Ph.D. and an M.Sc. in Physics, along with B.Ed. from Gauhati University. With over two decades **of academic experience**, Dr. Deb has made significant contributions to the field. Dr. Sujata Deb has several peer-reviewed research publications in indexed, high-impact international journals of repute to her credit. Dr. Deb is a life member of several professional bodies. As a **journal reviewer** and **published author**, she possesses a deep understanding of nanostructured materials **and devices**. In addition to her research, Dr. Deb is a dedicated educator to undergraduate, graduate students and **Ph.D. scholars**, helping to shape the researchers of tomorrow.

Chapter 12
Bio-nanomaterials and Composites in Artistry and Cultural Heritage Preservation

Raphael Kanyire Seidu, Emmanuel Abankwah Ofori, George Kwame Fobiri, Benjamin Tawiah, and Shouxiang Jiang

Abstract The preservation of artistry and cultural heritage provides unique values and a deep understanding that connects the people to their identities in history. This ensures cultural continuity and safeguards the transfer of generational knowledge. Bionanomaterials provide a promising approach for the restoration and preservation of the various artworks while preserving the natural environment without hurting fauna. The unique combination of nanomaterials and biological components produces materials and composites which are lightweight, biocompatible, have good mechanical properties and, most importantly, are eco-friendly. This provides great opportunities for application in restoring and preserving historical artworks from environmental contaminants that result in short to long-term damages. In this chapter, we discuss the application of bio-nanomaterials and composites in artistry and cultural heritage preservation under three (3) distinct areas: bio-nano and composite materials in artistry; methods of fabrication and applications, economic role in conservation and restoration, safety concerns, challenges and regulatory issues. We

R. K. Seidu (✉) · B. Tawiah · S. Jiang (✉)
School of Fashion and Textiles, The Hong Kong Polytechnic University, Kowloon, Hong Kong SAR, China
e-mail: seiduraphael@gmail.com; raphael-kanyire.seidu@connect.polyu.hk

S. Jiang
e-mail: kinor.j@polyu.edu.hk

E. A. Ofori
Department of Materials Science and Engineering, Hubei University of Automotive Technology, Shiyan, Hubei, China

G. K. Fobiri
Department of Fashion Design and Textiles Studies, Kumasi Technical University, Kumasi, Ghana

B. Tawiah
Department of Industrial Art (Textile and Fashion), Kwame Nkrumah University of Science and Technology, Kumasi, Ghana

S. Jiang
Research Centre of Textiles for Future Fashion, The Hong Kong Polytechnic University, Kowloon, Hong Kong SAR, China

S. C. George and B. Tawiah (eds.), *Advanced Bionanocomposite Materials*, Advanced Structured Materials 241, https://doi.org/10.1007/978-981-96-8225-6_12

present case studies of real-world applications of these bionanocomposite nanomaterials in the preservation of artworks and culture. We further discuss the role of bio-nanocomposites in cultural anthropology research and conclude with insights into the most promising areas for future studies on bionanocomposite applications in artistry and cultural heritage preservation with a focus on reducing environmental and societal impacts.

Keywords Bionanomaterials · Bionanocomposites · Eco-friendly · Artistry · Cultural heritage · Preservation

12.1 Introduction

Artistry and cultural heritage are considered unique and valuable, hence embodying the identity and traditions of a group of people in a community. These artefacts and heritage from ancient architecture, and manuscripts to artworks convey historical events and happenings on the achievements, beliefs and lives of their ancestors. They further represent the valuable and unique records of creativity and human history (Doni et al. 2024). For example, ancient architecture such as the Egyptian pyramids and hieroglyphs plays a critical role in the education of the history, identity, and early civilization of the Egyptians. Cultural heritage involves both tangible and intangible elements that are inherent in the culture or traditions or customs of the people (Król et al. 2019; Md Ramli et al. 2016). To help promote cultural identity, the preservation of artistry and cultural heritage provides unique values and deep understanding that connect people to their identities or roots in history. This ensures cultural continuity and safeguards the transfer of generational knowledge, which is at risk of being lost or destroyed from history. It is thus significant to preserve and protect this artistry and cultural heritage from deterioration and extinction.

Some conventional techniques have been used for monitoring to ensure the protection and preservation of cultural heritage artefacts. These include restoration techniques of reinforcing, retouching, and drawing out missing areas, cleaning with agents, and using protective enclosures to minimize environmental effects. Conventional methods for preserving this artistry and cultural heritage lead to changes that are irreversible, and invasive and limit the prolonged stability of the artefacts (Doni et al. 2024; Pérez-Gandarillas et al. 2024). Some conventional solvents, on the other hand, for cleaning purposes are harmful and toxic, hence posing some risks to conservators and the artefacts (Casini et al. 2023). This alters and leads to residue on the original artefacts that would require some invasive approaches to repair (Baglioni et al. 2021). With such negatives, advanced methods have been explored to ensure a significant restoration and preservation of artistry and cultural heritage that are valuable to the history and culture of the people. As such, nanotechnology poses a significant approach and innovative solutions to preserve and protect this artistry and cultural heritage (Rodica-Mariana et al. 2017; Daniele and Taglieri 2010; Baglioni et al. 2015a) and prevent the natural ageing of artefacts (Dei and Salvadori 2006).

Nanotechnology involves the manipulation of nanoparticles at the nanoscale from 1–100 nm, which exhibit unique properties (such as thermal, chemical stability and improved strength) (Tawiah et al. 2024). The application of nanotechnology has enhanced cleaning and preservation without deterioration. Most especially, bio-nanomaterials and bionanocomposites provide a promising approach for restoring and preserving various artworks with less impact on the natural environment. Bio-nanomaterials, also known as biological nanomaterials, are obtained from biological sources or origins which are biodegradable, non-toxic and biocompatible due to their physical, mechanical, chemical and biological properties (Karki et al. 2023). These properties, coupled with the small size of the nanomaterials, play a critical role in their unique functional purposes for different applications (Boverhof et al. 2015). Alternatively, bio-nanocomposites are composed of a reinforcing agent (dispersed phase) and a biopolymer matrix (continuous phase), with enhanced thermal, barrier, mechanical, biocompatible and biodegradable properties (Farooq and Sharma 2024). These properties, together with the antimicrobial abilities of bio-nanomaterials and bionanocomposites, have advanced their application for the long preservation of artistry and cultural heritage.

A trend analysis was conducted on publications regarding bionanocomposites in cultural heritage and artistry with data extracted from the Scopus online database. This was done to further understand case studies of bionanocomposite applications for cultural heritage conservation. Relevant data was obtained by entering keywords including nanomaterials, biocomposites, cultural, heritage and artworks. The trend analysis revealed that the first publication was recorded in 2003 with a document. A document was again recorded in 2005, 2006 and 2008, each with no publication found in 2004 and 2007. However, an increase in publication was observed in 2010, as 10 documents were found in the same year. The trend also indicates a decrease in 2010 with a document being recorded. From 2012 to 2015, the publication trend increased (2 studies in 2012, 7 in 2013 and 8 in 2014). With a continuous fluctuation in publications, a good number was recorded in 2019 (14 documents) with the highest number of publications being recorded in 2023 (19 documents). The current year (2025) has recorded 3 studies in February, with the hope that a possible increase would be seen in the same year following the trend line (as shown in Fig. 12.1).

From the trend analysis, it could be said that research on bionano application in culture and artistry has been gaining popularity recently after discovering the concept of nanotechnology a couple of decades ago. Thus, it is important to investigate and discuss any potential long-term benefits and applications of bio-nanomaterials and composites in artistry and cultural heritage preservation. The objective of the chapter is to provide a comprehensive understanding of the prospects and application of bio-nanomaterials and composites for cleaning and restoration in artistry, cultural anthropology, cultural heritage and artefacts. Further discussion on case studies of real-world applications and the safety and regulatory issues of applying bio-nanomaterials and composites are provided. Lastly, key research opportunities and directions are provided, which form part of efforts to overcome challenges and foster advanced sustainable nanomaterials and nanocomposite applications in restoring and

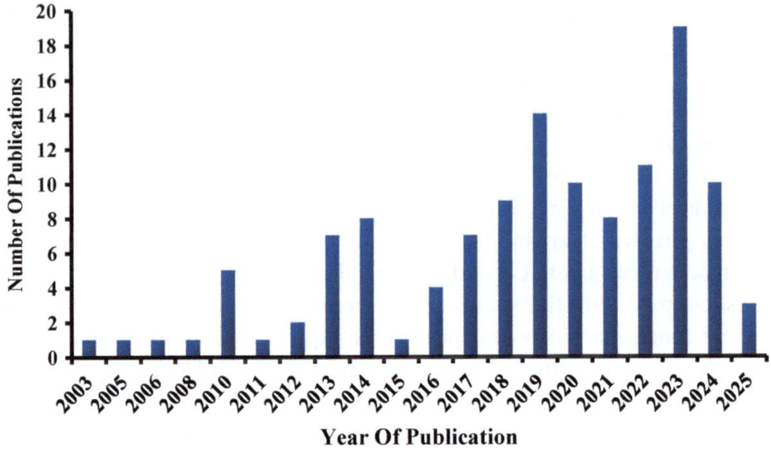

Fig. 12.1 Publication trend on bionanocomposite application in cultural heritage and artistry

preserving historical artworks, hence limiting environmental contaminants that result in short to long-term damages.

12.2 Bio-nanomaterials and Their Nanocomposites in Artistry

The application of bio-nanomaterials and nanocomposites in artistry holds significant importance due to their biodegradability, enhanced mechanical properties, and reversibility, which limits the degradation of artefacts, hence maintaining their historical value and authenticity. Due to the unique qualities of bio-nanomaterials and nanocomposites, they are used for the protection, preservation and restoration of historical documents, textiles, sculptures and paintings, among others. These artistry are considered very fragile and can easily deteriorate over time when exposed to or stored in rooms with unfavourable environmental conditions.

Historical papers are important ancient materials that contain key information and knowledge. These are exposed to frequent mechanical damages (such as deformation, cuts, cracks) and localized deterioration due to exogenous influences such as attacks from bacteria, humidity, light and air, among others (Fornari et al. 2022), hence, restoring and conserving these historical papers are imperative to prolong their life span. Studies have been advanced to apply bio-nanomaterials and nanocomposites to improve the resistance of this artistry to environmental conditions and repair or restore them from further degradation. As such, a study was conducted by Völkel et al. (Völkel et al. 2017) using nano-fibrillated cellulose (nanocellulose and phytate) suspensions to conserve historical papers. Herein, a film applicator (Fig. 12.2b) or brush was used to apply the suspensions onto the affected areas. Results show that the losses and cracks on the affected areas are consolidated for paper stabilization and conservation without any further damage. In a subsequent study, nano-fibrillated cellulose (nanocellulose and phytate) suspensions were applied on historical papers to prevent any degradation of iron gall ink (Völkel et al. 2020). Findings show the prevention of further damage; however, imperfections and cracks were sealed using the suspension as shown in Fig. 12.2c. This implies that the effective performance of these nanocellulose is influenced by the suspension medium, which influences the strength of the paper and the bonds. Confirming these results from an experimental study revealed a decrease in paper strength when nanocellulose was suspended in an ethanol solvent (Perdoch et al. 2022).

Nanocrystalline cellulose is another promising material for restoration treatments of historical papers due to its high crystallinity, optical transparency, thermal and mechanical properties, non-toxic and widely abundant (Zugenmaier 2001; Nogi et al. 2009; Nishino et al. 1995; Dufresne 2013). As such, cellulose nanocrystals were applied for paper conservation and restoration through reversible consolidation (Operamolla et al. 2021). Herein, the fabrication of neutral cellulose nanocrystals ensured the conservation of the sampled papers without compromising the mechanical and pH properties during ageing. This reversible treatment process can easily be removed by using gellan hydrogel without causing any damage to the papers (Fig. 12.3).

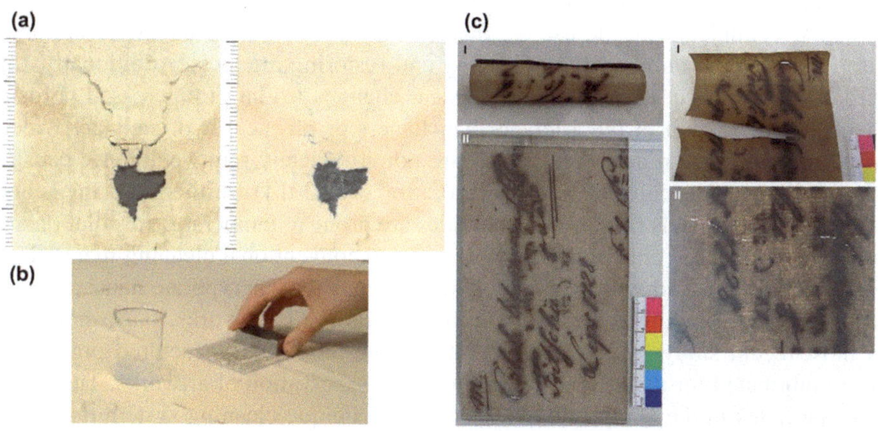

Fig. 12.2 **a** Losses and cracks after mechanical damages (left) and treated surface after nanocellulose application, **b** film applicator used to apply the nanocellulose (Völkel et al. 2017). (This is an open-access article that is distributed under the terms of the creative commons attribution 4.0 international licence). **c** Appearance of the samples before (I) and after (II) treatments (Völkel et al. 2020). (This is an open-access article that is distributed under the terms of the creative commons attribution 4.0 international licence)

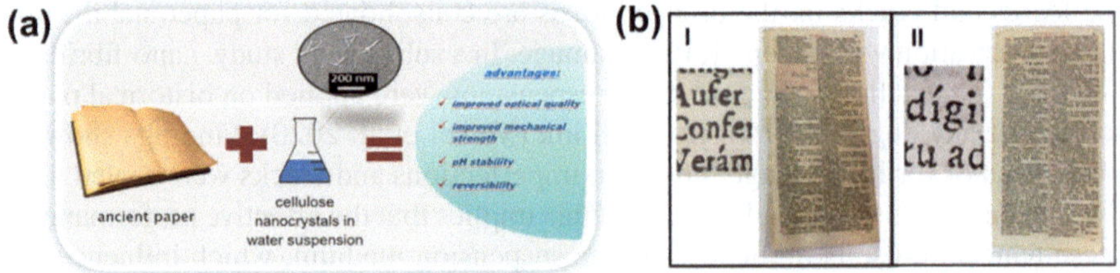

Fig. 12.3 **a** Illustration of the conservation approach, **b** photographs of treated details and pages with (I) water (pristine) and (II) N_CNC. (Operamolla et al. 2021). (This publication is licenced under CC-BY 4.0)

12.3 Bio-nanomaterials and Composites for Cleaning and Restoration

Cultural heritage works or artefacts exposed to weathering conditions over a long period experience deteriorating effects. This results in efforts to protect their surfaces using coatings, clean and remove unwanted pollutants and strengthen the artefacts from deterioration (Caruso et al. 2023; David et al. 2020). According to Ortega-Morales et al. (Ortega-Morales and Gaylarde 2021), this requires the adoption of compatible conservation methods for the artefacts. Physical or mechanical cleaning processes coupled with the use of solvents may pose harmful threats to the operator and the environment when cleaning and restoring artefacts. Additionally, this process may produce abrasions or physical changes leading to increased porosity hence, deterioration of the artefacts. Thickeners or physical hydrogels are widely used in conservation activities, which are not completely removed, hence causing the swelling and leaching of paint layers (Casoli et al. 2014). Furthermore, thickeners or physical hydrogels from polysaccharides are prone to temperature oscillation and microbial attacks (Feller and Wilt 1991). These conventional cleaning materials or solvents are replaced with biodegradable materials that incorporate nanostructures for enhanced cleaning and restoration of cultural heritage works. For example, in a most recent study by Poggi et al. (2023), bio-based nanostructured organogels were fabricated for the effective cleaning and conservation of artworks. The use of nanoparticles in the composite matrix creates a unique cleaning system that limits degradation processes, pollutants deposition and preserves the appearance of the heritage materials (David et al. 2020). It is thus important that these nanoparticles have good adaptability properties, low cost, non-toxic, chemically and biologically inert and thermally stable (Baglioni and Giorgi 2006). Aside from enhancing the mechanical properties of historical artefacts, the adoption of bio-nanomaterials and composites for cleaning and restoration further enhances the safety of the conservator to limit any health threats.

Due to the unique properties of bio-nanomaterials and composites, these innovative solutions have been applied in the cleaning and restoration of cultural heritage works. Cleaning entails a critical and delicate process of removing aged coatings, dust and dirt from an original artefact without destroying it. These advanced materials replaced try-and-error approaches such as rudimental soaps, ash or biological fluids (Stoner and Rushfield 2020). Nanocellulose composites, a notable example of bio-nanomaterials and composites that combine plant fibres and polymers, are used to conserve cultural heritage artefacts (Waked 2011). They play a critical role in restoring and cleaning historical papers, paintings and wood (Fornari et al. 2022). There is frequent discolouration of cultural heritage stones due to the corrosive nature of unprotected and exposed copper. To overcome this issue, ethylene diamine tetra acetic acid (EDTA) loaded in the bacterial nanocellulose hydrogel was fabricated as a cleaning agent (Sonaglia et al. 2024) to clean copper stains for about 120 min on the marble structure. The success of the findings presents an alternative cleaning agent, bacterial nanocellulose, as eco-friendly with high mechanical and crystallinity properties for the conservation of stone heritage. Other approaches utilized CuO nanoparticles from plant extracts such as orange peel, Lantana, Pongamia and neem leaves for preserving wood (Shiny et al. 2019). Findings revealed that the extracted CuO nanoparticles provided six-months termite protection, hence limiting biodeterioration of the wood in the environment.

For restoration, this process entails the use of effective repair and treatments to preserve the intangible and tangible aspects of cultural heritage and artefacts, hence ensuring longevity, authenticity and integrity. This process prevents or protects from any damage, decay and corrosion of cultural heritage and artefacts when exposed to environmental conditions. To ensure the longevity of wood, hence preventing any deterioration or decay, crystalline nanocellulose doped with siloxane derivatives (PDMS) or lignin was fabricated to reduce wood rot or decay. Results showed that after ten (10) impregnation cycles, there was an improvement in the stiffness of the treated wood. In a different approach, coatings made from photocurable resin/nanocellulose composite were fabricated to protect wood (Cataldi et al. 2017). Results showed an increased stiffness, glass transition temperature and dimensional and thermal stability when treated. Further results in Fig. 12.4a revealed improved hydrophobic properties of the wood treated with photocurable resin/nanocellulose composite, hence enhancing its water repellency (about 130° of contact angle) and limiting the growth of mould. There was also an improved surface quality and hardness even after the water sorption test, with no damages recorded for wood-treated with photocurable resin/nanocellulose composite (Fig. 12.4b). A protective coating from nanocellulose composite was fabricated using lignin nanoparticles, cellulose nanocrystals and cellulose nanofibrils (Fig. 12.4c), to conserve cellulosic artefacts such as fabric mock-ups, paper and wood (Camargos et al. 2022). Results show their reversible properties and abilities to block UV and prevent moist–heat ageing that keeps the surfaces from deterioration. The wettability test showed that the protective coatings prevented moisture leakage and humidity loading with the surfaces, hence making them hydrophobic. These properties are effective in protecting restored artefacts (Santos et al. 2015).

Subsequently, nanocellulose composites are used for the conservation of paper (Kryg et al. 2024). For example, the application of nanocellulose composites provides moisture resistance protection for paper, does not alter the paper but improves

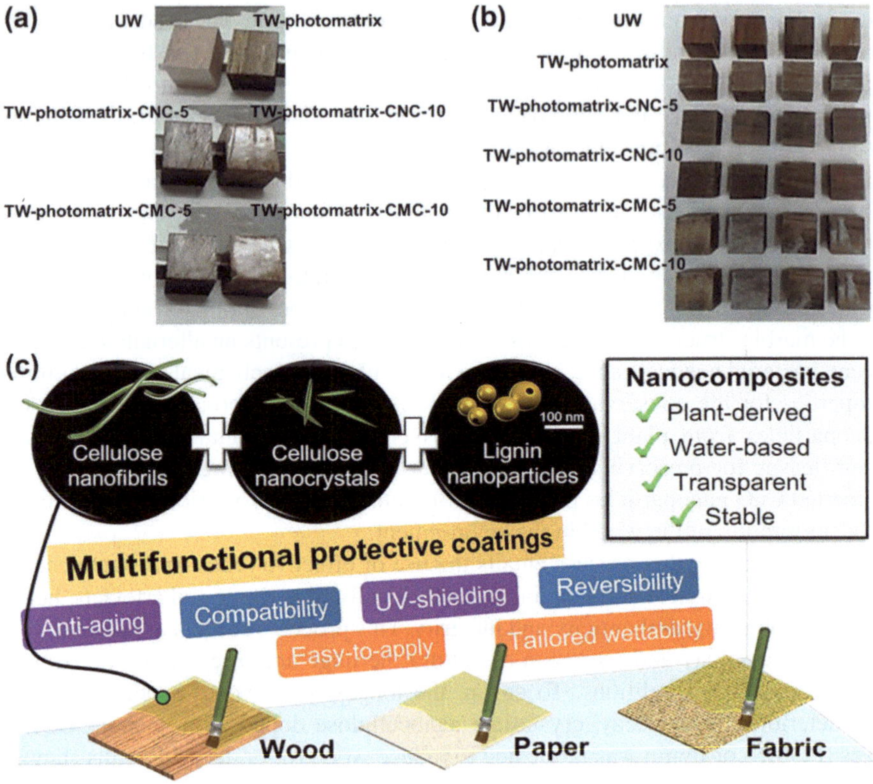

Fig. 12.4 **a** Hydrophobic test for the walnut wood samples, **b** water sorption test for the walnut wood samples. Reproduced with permission from Cataldi et al. (2017) Copyright 2017 Elsevier. **c** Reproduced with permission from Camargos et al. (2022) Copyright 2022 American Chemical Society

the mechanical properties. The fabrication of nanosized $Mg(OH)_2$, nano-TiO_2 and hydroxyethyl cellulose as nanocomposites with multifunctional abilities results in resistance to fungal growth, reinforcement of the mechanical properties and enhanced longevity of the paper (Wang et al. 2013). Other studies fabricated CNC-based nanocomposites (Camargos et al. 2017) and cellulose nanofibrils (Camargos et al. 2015) to improve the mechanical properties of paper consolidation. Aside from paper, artefacts made of stones or marble are also subjects of deterioration from external conditions over time. Studies fabricated materials with self-cleaning abilities for the protection of the surface of stones, such as using superhydrophobic nanocomposite film from TiO_2–ZnO (Speziale et al. 2020), bio-nanocomposites from SiO_2–PLA (Pedna et al. 2016), and PLA/montmorillonite clay (Ocak et al. 2015; Ocak 2013). These advancements limited the deterioration (surface erosion) of the surfaces that could disfigure them.

The preservation of heritage artefacts made of metal remains a critical task to limit pollutant accumulation, tarnishing and corrosion. Conventional processes for conserving historical artwork based on metals lead to damage and pose serious threats to the environment and the health of humans, hence the need for sustainable approaches (Passaretti et al. 2021). The advancement of bio-nanomaterials and composites provides potential solutions which ensure the stability and durability of the artefacts restored and protected from deterioration (Doni et al. 2024). For example, microbial derivatives contained in green gels were fabricated as a bio-nanomaterial for the removal of tarnish and corrosion on historical metal artworks (Passaretti 2023). The effectiveness of this approach will ensure the long-term preservation and integrity of archaeological artefacts.

12.4 Bionanocomposites for Cultural Heritage and Artefacts

Cultural heritage materials frequently present practitioners with several difficulties when tackling conservation issues. High selectivity, precise and controlled solvent administration, minimal exposure to hazardous chemicals and solvent access to extremely discrete locations are all necessary for the curative effect. The choice and use of conservation products that are safer for the artefacts and the conservators have also been critically examined as the green approach to conservation gains traction. Nanocomposites offer creative ways to protect and preserve materials, and they have demonstrated significant promise in the conservation of cultural heritage items (Kanth and Soni 2023; Paolino et al. 2024; Mohamed et al. 2024). For instance, the cleaning of wall paintings in the Brancacci Chapel in Florence, Italy, in the late 1980s marked the first instance of nanoscience being used for art conservation (Baglioni et al. 2015b). During restoration, wax stains were removed from the Brancacci Chapel paintings' surface using an oil-in-water microemulsion containing dodecane nanodroplets stabilized in water by a surfactant. In another discovery, a heavy layer of old acrylic coatings was removed off the surface of wall paintings in Venice's San Salvador cathedral using an hMHEC dispersion that contained an oil-in-water microemulsion (Tawiah et al. 2024; Baglioni et al. 2015c) (see Fig. 12.5a, b). Similarly, nanostructured calcium hydroxide ($Ca(OH)_2$) became a very useful restoration material towards the end of the twentieth century since it could be used to reinforce weak limestones and murals (Ambrosi et al. 2001), and neutralizing the acidity that degrades cellulose in wood and paper artefacts (Giorgi et al. 2009). One notable example of this practice is the consolidation of Maya wall paintings in Calakmul, Mexico (Fig. 12.5c) (Baglioni and Chelazzi 2013).

The use of bionanocomposites in the preservation of cultural heritage artefacts, however, offers a promising way to improve the longevity and protection of these priceless items due to advancements in materials science and history. Through the utilization of bio-based polymers and nanoparticles' special qualities, scientists are

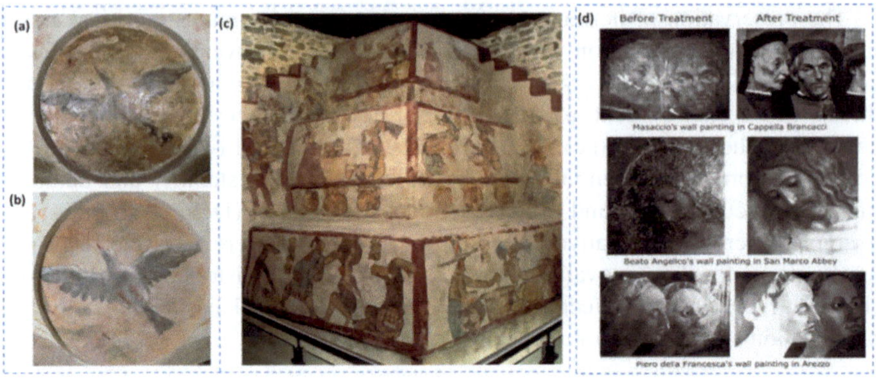

Fig. 12.5 **a, b** Removal of aged acrylic coatings from the wall paintings of the San Salvador church sacristy in Venice, Italy. **c** Maya wall paintings in the United Nations educational, scientific and cultural organization world heritage site of Calakmul, Mexico. Reproduced with permission from Baglioni et al. (2015c). Copyright 2015 Springer Nature Limited. **d** Difference between pre- and post-restoration of artistic heritages (Waked 2011). (This is an open-access publication)

opening the door to efficient and sustainable preservation techniques that can protect cultural heritage for the next generation. Gomez-Villalba (Gomez-Villalba et al. 2023a) reports that the University of Florence in Italy has restored damaged artwork due to the aggressive action of pollutants that have been deposited on its surface. The same organization utilized wood restoration to repair old galleons that had been damaged by chemicals and water. Similarly, effective outcomes include the preservation and restoration of archaeologically significant wall paintings that have been severely impacted over ages by extreme humidity and abrupt temperature fluctuations (Baglioni and Giorgi 2006). In a similar vein, Waked et al. (2011) discuss the use of nanomaterials to preserve wall paintings, such as those by Beato Angelico within Florence's San Marco Abbey and Piero della Francesca in Arezzo (see Fig. 12.5d). Recently, Ocak et al. (2015) investigated the use of poly L-lactide-based montmorillonite nano clay bio-nanocomposites as protective coatings for marble surfaces from pollution. The study findings report a significant improvement on the surface with an inhibition of the sulphation reaction when exposed to an acidic environment.

Egil et al. (2022) explored the application of calcium/chitosan nanoparticles (Ca/CS NPs) as a novel way to protect historical artefacts made of paper against fungus and bacteria. According to the results, when calcium was added to chitosan nanoparticles, the antibacterial activity of the nanoparticles against microbial strains like *E. coli and C. albicans* was significantly enhanced. According to antibacterial testing, Ca/CS NPs had a more potent inhibitory impact, particularly when it came to gram-positive bacteria like Bacillus subtilis and Bacillus megaterium. Because acidic environments speed up degradation and maintaining a neutral or slightly alkaline pH is crucial for the durability of paper artefacts, NPs' stability over time was an important factor. Additionally, studying the facade of Neb-Maat-Ra's palace gate from ancient Egypt, Ramadan et al. (2024) investigated the use of chitosan (Cs), chitosan nanoparticles (CsNPs), and silica-loaded chitosan nanoparticles (SiO_2@CsNPs) to improve the consolidation and protection of mural paintings. The results demonstrate that biopolymer-based materials have substantial advantages for mural conservation, particularly when combined with nanosilica, providing a sustainable substitute for conventional synthetic consolidants. This antimicrobial effect reduces biodeterioration risks, which are common in exposed cultural heritage artefacts.

12.5 Case Studies of Real-World Applications

The application of bio-nanocomposites in artistry and cultural heritage preservation is evident in everyday approaches to protecting cultural products, including ceramics, stone, wooden, and metallic artefacts, among others. Most bio-nanomaterials are applied to artefacts in the form of coating for preservation. The films or coats serve significant preservation to the artefacts by sealing dents and cracks on artefacts, serving as antimicrobials, UV protection, eco-friendly, self-healing capabilities, increasing lustre for aesthetic preservations, etc. (Kanth and Soni 2023; Mhd Haniffa et al. 2016). These significant properties of bio-nanocomposites have called for several reports from scholars on their application in cultural product preservation. To begin with, Ocak et al. (2015) report the preparation and use of a bio-nanocomposite comprising a biodegradable polymer (poly-L-lactide) and montmorillonite clay as a protective bio-nanocomposite material for marble surface protection. In this exploration, the poly-L-lactide prepared into a solution was added to the montmorillonite clay, which was also prepared into a solution, after which the composite was uniformly mixed and coated on the marble surface. Findings reveal an increased hydrophobic property of the marble surface with the coated film for effective protection from atmospheric pollutants. Again, a study investigating suitable bionanocomposite material for art preservation utilized elephant grass as biomass material to prepare a sustainable coating to be applied on cellulose artefacts for preservation (Camargos et al. 2022). The quest to go green in cultural heritage preservation is also observed in a study that investigates the manufacturing of sustainable materials for cultural heritage preservation using polymeric gels and oligomers to obtain a new class of chemicals for art preservation (Bandelli et al. 2024). It is reported that the bio-nanocomposite coating possesses excellent preservation abilities on paper, wood and fabric art pieces, coupled with good UV protection for the artefacts. It is also evident in previous studies that using colloids and nanomaterials science in cleaning and consolidation is a step towards conserving artefacts (Amornkitbamrung et al. 2020; Baglioni et al. 2013; Chelazzi and Baglioni 2023). For instance, Jiang et al. (2025) recently extensively reviewed nanocellulose composites for paper-based cultural heritage preservation and established their significant performance in UV protection, antimicrobial function, antioxidation, flame retardancy, self-cleaning, etc.

Moreover, Hafez and Biskos (2024) report a bionanocomposite material prepared from hydroxyapatite nanocrystals and polyelectrolyte multilayers applied in situ on a sample marble to investigate its performance characteristics. The manufactured nanostructured composite coating is reported to possess hydrophilicity, self-healing properties, improved adhesion and protection for stone artefacts. It is also observed that using traditional consolidants for consolidating mural paintings in hypogea has drawbacks of surface incompatibility, harsh microclimatic conditions, mechanical resistance and harmful effects of formulating solvents (Iafrate et al. 2023). To address these challenges, Iafrate et al. (2023) explored pure and crystalline calcium hydroxide nanoparticles dispersed in water as an innovative aqueous nano lime dispersion to be applied to mural paintings for conservation. Findings reveal good cohesion and the ability of the aqueous material to preserve the aesthetic features of mural paintings in hypogenic environments. Similarly, a methodology for applying eco-friendly nanocomposites on stone artworks has recently been reported where natural biocides, including oregano essential oil and eugenol, are used on the artefact through direct application in solution form and encapsulation technique with silica nanocapsules (Bartoli et al. 2024). The image of the silica nanocapsules in various states, including the absence of biocides, oregano oil-loaded, and eugenol-loaded, respectively, indicate almost identical morphologies of nano containers. Further results indicate less than 600 nm diameter distribution of the nanoparticles, with the densest area showing 200 nm and even less. This presupposes that a larger portion of the nanoparticles have ≤ 200 nm. Again, with a microscope of $20 \times$ magnification in visible light and ultraviolet green and red filters, the efficacy of biocides (eugenol-loaded and oregano oil-loaded) was tested at six intervals of time and presented the efficiency trend. With the red fluorescence indicating vitality, both the eugenol-loaded and oregano oil-loaded nanocapsules showed antimicrobial activity, demonstrating good efficacy for protecting cultural heritage through an environmentally friendly approach.

12.6 Bionanocomposites in Cultural Anthropology

Bio-nanocomposites are potential materials in cultural anthropology to promote and preserve cultural heritage. Artworks are invaluable cultural heritage worldwide and are worth keeping; hence, scientists, researchers and curators are searching for appropriate conservational and eco-friendly methodologies to promote and preserve such products (Fornari et al. 2022; Tortora 2020). As cultural anthropology involves the study of the present and past behaviours of humans and societies, research in this area is anchored on visual arts, and these have been employed by researchers for a deep understanding of a society, its values, and politics (Shiya 2023). For instance, visual arts such as paintings and drawings are good research tools in cultural anthropology (Soukup and McCallum 2020; Chakraborty et al. 2022). The multifaceted nature of nanotechnology makes it an effective technology in the cultural anthropology domain. The value of cultural artefacts is maintained through nanotechnology for the continuous education of the world on the way of life of various societies and people globally. It is observed that cultural heritage such as paintings, street art, textiles, sculptures, and monumental pieces are attacked by unfriendly weather conditions, eventually diminishing their aesthetic and philosophical functions (Awad and Baias 2020; Coletti 2024; Mascitelli et al. 2023; Pagnin et al. 2023; Tripa et al. 2023). Oil paintings, for instance, are prone to chemical deterioration and mechanical damage and demand innovative approaches to maintain their value effectively (Khalid et al. 2024). Following this, a non-invasive method of detecting deterioration in paintings is proposed, where a quantitative reflectance transformation imaging surface normal is used to identify the morphology of paintings for cultural heritage preservation (Manfredi et al. 2014). In the quest to go green in cultural heritage preservation, adopting bio-nanomaterials is an innovative and sustainable step to help advance the field of cultural anthropology. In this vein, Tripa et al. (2023) investigated biomaterials for cultural heritage preservation. They established that plant extracts and essential oils are potential eco-friendly materials for the biotreatment of cultural heritage since they possess strong antibacterial, antifungal, and insecticidal properties. It is noted that the surface area-to-volume ratio of a nanoparticle significantly influences its performance (Mekuye and Abera 2023), and applying materials at their nano state yields effective results. The exceptional properties of materials are shown at their nanometre scale, making them a preference in such states over their bulk state (Otero et al. 2023). Applying biomaterials at the nanoscale makes room for full exploration of their potential in conserving artworks in cultural anthropology.

In exploring bio-nanomaterials in cultural anthropology, a focus is placed on the sustainable use of preservative materials to protect the environment and the artefacts. In conventional preservative measures with composite materials, synthetic polymer materials have received critiques from several studies on their unsustainable nature, lack of compatibility with substrates, and insufficient preservative performance to artefacts (Doni et al. 2024; Baglioni et al. 2015b; Galvagno et al. 2024; Rodriguez-Navarro and Ruiz-Agudo 2018; Ziegenbalg et al. 2018). It is established that bio-based materials classified basically into polysaccharides, protein and polyester are sustainable alternatives to synthetic materials for cultural heritage preservation for subsequent advancement in cultural anthropology (Caruso et al. 2023). As such, bio-nanocomposites from various sustainable materials have been explored by scientists for conservators in cultural anthropology. Vizárová et al. (2011) explored gelatin obtained from fish and mammals for consolidating baroque oil paintings on canvas as a sustainable alternative to synthetic preservative chemicals. It is reported that the prepared recipes, apart from wax, are appropriate restoration media for oil paintings. A previous study on sustainable preservative materials also explored nanoscale consolidants for waterlogged archaeological wood preservation using cellulose and lignin (Antonelli 2020). The study utilized lignin nanoparticles, bacterial nanocellulose and cellulose nanocrystals as consolidants to be applied to the wooden cultural heritage. It is reported that the bacterial nanocellulose exhibited good penetration into the wooden artefacts compared to the cellulose nanocrystals and lignin nanoparticles, although all the consolidants are potential promoters of cultural anthropology. Another instance of the bio-nano application in cultural anthropology is found in a study in which bio-nanocomposite films were prepared using chitosan and pectin together with halloysite nanotubes acting as antioxidant molecule carriers and a compatibilizer for cultural heritage protection (Infurna et al. 2020). The bio-nanocomposite film is reported to be a potential protective material for cultural heritage.

To improve the performance of bio-based composites, they are combined with nanomaterials (Caruso et al. 2023). Enhancing cultural anthropology through nanotechnology revolves around the sustainable and effective preservation of cultural heritage with bio-nanocomposite materials. Cultural heritage produced with wood, canvas, paper, metals, concrete, etc., has been successfully restored and protected over the decades through nanotechnology, leading to cultural anthropology advancement (Doni et al. 2024; David et al. 2020; Kryg et al. 2024). It is worth noting that marrying nanotechnology with cultural anthropology, especially in the presence of biomaterials, is a strategic step in cultural and environmental preservation for sustainable development.

12.7 Safety Concerns, Challenges and Regulatory Issues

Artistry and cultural heritage preservation are critical to maintaining cultural traditions and identity, providing great insights into historical happenings and events, hence serving as educational resources for generations. This preservation is advanced using innovative solutions with bio-nanomaterials and composites, however, they pose some safety concerns, challenges and regulatory issues. Safety concerns associated with these innovative materials must be addressed to ensure their effective use for the conservation of historical artefacts. The potential toxicity of nanomaterials for bio-nanomaterials and composites is of great concern since the small particle size could easily enter the body and the environment. This could lead to serious threats to humans and the environment, a situation which requires appropriate protective equipment to limit such risks, especially for humans in the short to long term (Gomez-Villalba et al. 2023b). Even though the application of bio-nanomaterials and composites improves the conservation of historical and valuable materials for a long time, there are still some challenges that need to be addressed. The compatibility of these advanced materials with the cultural and historical artefacts remains a critical challenge to help prevent further deterioration. These cultural and historical artefacts are inherent with different material compositions that could be altered by the introduction of chemicals. Hence, the use of these bio-nanomaterials and composites should be compatible and not affect the appearance and properties of the cultural and historical artefacts. This implies that it is crucial to ensure the physical and chemical compatibility of bio-nanomaterials and composites to prevent any degradation reactions (Baglioni et al. 2021). The issue of reversibility of residues on artefacts for future cleaning and restoration treatment is of great importance. This is critical to ensure that the original appearance is not permanently damaged or destroyed. Hence, the fabrication of bio-nanomaterials and composites must adhere to this principle, where future treatments could easily remove and replace such materials without permanently altering the artefacts (Doni et al. 2024). This further leads to the issue of ageing and the stability of these treatments in the long term. It is important to ensure that their application would not be affected by external factors such as temperature, light and humidity to cause rapid degradation and discolouration of the original artefacts over time.

Regulatory issues remain a critical problem that influences the application of bio-nanomaterials and composite treatments on historical artefacts for longevity without any damage. There are limited universally standardized protocols or guidelines governing the safe application of these materials on historical artefacts. This situation could lead to varying practices that influence effectiveness and safety. Additionally, conservators must be effectively trained through standardized certification programmes that address core areas of regulatory and ethical consideration when handling bio-nanomaterials and composites. Alternatively, risk and safety assessment frameworks must effectively address the disposal, handling and production hazards potential associated with bio-nanomaterials and composites. For example, the control and monitoring of nanomaterials usage in the fabrication of bio-nanomaterials and composites is critical to limit any pollution to the environment when applied during cleaning, restoration or conservation practices (Mastrangelo et al. 2024).

12.8 Conclusions and Future Research Opportunities

The preservation of artistry and cultural heritage using bio-nanomaterials and composites presents significant innovations in the field. These materials are inherent with properties such as compatibility, stability and enhanced durability that limit further damage to these historical artefacts in the long term. These artistry and cultural artefacts are preserved without any alteration to their appearance that influences their value and originality. The use of bio-nanomaterials and composites has replaced the traditional methods of preserving artistry and cultural artefacts, which usually alter or cause some significant damage. Aside from these significant benefits, the issues of potential health and safety risks cannot be overlooked when using bio-nanomaterials and composites. This is due to the influence of the small nanoparticle sizes that can easily penetrate the body and environment, posing serious threats. Thus, material stability, compatibility, ethical, regulatory and environmental considerations must be paramount and strict ethical guidelines must be followed to ensure the protection of the artefacts, environment and conservators. Premise on these, future research opportunities can be directed in key areas like evaluation of material reversibility and compatibility, testing procedures on durability and stability, and development of regulatory and safety frameworks. Additionally, future studies could conduct research in other key areas like assessing the health risk and toxicity levels of bio-nanomaterials and composites. This would provide for an effective understanding of the health hazards of these materials, which still drive a significant engagement with stakeholders to develop ethical guidelines and frameworks. This approach will ensure the safe usage and application of bio-nanomaterials and composites for the conservation of artistry and cultural artefacts without compromising on their appearance, originality and integrity.

References

Ambrosi M et al (2001) Stable dispersions of $Ca(OH)_2$ in aliphatic alcohols: properties and application in cultural heritage conservation. Prog Colloid Polym Sci 118:68–72

Amornkitbamrung L et al (2020) Comparison of trimethylsilyl cellulose-stabilized carbonate and hydroxide nanoparticles for deacidification and strengthening of cellulose-based cultural heritage. ACS Omega 5(45):29243–29256

Antonelli F et al (2020) Cellulose and lignin nano-scale consolidants for waterlogged archaeological wood. Front Chem 8

Awad WM, Baias M (2020) How mobile NMR can help with the conservation of paintings. Magn Reson Chem 58(9):792–797

Baglioni P, Giorgi R (2006) Soft and hard nanomaterials for restoration and conservation of cultural heritage. Soft Matter 2(4):293–303

Baglioni P et al (2013) Colloid and materials science for the conservation of cultural heritage: cleaning, consolidation, and deacidification. Langmuir 29(17):5110–5122

Baglioni P, Chelazzi D (eds) (2013) Nanoscience for the conservation of works of art. The Royal Society of Chemistry

Baglioni P, Chelazzi D, Giorgi R (2015a) Innovative nanomaterials: principles, availability and scopes. In: Nanotechnologies in the conservation of cultural heritage: a compendium of materials and techniques. Springer Netherlands, Dordrecht, pp 1–14

Baglioni P, Chelazzi D, Giorgi R (2015b) Consolidation of wall paintings and stone. In: Nanotechnologies in the conservation of cultural heritage: a compendium of materials and techniques. Springer Netherlands, Dordrecht, pp 15–59

Baglioni P, Carretti E, Chelazzi D (2015c) Nanomaterials in art conservation. Nat Nanotechnol 10(4):287–290

Bandelli D et al (2024) New sustainable polymers and oligomers for cultural heritage conservation. Chem Sci 15(7):2443–2455

Baglioni M et al (2021) Advanced materials in cultural heritage conservation. Molecules 26(13):3967

Bartoli F et al (2024) In vitro viability tests of new ecofriendly nanosystems incorporating essential oils for long-lasting conservation of stone artworks. Gels 10(2):132

Boverhof DR et al (2015) Comparative assessment of nanomaterial definitions and safety evaluation considerations. Regul Toxicol Pharmacol 73(1):137–150

Camargos CHM, Figueiredo JCD, Pereira FV (2017) Cellulose nanocrystal-based composite for restoration of lacunae on damaged documents and artworks on paper. J Cult Herit 23:170–175

Camargos CHM et al (2022) Protective coatings based on cellulose nanofibrils, cellulose nanocrystals, and lignin nanoparticles for the conservation of cellulosic artifacts. ACS Appl Nano Mater 5(9):13245–13259

Camargos CHMD et al (2015) Cellulose nanofibril-based composites: performance evaluation and potential applications for restoration of cultural heritage, in 14. Brazil MRS meeting. Rio de Janeiro, RJ (Brazil)

Caruso MR et al (2023) A review on biopolymer-based treatments for consolidation and surface protection of cultural heritage materials. J Mater Sci 58(32):12954–12975

Casini A, Chelazzi D, Baglioni P (2023) Advanced methodologies for the cleaning of works of art. Sci China Technol Sci 66(8):2162–2182

Casoli A, Di Diego Z, Isca C (2014) Cleaning painted surfaces: evaluation of leaching phenomenon induced by solvents applied for the removal of gel residues. Environ Sci Pollut Res 21(23):13252–13263

Cataldi A et al (2017) Photocurable resin/nanocellulose composite coatings for wood protection. Prog Org Coat 106:128–136

Chakraborty M, Das K, Mukherjee K (2022) Visual anthropology: a systematic representation of ethnography. Athens J Soc Sci 10(1):95–110

Chelazzi D, Baglioni P (2023) From nanoparticles to gels: a breakthrough in art conservation science. Langmuir 39(31):10744–10755

Coletti C (2024) Climate change threats to stone cultural heritage: state of the art of quantitative damage functions and new challenges for a sustainable future. Heritage 7(6):3276–3290

Daniele V, Taglieri G (2010) Nanolime suspensions applied on natural lithotypes: the influence of concentration and residual water content on carbonatation process and on treatment effectiveness. J Cult Herit 11(1):102–106

David ME et al (2020) Nanomaterials used in conservation and restoration of cultural heritage: an up-to-date overview. Materials 13(9):2064

Dei L, Salvadori B (2006) Nanotechnology in cultural heritage conservation: nanometric slaked lime saves architectonic and artistic surfaces from decay. J Cult Herit 7(2):110–115

Doni M, Fierascu I, Fierascu RC (2024) Recent developments in materials science for the conservation and restoration of historic artifacts. Appl Sci 14(23):11363

Dufresne A (2013) Nanocellulose: a new ageless bionanomaterial. Mater Today 16(6):220–227

Egil AC et al (2022) Chitosan/calcium nanoparticles as advanced antimicrobial coating for paper documents. Int J Biol Macromol 215:521–530

Farooq K et al (2024) Bio-nanocomposites: fundamentals and recent advances. In: Sharma K et al (eds) Novel bio-nanocomposites for biomedical applications. Springer International Publishing, Cham, pp 1–35

Feller RL, Wilt MH (1991) Evaluation of cellulose ethers for conservation. Getty Publications

Fornari A et al (2022) A review of applications of nanocellulose to preserve and protect cultural heritage wood, paintings, and historical papers. Appl Sci 12(24):12846

Galvagno E et al (2024) Present status and perspectives of graphene and graphene-related materials in cultural heritage. Adv Func Mater 34(13):2313043

Giorgi R et al (2009) Nanoparticles of calcium hydroxide for wood deacidification: decreasing the emissions of organic acid vapors in church organ environments. J Cult Herit 10:206–213

Gomez-Villalba LS, Salcines C, Fort R (2023a) Application of inorganic nanomaterials in cultural heritage conservation, risk of toxicity, and preventive measures. Nanomaterials (Basel) 13(9)

Gomez-Villalba LS, Salcines C, Fort R (2023b) Application of inorganic nanomaterials in cultural heritage conservation, risk of toxicity, and preventive measures. Nanomaterials 13(9):1454

Hafez IT, Biskos G (2024) Bioinspired nanostructured hydroxyapatite-polyelectrolyte multilayers for stone conservation. J Colloid Interface Sci 674:459–473

Iafrate S et al (2023) New perspectives for the consolidation of mural paintings in hypogea with an innovative aqueous nanolime dispersion, characterized by compatible, sustainable, and eco-friendly features. Nanomaterials 13(2):317

Infurna G et al (2020) Bionanocomposite films containing halloysite nanotubes and natural antioxidants with enhanced performance and durability as promising materials for cultural heritage protection. Polymers 12(9):1973

Jiang M et al (2025) Recent advances in paper conservation using nanocellulose and its composites. Molecules 30(2):417

Kanth AP, Soni AK (2023) Application of nanocomposites for conservation of materials of cultural heritage. J Cult Herit 59:120–130

Karki Y, Verma SS, Naz F (2023) Biological nanomaterials and their development. In: Khan T et al (eds) Nanomaterials: the building blocks of modern technology: synthesis, properties and applications. Springer Nature Singapore, Singapore, pp 69–90

Khalid S et al (2024) A review on traditional and artificial intelligence-based preservation techniques for oil painting artworks. Gels 10(8):517
Król K, Kao R, Hernik J (2019) The Scarecrow as an indicator of changes in the cultural heritage of rural Poland. Sustainability 11(23):6857
Kryg P et al (2024) Challenges and prospects of applying nanocellulose for the conservation of wooden cultural heritage—a review. Forests 15(7):1174
Manfredi M et al (2014) A new quantitative method for the non-invasive documentation of morphological damage in paintings using RTI surface normals. Sensors 14(7):12271–12284
Mascitelli A et al (2023) Cultural heritage resilience in the face of extreme weather: lessons from the UNESCO site of alberobello. Sustainability 15(21):15556
Mastrangelo R, Chelazzi D, Baglioni P (2024) New horizons on advanced nanoscale materials for cultural heritage conservation. Nanoscale Horiz 9(4):566–579
Md Ramli A et al (2016) Determinants of food heritage towards food identity. Environ-Behaviour Proc J 1(1):207–216
Mekuye B, Abera B (2023) Nanomaterials: an overview of synthesis, classification, characterization, and applications. Nano Select 4(8):486–501
Mhd Haniffa MAC et al (2016) Review of bionanocomposite coating films and their applications. Polymers 8(7):246
Mohamed MG et al (2024) The potential of new eco-friendly formulations in enhancing the protection of ceramic artifacts. J Cult Herit 67:404–413
Nishino T, Takano K, Nakamae K (1995) Elastic modulus of the crystalline regions of cellulose polymorphs. J Polym Sci, Part B: Polym Phys 33(11):1647–1651
Nogi M et al (2009) Optically transparent nanofiber paper. Adv Mater 21(16):1595–1598
Ocak Y (2013) Use of bio-nano composites as protective coating on natural stone surfaces. In: Department of chemical engineering. Izmir Institute of Technology (Turkey), Turkey, p 136
Ocak Y et al (2015) Sustainable bio-nano composite coatings for the protection of marble surfaces. J Cult Herit 16(3):299–306
Operamolla A et al (2021) Toward a reversible consolidation of paper materials using cellulose nanocrystals. ACS Appl Mater Interfaces 13(37):44972–44982
Ortega-Morales BO, Gaylarde CC (2021) Bioconservation of historic stone buildings—an updated review. Appl Sci 11(12):5695
Otero J, Borsoi G, Monasterio-Guillot L (2023) The boom in nanomaterials for built heritage conservation: why does size matter? Materials 16(8):3277
Pagnin L et al (2023) Protecting street art from outdoor environmental threats: what are the challenges? Coatings 13(12):2044
Paolino B, Sorrentino MC, Pacifico S (2024) Greener solutions for biodeterioration of organic-media cultural heritage: where are we? Heritage Sci 12(1):334
Passaretti A et al (2021) Biologically derived gels for the cleaning of historical and artistic metal heritage. Appl Sci 11(8):3405
Passaretti A et al (2023) Biocleaning of historical metal artworks: innovative green gels amended with microbial derivatives. Front Mater 10
Pedna A et al (2016) Obtaining SiO_2—fluorinated PLA bionanocomposites with application as reversible and highly-hydrophobic coatings of buildings. Prog Org Coat 90:91–100
Perdoch W et al (2022) Influence of nanocellulose structure on paper reinforcement. Molecules 27(15):4696
Pérez-Gandarillas L et al (2024) Conservation and protection treatments for cultural heritage: insights and trends from a bibliometric analysis. Coatings 14(8):1027
Poggi G et al (2023) Nanostructured bio-based castor oil organogels for the cleaning of artworks. J Colloid Interface Sci 638:363–374
Ramadan S, El-Kholy SA, ElHagrassy AF (2024) Comparative study on the effect of chitosan, chitosan nanoparticles, and SiO_2 loaded chitosan for the consolidation of the painted stone. J Cult Herit 70:312–322

Rodica-Mariana I, Sanda-Maria D, Daniela Ţurcanu C (2017) Nanotechnologies in cultural heritage—materials and instruments for diagnosis and treatment. In: George ZK, Athanasios CM (eds) Novel nanomaterials. IntechOpen, Rijeka, p Ch. 10

Rodriguez-Navarro C, Ruiz-Agudo E (2018) Nanolimes: from synthesis to application. Pure Appl Chem 90(3):523–550

Santos SM et al (2015) Characterization of purified bacterial cellulose focused on its use on paper restoration. Carbohyd Polym 116:173–181

Shiny KS et al (2019) A new approach to wood protection: Preliminary study of biologically synthesized copper oxide nanoparticle formulation as an environmental friendly wood protectant against decay fungi and termites. Maderas. Ciencia Tecnología 21:347–356

Shiya R (2023) The role of anthropology in understanding culture and society. Int Res J Arts Soc Sci 11(3)

Sonaglia E et al (2024) Bacterial nanocellulose hydrogel for the green cleaning of copper stains from marble. Gels 10(2):150

Soukup M (2020) Visuality and anthropology: drawings, painting, photography, and movie. In: McCallum D (ed) The Palgrave handbook of the history of human sciences. Springer Nature Singapore, Singapore, pp 1–29

Speziale A et al (2020) Development of multifunctional coatings for protecting stones and lime mortars of the architectural heritage. Int J Architect Heritage 14(7):1008–1029

Stoner JH, Rushfield R (eds) (2020) Conservation of easel paintings, 2nd edn. Routledge, London

Tawiah B, Ofori EA, George SC (2024) Nanotechnology in societal development. In: George SC, Tawiah B (eds) Nanotechnology in societal development. Springer Nature Singapore, Singapore, pp 1–64

Tortora L et al (eds) (2020) Nanoscience and nanomaterials for the knowledge and conservation of cultural heritage. Front Mater 7

Tripa S et al (2023) A review on deterioration of textile cultural heritage objects and sustainable solutions to mitigate the degradation. Ind Textila 74(5):555–563

Vizárová K et al (2011) Stability studies of materials applied in the restoration of a baroque oil painting. J Cult Herit 12(2):190–195

Völkel L et al (2017) Nano meets the sheet: adhesive-free application of nanocellulosic suspensions in paper conservation. Heritage Sci 5(1):23

Völkel L, Prohaska T, Potthast A (2020) Combining phytate treatment and nanocellulose stabilization for mitigating iron gall ink damage in historic papers. Heritage Sci 8(1):86

Waked A (2011) Nano materials applications for conservation of cultural heritage. WIT Trans Built Environ 118:577–588

Waked AM (2011) Nano materials applications for conservation of cultural heritage. Modern academy for engineering and technology, Cairo, Egypt

Wang H et al (2013) Multifunctional nanocomposites for paper conservation. Stud Conserv 58(1):23–29

Ziegenbalg G et al (2018) Nanomaterials in architecture and art conservation. CRC Press

Zugenmaier P (2001) Conformation and packing of various crystalline cellulose fibers. Prog Polym Sci 26(9):1341–1417

Raphael Kanyire Seidu (Ph.D.) is a Research Assistant at the School of Fashion and Textiles at The Hong Kong Polytechnic University, Hong Kong. He holds a PhD in Fashion and Textiles Design from The Hong Kong Polytechnic University, Hong Kong SAR. He was also a Visiting Research Student at Nottingham Trent University (NTU) in the UK. He has several publications in leading journals in textile, design, technology, and material science. His recent paper was adjudged the "Second-Best Paper" presented at the 12th Sustainable Education and Development Research Conference in Somanya, Ghana. His research interest focuses on retro-reflective materials, smart textiles, cloth culture, weaving, and print technologies. He has several papers and has filed for a patent from his Ph.D. project which aims at developing smart interactive retro-reflective clothing for pedestrians' safety at night-time.

Emmanuel Abankwah Ofori is a graduate and researcher with a degree in BSc. Textile Design and Technology from the Industrial Art Department of Kwame Nkrumah University of Science and Technology, Kumasi. Emmanuel's research interest focuses on sustainable environmental practices, textile science, polymers, and energy efficiency. Recently, he co-authored a paper that was adjudged the "Second-Best Paper" presented at the 12th Sustainable Education and Development Research Conference in Somanya, Ghana, and has actively participated in several research projects, working with renowned authors to explore various facets of environmental sustainability, adaptation to climate change, green dyeing, and sustainable renewable energy consumption, thanks to Dr. Benjamin Tawiah's guidance. Emmanuel is dedicated to fostering the next generation of environmental and scientific leaders as a passionate textile scientist and educator. Through his writings and research, Emmanuel continues to motivate his mentees, readers of his works, and colleagues in the field, fostering greater awareness of the planet's vulnerability and the need to protect its ecological vibrancy in order to achieve a sustainable future at large.

George Kwame Fobiri is an Assistant Lecturer at the Department of Fashion Design and Textiles Studies, at Kumasi Technical University, Kumasi, Ghana. He holds a bachelor's degree in Industrial Art (Textiles) and an MFA in Textile Design from the Department of Industrial Art, Kwame Nkrumah University of Science and Technology (KNUST). He is currently pursuing a Ph.D. in Textile Design Technology at KNUST. His research interests cover textiles, fashion, computer-aided design (CAD), e-textiles, wearable textiles, textile printing, textile dyeing, etc.

Benjamin Tawiah (Ph.D) received his B.A. Industrial Art (Textiles Option) and MEng in Textile Engineering from the Kwame Nkrumah University of Science and Technology, Ghana, and Jiangnan University, China, respectively. Prior to his Master of Engineering degree, he earned a postgraduate Diploma in Management Information Systems from the Ghana Institute of Management and Public Administration (GIMPA) and then studied for his Ph.D at the Hong Kong Polytechnic University where he also worked as a postdoctoral Fellow. He also worked as a research associate at the City University of Hong Kong. He is a lecturer at the Kwame Nkrumah University of Science and Technology. His research interest focuses on flame retardant polymer composites, advanced energy electrochemical energy storage systems, and flexible fiber-based current collectors for zinc ion batteries.

Shouxiang Jiang is currently a Professor of fashion and textiles at The Hong Kong Polytechnic University. He teaches a broad range of subjects such as innovative textile design approach, advanced textile and fashion, and international fashion and textiles design. Prof. Jiang's research deals with innovative textile effects using physical and chemical treatment technologies. He is internationally renowned for his development of the metallised textiles. Leading a team that has developed a sputtering coating technology with research funding from the HKSAR government and sponsorships from the textile industries, Prof. Kinor Jiang has designed and launched an anhydrous system that uses metallising technologies to produce functional and decorative textiles for the production line. He holds 30 granted and filed patents based on the coating technology. Prof. Jiang has published more than 200 journal papers and more than 10 books. His designs and art projects have been exhibited worldwide and received more than 30 awards.

Chapter 13
Lightweight Bionanocomposites in Transportation Structures

Jahanvi Thakur, Karishma Rawat, Shreya, Peeyush Phogat, Sukhveer Singh, and Ranjana Jha

Abstract This chapter examines the transformative role of lightweight bionanocomposites in advancing sustainable transportation. Focusing on their integration in sectors like automotive, aerospace, marine, and rail, it highlights how these innovative materials, crafted from biopolymers, natural fibers, and nanoscale reinforcements, offer superior mechanical strength, durability, and weight reduction—crucial for improving fuel efficiency and reducing emissions. Within the transportation industry, bionanocomposites are increasingly applied to vehicle body panels, interiors, and battery housings, as well as aerospace fuselages and wings, where their lightweight properties help achieve significant energy savings. For marine applications, bionanocomposites provide durability and sustainability in hulls and decks, offering a competitive edge over traditional materials. The chapter also emphasizes their alignment with eco-friendly manufacturing practices and their potential for recyclability, addressing sustainability and environmental impact concerns. Moreover, it delves into challenges in scaling up production and meeting regulatory standards, alongside exploring future trends that leverage these composites for greener and more efficient transportation structures. With rising demand for sustainable materials, this chapter underscores the pivotal role of lightweight bionanocomposites in reducing transportation's carbon footprint and advancing clean energy goals.

Keywords Lightweight bionanocomposites · Sustainable transportation · Automotive · Aerospace · Marine · Biopolymers · Natural fibers · Emissions reduction

---

J. Thakur · K. Rawat · Shreya (✉) · P. Phogat · S. Singh · R. Jha
Research Lab for Energy Systems, Department of Physics, Netaji Subhas University of Technology, Dwarka, New Delhi, India
e-mail: shreyasharma.aug15@gmail.com

Shreya · P. Phogat
Council of Scientific and Industrial Research-National Institute of Science Communication and Policy Research, New Delhi, India

S. C. George and B. Tawiah (eds.), *Advanced Bionanocomposite Materials*, Advanced Structured Materials 241, https://doi.org/10.1007/978-981-96-8225-6_13

13.1 Introduction

The transportation sector is a major contributor to greenhouse gas emissions and energy consumption worldwide, driving the need for materials that can offer both lightweight properties and environmental benefits (Azhar et al. 2024). In this regard, lightweight bionanocomposites have garnered significant interest for their potential to replace conventional, heavier materials with sustainable alternatives that enhance efficiency, reduce emissions, and promote the use of renewable resources (Tripathy and Gupta 2024).These materials, composed of biopolymers, natural fibers, and nanoscale reinforcements, leverage both the environmental advantages of bio-based ingredients and the performance improvements that nanotechnology can deliver (Bilal et al. 2020). The result is a class of composites that offers strength, durability, and flexibility while remaining lightweight, opening new possibilities for applications in automotive, aerospace, marine, and public transportation sectors as shown in Fig. 13.1.

Lightweight bionanocomposites are materials designed to combine the best properties of bio-based and nanoscale materials. At their core, these composites typically involve a matrix made of biopolymers or other renewable substances, reinforced with nanoscale fibers or particles that enhance mechanical properties like strength, rigidity, and impact resistance. By utilizing natural ingredients such as cellulose (Njuguna et al. 2011), chitosan (Hosseini et al. 2015), and other plant-derived fibers

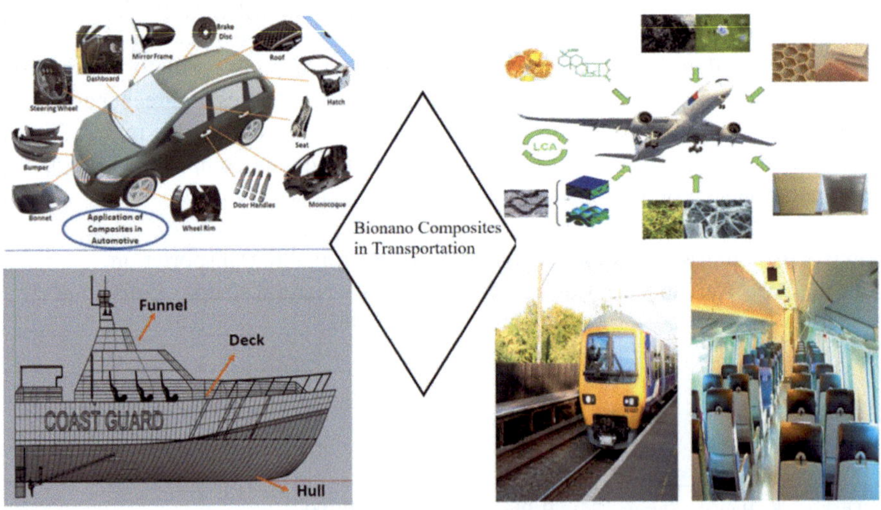

Fig. 13.1 Applications of bionanocomposites in automotive, aerospace, marine, and railway sectors to enhance sustainability through weight reduction, improved durability, and lower environmental impact. This image is reproduced with permission from Gupta et al. (2021) © 2021 by the authors. Licensee MDPI, Basel, Switzerland. Trimulyono et al. (2024) © 2024 by the authors. Licensee MDPI, Basel, Switzerland. Silva et al. (2017) © 2017 by the authors. Licensee MDPI, Basel, Switzerland

(Thapliyal et al. 2023), these composites are inherently biodegradable or recyclable, reducing the long-term environmental impact compared to traditional composites that rely on synthetic polymers and metals. The nanoscale components, often made of materials like nano-silica, carbon nanotubes, or graphene, are incorporated to achieve advanced properties without adding significant weight.

Nanotechnology enables the positioning of these materials in well-defined structures at the molecular level, which makes composites strong as well as lightweight. For example, it is possible to disperse nanofillers in the polymer matrix, thus making for a highly interconnected structure, giving improved load transfer and material cohesion (Guchait et al. 2022). Furthermore, nanotechnology can enhance the thermal, electrical, and chemical stability of these composites, making them suitable for diverse applications in transportation, where performance under various environmental conditions is crucial. The convergence of bio-based materials and nanotechnology not only yields functional benefits but also meets the rising demand for sustainable materials, as industries strive to minimize reliance on fossil fuels and reduce the carbon footprint associated with material production and disposal.

The application of lightweight bionanocomposites in transportation represents a significant step forward in sustainable design and engineering. Weight reduction in vehicles, whether cars, airplanes, or ships, directly contributes to fuel efficiency, reducing energy consumption and emissions (Gouripriya et al. 2024). For example, in the automotive industry, every 10% reduction in vehicle weight can lead to approximately 6–8% improvement in fuel efficiency (National Research Council 2010). For aircraft and other high-fuel-consuming vehicles, even minor weight reductions can lead to substantial fuel savings and emissions reductions over their operational lifespan.

Beyond fuel efficiency, lightweight bionanocomposites contribute to sustainability through the use of renewable materials that are often sourced from agricultural or natural waste, reducing dependency on non-renewable resources (Dalbanjan et al. 2024). By substituting synthetic materials with bio-based alternatives, the environmental impact of material extraction, processing, and disposal can be mitigated. Moreover, the biodegradability of many biopolymers means that end-of-life disposal has less environmental consequence compared to traditional materials, which often persist in landfills or release pollutants during incineration (Hottle et al. 2017).

Aside from environmentally related gains, bionanocomposites entail possible economic returns. Despite very high setup expenses to experience during their generation and utilization periods, they go a long way in proving worthwhile through long run cost efficiency gained in reduced amounts of fuel. Moreover, the lightweight and strong properties of these composites reduce wear and tear on transportation infrastructure, potentially lowering maintenance costs for vehicles and infrastructure alike (Ali and Hoque 2022).

This chapter explores the multifaceted role of lightweight bionanocomposites in advancing sustainable transportation. By combining bio-based materials like biopolymers, natural fibers, and nanofibers, these composites offer exceptional structural and functional properties suitable for automotive, aerospace, and marine applications. Through case studies and examples, the chapter highlights how these materials contribute to fuel efficiency, emission reduction, and enhanced performance in various transportation sectors. Additionally, this chapter examines the environmental and economic impacts of adopting lightweight bionanocomposites, focusing on their recyclability, biodegradability, and compatibility with circular economy principles. Key regulatory and safety considerations are discussed, emphasizing the need to meet industry standards while driving sustainability goals.

In covering both the technical aspects and broader implications, this chapter offers a comprehensive overview of the current state and future potential of bionanocomposites in transportation. These materials represent a promising shift toward environmentally responsible design, providing innovative solutions that align with global efforts to reduce the carbon footprint and resource dependency of transportation structures.

13.2 Fundamental Properties of Bionanocomposites

Bionanocomposites offer a compelling combination of properties that cater to sustainable engineering needs, particularly in sectors that demand lightweight, high-strength, and environmentally friendly materials. As seen in Table 13.1, these materials possess a unique set of characteristics, including a high strength-to-weight ratio, thermal stability, and impact resistance, which are particularly advantageous in transportation and biomedical applications.

These composites have lightweight properties based on their bio-based polymers and nanostructured reinforcements, and they can greatly reduce fuel consumption in vehicles with the required strength (Balakrishnan et al. 2016). Additionally, these materials' recyclability and biodegradability align with sustainability initiatives, addressing both lifecycle efficiency and end-of-life waste management. Their thermal stability also enables performance under high-temperature conditions, expanding their application in various transportation structures (Mousa et al. 2016). These composites thus present an opportunity to replace traditional materials while enhancing both environmental responsibility and operational performance.

13.2.1 Composition: Biopolymers, Natural Fibers, and Nanomaterials

The composition of bionanocomposites integrates biopolymers, natural fibers, and nanomaterials to create materials that are both functional and sustainable (Andrew and Dhakal 2022). This combination of bio-derived and nanoscale components

Table 13.1 Key properties and advantages of bionanocomposites in sustainable applications

Property	Description	Advantage	References
High strength-to-weight ratio	Superior mechanical properties with reduced weight	Improves fuel efficiency in vehicles	Mostafa et al. (2017)
Thermal stability	Maintains integrity under high temperatures	Suitable for extreme conditions	Mittal (2011)
Impact resistance	Enhanced ability to withstand physical stresses	Ideal for safety–critical applications	Nimbagal et al. (2021)
Recyclability	Designed for ease of material recovery and reuse	Supports circular economy practices	Sharma et al. (2023)
Biodegradability	Capable of breaking down naturally in the environment	Reduces long-term waste	Zhao et al. (2008)
Durability	High resistance to wear and degradation over time	Extends material lifespan	Muthukumar et al. (2022)

enhances the physical, mechanical, and environmental properties of composites, making them particularly well-suited for applications in fields like transportation and biomedical engineering (Sharma et al. 2018).

(A) **Biopolymers**, form the matrix of bionanocomposites, providing a flexible, biodegradable, and renewable alternative to synthetic polymers. Derived from natural sources such as starch, cellulose, polylactic acid (PLA), and polyhydroxyalkanoates (PHAs), biopolymers bring inherent environmental benefits (Kumari et al. 2022). Their biodegradable nature minimizes waste and reduces reliance on petroleum-based plastics, supporting a circular economy approach. Figure 13.2a shows various biopolymer sources, including proteins (gelatin, collagen, wheat gluten), plants (tapioca, corn, cotton), microorganisms (polyhydroxyalkanoates, polyhydroxybutyrate), and polysaccharides (anionic, neutral, and cationic types such as alginate, starch, cellulose, and chitosan).

(B) **Natural fibers**, such as flax, hemp, jute, and bamboo as shown in Fig. 13.2b serve as reinforcement agents within the composite, offering enhanced strength and stiffness while remaining lightweight (Al Mahmud et al. 2024). Natural fibers also have the advantage of being lighter than conventional materials, like glass or carbon fibers, which reduces the overall weight of the composite and, in transportation applications, leads to improved fuel efficiency and reduced emissions (Chauhan et al. 2022). However, because natural fibers may absorb moisture and degrade over time, they are often treated chemically or combined with other materials to improve durability and resistance to environmental factors (Ali et al. 2018).

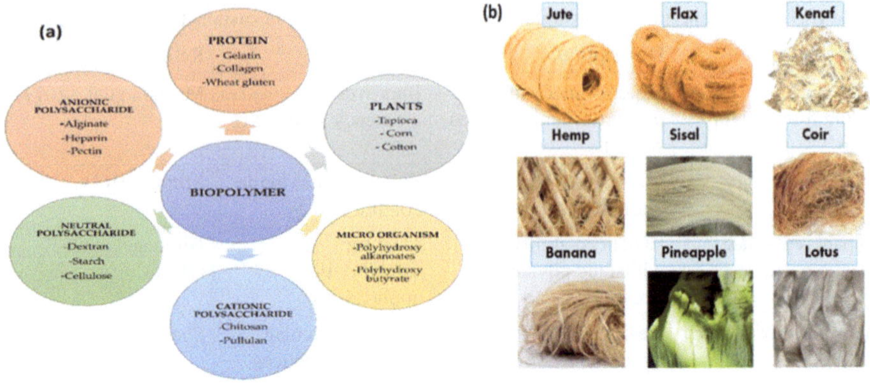

Fig. 13.2 **a** Sources of biopolymers. This image is reproduced with permission from Baranwal et al. (2022) © 2022 by the authors. Licensee MDPI, Basel, Switzerland. **b** Renewable and biodegradable natural fibers, including jute, flax, kenaf, hemp, sisal, coir, banana, pineapple, and lotus. This image is reproduced with permission from Jayaraman et al. (2024) © 2024. The Author(s)

(C) **Nanomaterials**, include nanoclay, carbon nanotubes (CNTs), and cellulose nanofibers. Carbon nanotubes, for instance, possess high tensile strength and electrical conductivity, which can contribute to improved structural integrity and multifunctional capabilities. Nanoclays add dimensional stability and enhance flame resistance, making the composites more suitable for high-temperature applications (Pandey et al. 2014). Additionally, their nanoscale dimensions tend to produce an enormous surface area; therefore, improving the matrix-to-fibers reinforcement interfacial areas should as expected, eventually distribute the applied load effectively in favor of much-improved mechanical performance.

13.2.2 Mechanical Properties: Strength, Flexibility, and Durability

Strength is one of the primary advantages of bionanocomposites. The inclusion of reinforcement materials like natural fibers (e.g., flax, hemp) and nanomaterials (e.g., carbon nanotubes, nanoclays) enhances the load-bearing capacity of the composite. These materials show a great tensile strength value, so it can be stressed at a higher level without deforming under conditions of high stress. In transportation applications, for example, this property helps ensure vehicle parts made from bionanocomposites that can endure impact and repetitive stress, contributing to structural integrity while remaining lightweight (Mohammed et al. 2024).

Flexibility is also crucial, particularly for applications that involve varying shapes, movements, or forces. Biopolymers such as PLA and PHA offer a degree of flexibility, allowing the composites to absorb impact and bend without breaking (Smith and

Verbeek 2018). The integration of nanomaterials can further enhance this property by enabling a more distributed stress transfer across the composite matrix. Flexibility is particularly beneficial in areas like biomedical engineering, where materials often need to conform to complex shapes or dynamic environments while maintaining performance.

Durability refers to the composite's ability to withstand environmental exposure, wear, and tear over time. Bionanocomposites are engineered to be resilient to both physical stress and environmental factors, such as temperature fluctuations and moisture. While natural fibers can be sensitive to environmental degradation, treatments and the addition of nanomaterials improve the composite resistance to moisture absorption, UV light, and microbial degradation (Al-Maharma and Al-Huniti 2019). This durability is essential for applications in which the composite must retain its properties over extended periods, such as in transportation components exposed to the elements or biomedical devices used within the body.

Bionanocomposites have very high versatility and reliability due to their mechanical properties, which include strength, flexibility, and durability. Optimization of these properties by material selection and design allows manufacturers to produce composites that are not only performing according to performance requirements but also provide sustainable and long-lasting solutions for industries.

13.2.3 Environmental Stability and Biodegradability

This development of bionanocomposites is more pertinent in transport, packaging, and biomedical sectors due to increased environmental stability and biodegradability (Kishore et al. 2023). These types of materials potentially contribute to minimalizing the effect of environmental influences and can, upon reaching their end-of-lifecycle, easily degrade or get recycled. Environmental stability in bionanocomposites is achieved by integrating materials that resist degradation under various conditions, such as exposure to sunlight, moisture, or temperature changes. This is particularly important in transportation and outdoor applications, where components may be exposed to the elements. For example, by incorporating nanomaterials like nanoclays or graphene, the composites can improve their resistance to UV radiation, moisture, and thermal stress. This is the case with stabilizing nanofillers combined with biopolymers such as PLA, which enhance the durability and resistance of the material to oxidation. These modifications help the composites maintain their mechanical properties and structural integrity over time, ensuring they can withstand rigorous environmental exposure while extending the lifecycle of the products made from them.

Biodegradability is one of the most compelling advantages of bionanocomposites. Unlike traditional composites that rely on synthetic, petroleum-based polymers, bionanocomposites often utilize biodegradable biopolymers and natural fibers. Materials like PLA, PHA, and cellulose are derived from renewable sources and can decompose naturally under specific environmental conditions, such as composting

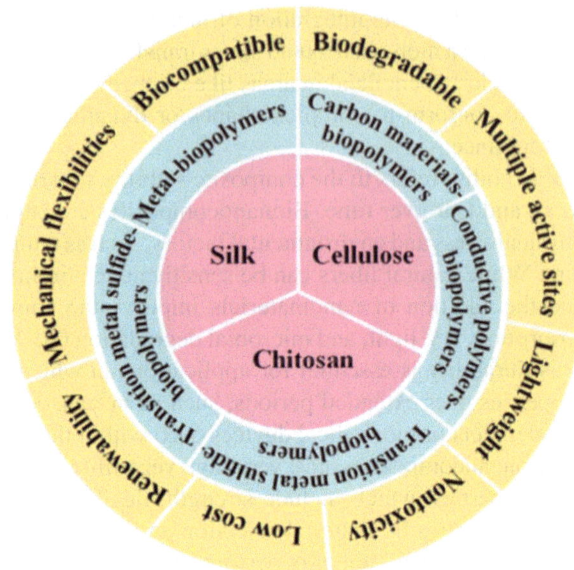

Fig. 13.3 Fundamental properties of bionanocomposites for potential applications in the transportation sector. This image is reproduced with permission from Wang et al. (2021) © 2021 by the authors. Licensee MDPI, Basel, Switzerland

(Agboola et al. 2016). The biodegradation process enables these composites to break down into natural byproducts, reducing waste accumulation and minimizing ecological harm. This property is particularly beneficial for use in products that are single-use or for medical implants, where the material can be formulated to degrade harmlessly once its function is accomplished, minimizing long-term environmental effects (Chatterjee and Dhar 2025). Figure 13.3 illustrates the fundamental properties of bionanocomposites for potential applications in the transportation sector.

Combining environmental stability with biodegradability in bionanocomposites creates a balanced approach that supports both product longevity and sustainability. Such designs allow for the stability of the materials in use, but the decomposability or recyclability after their useful life ensures contribution to a circular economy with minimal waste and proper management of material resources. Hence, bionanocomposites offer a high performance but are also sustainable and applicable in those industries that would want to minimize their footprint in the environment.

13.3 Synthesis and Fabrication Techniques

Producing lightweight bionanocomposites requires advanced synthesis and fabrication techniques that integrate nanoscale reinforcement materials into biopolymer matrices (Xiong et al. 2018). These methods enhance both mechanical and environmental properties, enabling the composites to meet the high-performance standards required in applications like transportation, aerospace, and biomedical industries.

This section details various techniques and the innovative approaches that facilitate both precision in performance and sustainability.

13.3.1 Fabrication of Lightweight Bionanocomposites

The fabrication of lightweight bionanocomposites is concerned with achieving materials that are strong yet light and environmentally friendly. Solution casting, melt blending, and electrospinning are three major techniques used for fabricating these composites, each of which provides specific advantages for any given application. A comparative summary of these fabrication techniques, along with their respective biomaterials, nanomaterials, characteristics, and transportation applications, is presented in Table 13.2.

Table 13.2 Fabrication techniques for lightweight bionanocomposites and their applications in transportation structures

Fabrication method	Biomaterial used	Type of nanomaterial	Characteristics	Application in transportation structures	References
Solution casting	PLA, Cellulose	Nanoclays, Carbon Nanotubes, Graphene	Lightweight, flexible, high mechanical strength	Used in lightweight automotive interior components, coatings for improved durability, and flexible electronic displays in vehicles	Giammaria et al. (2024)
Melt blending	Biodegradable thermoplastics (PLA, PHA)	Carbon Nanotubes, Nano-Silica, Nanoclays	High thermal stability, enhanced mechanical strength, solvent-free process	Applied in vehicle body panels, aerospace structural components, and lightweight under-the-hood automotive parts	Bagal-Kestwal et al. (2019)
Electrospinning	Biopolymer nanofibers (chitosan, gelatin, cellulose)	Carbon nanotubes, nanoclays, metal oxide nanoparticles	High surface area, porous, strong yet lightweight	Used in lightweight structural reinforcements, vibration-damping layers in aircraft, and protective coatings in transportation structures	Li and Shimizu (2009)

(A) Solution Casting: Solution casting is a widely used technique in the fabrication of biopolymer-based composites (Kumar and Sinha-Ray 2018). In this process, a polymer matrix, such as PLA or cellulose, is dissolved in a suitable solvent. Nanomaterials like nanoclays, carbon nanotubes, or graphene are then dispersed into the polymer solution. After achieving uniform dispersion, the solvent is evaporated, leaving behind a dense composite film. This approach is especially suitable for the manufacturing of ultra-thin, flexible films, with fillers homogeneously dispersed, which is vital for applications requiring light films with great mechanical properties. Due to its simplicity and scalability, solution casting is frequently used in packaging, biomedical films, and lightweight automotive components.

(B) Melt Blending: Melt blending, also known as melt compounding, involves heating a thermoplastic polymer until it melts and then mixing it with nanoscale fillers under high shear conditions (Scaffaro and Botta 2014). This technique does not require solvents, making it more environmentally friendly than solution casting. The heat and shear forces help disperse the fillers evenly throughout the polymer matrix. This method is particularly advantageous for fabricating bulkier, more structural components that require enhanced thermal stability and mechanical properties, such as vehicle body panels and aerospace components. By selecting specific biopolymers and fillers, manufacturers can achieve composites with tailored strength-to-weight ratios, suitable for both structural and semi-structural applications in industries where weight reduction is critical.

(C) Electrospinning: Electrospinning is a technique that produces nanofiber mats with unique properties such as high surface area and porosity (Xue et al. 2019). A polymer solution is drawn into ultrafine fibers under a high-voltage electric field, creating a mat that is lightweight yet strong. Nanofibers made from biopolymers and reinforced with nanomaterials (such as nanoclays or carbon nanotubes) can have mechanical strength and flexibility that makes them suitable for a range of applications, from medical dressings to filtration materials and lightweight protective equipment. This method is ideal for biomedical applications, as it allows for the incorporation of bioactive compounds or therapeutic agents into the composite material, opening up possibilities in tissue engineering and drug delivery.

Solution casting offers several advantages, such as high flexibility and scalability, making it ideal for applications like automotive interior components and flexible electronics. The process also allows for the uniform dispersion of nanomaterials, ensuring consistent reinforcement throughout the composite. However, challenges such as the use of solvents, limited thickness of the produced films, and long processing times exist. To address the solvent-related environmental concerns, steps like solvent recovery systems and the use of greener solvents are being explored. Additionally, for the limitation on thickness, hybrid composites and multi-layered casting techniques are being implemented to achieve the necessary strength while maintaining lightweight properties. Melt Blending is another commonly used technique, known for its solvent-free process, high mechanical strength, and thermal stability, making it suitable for structural components like vehicle body panels. This method also allows for customized strength-to-weight ratios by adjusting the filler and biopolymer combinations. However, challenges such as high energy requirements, difficulty in achieving uniform filler dispersion, and compatibility issues between the polymer matrix and fillers can arise. To resolve these issues, researchers are developing more energy-efficient processing methods, such as optimizing temperature profiles and using lower-melting-point biopolymers to reduce energy consumption. The uniform dispersion of fillers can be improved by fine-tuning shear force and temperature parameters or by employing advanced mixing technologies. Electrospinning offers unique advantages, including the production of nanofibers with high surface area and porosity, which are ideal for applications in lightweight structural reinforcements and vibration-damping layers. It also allows for the incorporation of functional agents, such as bioactive compounds, providing additional properties like conductivity or antibacterial effects. Despite these benefits, challenges like slow production rates, nanofiber alignment, and sensitivity to environmental conditions exist. To address the slow production rate, researchers are focusing on optimizing the electrospinning setup for higher throughput, such as using multi-needle electrospinning or enhancing the collection techniques. The challenge of nanofiber alignment is being resolved by employing techniques like electrospinning in a magnetic field or using templates to guide fiber orientation.

13.3.2 Nanostructuring and Hybrid Reinforcement

Nanostructuring and hybrid reinforcement techniques significantly enhance the properties of bionanocomposites by precisely controlling the size, shape, and distribution of the reinforcing nanomaterials within the polymer matrix (Aston et al. 2013). These approaches help in achieving high-performance characteristics essential for demanding applications.

(A) 1D and 2D Nanostructures: Nanostructuring at the 1D and 2D levels involves using nanomaterials such as nanotubes, nanofibers, and nanosheets (Liu et al. 2020). For instance, carbon nanotubes and nanofibers provide exceptional mechanical strength and elasticity, improving the load-bearing capacity of the composite. Meanwhile, 2D nanosheets like graphene and montmorillonite offer high surface areas and increase the thermal stability, electrical conductivity, and barrier properties of the composite. In the context of transportation, these nanostructures enhance the composite's impact resistance and durability, making it suitable for high-stress applications like body panels and structural frames. Furthermore, these materials can also enhance flame retardancy, which is crucial in both automotive and aerospace industries where safety is paramount.

(B) Hybrid Reinforcement: Hybrid composites combine multiple types of nanofillers, such as cellulose nanofibers and carbon nanotubes, within the same polymer matrix to achieve synergetic properties (Saba et al. 2014). By carefully selecting the type and concentration of each filler, hybrid reinforcement allows for the customization of properties, balancing mechanical strength with flexibility, or combining thermal conductivity with electrical insulation. For example, a hybrid composite used in battery housing can be designed to be lightweight, strong, and capable of dissipating heat effectively, while still being resistant to corrosion and environmental wear (Guo et al. 2024). In transportation structures, such as automotive body panels, aerospace components, and structural reinforcements in railway and marine applications, hybrid composites are highly beneficial for their ability to optimize strength-to-weight ratios, enhance durability, and improve overall performance (Shubham and Ray 2024). Studies have shown that these composites provide enhanced impact resistance, thermal stability, and flame retardancy, making them ideal for high-stress and safety–critical environments in the transportation sector (Broughton 2012).

1D and 2D nanostructures offer significant advantages such as exceptional mechanical strength, high surface area, and enhanced thermal and electrical properties. However, challenges like poor dispersion, difficulty in achieving uniformity, and the high cost of certain nanomaterials can hinder their widespread application. To overcome these challenges, researchers are focusing on optimizing dispersion techniques through surface modifications, using functionalized fillers, and employing advanced processing methods such as sonication or melt mixing to ensure uniform distribution. Additionally, the high cost of materials can be mitigated by exploring alternative, cost-effective nanofillers and scaling up production techniques to reduce overall costs. Hybrid Reinforcement combines multiple types of nanofillers, such as cellulose nanofibers and carbon nanotubes. This approach allows for balancing mechanical strength with flexibility, enhancing thermal conductivity while maintaining electrical insulation, making it suitable for complex applications like battery housings or multifunctional components. The main challenge with hybrid reinforcement lies in achieving optimal compatibility between the different nanofillers and the polymer matrix, which can affect the overall performance and mechanical properties. Additionally, the complexity of fabricating hybrid composites may result in higher manufacturing costs. To address these issues, efforts are being made to develop better compatibilizers and coupling agents that improve the interaction between the fillers and matrix, as well as refine processing techniques to streamline production. By selecting the appropriate nanomaterials and adjusting concentrations, the overall properties of the composite can be tailored more effectively for the desired application, reducing the complexity and improving the performance of the final product.

13.3.3 Innovations in Sustainable Manufacturing Techniques

In response to the growing demand for environmentally responsible materials, significant advancements have been made in sustainable manufacturing techniques for bionanocomposites. Table 13.3 summarizes key advancements in sustainable manufacturing techniques, highlighting their benefits and applications in reducing environmental impact while enhancing material performance. These innovations reduce the environmental impact of composite production while improving the quality and performance of the materials.

(A) Bio-based Solvents and Green Chemistry: Traditional composite manufacturing often relies on harsh, petrochemical-based solvents that can be toxic and environmentally damaging. Innovations in green chemistry have led to the development of bio-based solvents, derived from renewable sources like plants, that reduce the environmental footprint of the production process (Idoko et al.

Table 13.3 Innovations in sustainable manufacturing techniques for bionanocomposites

Innovation	Description	Benefits	Applications
Bio-based solvents and green chemistry	Utilization of renewable, plant-based solvents instead of traditional petrochemical solvents	– Reduces environmental footprint – Lowers emissions – Simplifies disposal and recycling – Minimizes hazardous substances and energy use	Biomedical devices, packaging
Additive manufacturing (3D printing)	Layer-by-layer fabrication using biopolymers infused with nanomaterials	– Minimizes material waste – Enables complex, lightweight structures – Enhances fuel efficiency in aerospace and automotive – Supports rapid prototyping and customization	Aerospace, automotive
Low-temperature processing	Manufacturing at room temperature or slightly elevated temperatures, e.g., UV curing	– Reduces energy consumption – Lowers greenhouse gas emissions – Improves manufacturing efficiency	Coatings, films, lightweight panels

2024). For example, using water-based or plant-based solvents in place of traditional solvents not only decreases emissions but also simplifies disposal and recycling processes (Aguiar et al. 2022). Additionally, green chemistry principles advocate for the reduction of hazardous substances, energy consumption, and waste throughout the manufacturing process, aligning with global sustainability goals. These eco-friendly approaches are especially beneficial in applications like biomedical devices and packaging, where material purity and environmental impact are key considerations.

(B) Additive Manufacturing (3D Printing): Additive manufacturing, or 3D printing, represents a groundbreaking approach to sustainable fabrication by enabling layer-by-layer construction of complex shapes with minimal material waste. Biopolymers infused with nanomaterials can be printed into lightweight, intricate structures that maximize strength while using only the necessary material (Veeman et al. 2021). This technique is particularly advantageous in aerospace and automotive sectors, where complex, weight-optimized parts are needed to improve fuel efficiency. Furthermore, 3D printing allows for rapid prototyping and customization, enabling manufacturers to produce parts with specific properties on demand, thereby minimizing overproduction and associated waste.

(C) Low-Temperature Processing: Traditional composite manufacturing processes often require high temperatures, which increase energy consumption and greenhouse gas emissions. Low-temperature processing techniques, such as UV curing, allow for the fabrication of composites at room temperature or slightly elevated temperatures, significantly reducing energy use (Xiao et al. 2023). For instance, UV curing involves exposing the composite material to ultraviolet light to initiate polymerization, creating a solid composite without the need for high heat. This method is particularly suitable for coatings, films, and lightweight panels, where rapid curing at lower temperatures improves manufacturing efficiency and reduces the environmental impact (Collinson et al. 2022).

Bio-based solvents and green chemistry offer significant environmental benefits by reducing the reliance on petrochemical-based solvents, which are toxic and harmful to the environment. However, challenges exist in finding suitable bio-based solvents with similar efficacy and cost-effectiveness as traditional solvents. Additionally, the limited availability and performance of bio-based solvents in specific applications can pose challenges to large-scale adoption. To overcome these issues, research has focused on improving the performance of bio-based solvents by refining extraction methods, enhancing solvent stability, and optimizing formulations to ensure they meet the necessary technical standards for transportation applications. Additive manufacturing (3D printing) provides a unique advantage by enabling the creation of complex, lightweight, and customized parts with minimal material waste. However, challenges include material limitations, particularly in terms of strength, durability, and the ability to incorporate diverse nanomaterials into the 3D printing process. The complexity of 3D printing processes also results in relatively slow production speeds and high equipment costs. To address these challenges, ongoing research focuses on developing new 3D-printable nanocomposites with enhanced mechanical properties, improved compatibility with various nanofillers, and faster print speeds. Low-temperature processing techniques, such as UV curing, significantly reduce energy consumption and greenhouse gas emissions by enabling the fabrication of composites at lower temperatures. However, challenges include the need for specialized UV light sources, limited material choices, and slower curing times compared to traditional high-temperature methods. To overcome these barriers, research is focused on developing new photoinitiators and optimizing UV curing processes to enhance the speed and versatility of the technique. Additionally, improving the material selection for low-temperature processing can help extend its applicability to a wider range of transportation components, ensuring that it can meet the performance requirements of various transportation structures.

13.4 Automotive Applications of Bionanocomposites

Bionanocomposites have gained significant traction in the automotive industry due to their lightweight, high-strength characteristics and sustainable nature. The use of these materials is transforming automotive design, especially in the push for fuel efficiency, emission reductions, and sustainable materials (Sapuan and Ilyas 2020). Bionanocomposites, developed from a blend of biopolymers, nanofibers, and natural fibers, provide both structural integrity and reduced weight, essential for advanced vehicle designs. This section explores the role of bionanocomposites in lightweight vehicle design, focusing on their applications in body panels, interiors, and battery housings.

13.4.1 Role in Lightweight Vehicle Design

The drive for lightweight vehicle design in the automotive industry is motivated by both regulatory pressures and consumer demand for more fuel-efficient, eco-friendly vehicles (Patel et al. 2024). With the implementation of stricter emission standards across the world and about climate change awareness, car manufacturers aim to make vehicles that are high-performance and less detrimental to the environment. Lightweight design plays a key role in this endeavor because reducing the weight of a vehicle directly influences the rate at which fuel and emissions are generated. For example, studies show that reducing a vehicle's weight by just 10% can improve fuel economy by 6–8% and reduce emissions by a comparable amount. Bionanocomposites, with their excellent strength-to-weight ratio, present a promising solution to meet these goals.

Traditional automotive materials, such as steel and aluminum, have long been the standard in vehicle manufacturing (Taub et al. 2007). While metals offer high strength and durability, they add considerable weight to vehicles, which can lower fuel efficiency and increase emissions. By contrast, bionanocomposites, which are made by combining biopolymers and natural fibers with reinforcing nanomaterials, provide comparable strength while being significantly lighter (Ogah et al. 2024). For example, nanocellulose has very high tensile strength but with an extremely low weight, ideal for automotive applications where enhancement of strength and reduction of weight are crucial. Through a clever use of these materials, car manufacturers can achieve further weight reduction without compromising durability and performance.

In addition to improving fuel efficiency, lightweight design enhances other performance aspects of the vehicle, such as acceleration, braking, and handling. Reduced vehicle weight allows for faster acceleration, shorter stopping distances, and more responsive handling, which is especially beneficial in electric vehicles (EVs) and high-performance cars. For electric vehicles, in particular, the use of bionanocomposites can help extend the driving range by reducing the energy needed to move the car, a crucial consideration as the automotive industry increasingly shifts toward electrification (Nugroho et al. 2024). Lower weight in EVs translates to longer battery life and increased range, making these composites an attractive option for electric vehicle manufacturers.

Another advantage of using bionanocomposites in lightweight vehicle design is their adaptability for different structural requirements. Due to their modular nature, bionanocomposites can be engineered with specific properties tailored to the needs of different automotive components. For example, materials reinforced with carbon nanofibers or nanocellulose can be adjusted to offer greater flexibility, impact resistance, or thermal stability, making them versatile for use in a wide range of applications within a vehicle, from structural components to intricate interior parts (Ilyas et al. 2021). This flexibility allows automakers to optimize each part of the vehicle for maximum efficiency, durability, and environmental impact.

The advantages associated with bionanocomposites are not limited to their lightweight properties. Many bionanocomposites derive from renewable resources, such as plant fibers and biopolymers, whose production has drastically reduced the footprints associated with the environment. The use of sustainable, bio-based materials aligns with the automotive industry's goals for a circular economy, where materials can be reused, recycled, and repurposed rather than contributing to waste (Bhagabati 2020). Additionally, the biodegradability of certain bionanocomposites offers a viable end-of-life solution, addressing the issue of automotive waste and further enhancing the sustainability of the industry.

Moreover, bionanocomposites are well-suited to meet the increasing demand for safer, more durable vehicles. The structural integrity and impact resistance of these materials provide a level of safety comparable to traditional metals. Many bionanocomposites are engineered to withstand extreme temperatures, environmental exposure, and mechanical stress, ensuring they perform reliably over the vehicle's lifetime (Gupta et al. 2022). The toughness of these materials contributes not only to the entire lifetime of a vehicle but also low-cost maintenance because the replacement parts of vehicles manufactured through bionanocomposites tend to be less wearable and corrosion-prone.

Bionanocomposites are enabling a transformative shift in lightweight vehicle design. By integrating these materials into automotive structures, manufacturers are creating vehicles that are lighter, more efficient, and environmentally responsible. These materials have the potential to revolutionize the industry, offering a high-performance alternative to traditional materials and aligning with the global push toward sustainability. The use of bionanocomposites in lightweight vehicle design illustrates how innovation in materials science can help the automotive industry achieve its performance, safety, and environmental goals.

13.4.2 Applications in Body Panels, Interiors, and Battery Housings

One of the most promising applications of bionanocomposites in automotive manufacturing is in body panels (Alam et al. 2021). Traditional metal body panels are not only heavy but also susceptible to corrosion, making them less ideal from both performance and environmental perspectives. Bionanocomposites offer a solution with their properties of high impact resistance, corrosion resistance, and lightweight. Such lightweight yet stronger than metal body panels can be achieved by using materials like nanocellulose and flax fibers integrated into biopolymer matrices (Ferreira et al. 2019). These panels also provide better vibration damping, contributing to a quieter cabin experience.

Case Study: Ford's Use of Bionanocomposites in the Mustang EcoBoost Model: In recent years, Ford Motor Company has pioneered the use of bionanocomposites in its vehicles (Ilyas 2019). One example is the Mustang EcoBoost model, where Ford implemented a bionanocomposite material made from soy-based polyurethane for seat cushions and natural fiber composites for interior panels (Mielewski et al. 2005). This approach reduced the vehicle's overall weight, contributing to improved fuel efficiency without sacrificing performance or safety. By replacing traditional materials with natural fiber-reinforced composites, Ford demonstrated that these materials can be used in high-performance vehicles to achieve sustainability goals.

Bionanocomposites are equally transformative in automotive interiors, where they are used for everything from dashboards to seat structures. Interior components require materials that are not only lightweight but also visually appealing and durable. Natural fiber composites, such as those reinforced with kenaf or hemp, are increasingly used in dashboards, door panels, and trim parts as shown in Fig. 13.4a. These materials are not only lightweight but also provide a natural aesthetic that aligns with the growing trend of sustainable design in the automotive industry.

The shift to bionanocomposites in interiors has the added benefit of reducing the levels of volatile organic compounds (VOCs) that are commonly associated with synthetic materials. This improvement is crucial for vehicle air quality and aligns with consumer demand for eco-friendly products. Additionally, the flexibility of bionanocomposites allows for unique design options, giving designers the ability to

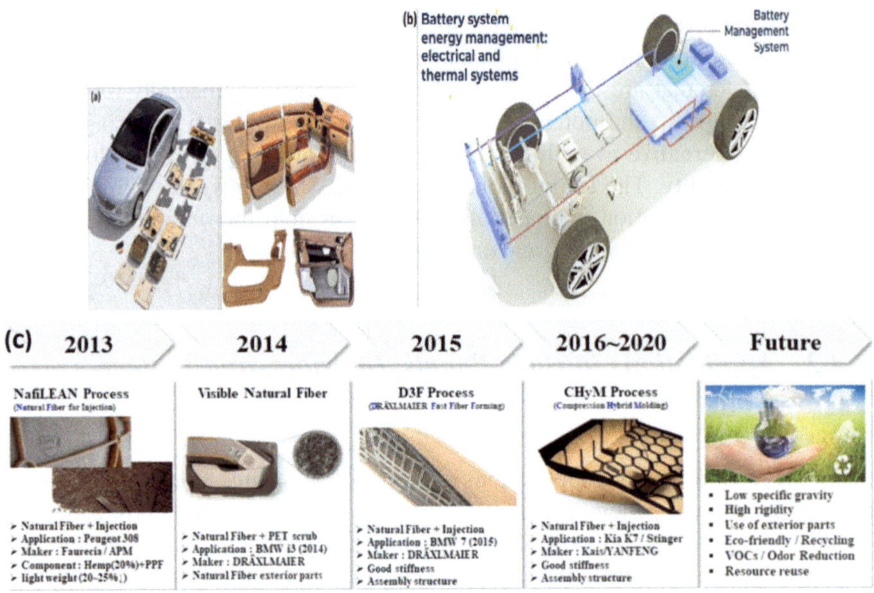

Fig. 13.4 Applications of lightweight bionanocomposite in **a** body panels and interior components of automobiles. This image is reproduced with permission from Teles et al. (2015) @ copyrights by authors. **b** Battery Housings, this image is reproduced with permission from Krishna et al. (2024) © 2024 by the authors. Licensee MDPI, Basel, Switzerland. **c** Global shift toward sustainable and lightweight automobile parts. This image is reproduced with permission from Lee et al. (2022). © 2022 by the authors. Licensee MDPI, Basel, Switzerland

create intricate shapes and textures that would be difficult or impossible to achieve with traditional materials (Al Azim Baiz 2022).

The demand for electric vehicles (EVs) has created a need for lightweight and durable materials that can safely house high-energy batteries. Battery housings must protect battery cells from impact, heat, and environmental factors, all while keeping the structure as lightweight as possible to extend the vehicle's range (Ibrahim et al. 2019). Bionanocomposites are particularly suitable for battery housings due to their high strength-to-weight ratios and thermal stability. Composites reinforced with nanomaterials, such as graphene or carbon nanotubes, provide excellent electrical and thermal conductivity, ensuring that the battery remains cool and performs efficiently (Scaffaro and Maio 2019).

In addition to their structural benefits, bionanocomposites offer opportunities for integrated functionalities within battery housings. For example, the thermal conductivity of certain nanomaterials can help dissipate heat generated by the battery cells, reducing the need for additional cooling systems as shown in Fig. 13.4b. Furthermore, the lightweight nature of these materials contributes directly to the overall range and efficiency of electric vehicles, making them a valuable addition to EV design. The global automobile industry is experiencing a significant shift toward eco-friendly and lightweight components as manufacturers strive to meet stricter environmental regulations and reduce fuel consumption as shown in Fig. 13.4c.

Case Study: BMW's i3 electric car with biocomposites in battery housing BMW's i3 model incorporates biocomposites in its battery housing to achieve a lightweight yet strong structure (Carlstedt and Asp 2020). By using bio-based polyamide reinforced with natural fibers, BMW reduced the weight of the battery housing while still meeting the strict safety and performance standards required for EV components. This approach not only enhanced the range of the i3 but also aligned with BMW's commitment to sustainable materials, as the bio-based housing material significantly reduced the vehicle's overall carbon footprint.

13.5 Aerospace Applications

Bionanocomposites have gained increasing importance within aerospace applications as they can also provide a high strength-to-weight ratio, better environmental stability, and improved toughness (Idumah and Hassan 2016). As the aerospace sector shifts its focus to achieving greater performance, reducing environmental impact, and adhering to stringent safety standards, bionanocomposites offer unique benefits in optimizing aircraft design. These materials leverage the properties of bio-based polymers, natural fibers, and advanced nanomaterials to provide durable, lightweight alternatives to traditional materials, paving the way for more sustainable aerospace solutions.

13.5.1 Benefits in Weight Reduction and Fuel Savings

Weight reduction is a primary focus in the aerospace industry due to its direct impact on fuel efficiency, environmental sustainability, and overall operational costs (Huang et al. 2016). Even small weight decreases in an airplane can result in substantial savings, as less fuel is required to maintain a lighter aircraft in the air. That means less fuel burned, resulting in lower levels of greenhouse gas emissions and less carbon footprint overall, which fits within the broader goals of sustainability for the industry. Bionanocomposites are emerging as a vital material in achieving this weight reduction. They offer comparable or even superior strength-to-weight ratios than traditional materials like aluminum and titanium, making them an ideal choice for aerospace applications where every gram saved counts (Nagaraju et al. 2023).

The lighter weight of bionanocomposites translates to more efficient fuel consumption, which directly benefits both the airline industry and the environment. The aviation industry is responsible for a substantial portion of global greenhouse gas emissions, and reducing aircraft weight is one of the most effective ways to minimize this impact (Shine et al. 2005). By switching to lightweight materials, aircraft can reduce fuel use, cutting down on emissions per flight and potentially extending the range of aircraft without requiring additional fuel. For example, replacing a portion of an aircraft's aluminum components with bionanocomposites could reduce overall weight by 20–30%, resulting in significant fuel savings and decreased environmental impact (Adekomaya and Majozi 2024).

The ability to enhance fuel efficiency through lightweighting is especially relevant for commercial airlines, which operate flights over long distances and face increasing pressure to reduce costs. Slightly higher fuel efficiency improvements from the bionanocomposites would translate into tremendous saving through a fleet's operating life during long-haul flights. In long-haul flights, even slight improvements in fuel efficiency can lead to substantial savings over a fleet's operational life. These savings make bionanocomposites not only an environmentally sustainable choice but also an economically viable one for the aerospace industry (Bhat et al. 2021). Fuel is one of the largest operational costs for airlines, and as fuel prices fluctuate, the demand for lightweight materials that reduce consumption and cost becomes even more critical. Furthermore, reducing fuel consumption extends aircraft range, allowing airlines to consider new, longer routes that were previously unfeasible due to fuel limitations.

Bionanocomposites also provide unique benefits in terms of the structural integrity and safety of aircraft. They often consist of bio-based polymers reinforced with nanomaterials such as cellulose nanofibers, carbon nanotubes, or other nanoparticles, which provide exceptional strength while remaining light. The inclusion of nanostructures enhances the mechanical properties of these composites, such as tensile strength and rigidity, without adding extra weight. The high stresses and harsh environmental conditions imposed on materials within aerospace applications often require that those materials not alter the overall aircraft weight or the fuel efficiency that the aircraft produces. Bionanocomposites maintain strength, even at decreased weights, thereby allowing them to be used on critical parts of the aircraft—wings, fuselage, or internal components—to enhance the entire vehicle's performance and durability.

Another important benefit of bionanocomposites in weight reduction is their potential for multi-functionality. Beyond their strength and lightweight properties, these composites can also be engineered to possess thermal resistance, electrical conductivity, and other functional attributes that may eliminate the need for additional materials. For example, composites embedded with conductive nanomaterials can be used in areas where both structural support and electronic functionality are needed, further reducing the weight by eliminating separate wiring or conductive materials. This level of integration opens new design possibilities and enhances the efficiency of aerospace components.

13.5.2 Use in Aircraft Structures: Wings, Fuselages, and Interiors

Bionanocomposites are finding their way into various parts of aircraft, including wings, fuselages, and cabin interiors as shown in Fig. 13.5a. Each of these components demands a unique combination of strength, flexibility, and durability, which bionanocomposites can provide due to their customizable nature as shown in Fig. 13.5a.

(A) Wings: Aircraft wings experience continuous aerodynamic forces that demand materials capable of handling both flexibility and resilience. Bionanocomposites, particularly those reinforced with carbon nanofibers or nanotubes, are able to meet these requirements effectively (Fonseca-Florido et al. 2024). These materials allow wings to maintain a stable shape, resist deformation, and absorb vibrations during flight, all while minimizing weight. By optimizing the strength-to-weight ratio, bionanocomposites help improve fuel efficiency by enhancing aerodynamic performance and reducing drag. The durability of these materials also helps to reduce maintenance needs and extends the lifespan of the wings, which is vital for both safety and cost management in aviation. Researchers are exploring bio-based fibers (flax, ramie) and recycled carbon fibers as sustainable alternatives for aerospace composites. The ECO-COMPASS project has developed bio-based resins from rosin to replace synthetic materials in aircraft structures like interior panels and winglets (https://projects.research-and-innovation.ec.europa.eu/en/projects/success-stories/all/bio-based-materials-aircraft).

(B) Fuselages: The fuselage is the main body of the aircraft and houses passengers and cargo. It must be impact-resistant and capable of withstanding both internal and external forces. Bionanocomposites reinforced with natural fibers

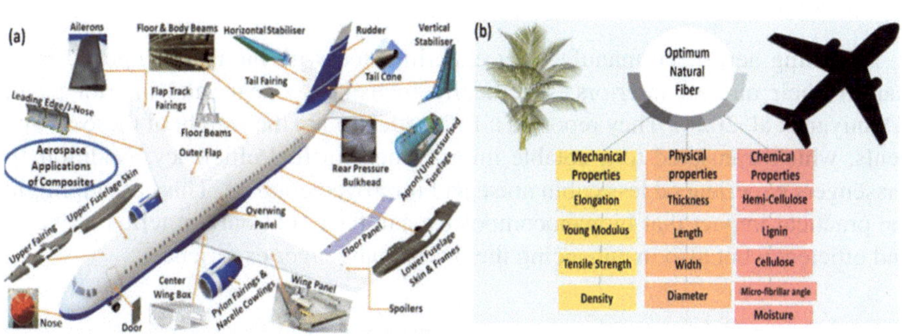

Fig. 13.5 **a** Applications of bionanocomposites in aircraft: Enhancing performance in Wings, Fuselages, and Interiors, This image is reproduced with permission from Bachmann et al. (2021) © 2021 by the authors. Licensee MDPI, Basel, Switzerland. **b** Customizable properties of Bionanocomposites for aircraft components. This image is reproduced with permission from Balo and Sua (2024) © 2024 by the authors. Licensee MDPI, Basel, Switzerland

or nanocellulose offer the ideal blend of impact resistance and weight reduction needed for fuselage construction (Mehdinia et al. 2024). These materials help decrease the overall weight of the fuselage while still offering high durability, essential for passenger safety. Additionally, their sound-dampening properties enhance the cabin environment by reducing noise and vibrations, creating a quieter and more comfortable experience for passengers. A bio-active living coating has been designed to protect surfaces like concrete, plastic, and metal. This self-healing biofilm enhances durability, improves air quality, and promotes eco-friendly, cost-effective building materials. These innovations contribute to a more sustainable future in aerospace and architecture (https://projects.research-and-innovation.ec.europa.eu/en/horizon-magazine/new-aerospace-and-building-materials-could-repair-themselves-thanks-fungi-and-bacteria).

(C) Interiors: Aircraft interiors include seats, overhead bins, and panels, all of which must be lightweight and durable. Bionanocomposites allow manufacturers to use sustainable materials while maintaining a high level of strength and flexibility, ideal for interior components that are subject to frequent wear and tear (Arockiam et al. 2018). Using bionanocomposites in these areas not only reduces the overall aircraft weight but also minimizes the reliance on petroleum-based plastics, enhancing sustainability. Natural fibers and biopolymers are particularly suitable for these applications as they are both aesthetically pleasing and capable of withstanding the operational demands of aerospace interiors. Researchers have developed 3D-printed fungal composites by combining mycelium and wood-based hydrogels. Mycelium forms a natural sensing network, enabling self-repair and enhanced material performance. These lightweight, sustainable bio-composites could replace plastics and metals in aircraft interiors, making components biodegradable and easy to recycle (https://projects.research-and-innovation.ec.europa.eu/en/horizon-magazine/new-aerospace-and-building-materials-could-repair-themselves-thanks-fungi-and-bacteria).

A leading aerospace manufacturer experimented with bionanocomposite materials in their aircraft interiors, specifically focusing on seats and interior panels (Olaniyan et al. 2024). They reported a 15% reduction in the weight of these components, which translated to a notable improvement in fuel efficiency. Additionally, passengers experienced less cabin noise and improved comfort. This case highlights the practical impact that bionanocomposites can have, not only in terms of weight and efficiency but also in enhancing the overall passenger experience.

13.5.3 Role of Bionanocomposites in Space Mission

Bionanocomposites have gained significant attention in the aerospace industry, particularly due to their potential to enhance the performance and efficiency of materials used in space missions. The inherent properties of bionanocomposites, such as their lightweight nature, mechanical strength, and thermal stability, make them ideal candidates for use in aerospace and space exploration applications. In the aerospace industry, companies like Airbus and Boeing have explored and integrated bionanocomposites into their designs (Muhammad et al. 2021). For instance, Airbus has utilized hybrid composites, combining materials like cellulose nanofibers and carbon nanotubes within polymer matrices, to improve the strength-to-weight ratio in aircraft body panels and structural reinforcements. These materials help reduce fuel consumption and carbon emissions without compromising the durability or safety of the aircraft. Boeing, similarly, has incorporated bionanocomposites into wings and fuselage sections, achieving significant weight reduction while enhancing impact resistance, crucial for long-distance air travel (Epstein et al. 2020). In the context of space missions, bionanocomposites offer promising solutions for spacecraft structures, insulation, and storage systems. Their lightweight nature is essential in space travel, where weight constraints are critical, while their mechanical strength ensures they can withstand the stresses of launch and space conditions. Additionally, their excellent thermal properties make them suitable for thermal protection and insulation systems, shielding spacecraft from extreme temperature fluctuations and radiation exposure in space. Furthermore, their resistance to corrosion and environmental wear makes them ideal for storage tanks and compartments that must maintain integrity over long periods. As research and development in this area continue, the integration of bionanocomposites into both aerospace and space mission applications offers promising advancements in material performance while contributing to sustainability and efficiency.

Plant fiber-inspired nanocomposites, derived from nanocellulose, hemp, jute, and flax, offer high strength, flexibility, and biodegradability, making them ideal for space applications (Bilal et al. 2020). Their lightweight structure enhances fuel efficiency and increases payload capacity, making them suitable for structural components like spacecraft panels, chassis, and frameworks. These nanocomposites also provide thermal insulation, with nanocellulose-based aerogels protecting spacecraft from extreme temperature variations (Miyashiro et al. 2020). Their integration with radiation-absorbing materials like graphene oxide and boron nanoparticles enhances radiation shielding, ensuring astronaut safety. Beyond structure and protection, these materials support life-support systems, with nanocellulose membranes filtering water contaminants and capturing CO_2 for air recycling (Fang et al. 2019). Additionally, self-healing bionanocomposites can autonomously repair microcracks, increasing spacecraft durability and reducing maintenance needs.

13.5.4 Standards and Certifications for Aerospace Adoption

The aerospace industry boasts some of the most rigid standards for safety and quality, both in its manufacture and elsewhere. The attainment of such requirements ensures that passenger safety, longevity of aircraft, and environmental factors are all safeguarded through compliance with regulations, testing, and industry acceptance. Innovative materials like bionanocomposites achieve the appropriate certification through meticulous processes of testing, regulatory compliance, and industry acceptance. The unique composition of bionanocomposites means that these biopolymers are combined with nanomaterials, making this a highly complex process. This implies that standards must validate not only the material strength and resilience but its long-term stability, as well as its sustainability in various aerospace applications.

Aerospace materials must demonstrate exceptional structural integrity to withstand the extreme environmental conditions encountered during flight. These materials are required to meet strict guidelines set by regulatory bodies like the Federal Aviation Administration (FAA) in the U.S. (McIntyre 2002) and the European Union Aviation Safety Agency (EASA) (Groenleer et al. 2010). Standards include testing for load-bearing capacity, tensile and compressive strength, resistance to impact, and fatigue. For bionanocomposites, these tests ensure that the materials can sustain repeated stress without losing structural integrity or becoming prone to cracks or deformation.

One of the prime requirements for aerospace applications is the resistance to fatigue, as the aircraft components sustain cyclic loading for many years during their operation. Bionanocomposites go through accelerated testing to simulate the long-term conditions and thus will help in establishing any weakness of the material beforehand. For example, a composite material may have to undergo millions of load cycles to simulate stresses it would go through over an aircraft's lifetime. Ensuring that bionanocomposites can meet these structural standards is vital to gaining industry acceptance and achieving the certifications necessary for use in critical structural components.

Aircraft are also subject to stringent fire safety standards to prevent and contain fires, which are critical for passenger safety. Materials used in aircraft interiors, for instance, must comply with specific flammability requirements, including the rate of flame spread, smoke density, and toxicity of combustion gases. Bionanocomposites need to demonstrate fire resistance that equals or exceeds that of traditional materials like aluminum alloys and thermoset composites.

Fire resistance standards can be achieved with nanomaterial reinforcement within bionanocomposites. Nanoclays or certain metal oxide nanoparticles may be incorporated into bionanocomposites to lower their flammability (Satdive et al. 2022), increasing their suitability for aerospace interiors and other sensitive components.

In this manner, they may find increased utility in aerospace interior and other sensitive applications. The key for the usage of bionanocomposites in aircraft cabins is the successful passing of fire safety tests because they need to guarantee safety to passengers during emergency conditions. Such testing is overseen by agencies like the FAA and EASA, which assess these materials for flame resistance and overall safety performance.

Another crucial certification involves environmental resilience, as aircraft components must withstand a range of temperatures, humidity levels, and atmospheric pressures. Bionanocomposites must demonstrate durability against these environmental factors without deteriorating, swelling, or warping. In the aerospace context, materials are exposed to varying weather conditions and altitudes, which could range from intense UV radiation at high altitudes to freezing temperatures in various climates.

Certification standards for environmental durability include tests for UV stability, thermal stability, and moisture resistance. For bionanocomposites, meeting these standards may involve incorporating UV-blocking agents or moisture-resistant nanomaterials to ensure the material does not degrade over time. This testing ensures that bionanocomposites can withstand the unique environmental stresses of aerospace applications, such as the cyclic temperature changes experienced during takeoff, cruising, and landing.

Increasingly, aerospace regulatory bodies are also emphasizing sustainability, encouraging the adoption of materials with a reduced carbon footprint. Certifications related to sustainability assess the production, lifecycle, and disposal methods of materials, making sure they align with environmental standards and goals. Bionanocomposites, due to their bio-based components and potential for recyclability or biodegradability, may find favorable consideration in this area. However, they must still pass strict environmental compliance tests to confirm they do not release harmful substances or degrade in a way that would impact the environment.

This sustainability aspect is particularly relevant for newer certifications and standards developed in alignment with global environmental goals, like the European Union's REACH (Registration, Evaluation, Authorization, and Restriction of Chemicals) regulation (Williams et al. 2009) and the EPA's Safer Choice standards. Complying with these certifications helps ensure that bionanocomposites offer not only functional advantages but also align with the industry's shift toward more sustainable practices.

Beyond regulatory certifications, many aerospace companies have proprietary approval processes for materials. These processes involve rigorous in-house testing, where manufacturers of bionanocomposites must meet specifications unique to each aircraft maker. Industry-leading companies like Boeing and Airbus, for instance, maintain internal databases of approved materials, each of which has gone through years of testing and analysis before being qualified for use. For bionanocomposites to gain entry into these databases, they must consistently demonstrate performance reliability, cost-effectiveness, and compatibility with existing manufacturing processes.

To achieve industry adoption, bionanocomposite manufacturers often work directly with aerospace engineers to test and refine the materials according to specific aircraft design needs. This partnership approach can expedite the certification process by aligning the development of bionanocomposites with specific industry demands and engineering standards. Manufacturers that meet these proprietary standards can secure preferred supplier status, opening doors for widespread adoption across multiple aircraft models.

13.6 Marine and Rail Applications

As industries strive toward sustainability and efficiency, bionanocomposites offer transformative potential in both marine and rail sectors. By combining the strength, lightweight, and durability of conventional materials with eco-friendly properties, bionanocomposites are gaining traction as valuable alternatives in environments that require high performance under challenging conditions. Traditional materials, including metals, fiberglass, and standard composites, have been the mainstay of marine and rail structures, yet they come with drawbacks related to weight, environmental impact, and maintenance needs. Bionanocomposites offer solutions that address these challenges while aligning with industry goals for reducing fuel consumption, carbon emissions, and lifecycle costs.

13.6.1 Marine Applications: Hulls, Decks, and Interior Components

In the marine sector, lightweight bionanocomposites offer promising advancements in material performance, particularly in terms of weight reduction, corrosion resistance, and durability in harsh marine environments (Yang et al. 2018). Traditional materials like steel, aluminum, and fiberglass have long dominated marine applications due to their strength and structural stability, but they come with significant drawbacks, such as vulnerability to corrosion, biofouling, and high weight, which affects fuel efficiency (Hosseinabadi and Khedmati 2021). Bionanocomposites, which incorporate bio-based polymers and natural fibers with nanomaterial reinforcements, present a sustainable and high-performance alternative.

The hull is the core structure of a vessel, and its material directly impacts buoyancy, stability, and fuel consumption. Bionanocomposites in hull construction help to reduce overall vessel weight, enabling more efficient propulsion and fuel savings. Nanostructured additives within these composites, such as graphene or nanoclays, enhance the material's resistance to water penetration and marine biofouling—a critical factor for ships operating in oceanic environments. This biofouling, which occurs when microorganisms adhere to hull surfaces, creates drag that reduces fuel efficiency. Enhanced bionanocomposites can prevent this accumulation, resulting in lower maintenance requirements, improved hydrodynamics, and extended operational lifespan.

Bionanocomposites are also highly effective in decks and interior components, where resistance to corrosion, strength, and lightweight properties are essential. For decks, which are constantly exposed to moisture, UV light, and physical stress, bionanocomposites reinforced with natural fibers like hemp or flax offer a lightweight and robust solution. These composites resist water absorption and are less likely to degrade under UV exposure, maintaining their structural integrity over time (Kushwaha et al. 2015).

In interior applications, bionanocomposites are used in panels, furnishings, and other structural elements, providing a more sustainable, lightweight alternative to conventional materials like wood or metal. This weight reduction contributes to a vessel's overall fuel efficiency, as lighter vessels require less energy to operate. Furthermore, bionanocomposites are less susceptible to temperature and humidity fluctuations, maintaining dimensional stability and reducing the likelihood of warping or cracking—common issues with traditional materials in marine environments.

A notable example is the integration of natural fiber-reinforced bionanocomposites in racing yachts. By replacing heavier materials, designers achieved a reduction in hull weight, significantly improving speed and fuel efficiency. With the added durability and reduced maintenance needs due to the composite's resistance to saltwater and UV exposure, these yachts perform well under rigorous marine conditions, demonstrating the effectiveness of bionanocomposites in enhancing performance and sustainability.

The polymer-marine biomass-based bionanocomposite is a promising material for marine water treatment and pollution control. With a high adsorption capacity (384.615 mg/g) and rapid removal efficiency (99.79%), it effectively captures organic contaminants like malachite green (MG) dye from marine wastewater (Sarojini et al. 2022). Its spontaneous and exothermic adsorption behavior makes it highly effective in marine environments where temperature fluctuations occur. Given its ability to remove toxic dyes and contaminants, this bionanocomposite can be utilized in coastal wastewater treatment, oil spill cleanups, and marine aquaculture systems, promoting cleaner and more sustainable ocean ecosystems.

Bionanocomposites provide the marine industry with a sustainable, high-performance alternative to traditional materials. Their properties contribute to improved fuel efficiency, reduced maintenance, and lower environmental impact, paving the way for eco-friendly innovation in marine engineering.

13.6.2 Rail Applications: Lightweighting and Structural Reinforcement

The use of bionanocomposites in the rail industry is gaining traction as a response to the growing demand for energy efficiency, enhanced durability, and environmental sustainability in transportation (Atodariya et al. 2024).

Bionanocomposites offer benefits such as lightweighting and structural reinforcement and provide a sustainable, high-performance alternative to traditional materials like steel and aluminum. These composites, made from biopolymers reinforced with nanomaterials and natural fibers, align with the rail industry's commitment to reducing energy consumption and emissions while maintaining high standards of safety and durability.

Reducing the weight of rail vehicles is a top priority in the industry, as lighter vehicles consume less energy, require less power for acceleration, and exert lower track wear and tear, ultimately reducing operating costs and environmental impact. Bionanocomposites enable substantial weight reductions without compromising structural strength, which is critical for rail applications. For instance, components such as body panels, flooring, and seating can be made from bionanocomposites instead of heavier metals. By replacing traditional materials with bionanocomposites, weight savings of up to 30–40% are achievable in these non-structural components (Patel and Mahajan 2014).

Rail vehicle interiors benefit greatly from lightweight bionanocomposites, where natural fibers like flax, jute, or hemp provide sufficient strength while keeping the components light (Ramli et al. 2018). Thus, it reduces overall weight but also improves passenger comfort by creating a quieter, softer interior. In addition, bionanocomposites can be customized to any design requirement that may include intricate geometries and shapes which cannot be easily manufactured with metals; hence, there is flexibility for manufacturers in designing the interior, while the vehicles remain light in weight and energy-efficient. Structural integrity and durability are paramount in rail applications due to the high loads and vibrations experienced by rail vehicles over their long operational lifespans. Bionanocomposites reinforced with nanomaterials, such as carbon nanotubes or nanoclays, enhance strength and stiffness, making them suitable for reinforcing load-bearing components. These composites offer improved resistance to fatigue and impact, reducing the risk of material degradation or failure over time.

For structural applications, bionanocomposites are often used in hybrid formulations, where layers of nanomaterial-reinforced composites are combined with traditional metals to maximize strength while reducing weight (Kottappara et al. 2021). For instance, composite sandwich panels can reinforce the structural framework of rail cars, providing high stiffness and load-bearing capacity with minimal added weight. This hybrid approach improves the resilience of rail cars to dynamic stresses and impacts, particularly important in high-speed rail, where components are subjected to intense loads and vibrations.

Beyond mechanical benefits, bionanocomposites offer additional environmental and safety advantages for the rail industry. As they are derived from renewable resources, these composites contribute to reduced carbon emissions, aligning with sustainability goals in rail transportation. Bionanocomposites are often biodegradable or recyclable, supporting circular economy practices in the industry. Furthermore, the ability to source materials locally, such as natural fibers, reduces the environmental footprint associated with transportation and processing of raw materials.

From a safety perspective, bionanocomposites can be engineered to improve fire resistance, which is critical in rail applications. Innovations in nanomaterial integration have enabled the development of composites that are less prone to catching fire, providing increased safety for passengers and aligning with stringent industry regulations. For example, flame-retardant additives can be incorporated into bionanocomposites used in flooring or wall panels, preventing fire propagation and increasing passenger protection in case of emergencies.

One successful application of bionanocomposites in the rail industry is in passenger seating (Jayalakshmy 2017). Composite seats are in the best embodiment of the potential that bionanocomposites have for the modernization of rail interiors in a sustainable way. These seats reduce the vehicle's overall weight and allow for increased passenger capacity or fuel savings without compromising structural performance. With their reduced carbon footprint and long-term durability, composite seats represent the potential of bionanocomposites to modernize rail interiors sustainably.

These materials offer a paradigm shift in rail transportation toward greener and more efficient solutions through the reduction of energy consumption, durability enhancement, and support of sustainability goals. With the pace of innovation, it is very likely that bionanocomposites will become the cornerstone in the design and manufacture of the next generation of rail vehicles.

13.6.3 Comparison with Traditional Composite Materials

While both traditional composites and bionanocomposites are known for their strength and versatility, bionanocomposites offer distinct advantages that make them especially valuable for applications in industries like marine and rail. Table 13.4 outlines key differences.

Bionanocomposites present a sustainable edge and significant environmental benefits compared to traditional composites. Derived from renewable resources, bionanocomposites typically feature lower carbon footprints, aligning with global sustainability goals. Their recyclability and, in some cases, biodegradability allows

Table 13.4 Comparison of properties between traditional composites and bionanocomposites

Property	Traditional composites	Bionanocomposites
Weight	Heavier due to synthetic components	Lighter, especially with natural fibers
Strength-to-weight ratio	High, especially with carbon fibers	Comparable or better with nanomaterial reinforcement
Corrosion resistance	Susceptible to corrosion, especially in marine environments	High natural resistance, enhanced with treatments
Biodegradability	Limited, challenging to recycle	Often biodegradable or easier to recycle
Environmental impact	High energy use in production	Reduced impact with bio-based content
Cost	Often higher due to complex production processes	Potentially more cost-effective

for better end-of-life management, which is particularly important in industries like marine and rail where environmental regulations are tightening.

Traditional composites, such as fiberglass, have been reliable for decades, but they are often energy-intensive in production and may not provide the same level of environmental benefits as bionanocomposites. Furthermore, the ability of bionanocomposites to reduce weight while maintaining or even improving strength makes them an excellent choice for transportation applications. In such fields, weight reduction directly translates into fuel savings and reduced emissions, offering a more sustainable alternative to traditional materials.

13.7 Environmental and Economic Benefits

As global environmental concerns become more pronounced, industries worldwide are striving to reduce their ecological impact. The transportation sector, a significant contributor to greenhouse gas emissions, is under increasing pressure to adopt more sustainable practices. Bionanocomposites materials that integrate biological and nanomaterial components offer a promising solution. These materials are not only environmentally friendly but also provide economic advantages, making them key players in the future of sustainable transportation. In this section, we will explore the major environmental and economic benefits that bionanocomposites bring to various industries, particularly transportation, focusing on sustainability, carbon footprint reduction, lifecycle cost efficiency, and recyclability.

13.7.1 Sustainability and Carbon Footprint Reduction

As concerns over climate change and environmental degradation continue to rise, industries across the globe are facing increasing pressure to adopt more sustainable practices. The transportation sector, in particular, is a major contributor to greenhouse gas emissions, with cars, trucks, airplanes, and ships relying on fossil fuels that release carbon dioxide (CO_2) into the atmosphere (Samaras and Meisterling 2008). As a requirement for even greener options escalates, there has emerged hope in bionanocomposites, with more promising outlook beyond just increased functionality, and decreased environmental impacts such as what has been shown in Fig. 13.6.

Bionanocomposites generally contain the core material in their structure being either biopolymer or a variety of natural fibers originating from renewable biodegradable origins for instance crops or crop residues. Biopolymers, such as PLA and PHA, are produced from renewable agricultural resources, such as corn, sugarcane, and vegetable oils, and are inherently more sustainable than traditional petroleum-based polymers (Mori 2023). Unlike fossil fuels, which are finite and environmentally

Fig. 13.6 Sustainability and carbon footprint reduction in transportation through the use of bionanocomposites. This is an AI-generated image taken from ChatGPT.com

damaging to extract and refine, these materials can be replenished annually, reducing the reliance on non-renewable resources.

Natural fibers used in bionanocomposites—such as hemp, flax, and jute—also play a significant role in reducing the environmental footprint of these materials. These fibers are grown with minimal resource consumption (requiring little water or pesticides) and have a lower environmental impact during production compared to synthetic fibers made from petroleum. Moreover, natural fibers are biodegradable, which reduces waste accumulation in landfills (Khalid et al. 2021).

In addition to their renewable sourcing, bionanocomposites also have a lower environmental impact during the manufacturing process. The production of traditional composites, such as those made from fiberglass or carbon fiber, involves energy-intensive processes, including the extraction and refinement of raw materials, which results in high emissions of CO_2 and other pollutants. On the other hand, the production of bionanocomposites requires less energy, as many of these materials can be processed at lower temperatures and with less chemical input. This results in a lower carbon footprint during manufacturing.

Furthermore, the use of nanomaterials in bionanocomposites can further reduce the amount of material needed to achieve desired strength and durability. For example, the addition of carbon nanotubes or nanocellulose enhances the mechanical properties of bionanocomposites, allowing manufacturers to use smaller quantities of material (Miyashiro et al. 2020). This not only decreases resource consumption but also results in lighter-weight products that require less energy during use.

The lightweight nature of bionanocomposites also contributes significantly to carbon footprint reduction during the operational phase of transportation (Correa et al. 2019). By replacing traditional metals (such as steel and aluminum) with lightweight composite materials in vehicles, aircraft, and ships, manufacturers can reduce the overall weight of these transportation modes (Faruk et al. 2017). Lighter vehicles require less energy to move, whether it's gasoline for cars, jet fuel for planes, or marine fuel for ships. This results in lower carbon emissions during operation.

For example, substituting steel body panels with bionanocomposites can help to enhance the fuel efficiency by up to 30% depending upon the percentage weight reduction in automotive applications. The same thing goes for aerospace applications; by reducing the weight of aircraft structures, considerable savings can be obtained, especially in long-haul flights. Being the major emitter, the aviation industry would largely benefit from the wide adoption of lightweight composites.

Another unique environmental benefit of bionanocomposites is their ability to act as carbon sinks during their lifecycle. Natural fibers used in composites, such as hemp and flax, capture and store carbon during their growth, thus helping to offset carbon emissions during the material's lifespan. In fact, it has been estimated that the carbon absorbed by these fibers during growth can offset a significant portion of the carbon emissions produced during the composite's manufacturing and use.

At the end of their lifecycle, bionanocomposites can be composted or biodegraded, reducing their contribution to landfills and the associated methane emissions. Many biopolymers, such as PLA, are fully compostable, breaking down into water and carbon dioxide in appropriate conditions. This biodegradability contrasts with conventional petroleum-based plastics, which can persist in the environment for hundreds of years, contributing to long-term pollution.

13.7.2 Lifecycle and Cost Efficiency of Bionanocomposites

While sustainability is a critical factor in the widespread adoption of bionanocomposites, it is equally important to consider their economic viability and lifecycle cost efficiency. Bionanocomposites offer several advantages over traditional materials, both in terms of production cost and operational efficiency. In the context of transportation, where reducing costs while improving performance is key, bionanocomposites offer a competitive advantage.

The use of renewable, plant-based fibers and biopolymers in the production of bionanocomposites can significantly lower raw material costs. Traditional composites, such as fiberglass or carbon fiber, often require expensive, non-renewable raw materials, including petroleum-based polymers and synthetic fibers, which can be costly to produce. Bionanocomposites, on the other hand, use bio-based materials that are more abundant and generally less expensive to source, making them an economically attractive alternative (Riaz et al. 2024).

Additionally, the manufacturing process for bionanocomposites typically requires less energy, leading to reduced operational costs. Traditional composite materials often require high temperatures and specialized equipment for manufacturing, whereas bionanocomposites can be processed using simpler, more energy-efficient techniques. For example, many biopolymers used in bionanocomposites can be processed at lower temperatures compared to petroleum-based polymers, leading to lower energy consumption during production (Pradhan 2022).

One of the most significant economic benefits of bionanocomposites in transportation is their ability to reduce fuel consumption. Lighter vehicles require less energy to operate, which directly translates to reduced fuel costs. In the automotive industry, the adoption of lightweight composites can lead to significant fuel savings over the lifetime of a vehicle. For example, a vehicle that is 10–20% lighter can improve fuel efficiency by up to 10%, depending on the type of composite used. In addition, reducing weight in aircraft and marine vessels can lead to similar fuel savings, particularly for long-haul flights or ocean voyages.

These fuel savings not only result in lower operational costs for businesses but also have long-term economic benefits by extending the lifespan of transportation assets. By reducing fuel consumption, bionanocomposites contribute to a reduction in wear and tear on engines and other mechanical components, thereby lowering maintenance and repair costs.

Another economic advantage of bionanocomposites is their superior durability compared to traditional materials. Many bionanocomposites are highly resistant to corrosion, wear, and fatigue, which extends the life of transportation components. For example, the use of natural fibers in automotive body panels or aircraft fuselages can result in components that are less prone to corrosion, reducing the need for frequent repairs and replacements. This can lead to significant long-term cost savings, particularly in industries where maintenance is a major operational expense, such as aviation and shipping.

Furthermore, bionanocomposites often have higher strength-to-weight ratios than conventional materials, meaning that thinner, lighter components can be used without sacrificing strength. This reduces the material costs associated with manufacturing, as less material is needed to achieve the desired performance.

The scalability of bionanocomposite production is another important factor in their economic viability. As demand for sustainable materials increases, manufacturers are investing in technologies to scale up the production of bionanocomposites. These materials can be produced using existing infrastructure and manufacturing processes, reducing the need for expensive new facilities. Additionally, as research into bionanocomposites continues to progress, the cost of production is expected to decrease, further improving their cost-effectiveness.

13.7.3 Recyclability and End-of-Life Considerations

As worldwide efforts to minimize waste and environmental pollution heighten, the recyclability of materials and end-of-life (EOL) management have become crucial aspects of sustainable product design (Kupolati et al. 2022). The wreckage of the composites of such industries as automobiles, aviation, and construction is getting greater due to the fact the majority of the conventional composite materials are difficult to recycle and usually are dumped in landfills which becomes the leading cause of permanent environmental pollution. On the other hand, Bionanocomposites offer a great solution to recycling challenges because of their renewable, biodegradable components and the emerging recycling technologies that are being developed (Dwivedi et al. 2022). The focus of these materials is sustainability and the cycle of recycling. Therefore, they provide many more benefits than the traditional synthetic composites that are currently mostly used in the industry.

The biodegradability of bionanocomposites is one of their most notable advantages, particularly in the context of end-of-life disposal. Many conventional composites, particularly those made from synthetic fibers like carbon fiber or glass, are not biodegradable (Egan and Salmon 2022). These materials can take hundreds of years to decompose, leading to long-lasting pollution in landfills. In contrast, bionanocomposites often incorporate natural fibers (such as hemp, jute, or flax) and biopolymers (such as polylactic acid, PLA, or polyhydroxyalkanoates, PHA) that break down through natural processes after they have reached the end of their useful life (Shrivastava and Dondapati 2021). This biodegradability significantly reduces the environmental burden of these materials at the end of their life cycle, as they do not persist in the environment for extended periods.

Renewable natural fibers, which are biodegradable, are the chief contributors to the safe return of bionanocomposites to nature. A case-in-point, when bionanocomposites are dumped in composting conditions, the fibers and biopolymers biodegrade naturally and turn into organic matter, water, and carbon dioxide (Ragitha and Edison 2022). This is a much faster process compared to the natural decomposition of conventional composites that involve chemicals or high temperatures to break down the materials. As such, the use of bionanocomposites in products like automotive parts, construction materials, and packaging can help mitigate the growing issue of waste accumulation.

While biodegradability is a significant advantage, the ability to recycle bionanocomposites also plays an important role in their sustainability. Recycling allows for the recovery of valuable materials, reducing the need for new raw materials and helping to conserve resources. The recycling of traditional composites, especially those involving synthetic resins and fibers, is often challenging due to the complex mixture of materials that are difficult to separate. In contrast, bionanocomposites, particularly those using natural fibers and biopolymers, are easier to process and recycle.

For instance, natural fibers like hemp, flax, and jute are easier to separate from the matrix material, allowing them to be reused in new composite products. Biopolymers, which are also used in bionanocomposites, are often more amenable to recycling processes compared to petroleum-based plastics. Polylactic acid (PLA), one of the most common biopolymers used in bionanocomposites, can be broken down and reprocessed into new products at the end of its life cycle (Ranakoti et al. 2022). Some companies have already developed systems for recycling PLA-based composites, which further enhances their environmental credentials.

However, the recycling of bionanocomposites is still an area of active research. In many cases, existing recycling systems are not fully equipped to handle the specific types of materials found in these composites. For instance, the presence of nanomaterials, such as nanocellulose or carbon nanotubes, in bionanocomposites can complicate recycling efforts. These nanomaterials, whilst offering performance gains in terms of strength and resilience, may make recovery and ultimate reuse dependent upon specialized equipment that is still emerging. Researchers are currently developing new recycling techniques to handle composites containing nanomaterials, which will be crucial to making bionanocomposites fully recyclable and a more sustainable option in the long term.

Another essential feature of bionanocomposites is their safety decomposability if in service. Unlike conventional composites that need landfilling or incineration, biodegradation through composting in the case of bionanocomposites ensures that their environmental footprint on disposal is highly reduced. Biodegradable biopolymers like PLA or PHA are known to break down under natural conditions when exposed to microorganisms, heat, and moisture (Burzic et al. 2019). These materials do not produce toxic byproducts during decomposition, making them safe for the environment.

For example, automotive parts made from bionanocomposites could, at the end of their life, be composted or subjected to biodegradation in an industrial composting facility. The natural fibers in the composite would break down into organic matter, enriching the soil without leaving harmful residues. This is particularly valuable in industries like packaging and agriculture, where products made from bionanocomposites can be disposed of without posing long-term environmental risks.

However, it is important to note that the rate of biodegradation can depend on factors such as temperature, moisture, and microbial activity, which may not always be present in landfills or other disposal sites. As such, it is essential that proper disposal and waste management systems be developed to ensure that bionanocomposites can degrade effectively at the end of their life.

Despite the promising potential of bionanocomposites, there are still several challenges that need to be addressed in terms of scaling up recycling technologies. One major hurdle is the need for infrastructure capable of processing bionanocomposites on a large scale. Recycling systems for conventional materials, such as metals and plastics, are already well-established, but the systems for recycling bionanocomposites are still in the early stages of development.

Additionally, as bionanocomposites often incorporate a variety of natural fibers, biopolymers, and nanomaterials, it can be difficult to standardize recycling processes for all types of composites. This requires research into developing universal recycling protocols and technologies that can effectively handle different types of bionanocomposite materials (Botlhoko et al. 2024). The use of nanomaterials, in particular, may complicate the separation and recovery of fibers, as the small size and high surface area of nanomaterials can affect their behavior during recycling processes.

Another economic issue linked to the recycling of bionanocomposites is that related to cost effectiveness. The fact that recycling processes are environmental friendly does not discount the high price tag on developing and sustaining recycling infrastructure and technology. Thus, the economic merits of recycled materials need to be weighed against the cost of recycling bionanocomposites; unless recycling techniques become more effective, then its large-scale use will face various economic difficulties.

As more industries seek more sustainable options, bionanocomposites present huge potential for both minimizing waste and pollution but also, ultimately toward the achievement of a circular economy. By developing new recycling technologies, enhancing biodegradability, and improving disposal methods, bionanocomposites can play a key role in reducing the environmental footprint of industries that rely heavily on composite materials. The future of bionanocomposites lies in their ability to be reused, recycled, and responsibly disposed of, making them a critical component of a sustainable, circular economy.

13.8 Challenges and Future Directions

This implies that the scale up, market acceptance, production challenges, and regulatory compliance become more significant factors that need to be addressed while there is still significant growth of adoption in industries across the world. The environmental sustainability, lightness, and exceptional mechanical properties provide a platform to overcome most obstacles that arise for the full potential realization of bionanocomposites. The future of these materials depends on overcoming these challenges through technological advancements, market education, and regulatory evolution as shown in Fig. 13.7.

13.8.1 Production Scale-Up and Cost Challenges

One of the most significant challenges in advancing bionanocomposites from the research stage to large-scale industrial use is the scaling up of production processes while maintaining cost-effectiveness (Reshmy et al. 2022). While the lab-scale synthesis of bionanocomposites is relatively straightforward, transferring this process to large-scale manufacturing involves complexities related to material sourcing, production rates, and the need for specialized equipment.

The production of bionanocomposites often relies on natural fibers and biopolymers, which may not always be available in sufficient quantities to meet the demands

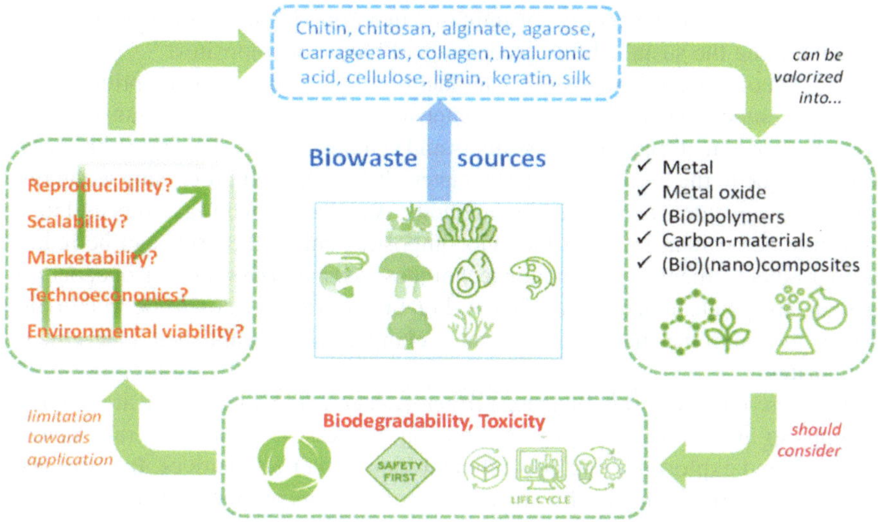

Fig. 13.7 Challenges and future directions in the use of bionanocomposites for enhancing sustainability and performance in the transportation sector. This image is reproduced with permission from Lizundia et al. (2022) © The Royal Society of Chemistry 2022

of large-scale industrial applications. While materials like hemp, flax, and jute have gained popularity in various sectors, securing a consistent and sustainable supply of these natural fibers is a critical challenge. Additionally, the cost of extracting and processing biopolymers, especially biodegradable polymers like PLA or PHA can be more expensive than synthetic polymers derived from petroleum. The cost of raw materials can, therefore, be a significant factor influencing the final cost of bionanocomposites.

The fabrication of bionanocomposites requires specialized manufacturing techniques that may not be as widely available as those used for conventional composites, such as thermoset or thermoplastic composites. Processes such as fiber impregnation, molding, and curing need to be adjusted for the unique properties of natural fibers and biopolymers. Moreover, nanocomposites containing carbon nanotubes or nanocellulose require additional manufacturing efforts because their behavior on a nanometer scale and the tendency for agglomeration or poor dispersion in the matrix add complexity (Dhar and Katiyar 2020). This requires advanced equipment and precision handling to ensure uniform distribution of the nanofillers, which can be more costly than traditional methods used in manufacturing conventional composites.

For large-scale bionanocomposite production to be sustainable the, scale and energy efficiency during this entire production process must also be considered. Although composites are supposed to be more sustainable, the energy consumed in processing natural fibers and biopolymers, particularly at larger volumes, could offset part of those benefits. Therefore, reducing the environmental impact of the production process and minimizing energy consumption in optimizing its production will be key to maintaining the sustainability credentials of bionanocomposites.

The cost of bionanocomposites is another barrier to widespread adoption. Despite their environmental benefits, bionanocomposites often come at a premium compared to conventional materials like glass fibers and synthetic resins. This price disparity is primarily due to the higher cost of raw materials, specialized processing techniques, and the energy-intensive production methods. As a result, manufacturers and industries may be hesitant to adopt bionanocomposites unless the price can be reduced through advancements in production techniques, raw material sourcing, or economies of scale (Nagalakshmaiah et al. 2019). For bionanocomposites to be viable alternatives to traditional materials, they must compete with established, cost-effective composites.

To overcome these challenges, further research and innovation are required. Scaling up production involves optimizing processes, improving raw material sourcing, and exploring more cost-efficient biopolymer alternatives. Innovations in nanomaterial processing, such as using greener solvents or more efficient dispersion techniques, will be necessary to reduce costs and increase the commercial feasibility of bionanocomposites (Lonkar et al. 2011). Additionally, establishing industrial partnerships and collaborations with raw material suppliers, as well as exploring the development of more sustainable production methods, can help drive down costs and make bionanocomposites more competitive in the market.

13.8.2 Market Acceptance and Regulatory Compliance

As bionanocomposites become more widely researched and developed, it is crucial for industries and consumers to embrace these materials for their potential advantages (Hasan et al. 2020). However, market acceptance of bionanocomposites faces challenges, such as the inertia of established industries that rely on conventional materials, as well as concerns over the performance, reliability, and cost-effectiveness of bionanocomposites in comparison to their synthetic counterparts.

However, the lack of awareness and understanding of the advantages of bionanocomposites by industry actors and consumers limits the adoption of bionanocomposites, in spite of the increasing interests towards the sustainability. The industries (e.g., automotive and aerospace) have been utilizing the classical composite materials (i.e., carbon fiber and glass fiber) by which they are characterized in terms of performance and production process (Tripathi et al. 2020). Introducing new materials, even those that promise improved environmental benefits, requires overcoming the skepticism of manufacturers who may be hesitant to change their long-standing practices. Additionally, consumers and end-users may not yet fully recognize the advantages of bionanocomposites, especially in industries that require high-performance materials.

To overcome this challenge, it is imperative to provide training to manufacturers, engineers and end-users on the possible advantages of and application of bionanocomposites. Industrial conferences, seminars, and collaborative research teams could be used to advance the relative advantages of bionanocomposites and their applicability to a variety of applications. Furthermore, showcasing real-world case studies that highlight the successful use of bionanocomposites in products—such as lightweight vehicle parts or sustainable building materials—can help build confidence in the technology.

For bionanocomposites to gain acceptance, they must demonstrate performance and durability equivalent to or better than traditional materials. Industries like automotive, aerospace, and construction demand high standards in terms of strength, stiffness, fire resistance, and impact resistance. While bionanocomposites show promise in these areas, they must undergo extensive testing and validation to prove their reliability. Ensuring that bionanocomposites meet the stringent performance requirements of these sectors will be key to their market acceptance (Anton et al. 2015).

Apart from performance, concerns about the long-term stability of bionanocomposites remain to be resolved. For instance, although certain natural fibers are very strong, they can be more susceptible to damage from moisture or UV exposure with the passage of time. The matrix materials of the bionanocomposites (biopolymers may also be exposed to the environmental conditions such as temperature changes, which can influence the stability of the matrix materials). In order that these materials find market success, design and testing under severe conditions for long-term stability are needed.

Compliance with regulatory standards is another important challenge for the commercialization of bionanocomposites. Every industry, particularly the automotive, aerospace, and construction sectors, is governed by strict regulatory frameworks to ensure the safety, performance, and environmental impact of materials. For example, materials used in automotive or aircraft manufacturing must meet stringent safety regulations, such as crash tests, fire resistance, and long-term durability tests (Pinkse et al. 2014). Similarly, construction materials must comply with building codes and environmental standards.

Bionanocomposites need to adhere to these regulatory standards to gain acceptance in these sectors. However, since bionanocomposites are relatively new, they often lack established standards or certifications. Regulatory bodies and industry organizations need to develop and implement specific testing protocols and performance standards for these materials. Establishing clear, internationally recognized guidelines will help ensure the safe and reliable use of bionanocomposites across different industries.

Furthermore, standardization of certification procedures for the environmental impact of bionanocomposites, i.e., life cycle assessment (LCA), carbon footprint and recyclability, is required. Since these materials are marketed as eco-friendly substitutes, they must be assessed for their complete environmental footprint in order to validate their eco friendliness. This will necessitate a partnership amongst industry practitioners, regulators, and research organizations, in order to objectively and transparently evaluate bionanocomposites.

Addressing the challenges of market acceptance and regulatory compliance requires coordinated efforts across multiple fronts. Policymakers can play a vital role by offering incentives and support for research and development in the bionanocomposites sector. By promoting sustainable practices and providing funding for technology advancements, governments can encourage the adoption of bionanocomposites in key industries.

Furthermore, industry collaboration is essential in overcoming regulatory hurdles. Standardization organizations, such as ASTM International or ISO, need to develop frameworks and testing methods for bionanocomposites to ensure they meet industry-specific requirements (Klaessig 2017). By working together, industry leaders and regulatory bodies can facilitate the smooth integration of bionanocomposites into mainstream markets.

The widespread adoption of bionanocomposites faces challenges related to production scale-up, cost, market acceptance, and regulatory compliance. Overcoming these hurdles will require continuous innovation, investment in research, and collaboration across industries and regulatory bodies. However, by addressing these challenges effectively, bionanocomposites have the potential to revolutionize industries and contribute significantly to a more sustainable, eco-friendly future.

13.9 Conclusion

Bionanocomposites are poised to transform industries by offering a sustainable alternative to conventional materials. Their lightweight properties are particularly advantageous in sectors like automotive, aerospace, marine, and rail transportation, where reducing weight leads to improved fuel efficiency, lower emissions, and reduced environmental impact. These composites are not only mechanically strong and durable but also possess the added benefit of recyclability and biodegradability, supporting a circular economy and reducing waste. For instance, in automotive applications, bionanocomposites contribute to lighter body panels, interiors, and battery housings, improving the efficiency of both traditional and electric vehicles. Similarly, their high strength-to-weight ratio benefits the aerospace sector by enabling lower fuel consumption in aircraft.

Despite their clear advantages, widespread adoption faces challenges, including high production costs, complex manufacturing processes, and the need for compliance with strict regulatory standards. Addressing these challenges through innovations in fabrication technologies and process optimizations is crucial to making bionanocomposites more economically viable and accessible for large-scale use. Looking ahead, ongoing research and development must focus on enhancing the properties of these materials and improving cost-efficiency. By advancing nanostructuring, hybrid reinforcement techniques, and sustainable manufacturing practices, bionanocomposites can play a key role in reshaping the future of materials, contributing to global sustainability goals, and driving greener, more efficient industrial practices.

References

Adekomaya O, Majozi T (2024) Sustainable reclamation of synthetic materials as automotive parts replacement: effects of environmental response on natural fiber vulnerabilities. Environ Sci Pollut Res 31(12):18396–18411. https://doi.org/10.1007/s11356-024-32436-5

Agboola O, Sadiku R, Mokrani T, Amer I, Moropeng ML, Ramakokovhu MM (2016) Polyethylene/polyhydroxyalkanoates-based biocomposites and bionanocomposites. In: Polyethylene-based biocomposites and bionanocomposites, pp 201–278

Aguiar JB, Martins AM, Almeida C, Ribeiro HM, Marto J (2022) Water sustainability: a waterless life cycle for cosmetic products. Sustain Prod Consum 32:35–51

Al Azim Faiz BINZ (2022) Design of kenaf fiber polymer composite for interior car door panel application. Universiti Sains Malaysia

Al Mahmud MZ, Rabbi SMF, Islam MD, Hossain N (2024) Synthesis and applications of natural fiber-reinforced epoxy composites: a comprehensive review. SPE Polym

Alam MA, Sapuan SM, Ya HH, Hussain PB, Azeem M, Ilyas RA (2021) Application of biocomposites in automotive components: a review. In: Biocomposite and synthetic composites for automotive applications, pp 1–17

Ali A et al (2018) Hydrophobic treatment of natural fibers and their composites—a review. J Ind Text 47(8):2153–2183

Ali N, Hoque ME (2022) Bionanocomposites in the automotive and aerospace applications. In: Muthukumar C, Thiagamani SMK, Krishnasamy S, Nagarajan R, Siengchin S (eds) Polymer based bio-nanocomposites: properties, durability and applications. Springer, Singapore, pp 237–253. https://doi.org/10.1007/978-981-16-8578-1_13

Al-Maharma AY, Al-Huniti N (2019) Critical review of the parameters affecting the effectiveness of moisture absorption treatments used for natural composites. J Compos Sci 3(1):27

Andrew JJ, Dhakal HN (2022) Sustainable biobased composites for advanced applications: recent trends and future opportunities—a critical review. Compos C Open Access 7:100220

Anton PS, Silberglitt R, Schneider J (2001) The global technology revolution: bio/nano/materials trends and their synergies with information technology by 2015. Rand Corporation

Arockiam NJ, Jawaid M, Saba N (2018) 6—Sustainable bio composites for aircraft components. In: Jawaid M, Thariq M (eds) Woodhead Publishing series in composites science and engineering. Woodhead Publishing, pp 109–123. https://doi.org/10.1016/B978-0-08-102131-6.00006-2

Aston DE, Bow JR, Gangadean DN (2013) Mechanical properties of selected nanostructured materials and complex bio-nano, hybrid and hierarchical systems. Int Mater Rev 58(3):167–202

Atodariya J, Agrawal M, Singh A, Patni N (2024) "Advancements in green nanocomposites: a comprehensive review on cellulose-based materials in biocomposites and bionanocomposites. In: Visakh (ed) Rubber based bionanocomposites: characterisation. Springer Nature, Singapore, pp 213–239. https://doi.org/10.1007/978-981-10-2978-3_9

Azhar U, Yaseen S, Arif M, Babar M, Sagir M (2024) Chapter Seven—Emission of greenhouse gases from transportation. In: Rahimpour MR, Makarem MA, Meshksar M (eds) Advances and technology development in greenhouse gases: emission, capture and conversion. Elsevier, pp 147–163. https://doi.org/10.1016/B978-0-443-19231-9.00016-8

Bachmann J et al (2021) Towards a circular economy in the aviation sector using eco-composites for interior and secondary structures. Aerospace. https://doi.org/10.3390/aerospace8050131

Bagal-Kestwal DR, Pan MH, Chiang BH (2019) Processing methods for bionanocomposites. In: Bio monomers for green polymeric composite materials, pp 25–55. https://doi.org/10.1002/978 1119301714.CH2

Balakrishnan P, John MJ, Pothen L, Sreekala MS, Thomas S (2016) 12—Natural fibre and polymer matrix composites and their applications in aerospace engineering. In: Rana S, Fangueiro R (eds) Advanced composite materials for aerospace engineering. Woodhead Publishing, pp 365–383. https://doi.org/10.1016/B978-0-08-100037-3.00012-2

Balo F, Sua L (2024) Green composites in aviation: optimizing natural fiber and polymer selection for sustainable aircraft cabin material. Textiles. https://doi.org/10.3390/textiles4040033

Baranwal J, Barse B, Fais A, Delogu GL, Kumar A (2022) Biopolymer: a sustainable material for food and medical applications. Polymers. https://doi.org/10.3390/polym14050983

BhagabatiP (2020) Biopolymers and biocomposites-mediated sustainable high-performance materials for automobile applications. In: Sustainable nanocellulose and nanohydrogels from natural sources. Elsevier, pp 197–216

Bhat A et al (2021) Review on nanocomposites based on aerospace applications. Nanotech Rev 10(1):237–253. https://doi.org/10.1515/ntrev-2021-0018

Bilal M, Rasheed T, Nabeel F, Iqbal HMN (2020) Bionanocomposites from biofibers and biopolymers. In: Khan A, Mavinkere Rangappa S, Siengchin S, Asiri AM (eds) Biofibers and biopolymers for biocomposites: synthesis, characterization and properties. Springer International Publishing, Cham, pp 135–157. https://doi.org/10.1007/978-3-030-40301-0_6

Bio-based materials for aircraft | Research and Innovation. Accessed: 30 Jan 2025. [Online]. Available: https://projects.research-and-innovation.ec.europa.eu/en/projects/success-stories/all/bio-based-materials-aircraft

Botlhoko OJ, Lekalakala R, Muniyasamy S (2024) Bioplastics and its bio-nanocomposites for a circular economy: standards, labelling and regulations. In: Biodegradable polymers, blends and biocomposites. CRC Press, pp 257–270

Broughton W (2012) Environmental induced failure in fibre-reinforced plastics. In: Failure mechanisms in polymer matrix composites, pp 393–440. https://doi.org/10.1533/9780857095329.2.393

Burzic I et al (2019) Impact modification of PLA using biobased biodegradable PHA biopolymers. Eur Polym J 114:32–38

Carlstedt D, Asp LE (2020) Performance analysis framework for structural battery composites in electric vehicles. Compos B Eng 186:107822. https://doi.org/10.1016/j.compositesb.2020.107822

Chatterjee N, Dhar P (2025) Bio-nanocomposites: curse or miracle for the society? In: Characterization techniques in bionanocomposites. Elsevier, pp 3–39

Chauhan V, Kärki T, Varis J (2022) Review of natural fiber-reinforced engineering plastic composites, their applications in the transportation sector and processing techniques. J Thermoplast Compos Mater 35(8):1169–1209

Collinson MG, Bower MP, Swait TJ, Atkins CP, Hayes SA, Nuhiji B (2022) Novel composite curing methods for sustainable manufacture: a review. Compos C Open Access 9:100293

Correa JP, Montalvo-Navarrete JM, Hidalgo-Salazar MA (2019) Carbon footprint considerations for biocomposite materials for sustainable products: A review. J Clean Prod 208:785–794

Dalbanjan NP et al (2024) Descriptive review on conversion of waste residues into valuable bionanocomposites for a circular bioeconomy. Nano-Structures & Nano-Objects 39:101265. https://doi.org/10.1016/j.nanoso.2024.101265

Dhar P, Katiyar V (2020) 3. Benchmarking nanocellulose production: scale-up strategies and life-cycle assessment. In: Cellulose nanocrystals, pp 49–80

Dwivedi N, Dwivedi S, Sharma D (2022) Eco-design and modification study of bionanoparticles and life cycle assessment: keys to sustainability. In: Bionanotechnology towards sustainable management of environmental pollution. CRC Press, pp 119–140

Egan J, Salmon S (2022) Strategies and progress in synthetic textile fiber biodegradability. SN Appl Sci 4:1–36

Epstein AH et al (2025) Some steps toward autonomy in aeronautics. Accessed: 30 Jan 2025. [Online]. Available: https://www.researchgate.net/profile/Nicholas-Margiewicz/post/Improvements_in_Synthetic_Fuel/attachment/616c839df5675b211b099eb8/AS%3A1079064483897346%401634280429430/download/032841NAEBridgev50n2-Summer2020TXTstitchedhyperlinks.pdf#page=45

Fang Z, Hou G, Chen C, Hu L (2019) Nanocellulose-based films and their emerging applications. Curr Opin Solid State Mater Sci 23(4):100764. https://doi.org/10.1016/J.COSSMS.2019.07.003

Faruk O, Tjong J, Sain M (2017) Lightweight and sustainable materials for automotive applications. CRC Press, Boca Raton

Ferreira FV, Pinheiro IF, de Souza SF, Mei LHI, Lona LMF (2019) Polymer composites reinforced with natural fibers and nanocellulose in the automotive industry: a short review. J Compos Sci 3(2):51

Fonseca-Florido HA, Yañez-Macías R, Ávila-Orta CA, Cruz-Delgado VJ, Covarrubias-Gordillo CA (2024) Starch/carbon nanofibers bionanocomposites via melt mixing: Effect of dispersion and compatibility on conductivity and mechanical properties. Int J Biol Macromol 256:128519. https://doi.org/10.1016/j.ijbiomac.2023.128519

Giammaria V, Capretti M, Del Bianco G, Boria S, Santulli C (2024) Application of poly(lactic acid) composites in the automotive sector: a critical review. Polymers 16(21):3059. https://doi.org/10.3390/POLYM16213059

Gouripriya DA, Debnath P, Saha P (2024) Structure, Properties, and applications of plant-based and bacterial nanocellulose in tissue engineering. In: Sreedharan M, Thomas S, Kalarikkal N, Vijayamma R, Grohens Y, Yang G (eds) Nanocellulose-based hybrid systems for tissue engineering, vol 18. Royal Society of Chemistry. https://doi.org/10.1039/9781837673094-00016

Groenleer M, Kaeding M, Versluis E (2010) Regulatory governance through agencies of the European Union? The role of the European agencies for maritime and aviation safety in the implementation of European transport legislation. J Eur Public Policy 17(8):1212–1230. https://doi.org/10.1080/13501763.2010.513577

Guchait A, Saxena A, Chattopadhyay S, Mondal T (2022) Influence of nanofillers on adhesion properties of polymeric composites. ACS Omega 7(5):3844–3859. https://doi.org/10.1021/acsomega.1c05448

Guo H, Zhou X, Liu Z (2024) Advanced lightweight structural materials for automobiles: properties, manipulation, and perspective. Sci Adv Mater 16(5):563–580. https://doi.org/10.1166/SAM.2024.4686

Gupta P, Toksha B, Patel B, Rushiya Y, Das P, Rahaman M (2022) Recent developments and research avenues for polymers in electric vehicles. Chem Rec 22(11):e202200186

Gupta R et al (2021) A review of sensing technologies for non-destructive evaluation of structural composite materials. https://doi.org/10.3390/jcs5120319

Hasan KMF, Horváth PG, Alpár T (2020) Potential natural fiber polymeric nanobiocomposites: a review. Polymers (Basel) 12(5):1072

Hosseinabadi OF, Khedmati MR (2021) A review on ultimate strength of aluminium structural elements and systems for marine applications. Ocean Eng 232:109153. https://doi.org/10.1016/j.oceaneng.2021.109153

Hosseini SF, Rezaei M, Zandi M, Farahmandghavi F (2015) Fabrication of bio-nanocomposite films based on fish gelatin reinforced with chitosan nanoparticles. Food Hydrocoll 44:172–182. https://doi.org/10.1016/j.foodhyd.2014.09.004

Hottle TA, Bilec MM, Landis AE (2017) Biopolymer production and end of life comparisons using life cycle assessment. Resour Conserv Recycl 122:295–306. https://doi.org/10.1016/j.resconrec.2017.03.002

Huang R et al (2016) Energy and emissions saving potential of additive manufacturing: the case of lightweight aircraft components. J Clean Prod 135:1559–1570. https://doi.org/10.1016/j.jclepro.2015.04.109

Ibrahim ID et al (2019) Biopolymer composites and bionanocomposites for energy applications. In: Gnanasekaran D (ed) Green biopolymers and their nanocomposites. Springer, Singapore, pp 313–341. https://doi.org/10.1007/978-981-13-8063-1_14

Idoko FA, Ezeamii GC, Ojochogwu OJ (2024) Green chemistry in manufacturing: innovations in reducing environmental impact. World J Adv Res Rev 23(3):2826–2841

Idumah CI, Hassan A (2016) Emerging trends in flame retardancy of biofibers, biopolymers, biocomposites, and bionanocomposites. Rev Chem Eng 32(1):115–148. https://doi.org/10.1515/revce-2015-0017

Ilyas RA et al (2019) Potential of natural fibre composites for transport industry: a review. In: Prosiding Seminar Enau Kebangsaan, Persatuan Pembangunan dan Industri Enau Malaysia (PPIEM) Bahau, Negeri, pp 2–11

Ilyas RA et al (2021) Macro to nanoscale natural fiber composites for automotive components: research, development, and application. In: Biocomposite and synthetic composites for automotive applications, pp 51–105

Jayalakshmy MS (2017) Applications of rubber based biocomposites and bionanocomposites. In: Visakh PM (ed) Rubber based bionanocomposites: preparation. Springer International Publishing, Cham, pp 167–176. https://doi.org/10.1007/978-3-319-48806-6_8

Jayaraman P, Pai A, Rodriguez-Millan M, Shenoy S, Chethan KN, Hegde S (2024) Exploring acoustic properties of banana fiber composites with elastomeric filler through a computational approach. Mater Res Express 11(1):15508

Khalid MY et al (2021) Developments in chemical treatments, manufacturing techniques and potential applications of natural-fibers-based biodegradable composites. Coatings 11(3):293

Kishore A, Mithul Aravind S, Singh A (2023) Bionanocomposites for active and smart food packaging: a review on its application, safety, and health aspects. J Food Process Eng 46(5):e14320

Klaessig FC (2017) Nanotechnology definitions at ISO and ASTM International: origin, usage, and relationship to nomenclature and regulatory and metrology activities. In: Metrology and standardization of nanotechnology: protocols and industrial innovations, pp 15–48

Kottappara R, Pillai SC, Vijayan BK (2021) copper- and iron-based bio-nanocomposites for green applications. In: Sarma H, Joshi SJ, Prasad R, Jampilek J (eds) Biobased nanotechnology for green applications. Springer International Publishing, Cham, pp 41–72. https://doi.org/10.1007/978-3-030-61985-5_2

Krishna TNV, Kumar SVSVPD, Srinivasa Rao S, Chang L (2024) Powering the future: advanced battery management systems (BMS) for electric vehicles. https://doi.org/10.3390/en17143360

Kumar A, Sinha-Ray S (2018) A review on biopolymer-based fibers via electrospinning and solution blowing and their applications. Fibers 6(3):45

Kumari SVG, Pakshirajan K, Pugazhenthi G (2022) Recent advances and future prospects of cellulose, starch, chitosan, polylactic acid and polyhydroxyalkanoates for sustainable food packaging applications. Int J Biol Macromol 221:163–182

Kupolati WK, Sadiku ER, Eze AA, Ibrahim ID, Agboola O (2022) Life cycle assessment, recycling and re-use of the bionanocomposites. In: Polymer based bio-nanocomposites: properties, durability and applications. Springer, pp 205–216

Kushwaha OS, Avadhani CV, Singh RP (2015) Preparation and characterization of self-photostabilizing UV-durable bionanocomposite membranes for outdoor applications. Carbohydr Polym 123:164–173. https://doi.org/10.1016/j.carbpol.2014.12.062

Lee J-E, Shim S-B, Park J-H, Chung I (2022) Interfacial properties and melt processability of cellulose acetate propionate composites by melt blending of biofillers. Polymers. https://doi.org/10.3390/polym14204286

Li Y, Shimizu H (2009) Fabrication of a stretchable, elastic and electrically conductive nanocomposite by high-shear processing. Accessed: 30 Jan 2025. [Online]. Available: https://citeseerx.ist.psu.edu/document?repid=rep1&type=pdf&doi=4e232898dc21bbfa4450651b9211db6d807a316f#page=32

Liu J et al (2020) Synergistic engineering of 1D electrospun nanofibers and 2D nanosheets for sustainable applications. Sustain Mater Technol 26:e00214

Lizundia E, Luzi F, Puglia D (2022) Organic waste valorisation towards circular and sustainable biocomposites. Green Chem 24(14):5429–5459

Lonkar PS, Kumar AP, Singh RP (2011) Commercial aspects associated with bio-nanocomposites. In: Nanocomposites with biodegradable polymers: synthesis properties and future perspectives. Oxford University Press, Oxford, pp 400–421

McIntyre GR (2002) The application of system safety engineering and management techniques at the US Federal Aviation Administration (FAA). Saf Sci 40(1):325–335. https://doi.org/10.1016/S0925-7535(01)00052-2

Mehdinia M, Farajollah Pour M, Yousefi H, Dorieh A, Lamanna AJ, Fini E (2024) "Developing bio-nano composites using cellulose-nanofiber-reinforced epoxy. https://doi.org/10.3390/jcs8070250

Mielewski DF, Flanigan CM, Perry C, Zaluzec MJ, Killgoar PC (2005) Soybean oil auto applications: developing flexible polyurethane foam formulations containing functionalized soybean oil for automotive applications. Ind Biotechnol 1(1):32–34

Mittal V (2011) Bio-nanocomposites: future high-value materials. In: Bionanocomposites: future high-value materials, pp 1–27

Miyashiro D, Hamano R, Umemura K (2020) A review of applications using mixed materials of cellulose, nanocellulose and carbon nanotubes. Nanomaterials 10(2):186

Mohammed M et al (2024) A review on the advancement of renewable natural fiber hybrid composites: prospects, challenges, and industrial applications. J Renew Mater 12(7)

Mori R (2023) Replacing all petroleum-based chemical products with natural biomass-based chemical products: a tutorial review. RSC Sustain 1(2):179–212

Mostafa NH, Ismarrubie ZN, Sapuan SM, Sultan MTH (2017) Fibre prestressed polymer-matrix composites: a review. J Compos Mater 51(1):39–66

Mousa MH, Dong Y, Davies IJ (2016) Recent advances in bionanocomposites: preparation, properties, and applications. Int J Polym Mater Polym Biomater 65(5):225–254

Muhammad A, Rahman MR, Baini R, Bin Bakri MK (2021) Applications of sustainable polymer composites in automobile and aerospace industry. In: Advances in sustainable polymer composites, pp 185–207. https://doi.org/10.1016/B978-0-12-820338-5.00008-4

Muthukumar C, Thiagamani SMK, Krishnasamy S, Nagarajan R, Siengchin S (2022) Polymer based bio-nanocomposites: properties, durability and applications. Springer

Nagalakshmaiah M et al (2019) Biocomposites: present trends and challenges for the future. In: Green composites for automotive applications, pp 197–215

Nagaraju SB, Priya HC, Girijappa YGT, Puttegowda M (2023) 9—Lightweight and sustainable materials for aerospace applications. In: Rangappa SM, Doddamani SM, Siengchin S, Doddamani M (eds) Lightweight and sustainable composite materials. Woodhead Publishing, pp 157–178. https://doi.org/10.1016/B978-0-323-95189-0.00007-X

National Research Council (2010) Technologies and approaches to reducing the fuel consumption of medium-and heavy-duty vehicles. National Academies Press

New aerospace and building materials could repair themselves thanks to fungi and bacteria | Horizon Magazine. Accessed: 30 Jan 2025. [Online]. Available: https://projects.research-and-innovation.ec.europa.eu/en/horizon-magazine/new-aerospace-and-building-materials-could-repair-themselves-thanks-fungi-and-bacteria

Nimbagal V, Banapurmath NR, Sajjan AM, Patil AY, Ganachari SV (2021) Studies on hybrid bio-nanocomposites for structural applications. J Mater Eng Perform 30(9):6461–6480

Njuguna J, Wambua P, Pielichowski K, Kayvantash K (2011) Natural fibre-reinforced polymer composites and nanocomposites for automotive applications. In: Kalia S, Kaith BS, Kaur I (eds) Cellulose fibers: bio- and nano-polymer composites: green chemistry and technology. Springer, Berlin, Heidelberg, pp 661–700. https://doi.org/10.1007/978-3-642-17370-7_23

Nugroho A et al (2024) Enhancing tribological performance of electric vehicle lubricants: nanoparticle-enriched palm oil biolubricants for wear resistance. Heliyon 10(22)

Ogah OA, Archibong FN, Pauline OU, Mbam NJ, Yudhanto F (2024) Investigating natural fiber reinforced polymer composites, biocomposites, bionanocomposites thermo-mechanical attributes emerging implementations. J Fibers Polym Compos 3(2):119–157

OlaniyanOT, Adeyomoye O, Adetunji CO, Oloruntoba A (2024) 10—Biocomposites for aerospace engineering applications. In: Karak N (ed) Woodhead Publishing series in composites science and engineering. Woodhead Publishing, pp 285–298. https://doi.org/10.1016/B978-0-443-19074-2.00010-1

Pandey P, Mohanty S, Nayak SK (2014) Improved flame retardancy and thermal stability of polymer/clay nanocomposites, with the incorporation of multiwalled carbon nanotube as secondary filler: evaluation of hybrid effect of nanofillers. High Perform Polym 26(7):826–836

Patel VK, Patel N, Patel MD, Shukla HA, Patel AB, Parmar NJ (2024) Innovations in lightweight materials for automotive engineering. J Electr Syst 20(10s):2121–2133

Patel V, Mahajan Y (2014) Polymer nanocomposites: emerging growth driver for the global automotive industry. In: Pandey JK, Reddy KR, Mohanty AK, Misra M (eds) Handbook of polymer-nanocomposites. Processing, performance and application: volume a: layered silicates. Springer, Berlin, Heidelberg, pp 511–538. https://doi.org/10.1007/978-3-642-38649-7_23

Pinkse J, Bohnsack R, Kolk A (2014) The role of public and private protection in disruptive innovation: the automotive industry and the emergence of low-emission vehicles. J Prod Innov Manag 31(1):43–60

Pradhan RA (2022) Life-cycle assessment and circular economy concepts for bionanocomposites. In: Bionanocomposites for food packaging applications. Elsevier, pp 357–369

Ragitha VM, Edison LK (2022) Safety issues, environmental impacts, and health effects of biopolymers. In: Handbook of biopolymers. Springer, pp 1–27

Ramli N et al (2018) Natural fiber for green technology in automotive industry: a brief review. IOP Conf Ser Mater Sci Eng 368(1):12012. https://doi.org/10.1088/1757-899X/368/1/012012

Ranakoti L et al (2022) Critical review on polylactic acid: properties, structure, processing, biocomposites, and nanocomposites. Materials 15(12):4312

Reshmy R et al (2022) Biorefinery aspects for cost-effective production of nanocellulose and high value-added biocomposites. Fuel 311:122575

Riaz F et al (2024) Environmental impact, health implications, and life cycle assessment of bionanocomposites. In: Advances in bionanocomposites. Elsevier, pp 387–406

Saba N, Md Tahir P, Jawaid M (2014) A review on potentiality of nano filler/natural fiber filled polymer hybrid composites. Polymers (Basel) 6(8):2247–2273

Samaras C, Meisterling K (2008) Life cycle assessment of greenhouse gas emissions from plug-in hybrid vehicles: implications for policy. Environ Sci Technol 42(9):3170–3176. https://doi.org/10.1021/es702178s

Sapuan SM, Ilyas RA (2020) Biocomposite and synthetic composites for automotive applications. Woodhead Publishing

Sarojini G, Venkatesh Babu S, Rajamohan N, Rajasimman M (2022) Performance evaluation of polymer-marine biomass based bionanocomposite for the adsorptive removal of malachite green from synthetic wastewater. Environ Res 204:112132. https://doi.org/10.1016/J.ENVRES.2021.112132

Satdive A, Tayde S, Chatterjee A (2022) Flammability properties of the bionanocomposites reinforced with fire retardant filler. In: Muthukumar C, Thiagamani SMK, Krishnasamy S, Nagarajan R, Siengchin S (eds) Polymer based bio-nanocomposites: properties, durability and applications. Springer, Singapore, pp 69–86. https://doi.org/10.1007/978-981-16-8578-1_4

Scaffaro R, Botta L (2014) Nanofilled thermoplastic–thermoplastic polymer blends. In: Nanostructured polymer blends. Elsevier, pp 133–160

Scaffaro R, Maio A (2019) Integrated ternary bionanocomposites with superior mechanical performance via the synergistic role of graphene and plasma treated carbon nanotubes. Compos B Eng 168:550–559. https://doi.org/10.1016/j.compositesb.2019.03.076

Sharma G et al (2018) Applications of nanocomposite hydrogels for biomedical engineering and environmental protection. Environ Chem Lett 16:113–146

Sharma B, Thomas S, Bajpai PK, Ghosal K, Shekhar S (2023) Advances in bionanocomposites: materials, applications, and life cycle. Elsevier

Shine KP, Fuglestvedt JS, Hailemariam K, Stuber N (2005) Alternatives to the global warming potential for comparing climate impacts of emissions of greenhouse gases. Clim Change 68(3):281–302. https://doi.org/10.1007/s10584-005-1146-9

Shrivastava A, Dondapati S (2021) Biodegradable composites based on biopolymers and natural bast fibres: a review. Mater Today Proc 46:1420–1428

Shubham, Ray BC (2024) Introduction to composite materials. Eng Mater Part F2264:1–20. https://doi.org/10.1007/978-981-99-9746-6_1
Silva RF, Kaewunruen S (2017) Recycling of rolling stocks. https://doi.org/10.3390/environments4020039
Smith MJ, Verbeek CJR (2018) Energy absorption mechanisms and impact strength modification in multiphase biopolymer systems. Recent Patents Mater Sci 11(1):2–18
Taub AI, Krajewski PE, Luo AA, Owens JN (2007) The evolution of technology for materials processing over the last 50 years: the automotive example. JOM 59:48–57
Teles MCA et al (2015) Evaluation of the diameter influence on the tensile strength of pineapple leaf fibers (PALF) by Weibull method. Mater Res 18:185–192
Thapliyal D et al (2023) Natural fibers composites: origin, importance, consumption pattern, and challenges. https://doi.org/10.3390/jcs7120506
Trimulyono A et al (2024) Numerical investigations of deckhouse height to the self-righting moment of the patrol boat. https://doi.org/10.3390/designs8050086
Tripathi DK, Sarkar S, Meher MK, Poluri KM, Limitations in commercialization of green polymeric nanocomposites and avenues for rectification. In: Green polymeric nanocomposites. CRC Press, pp 281–312
Tripathy DB, Gupta A (2024) Nanocomposites as sustainable smart materials: a review. J Reinforced Plastics Compos, p. 07316844241233162. https://doi.org/10.1177/07316844241233162
Veeman D et al (2021) Additive manufacturing of biopolymers for tissue engineering and regenerative medicine: an overview, potential applications, advancements, and trends. Int J Polym Sci 2021(1):4907027
Wang Z et al (2021) Recent advances in natural functional biopolymers and their applications of electronic skins and flexible strain sensors. Polymers. https://doi.org/10.3390/polym13050813
Williams ES, Panko J, Paustenbach DJ (2009) The European Union's REACH regulation: a review of its history and requirements. Crit Rev Toxicol 39(7):553–575. https://doi.org/10.1080/10408440903036056
Xiao T, Geng L, Dai Y, Zhao J, Liu C (2023) UV-cured polymer aided phase change thermal energy storage: preparation, mechanism and prospects. J Energy Storage 64:107066
Xiong R, Grant AM, Ma R, Zhang S, Tsukruk VV (2018) Naturally-derived biopolymer nanocomposites: interfacial design, properties and emerging applications. Mater Sci Eng R Rep 125:1–41
Xue J, Wu T, Dai Y, Xia Y (2019) Electrospinning and electrospun nanofibers: methods, materials, and applications. Chem Rev 119(8):5298–5415
Yang MFM, Hamid H, Abdullah AM (2018) Potential use of cellulose fibre composites in marine environment—a review. In: Öchsner A (ed) Engineering applications for new materials and technologies. Springer International Publishing, Cham, pp 25–55. https://doi.org/10.1007/978-3-319-72697-7_3
Zhao R, Torley P, Halley PJ (2008) Emerging biodegradable materials: starch-and protein-based bio-nanocomposites. J Mater Sci 43(9):3058–3071

Ms. Jahanvi Thakur, a Physics graduate from University of Delhi is currently pursuing her M.Sc. in Physics at Netaji Subhas University of Technology where she is pursuing research on nanomaterials for renewable energy, particularly 2D transition metal dichalcogenides. Her focus lies on the research for highly efficient solar cells, supercapacitors, and photodetectors, driven by a commitment to sustainable energy solutions. She has co-authored several publications, including, journal papers, conference proceedings, and book chapters.

Ms. Karishma Rawat, a graduate from University of Delhi, is currently pursuing her M.Sc. in Physics at Netaji Subhas University of Technology. Her research journey is centered on exploring advanced nanomaterials, with a particular focus on 2D materials like reduced graphene oxide and metal sulfide nanocomposites. She is deeply engaged in developing innovative nanocomposites aimed at enhancing the efficiency and functionality of energy storage and optoelectronic devices, including solar cells, supercapacitors, and photodetectors. With a strong emphasis on sustainability, she is driven by the vision of creating impactful solutions for renewable energy challenges. Ms. Rawat has co-authored multiple research works, conference papers, and book chapters, showcasing her significant contributions to the field.

Dr. Shreya is a physicist who earned her Ph.D. in Physics from Netaji Subhas University of Technology, where her research was centered on the exploration of exceptional properties exhibited by nanomaterials and their potential contributions to advancing renewable energy technologies. Her primary focus lies in investigating the unique properties of 2D materials and their electrochemical study, specifically transition metal dichalcogenides, with the aim of applying these findings to enhance solar cells, supercapacitors, photodetectors, hydrogen production, and electrochemical sensing. Her dedication to leveraging advanced materials for sustainable energy solutions reflects her vision of contributing to societal progress through innovative technology. Currently, she is working to identify sustainable energy resources, with a specific focus on blue energy foresight. She is also the author and co-editor of multiple scientific books and papers in esteemed journals. Driven by a deep commitment to environmental sustainability, Dr. Shreya's work embodies her passion for scientific advancement in service to future energy needs.

Dr. Peeyush Phogat, completed his Ph.D. in the Department of Physics at Netaji Subhas University of Technology. His current area of research is synthesis and characterization of materials, focusing particularly on applications in solar energy and capacitors. Holding a master's degree with specialization in condensed matter physics, He is driven by the pursuit of uncovering novel materials and their distinct properties, meticulously tailored for effective solar energy capture and the design of efficient capacitors. His scholarly endeavors have resulted in significant contributions towards the development of new and advanced materials for solar applications, which are evidenced by his 35 publications in esteemed SCI/SCIE journals and authorship of 14 book chapters within the renowned book series. His expertise extends comprehensively into the domain of capacitors, where his research showcases a remarkable ability to engineer materials that optimize energy storage and discharge capabilities.

Prof. Sukhveer Singh is presently working with Department of Nanotechnology, Delhi Skill and Entrepreneurship University, New Delhi, 110077, India. He worked as Professor Emeritus at Department of Physics, Netaji Subhash University of Technology, Dwarka, New Delhi, India. Previously, he worked as Ex. Emeritus Scientist at CSIR-National Physical Laboratory, New Delhi India, Ex. Sr. Principal Scientist, and Head Electron and Ion Microscopy Group, CSIR National Physical Laboratory, New Delhi India. Dr. Sukhveer Singh is Ph. D in physics from Jamia Millia Islamia, New Delhi, India. Synthesis and characterization of bulk and nanostructured thin films of compound semiconductors, metal oxides, nanocomposites and investigation of their structural, optical, electronics, and electrical properties in order to ascertain their application for various sensors and device applications are the main areas of his research. The development of standard reference materials (SRM) for the calibration of SEM and TEM is another area of his specialized research under him. He has also established a Centre for Nano Scale Science at CSIR National Physical Laboratory, New Delhi where highly sophisticated and state-of-the-art advance equipment, such as XRD/HRXRD, TEM/HRTEM, SEM/EDS, AFM, TOF-SIMS, PL spectrometer and highly sophisticated specimen preparation tools for TEM and SEM are installed and serving the nation in various areas of the current research. He has published more than 150 research papers in SCI journals of high repute. He has three no. of US, South Africa and Asia patents granted in his name.

Prof. Ranjana Jha is currently serving as Vice Chancellor, IGDTUW, Delhi, India. Previously, she worked as a professor and head of Department of Physics, Netaji Subhas University of Technology. She has been actively involved in research and teaching in the field of Solar Energy Materials and Applied Physics. Received Lifetime Achievement Award in Science and several International and National awards for academics and research in Solar Energy Utilization, Materials and Device fabrication. She is the founder and in-charge of Research Lab for Energy Systems at Department of Physics. Her research interests are Solar Energy Materials and Solar Energy Utilization, Single Crystal Formation, Development of Nano-structured thin film Solar cells, and Characterization of Energy Materials for Device Applications. Published 67 manuscripts in reputed International/National Journals. Attended/presented papers as an author, co-author, and corresponding author in 123 International and National conferences.

Chapter 14
Bionanocomposites in Nanostructured Coatings and Wearable Textile Applications

Raphael Kanyire Seidu, Emmanuel Abankwah Ofori, Benjamin Tawiah, Shouxiang Jiang, and Bin Fei

Abstract The use of nanoscale materials provides a promising opportunity for advancement in different applications, most especially for coatings and wearable technology. Biological and nanoscale composites nanostructured coatings are a promising approach for fabricating thin films with enhanced corrosion resistance, durability and self-cleaning, amongst others. Their applications have been demonstrated to prolong the lifespan and performance of surfaces, offering a sustainable approach for multifunctional applications. Additionally, the application of bionanocomposites for wearable textiles presents a unique opportunity to achieve durability, comfort, breathability, conductivity and a moisture-wicking effect besides their excellent biocompatibility. These properties expand the applications of bionanocomposite materials in rehabilitation, sports, fashion and healthcare. Despite

---

R. K. Seidu · B. Tawiah · S. Jiang · B. Fei (✉)
School of Fashion and Textiles, The Hong Kong Polytechnic University, Kowloon, Hong Kong SAR, China
e-mail: bin.fei@polyu.edu.hk

R. K. Seidu
e-mail: raphael-kanyire.seidu@connect.polyu.hk

B. Tawiah
e-mail: benjamin.tawiah@knust.edu.gh

S. Jiang
e-mail: kinor.j@polyu.edu.hk

E. A. Ofori
Department of Materials Science and Engineering, Hubei University of Automotive Technology, Shiyan, Hubei, China

B. Tawiah
Department of Industrial Art, Kwame Nkrumah University of Science and Technology, Kumasi, Ghana

S. Jiang
Research Centre of Textiles for Future Fashion, The Hong Kong Polytechnic University, Kowloon, Hong Kong SAR, China

S. C. George and B. Tawiah (eds.), *Advanced Bionanocomposite Materials*, Advanced Structured Materials 241, https://doi.org/10.1007/978-981-96-8225-6_14

the wide applications, their safety in humans and the environment is not well understood, hence the need for a critical assessment of the safety and consumer acceptance of bionanocomposite-based nanostructured coatings and wearable textiles, consistent with the UNDP SDGs. In this chapter, we discuss the different types of bionanocomposites used for nanostructured coatings and wearable textile applications, properties and mechanisms for enhancing the performance and functionality of surfaces and wearable textiles. We further discuss the possible safety issues with the application of bionanocomposite materials to understand the regulatory, scalability, user experience and integration issues in textile electronics. The chapter finally concludes with some insights into promising areas for future wearable e-textile applications and nanostructured coatings for limited environmental impact.

Keywords Bio-nanocomposites · Biocompatibility · Nanostructured coatings · Protective coatings · Wearable technology · Smart textiles

14.1 Introduction

Bionanocomposites are advanced materials composed of biological substances and unique nanomaterials (ranging from 1 to 100 nm) with good properties (Farooq et al. 2024). Herein, nanomaterials such as graphene or carbon nanotubes amongst others and biological materials such as natural, polysaccharides and protein fibres are fabricated for unique thermal, electrical, mechanical, biodegradability and biocompatibility properties. These composite materials have emerged and have gained potential but significant attention for applications in nanostructured coatings, offering novel solutions for enhanced performance, sustainability and functionality. It is, however, not surprising to note the continuous research trend in this area (Fig. 14.1a) with data extracted from the Scopus database. It was revealed that key countries like India, Iran, China and the United States (Fig. 14.1b) lead the chart in advancing research output in the bio-nanocomposites field.

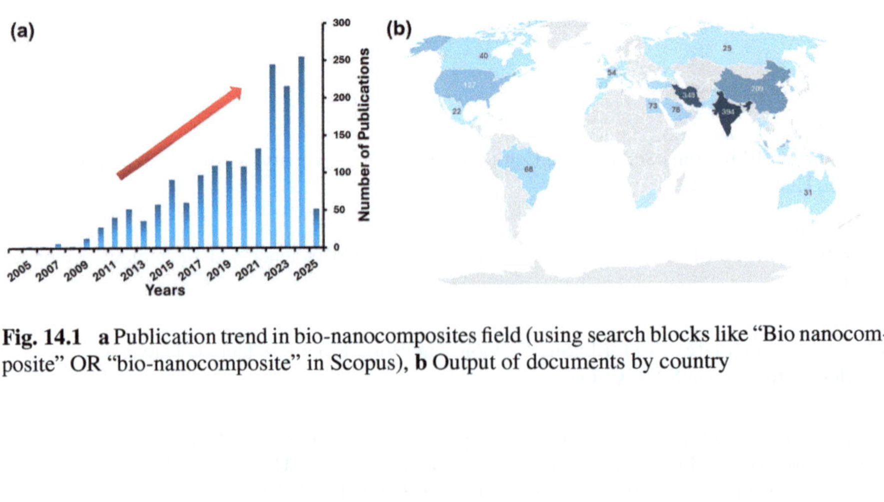

Fig. 14.1 **a** Publication trend in bio-nanocomposites field (using search blocks like "Bio nanocomposite" OR "bio-nanocomposite" in Scopus), **b** Output of documents by country

Conventional coatings have several challenges, such as lack of sustainability, insufficient mechanical properties, thermal stability and lack of functional properties such as anti-microbial and self-cleaning amongst others, which are solved by using bionanocomposites (Mhd Haniffa et al. 2016a). These materials have better mechanical strength, fire resistance, antibacterial, optical, thermal and gas barrier qualities, as well as improved shelf life and a smaller environmental impact (Senthil Kumar and Gunasundari 2018). Hence, nanostructured coatings made from bionanocomposites exhibit significant properties in durability, flexibility and mechanical strength for various applications in aerospace, electronics, automotive and healthcare. For wearable textiles that integrate electronic technology into clothing for effective sensing, tracking and monitoring (Meddeb 2024), the incorporation of bionanocomposites expands the development possibilities of smart fabrics. Significant properties of bionanocomposites, such as biodegradability, environmentally friendly, enhanced mechanical, flame retardants, thermal and electrical conductivity, biocompatibility and non-toxic, have improved functional, structural and physical properties (Azadi et al. 2024) and have advanced applications for wearable textiles. As such, conductive fibres, yarns and fabrics fabricated from these composite materials can be easily integrated into clothing with embedded energy, actuator and sensor devices for effective real-time monitoring and other functional purposes. Aside from these properties, the use of bionanocomposites in nanostructured coatings and wearables addresses the need for sustainable materials. These high-performing bionanocomposites are environmentally friendly, resulting in limited effects and impacts for a minimized carbon footprint. Furthermore, by advancing these materials, key challenges can be overcome for their full utilization in nanostructured coatings and wearable textiles. The objectives of this chapter, therefore, highlight and discuss the fabrication methods, properties and applications with the use of bionanocomposites for nanostructured coatings and wearable textiles. Key safety concerns, challenges and regulatory issues are further discussed to provide an effective understanding for rapid advancement in the field for future research.

14.2 Fundamentals of Nanostructured Coatings and Wearable Textiles

The advancement of nanostructured coatings offers a significant material with unique properties for various applications. These are coatings produced from nanomaterials which have their thickness from 1 to 100 nm (Farooq et al. 2022). These composing materials have unique properties that influence the fabrication of nanostructure coatings (Fig. 14.2a–g) with enhanced properties. Nanoscale and interface effects linked to the structural characteristics of > 100 nm crystalline size and an extremely high interfacial phase (Ovid'ko 2006) influence the properties of nanostructured coatings. These structural characteristics, for example, the nanoscale effect, arise due to the several fundamental processes in solids linked with less nanometre length

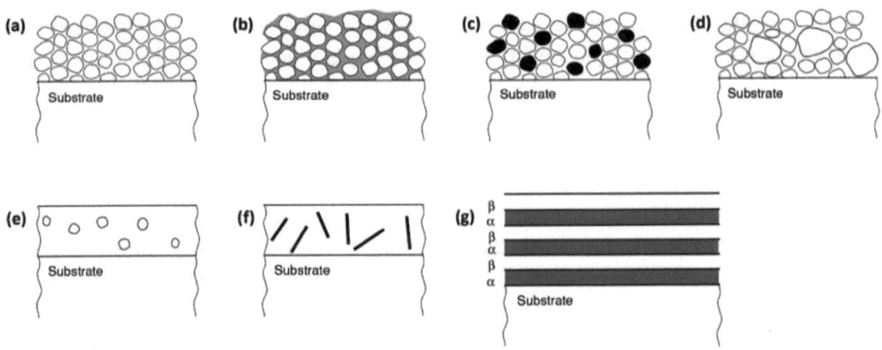

Fig. 14.2 Examples of nanostructured coatings, **a** nano single-phase coating, **b** nanolayer composite coating, **c** nanocomposite coating, **d** nano-micro coating, **e** nanoparticles and **f** nanotubes (nanofibres). Reproduced with permission from Ovid'ko (2006) Copyright 2006, Springer science business media, LLC

scales. The interfacial effects arise due to the factor that, at the interface level, there is a larger fraction of atoms. The nanoscale and interfacial effects are the contributory factors for the unique properties of the nanostructured coatings. Nanostructured coatings have exceptional functional, chemical and mechanical characteristics and surface protection properties that advance their application in food packaging, medicine and electronics, amongst others (Gu et al. 2020). Alternatively, they exhibit enhanced tribological and super hardness characteristics (Barnett et al. 2003; Münz 2003). Other significant properties include an increase in decadence, abrasion, wear strength and hardness, as compared to the conventional coatings (Aliofkhazraei 2011). Based on their unique characteristics, the flame retardancy, wear resistance, anti-reflection properties, IR shielding effect, corrosion, chemical and abrasion resistance of products are improved for different applications (Sharma and Aliofkhazraei 2015).

Key parameters such as thickness, microstructure, grain size, nanoparticle type, substrate material and synthesis method must be considered during the formation of nanostructured coatings for effective functionality and application (Farooq et al. 2022). By adopting these parameters, nanostructured coatings can be fabricated using physical and chemical vapour deposition, sol–gel method, spray coating method, electrodeposition and laser cladding methods (Farooq et al. 2022; Gu et al. 2020). Others are self-assembly coatings, electroless deposition and layer-by-layer assembly (Tettey et al. 2010; Nguyen-Tri et al. 2018). Alternatively, factors such roughness, surface defects and the cleanliness of the substrate influence the quality of nanostructured coatings during applications (Aliofkhazraei 2011). Together with cost-related issues, scalability and maintaining its durability and stability in the long term has remained a critical problem.

Wearable textiles are electronic textiles or smart textiles that combine and integrate electronic technology into clothing for effective sensing, tracking and monitoring

(Meddeb 2024). These innovative materials are integrated with electronic components like microcontrollers, actuators and sensors into the textiles, which expands their applications. They can "sense, react and adapt" to conditions externally (Kan and Lam 2021). This opens possibilities for wearable textiles to perform key functions like energy generation, feedback provision and monitoring the vitals or signals from the human body, amongst others. The integration of key technologies (Table 14.1) transforms conventional textiles into functional and intelligent textiles (Meddeb 2024; Ruckdashel et al. 2022). The working principle of wearable textiles is that the electronic feedback process is received from stimuli with or sensing (concerning light, sound, heat and touch) from the physical environment, where this detected stimulation is processed and converted into quantitative data for effective analysis (Shuvo et al. 2022). In principle, wearable textiles sense physical changes from the environment and human motions, collect and transmit the stored data for an appropriate feedback (Meena et al. 2023). Based on these unique principles and components, wearable textiles have several advantages that are significant for users to fully explore their functionality for various applications. Continuous monitoring of physiological changes and vital signals is a significant advantage of wearable textiles, which finds applications in healthcare. This approach has greatly transformed disease management and patient care, since these wearable textiles equipped with electronic components monitor, track and record data in real-time on critical vital signs from the human body. These data are then transmitted to their respective healthcare provider for quick detection and timely interventions to save the lives of users. Key examples include an integrated belt senso system that was developed to measure respiratory cycle for easy detection of hypopnea/apnoea (Adnane et al. 2009), a textile-based humidity and temperature sensor was developed (Soukup et al. 2014) and a smart glove system with a transcutaneous electrical nerve stimulator was developed to stimulate the meridian point on the palm for treating hypertensive disease.

14.3 Bionanocomposites for Nanostructured Coatings

Nanotechnology represents the development and application of materials with a size < 100 nm (Saxena et al. 2012; Joshi and Bhattacharyya 2011). Nanomaterials research, development and application have advanced rapidly since the early 1980s (Gleiter 1991), when nanotechnology first emerged. In three-dimensional space, a nanomaterial is defined as a material with at least one dimension in the nanometre scale (Tripathi and Chung 2019; Khan et al. 2019; Mekuye and Abera 2023) and a spherical surface area to volume ratio greater than $60\frac{m^2}{cm^3}$ (Santos et al. 2015). Materials that integrate nanosized particles into a conventional material matrix are referred to as nanocomposites (Onyeaka et al. 2022; Bharathi and Jayas 2014). The constituent materials of a composite material, such as inorganic or organic materials, nanoscale minerals, and natural or manmade biodegradable polymers, determine its qualities (Mhd Haniffa et al. 2016b). Nanocomposite coatings have garnered

Table 14.1 Technologies used in wearable textiles

Technology	Description	Ref.
Sensors	These are electronic components that can monitor, detect and measure different physical changes like motion, humidity, pressure and temperature. Herein, electrical signals are generated from these detections. For example, measurable electrical signals are generated from mechanical deformation by strain sensors. These electrical signals are then transferred to a control unit for subsequent actions	Meddeb (2024); Yang et al. (2019); Takebira et al. (2024)
Actuators	They produce physical actions from converted electrical signals based on their response to physical and environmental changes. This implies that actuators are designed to perform the needed physical actions based on the sensor's input signals that are processed by a processor. Examples of actuators include piezoelectric materials, electroactive polymers and shape memory alloys	Meddeb (2024); Çelikel (2020)
Data processing	This process is relevant data collected from sensors. Based on this, the appropriate processor, e.g. microcontrollers is needed to actively process these data for effective evaluation. This means that the type of analysis needed will determine the suitable processor	Meddeb (2024); Çelikel (2020)
Communication	Wearable textiles can effectively communicate wirelessly with other surrounding devices to ensure data is collected in real-time for processing and analysis. Key electronic components or elements are embedded in the garment to communicate to users. Herein, different forms of communication protocols like optical fibres, and wireless communication via antennas are advanced	Meddeb (2024); Takebira et al. (2024)

(continued)

Table 14.1 (continued)

Technology	Description	Ref.
Storage	Storage in wearable textiles comes in two folds; storing of collected data and storing of appropriate energy. Herein, the ability of the wearable to store the need energy for self-use is important. Energy harvesters, generators and convention batteries are widely used to generate and storage energy to power the various electronics on the wearables. These energy harvesters can easily harvest energy directly from human motions when the textile is in contact with the human body	Çelikel (2020); Meyer et al. (2006); Engin et al. (2005); Fernández-Caramés and Fraga-Lamas (2018)
Conductive threads	These are threads coated with materials such as copper or silver for effective electrical conductivity. Conductive threads are integrated into the clothing system to transmit electrical signals amongst the electronic components	Meddeb (2024)

attention in recent years from a variety of key industries, including electronics, automotive, aerospace and petroleum. These nanocomposite coatings are fabricated, offering surface coatings with exceptional antifogging, antibacterial, anti-corrosion and adhesive qualities that are both aesthetically pleasing and cost-effective (Abu-Thabit et al. 2015). As illustrated in Fig. 14.3, nanocomposites are categorized based on their application, structural configuration, synthesis, reinforcing properties and matrix material (Senthil Kumar and Gunasundari 2018; Bharathi and Jayas 2014; Okpala 2013). However, recent advances have seen the fabrication of green composites, bio-composites, or bio-nanocomposites (Zafar et al. 2016; Pande and Sanklecha 2017). These materials have better mechanical strength, fire resistance, antibacterial, optical, thermal and gas barrier qualities, as well as improved shelf life and a smaller environmental impact (Senthil Kumar and Gunasundari 2018).

Bionano interactions have been studied using a variety of polymeric nanomaterials, including cellulose, silk, chitosan, poly(methyl methacrylate), poly(alkyl cyanoacrylate), polyurethane, poly(ε-caprolactone), poly(D,L-lactide) and poly(D,L-lactide-co-glycolide) (Mailänder and Landfester 2009). Gold and silver nanoparticles and their nanocomposites are also the most extensively researched nanocomponents or reinforcers because of their distinct optoelectronic characteristics and prospective applications in imaging, drug delivery and therapies (Bardhan et al. 2011; Brown et al. 2010). Other significant metal-based nanostructural materials are metal oxides, including copper, zinc, titanium and iron oxides. They have explored their bio-nano interfaces and employed their nanocomposites in biomedical fields (Etoc et al. 2013; Edelmann et al. 2014). Conversely, emerging nanomaterials like mesoporous silica nanoparticles and nanocomposites are utilized for a variety of applications amongst non-metals because of their uniform porosity, stable high pore volume, water dispersion, in vivo biodegradability and outstanding biocompatibility (Park et al. 2009; Nandhini et al. 2024). Phyllosilicates of the 2:1

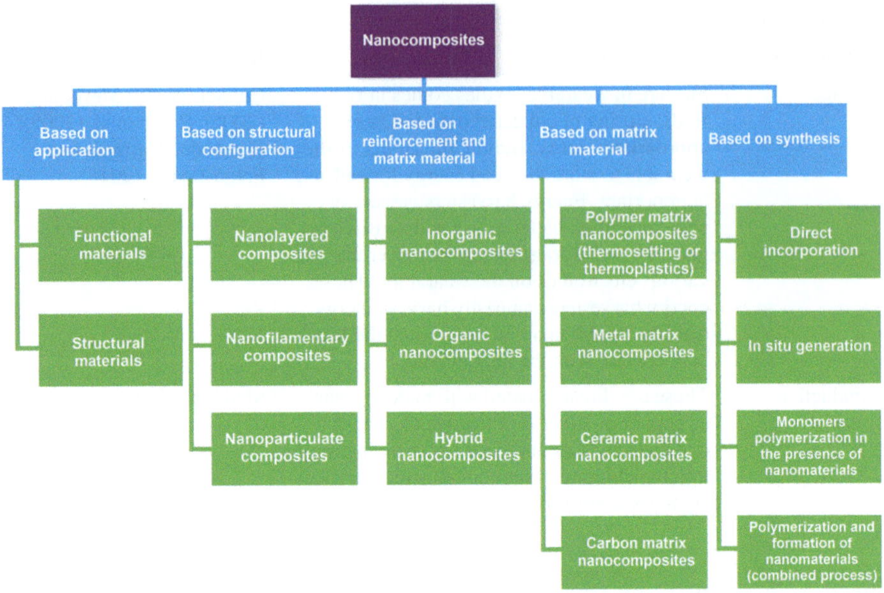

Fig. 14.3 Classification of nanocomposites (Bharathi and Jayas 2014). This article is an open-access article distributed under the terms and conditions of the Creative Commons Attribution (CC BY) license

type, such as hectorite, saponite and montmorillonite, are layered silicates or plate-like clay minerals with a large aspect ratio due to their extremely low thickness (w1 nm) and high sheet length (mm). These minerals are widely used as sustainable nanomaterials in various polymer nanocomposites. Additionally, carbon-based nanomaterials, including carbon nanotubes (CNTs), graphene and carbon quantum dots (CQDs), are three of the most well-known materials in this group used as reinforcers for biopolymers in bio-nanocomposite production (Sajid 2022).

According to studies, bionanocomposites are used to create nanostructured coatings. Waterborne hyperbranched polyester/carbon dot@TiO_2 nanocomposites based on renewable resources, for instance, demonstrated promise as antifogging, anti-icing and self-cleaning surface coating materials with excellent chemical, thermal and mechanical resistance qualities (Hazarika and Karak 2018). For paints and coatings of active surfaces, a thermosetting nanocomposite based on organically modified montmorillonite clay and hyperbranched epoxy/neem oil showed tremendous promise as a durable, biodegradable and antibacterial binder (De et al. 2015). The use of poly L-lactide-based montmorillonite nano clay bio-nanocomposites as pollution-protective coatings for marble surfaces was examined in Ocak et al. (Ocak et al. 2015). The results of the study show a notable improvement in the surface with an inhibition of the sulphation reaction when exposed to an acidic environment. Similarly, coating agents of low and high molecular weight polyhydroxybutyrate (PHB) and poly-L-lactide (PLA) were applied to marble surfaces to slow down the SO_2-calcite reaction (Ocak et al. 2009). The application of high molecular weight PLA polymers to marble surfaces, according to the researchers, significantly reduced SO_2 and water vapour exposure for 90 days by up to 60%. Consequently, bionanocomposites have been applied in several applications based on their properties. The fabrication, characterization, properties and their eventual applications are discussed below.

14.4 Fabrication, Characterization, Properties and Applications

Nanocoatings are developed based on the application areas using different fabrication methods. These include physical and chemical vapour deposition, laser cladding, electrodeposition, sol–gel and spray coating methods (Farooq et al. 2022; Gu et al. 2020). These fabrication processes are discussed in the next sub-headings.

14.4.1 Fabrication

Chemical vapour deposition (CVD) is the process of depositing and forming thin solid films onto the surfaces of materials because of multiple reactants undertaking chemical reactions (Cheng et al. 2023; Urrutia et al. 2021). Alternatively, this method segregates vaporous materials via chemical reactions where coatings are deposited on heated surfaces (Farooq et al. 2022). This technology produces high-quality and high-performance thin films and coatings. Figure 14.4 shows the general procedure in a typical chemical vapour deposition process. Herein, crystalline coatings are deposited using higher temperatures directly onto surfaces (Urrutia et al. 2021). Through a two-step process, the working principle of CVD comprises the delivery of reactants and the deposition of coatings onto the surface (Sun et al. 2021). Reactants of liquid or gas form are first introduced into the reaction chamber where highly active intermediates are produced under higher temperatures when reactants undertake a chemical reaction. Deposition thus begins when the intermediates are transported to the surface of the substrate. Solid deposits are then formed via "adsorption and diffusion" on the surface with the development of "the first nanolayer for the individual coating" (Cheng et al. 2023). The characteristics of these coatings are influenced by the vapour flow, composition, substrate and temperature (Dhiman and Singla 2024).

Physical vapour deposition (PVD) is a coating process where, at the atomic scale level, materials are vaporized and transferred (Faraji et al. 2018) (Fig. 14.5). Unlike CVD, PVD converts a material into a gaseous phase, which is subsequently deposited on the surface of the substrate (Farooq et al. 2022). Herein, resistive heating, electron beam, laser energy and thermal energy are used in the conversion process (Lente and Til 2008). PVD undergoes three key steps: firstly, the vaporization of the target material via physical means, the transport of the vaporized materials to the surface

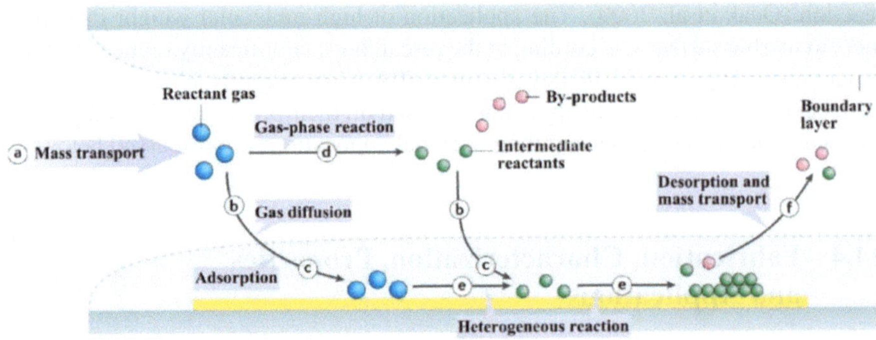

Fig. 14.4 General procedure in chemical vapour deposition. Reproduced with permission from Sun et al. (2021) Copyright 2021, Springer Nature Limited

of the substrate which is of a lower pressure and lastly, the deposition of the materials to form thin films on the surface of the substrates (Faraji et al. 2018; Baptista et al. 2018). This method of deposition distinguishes PVD from CVD (Gu et al. 2020). This coating technique is used to form free-standing structures, deposits that are very thick, multilayer coatings and thin films with a thickness of "a few manometers to thousands of manometers" (Faraji et al. 2018). Laser pulse, magnetron sputtering, cathodic arc and thermal evaporation are key methods of depositions (Nguyen-Tri et al. 2018; Shishkovsky et al. 2011). PVD has several advantages such as protection of the environment (meaning no pollution), flexibility under required temperatures, no restriction on the substrate, complex shapes that can be coated, repeatability is high and uniform coatings with desired thickness can be produced (Helmersson et al. 2006).

Sol–gel method (Fig. 14.6) is a process of producing a sol system where the precursor (an active chemical component) contained in a solvent undergoes hydrolysis and polycondensation (Farooq et al. 2022; Gu et al. 2020). The formed sol is then deposited onto the substrate's surface, where heat and dry treatments are applied to form the coating (Singh et al. 2014; Sun et al. 2019). Uniform physical and chemical properties of coatings are obtained from the process. Subsequently, key factors such as pH, thickness, temperature and materials influence the production of the coating. These factors further affect the coating density and the hydrolysis speed (Farooq et al. 2022). The key advantages of the sol–gel process are simple procedures, easy to control process with the formation of large-area film, operating under low temperatures and the ability to fabricate homogenous coating (Asri et al. 2016). Water-resistant coatings are easily produced via sol–gel methods (Jiang et al. 2019; Claire et al. 2016).

Spray coating method is a simple method of depositing coats onto the surfaces by passing the substrate across a nozzle (Yilmaz et al. 2019). The coated surface is then exposed to heat at a controlled temperature and time for drying. A coated layer formed over the surface could be either softened, molten or liquid particles (Makhlouf et al.

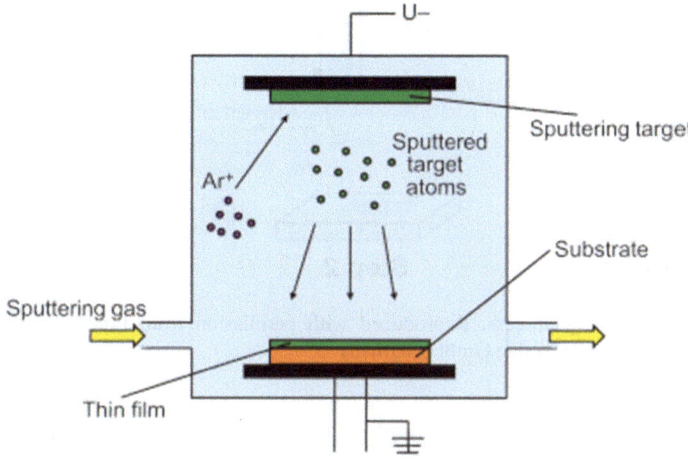

Fig. 14.5 Diagram of a physical vapour deposition process. Reproduced with permission from Faraji et al. (2018) Copyright 2018, Elsevier Inc

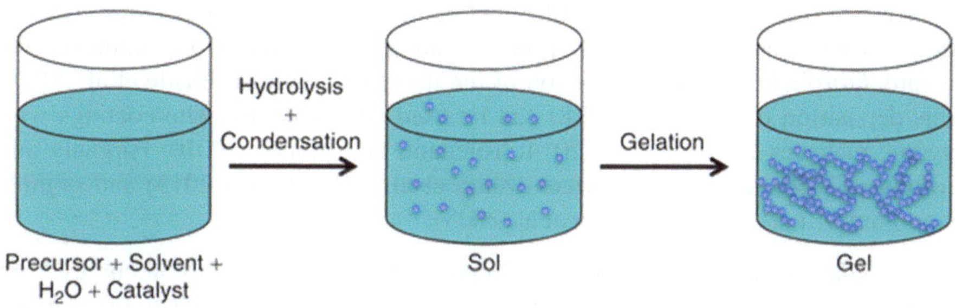

Fig. 14.6 Sol–gel process. Reproduced with permission from Dave et al. (2012) Copyright 2012, Springer Science Business Media B.V

2011). This technique is widely used to coat leather, textiles, wood, metal and glass surfaces (Kuroda 2008). The required thickness of the coatings on the surface can be achieved by repeating the process. Figure 14.7 shows a representation diagram of the spray coating method. Key factors such as the solution and the spray device (Frederichi et al. 2021), spray velocity, the distance between the substrate and spray gun and size distribution of coating materials (Farooq et al. 2022), could influence the quality of coats on the surfaces of substrates. As such, different variations of spray coating methods have been developed: powder spray coating (Li et al. 2020), thermal spray coating (Farooq et al. 2022; Gardon and Guilemany 2014), plasma spray coating (Zhai et al. 2020) and cold spray coating (Nguyen-Tri et al. 2018). For example, the thermal coating process utilizes a chemical or electrical means to heat or meat the coating particles, which are then sprayed onto the surface of the substrate (Leroux et al. 2008).

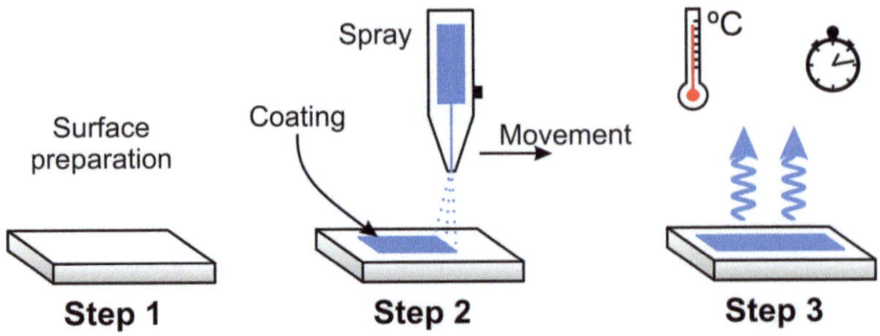

Fig. 14.7 Spray coating process. Reproduced with permission from Frederichi et al. (2021) Copyright 2020, Springer-Verlag GmbH Germany

Electrodeposition is a deposition method where alloy or metal coating via electrolysis (electric current) is applied to the conductive surface of a substrate that is immersed in a bath (electrolyte) which contains a salt of an alloy or a metal (Farooq et al. 2022). This electrochemical process produces thin films with improved heat tolerance, reduced friction and wear, corrosion and electrical resistance finish on surfaces (Souad 2015; Ho et al. 2018). Herein, the solution composition, deposit time, current density and applied potential are key parameters that influence the texture, morphology and composition of the thin film coatings (Basile et al. 2010). This deposition method (Fig. 14.8) can be used for nanocomposite coatings with organic matrices (Weaver et al. 2014; Toledano and Mandler 2010; Farghaly and Collinson 2016), inorganic matrices (Zhao et al. 2014; Benea 2013) and organic nanofillers (Li et al. 2005; Koleva et al. 2005).

Laser cladding method (Fig. 14.9) is a deposition method that uses an energy laser of high-power to melt the wire or powdered feedstock onto a substrate's surface (Gu et al. 2020). Herein, the surface of the substrate is first supplied with a powder system, which is then heated or melted using a high-power laser. This process forms a cladding layer after the melted part is cooled. The coating quality is influenced by key parameters such as the scan speed, power feed and laser power (Zou et al. 2020; Rahman Rashid et al. 2019). The laser cladding method is achieved in either a one-step (continuous supply of materials in the laser zone with fine powder) or two-step process (deposition of a thin layer onto the substrate, which is subsequently melted with the laser beam) (More et al. 2017). Key advantages of the laser cladding method are easy automation, minimal dilution and increased cooling rates (Farooq et al. 2022).

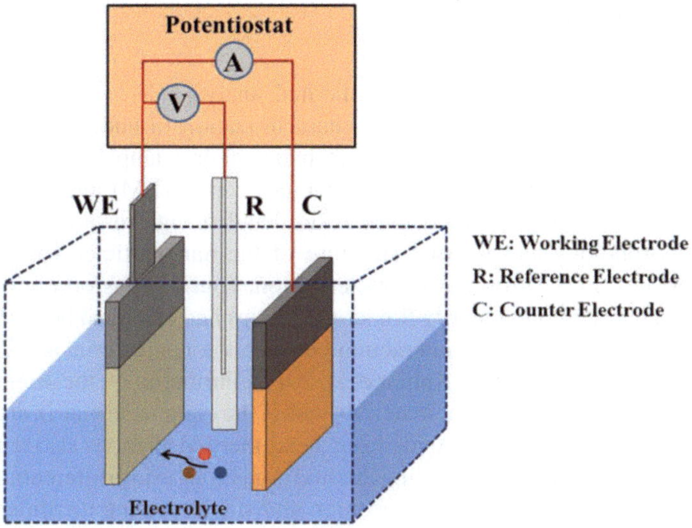

Fig. 14.8 Diagram of the electrodeposition coating process (Ho et al. 2018). (This is an open-access paper licensed under a Creative Commons Attribution License)

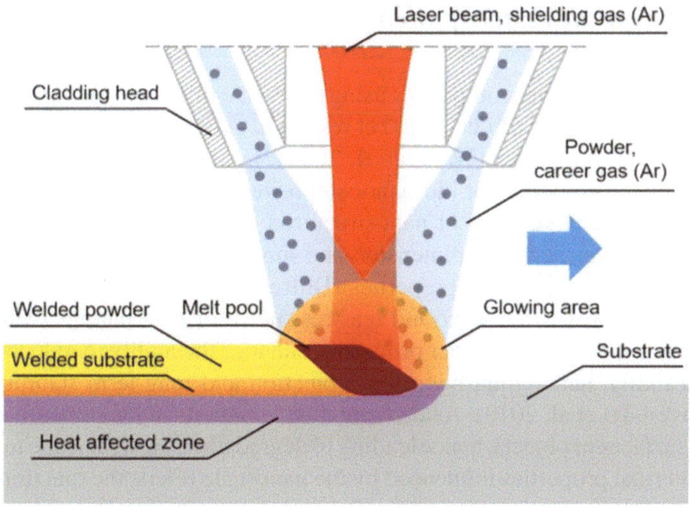

Fig. 14.9 Diagram of laser cladding coating process (Chkalov and Chkalova 2020) (This is an open-access paper published under the terms of the creative commons attribution 3.0 licence)

14.4.2 Characterization

Several advanced techniques, such as thermal analysis, X-ray diffraction, spectroscopy and microscopy, are used for characterization to understand their properties and structure for their intended performance. For example, microscopy characterization such as transmission electron microscopy (TEM) and scanning electron microscopy (SEM) are used to provide detailed high-resolution imagery of the size, morphology and internal structure of the nanoparticles within the bio-nanocomposites. In spectroscopy characterization, Fourier-transform infrared spectroscopy (FTIF) is used to identify functional groups and chemical bonds within the composite material; hence, the interaction between the nonmaterial and the biological components is understood (Kumar et al. 2014). Furthermore, the orientation and degree of crystallinity of the nanomaterials within the composite matrix are analysed using X-ray diffraction (XRD). Aside from these, thermal analyses also remain a key characterization tool for understanding thermal transitions using differential scanning calorimetry (DSC) and thermal composition and stability using thermogravimetric analyses (TGA).

14.4.3 Properties and Applications

The formation of bio-nanocomposites using nano-size fillers and biological or natural polymer components (Karki et al. 2021) results in enhanced performance, mechanical and thermal properties (Wang et al. 2005). Enhanced mechanical property is one of the major properties of bio-nanocomposites, which are crucial for the durability, toughness and strength of nanostructured coatings. Key factors such as the percolating whisker network, dimension and morphology of the component materials influence the mechanical properties (Dufresne 2008). The enhanced mechanical property improves the wear and tear resistance of the coatings (Mhd Haniffa et al. 2016a). Furthermore, barrier properties enhance the ability to reduce water and gas permeability, hence improving the protection and long-term stability of coatings (Nguyen-Tri et al. 2018). Aside from this, variations in temperature conditions affect the surfaces of objects, hence leading to degradation. However, due to increased inherent thermal properties influenced by the nanoscale fillers, the functional protective ability of these coatings can withstand thermal conditions or resist temperature variations in the environment for longer periods (Sumrith et al. 2019). Other properties such as electrical conductivity, chemical resistance and flame retardancy are key properties that prevent the degradation from chemicals and reduce the flammability of the coatings. Further engineering of bio-nanocomposites enhances functional properties such as self-cleaning and anti-microbial abilities for nanostructured coatings.

These unique properties of anti-microbial, thermal stability and mechanical strength, amongst others, have advanced the potential application of nanostructured coatings for improved surface functionality and protection in the medical, automotive, packaging, aerospace, construction, energy and electronics. For example, the anti-microbial ability of the coatings advances their application for medical devices to limit the risk of infections; the mechanical properties improve the resistance to temperature, corrosion and wear for the longevity of materials in the aerospace and automotive industries.

14.5 Bionanocomposites for Wearable Textiles

For wearable electronics, due to the breathability and outstanding deformability of textiles, they are widely considered (Bairagi et al. 2023). The next generation of "smart" textiles will depend on the synergistic combination of components like enzymes, nanoparticles and synthetic chemical complexes. Most importantly, the development of bio-nanocomposites for wearable textiles has been a topic of interest in recent years. Key application areas of nanotechnology for wearables have been reported extensively by Shah et al. (2022) for sensors, energy storage, anti-static, UV resistance and hydrophobicity, amongst others. Wearable technology is helping to move towards seamless connectivity, and applications of wearable technology and functional "smart" fabrics, more generally, are improving lives every day. One path that scientists are pursuing is the fusion of biotechnology and nanomaterial science, which will create advanced materials with enhanced functionalities, ranging from antibacterial activity to sun protection. This is due to their potential to sense and harvest energy from the environment (Berger 2008; Sengupta and Kottapalli 2024). For example, Gheibi et al. (2014) focused on fabricating a nanogenerator based on electrospun nanofibrous materials, specifically Poly (vinylidene fluoride) (PVDF), for self-powering wearable electronic textiles. This research highlights the potential of nanofibrous structures in enhancing the functionality of wearable textiles. In a similar vein, Dong et al. (2016) proposed a novel strategy for the simultaneous production of high-performance flexible textile electrodes and fibre electrodes for wearable energy storage using activated carbon fibre cloth (ACFC)/carbon nanotubes (CNT). The results of the investigation showed a power density of 16,287 $\mu W\ cm^{-2}$, an energy density of 56.9 $\mu Wh\ cm^{-2}$, and an areal capacitance of 2542 $mF\ cm^{-2}$. Additionally, the produced composite textiles' excellent electrical conductivity and porous nature made it possible to assemble thicker textile electrodes with an energy output of up to 88.5 $\mu Wh\ cm^{-2}$. To incorporate nanofibrous materials into textiles for wearable and smart electronics applications, researchers have been investigating several methods.

For example, Naidonov et al. (2023) synthesized a nanocellulose polyvinyl alcohol bio-nanocomposite to be used as a sensor substrate for analysing human muscle activities. The obtained bend sensors achieved a strain sensitivity coefficient of 7.52, with improved optical and mechanical properties upon the addition of PVA to NC. This found application in medicine, sports and rehabilitation. Also, Koç et al. (2020) created an NF-TENG utilizing a nanocomposite fabric made of PVDF/cellulose nanocrystal (CNC) nanocomposite yarns. The mechanical strength of the nanocomposite yarns was greater than that of the pure PVDF yarns. The NF-TENG showed that, at 9.8 kPa of pressure, its output voltage, current density, and power density were roughly 2 V, -155 nA/cm^2 and 219 nW/cm^2, respectively. To assess the viability of using CNC1 fibres in a smart sensing textile, the NF-TENG was used as a real-time pressure sensor. It is interesting to note that the triboelectric pressure sensor was found to be able to detect pressures between 0.98 and 98 kPa (with a detecting surface of 10×10 mm^2). Chen et al. (2023) reported on an all-day PTM, which is made up of silica microparticles (SiO_2) and chitosan (CS) fibres. Due to fibre porosity, there are exceptional properties of high infrared emissivity and high reflectivity for the woven CS/SiO_2 textiles. Subsequently, the fibres' physical properties enhance textile comfortability with good strength, breathability and flexibility. It was also revealed that the cooling power was 103.3 W m^{-2} (for nighttime) and 95.7 W m^{-2} (for daytime). Qian et al. (2022) fabricate a wearable electronic textile with higher antibacterial and thermal stability properties from bio-based chitosan silver nanoparticle composites via the in situ technique. The as-produced substrate had an electrical resistance of 0.93 V/sq and 0.20 V/sq after one and four cycles, respectively.

In a similar vein, Qin et al. (2018) manufactured a novel wearable electrode for smart garments using chitosan biopolymers composited with silver nanoparticles. According to the study findings, the authors report an electrical resistance of 0.0332 $\pm$ 0.0041 Ω/sq due to the reactivity between amine and silver ions. However, the smart textile was functional for electrocardiogram signals capturing from the body in jogging and static motion, by displaying good data acquisition and sensitivity. An NF-TENG nanocomposite with Ecoflex/CNT-modified conductive fabric has been described. It was found that the NF-TENG can generate a high output voltage of ~490 V, a current of 43 μA, and a power density of ~1.6 mW/cm^2, respectively (Bai et al. 2019). Mokhtari et al. (2020) investigated the piezoelectric characteristics of a piezoelectric nanogenerator based on nanocomposite fabric and constructed a nanogenerator utilizing PVDF/barium titanate nanocomposite filament. Because the filler and polymer interacted more when the loading of barium titanate nanoparticles increased, PVDF's electroactive phase content rose as well. After that, the fibres were braided in a triaxial pattern to create energy harvesters. To create wearable energy harvesters, the braided fibres were subsequently knitted into structures. These energy harvesters produced output voltages of up to 4 V, had a power density of 87 μWcm^{-3}, and had outstanding sensitivity. Another study by Parandeh et al. (2019) fabricated a hybrid triboelectric nanogenerator with graphene oxide (GO), which is integrated with cellulose paper and polycaprolactone (PCL). Herein, PCL electrospun nanocomposite web infused with GO and cellulose paper as negative and positive triboelectric layers, respectively. The electrospun PCL/GO nanocomposite was fabricated with GO of different concentrations. The electrical performance of the NEW-TENG is increased with increasing GO concentration up to 4%. It was found that the output voltage, output current, and power were increased from 0.473 V to 22.73 V, 4–165 nA, and 0.473 nW to 938 nW, respectively, at 8 N applied load and 2 Hz frequency.

In another study, Bayan et al. (2022) combined carbon cloth fibres with silver nanoparticles loaded on graphitic carbon nitride (g–C_3N_4) nanosheets to create a textile-based triboelectric nanogenerator (T-TENG). They accomplished charge trapping and interface modification by sandwiching a nylon layer between the nanosheets and carbon fibres, which greatly improved the T-TENG's triboelectric characteristics. Furthermore, temperature-dependent investigations demonstrated that the Ag/CN/nylon-Teflon-based bi-layers T-TENG have a high-power conversion efficiency of up to 65 °C and maintain their stability at higher temperatures. This device is positioned as a promising power source for wearable electronics since it can charge Teflon material under mechanical agitation and provide an open-circuit voltage (up to 200 V). Due to the biocompatibility of zinc oxide (ZnO), it was reported as an alternative to zirconate titanate or other filler materials (Bairagi et al. 2023), which is widely used for piezoelectric fabrics based on ZnO nanocomposites. Rafique et al. (2020) reported on a textile-based piezoelectric nanogenerator that was created by cultivating silver-doped ZnO nanorods on cotton fabric. The piezoelectric performance of the developed piezoelectric nanogenerator was tested under a cyclic load of 3 kgf. The Ag–ZnO/cotton-based piezoelectric nanogenerator demonstrated a maximum power density of 1.45 mW/cm^2 at a load resistance of 31 MΩ, and they also discovered that capacitive reactance influences the PENG's charging capacity.

Aside from the above, the advancement of nanostructure for bionanocomposite coatings for wearable textiles has widened the frontiers for enhanced durability and functionality for effective applications. Due to the biocompatibility, chemical reactivity, mechanical strength and high surface area, their application as coatings in textiles ensures self-cleaning abilities, UV protection and anti-microbial abilities (Pakdel et al. 2018). For example, the integration of zinc oxide and titanium dioxide for bio-nanocomposite coatings for textiles improves the anti-microbial properties and UV protection properties from the harmful radiation of the Sun (Antony 2024; Ielo et al. 2021). These materials replace conventional coatings, which have less strength, durability and abrasion resistance for effective functionality (Nguyen-Tri et al. 2018; Saleem and Zaidi 2020). Another study by Babaeipour et al. (2024) fabricated bio-based coatings on CNF through a layer-by-layer method for textiles. Herein, results showed unique properties of hydrophobicity, water contact angle, antibacterial and UV-shielding abilities, providing their application for outdoor clothing, medical and protective textiles, respectively. Also, ZnO nanoparticles were biosynthesized using bioactive macromolecules for effective UV and antibacterial protection activity for medical textiles (Fouda et al. 2018). These innovative materials provide different functional performances but ensure sustainability to advance the next generation of wearable textiles.

14.6 Fabrication Methods, Properties and Applications

14.6.1 Fabrication Methods

Bionanocomposites are developed for various wearable applications using different fabrication methods. These conductive materials are either fabricated into a fibre or yarn, which is then made into a textile or applied directly onto an existing textile material (Azani and Hassanpour 2024). Table 14.2 provides a comprehensive illustration of the key fabrication methods adopted. However, common methods like screen printing, electrochemical depositions and spinning techniques like electrospinning, wet spinning and dry spinning, amongst others, are used to deposit conductive materials from bio-nanocomposites onto fibres or fabrics for wearable textiles.

Spinning techniques such as melt, solvent, wet, electro and dry spinning methods are widely adopted to produce conductive nanofibers. For example, in the electrospinning process (Fig. 14.10a), the polymer melt is applied with a high voltage, transforming it into a charged jet that is extruded through a needle tip and solidifies to form nanofibres (Azani and Hassanpour 2024; Jaeger et al. 1996). The extruded nanofibres are then collected onto a collector, which could be in the form of a dual ring, steel sheet, mandrel, edge type, grid type, rotary drum and plate (Dalton et al. 2005; Zucchelli et al. 2009; Teo and Ramakrishna 2006; Theron et al. 2001; Zhang et al. 2009; Pedicini and Farris 2003). The resultant fibres have nano to micro-size diameters. Key parameters such as the ambient, process and solution must affect the

Table 14.2 Key fabrication methods adopted for fibres and fabrics

Materials	Main category	Sub-divisions	Methods
Fibre	Fabrication and incorporation of conductive yarns into textile materials	Textile structures incorporated with solid metal wires	Electroforming, casting, rolling, extrusion and wire drawing
		Conductive materials coated on yarns	Layer-by-layer deposition, electrochemical deposition and vapour deposition
			wet coating techniques such as in-situ polymerization, dip coating
		Conductive fibres are spun directly	Spinning techniques
Fabrics	Coating of conductive materials directly onto the textiles	Conductive materials are coated onto fibres	Layer-by-layer deposition, electrochemical deposition and vapour deposition
			Wet coating techniques such as in situ polymerization, roll coating, spray coating, dip coating
		Printing methods	Inkjet printing, screen printing

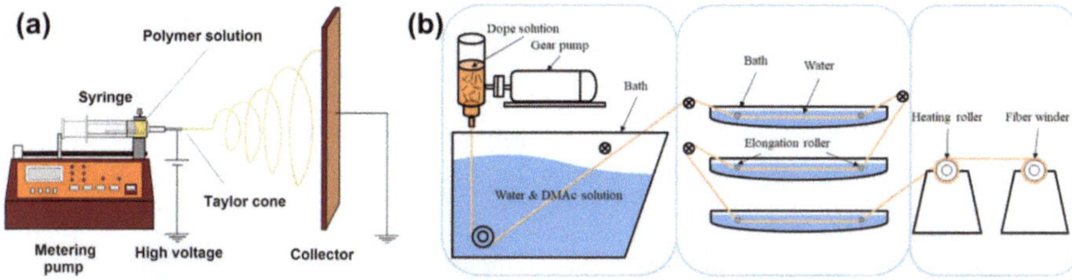

Fig. 14.10 Diagram of **a** electrospinning process. Reproduced with permission from Riazi et al. (2016) Copyright 2016 Elsevier Ltd and **b** wet spinning process (Kang et al. 2021) (This article is an open-access article distributed under the terms and conditions of the Creative Commons Attribution (CC BY) license)

diameter and morphology of the electrospun nanofibres (Li and Xia 2004). In others like wet spinning (Fig. 14.10b), the extruded charged jets are solidified through a coagulation bath, whereas for dry spinning, solvents (used in the preparation process) are evaporated using gas in a hot chamber to solidify the nanofibres. Based on these approaches, bio-nanocomposites can be incorporated into the solution mixture and extruded as conductive nanofibres with good inherent mechanical and performance properties.

Electrochemical depositions ensure a controlled and precise deposition of conductive materials via electrochemical reactions onto textile materials (Azani and Hassanpour 2024). Electroplating (which uses electric charge) and electroless plating (which uses a reducing agent and metal ions contained in a chemical bath) deposit thin films onto the surfaces of the textile materials.

Coating techniques such as in situ polymerization and dip coating are commonly used to incorporate conductive materials onto fibres, whereas dip coating, spray coating, roll coating and in situ polymerization for textiles (Azani and Hassanpour 2024). For example, in dip coating (Fig. 14.11a), prepared textile materials are immersed in the prepared solution with conductive materials. This process disperses and integrates the conductive materials onto both sides of the textiles to form a uniform coating after drying. Due to the simple nature of the process, achieving an appropriate thin layer on the textile can be somewhat challenging (Azani and Hassanpour 2024). The speed rate used to withdraw the coated materials could influence the coated layer on the textile surface. For in-situ polymerization, conductive polymers are directly grown on the fibres via a chemical reaction resulting in the linking between monomers. For spray coating (Fig. 14.11b), a special spraying device is used to deposition by spraying conductive materials onto the surface of a textile material, which is then cured under controlled heat or temperature (Frederichi et al. 2021).

Printing techniques such as direct, screen, inkjet, gravure and flexographic processes are widely applied to deposit conductive materials (from bio-nanocomposites) onto the surface of a textile material. For screen printing, conductive materials are applied onto the textile surface by squeezing. Herein, these conductive materials are passed

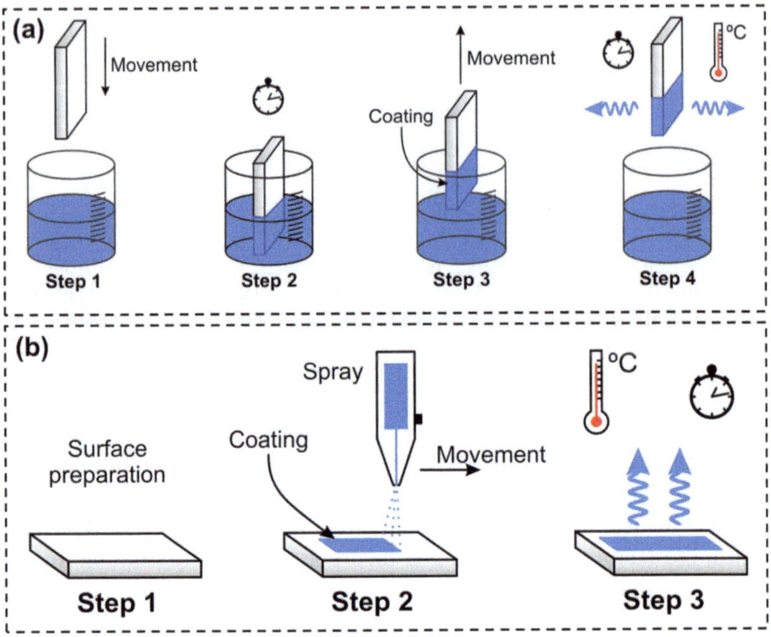

Fig. 14.11 Diagram of **a** dip coating process and **b** spray coating process. Reproduced with permission from Frederichi et al. (2021) Copyright 2020, Springer-Verlag GmbH Germany

through the open spaces in a mesh screen with stencil designs to form conductive pathways directly to the textile surface. In the inkjet printing process, conductive materials are deposited in a precise and controlled manner onto the surface of the textile material through a jet or nozzle. The choice of the inkjet printing technique is influenced by the desired functionality and the type of fabric (Azani and Hassanpour 2024).

14.6.2 Properties and Applications

Bionanocomposites are formed when nanoparticles enhance the natural polymer substrate in one or more small nano-size dimensions (Karki et al. 2021). As such, these are potential materials for wearable textiles due to their significant properties that advance their use in various applications. Herein, enhanced mechanical, flame retardants, thermal and electrical conductivity, biodegradable, environmentally friendly, biocompatibility and non-toxic, have improved functional, structural and physical properties (Azadi et al. 2024) are some properties of bionanocomposites. Integrating nanomaterials into the biopolymer matrix enhances the dimensional stability and strength associated with the mechanical properties. For example, temperature and moisture resistance, enhanced modulus and tensile strength were achieved by fabricating a bionanocomposite from bacterial cellulose nanofibres/PVA (Abral et al. 2019). Other key properties, such as enhanced mechanical, flame retardant, thermal and electrical properties are realized when composite fibres (Fig. 14.12) are fabricated from *Moringa oleifera* fibres in functionalized carbon nanotube ink (Kalam et al. 2025). High performance, high cyclic stability, specific capacitance and robust mechanical properties are obtained from the fabrication of hybrid nanocomposites (composed of carbon nanotubes, reduced graphene and PEDOT: PSS) on the nanofibres from cellulose origin (Carrascosa et al. 2024). Subsequently, the integration of nanomaterials in the biopolymer matrix enhances the anti-microbial property and chemical resistance to bacteria and chemicals or solvents, respectively (Krifa and Prichard 2020). For example, in a relevant study, polyester/copper nanocomposites were fabricated via melt-spinning processes (González-Sánchez et al. 2024). Herein, results revealed a 99% growth inhibition of pathogens, hence offering great protection. Other studies, such as Xue et al. (2012), Aravind et al. (2019) and Novi et al. (2022), deposited or coated textile materials with nanosized particles for antibacterial or anti-microbial properties. These properties make these bio-nanocomposites very suitable and versatile for wearable textile applications, which are discussed below.

Based on the unique properties of these bionanocomposites, they find applications in key areas such as healthcare, fitness and sportswear, protective clothing, energy harvesting and monitoring of the environment. In the healthcare sector, for example, fabricated materials are used for drug delivery, wound dressings and health monitoring. For example, chitosan/ZnO nanoparticles solution was coated on a cotton fabric (Alharbi et al. 2022) for enhanced antibacterial and antioxidant effects, which could be used effectively as medical bandages, virus masks, bed liners and in wound treatment. Due to the antibacterial properties of the fabricated materials when

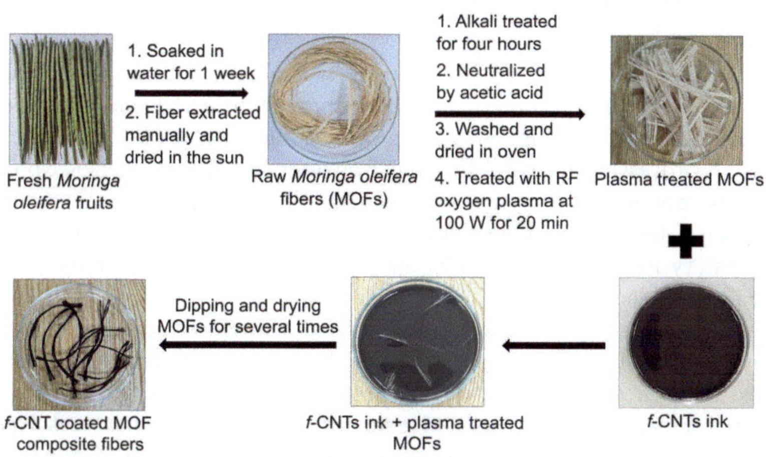

Fig. 14.12 Fabrication procedure for Moringa oleifera fibres in functionalized carbon nanotube ink (Kalam et al. 2025). (This is an open-access article distributed under the terms of the Creative Commons CC BY license)

nanoparticles are integrated into fibrous materials, relevant studies have advanced this approach for wound dressing applications; silk fibroin (SF) and chitosan composite nanofibrous membrane were fabricated via electrospinning (Cai et al. 2010), SF/ TiO_2 nanofibrous mats (Jao et al. 2012), and SF incorporated with silver nanoparticles (Kang et al. 2007). Alternatively, bionanocomposites are further applied as sensors due to their high sensitivity, biocompatibility and flexibility in wearable textiles to effectively sense and monitor the environment and human health, respectively. These sensory materials are fabricated either by printing, coating, spinning, embedding or attaching composite materials into the fibre or textile materials (Li et al. 2023). Key sensors such as pressure, temperature, sweat and strain are fabricated for effective sensing, monitoring and recording of critical vitals of the human body. In a relevant study, for example, a carbon nanotube/silk non-woven fabric was fabricated as a strain/pressure sensor (He et al. 2022). Results show these sensing materials can sense and record 10 Pa ultralow pressure and 0.05% low strain, which are vital for monitoring key health signals and human movements. Thus, they can monitor and record at very low limits under use. The strain/pressure sensor further exhibited good durability, reliability and reproducibility. Others, such as Xu et al. (2021), Correia et al. (2019) and Gan et al. (2015), have also advanced the incorporation of carbon-based materials into natural matrix to fabricate flexible strain sensors for monitoring human health. Subsequently, these strain sensors are seamlessly integrated into textiles for sports and fitness wear to monitor, detect and record human movements or physical activities. This is made possible due to their flexible nature, high sensitivity and real-time tracking performance.

14.7 Safety Concerns, Challenges and Regulatory Issues

The significant growth and advancements in wearable textiles and nanostructured coatings offer several advantages, such as improved properties and functionalities that widen their applications. For example, wearable textiles play a critical role in monitoring, processing and recording data from users for healthcare, hence providing a non-invasive approach (Sungmee and Jayaraman 2003; Park et al. 2002). Aside from these benefits, their use and development raise serious safety concerns, challenges and regulatory issues that need to be addressed to foster their effective usage and safety in real-life applications.

Safety remains a significant concern associated with nanostructured coatings since they involve nanomaterial incorporation, which poses serious threats to the environment and human health. These safety concerns centre on storage and handling, impact on the environment, health threats, and regulatory and preventive measures. Nanoparticles disposed into the environment through products containing nanoparticles or through the discharge of wastewater pose a significant threat to wildlife and the ecosystem. These ecological consequences are influenced by the high reactivity and small size (Ma et al. 2024). Relevant studies have revealed the disruptive and high toxicity nature of nanomaterials on biodiversity, ecosystems, terrestrial and aquatic organisms (Bobori et al. 2020). For example, there is surface dissolution and transformation, which results in an environmentally aged state when nanomaterials enter the water bodies (Spurgeon et al. 2020). This comprehensive risk assessment should cover aged forms of nanomaterials together with the new nanomaterials (Ma et al. 2024). Subsequently, the handling and storage of nanomaterials is another safety concern for workers. Lack of sufficient information on the handling and storage of the nanomaterials puts these workers at great risk of exposure. Nanoparticles enter the health system by ingestion, through the skin and by inhalation. For the latter, without proper protective equipment, nanoparticles are inhaled into the human body and deposited into various systems or regions such as the alveolar, tracheobronchial (Castillo 2013) and in some cases the brain structure (Oberdörster et al. 2005). Addressing these threats will require some regulatory and preventive measures. Effective labelling, safety assessment and lack of standardization are some key regulatory issues for nanostructured coatings. Disclosing and labelling of products with nanoparticles such as coatings is required by regulatory bodies to provide information to users for an informed decision to limit exposures. Also, standardized testing protocols by regulatory agencies are imperative to assess the health impacts of these products. This is because current procedures vary from regulatory body to regulatory bodies and hence may cause inconsistency in the requirements and compliance by manufacturers for products.

Several challenges affect the design and fabrication of nanostructured coatings, which influences the eventual performance. These challenges include the suitability of the fillers or reinforcements, adhesion strength and selecting the appropriate nanocoating. One is the key challenges for the fabrication of nanostructured coatings is material selection and compatibility which ensures effective bonding and functionality of coatings. Selecting the appropriate materials like fillers or reinforcements, is important to prevent any mismatch that could affect the development process and the functionality of the nanocoating (Farooq et al. 2022). For example, the composition materials have different thermal and mechanical characteristics, hence selecting the right materials that are compatible is appropriate. Even though nanostructured coating exhibits unique tribological performance and corrosion behaviours, they are greatly influenced by changes in contact conditions, filler, operating environment, size of grain, parameters for processing, methods of synthesis, composition and coating materials. This calls for effective monitoring and control systems for materials. Any changes during mixing could produce undesirable properties that lead to layer complexity (Ganeshkumar et al. 2023). Alternatively, the dispersion of nanoparticles in coatings and high viscosity influenced by dispersing fine nanoparticles are challenges that need to be solved (Fernando 1008). Furthermore, this mismatch of material compatibility could affect the adhesion strength of the nanocoating. The adhesion strength is another challenge since there are undesirable reactions between the substrate and the nanostructured coating (Sasmita et al. 2024). This affects the effectiveness and durability of the coatings where failures could occur under operational use. The fabrication and deposition process remains another challenge that influences uniformity within the nanostructured coating. Ensuring stability and uniformity of nanoparticles for coatings during the fabrication process is important to prevent any agglomeration which could affect the properties and functionality of the coating. The time and cost involved in the production of coating is another challenge that affects their scalability for commercial use (Farooq et al. 2022).

Wearable textiles are electronic textiles or smart textiles that combine and integrate electronic technology into clothing for effective sensing, tracking and monitoring (Meddeb 2024). However, with the benefits derived from these innovations come critical safety concerns, challenges and regulatory issues that are associated with wearable textiles. These are capable of monitoring, collecting, processing, storing and sharing large data sets for applications in healthcare. This integration raises serious safety and ethical challenges to data privacy (Bouderhem 2023). Health data is very confidential and sensitive (Silva 2023), hence a breach or access to such information by third parties poses serious threats to the patients, which could lead to trust issues and misuse of data. Compliance with strict regulations and privacy policies, robust data storage and encryption of user data will help build confidence and ensure data protection. Such regulatory frameworks must include options for users to opt out or opt-in on the collection of data. This raises their awareness and informs them on how their personal data is controlled. Patent of innovations for intellectual property is critical to safeguard their innovations and developments of wearable textiles for various applications. As such, regulatory bodies must develop guidelines to safeguard these innovations. Wearable textiles are made of electronic components for effective monitoring, tracking and data transfer. However, the issue of safety, impact on the environment and standards of these wearable textiles are critical. Since these wearable textiles will be worn by the user, their safety is important to prevent overheating and any electric shocks that are harmful to human health. For effective compliance, key standards by industry associations and the International Organization for Standardization (ISO) have been developed to provide guidelines and ensure the safety and consistency of product quality (Meddeb 2024).

Subsequently, there are several challenges associated with wearable textiles such as data reliability and accuracy, system durability, power efficiency and user-friendliness (Kan and Lam 2021). Textiles are integrated with electronics, which makes it difficult to wash unless these electronic components are separated. This makes them not user-friendly unless waterproof enclosures are developed for these electronics to address the issue (Kubicek et al. 2022). The electronics used most especially batteries, make the wearables bulky in nature and increase weight (Dolez et al. 2020), making them not comfortable to wear, hence not user-friendly. The issue of the charging process for batteries is also another challenge for wearable textiles which affects their power efficiency. It is thus important to develop a flexible energy device that can be easily integrated into the garment without compromising the comfort and weight (Pu et al. 2016). Wearable textiles are subject to mechanical agitation, stretching and bending during washing and usage. These actions affect the integrated electronics which questions the durability of the systems and cyclic stability (Kan and Lam 2021; Islam et al. 2020; Merhi et al. 2020). Lastly, data reliability and accuracy are important for reproducibility, repeatability and trust in the use of these wearable textiles for different applications (Dolez et al. 2020). Key external and internal conditions such as rain and body sweat could affect the effective performance of the wearable textiles (Firšt Rogale et al. 2018), the accuracy and reliability of the data recorded is questionable. Furthermore, the lack of required regulatory frameworks and standard testing methods for wearable textiles limits a complete understanding of their behaviour, safety and operation, which makes be of great risk to users (Kan and Lam 2021; Dolez et al. 2020).

In conclusion, advancements in wearable textiles and nanostructured coatings offer several advantages for different applications; however, addressing related safety concerns, challenges and regulatory issues will foster their effective usage and safety. It is thus suggested that effective collaborations between researchers, regulator bodies and the industry to advance the frontiers for a responsible technology for the benefit of all.

14.8 Conclusions, Future Research Trends and Opportunities

The unique properties of bionanocomposites have advanced their applications in nanostructured coatings and wearable textiles. For nanostructured coatings such as exceptional functional, chemical and mechanical characteristics, and surface protection properties that advance their application in food packaging, medicine and electronics amongst others (Gu et al. 2020). However, future research trends and opportunities should focus on improving the functionality, mechanical properties and scalability of bionanocomposites for nanostructured coatings. For example, developing and improving the interaction ability and compatibility with other substrates, hence quality novel coating films, could be advanced in future research (Mhd Haniffa et al. 2016a). Furthermore, future research of these coatings could be enhanced to have multifunctionality such as enhanced strength, self-cleaning and healing abilities, and anti-microbial properties. Cost-effective manufacturing procedures could be advanced to address scalability issues of bio-nanocomposites in the future.

Wearable textiles which integrate electronic components into textiles are potential innovations with unique functionality for different applications. Future research trends and opportunities should focus on core areas like user comfort, durability, functionality, effective integration technologies and flexible energy harvesting materials for easy integration. The issue of the weight of the components integrated into wearable textiles influences their wear comfort by users. Adopting flexible electronic components that are durable with less weight will ensure their wear comfort. Alternatively, the durability of the textile materials coupled with the electronics is also a serious challenge since they are not resistant to washing and other mechanical actions. Future studies should innovate effective encapsulation approaches to protect the electronics from damage by rain and wash. The functionality of these wearable textiles is important to ensure users fully benefit. Further incorporation of sensing properties into the textiles using appropriate materials will advance their applications in key areas like healthcare for real-time monitoring of vitals from the human body. Lastly, the design and development of flexible energy harvesting materials are ideal for alternative energy supply for wearable textiles. Alternative rechargeable batteries can be developed to power wearable textiles, enabling storage device solutions with less weight (Islam et al. 2020).

References

Abral H, Kadriadi, Mahardika M, Handayani D, Sugiarti E, Muslimin AN (2019) Characterization of disintegrated bacterial cellulose nanofibers/PVA bionanocomposites prepared via ultrasonication. Int J Biol Macromol 135:591–9. https://doi.org/10.1016/j.ijbiomac.2019.05.178

Abu-Thabit NY, Makhlouf ASH (2015) Chapter 24—recent advances in nanocomposite coatings for corrosion protection applications. In: Makhlouf ASH, Scharnweber D (eds) Handbook of nanoceramic and nanocomposite coatings and materials. Butterworth-Heinemann, pp 515–549

Adnane M, Jiang Z, Choi S, Jang H (2009) Detecting specific health-related events using an integrated sensor system for vital sign monitoring. Sensors 9(9):6897–6912

Alharbi ND, Amer HH, El-Zaher NA, Guirguis OW (2022) Development and characterization of cotton fabrics by dipping in solutions of chitosan and ZnO-nanoparticles as promising environmentally friendly reinforcements for polymer composites. Polym Polym Compos 30:09673911221148826. https://doi.org/10.1177/09673911221148826

Aliofkhazraei M (2011) Synthesis, processing and application of nanostructured coatings. Nanocoatings: size effect in nanostructured films. Springer Berlin Heidelberg, Berlin, Heidelberg, pp 1–28

Antony ASMC (2024) Effects of nano-coating, nanocomposites, biocomposites to develop UV protective knit fabrics: a synergistic review. Open J Compos Mater 15(1):1–30

Aravind HP, Jadhav SA, More VB, Sonawane KD, Patil PS (2019) Novel one step sonosynthesis and deposition technique to prepare silver nanoparticles coated cotton textile with antibacterial properties. Colloid J 81(6):720–727. https://doi.org/10.1134/S1061933X19070019

Asri RIM, Harun WSW, Hassan MA, Ghani SAC, Buyong Z (2016) A review of hydroxyapatite-based coating techniques: sol–gel and electrochemical depositions on biocompatible metals. J Mech Behav Biomed Mater 57:95–108. https://doi.org/10.1016/j.jmbbm.2015.11.031

Azadi E, Dinari M, Derakhshani M, Reid KR, Karimi B (2024) Sources and extraction of biopolymers and manufacturing of bio-based nanocomposites for different applications. Molecules 29(18):4406

Azani M-R, Hassanpour A (2024) Electronic textiles (E-Textiles): types, fabrication methods, and recent strategies to overcome durability challenges (washability and flexibility). J Mater Sci: Mater Electron 35(29):1897. https://doi.org/10.1007/s10854-024-13347-0

Babaeipour S, Nousiainen P, Kimiaei E, Tienaho J, Kohlhuber N, Korpinen R et al (2024) Thin multifunctional coatings for textiles based on the layer-by-layer application of polyaromatic hybrid nanoparticles. Mater Adv 5(15):6114–6131. https://doi.org/10.1039/D4MA00085D

Bai Z, Zhang Z, Li J, Guo J (2019) Textile-based triboelectric nanogenerators with high-performance via optimized functional elastomer composited tribomaterials as wearable power source. Nano Energy 65:104012. https://doi.org/10.1016/j.nanoen.2019.104012

Bairagi S, Shahid ul I, Kumar C, Babu A, Aliyana AK, Stylios G et al (2023) Wearable nanocomposite textile-based piezoelectric and triboelectric nanogenerators: progress and perspectives. Nano Energy 118:108962. https://doi.org/10.1016/j.nanoen.2023.108962

Baptista A, Silva FJG, Porteiro J, Míguez JL, Pinto G, Fernandes L (2018) On the physical vapour deposition (PVD): evolution of magnetron sputtering processes for industrial applications. Procedia Manuf 17:746–757. https://doi.org/10.1016/j.promfg.2018.10.125

Bardhan R, Lal S, Joshi A, Halas NJ (2011) Theranostic nanoshells: from probe design to imaging and treatment of cancer. Acc Chem Res 44(10):936–946. https://doi.org/10.1021/ar200023x

Barnett SA, Madan A, Kim I, Martin K (2003) Stability of nanometer-thick layers in hard coatings. MRS Bull 28(3):169–172. https://doi.org/10.1557/mrs2003.57

Basile F, Benito P, Fornasari G, Monti M, Scavetta E, Tonelli D et al (2010) A novel electrochemical route for the catalytic coating of metallic supports. In: Gaigneaux EM, Devillers M, Hermans S, Jacobs PA, Martens JA, Ruiz P (eds) Studies in surface science and catalysis. Elsevier, pp 51–58

Bayan S, Pal S, Ray SK (2022) Interface engineered silver nanoparticles decorated g-C_3N_4 nanosheets for textile based triboelectric nanogenerators as wearable power sources. Nano Energy 94:106928

Benea L (2013) Electrochemical impedance spectroscopy and corrosion behavior of Co/CeO_2 nanocomposite coatings in simulating body fluid solution. Metall Mater Trans A 44(2):1114–1122. https://doi.org/10.1007/s11661-012-1422-z

Berger M (2008) Nanotechnology e-textiles for bio-monitoring and wearable electronics

Bharathi VSK, Jayas DS (2024) Evolution of bionanocomposites: innovations and applications in food packaging. Foods 13(23). https://doi.org/10.3390/foods13233787

Bobori D, Dimitriadi A, Karasiali S, Tsoumaki-Tsouroufli P, Mastora M, Kastrinaki G et al (2020) Common mechanisms activated in the tissues of aquatic and terrestrial animal models after TiO_2 nanoparticles exposure. Environ Int 138:105611. https://doi.org/10.1016/j.envint.2020.105611

Bouderhem R (2023) Privacy and regulatory issues in wearable health technology. Eng Proc 58(1):87

Brown SD, Nativo P, Smith J-A, Stirling D, Edwards PR, Venugopal B et al (2010) Gold nanoparticles for the improved anticancer drug delivery of the active component of oxaliplatin. J Am Chem Soc 132(13):4678–4684. https://doi.org/10.1021/ja908117a

Cai Z-x, Mo X-m, Zhang K-h, Fan L-p, Yin A-l, He C-l et al (2010) Fabrication of chitosan/silk fibroin composite nanofibers for wound-dressing applications. Int J Mol Sci 11(9):3529–3539

Carrascosa A, Sánchez JS, Morán-Aguilar MG, Gabriel G, Vilaseca F (2024) Advanced flexible wearable electronics from hybrid nanocomposites based on cellulose nanofibers, PEDOT:PSS and reduced graphene oxide. Polymers 16(21):3035

Castillo AMPD (2013) Nanomaterials and workplace health and safety. What are the issues for workers?, pp 1–44

Çelikel DC (2020) Smart e-textile materials. In: Tasaltin N, Nnamchi PS, Al-Humairi SNS (eds) Advanced functional materials. IntechOpen, Rijeka

Chen C, Jia X, Li X, Shi M, Hu J, Song M et al (2023) Scalable wet-spinning of wearable chitosan-silica textile for all-day radiative cooling. Chem Eng J 475:146307. https://doi.org/10.1016/j.cej.2023.146307

Cheng W, Wang J, Ma X, Liu P, Liaw PK, Li W (2023) A review on microstructures and mechanical properties of protective nano-multilayered films or coatings. J Market Res 27:2413–2442. https://doi.org/10.1016/j.jmrt.2023.10.012

Chkalov RV, Chkalova DG (2020) Laser powder cladding complex: principles of advanced automated control. IOP Conf Ser: Mater Sci Eng 969(1):012093. https://doi.org/10.1088/1757-899X/969/1/012093

Claire L, Marie G, Julien G, Jean-Michel S, Jean R, Marie-Joëlle M et al (2016) New architectured hybrid sol-gel coatings for wear and corrosion protection of low-carbon steel. Prog Org Coat 99:337–345. https://doi.org/10.1016/j.porgcoat.2016.06.015

Correia DM, Ribeiro S, da Costa A, Ribeiro C, Casal M, Lanceros-Mendez S et al (2019) Development of bio-hybrid piezoresistive nanocomposites using silk-elastin protein copolymers. Compos Sci Technol 172:134–142. https://doi.org/10.1016/j.compscitech.2019.01.017

da Silva JP (2023) Privacy data ethics of wearable digital health technology. https://cdh.brown.edu/news/2023-05-04/ethics-wearables. Accessed 29 Jan 2025

Dalton PD, Klee D, Möller M (2005) Electrospinning with dual collection rings. Polymer 46(3):611–614. https://doi.org/10.1016/j.polymer.2004.11.075

Dave BC, Lockwood SB (2012) Sol-Gel Method. In: Bhushan B (ed) Encyclopedia of nanotechnology. Springer, Netherlands, Dordrecht, pp 2459–2470

De B, Gupta K, Mandal M, Karak N (2015) Tough hyperbranched epoxy/neem-oil-modified OMMT thermosetting nanocomposite with an antimicrobial attribute. New J Chem 39(1):595–603. https://doi.org/10.1039/C4NJ01558D

Dhiman N, Singla N (2024) Smart nanocoating—an innovative solution to create intelligent functionality on surface. ChemistrySelect 9(41):e202403038. https://doi.org/10.1002/slct.202403038

Dolez PI, Decaens J, Buns T, Lachapelle D, Vermeersch O (2020) Applications of smart textiles in occupational health and safety. IOP Conf Ser: Mater Sci Eng 827(1):012014. https://doi.org/10.1088/1757-899X/827/1/012014

Dong L, Xu C, Li Y, Wu C, Jiang B, Yang Q et al (2016) Simultaneous production of high-performance flexible textile electrodes and fiber electrodes for wearable energy storage. Adv Mater 28(8):1675–1681. https://doi.org/10.1002/adma.201504747

Dufresne A (2008) Polysaccharide nano crystal reinforced nanocomposites. Can J Chem 86(6):484–494. https://doi.org/10.1139/v07-152

Edelmann MJ, Shack LA, Naske CD, Walters KB, Nanduri B (2014) SILAC-based quantitative proteomic analysis of human lung cell response to copper oxide nanoparticles. PLoS ONE 9(12):e114390. https://doi.org/10.1371/journal.pone.0114390

Engin M, Demirel A, Engin EZ, Fedakar M (2005) Recent developments and trends in biomedical sensors. Measurement 37(2):173–188. https://doi.org/10.1016/j.measurement.2004.11.002

Etoc F, Lisse D, Bellaiche Y, Piehler J, Coppey M, Dahan M (2013) Subcellular control of Rac-GTPase signalling by magnetogenetic manipulation inside living cells. Nat Nanotechnol 8(3):193–198. https://doi.org/10.1038/nnano.2013.23

Faraji G, Kim HS, Kashi HT (2018) Introduction. In: Faraji G, Kim HS, Kashi HT (eds) Severe plastic deformation. Elsevier, pp 1–17

Farghaly AA, Collinson MM (2016) Mesoporous hybrid polypyrrole-silica nanocomposite films with a strata-like structure. Langmuir 32(23):5925–5936. https://doi.org/10.1021/acs.langmuir.6b00872

Farooq K, Kumawat YK, Sharma K, Kumar V (2024) Bio-nanocomposites: fundamentals and recent advances. In: Sharma K, Tiwari SK, Kumar V, Kalia S (eds) Novel bio-nanocomposites for biomedical applications. Springer International Publishing, Cham, pp 1–35

Farooq SA, Raina A, Mohan S, Arvind Singh R, Jayalakshmi S, Irfan Ul Haq M (2022) Nanostructured coatings: review on processing techniques, corrosion behaviour and tribological performance. Nanomater 12(8):1323

Fernández-Caramés TM, Fraga-Lamas P (2018) Towards the internet of smart clothing: a review on iot wearables and garments for creating intelligent connected e-textiles. Electronics 7(12):405

Fernando RH (2009) Nanocomposite and nanostructured coatings: recent advancements. Nanotechnology applications in coatings. ACS symposium series, vol 1008. American Chemical Society, pp 2–21

Firšt Rogale S, Rogale D, Nikolić G (2018) Intelligent clothing: first and second generation clothing with adaptive thermal insulation properties. Text Res J 88(19):2214–2233. https://doi.org/10.1177/0040517517718190

Fouda A, El-Din Hassan S, Salem SS, Shaheen TI (2018) In-vitro cytotoxicity, antibacterial, and UV protection properties of the biosynthesized zinc oxide nanoparticles for medical textile applications. Microb Pathog 125:252–261. https://doi.org/10.1016/j.micpath.2018.09.030

Frederichi D, Scaliante MHNO, Bergamasco R (2021) Structured photocatalytic systems: photocatalytic coatings on low-cost structures for treatment of water contaminated with micropollutants—a short review. Environ Sci Pollut Res 28(19):23610–23633. https://doi.org/10.1007/s11356-020-10022-9

Gan L, Shang S, Yuen CWM, Jiang S-X (2015) Graphene nanoribbon coated flexible and conductive cotton fabric. Compos Sci Technol 117:208–14. https://doi.org/10.1016/j.compscitech.2015.06.019

Ganeshkumar S, Kumar A, Maniraj J, Suresh Babu Y, Ansu AK, Goyal A et al (2023) Exploring the potential of nano technology: a assessment of nano-scale multi-layered-composite coatings for cutting tool performance. Arab J Chem 16(10):105173. https://doi.org/10.1016/j.arabjc.2023.105173

Gardon M, Guilemany JM (2014) Milestones in functional titanium dioxide thermal spray coatings: a review. J Therm Spray Technol 23(4):577–595. https://doi.org/10.1007/s11666-014-0066-5

Gheibi A, Latifi M, Merati AA, Bagherzadeh R (2014) Piezoelectric electrospun nanofibrous materials for self-powering wearable electronic textiles applications. J Polym Res 21:469. https://doi.org/10.1007/s10965-014-0469-5

Gleiter H (1991) Nanocrystalline materials. Advanced structural and functional materials: proceedings of an international seminar organized by Deutsche Forschungsanstalt für Luft-und Raumfahrt (DLR), Köln, June 1991. Springer, pp 1–37

González-Sánchez A, Rosas-Macías R, Hernández-Bautista JE, Valdez-Garza JA, Rodríguez-Fuentes N, Soriano-Corral F et al (2024) Antimicrobial properties of polyester/copper nanocomposites by melt-spinning and melt-blowing techniques. Textiles 4(1):1–16

Gu Y, Xia K, Wu D, Mou J, Zheng S (2020) Technical characteristics and wear-resistant mechanism of nano coatings: a review. Coatings 10(3):233

Hazarika D, Karak N (2018) Unprecedented influence of carbon Dot@TiO(2) nanohybrid on multifaceted attributes of waterborne hyperbranched polyester nanocomposite. ACS Omega 3(2):1757–1769. https://doi.org/10.1021/acsomega.7b02079

He Y, Zhou M, Mahmoud MHH, Lu X, He G, Zhang L et al (2022) Multifunctional wearable strain/pressure sensor based on conductive carbon nanotubes/silk nonwoven fabric with high durability and low detection limit. Adv Compos Hybrid Mater 5(3):1939–1950. https://doi.org/10.1007/s42114-022-00525-z

Helmersson U, Lattemann M, Bohlmark J, Ehiasarian AP, Gudmundsson JT (2006) Ionized physical vapor deposition (IPVD): a review of technology and applications. Thin Solid Films 513(1):1–24. https://doi.org/10.1016/j.tsf.2006.03.033

Ho SM, SA V, Ahmed G, Vidya NS (2018) A review of nanostructured thin films for gas sensing and corrosion protection. Mediterr J Chem 7(6). https://doi.org/10.13171/mjc7618111916hs

Ielo I, Giacobello F, Sfameni S, Rando G, Galletta M, Trovato V et al (2021) Nanostructured surface finishing and coatings: functional properties and applications. Materials 14(11):2733

Islam GMN, Ali A, Collie S (2020) Textile sensors for wearable applications: a comprehensive review. Cellulose 27(11):6103–6131. https://doi.org/10.1007/s10570-020-03215-5

Jaeger R, Schönherr H, Vancso GJ (1996) Chain packing in electro-spun poly(ethylene oxide) visualized by atomic force microscopy. Macromolecules 29(23):7634–7636. https://doi.org/10.1021/ma9610673

Jao W-C, Yang M-C, Lin C-H, Hsu C-C (2012) Fabrication and characterization of electrospun silk fibroin/TiO_2 nanofibrous mats for wound dressings. Polym Adv Technol 23(7):1066–1076. https://doi.org/10.1002/pat.2014

Jiang Y, Shi K, Tang H, Wang Y (2019) Enhanced wettability and wear resistance on TiO_2/PDA thin films prepared by sol-gel dip coating. Surf Coat Technol 375:334–340. https://doi.org/10.1016/j.surfcoat.2019.07.051

Joshi M, Bhattacharyya A (2011) Nanotechnology–a new route to high-performance functional textiles. Text Prog 43(3):155–233

Kalam MA, Mustakim S, Rahman MJ, Uddin MT, Alam MA, Rahman MM (2025) Bio-degradable smart nanocomposite fiber from Moringa oleifera fiber reinforced with safely functionalized carbon nanotubes. AIP Adv 15(1). https://doi.org/10.1063/5.0238023

Kan C-W, Lam Y-L (2021) Future trend in wearable electronics in the textile industry. Appl Sci 11(9):3914

Kang M, Jung R, Kim H-S, Youk JH, Jin H-J (2007) Silver nanoparticles incorporated electrospun silk fibers. J Nanosci Nanotechnol 7(11):3888–3891. https://doi.org/10.1166/jnn.2007.056

Kang K-W, Choi C-W, Jin J-W (2021) A wet-spinning process for producing carbon nanotube/polyvinylidene fluoride fibers having highly consistent electrical and mechanical properties. Polymers 13(22):4048

Karki S, Gohain MB, Yadav D, Ingole PG (2021) Nanocomposite and bio-nanocomposite polymeric materials/membranes development in energy and medical sector: a review. Int J Biol Macromol 193:2121–2139. https://doi.org/10.1016/j.ijbiomac.2021.11.044

Khan I, Saeed K, Khan I (2019) Nanoparticles: properties, applications and toxicities. Arab J Chem 12(7):908–931. https://doi.org/10.1016/j.arabjc.2017.05.011

Koç M, Paralı L, Şan O (2020) Fabrication and vibrational energy harvesting characterization of flexible piezoelectric nanogenerator (PEN) based on PVDF/PZT. Polym Testing 90:106695. https://doi.org/10.1016/j.polymertesting.2020.106695

Koleva D, Boshkov N, Raichevski G, Veleva L (2005) Electrochemical corrosion behaviour and surface morphology of electrodeposited zinc, zinc–cobalt and their composite coatings. Trans IMF 83(4):188–193. https://doi.org/10.1179/002029605X61676

Krifa M, Prichard C (2020) Nanotechnology in textile and apparel research—an overview of technologies and processes. J Text Inst 111(12):1778–1793. https://doi.org/10.1080/00405000.2020.1721696

Kubicek J, Fiedorova K, Vilimek D, Cerny M, Penhaker M, Janura M et al (2022) Recent trends, construction, and applications of smart textiles and clothing for monitoring of health activity: a comprehensive multidisciplinary review. IEEE Rev Biomed Eng 15:36–60. https://doi.org/10.1109/RBME.2020.3043623

Kumar P, Sandeep K, Alavi S, Truong V (2014) Analytical techniques for structural characterization of biopolymer-based nanocomposites. Polymers for packaging applications Apple Academic Press Inc., New Jersey, pp 307–333

Kuroda A (2008) Concurrent but non-integrable currency circuits: complementary relationships among monies in modern China and other regions. Financ Hist Rev 15(1):17–36. https://doi.org/10.1017/S0968565008000036

Leroux F, Campagne C, Perwuelz A, Gengembre L (2008) Fluorocarbon nano-coating of polyester fabrics by atmospheric air plasma with aerosol. Appl Surf Sci 254(13):3902–3908. https://doi.org/10.1016/j.apsusc.2007.12.037

Li D, Xia Y (2004) Electrospinning of nanofibers: reinventing the wheel? Adv Mater 16(14):1151–1170. https://doi.org/10.1002/adma.200400719

Li J, Sun Y, Sun X, Qiao J (2005) Mechanical and corrosion-resistance performance of electrodeposited titania–nickel nanocomposite coatings. Surf Coat Technol 192(2):331–335. https://doi.org/10.1016/j.surfcoat.2004.04.082

Li F, Lan X, Wang L, Kong X, Xu P, Tai Y et al (2020) An efficient photocatalyst coating strategy for intimately coupled photocatalysis and biodegradation (ICPB): powder spraying method. Chem Eng J 383:123092. https://doi.org/10.1016/j.cej.2019.123092

Li S, Li H, Lu Y, Zhou M, Jiang S, Du X et al (2023) Advanced textile-based wearable biosensors for healthcare monitoring. Biosensors 13(10):909

Ma X, Tian Y, Yang R, Wang H, Allahou LW, Chang J et al (2024) Nanotechnology in healthcare, and its safety and environmental risks. J Nanobiotechnol 22(1):715. https://doi.org/10.1186/s12951-024-02901-x

Mailänder V, Landfester K (2009) Interaction of nanoparticles with cells. Biomacromol 10(9):2379–2400. https://doi.org/10.1021/bm900266r

Makhlouf ASH (2011) 1—current and advanced coating technologies for industrial applications. In: Makhlouf ASH, Tiginyanu I (eds) Nanocoatings and ultra-thin films. Woodhead Publishing, pp 3–23

Meddeb A (2024) Smart textile industry: threats and opportunities. Springer Nature Singapore, Singapore, pp 245–64

Meena JS, Choi SB, Jung S-B, Kim J-W (2023) Electronic textiles: new age of wearable technology for healthcare and fitness solutions. Mater Today Bio 19:100565. https://doi.org/10.1016/j.mtbio.2023.100565

Mekuye B, Abera B (2023) Nanomaterials: an overview of synthesis, classification, characterization, and applications. Nano Select 4(8):486–501. https://doi.org/10.1002/nano.202300038
Merhi Y, Mikkelsen PH, Suetta C, Nygaard JV, Agarwala S (2020) Mechanical performance of electronically functional smart textiles. Trans Add Manuf Meets Med 2(1). https://doi.org/10.18416/AMMM.2020.2009012
Meyer J, Lukowicz P, Troster G (2006) Textile pressure sensor for muscle activity and motion detection. 2006 10th IEEE international symposium on wearable computers, pp 69–72
Mhd Haniffa MAC, Ching YC, Abdullah LC, Poh SC, Chuah CH (2016a) Review of bionanocomposite coating films and their applications. Polymers 8(7):246
Mhd Haniffa MA, Ching YC, Abdullah LC, Poh SC, Chuah CH (2016b) Review of bionanocomposite coating films and their applications. Polymers. https://doi.org/10.3390/polym8070246
Mokhtari F, Spinks GM, Fay C, Cheng Z, Raad R, Xi J et al (2020) Wearable electronic textiles from nanostructured piezoelectric fibers. Adv Mater Technol 5(4):1900900. https://doi.org/10.1002/admt.201900900
More SR, Bhatt DV, Menghani JV (2017) Resent research status on laser cladding as erosion resistance technique—an overview. Mater Today: Proc 4(9):9902–9908. https://doi.org/10.1016/j.matpr.2017.06.291
Münz WD (2003) Large-scale manufacturing of nanoscale multilayered hard coatings deposited by cathodic arc/unbalanced magnetron sputtering. MRS Bull 28(3):173–179. https://doi.org/10.1557/mrs2003.58
Naidonov AO, Koval VM, Dusheiko MH, Barbash VA, Yashchenko OV (2023) Bend sensors based on nanocellulose composites: influence of configuration of resistive strain gauge. Radioelectron Commun Syst 66(10):556–562. https://doi.org/10.3103/S0735272723090042
Nandhini J, Karthikeyan E, Rajeshkumar S (2024) Eco-friendly bio-nanocomposites: pioneering sustainable biomedical advancements in engineering. Discov Nano 19(1):86. https://doi.org/10.1186/s11671-024-04007-7
Nguyen-Tri P, Nguyen TA, Carriere P, Ngo XC (2018) Nanocomposite coatings: preparation, characterization, properties, and applications. Int J Corros 2018(1):4749501. https://doi.org/10.1155/2018/4749501
Novi VT, Gonzalez A, Brockgreitens J, Abbas A (2022) Highly efficient and durable antimicrobial nanocomposite textiles. Sci Rep 12(1):17332. https://doi.org/10.1038/s41598-022-22370-2
Oberdörster G, Maynard A, Donaldson K, Castranova V, Fitzpatrick J, Ausman K et al (2005) Principles for characterizing the potential human health effects from exposure to nanomaterials: elements of a screening strategy. Part Fibre Toxicol 2(1):8. https://doi.org/10.1186/1743-8977-2-8
Ocak Y, Sofuoglu A, Tihminlioglu F, Böke H (2009) Protection of marble surfaces by using biodegradable polymers as coating agent. Prog Org Coat 66:213–220. https://doi.org/10.1016/j.porgcoat.2009.07.007
Ocak Y, Sofuoglu A, Tihminlioglu F, Böke H (2015) Sustainable bio-nano composite coatings for the protection of marble surfaces. J Cult Herit 16(3):299–306. https://doi.org/10.1016/j.culher.2014.07.004
Okpala CC (2013) Nanocomposites—an overview. Int J Eng Res Dev 8(11):17–23
Onyeaka HN, Nwabor OF (2022) Chapter 18—nanotechnology and nanomaterial delivery in the food system. In: Onyeaka HN, Nwabor OF (eds) Food preservation and safety of natural products. Academic Press, pp 247–257
Ovid'ko IA (2006) Defects and deformation mechanisms in nanostructured coatings. In: Cavaleiro A, De Hosson JTM (eds) Nanostructured coatings. Springer New York, New York, NY, pp 78–108
Pakdel E, Fang J, Fang J, Sun L, Wang X, Wang X (2018) Nanocoatings for smart textiles. Smart textiles, pp 247–300
Pande VV, Sanklecha VM (2017) Bionanocomposite: a review. Nanomed Nanotechnol

Parandeh S, Kharaziha M, Karimzadeh F (2019) An eco-friendly triboelectric hybrid nanogenerators based on graphene oxide incorporated polycaprolactone fibers and cellulose paper. Nano Energy 59:412–421. https://doi.org/10.1016/j.nanoen.2019.02.058

Park JH, Gu L, von Maltzahn G, Ruoslahti E, Bhatia SN, Sailor MJ (2009) Biodegradable luminescent porous silicon nanoparticles for in vivo applications. Nat Mater 8(4):331–336. https://doi.org/10.1038/nmat2398

Park S, Mackenzie K, Jayaraman S (2002) The wearable motherboard: a framework for personalized mobile information processing (PMIP). Proceedings of the 39th annual design automation conference. Association for Computing Machinery, New Orleans, Louisiana, USA, pp 170–4

Pedicini A, Farris RJ (2003) Mechanical behavior of electrospun polyurethane. Polymer 44(22):6857–6862. https://doi.org/10.1016/j.polymer.2003.08.040

Pu X, Liu M, Li L, Han S, Li X, Jiang C et al (2016) Wearable textile-based in-plane microsupercapacitors. Adv Energy Mater 6(24):1601254. https://doi.org/10.1002/aenm.201601254

Qian L-Y, He D, Qin H, Cao X, Huang J, Li J (2022) Chitosan fabrics with synergy of silver nanoparticles and silver nanowires for enhanced conductivity and antibacterial activity. J Ind Text 51:152808372211016. https://doi.org/10.1177/15280837221101650

Qin H, Li J, He B, Sun J, Li L, Qian L (2018) Novel wearable electrodes based on conductive chitosan fabrics and their application in smart garments. Materials (Basel) 11(3). https://doi.org/10.3390/ma11030370.

Rafique S, Kasi AK, Kasi JK, Aminullah, Bokhari M, Shakoor Z (2020) Fabrication of silver-doped zinc oxide nanorods piezoelectric nanogenerator on cotton fabric to utilize and optimize the charging system. Nanomater Nanotechnol 10:1847980419895741. https://doi.org/10.1177/1847980419895741

Rahman Rashid RA, Nazari KA, Barr C, Palanisamy S, Orchowski N, Matthews N et al (2019) Effect of laser reheat post-treatment on the microstructural characteristics of laser-cladded ultra-high strength steel. Surf Coat Technol 372:93–102. https://doi.org/10.1016/j.surfcoat.2019.05.021

Riazi K, Kübel J, Abbasi M, Bachtin K, Indris S, Ehrenberg H et al (2016) Polystyrene comb architectures as model systems for the optimized solution electrospinning of branched polymers. Polymer 104:240–250. https://doi.org/10.1016/j.polymer.2016.05.032

Ruckdashel RR, Khadse N, Park JH (2022) Smart e-textiles: overview of components and outlook. Sensors 22(16):6055

Sajid M (2022) Nanomaterials: types, properties, recent advances, and toxicity concerns. Curr Opin Environ Sci Health 25:100319. https://doi.org/10.1016/j.coesh.2021.100319

Saleem H, Zaidi SJ (2020) Sustainable use of nanomaterials in textiles and their environmental impact. Materials 13(22):5134

Santos CS, Gabriel B, Blanchy M, Menes O, García D, Blanco M et al (2015) Industrial applications of nanoparticles—a prospective overview. Mater Today: Proc 2(1):456–465

Sasmita B, Rani N, Hasan T, Nelson D, Mukesh S, Manikandan S (2024) Emerging trends in nanostructured coatings: unraveling processing techniques, corrosion mechanisms, and tribological performance. J Nano- Electron Phys 16(6). https://doi.org/10.21272/jnep.16(6).06003

Saxena A, Tripathi R, Zafar F, Singh P (2012) Green synthesis of silver nanoparticles using aqueous solution of Ficus benghalensis leaf extract and characterization of their antibacterial activity. Mater Lett 67(1):91–94

Sengupta D, Kottapalli AGP (2024) Nanomaterials-based bioinspired next generation wearable sensors: a state-of-the-art review. Adv Electron Mater 10(1):2300436. https://doi.org/10.1002/aelm.202300436

Senthil Kumar P, Gunasundari E (2018) Nanocomposites: recent trends and engineering applications. Nano Hybrids Compos 20:65–80

Shah MA, Pirzada BM, Price G, Shibiru AL, Qurashi A (2022) Applications of nanotechnology in smart textile industry: a critical review. J Adv Res 38:55–75. https://doi.org/10.1016/j.jare.2022.01.008

Sharma KR (2015) 6 - Process considerations for nanostructured coatings. In: Aliofkhazraei M (ed) Anti-abrasive nanocoatings. Woodhead Publishing, pp 137–153
Shishkovsky IV, Lebedev PN (2011) 3—chemical and physical vapor deposition methods for nanocoatings. In: Makhlouf ASH, Tiginyanu I (eds) Nanocoatings and ultra-thin films. Woodhead Publishing, pp 57–77
Shuvo II, Shah A, Dagdeviren C (2022) Electronic textile sensors for decoding vital body signals: state-of-the-art review on characterizations and recommendations. Adv Intell Syst 4(4):2100223. https://doi.org/10.1002/aisy.202100223
Singh LP, Bhattacharyya SK, Kumar R, Mishra G, Sharma U, Singh G et al (2014) Sol-Gel processing of silica nanoparticles and their applications. Adv Coll Interface Sci 214:17–37. https://doi.org/10.1016/j.cis.2014.10.007
Souad M (2015) Electrodeposition of nanostructure materials. In: Aliofkhazraei M (ed) Electroplating of nanostructures. IntechOpen, Rijeka
Soukup R, Hamacek A, Mracek L, Reboun J (2014) Textile based temperature and humidity sensor elements for healthcare applications. Proceedings of the 2014 37th international spring seminar on electronics technology, pp 407–11
Spurgeon DJ, Lahive E, Schultz CL (2020) Nanomaterial transformations in the environment: effects of changing exposure forms on bioaccumulation and toxicity. Small 16(36):2000618. https://doi.org/10.1002/smll.202000618
Sumrith N, Rangappa SM, Dangtungee R, Siengchin S, Jawaid M, Pruncu CI (2019) Biopolymers-based nanocomposites: properties and applications. In: Sanyang ML, Jawaid M (eds) Bio-based polymers and nanocomposites: preparation, processing, properties and performance. Springer International Publishing, Cham, pp 255–272
Sun Z, Yang L, Zhang D, Song W (2019) High performance, flexible and renewable nano-biocomposite artificial muscle based on mesoporous cellulose/ionic liquid electrolyte membrane. Sens Actuators, B Chem 283:579–589. https://doi.org/10.1016/j.snb.2018.12.073
Sun L, Yuan G, Gao L, Yang J, Chhowalla M, Gharahcheshmeh MH et al (2021) Chemical vapour deposition. Nat Rev Methods Primers 1(1):5. https://doi.org/10.1038/s43586-020-00005-y
Sungmee P, Jayaraman S (2003) Enhancing the quality of life through wearable technology. IEEE Eng Med Biol Mag 22(3):41–48. https://doi.org/10.1109/MEMB.2003.1213625
Takebira UM, Mohibullah ATM, Mehbub A, Islam MM, Pervez MN, Mondal MIH (2024) Chapter 13—wearable smart textiles—recent development and applications. In: Mondal MIH (ed) Smart textiles from natural resources. Woodhead Publishing, pp 395–442
Teo WE, Ramakrishna S (2006) A review on electrospinning design and nanofibre assemblies. Nanotechnology 17(14):R89. https://doi.org/10.1088/0957-4484/17/14/R01
Tettey KE, Yee MQ, Lee D (2010) Photocatalytic and conductive MWCNT/TiO_2 nanocomposite thin films. ACS Appl Mater Interfaces 2(9):2646–2652. https://doi.org/10.1021/am1004656
Theron A, Zussman E, Yarin AL (2001) Electrostatic field-assisted alignment of electrospun nanofibres. Nanotechnology 12(3):384. https://doi.org/10.1088/0957-4484/12/3/329
Toledano R, Mandler D (2010) Electrochemical codeposition of thin gold nanoparticles/sol−gel nanocomposite films. Chem Mater 22(13):3943–3951. https://doi.org/10.1021/cm1005295
Tripathi R, Chung SJ (2019) Biogenic nanomaterials: synthesis, characterization, growth mechanism, and biomedical applications. J Microbiol Methods 157:65–80
Urrutia A, Rivero PJ, Goicoechea J, Arregui FJ (2021) Chapter 23—micro/nanodeposition techniques for enhanced optical fiber sensors. In: Hussain CM, Kailasa SK (eds) Handbook of nanomaterials for sensing applications. Elsevier, pp 531–573
van Lente H, van Til JI (2008) Articulation of sustainability in the emerging field of nanocoatings. J Clean Prod 16(8):967–976. https://doi.org/10.1016/j.jclepro.2007.04.020
Wang SF, Shen L, Tong YJ, Chen L, Phang IY, Lim PQ et al (2005) Biopolymer chitosan/montmorillonite nanocomposites: preparation and characterization. Polym Degrad Stab 90(1):123–131. https://doi.org/10.1016/j.polymdegradstab.2005.03.001
Weaver CL, LaRosa JM, Luo X, Cui XT (2014) Electrically controlled drug delivery from graphene oxide nanocomposite films. ACS Nano 8(2):1834–1843. https://doi.org/10.1021/nn406223e

Xu X, Wu S, Cui J, Yang L, Wu K, Chen X et al (2021) Highly stretchable and sensitive strain sensor based on polypyrrole coated bacterial cellulose fibrous network for human motion detection. Compos Part B: Eng 211:108665. https://doi.org/10.1016/j.compositesb.2021.108665

Xue C-H, Chen J, Yin W, Jia S-T, Ma J-Z (2012) Superhydrophobic conductive textiles with antibacterial property by coating fibers with silver nanoparticles. Appl Surf Sci 258(7):2468–2472. https://doi.org/10.1016/j.apsusc.2011.10.074

Yang K, Isaia B, Brown LJE, Beeby S (2019) E-textiles for healthy ageing. Sensors 19(20):4463

Yilmaz O, Yorgancioglu A (2019) Chapter 8—nanocoatings: preparation, properties, and biomedical applications. In: Vasile C (ed) Polymeric nanomaterials in nanotherapeutics. Elsevier, pp 299–331

Zafar R, Zia KM, Tabasum S, Jabeen F, Noreen A, Zuber M (2016) Polysaccharide based bionanocomposites, properties and applications: a review. Int J Biol Macromol 92:1012–1024. https://doi.org/10.1016/j.ijbiomac.2016.07.102

Zhai M, Liu Y, Huang J, Hou W, Wu S, Zhang B et al (2020) Fabrication of TiO_2–$SrCO_3$ composite coatings by suspension plasma spraying: microstructure and enhanced visible light photocatalytic performances. J Therm Spray Technol 29(5):1172–1182. https://doi.org/10.1007/s11666-020-01022-9

Zhang K, Wang X, Jing D, Yang Y, Zhu M (2009) Bionic electrospun ultrafine fibrous poly(L-lactic acid) scaffolds with a multi-scale structure. Biomed Mater 4(3):035004. https://doi.org/10.1088/1748-6041/4/3/035004

Zhao C, Yao Y, He L (2014) Electrodeposition and characterization of Ni-W/ZrO_2 nanocomposite coatings. Bull Mater Sci 37(5):1053–1058. https://doi.org/10.1007/s12034-014-0044-z

Zou Y, Ma B, Cui H, Lu F, Xu P (2020) Microstructure, wear, and oxidation resistance of nanostructured carbide-strengthened cobalt-based composite coatings on Invar alloys by laser cladding. Surf Coat Technol 381:125188. https://doi.org/10.1016/j.surfcoat.2019.125188

Zucchelli A, Fabiani D, Gualandi C, Focarete ML (2009) An innovative and versatile approach to design highly porous, patterned, nanofibrous polymeric materials. J Mater Sci 44(18):4969–4975. https://doi.org/10.1007/s10853-009-3759-2

Raphael Kanyire Seidu (Ph.D) holds a Ph.D in Fashion and Textiles Design from the School of Fashion and Textiles at The Hong Kong Polytechnic University, Hong Kong. Before this, he gained both his MFA degree in Textile Design and BA. Degree in Industrial Art (Textile option) from Kwame Nkrumah University of Science and Technology, Ghana. He was also a Visiting Research Student at Nottingham Trent University (NTU) in the UK. He has several publications in leading journals in textile, design, technology, and material science. His recent paper was adjudged the "Second-Best Paper" presented at the 12th Sustainable Education and Development Research Conference in Somanya, Ghana. His research interest focuses on retro-reflective materials, smart textiles, cloth culture, weaving, and print technologies. He has several papers and has an approved patent from his Ph.D. project which aims at developing smart interactive retro-reflective clothing for pedestrians' safety at night-time.

Emmanuel Abankwah Ofori is a graduate and researcher with a degree in BSc. Textile Design and Technology from the Industrial Art Department of Kwame Nkrumah University of Science and Technology, Kumasi. Emmanuel's research interest focuses on sustainable environmental practices, textile science, polymers, and energy efficiency. Recently, he co-authored a paper that was adjudged the "Second-Best Paper" presented at the 12th Sustainable Education and Development Research Conference in Somanya, Ghana, and has actively participated in several research projects, working with renowned authors to explore various facets of environmental sustainability, adaptation to climate change, green dyeing, and sustainable renewable energy consumption, thanks to Dr. Benjamin Tawiah's guidance. Emmanuel is dedicated to fostering the next generation of environmental and scientific leaders as a passionate textile scientist and educator. Through his writings and research, Emmanuel continues to motivate his mentees, readers of his works, and colleagues in the field, fostering greater awareness of the planet's vulnerability and the need to protect its ecological vibrancy in order to achieve a sustainable future at large.

Benjamin Tawiah (Ph.D.) received his B.A. Industrial Art (Textiles Option) and MEng in Textile Engineering from the Kwame Nkrumah University of Science and Technology, Ghana, and Jiangnan University, China, respectively. Prior to his Master of Engineering degree, he earned a postgraduate Diploma in Management Information Systems from the Ghana Institute of Management and Public Administration (GIMPA) and then studied for his Ph.D. at the Hong Kong Polytechnic University where he also worked as a postdoctoral Fellow. He also worked as a research associate at the City University of Hong Kong. He is a senior lecturer at the Kwame Nkrumah University of Science and Technology. His research interest focuses on flame retardant polymer composites, advanced energy electrochemical energy storage systems, and flexible fiber-based current collectors for zinc ion batteries. He is listed among the top 2% scientist globally by Stanford University/ Elsevier ranking 2025.

Shouxiang Jiang is currently a Professor of fashion and textiles at The Hong Kong Polytechnic University. He teaches a broad range of subjects such as innovative textile design approach, advanced textile and fashion, and international fashion and textiles design. Prof. Jiang's research deals with innovative textile effects using physical and chemical treatment technologies. He is internationally renowned for his development of the metallised textiles. Leading a team that has developed a sputtering coating technology with research funding from the HKSAR government and sponsorships from the textile industries, Prof. Kinor Jiang has designed and launched an anhydrous system that uses metallising technologies to produce functional and decorative textiles for the production line. He holds 30 granted and filed patents based on the coating technology. Prof. Jiang has published more than 200 journal papers and more than 10 books. His designs and art projects have been exhibited worldwide and received more than 30 awards.

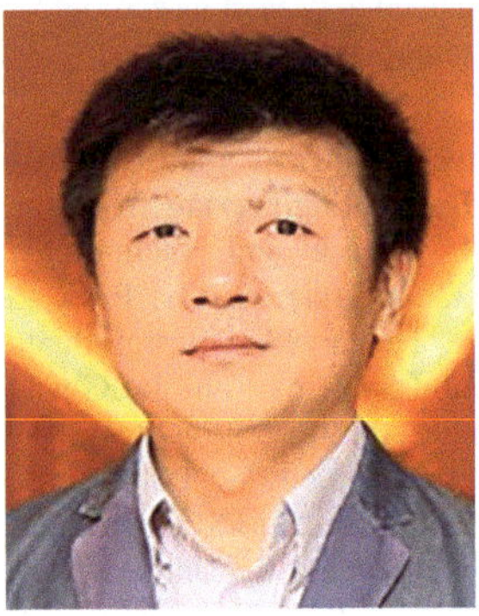

Bin Fei is currently a Professor of fashion and textiles at School of Fashion and Textiles (SFT), The Hong Kong Polytechnic University. He obtained a BSc. in Chemistry from the Department of Applied Chemistry, University of Science and Technology (USTC), China, in 1996 and a PhD from the Changchun Institute of Applied Chemistry, Chinese Academy of Sciences, in 2003. With more than two decades of experience in polymer and catalysis research, Prof. Fei has been engaged in fundamental and applied research on health care, energy devices and sustainable fashion.

Chapter 15
Bionanocomposites in Structural Engineering, Building and Construction

Festus Ben

Abstract This chapter explores the role of bionanocomposites, advanced materials that combine renewable, bio-based polymers with nanoscale reinforcements in transforming structural engineering, building, and construction. It reviews the composition and various types of bionanocomposites, highlighting their potential to deliver superior mechanical, thermal, and environmental performance while supporting sustainable green building practices. Bionanocomposite materials hold promises for specific applications in the building sector, including lightweight concrete, thermal insulators, and retrofitting solutions for commercial housing and infrastructure projects, aligning with sustainable development goals 9, 11, and 12. The chapter also examines different innovative manufacturing techniques with a focus on green methods, such as sonochemical processing, in-situ polymerization, layer-by-layer assembly, electrospinning, melt compounding, and the sol–gel process, and assesses their advantages and shortcomings. Furthermore, the chapter explores diverse applications ranging from enhanced structural components, improved insulation and thermal regulation, and advanced coatings and finishes to self-healing materials and smart structural monitoring systems that collectively contribute to energy efficiency, durability, and reduced environmental impact. By integrating cutting-edge research with practical case studies, this chapter elucidates the potential of bionanocomposites to play a pivotal role in the development of eco-friendly, resilient infrastructure aligned with sustainable development goals.

Keywords Nanocomposites · Structural engineering · Building · Construction materials · Biomaterials

F. Ben (✉)
Centre for Nanoengineering and Advanced Materials, Department of Metallurgy, University of Johannesburg, Johannesburg, South Africa
e-mail: festusb@uj.ac.za; benfestus@federalpolyede.edu.ng

Centre for Advanced Materials Research and Development, Department of Physics, Federal Polytechnic Ede, Ede, Nigeria

S. C. George and B. Tawiah (eds.), *Advanced Bionanocomposite Materials*, Advanced Structured Materials 241, https://doi.org/10.1007/978-981-96-8225-6_15

15.1 Introduction

Rising costs of building materials have made life increasingly difficult for an estimated 1.6 billion people, with 1.1 billion urban residents currently living in slums and projections indicating an additional 2 billion will join them over the next 30 years (UN 2023). According to Sustainable Development Goal (SDG) 11, cities are expected to become safe, inclusive, sustainable, and resilient by 2030, with upgraded slums and universally accessible housing and essential services. However, escalating carbon emissions, rapid industrialization, and expanding populations are making these targets more challenging while infrastructure struggles to keep pace with urban growth and disaster-related disruptions.

Structural engineering, building, and construction depend on diverse materials, which are selected for their distinct properties, including maintenance, environmental, and mechanical. Traditional building materials such as concrete and steel remain prevalent, despite their significant contribution to CO_2 emissions, which impacts negatively on the ozone layer (Oh et al. 2016). The increasing demand for sustainable and eco-friendly construction practices underscores the need for green materials, which reduce carbon emissions and ensure durability, safety, and efficiency amid changing climatic conditions. Research efforts on the use of green materials as an eco-friendly, renewable, and cost-effective substitute have increased in recent years in response to the United Nations 2030 carbon footprint and net-zero reduction. Green materials such as bamboo-based composites (Chen et al. 2021), plant fibers (Ben and Olubambi 2025a), recycled wastes (Ricciardi et al. 2020), agro-waste residues (Ben 2024, Ben and Olubambi 2025b), and innovative materials like shape memory alloys, fiber optics, and piezoelectric materials (Patil et al. 2022) have been extensively explored in the construction industry.

The use of bionanocomposites as an emerging construction material is innovative, owing to their distinct properties. Integrating sustainable bio-based materials with nanomaterials results in bionanocomposites, which offer multifunctional properties, including superior thermal regulation, excellent mechanical strength, and enhanced chemical durability, while mitigating environmental impact (Nepal et al. 2022; Sanmugam et al. 2023). Bionanocomposites possess superior strength-to-weight ratio and inherent sustainability that make them particularly suited for modern construction applications, which are lighter, stronger, and eco-friendly, aligning with contemporary infrastructure requirements.

These unique properties of bionanocomposites enhance their potential in the development of innovative building and construction structures. This chapter contributes to existing knowledge by exploring innovative applications of bionanocomposites and investigating their thermal and mechanical properties within the structural engineering and construction sector. This chapter underscores the potential of bionanocomposites to address global housing shortages and mitigate the environmental impacts of traditional materials, paving the way for the development of lightweight concrete, thermal insulators, and retrofitting solutions for sustainable commercial housing and infrastructure projects.

15.2 Bionanocomposites: Composition and Applications

Bionanocomposites are advanced materials that consist of biopolymers reinforced with nanofillers. These composites, prevalent in ceramics and polymer material science, feature at least one dimension within the 1-to-100 nm range. The compositions and applications of bionanocomposites across various industries are illustrated in Fig. 15.1.

Typical examples of biopolymers include chitosan (gotten from chitin), cellulose (obtained from plant fibers), lignin (a paper residue of the pulp industry), polylactic acid (sourced from sugarcane or corn starch), and polycaprolactone. Chitosan is valued for its antimicrobial, biocompatible, and biodegradable properties (Kreuz et al. 2024), while cellulose is a renowned good reinforcing material that improves the mechanical properties of the composite (Zidi and Miraoui 2024). Conversely, lignin biopolymers provide rigidity and thermal stability for the composite (Zor et al. 2023), with polylactic acid offering excellent strength, clarity, and biocompatibility, and polycaprolactone showing properties such as low melting point and good flexibility (Raza et al. 2024; Cozzani et al. 2024).

Nanofillers that can be used with biopolymers include silver nanoparticles (AgNPs), montmorillonite (MMT), carbon nanotubes (CNTs), nano-silica, nanoclay, kaolinite, graphene, zinc oxide nanoparticles (ZnO-NPs), and titanium dioxide nanoparticles (TiO_2-NPs). Nanoclay comprises of a layered structure, which enhances barrier properties and thermal stability. Nano-silica demonstrates excellent strength and durability, which impacts positively on the tribological performance

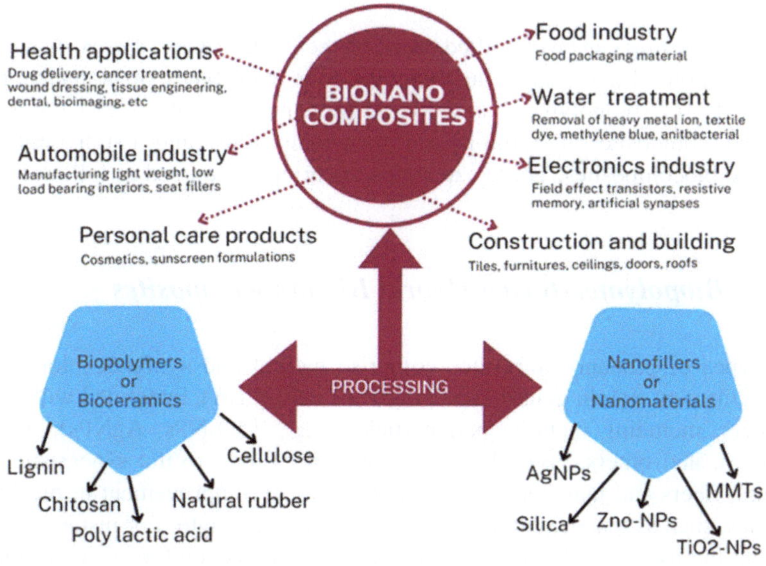

Fig. 15.1 Composition and applications of Bionanocomposites

of the composite. Graphene particles are renowned for their exceptional strength, flexibility, and electrical conductivity, while CNTs show a remarkable tendency to enhance mechanical strength, thermal conductivity, and electrical properties even at reduced concentrations. AgNPs are well-suited for improving both durability and hygienic performance owing to their potent antimicrobial properties. The ZnO-NPs and TiO_2-NPs both provide excellent antimicrobial and ultraviolet-blocking properties, imparting photocatalytic activity and increasing thermal stability and mechanical performance (Zhou et al. 2024). Kaolinite improves thermal stability and mechanical reinforcement, enhancing structural integrity without significantly increasing material density (Cheng et al. 2024).

Owing to their high surface-to-volume ratio, these nanofillers enhance polymer filler and matrix interactions, significantly enhancing the barrier and thermal properties of the composites even at reduced filler content, a distinct advantage over conventional composites. By leveraging the strengths of both natural and synthetic components, bionanocomposites exhibit enhanced barrier, mechanical, and thermal properties, rendering them ideal for applications in structural engineering, construction and building, biomedical devices, food packaging, and environmental remediation. Their biodegradability further broadens their appeal, making them valuable in sectors including water purification and treatment, healthcare, automotive, cosmetics, food, textiles, and electronics (Joseph et al. 2020; Fura et al. 2024). Table 15.1 details examples of bionanocomposite materials and their areas of application.

15.3 Types of Bionanocomposites

Bionanocomposites can be classified based on their constituents, synthesis methods, and applications. Figure 15.2 illustrates the four main types of bionanocomposites, namely: (1) biopolymer-based hybrid bionanocomposites, (2) natural fiber-reinforced composites, (3) bio-hybrid nanocomposites, and (4) natural filler-reinforced biocomposites. These categories are briefly discussed as follows:

15.3.1 Biopolymer-Based Hybrid Bionanocomposites

These types of bionanocomposites comprise natural and synthetic biopolymers, such as chitosan, starch, gelatin, and polylactic acid (PLA), integrated with various nanofillers, including metallic nanoparticles (e.g., ZnO-NPs, AgNPs), graphene derivatives, and others. Hybrid bionanocomposites, due to the synergistic effect of natural fibers and nanoparticles, exhibit significant improvement in mechanical properties, such as tensile, compressive, and flexural strength. For instance, a study involving the use of PLA and epoxy resin-based matrix reinforced with graphene nanofillers and multiwalled CNTs showed a significant increase in tensile and flexural

Table 15.1 Bionanocomposite materials and their applications

Biopolymer	Nanofiller	Remarks	Application	Source
Hydroxypropyl methylcellulose	AgNPs	Bactericidal films with excellent barrier and mechanical properties exhibit high tensile strength	Antimicrobial coatings in the food packaging industry	Khater et al. (2023)
Polylactic acid	Nano-silica	Significant enhancement in hybrid films of gas barrier properties	Elongating food products' shelf life by maintaining quality and preventing spoilage	Bang and Kim (2012)
Chitosan	Reduced graphene oxide and cerium oxide	Excellent biocompatibility and antimicrobial efficacy against pathogens like *Staphylococcus aureus* and *Escherichia coli*	Drug delivery systems, antimicrobial agents, biosensors, wound dressings, and energy storage solutions	Sanmugam et al. (2023)
Chitosan	Graphene oxide and superparamagnetic iron oxide nanoparticles	Theragnostic nanosystems that enhance drug delivery and imaging properties	Biomedical applications include drug delivery and diagnostic imaging with real-time treatment tracking	Aliabadi et al. (2017)
Chitosan and sodium alginate	Graphene oxide	Nanocomposites exhibit significant pH-dependent drug release, enhanced solubility, and reduced non-specific protein adsorption	Targeted delivery of anti-cancer drugs in cancer therapy	Lei et al. (2016)
Agro-waste residues	ZnO-NPs AgNPs	ZnO nanoparticles enhance mechanical properties and antimicrobial activity, while AgNPs provide potent antimicrobial effects	Active food packaging, tissue engineering, and wound dressings, leveraging the distinct properties of biobased composites	Ortega et al. (2021)

(continued)

Table 15.1 (continued)

Biopolymer	Nanofiller	Remarks	Application	Source
Chitosan	Bio-silica	Nanocomposites demonstrate effective adsorption of anionic dyes from aqueous solutions	Wastewater treatment, particularly in removing toxic organic dyes from industrial effluents	Darvishi Cheshmeh Soltani et al. (2013)
Natural rubber	Graphene and zinc oxide nanoparticles	Significant increase in physical crosslink density, tear strength, and energy dissipation	Rubber additives and products require enhanced mechanical properties and durability	Lin et al. (2015)
Acrylated epoxidized soybean oil	Cellulose fibers	Structures exhibit improved mechanical performance and address moisture issues in recycled paper, enhancing the curing process	Roof structures due to their strength and stiffness	Dweib et al. (2006)
Amyloidogenic protein alpha-synuclein	AuNPs	Enables production of nanoparticle organic memory field-effect transistors (MOSFETs) that mimic biological synaptic behavior and support programmable synaptic functions	Advanced electronic devices, particularly neuromorphic computing and memory storage technologies	Kim et al. (2016)

strength, with SEM micrographs revealing homogenous dispersion of the nanofillers in the epoxy-PLA matrix (Nimbagal et al. 2021).

Natural reinforcing materials like agro-waste residues have been shown to result in the production of lightweight composites owing to their reduced density compared to their synthetic counterparts (Ben and Olubambi 2024a; Ben et al. 2023). This makes the hybrid bionanocomposites suitable for producing lightweight concrete and building structures like door and window frames, roofing materials, and door handles, with improved mechanical and tribological performance. Additionally, these composites are particularly useful in civil engineering for repairing and retrofitting structures, including developing insulation materials. Despite the numerous benefits associated with using hybrid bionanocomposites, limitations such as elevated moisture absorption, reduced mechanical performance compared to synthetic fibers, and compatibility with polymer matrices should be addressed. Factors such as (a)

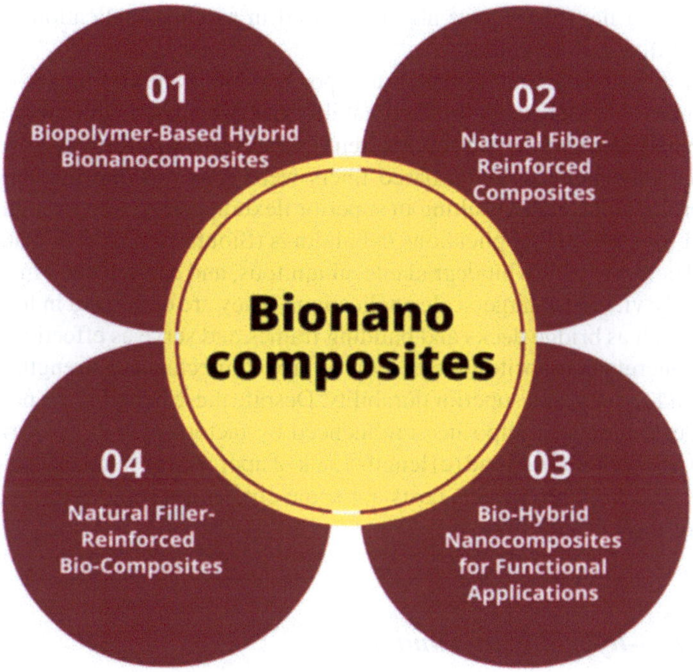

Fig. 15.2 Types of bionanocomposites

fiber orientation, (b) weight fraction, and (c) stacking sequence must also be considered when implementing hybrid bionanocomposites to optimize their properties for improved performance effectively.

15.3.2 Natural Fiber-Reinforced Composites

This category of bionanocomposites encompasses materials that combine wood, natural biofibers, or agro-waste byproducts such as plant-based and animal-based fibers, with polymer matrices to create sustainable, renewable, and recyclable composites. Natural fibers can be grouped into four, namely (1) stalk fibers, including bamboo, straw, and wood; (2) bast fibers, including flax, jute, hemp, ramie, and kenaf; (3) leaf fibers, including abaca, pineapple, agave, and sisal; and (4) seed fibers including kapok, coir, and cotton. These composites are engineered for superior performance in terms of durability, dimensional stability, and, in some cases, improved fire retardancy, making them suitable for structural and transportation applications. For example, bionanocomposites made from agro-waste residue like coconut husk have been developed into insulation panels with excellent thermal properties and mold resistance essential for energy-efficient buildings (Shamindi et al.

2024). Jute and flax fibers have also been used in roofing applications with good thermal stability and durability.

Natural fiber-reinforced polymer composites use natural fibers to reinforce polymer matrices, addressing the flammability risk associated with these materials. Natural reinforcers include flax, abaca, hemp, kenaf, rice straw, bamboo, and cotton stalk fibers. For instance, hemp-based fibers were found to improve the mechanical properties of mortars, resulting in superior flexural and tensile strength, making them ideal for structural applications in buildings (Siouta et al. 2024). Natural fiber-reinforced composites are biodegradable, ubiquitous, and eco-friendly, making them economically viable for large-scale applications. They are employed in load-bearing structures such as bridge decks and building frames and serve as effective reinforcements in concrete composites due to their improved mechanical strength, enhanced thermal conductivity, and superior durability. Despite these benefits, the performance of this type of bionanocomposite is influenced by factors such as (a) surface treatments, (b) fiber orientation, and (c) length. Lack of appropriate standardization testing techniques and high production costs are some challenges limiting the widespread adoption of these materials.

15.3.3 Bio-hybrid Nanocomposites

Bio-hybrid nanocomposites represent an emerging category of bionanocomposites that integrate biological components or natural fibers with nanofillers to create stretchable, flexible, and responsive systems for structural engineering, building, and construction applications. These materials are ideally suited for green buildings and sustainable construction practices, serving as energy storage materials, components in lightweight construction, and solutions for repairing or retrofitting existing structures. They offer distinct advantages, including improved durability, biodegradability, and moisture resistance, which are essential for long-term structural applications. For example, a study demonstrated that *Moringa oleifera* reinforced with epoxy nanofillers showed a significant increase in thermal stability, along with improvements in bending and tensile moduli by 38.2% and 31.4%, respectively (Ayrilmis et al. 2018). Similarly, the incorporation of graphene nanoplatelets with biocarbon enhanced both the impact strength and tensile modulus by approximately 120%, while nanoclay fillers reinforced in a kenaf-PLA matrix led to a 19.1% and 98.7% increase in tensile and impact strength, respectively.

Moreover, bio-hybrid nanocomposites have advanced sensing capabilities; they can detect large and subtle movements, making them highly promising for smart building applications. For instance, strain sensors developed from polyacrylamide and oxidized multiwalled CNTs (a conductive nanocomposite hydrogel) achieved a 90% recovery efficiency, 0.71 MPa tensile strength, 700% sensing range, and excellent sensing and stretching capabilities (Sun et al. 2020). In another study, Zhang et al. developed lepidolite-inspired sensors incorporating multiwalled CNTs, graphene, neodymium iron boron, and polydimethylsiloxane, which detected tensile strain and human movements with a linearity of 0.99 (Zhang et al. 2024). Additionally, Omar et al. developed a dual-sensing nanocomposite based on a high-frequency sensing piezoelectric polymer combined with carbon black, capable of detecting both slow and fast pressure fluctuations (Omar et al. 2024). By leveraging advanced sensing technologies to provide stable and repeatable electrical signals, bio-hybrid nanocomposites are auspicious materials for monitoring vital signs and various movements in a smart building.

15.3.4 Natural Filler-Reinforced Biocomposites

This type of bionanocomposites incorporates natural organic and inorganic fillers. Organic fillers include agro-waste residues from plant and animal fibers such as chitosan, wheat gluten, and bioceramics, while inorganic fillers include TiO_2, $CaCO_3$, ZnO, clay nanofillers (montmorillonite, kaolinite), nano-silica, graphene, CNTs, zirconia, magnesia, and alumina. These fillers enhance mechanical and thermal properties while maintaining biodegradability, resulting in lightweight, cost-effective, and environmentally friendly materials suitable for structural engineering applications. For instance, biowaste fillers from agro-waste residues such as coconut, pineapple, banana, have been used to produce ash-based fillers reinforced with epoxy matrix showed good adhesion properties with enhanced mechanical properties—flexural (22.11%) and impact (21.77%) (Sumesh et al. 2021). The addition of natural fillers to metal matrix composites has also been observed to improve the surface finish of these biocomposites mitigating the formation of coarse structure (Ben and Olubambi 2024b). Incorporating inorganic fillers into polymer composites enhances thermal stability, improves tribological performance, fosters better interfacial bonding, and boosts mechanical properties. Factors affecting the influence of fillers on biocomposite properties include (a) filler shape, (b) filler size, (c) filler surface area, (d) filler aspect ratio, and (e) filler dispersion.

Fillers with smaller particle sizes exhibit a high surface area per unit weight, resulting in enhanced interfacial bonding between the filler and the polymer. The surface area-to-volume ratio of fillers plays a critical role in enhancing composite properties, with higher ratios leading to more significant improvements, while reductions in filler size further amplify these benefits. Fillers with high aspect ratios and low widths—such as nanofillers—exhibit improved mechanical, physical, thermal, and optical properties in composite materials. The homogenous dispersion of nanofillers within the matrix results in reduced water adsorption and superior interfacial adhesion in the fiber-reinforced biocomposites allowing for effective stress transfer between fillers and the matrix during structural loading applications. Filler addition also mitigates the formation of pores or voids within the matrix leading to enhanced composite stiffness (Ben and Olubambi, 2025c), however, this must be done with caution, as it has equally been observed that when the nanofillers are increased beyond certain limits in a matrix, agglomeration occurs as micro-voids between the fillers expands, thus reducing the bond strength between the matrix and fibers. Natural fillers are finding applications in architectural moldings, interior roofs, doors, rail and decking products, and head liners.

15.4 Bionanocomposites Manufacturing Methods

Bionanocomposites are manufactured through various innovative methods that leverage biopolymers and nanomaterials. These approaches often utilize environmentally friendly techniques, such as green chemistry, to create composites with enhanced properties. Key green manufacturing methods include (a) sonochemical method, (b) in-situ polymerization, (c) layer-by-layer assembly, (d) electrospinning, and (e) melt compounding and sol–gel.

15.4.1 Sonochemical Method

Sonochemical method is an innovative green manufacturing technique that utilizes ultrasonic radiation, typically between 20 kHz and 10 MHz, to induce redox polymerization and modify nanostructures through a process known as acoustic cavitation. In this process, ultrasonic irradiation creates microbubbles in a liquid medium that collapse violently, forming localized hotspots with temperatures up to 5000 K and pressures around 1000 atm for transient moments. These extreme, transient conditions facilitate rapid chemical reactions and the formation of nanostructured materials. This method of producing bionanocomposites often employs natural extracts such as Aloe vera gel (a reducing agent) and pomegranate extract (a stabilizing agent) (Yan et al. 2022), to mitigate the presence of toxic chemicals in the production process. Studies have reported using ultrasonic techniques to extract nanofibers from bamboo, wood, flax fibers, and wheat straws and produce dense multilayered polymer particles with enhanced thermal and mechanical properties (Cai et al. 2022).

Figure 15.3 illustrates the block diagram of a sonochemical method. At the heart of the system is an ultrasonic generator that supplies electrical energy converted into mechanical vibrations by a piezoelectric transducer. These vibrations are then amplified by a titanium probe or horn, with its tip immersed directly in the sample. The probe transmits ultrasonic waves and vibrational energy into the sample solution, which is held in plastic or glass beakers (typically ranging from 50 to 500 mL) designed to withstand the ultrasonic energy. A circulatory water bath cooling system maintains a controlled sonication temperature. To minimize high-frequency noise emissions and protect the surrounding environment from potential ultrasonic damage, the sample holder and probe are enclosed within an acoustic cabinet. A pulse-driven potentiostat provides precise control over sonication parameters such as pulse mode, duration, and amplitude, ensuring that over-sonication and sample degradation are avoided. Regular probe cleaning is essential to prevent sample contamination as part of preventive maintenance.

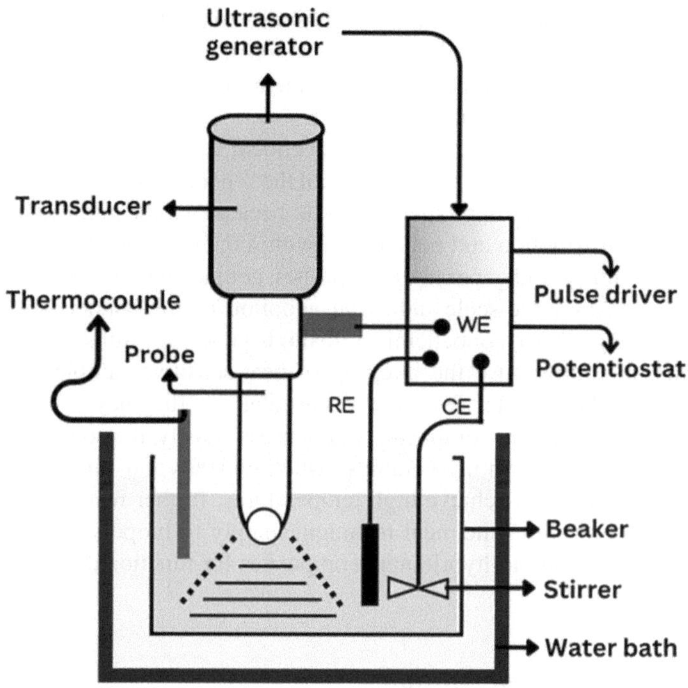

Fig. 15.3 Schematic diagram demonstrating the working principle of an ultrasonic probe technique

Sonochemical methods have been successfully utilized to produce various bionanocomposites. For example, Hatami et al. demonstrated the production of a chitosan-based composite in which cerium oxide nanoparticles, modified with polydopamine, were homogenously dispersed within the polymer matrix, enhancing the material's overall properties (Hatami et al. 2024). Similarly, sonication has been utilized to disperse nanoclay into cellulose matrices, yielding biocomposites with significantly enhanced water barrier and mechanical attributes that are ideal for sustainable structural solutions (Jali et al. 2024). Additionally, bio-hybrid nanocomposites incorporating carbon-based nanomaterials for flexible strain sensors have been synthesized using this method, resulting in devices with high sensitivity and rapid response suitable for wearable applications (Díaz-Mena et al. 2024).

Sonochemical methods of producing bionanocomposites have some advantages, which include (a) enhanced dispersion of nanofillers within the biopolymer matrix for efficient composite homogeneity, (b) precise control over reaction conditions for rapid synthesis and fine-tuning of material properties to meet specific application requirements, (c) high energy efficiency owing through reduction in ambient pressure and temperature, (d) elimination of harmful chemicals to make the method safer and eco-friendly, (e) a cost-effective and sustainable alternative to traditional bionanocomposite fabrication techniques, (f) ability to produce polymers with well-defined structural properties and molecular weights, and (g) versatility in synthesizing a wide range of nanomaterials such as metal oxides, carbon-based materials, semiconductors, and metallic nanoparticles.

Despite these promising benefits, sonochemical methods have shown several shortcomings. For instance, the effectiveness of the sonochemical method is primarily restricted to nanoscale reinforcements, this is because larger particles can disrupt uniform energy distribution and penetration within the reaction medium. Secondly, the high cost of ultrasound generators and probes, coupled with the need for frequent maintenance, makes large-scale industrial adoption of the sonochemical method highly challenging. Third, sonochemical methods generate a substantial amount of heat, which can compromise the integrity of heat-sensitive samples. In addition, sonochemical methods suffer from low encapsulation efficiency, which may limit their applicability in certain composite formulations. Lastly, the sonochemical technique is constrained by working volumes, which typically must not exceed 50 mL, along with its inability to achieve high temperatures, further restricting scalability for mass production. Sonochemical techniques apply to biopolymer-based hybrid bionanocomposites and bio-hybrid nanocomposites for functional applications.

15.4.2 In-Situ Polymerization

In-situ polymerization is a green manufacturing procedure that involves chemical encapsulation, with multiple polymerization steps carried out in the presence of nanofillers. In this method, monomers and pre-dispersed nanofillers are mixed under controlled conditions, and polymerization is initiated, often via free radical mechanisms or other chemical initiators (Fig. 15.4). As polymerization proceeds, the growing polymer chains form a matrix around the nanofillers, resulting in enhanced interfacial bonding and a uniform distribution of the reinforcement throughout the composite.

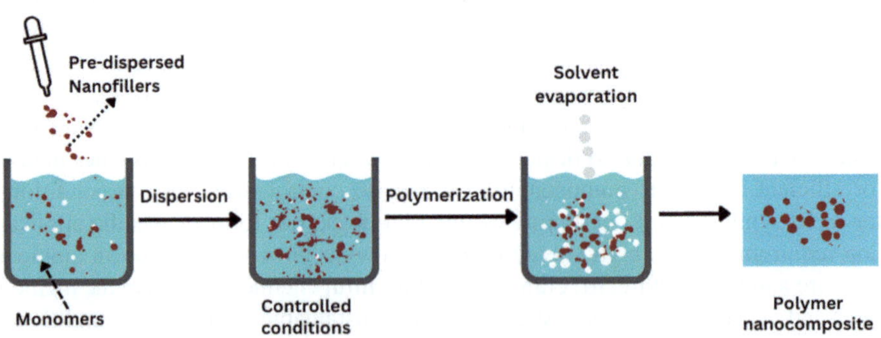

Fig. 15.4 Schematic diagram illustrating the in-situ polymerization technique

The process typically begins by creating a homogeneous, low-viscosity admixture of nanofillers in a solution. An initiator is introduced, and the mixture is then heated—using light or a furnace—to initiate the polymerization process. This leads to the creation of a cured polymer nanocomposite, in which the polymer chains are networked through covalent bonds, forming a perpetual secondary phase within the matrix. The thermosetting process, often facilitated by a hardener or catalyst, further promotes crosslinking, ultimately yielding a hybrid material that integrates the nanoparticles with the polymer molecules (Masoumi et al. 2022).

In-situ polymerization has been widely utilized in fabricating bionanocomposites for applications in structural engineering, building, and construction. They can enable the creation of multifunctional composites that exhibit properties like shape memory, making them suitable for integration into smart building applications. This method has been found suitable for enhancing the properties of cementitious composites. For instance, the addition of monomers into cement particulates has been shown by several studies to significantly enhance properties such as compressive, tensile, and flexural strengths, making them suitable for structural load-bearing applications. The continuous interpenetrating crosslink accounts for the superior flexural strength properties, further driving crack initiation and mitigating crack progression to a visible scale in structural elements. An in-situ polymeric layer created using carbon black (1–3%) and photoinitiator (1–5%) showed good permeability and enhanced tensile strength for use as a temporary polymeric barrier in construction sites (Li et al. 2021). Additionally, polymer matrix reinforced with CNT nanofillers using in-situ polymerization demonstrated good molecular adhesion resulting in remarkable tensile strength (776 MPa) and elastic modulus (31 GPa) (Chazot et al. 2020). Other research has combined in-situ polymerization with sonication to produce high-quality nanostructures such as nanorods, nanofibers, and capacitor electrodes (Yang et al. 2023; Manopriya and Hareesh 2021). Moreover, chitosan-based composites synthesized in-situ with graphene oxide have demonstrated markedly enhanced mechanical strength and antibacterial properties, while the in-situ polymerization of polylactic acid with nanoclay has yielded composites with superior gas barrier performance (Pradhan et al. 2025).

The advantages of in-situ polymerization include excellent nanofiller dispersion within the polymer matrix, precise control over polymerization conditions, excellent nanofiller interfacial adhesion, and promotion of thermodynamic compatibility at the matrix-reinforcement interface. This method also yields reinforcement surfaces that are free of contamination, thereby enhancing the overall properties of the composite. Shortcomings of the in-situ polymerization method include high temperatures during polymerization, resulting in the decomposition of polymer chains, ease of agglomeration, and the need for strict control over reaction parameters to avoid issues such as residual monomer content. Additionally, scaling up for industrial production can be difficult, and the complexity of the reaction system may limit the range of compatible monomers and nanofillers.

In-situ polymerization is particularly applicable to biopolymer-based hybrid nanocomposites and bio-hybrid nanocomposites for functional applications, where homogeneous dispersion and strong interfacial bonding at the nanoscale are critical. While it can also be employed for natural filler-reinforced biocomposites when the fillers are in nanoparticulate form, it is less suited to natural fiber-reinforced composites, which typically involve larger, continuous fibers.

15.4.3 Layer-By-Layer Assembly

Layer-by-layer (LBL) assembly is an established green manufacturing method for producing bionanocomposites that rely on the successive deposition of oppositely charged nanomaterials onto a substratum through various intermolecular interactions. This technique facilitates the addition of various chemicals by harnessing electrostatic forces, hydrogen bonding, and other attractive interactions, allowing for precise control of the structure and composition of the resulting composite. Additionally, it offers exceptional restraint over critical film properties such as thickness, mass, and porosity while also allowing for the integration of numerous components within the multilayer construct (Santos et al. 2024).

The basic formation mechanism of the LBL assembly process is illustrated in Fig. 15.5 and involves alternately exposing a negatively charged substrate to solutions containing positively and negatively charged species, such as polyelectrolytes and nanoparticles. In step (1), the substrate adsorbs a polycation, followed by rinsing with deionized water (step 2) to remove loosely bound material. It is then immersed in a polyanion solution (step 3) and rinsed again with deionized water (step 4). These steps are repeated to build multilayered films with controlled thickness and composition. By precisely regulating the number of deposition cycles, the method allows fine-tuning of the film's structure, ensuring uniformity, strong adhesion, and tailored surface functionalities for various applications. The LBL process is influenced by several factors, including cross-linkers, coating duration, ionic strength, and pH of the coating solution, and the choice of substrate and multilayer materials. The LBL method is desirable for applications that demand specific surface functionalities and precise material compositions.

The LBL technique has had a transformative impact on shape memory materials, radiation protection composites, advanced dielectric composites, and vibration-damping composites. Multilayered polymer composites produced via the LBL assembly technique can regulate micro-crack propagation, stress transfer, and stress distribution, with potential application in electromagnetic shielding, interfacial heat dissipation, optical films, and planar coatings. LBL-assembled films are employed in sensor technologies, where their tailored multilayer structures contribute to high sensitivity and selectivity. A study by Bonderer et al. demonstrated that shell-like ceramic/chitosan multilayered composites, prepared using high-strength submicron $CaCO_3$ and aluminum oxide templates along with chitosan, exhibited superior toughness and strength (Bonderer et al. 2008). Additionally, a single-layer and double-layer membrane structure was employed in the synthesis of modified asphalts using modified polyacrylonitrile (PAN) fibers. The integration of these PAN fibers into

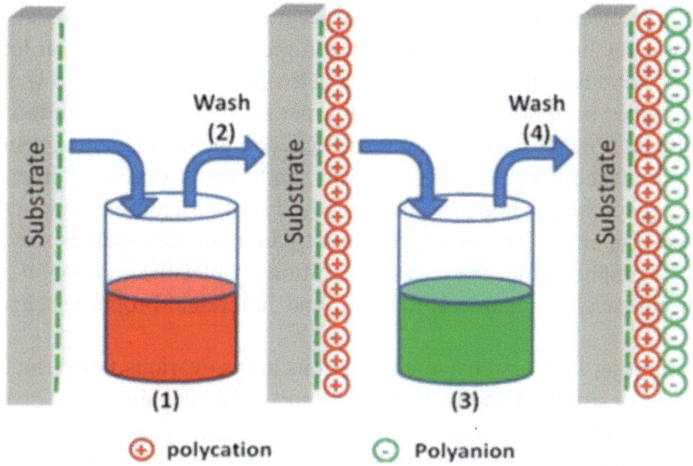

Fig. 15.5 Schematic diagram illustrating the LBL assembly deposition process (Qi et al. 2015). Reproduced with permission from Springer Nature

the asphalt mix during production altered the surface characteristics, significantly enhancing adhesion properties. This modification renders the asphalt more suitable for road construction and pavement materials by improving its durability and resistance to environmental stress (Ren et al. 2022).

Among its many advantages, LBL assembly is simple and cost-effective, offering a high degree of control over film thickness, composition, and morphology. It is compatible with a large range of nanofillers and bioactive compounds, enabling the design of composites with tailored functionalities. In addition, the method can be added to diverse substrates, including complex and flexible surfaces, further expanding its utility in advanced material applications. Despite these advantages, the LBL assembly technique has some shortcomings, including low efficiency, long preparation times, difficulty in precisely controlling the thickness of each layer in the final product, and the tendency of the polymer to approach a near-melting state.

15.4.4 Electrospinning

Electrospinning is a versatile green manufacturing technique for producing bionanocomposites, particularly valued for its ability to integrate eco-friendly materials, such as myrrh, to generate continuous nanofibers with high surface area-to-volume ratios. This process reduces dependence on toxic solvents, promoting environmental sustainability. The principle of electrospinning revolves around the interplay between electrostatic forces and surface tension. When a high voltage is applied to a polymer solution or melt containing dispersed nanofillers, the electric field overcomes the surface tension at the liquid surface, forming a Taylor cone from which a charged jet is ejected (Fig. 15.6). As the jet travels toward a grounded collector, it undergoes stretching and thinning, while rapid solvent evaporation leads to the deposition of solid nanofibers. The green electrospinning process allows for precise control over fiber morphology by adjusting parameters such as solution viscosity, applied voltage, flow rate, and the distance between the needle and collector. These parameters influence fiber diameter, porosity, and overall structural characteristics, making electrospinning a highly tunable method for fabricating bionanocomposites with tailored properties. The incorporation of nanomaterials such as nanoparticles, nanotubes, or graphene into the polymer solution enables the creation of composite fibers with enhanced mechanical, thermal, and functional properties (Princy et al. 2024).

Electrospinning finds extensive applications across various fields, including the building and construction industry. For instance, the technique has been used to produce fibers with various morphologies, such as hollow, porous, and core–shell structures, as well as three-dimensional structures with controlled fiber density, spacing, and layer thickness, all of which are crucial for applications in filtration and insulation. Superhydrophobic surfaces needed for water-repellent coatings have also been produced using electrospun fibers. Additionally, incorporating electrospun cellulose acetate polymer microfibers into cementitious materials used in construction applications has significantly enhanced compressive strength, with visual inspection confirming effective fiber blending (Daniel et al. 2025). Overall, electrospinning is one of the most widely used techniques for producing polymer nanofibers, yielding

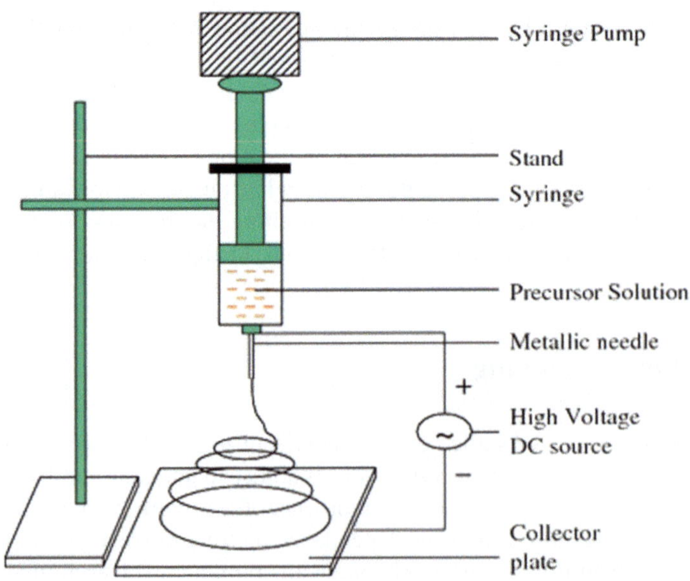

Fig. 15.6 Schematic representation of an electrospinning process (Panda and Ramakrishna 2007). Reproduced with permission from Springer Nature

nanostructured polymers with excellent mechanical and structural properties that make them highly suitable for building applications.

Among the four categories of bionanocomposites, electrospinning is particularly applicable to biopolymer-based hybrid bionanocomposites and bio-hybrid nanocomposites for functional applications. In these systems, the technique's ability to produce nanofibers with high control over fiber diameter, porosity, and nanofiller dispersion is crucial for enhancing mechanical, thermal, and electrical properties. While electrospinning can be adapted for natural filler-reinforced biocomposites when the fillers are in nanoparticulate form, it is less suited to natural fiber-reinforced composites that rely on larger, continuous fibers. Electrospinning is a powerful tool for fabricating advanced bionanocomposites tailored for high-performance applications across various industries.

The advantages of electrospinning are numerous. It offers a straightforward, cost-effective approach to fabricating nanofibers with controlled morphology and high uniformity. The method allows for directly incorporating nanofillers into the polymer matrix, ensuring a homogeneous distribution that significantly enhances the composite's overall properties. However, electrospinning also presents challenges. The process is susceptible to various parameters, and maintaining consistent fiber production can be difficult. Achieving high loadings of nanofillers without agglomeration remains a significant hurdle, and the reliance on organic solvents may pose environmental and safety concerns.

15.4.5 Melt Compounding and Sol–Gel

Melt compounding is a solvent-free green manufacturing technique commonly employed in producing bionanocomposites. In this technique, the biopolymer is heated to its molten state and mixed with nanofillers using high-shear equipment, such as twin-screw extruders or internal mixers. The vigorous mechanical mixing promotes the dispersion of nanofillers throughout the polymer matrix, leading to composites with excellent mechanical, thermal, and barrier performance. Figure 15.7 illustrates a simple melt compounding schematic used in manufacturing polymer/graphene nanocomposites. Advanced melt compounding structures incorporate adjustable mixing screws and chambers, spiral sitting blades, and arc-shaped scrapers to optimize the mixing process, further improving the material blending efficiency and effectiveness.

Melt compounding is an eco-friendly manufacturing method that utilizes renewable resources, such as cellulose nanocrystals derived from waste streams, thereby reducing reliance on fossil fuels and mitigating environmental impacts. Studies have demonstrated that melt compounding can significantly enhance the mechanical and structural properties of composites. For example, this technique has been employed to integrate kaolin fillers into polyethylene matrices, with the addition of vegetable lipids to improve miscibility, resulting in a marked increase in the

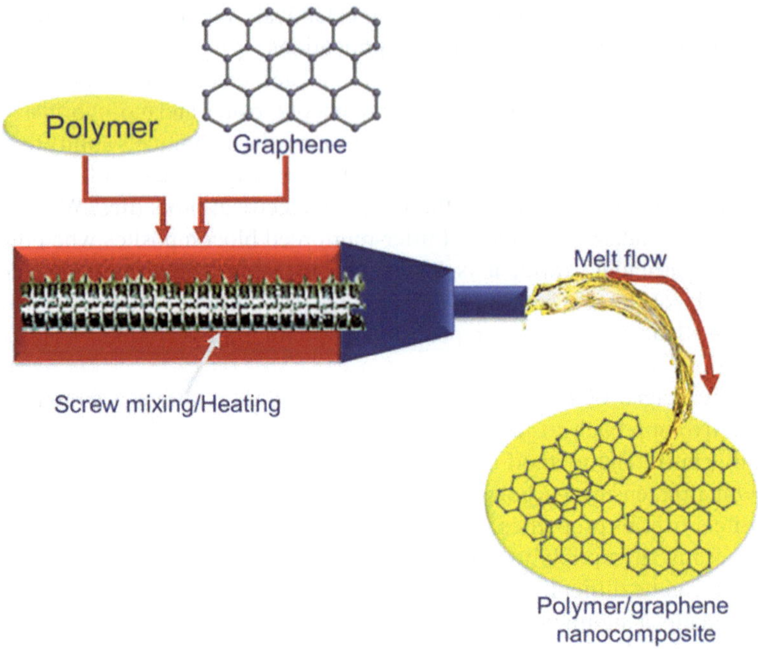

Fig. 15.7 Schematic diagram of a basic melt compounding setup (Kausar et al. 2023). Reproduced with permission from MPDI under the CC-BY license

composite's compressive strength (Sessini et al. 2021). Another typical application of melt compounding is in the development of fire-retardant cementitious composites. To address the weakening effects of high temperatures on concrete, a bio-inspired composite was prepared using low-density polyethylene as the base material and reinforced with melamine, pentaerythritol, and ammonium polyphosphate. Figure 15.8 presents optical micrographs of the internal pore structure of this innovative composite after exposure to high melting and charring temperatures. The composite exhibited excellent adhesive attraction and high heat stability, forming a steady intumescent charring layer upon flame exposure. This property makes the material particularly effective in enhancing the fire resistance of structural components (Zhang et al. 2019).

Advantages of the melt compounding technique include improved interfacial adhesion, which leads to efficient load distribution and improved mechanical properties such as tensile, flexural, and compressive strength of the composites. This technique provides a sustainable approach by mitigating plastic pollution and reducing the depletion of fossil-based resources through the use of bio-sourced alternatives like cellulose fibers. Additionally, melt compounding enhances the thermal stability of composites and improves their rheological properties, making processing easier and tailoring the material characteristics to meet specific construction needs. Its

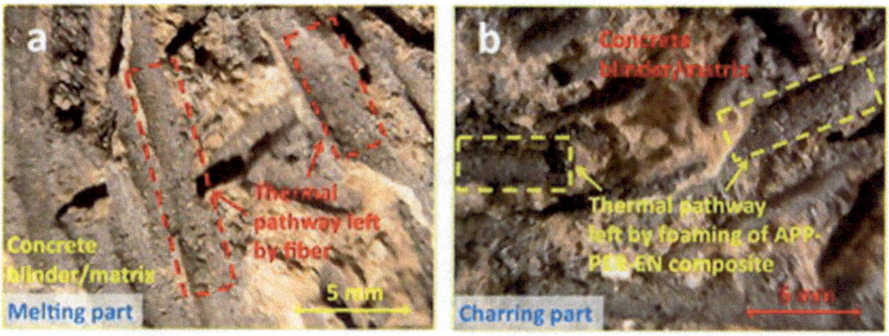

Fig. 15.8 Morphological properties showing internal pore structure of the **a** melting and **b** charring parts of a fire retardant composite (Zhang et al. 2019). Reproduced with permission from Elsevier

versatility allows for easy production and customization of composites with specific properties suited for various applications in the construction and building sectors.

Despite its advantages, melt compounding also presents several challenges. The high processing temperatures required can lead to the thermal degradation of sensitive biopolymers, and achieving uniform dispersion of nanofillers is often problematic, as the fillers tend to agglomerate. Scaling up the technique for large-scale construction applications can be challenging due to the high cost of specialized machinery and the significant energy input required. Furthermore, environmental and health concerns arise from the potential release of toxic chemicals during melt processing, particularly when these additives are used in large quantities in construction. The absence of standardized guidelines for this technique also raises serious quality control and assurance concerns.

The sol–gel method offers an alternative approach based on chemical processing rather than mechanical mixing. Figure 15.9 illustrates a simple experimental sol–gel setup that can be used to produce diverse and cost-effective materials such as xerogel, nanoparticles, glass, and ceramics. This technique prepares a colloidal suspension (or sol) of precursor materials, typically using metal alkoxides or other inorganic precursors, which undergo controlled hydrolysis and condensation reactions to form a continuous three-dimensional network (or gel). When applied to bionanocomposites, the sol–gel process allows inorganic nanofillers, such as TiO_2, SiO_2, or other metal oxides, to be incorporated into a biopolymer matrix at relatively low temperatures. This process provides exceptional control over the chemical composition and microstructure of the resulting composite material. Sol–gel-derived bionanocomposites have found applications in areas such as protective coatings, optical materials, sensors, and biomedical devices (Mondal et al. 2020; Shaban et al. 2023; Mathew Simon et al. 2021), where properties such as high thermal stability, excellent adhesion, and tailored porosity are desirable. Research has shown that sol–gel processing can effectively integrate TiO_2 nanoparticles into chitosan matrices (Anaya-Esparza et al. 2020), resulting in composites with improved photocatalytic and antibacterial properties.

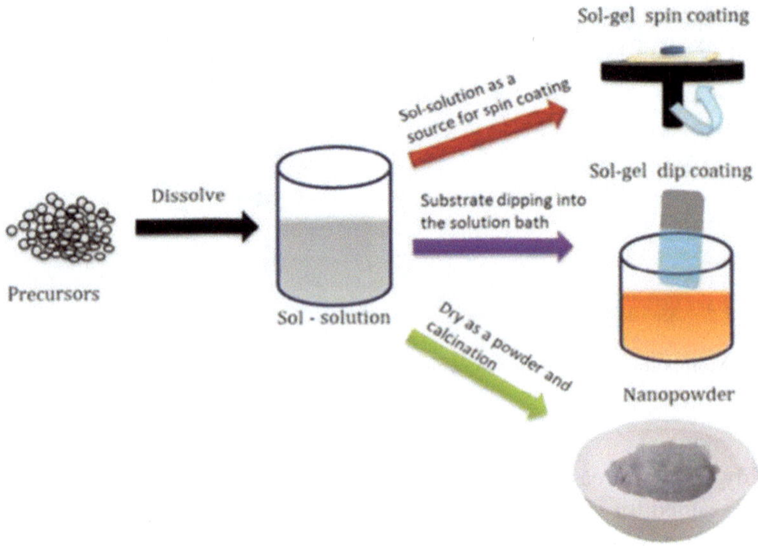

Fig. 15.9 Schematic diagram of a simple sol–gel process for the production of nanocomposites (Vikraman et al. 2017; Nwanna et al. 2020). Reproduced with permission from Springer Nature

The sol–gel process offers several advantages, including low-temperature chemistry that minimizes defects in the resulting materials. Its high surface-to-volume ratio makes it especially suitable for catalytic applications, while also yielding products of high quality and purity with excellent physical properties. Despite these advantages, the sol–gel method has some notable shortcomings. For instance, the method is time-consuming due to extended reaction and drying periods. It also suffers from shrinkage and cracking during the transition from sol to gel. Moreover, scaling up the sol–gel process for industrial applications remain a significant challenge.

When considering the four types of bionanocomposites, melt compounding is most applicable to biopolymer-based hybrid bionanocomposites and natural filler-reinforced bio-composites, where the reinforcements are in nanoscale or finely dispersed particulate form. Its high-shear mixing is particularly effective for these systems. However, it may also be adapted for specific natural fiber-reinforced composites if the fibers are thermally stable and can withstand the processing conditions. Conversely, the sol–gel method is ideally suited for biopolymer-based hybrid bionanocomposites and natural filler-reinforced bio-composites incorporating inorganic nanofillers. The low-temperature nature of the sol–gel method preserves the integrity of the biopolymer matrix while enabling the formation of highly uniform and well-integrated nanostructured networks. While both methods offer unique advantages, their selection is based on the actual requirements of the composite system and the intended application.

15.5 Innovative Applications of Bionanocomposites in Structural Engineering and Construction

Bionanocomposites have emerged as promising materials in structural engineering, building, and construction due to their unique properties and potential to enhance performance. By combining biological components with nanomaterials, these advanced composites offer improved mechanical, thermal, and chemical properties that significantly boost building performance and sustainability. They reinforce core structural elements, enhance insulation systems, refine surface finishes, and enable structures to self-repair. Figure 15.10 illustrates several innovative applications of bionanocomposites in this field, including structural components, insulation and thermal regulation, coatings and finishes, self-healing materials, and innovative structural monitoring systems. This section reviews various case studies investigating the use of bionanocomposites across these diverse applications.

15.5.1 Structural Components

Structural components such as beams, columns, and panels form the backbone of any building, bearing loads, ensuring stability, and maintaining overall structural integrity. Recent advances in bionanocomposites have introduced innovative alternatives to traditional construction materials by merging the sustainability of bio-polymers with the improved mechanical performance of nanomaterials. These

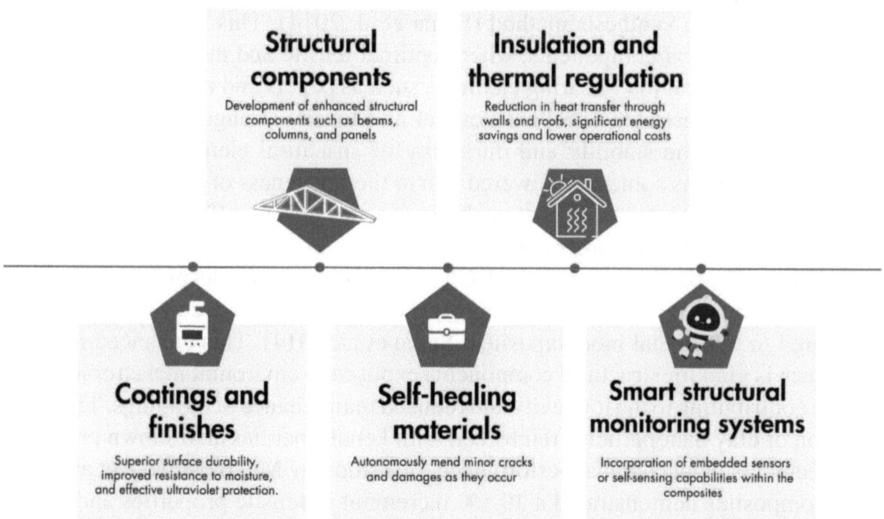

Fig. 15.10 Bionanocomposite applications in structural engineering, building, and construction

composites boast an excellent strength-to-weight ratio, increased flexibility, and improved durability, contributing to more efficient and resilient structural elements.

Incorporating nanofillers, such as CNTs, graphene, or nanoclay, into biopolymer matrices results in composite materials that withstand high loads and reduce overall weight. This capability is particularly advantageous in seismic zones, where lighter structures can lead to reduced inertial forces during an earthquake (Yenidogan 2021). Moreover, the superior mechanical properties of these composites allow for longer spans and thinner structural elements, paving the way for innovative architectural designs and potentially lowering material consumption. Several case studies have demonstrated improvements in load-bearing capacity using bionanocomposites, offering insights into their relative strengths and applications.

One such notable case study is the fabrication of a bionanocomposite reinforced with reduced graphene oxide, which significantly enhanced its mechanical performance by increasing the flexural strength by 43.97%, the compression strength by 58.08%, and the tensile strength by 39.84% compared to neat epoxy (Dubey et al. 2024). These improvements are particularly relevant for structural applications, as they translate into components that are not only stronger but also more resilient under various loading conditions. Additionally, the composite exhibited improved thermal stability and a 23.38% reduction in combustion rate, which are critical attributes for materials used in green building structures that must resist fire and maintain integrity under elevated temperatures.

Similar research on polylactic acid matrix reinforced with nanoclay also demonstrated significant improvements in tensile strength and stiffness, suggesting that such composites could serve as viable alternatives to conventional reinforced concrete or steel reinforcements. For instance, Sajna et al. fabricated green bionanocomposites by blending PLA with banana fiber and reinforcing the blend with nanoclay using the melt compounding synthesis method (Sajna et al. 2014). This study is particularly relevant to structural components, where optimal tensile and flexural properties are critical for structural load-bearing elements such as panels and beams. The resulting composites demonstrated excellent flexural and tensile strength, which are essential for ensuring the stability and durability of structural elements. Although the impact strength was somewhat lowered due to the brittleness of the PLA matrix, the overall performance suggests that such composites can be effectively used where high stiffness and rigidity are prioritized.

Additionally, the incorporation of Closite 30B nanoclay, despite causing some thermal degradation, significantly improved the water resistance of the composites compared to traditional biocomposites (Sajna et al. 2014). This enhanced moisture resistance is vital for structural components exposed to environmental stressors, ultimately contributing to the longevity and reduced maintenance of buildings. The incorporation of clay nanoparticles reinforced with kenaf fiber has also shown promising enhancements in mechanical performance. In a study by Moyo (Mondal et al. 2020), such composites demonstrated a 19.1% increment in tensile properties and a 9.8% improvement in flexural performance, while impact resistance was markedly boosted, with energy absorption capability and impact strength rising by 109% and 98.7%, respectively. These improvements are crucial for structural elements like beams, columns, and panels, where enhanced impact resistance and overall durability are imperative for withstanding unexpected loads and shocks.

Chitosan-based nanocomposites incorporating carbon-based nanofillers have also exhibited remarkable gains in both strength and flexibility, making them ideal candidates for load-bearing panels and beams. These enhancements, primarily due to the synergistic effects of chitosan's inherent properties are often attributed to the homogenous dispersion of the nanomaterials within the biopolymer matrix, resulting in optimized interfacial bonding and stress transfer. The incorporation of carbon nanotubes not only improves mechanical performance but also enhances thermal and electrical conductivity, offering multifunctional advantages. These characteristics are particularly beneficial in modern infrastructure, where smart materials with self-monitoring capabilities are increasingly being integrated.

The incorporation of graphene oxide and carbon nanotubes into chitosan matrices has been observed to significantly improve the mechanical properties. For instance, the Young's modulus and tensile strength of chitosan composites were enhanced by graphene oxide, increasing the elastic modulus and toughness of the material (Khachatryan et al. 2023). These improvements are crucial for structural components, where high tensile strength and flexibility are required to withstand varying loads and dynamic stresses. The enhanced toughness also contributes to the material's ability to resist crack propagation, improving the long-term durability of structural elements such as panels, beams, and support frames.

Further advancements have been observed with the development of a "rigid-soft" structure by grafting chitosan and carbon nanotubes onto carbon fibers. This innovative approach significantly increased the interfacial tensile and shear strengths of the composite (Cankaya and Yalcin 2022). The improved interfacial bonding is particularly valuable in load-bearing applications, where strong adhesion between reinforcement and matrix materials is critical for maintaining structural integrity. This enhanced bonding minimizes the risk of delamination or failure under mechanical stress, making the composite suitable for high-performance structural components, including bridge reinforcements, aerospace structures, and high-rise construction materials.

Cellulose nanofibers (CNFs) are another class of biopolymers that are increasingly used to develop structural components such as mirror housings, roofs, door handles, and panels. Due to its versatility, cellulose can be processed into derivatives like cellulose ethers and esters, facilitating the fabrication of components with high modulus and specific strength. In one study by Meysam et al., CNF was dispersed in an epoxy matrix using a dual-step alcohol/acetone treatment to boost dispersion and improve interfacial properties (Mehdinia et al. 2024). The resulting epoxy nanocomposite showed dramatic improvements in mechanical properties, with the flexural modulus increasing by 163% (from 1.4 to 3.7 GPa) and flexural strength by 197%, while tensile modulus and strength rose by 155% and 466%, respectively. These gains indicate that CNF/epoxy composites can withstand greater forces and impacts, ensuring long-term structural integrity and offering potential applications in lightweight load-bearing structures.

Kwon et al. demonstrated the structural viability of CNF reinforced with polybutylene succinate, a heterocyclic dispersing agent, by reporting significant improvements in mechanical properties, such as an 18.6% increment in tensile strength, a 77.1% boost in toughness, and a remarkable 33.3% rise in elongation at break (Kwon et al. 2024). These enhancements are particularly relevant for structural components where flexibility, toughness, and load distribution are critical, such as in high-performance composite panels, impact-resistant cladding, and earthquake-resistant building materials. The increased toughness ensures superior energy absorption, reducing the likelihood of catastrophic failure under dynamic or cyclic loading conditions. Furthermore, the improved elongation at break suggests a higher strain tolerance, making the material suitable for applications where ductility and resilience are necessary, such as in bridge decks and seismic reinforcements.

Joshi et al. employed an innovative strategy to fabricate platinum nanopillars for soft structures by integrating peptide self-assembly with advanced milling and deposition techniques (Joshi et al. 2009). This approach resulted in materials with high structural integrity, offering valuable insights into optimizing beam design for superior load-bearing capacity and stiffness. The ability of these nanopillars to provide both strength and flexibility highlights their potential in developing innovative adaptive materials for structural applications, including high-rise buildings and aerospace structures. The controlled self-assembly process ensures precise material architecture, leading to composites with tailored mechanical properties that balance rigidity and adaptability, crucial for mitigating stress concentrations and improving the long-term performance of structural elements.

15.5.2 Insulation and Thermal Regulation

Buildings contribute approximately 40% of global energy usage and 36% of greenhouse gas emissions, largely due to the significant loads required for heating and cooling to maintain thermal comfort. As the demand for energy-efficient designs grows, efficient thermal regulation and effective insulation have become critical components in near-zero energy (NZE) buildings. One promising solution in this regard is the development of bionanocomposites, utilizing innovative materials that merge sustainable, bio-based polymers with nanofillers to reduce thermal conductivity and enhance heat retention.

Derived from agricultural waste or other bio-resources, bionanocomposites not only offer improved thermal properties compared to traditional materials but also lower environmental impact. The thermal performance of these biocomposites is influenced by the type and proportion of both the polymer matrix and the nanofiller. By integrating nanoclay, carbon nanotubes, and graphene into polymer matrices, these composites form a complex microstructure in which dispersed nanofillers create a labyrinthine barrier (Akinyi and Iroh 2023). This barrier effectively disrupts heat flow, slowing thermal conduction and minimizing unwanted heat transfer. As a result, building components fabricated from these composites require less energy for heating in cold climates and cooling in warm environments, thereby reducing operational costs and the overall carbon footprint.

The enhanced insulation performance of bionanocomposites is due to their reduced thermal conductivity and ability to maintain temperature stability. The intrinsic insulating properties of bio-based polymers, combined with the reflective and heat-dissipating characteristics of the nanofillers, yield materials that moderate indoor temperatures effectively (Raja et al. 2023). This makes them ideal for various applications, from insulating panels and roof coatings to window frames, where consistent thermal performance is essential.

Research has also explored integrating PCMs into bionanocomposites for storing thermal energy. In a study by Selvasofia et al., an organic bio-nano-doped PCM was synthesized with paraffin wax, eggshell nanoparticles, and other bio-based additives (Anitha Selvasofia et al. 2024). Their work demonstrated that the bio-nano particles were uniformly distributed within the PCM matrix at lower mass fractions, as shown in Fig. 15.11a. Thermogravimetric analysis, depicted in Fig. 15.11b, revealed that incorporating these particles enhanced the thermal storage performance by 53.33%. This improvement was achieved without compromising latent heat stability, as the bio-nanoparticles effectively suppressed supercooling and increased thermal conductivity. The study found that an optimal mass fraction of 1.0% bio-nanoparticles yielded the best performance, striking a balance between enhanced thermal regulation and preserving latent heat properties.

Arpita et al. investigated the production of a cellulose-filled silica aerogel composite derived from a combination of silica aerogel and wheat straw, and incorporating sodium bicarbonate ($NaHCO_3$) for its flame retarding properties (Sarkar et al. 2024). The incorporation of $NaHCO_3$ significantly enhanced the microporosity of the aerogel, which in turn improved the thermal insulation and flame retarding properties of the composite. The resulting bionano material exhibited an

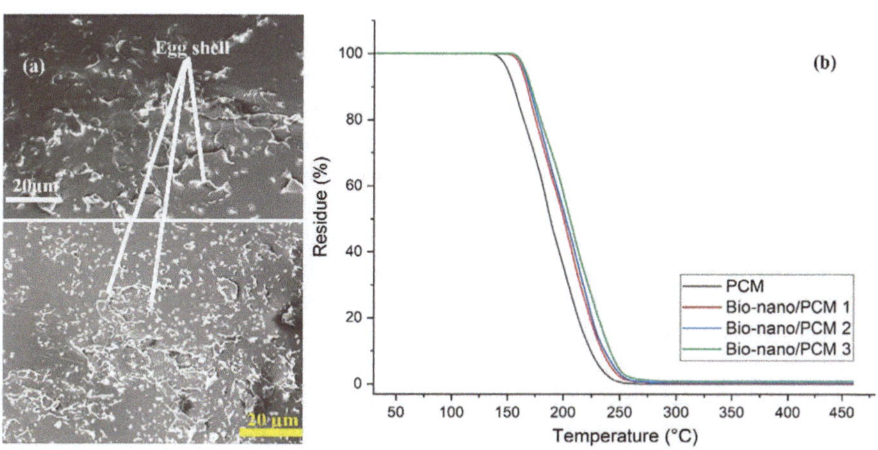

Fig. 15.11 Surface morphology **a** and thermogravimetric analysis **b** of bio-nano PCM samples. Modified with permissions from Anitha Selvasofia et al. (2024)

impressively reduced thermal conductivity (24.1 mW/mK), a high flexural modulus (736 MPa), and hydrophobic characteristics with a 110.42° water contact angle, making it a promising candidate for applications where both fire resistance and moisture management are essential.

In a complementary study, Mihai et al. explored bionanocomposites fabricated from cellulose pulp reinforced with natural fillers such as beeswax, fir resin, fir needles, rice flour, and horsetail (Pop et al. 2024). These composites were specifically designed for thermal and acoustic insulation. The study reported that the addition of these natural fillers not only enhanced thermal resistance, evidenced by an increase in oxidation temperatures by 60–70 °C, and improved sound-dampening properties, with acoustic absorption values ranging from 0.15 to 0.78. These improvements suggest that such composites can effectively regulate indoor temperatures while reducing noise, contributing to more comfortable and energy-efficient building environments. Table 15.2 provides a summary of additional studies that explored the use of bionanocomposites for insulation and thermal regulation in green building applications.

Together, these studies illustrate the significant potential of bionanocomposites in building insulation and thermal regulation, making them a key component in buildings that are energy-efficient and sustainable. Their ability to control heat flow, combined with enhanced thermal stability and energy storage capabilities, positions them as a critical material for reducing energy consumption and environmental impact in modern construction. However, challenges such as managing moisture and ensuring long-term durability remain essential considerations for their broader application in the building industry.

15.5.3 Coatings and Finishes

In response to escalating demands for sustainability and multifunctionality in the built environment, researchers are developing bionanocomposite coatings and finishes that protect building surfaces and enhance esthetic appeal and performance. These advanced coatings leverage bio-based polymers integrated with nanoscale materials, such as nanocellulose, graphene oxide, nanoclays, TiO_2 nanoparticles, and carbon nanotubes, to create films with superior barrier properties, adequate moisture and UV resistance, and multifunctional features including self-cleaning and antimicrobial activity. For example, emerging nanocellulose-based antiviral coatings offer promising solutions for maintaining hygienic surfaces while contributing to energy efficiency (Bakhshi et al. 2024).

The properties of bionanocomposite coatings depend on several factors namely, the type and characteristics of the nanomaterials, the stoichiometric balance with the biopolymer matrix, and the degree of cross-linking between the constituents. These parameters collectively determine the coating's mechanical strength, adhesion, and microstructural features, critical for long-term durability and environmental resilience. By carefully engineering the composite microstructure, the incorporated

Table 15.2 Typical studies exploring the use of bionanocomposites for insulation and thermal regulation of green buildings

Composite type (Matrix and nanofiller)	Key findings	Potential applications	References
Oil palm trunk particles and ramie fiber, with polylactic acid as the matrix and tapioca starch serving as a bio-binder	All the produced hybrid bionanocomposites exhibited density ranging from 732 to 801 kg/m^3 and thermal conductivity ranging from 0.0723 to 0.1344 W/mK. Additionally, co-reinforcement was found to positively influence water absorption and reduce voids in the composite structure	Suitable for thermal insulation in buildings	Mawardi et al. (2022)
Hemp shives and Lagarosiphon major fiber are separately used as a matrix, with mycelium as the binding agent	Both bionanocomposites exhibited thermal conductivity ~ 0.05 W/mK, with hemp-based composites showing higher water absorption than lake plant fiber composites. Both composites fall within the "WS 3" classification for moisture regulation under DIN 18,947	Suitable for moisture regulation in building applications. These materials could also play a significant role in renovating existing buildings with eco-friendly solutions	Pittau et al. (2022)
Starch -g-poly (N-isopropylacrylamide) as matrix, with eggshell powder as nanofiller	The incorporation of eggshell powder bio-filler significantly resulted in strong interfacial adhesion between the matrix and fillers, enhancing stress transfer, resulting in improved thermal stability, oxygen barrier properties, and fire retardancy	Suitable for creating biodegradable materials with enhanced thermal properties	Jena and Sahoo (2019)
Chitosan-grafted polyvinyl alcohol (CS-g-PVA) is used as the matrix material, with bio-based calcium carbonate serving as the nanofiller	Improved thermal stability with high char yield was observed. Fire retardant test shows a suppressed peak heat release rate at 44.96% and 32.54% reduction in total heat release	Suitable for fire-retardant insulation	Jena and Sahoo (2022)

(continued)

Table 15.2 (continued)

Composite type (Matrix and nanofiller)	Key findings	Potential applications	References
Octadecane PCM-based matrix combined with reduced graphene oxide as nanofiller	The bilayer phase change composite showed excellent thermal insulation and the ability to maintain a warm microclimate in cold environments (down to −5 °C). Thermal buffering to prevent overheating in hot conditions was also observed, showcasing its dual functionality in temperature regulation	Suitable for self-adaptive personal thermal management applications, with potential use in clothing and outdoor gear for harsh environments	Luo et al. (2023)
A tri-substituted cardanol-based benzoxazine (CBz) matrix, with functionalized rice husk ash (FRHA) nanofiller	Curing temperature was lower at 229 °C, while a higher glass transition temperature of 145 °C and char yield was observed with 20 wt.% FRHA-reinforced CBzFRHA	Suitable as adhesives, sealants, and encapsulants	Krishnadevi et al. (2020)

nanofillers form a complex barrier that disrupts heat flow and prevents moisture ingress, thereby maintaining stable surface conditions under harsh environmental exposure.

Various synthesis techniques, such as hot pressing, water evaporation, suspension-casting, and the more efficient pressure extrusion method, have been employed to produce these coatings (Jiang et al. 2015). Pressure extrusion has proven effective in minimizing issues associated with excessive water evaporation, thereby reducing processing time and ensuring a uniform nanofiller dispersion. For instance, graphene oxide enhances surface hardness and UV stability, while nanoclays significantly lower permeability to water and gases (Imoni et al. 2023). Moreover, TiO_2 nanoparticles enhance the performance of composite coatings by improving UV resistance and durability (Li et al. 2022).

Case studies in this area have demonstrated the considerable potential of bionanocomposite coatings for building applications, particularly in enhancing insulation, durability, and environmental resilience. For example, cellulose nanofiber-based composites have been explored as coatings for paints and adhesives, capitalizing on their three-dimensional network structure, high shear tolerance, and low environmental impact (Sharma et al. 2019). Similarly, nitrocellulose, a chemically modified cellulose, has been employed in the paint industry as potential substitute for conventional lacquers owing to its raw material abundance, industrial maturity, and biodegradability, making it a sustainable option for coatings (Mhd Haniffa et al. 2016). In addition, nanocellulose-reinforced polymer composites offer unique advantages due to their high optical transparency and robust reinforcement capabilities (Soeta et al. 2017).

The anticorrosion properties of coatings made from epoxy/nanocellulose applied on mild steel have been investigated (Ma et al. 2017). Their work revealed that even low loadings of nanocellulose significantly influenced the chemical and physical properties of the epoxy-diamine pre-polymer, with a 1 wt.% nanocellulose coating system showing the most pronounced anticorrosion performance over a 30-day electrochemical impedance study. This enhanced barrier effect not only prolongs the service life of metal substrates by mitigating corrosion but also contributes to improved thermal regulation by reducing the thermal conductivity of the coating.

Further innovations include the development of transparent films of chitosan reinforced with substantial amounts of nanofibrillated cellulose, as reported by Fernandes et al., which exhibit promising optical and barrier properties for coating applications (Fernandes et al. 2010). Additionally, Cecilia et al. formulated a bionanocomposite using polyetherimide resin modified with *Albizia benth* oil (Akintayo et al. 2011). The urethanation process in this composite led to lower curing temperatures, increased glass transition temperatures, and improved resistance to chemical corrosion, alongside enhanced scratch hardness and flexibility with higher toluene 2,4-diisocyanate loadings. These properties are essential for protective coatings in building envelopes where both aesthetic and functional durability are critical.

Beyond physical protection, bionanocomposite coatings also offer aesthetic enhancements. Their ability to be engineered at the nanoscale allows for precise control over surface texture, gloss, and color, providing designers with a versatile tool to achieve functional and visually appealing finishes. Such coatings can be formulated to exhibit self-cleaning properties, reducing the accumulation of dirt and pollutants and maintaining a pristine appearance over extended periods. For instance, utilizing inorganic and organic nanomaterials in polyurethane (PU)-based composites has resulted in coatings that combine biodegradability with a broad range of other properties including good toughness, superior corrosion and chemical resistance, mechanical strengths, low-temperature flexibility, and excellent adhesion (Ismail et al. 2011). Polyester-based polyurethanes have also been widely adopted in the coatings industry for similar reasons, providing reliable performance in demanding environmental conditions.

Despite these advantages, challenges remain in the development and large-scale implementation of bionanocomposite coatings. Ensuring uniform nanofiller dispersion, optimizing adhesion to diverse substrates, and addressing potential environmental and health impacts associated with nanomaterial release during application and disposal are areas that require ongoing research. As research continues to refine their formulation and processing, bionanocomposite coatings are poised to become a key component in the next generation of building envelope systems, contributing to safer, more durable, and visually appealing structures (Kolya and Kang 2024).

15.5.4 Self-Healing Materials

Bionanocomposites with self-healing potential present a major breakthrough in construction materials by autonomously repairing damage such as microcracks and delamination, thereby extending structural elements' service life and durability while reducing maintenance costs. Mimicking the natural self-repair mechanisms observed in living organisms, these innovative materials integrate healing functionalities into bio-based polymer matrices reinforced with nanomaterials.

Self-healing mechanisms in bionanocomposites can be broadly classified into (a) intrinsic (autonomic) and (b) extrinsic (non-autonomic) systems. Intrinsic self-healing leverages reversible bonds, such as hydrogen or dynamic covalent bonds, within the polymer matrix that can reform when triggered by external stimuli like heat or light. Conversely, in extrinsic self-healing, the healing agents are stored in encapsulated tiny vesicles or capillary networks. When damage occurs, these agents are released to fill and repair the crack, restoring the material's integrity.

Several case studies underscore the potential of self-healing bionanocomposites in structural applications. For example, modifying the surface of cellulose nanocrystals using a combinable dynamic covalent mechanophore enabled the nanocrystals to reinforce as a regenerative polymer through reversible covalent bonds (Imato et al. 2017). This approach allowed the composite to regain much of its original mechanical strength after damage. Complementary studies have demonstrated that localized repair at nanoscale cracks can restore 75–100% of the undamaged material's properties, while molecular dynamics simulations have shown that carbon nanotubes can serve both as reinforcement and as self-healing containers (Lanzara et al. 2009).

Self-healing strategies have also been applied to cementitious materials, a critical area given concrete's vulnerability to microcracking. Self-healing agents like bacteria can be embedded within the concrete in such systems. When water infiltrates a crack through ureolytic action, these bacteria metabolize and precipitate calcium carbonate to seal the damage. This method has proven effective even in cracks up to 200–300 μm wide (Kordas 2020). Other approaches use natural hollow fibers to store repair agents that release upon crack formation, mimicking the circulatory systems found in living organisms. Figure 15.12 illustrates the self-healing process of hollow fibers monitored using fluorescent dye.

Nanoparticles further enhance self-healing capabilities. For instance, iron oxide nanoparticles can be activated by light sources to facilitate rapid curing and repair. In another innovative study, Rajimol et al. encapsulated oxyresveratrol extracted from coconut shell waste in mesoporous silica nanocontainers (Rajimol et al. 2023). The resulting coating completely self-healed within 48 h, inhibiting the expansion of Gram-negative and positive bacteria. TiO_2 nanoparticles have also been observed to impart photocatalytic properties, enabling self-cleaning surfaces that actively decompose organic pollutants when subjected to sunlight (Zhou et al. 2024).

Polymer-based bionanocomposites inherently offer superior stiffness, thermal stability, lightweight properties, and strength; however, they can suffer from mesoscopic damage, degradation, and surface scratches over time. Integrating self-healing

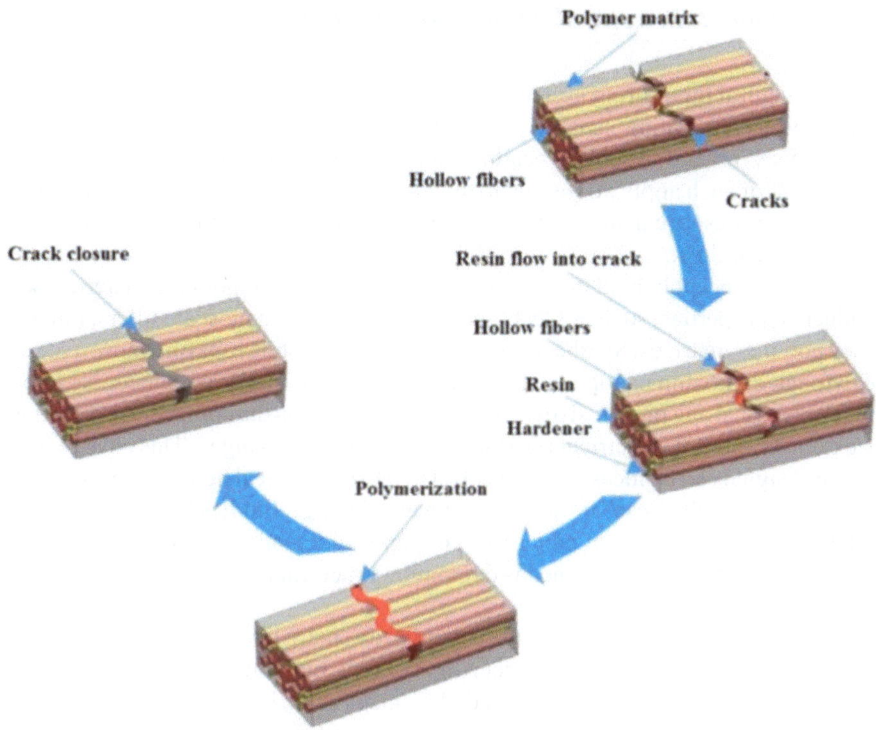

Fig. 15.12 Self-healing process of hollow fibers reinforced bionanocomposite (Ghosh 2008). Reproduced with permission from Elsevier

functionalities addresses these issues by enabling continuous repair and restoration of the material's properties, ensuring the structural integrity remains intact even after multiple damage events.

Applying self-healing bionanocomposites is particularly valuable for critical structural components such as beams, columns, and panels, where maintaining load-bearing capacity and durability is paramount. By autonomously repairing minor defects before they evolve into significant failures, these materials improve safety and performance and contribute to sustainability by reducing maintenance costs and material waste.

Incorporating self-healing mechanisms into bionanocomposites offers a promising pathway toward next-generation building materials. Whether through intrinsic reversible bonds or extrinsic healing agents, these composites are designed to repair themselves multiple times, handling defects of various sizes and ensuring optimal long-term performance. As research continues to refine these technologies, self-healing bionanocomposites are poised to become integral to sustainable and resilient construction practices.

15.5.5 Smart Structural Monitoring Systems

Smart structural monitoring systems represent a transformative approach in modern construction, integrating sensing capabilities directly within building materials to continuously assess structural health. Bionanocomposites are emerging as ideal candidates for such applications due to their ability to serve as load-bearing elements and as active sensors capable of detecting stress, strain, temperature fluctuations, and damage in real time.

The core concept behind these systems is the incorporation of nanomaterials with conductive properties, such as CNTs, graphene, and other nanoscale fillers, into bio-based polymer matrices. Synthesis of graphene from graphite rods on a mesoscale can be exfoliated into thin nanoplatelets (Festus et al. 2022). These nanomaterials impart unique electrical and piezoresistive properties so that when the composite is deformed by external forces, its electrical resistance or capacitance changes. This change in the electrical signal can be measured and correlated to mechanical deformation, enabling the material to "self-sense" its state of strain and damage. Studies have demonstrated that even a small addition of nanomaterials can significantly enhance a composite's sensitivity, offering a reliable means to monitor structural integrity over time (Zheng et al. 2019).

Case studies have reinforced the potential of smart bionanocomposite sensors in various applications. For instance, lightweight bio-composites with unidirectional flax fiber-epoxy matrices embedded in quantum resistive sensors (sQRS) have been shown to accurately monitor damage and predict failure through piezoresistive responses under different loading conditions (Tripathi et al. 2016). These sQRS signals align well with the mechanical behavior of the host material, providing continuous, real-time feedback on structural performance. Similarly, research has demonstrated that smart hybrid bionanocomposite sensors, created by integrating carbon fiber, multiwall carbon nanotubes, and graphene into a cement matrix, can convert applied strain into measurable changes in electrical resistance. Roopa et al. developed such sensors by embedding different percentages of these nano-additives into cement mortar (Roopa et al. 2020). Their tests indicated that the sensors improved flexural strength by approximately 30.08% with a 0.5% nanofiller combination and provided reliable strain sensing under both flexural and compressive loading conditions. This self-detecting capability is invaluable for real-time monitoring of structural components and initiating timely repairs.

The concept of self-sensing is not limited to polymer-based systems. Smart concrete nanocomposites, for instance, employ piezoresistivity to monitor damage under mechanical stress. In several studies, cementitious composites incorporated with CNTs and carbon fibers have shown a significant increment in electrical resistivity under load. D'Alessandro et al. (2015) and Castañeda-Saldarriaga et al. (2021) demonstrated that embedding nanometric conductive carbon-based fillers into concrete could generate electrical resistance maps correlating to strain and crack formation. These maps, often obtained through techniques like electrical impedance tomography (EIT), pinpoint the locations and severities of damage within a structure. Additional research by Taheri et al. (2022) explored the use of multiscale carbon nanofillers and lignocellulosic fibers, further validating the capability of these smart materials to detect early signs of structural degradation.

Another innovative approach involves using optical fiber sensors, shape memory alloys, and self-diagnosing polymers, which complement the smart capabilities of bionanocomposites. For example, Fig. 15.13a depicts a schematic diagram illustrating how electrical impedance tomography, combined with CNT-latex thin films coated on large aggregates, can yield spatial resistivity maps that accurately indicate damage in concrete structures. Similarly, fractional changes in electrical resistivity ($\Delta\rho_s/\rho_s$) have been correlated with external forces, as shown in Fig. 15.13b, demonstrating the strong link between mechanical stress and the electrical behavior of the biocomposites.

Despite these promising developments, challenges remain. Ensuring uniform dispersion of nanomaterials, maintaining long-term signal stability, and integrating data from smart sensors into existing building information systems are critical hurdles. Moreover, the environmental effects and possible health hazards associated with nanomaterial release during the life cycle of these composites must be carefully managed.

Smart structural monitoring systems based on bionanocomposites offer a sustainable and efficient solution for real-time health monitoring of civil infrastructures, including buildings, bridges, and roads. By integrating advanced nanomaterials into bio-based matrices, these composites enhance structural performance and provide invaluable diagnostic data, enabling predictive maintenance and significantly reducing the risk of catastrophic failures. As research in nanomaterial dispersion, wireless communication, and data analytics advances, the role of smart bionanocomposite sensors in the construction industry is poised to expand, heralding a new era of self-monitoring and resilient infrastructure.

15.6 Bionanocomposite Prospects in Civil and the Built Environment

Bionanocomposite exhibits exceptional mechanical, thermal, and functional properties, making it an ideal material for structural reinforcement in civil engineering. Their superior strength-to-weight ratio favors their use in high-rise buildings, bridges, and earthquake-resistant structures. For instance, the performance of asphalt modified with polyacrylonitrile fiber produced using a two-step LBL technique was found suitable for road and pavement constructions, demonstrating properties such as reduced temperature fracture resistance, high-temperature bonding, and excellent viscoelastic properties (Ren et al. 2022).

Bionanocomposite is also gaining attention in the built environment, enhancing the functionality of building materials. Their ability to improve moisture resistance,

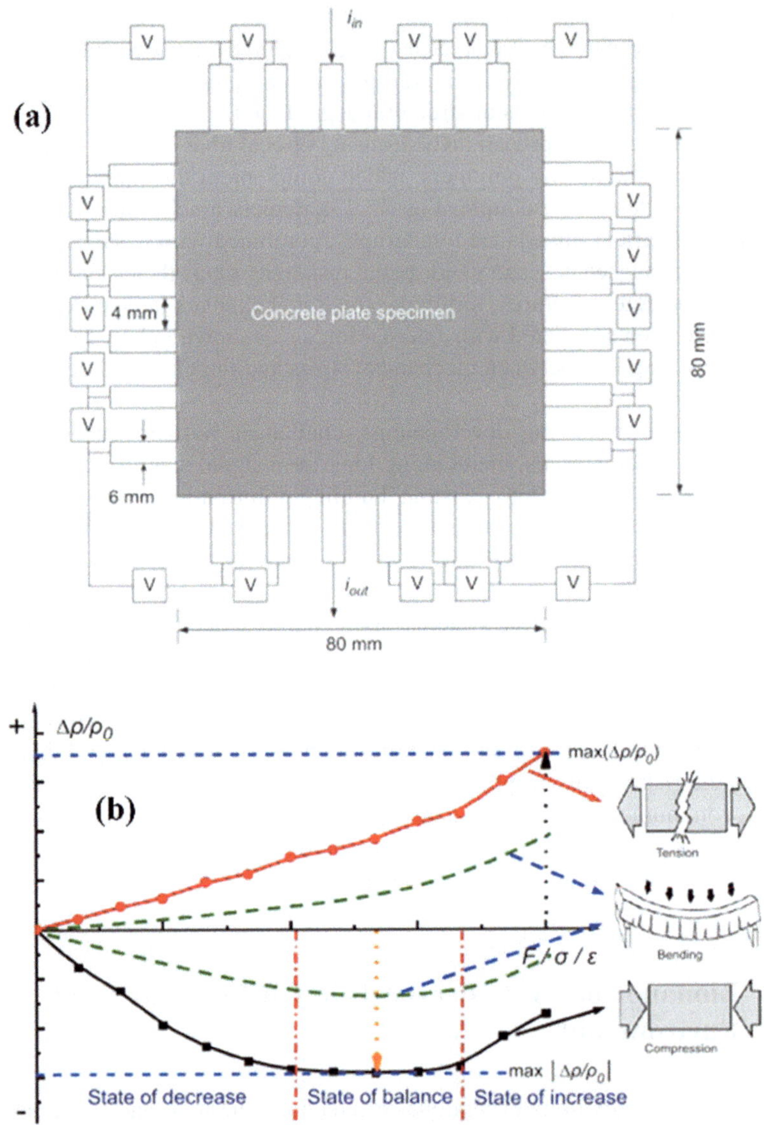

Fig. 15.13 Block diagram of an EIT **a** and typical sensing performance of nanocementitious composites under loading **b**. Reproduced from Zheng et al. (2019) with permission from Elsevier

fire resistance, thermal conduction, and antimicrobial properties presents them as sustainable biogenic materials, contributing to safer and more hygienic indoor environments. For instance, an experimental investigation of cellulose aerogel modified with sodium bicarbonate was observed to yield excellent fire resistance and improve thermal insulation (Sarkar et al. 2024). Additionally, low-density polyethylene modified with melamine, pentaerythritol, and ammonium polyphosphate produced using the melt compounding method was found to exhibit excellent fire-resistant properties, which can be thermally triggered and suitable for engineering structures (Zhang et al. 2019). Bio-based silver nanoparticles have been found to offer antimicrobial properties inhibiting the growth of pathogenic fungi and antibiotic-resistant strains, making them suitable for integration into surface coatings in buildings to combat infections and enhance indoor air quality. A typical commercial variant is the EcoClean, a functional surface coating that breaks down toxic airborne pollutants (such as viruses, mold, and bacteria) that come into contact with the coated surface, thus improving indoor air quality and reducing the spread of infection (EcoClean 2024). EcoClean enhances building wellness by mitigating sick building syndrome, and it can be applied to indoor walls, floors, and ceiling surfaces.

Bionanocomposites also extend to smart infrastructure with the integration of nanosensors and self-sensing materials to enable real-time structural health monitoring, stress and strain detection, and material failure prevention. This innovative application supports predictive maintenance strategies, mitigating the devastating risk of failures in tunnels, bridges, buildings, and large-scale infrastructure projects. Bionanocomposites also present a sustainable alternative for retrofitting and rehabilitating aging structures. Their ability to enhance the mechanical properties of existing materials gives them potential prospects for preserving historic buildings and structures, further improving their longevity. By utilizing biodegradable materials, bionanocomposites are able to reduce the carbon footprint of construction, promoting sustainable, greener building practices.

15.7 Conclusion

In this chapter, the transformative potential of bionanocomposites in building, construction, and structural engineering applications has been explored. Bionanocomposite materials have demonstrated remarkable performance resulting in enhanced mechanical properties such as tensile, flexural, and compressive strength, for use in structural components. They have also been utilized to produce lighter and more robust beams, columns, and panels. Bionanocomposites have also demonstrated superior insulation and barrier properties resulting in improved energy efficiency and thermal regulation, with advanced coatings and finishes delivering durable, attractive surfaces, enhanced UV resistance, and anti-corrosion performance. Furthermore, the self-healing capabilities and smart structural monitoring potential of bionanocomposites contribute to extending the lifespan of structures and enabling real-time health diagnostics.

The case studies presented in this chapter demonstrate that bionanocomposite offers a promising path toward energy-efficient, resilient, and low-maintenance infrastructures, despite challenges such as scalability, long-term durability, and environmental impact. Ongoing improvements in green manufacturing methods and application strategies suggest that these innovative materials are poised to play a critical role in shaping the next generation of green building systems and advancing the construction sector.

References

Akintayo CO, Akintayo ET, Ziegler T (2011) Studies on newly developed urethane modified polyetheramide coatings from Albizia benth oil. Prog Org Coat 71:89–97. https://doi.org/10.1016/j.porgcoat.2011.01.003

Akinyi C, Iroh JO (2023) Thermal decomposition and stability of hybrid graphene–clay/polyimide nanocomposites. Polymers (Basel) 15. https://doi.org/10.3390/polym15020299

Aliabadi M, Shagholani H, Yunessnia lehi A (2017) Synthesis of a novel biocompatible nanocomposite of graphene oxide and magnetic nanoparticles for drug delivery. Int J Biol Macromol 98:287–291. https://doi.org/10.1016/j.ijbiomac.2017.02.012

Anaya-Esparza LM, Ruvalcaba-Gómez JM, Maytorena-Verdugo CI, González-Silva N, Romero-Toledo R, Aguilera-Aguirre S, Pérez-Larios A, Montalvo-González E (2020) Chitosan-TiO_2: a versatile hybrid composite. Mater (Basel) 13. https://doi.org/10.3390/ma13040811

Anitha Selvasofia SD, Deepa Shri S, Manoj Kumar P, Prakash KB (2024) Characterization of bio-nano doped phase change material (bio-nano/PCM) for building energy management. Energy Sour, Part A Recover. Util Environ Eff 46:3760–3778. https://doi.org/10.1080/15567036.2024.2322690

Ayrilmis N, Ozdemir F, Nazarenko OB, Visakh PM (2018) Mechanical and thermal properties of Moringa oleifera cellulose-based epoxy nanocomposites. J Compos Mater 53:669–675. https://doi.org/10.1177/0021998318789732

Bakhshi A, Naghib SM, Rabiee N (2024) Antibacterial and antiviral nanofibrous membranes. In: Antibacterial and antiviral functional materials, vol 2. American Chemical Society, pp 2–47. https://doi.org/10.1021/bk-2024-1472.ch002

Bang G, Kim SW (2012) Biodegradable poly(lactic acid)-based hybrid coating materials for food packaging films with gas barrier properties. J Ind Eng Chem 18:1063–1068. https://doi.org/10.1016/j.jiec.2011.12.004

Ben F (2024) Valorization of Manihot esculenta peel from environmental pollutant to sustainable engineering solutions for a cleaner future. Environ Sci Pollut Res 32:1–27. https://doi.org/10.1007/s11356-024-35621-8

Ben F, Olubambi PA (2024a) Agro waste reinforcement of metal matrix composites, a veritable sustainable engineering achievement, or an effort in futility? A critical review Agro waste reinforcement of metal matrix composites, a veritable sustainable engineering achievement. Mater Res Express 11:1–37. https://doi.org/10.1088/2053-1591/ad5642

Ben F, Olubambi PA (2024b) Investigating the tribological behavior of bioinspired surfaces in agro-waste and alumina reinforced AA6063 matrix composites. Vacuum 230:1–17. https://doi.org/10.1016/j.vacuum.2024.113687

Ben F, Olubambi PA (2025a) Investigating the mechanical performance of innovative bean pod ash particulates reinforced in AA6063 alloy for construction and building applications. MRS Adv. https://doi.org/10.1557/s43580-025-01168-0

Ben F, Olubambi PA (2025b) Synergistic effects of agro—waste ash and alumina on the surface texture and tribological properties of hybrid AMCs. Discov Mater. https://doi.org/10.1007/s43939-025-00228-4

Ben F, Olubambi PA (2025c) Comparative study on mechanical performance of eco-friendly ceramic composites reinforced with polymer and sawdust additives. Discov Mater 5:1-24. https://doi.org/10.1007/s43939-025-00367-8

Ben F, Amodu FR, Olubambi PA (2023) Influence of green plantain peel ash and alumina reinforcement on the physio-mechanical properties of aluminium matrix hybrid composites. In: Materials science forum, pp 33–45. https://doi.org/10.4028/p-5CdJMw

Bonderer LJ, Studart AR, Gauckler LJ (2008) Bioinspired design and assembly of platelet reinforced polymer films. Science (80-.). 319:1069–1073. https://doi.org/10.1126/science.1148726

Cai B, Mazahreh J, Ma Q, Wang F, Hu X (2022) Ultrasound-assisted fabrication of biopolymer materials: a review. Int J Biol Macromol 209:1613–1628. https://doi.org/10.1016/j.ijbiomac.2022.04.055

Cankaya N, Yalcin S (2022) Chitosan-based nanocomposites for biological applications. In: Oueslati W (ed) IntechOpen. Rijeka. https://doi.org/10.5772/intechopen.106379

Castañeda-Saldarriaga DL, Alvarez-Montoya J, Martínez-Tejada V, Sierra-Pérez J (2021) Toward structural health monitoring of civil structures based on self-sensing concrete nanocomposites: a validation in a reinforced-concrete beam. Int J Concr Struct Mater 15:3. https://doi.org/10.1186/s40069-020-00451-8

Chazot CAC, Jons CK, Hart AJ (2020) In situ interfacial polymerization: a technique for rapid formation of highly loaded carbon nanotube-polymer composites. Adv Funct Mater 30:2005499. https://doi.org/10.1002/adfm.202005499

Chen Z, Ma R, Du Y, Wang X (2021) State-of-the-art review on research and application of original bamboo-based composite components in structural engineering. Structures 35:1010–1029. https://doi.org/10.1016/j.istruc.2021.11.059

Cheng H, Zhou Y, Xu P, Zhang M, Sun L (2024) Kaolinite-based form-stable phase change materials for thermal energy storage. J Energy Storage 87:111349. https://doi.org/10.1016/j.est.2024.111349

Cozzani M, Ferrari PF, Damonte G, Pellis A, Monticelli O (2024) On the development of polylactic acid/polycaprolactone blended films with high retention capacity. Macromol Biosci 24:2400272. https://doi.org/10.1002/mabi.202400272

D'Alessandro A, Rallini M, Ubertini F, Materazzi AL, Kenny JM, Laflamme S (2015) A comparative study between carbon nanotubes and carbon nanofibers as nanoinclusions in self-sensing concrete. In: 2015 IEEE 15th international conference nanotechnology, pp 698–701. https://doi.org/10.1109/NANO.2015.7388702

Daniel H, Omar OM, Kioumarsi M, Pilehvar S, Borrajo-Pelaez R (2025) Mechanical reinforcement of building materials with microfibers produced by electrospinning. Lect Notes Civ Eng 237:403–414. https://doi.org/10.1007/978-3-031-69626-8_34

Darvishi Cheshmeh Soltani R, Khataee AR, Safari M, Joo SW (2013) Preparation of bio-silica/chitosan nanocomposite for adsorption of a textile dye in aqueous solutions. Int Biodeterior Biodegradation 85:383–391. https://doi.org/10.1016/j.ibiod.2013.09.004

dos Santos VLS, Araújo RC, Lisboa ES, Lopes AM, de Albuquerque-Júnior RL, Cardoso JC, Blanco-Llamero C, Deshpande TA, Anderson HOW, Priefer R, Souto EB, Severino P (2024) Layer-by-layer assembly: a versatile approach for tailored biomedical films and drug delivery. J Drug Deliv Sci Technol 91:105243. https://doi.org/10.1016/j.jddst.2023.105243

Díaz-Mena V, Sánchez-Romate XF, Sánchez M, Ureña A (2024) Effect of sonication batch on electrical properties of graphitic-based PVDF-HFP strain sensors for use in health monitoring. Sensors 24. https://doi.org/10.3390/s24062007

Dubey U, Panneerselvam K (2024) Low-combustible, high-strength, and thermally stable bio-blended epoxy-based bio-nanocomposite using reduced graphene oxide as a strengthening agent. J Reinf Plast Compos 07316844241279013. https://doi.org/10.1177/07316844241279013

Dweib MA, Hu B, Shenton HW, Wool RP (2006) Bio-based composite roof structure: manufacturing and processing issues. Compos Struct 74:379–388. https://doi.org/10.1016/j.compstruct.2005.04.018

EcoClean (2024) Protecting indoor surfaces. https://ecocleancoatings.com/surfaces/indoor-surfaces/. Indoor Air Quality (IAQ). Accessed 13 March 2025

Fernandes SCM, Freire CSR, Silvestre AJD, Pascoal Neto C, Gandini A, Berglund LA, Salmén L (2010) Transparent chitosan films reinforced with a high content of nanofibrillated cellulose. Carbohydr Polym 81:394–401. https://doi.org/10.1016/j.carbpol.2010.02.037

Festus B, Enwerem GC, Faremi AA, Olubambi PA (2022) Characterisation of graphite powder processed from waste dry cell batteries. https://www.ajol.info/index.php/pngas/article/view/261186

Fura S, Hakim (Assala) Q, Toledo D, Agatemor C (2024) Chapter 15—biomedical applications of bionanocomposites. In: Sharma B, Thomas S, Kumar Bajpai P, Ghosal KSBT-A, Shekhar B (eds) Micro nano technology. Elsevier, pp 351–364. https://doi.org/10.1016/B978-0-323-91764-3.00015-2

Ghosh SK (2008) Self-healing materials: fundamentals, design strategies, and applications. Wiley. https://doi.org/10.1002/9783527625376

Hatami M, Jouybari FN, Djafarzadeh M, Qandalee M, Alikarami M, Panah MY, López-Coca IM, Durán-Valle CJ (2024) Bionanocomposites preparation by design: planning based on the modification of ceria nanoparticles by polydopamine interleaved with the chitosan biopolymer. Polymer (Guildf) 304:127140. https://doi.org/10.1016/j.polymer.2024.127140

Imato K, Natterodt JC, Sapkota J, Goseki R, Weder C, Takahara A, Otsuka H (2017) Dynamic covalent diarylbibenzofuranone-modified nanocellulose: mechanochromic behaviour and application in self-healing polymer composites. Polym Chem 8:2115–2122. https://doi.org/10.1039/C7PY00074J

Imoni S, Tiza MT, Onyebuchi M, Akande EO (2023) The use of nanomaterials for road construction. Bincang Sains Dan Teknol 2:108–117. https://doi.org/10.56741/bst.v2i03.435

Ismail EA, Motawie AM, Sadek EM (2011) Synthesis and characterization of polyurethane coatings based on soybean oil–polyester polyols. Egypt J Pet 20:1–8. https://doi.org/10.1016/j.ejpe.2011.06.009

Jali S, Mohan TP, Mwangi FM, Kanny K (2024) A review on barrier properties of cellulose/clay nanocomposite polymers for packaging applications. Polymers (Basel) 16. https://doi.org/10.3390/polym16010051

Jena DK, Sahoo PK (2019) New strategies for the construction of eggshell powder reinforced starch based fire hazard suppression biomaterials with tailorable thermal, mechanical and oxygen barrier properties. Int J Biol Macromol 140:496–504. https://doi.org/10.1016/j.ijbiomac.2019.08.156

Jena DK, Sahoo PK (2022) A facile approach for synthesising and studying (chitosan-g-poly vinyl alcohol)/bio-$CaCO_3$ fire retardant bionanocomposites. Polym Bull 79:6673–6691. https://doi.org/10.1007/s00289-021-03834-3

Jiang Y, Song Y, Miao M, Cao S, Feng X, Fang J, Shi L (2015) Transparent nanocellulose hybrid films functionalized with ZnO nanostructures for UV-blocking. J Mater Chem C 3:6717–6724. https://doi.org/10.1039/C5TC00812C

Joseph B, Krishnan S, Sagarika VK, Tharayil A, Kalarikkal N, Thomas S (2020) Bionanocomposites as industrial materials, current and future perspectives: a review. Emergent Mater 3:711–725. https://doi.org/10.1007/s42247-020-00133-x

Joshi KB, Singh P, Verma S (2009) Fabrication of platinum nanopillars on peptide-based soft structures using a focused ion beam. Biofabrication 1:25002. https://doi.org/10.1088/1758-5082/1/2/025002

Kausar A, Ahmad I, Eisa MH, Maaza M, Khan H (2023) Manufacturing strategies for graphene derivative nanocomposites—current status and fruitions. Nanomanufacturing 3:1–19. https://doi.org/10.3390/nanomanufacturing3010001

Khachatryan G, Khachatryan K, Szczepankowska J, Krzan M, Krystyjan M (2023) Design of carbon nanocomposites based on sodium alginate/chitosan reinforced with graphene oxide and carbon nanotubes. Polymers (Basel). 15. https://doi.org/10.3390/polym15040925

Khater E-S, Bahnasawy A, Gabal BA, Abbas W, Morsy O (2023) Effect of adding nano-materials on the properties of hydroxypropyl methylcellulose (HPMC) edible films. Sci Rep 13:5063. https://doi.org/10.1038/s41598-023-32218-y

Kim C-H, Bhak G, Lee J, Sung S, Park S, Paik SR, Yoon M-H (2016) Controlled charge trapping and retention in large-area monodisperse protein metal-nanoparticle conjugates. ACS Appl Mater Interfaces 8:11898–11903. https://doi.org/10.1021/acsami.6b02268

Kolya H, Kang C-W (2024), Eco-friendly polymer nanocomposite coatings for next-generation fire retardants for building materials. Polymers (Basel) 16. https://doi.org/10.3390/polym16142045

Kordas G (2020) Self-healing of concrete through ceramic nanocontainers loaded with corrosion inhibitors and microorganisms. In: Mhadhbi M (ed) IntechOpen. Rijeka. https://doi.org/10.5772/intechopen.93514

Kreuz A, Cadorin L, da Silva DB, de Oliveira VB, Malschitzky MET, Zimmermann LM, Andreaus J, Lukasik RM (2024) Chitosan: a biobased sustainable textile auxiliary. In: Advances in renewable natural materials for textile sustainability, 1st edn. CRC Press, p 17

Krishnadevi K, Devaraju S, Sriharshitha S, Alagar M, Keerthi Priya Y (2020) Environmentally sustainable rice husk ash reinforced cardanol based polybenzoxazine bio-composites for insulation applications. Polym Bull 77:2501–2520. https://doi.org/10.1007/s00289-019-02854-4

Kwon H, Kim Y, Bae J, Bae SH, Park JH, Choi K, Hong PH, Ko MJ, Ryu J, Jung HW, Hong SW (2024) A bionanocomposite based on cellulose nanofibers modified by a sustainable heterocyclic dispersing agent with outstanding mechanical properties. Chem Eng J 487:150360. https://doi.org/10.1016/j.cej.2024.150360

Lanzara G, Yoon Y, Liu H, Peng S, Lee W-I (2009) Carbon nanotube reservoirs for self-healing materials. Nanotechnology 20:335704. https://doi.org/10.1088/0957-4484/20/33/335704

Lei H, Xie M, Zhao Y, Zhang F, Xu Y, Xie J (2016) Chitosan/sodium alginate modificated graphene oxide-based nanocomposite as a carrier for drug delivery. Ceram Int 42:17798–17805. https://doi.org/10.1016/j.ceramint.2016.08.108

Li L, Geng Z, Huang J, Xiao H, Pei Y, Liu Y (2021) Creating in-situ polymeric layer based on photocuring for various construction applications. Constr Build Mater 273:121732. https://doi.org/10.1016/j.conbuildmat.2020.121732

Li H, Lai J, Yao M, Leng Y, Wu Z, Zhang J, Peng H, Qiu Z (2022) Development of high performance superhydrophobic coating with excellent corrosion resistance, durability, and self-cleaning properties using M-TiO_2@EP composites. Appl Surf Sci 601:154109. https://doi.org/10.1016/j.apsusc.2022.154109

Lin Y, Chen Y, Zeng Z, Zhu J, Wei Y, Li F, Liu L (2015) Effect of ZnO nanoparticles doped graphene on static and dynamic mechanical properties of natural rubber composites. Compos Part A Appl Sci Manuf 70:35–44. https://doi.org/10.1016/j.compositesa.2014.12.008

Luo Z, Yang D, Liu J, Zhao H-Y, Zhao T, Li B-X, Yang W-G, Yu Z-Z (2023) Nature-inspired solar-thermal gradient reduced graphene oxide aerogel-based bilayer phase change composites for self-adaptive personal thermal management. Adv Funct Mater 33:2212032. https://doi.org/10.1002/adfm.202212032

Ma IAW, Shafaamri A, Kasi R, Zaini FN, Balakrishnan V, Subramaniam R, Arof AK (2017) Anticorrosion properties of epoxy/nanocellulose nanocomposite coating. BioResources 12:2912–2929. https://doi.org/10.15376/biores.12.2.2912-2929

Manopriya S, Hareesh K (2021) The prospects and challenges of solar electrochemical capacitors. J Energy Storage 35:102294. https://doi.org/10.1016/j.est.2021.102294

Masoumi S, O'Shaughnessy S, Pakdel A (2022) Organic-based flexible thermoelectric generators: from materials to devices. Nano Energy 92:106774. https://doi.org/10.1016/j.nanoen.2021.106774

Mathew Simon S et al (2021) Recent advancements in multifunctional applications of sol-gel derived polymer incorporated TiO_2–ZrO_2 composite coatings: a comprehensive review. Appl Surf Sci Adv 6:100173. https://doi.org/10.1016/j.apsadv.2021.100173

Mawardi I, Aprilia S, Faisal M, Rizal S (2022) Investigation of thermal conductivity and physical properties of oil palm trunks/ramie fiber reinforced biopolymer hybrid composites as building bio-insulation. Mater Today Proc 60:373–377. https://doi.org/10.1016/j.matpr.2022.01.249

Mehdinia M, Farajollah Pour M, Yousefi H, Dorieh A, Lamanna AJ, Fini E (2024) Developing bio-nano composites using cellulose-nanofiber-reinforced epoxy. J Compos Sci 8. https://doi.org/10.3390/jcs8070250

Mhd Haniffa MAC, Ching YC, Abdullah LC, Poh SC, Chuah CH (2016) Review of bionanocomposite coating films and their applications. Polymers (Basel) 8. https://doi.org/10.3390/polym8070246

Mondal D, Zaharia A, Mequanint K, Rizkalla AS (2020) Sol-gel derived tertiary bioactive glass–ceramic nanorods prepared via hydrothermal process and their composites with poly(Vinylpyrrolidone-Co-Vinylsilane). J Funct Biomater 11. https://doi.org/10.3390/jfb11020035

Nepal D, Tsukruk VV, Keten S, Brinson LC, Dayal K, Vignolini S, Buehler MJ, Wang Y, Bockstaller MR, Kaplan DL, Coveney PV, Heinz H, Kanhaiya K, Schatz GC, Kotov NA, Henderson LC, Yakobson BI, Kang S, Vollrath F, Adstedt K, El-Awady J (2022) Hierarchically structured bioinspired nanocomposites. Nat Mater 22:18–35. https://doi.org/10.1038/s41563-022-01384-1

Nimbagal V, Banapurmath NR, Sajjan AM, Patil AY, Ganachari SV (2021) Studies on hybrid bionanocomposites for structural applications. J Mater Eng Perform 30:6461–6480. https://doi.org/10.1007/s11665-021-05843-9

Nwanna EC, Imoisili PE, Jen T-C (2020) Fabrication and synthesis of SnOX thin films: a review. Int J Adv Manuf Technol 111:2809–2831. https://doi.org/10.1007/s00170-020-06223-8

Oh BK, Choi SW, Park HS (2016) Influence of variations in CO_2 emission data upon environmental impact of building construction. J Clean Prod 140:1194–1203. https://doi.org/10.1016/j.jclepro.2016.10.041

Omar AM, Hassan MH, Daskalakis E, Smith A, Donoghue J, Mirihanage W, Bartolo PJDS (2024) Biomimetic dual sensing polymer nanocomposite for biomedical applications. Front Bioeng Biotechnol 12. https://doi.org/10.3389/fbioe.2024.1322753

Ortega F, Versino F, López OV, García MA (2021) Biobased composites from agro-industrial wastes and by-products. Emergent Mater 53(5):873–921. https://doi.org/10.1007/S42247-021-00319-X

Panda PK, Ramakrishna S (2007) Electrospinning of alumina nanofibers using different precursors. J Mater Sci 42:2189–2193. https://doi.org/10.1007/s10853-007-1581-2

Patil M, Boraste S, Minde P (2022) A comprehensive review on emerging trends in smart green building technologies and sustainable materials. Mater Today Proc 65:1813–1822. https://doi.org/10.1016/j.matpr.2022.04.866

Pittau F, Carcassi OG, Servalli M, Pellegrini S, Claude S (2022) Hygrothermal characterization of bio-based thermal insulation made of fibres from invasive alien lake plants bounded with mycelium. IOP Conf Ser Earth Environ Sci 1078:12069. https://doi.org/10.1088/1755-1315/1078/1/012069

Pop MA, Croitoru C, Matei S, Zaharia S-M, Coşniţă M, Spîrchez C (2024) Thermal and sound insulation properties of organic biocomposite mixtures. Polymers (Basel). 16. https://doi.org/10.3390/polym16050672

Pradhan S et al (2025) Augmented barrier properties of poly (lactic acid) composite packaging film reinforced by antibacterial hybrid bio-filler: extending the shelf life of fruits. Polym Compos 46:1753–1767. https://doi.org/10.1002/pc.29071

Princy S, Gupta S, Sharma AK (2024) Chapter 10—electrospinning of bionanocomposites: properties and applications. In: Sharma B, Thomas S, Kumar Bajpai P, Ghosal KSBT-A, Shekhar B (eds) Micro nano technology. Elsevier, pp 227–246. https://doi.org/10.1016/B978-0-323-91764-3.00007-3

Qi S, Tang CY (2015) Layer-by-layer (LbL) method. In: Drioli E, Giorno L (eds) Encyclopedia of membranes. Springer Berlin Heidelberg, Berlin, Heidelberg, pp 1–2. https://doi.org/10.1007/978-3-642-40872-4_1487-1

Raja P, Murugan V, Ravichandran S, Behera L, Mensah RA, Mani S, Kasi AK, Balasubramanian KBN, Sas G, Vahabi H, Das O (2023) A review of sustainable bio-based insulation materials for energy-efficient buildings. Macromol Mater Eng 308:2300086. https://doi.org/10.1002/mame.202300086

Rajimol PR et al (2023) Development of an oxyresveratrol incorporated bio-based smart nanocomposite coating with anti-corrosive, self-healing, and anti-microbial properties. Green Chem 25:7189–7215. https://doi.org/10.1039/D3GC00773A

Raza MA, Sharma MK, Nagori K, Jain P, Ghosh V, Gupta U, Ajazuddin (2024) Recent trends on polycaprolactone as sustainable polymer-based drug delivery system in the treatment of cancer: biomedical applications and nanomedicine. Int J Pharm 666:124734. https://doi.org/10.1016/j.ijpharm.2024.124734

Ren D, Liu Y, Muhammad Y, Li X, Mao C, Zhao Z, Pei R, Li J (2022) Preparation of double-layer film modified polyacrylonitrile fiber modifier for designing and performance evaluation of PAN/SBS composite asphalt. Constr Build Mater 350:128920. https://doi.org/10.1016/j.conbuildmat.2022.128920

Ricciardi P, Cillari G, Miino MC, Collivignarelli MC (2020) Valorization of agro-industry residues in the building and environmental sector: a review. WASTE Manag Res 38:487–513. https://doi.org/10.1177/0734242X20904426

Roopa AK, Hunashyal AM, Venkaraddiyavar P, Ganachari SV (2020) Smart hybrid nano composite concrete embedded sensors for structural health monitoring. Mater Today Proc 27:603–609. https://doi.org/10.1016/j.matpr.2019.12.071

Sajna VP, Mohanty S, Nayak SK (2014) Hybrid green nanocomposites of poly(lactic acid) reinforced with banana fibre and nanoclay. J Reinf Plast Compos 33:1717–1732. https://doi.org/10.1177/0731684414542992

Sanmugam A, Abbishek S, Kumar SL, Sairam AB, Palem VV, Kumar RS, Almansour AI, Arumugam N, Vikraman D (2023) Synthesis of chitosan based reduced graphene oxide-CeO_2 nanocomposites for drug delivery and antibacterial applications. J Mech Behav Biomed Mater 145:106033. https://doi.org/10.1016/j.jmbbm.2023.106033

Sarkar A, Islam A, Zhu L, Ren S (2024) Flame-retardant cellulose-aerogel composite from agriculture waste for building insulation. Appl Mater Today 36:102080. https://doi.org/10.1016/j.apmt.2024.102080

Sessini V, Haseeb B, Boldizar A, Lo Re G (2021) Sustainable pathway towards large scale melt processing of the new generation of renewable cellulose–polyamide composites. RSC Adv 11:637–656. https://doi.org/10.1039/D0RA07141B

Shaban M, Hassen A, El Sayed AM, Al-Ghamdi A (2023) Synthesis of some functional oxides and their composites using sol-gel method. In: Singh JP, Acharya SS, Kumar S, Dixit SK (eds) IntechOpen. Rijeka. https://doi.org/10.5772/intechopen.111384

Shamindi A, Kudaligama A, Udawattha C (2024) Sustainable composite insulation panels: investigating diverse alternatives for insulating applications utilizing coconut husk and agro-industrial residue composites. FARU J 11:105–114. https://doi.org/10.4038/faruj.v11i2.327

Sharma A, Thakur M, Bhattacharya M, Mandal T, Goswami S (2019) Commercial application of cellulose nano-composites—a review. Biotechnol Rep 21:e00316. https://doi.org/10.1016/j.btre.2019.e00316

Siouta L, Apostolopoulou M, Bakolas A (2024) Natural fibers in composite materials for sustainable building: a state-of-the-art review on treated hemp fibers and hurds in mortars. Sustainability 16. https://doi.org/10.3390/su162310368

Soeta H, Fujisawa S, Saito T, Isogai A (2017) Interfacial layer thickness design for exploiting the reinforcement potential of nanocellulose in cellulose triacetate matrix. Compos Sci Technol 147:100–106. https://doi.org/10.1016/j.compscitech.2017.05.010

Sumesh KR, Kavimani V, Rajeshkumar G, Indran S, Saikrishnan G (2021) Effect of banana, pineapple and coir fly ash filled with hybrid fiber epoxy based composites for mechanical and morphological study. J Mater Cycles Waste Manag 23:1277–1288. https://doi.org/10.1007/s10163-021-01196-6

Sun X, Qin Z, Ye L, Zhang H, Yu Q, Wu X, Li J, Yao F (2020) Carbon nanotubes reinforced hydrogel as flexible strain sensor with high stretchability and mechanically toughness. Chem Eng J 382:122832. https://doi.org/10.1016/j.cej.2019.122832

Taheri S, Georgaklis J, Ams M, Patabendigedara S, Belford A, Wu S (2022) Smart self-sensing concrete: the use of multiscale carbon fillers. J Mater Sci 57:2667–2682. https://doi.org/10.1007/s10853-021-06732-1

Tripathi KM, Vincent F, Castro M, Feller JF (2016) Flax fibers—epoxy with embedded nanocomposite sensors to design lightweight smart bio-composites. Nanocomposites 2:125–134. https://doi.org/10.1080/20550324.2016.1227546

UN (2023) Make cities and human settlements inclusive, safe, resilient and sustainable. Department of economics social affiliation sustainable development. https://sdgs.un.org/goals/goal11. Accessed 14 Feb 2025

Vikraman D, Thiagarajan S, Sanmugam A (2017) Facile methodology of sol-gel synthesis for metal oxide nanostructures. In: Chandra U (ed) IntechOpen. Rijeka. https://doi.org/10.5772/intechopen.68708

Yan J, Karmakar B, Zaki MSA, Osman R, Abdalla AM, Shati AA, El-kott AF, Yousef FM, Aggad SS, Batiha GE-S (2022) Introducing a bionanocomposite (Ultrasound-assisted synthesis of Ag nanoparticles embedded Aloe vera gel) for the treatment of cervical carcinoma. J Exp Nanosci 17:617–630. https://doi.org/10.1080/17458080.2022.2134562

Yang X, Chen Y, Zhang C, Duan G, Jiang S (2023) Electrospun carbon nanofibers and their reinforced composites: preparation, modification, applications, and perspectives. Compos Part B Eng 249:110386. https://doi.org/10.1016/j.compositesb.2022.110386

Yenidogan C (2021) Earthquake-resilient design of seismically isolated buildings: a review of technology. Vibration 4:602–647. https://doi.org/10.3390/vibration4030035

Zhang H, Yan Z, Zhang T, Wang J, Wang X, Chen Y, Zhu S, Li Z, Chen Y, Hong W, Zhao Y, Chen S, Hong Q, Xu Y, Guo X (2024) Bioinspired high-linearity, wide-sensing-range flexible stretchable bioelectronics based on MWCNTs/GR/$Nd_2Fe_{14}B$/PDMS nanocomposites for human-computer interaction and biomechanics detection. ACS Sens 9:3947–3957. https://doi.org/10.1021/acssensors.4c00664

Zhang T, Zhang Y, Xiao Z, Yang Z, Zhu H, Ju JW, Yan Z (2019) Development of a novel bio-inspired cement-based composite material to improve the fire resistance of engineering structures. Constr Build Mater 225:99–111. https://doi.org/10.1016/j.conbuildmat.2019.07.121

Zheng Q, Han B, Ou J (2019) NanoComposites for structural health monitoring. In: Pacheco-Torgal F, Diamanti MV, Nazari A, Granqvist CG, Pruna ASBT-NEC; Amirkhanian E (eds) Woodhead Publishing service civil structure engineering, Woodhead Publishing, pp 227–259. https://doi.org/10.1016/B978-0-08-102641-0.00011-6

Zhou M, Xiang L, Quan Y, Li L, Zhang X (2024) Development of superhydrophilic ZnOX/TiO_2 with visible light-induced self-cleaning and antibacterial properties. Surf Interfaces 52:104933. https://doi.org/10.1016/j.surfin.2024.104933

Zidi S, Miraoui I (2024) Mechanical and morphological analysis of cellulose extracted from sisal fibers and their effect on bio-based composites mechanical properties. Funct Compos Struct 6:15008. https://doi.org/10.1088/2631-6331/ad2fe5

Zor M, Şen F, Yazıcı H, Candan Z (2023) Thermal, mechanical and morphological properties of cellulose/lignin nanocomposites. Forests 14. https://doi.org/10.3390/f14091715

Dr. Festus Ben is a Postdoctoral Research Fellow at the Department of Metallurgy, University of Johannesburg, South Africa. He specializes in Electronics and Instrumentation, Material Science, and Artificial Intelligence. His research focuses on sustainable waste management techniques, addressing challenges, opportunities, health implications, control systems engineering, and renewable energy systems. Dr. Festus holds a patent for developing the first smart cooling device. As Director of Research and Development at the Federal Polytechnic Ede Osun State, Dr. Festus has contributed extensively to research publications and led impactful projects in environmental waste management and recycling. His work is widely recognized in academic circles, with citations highlighting his significant contributions to the field.

Chapter 16
Life Cycle Assessment of Bionanocomposite Materials: Assessing the Ecological Footprints

Jakpa Wizi, Lixiao Ni, and Benjamin Tawiah

Abstract The growing interest in sustainable materials has led to a rise in demand for bionanocomposite materials, which integrate biodegradable polymers with nanoscale fillers. These composites offer innovative solutions for various applications but also pose significant questions regarding their environmental impacts throughout their life cycles. This chapter examines the life cycle assessment (LCA) of bionanocomposite materials, focusing on their ecological footprints throughout various stages of production, use, and disposal. The chapter begins by discussing the concept of bionanocomposites and the pressing need to evaluate their environmental impacts in the context of global sustainability challenges consistent with the UNDP sustainable development goals. The current methodologies for LCA are discussed taking into consideration resource extraction, application, and disposal with their associated water and energy consumptions. The challenges and opportunities in conducting comprehensive LCA for bionanocomposites are thoroughly discussed, and recommendations for achieving sustainability in bionanocomposites production and applications are recommended. Also, the comparative environmental benefits and drawbacks of bionanocomposites relative to conventional materials are examined. This chapter will serve as a valuable resource for researchers and practitioners of LCA and ESG professionals in understanding the potential of bionanocomposite materials in contributing to a more sustainable future. Ultimately, this chapter provides insights that can guide future research and policy initiatives aimed at promoting the adoption of environmentally friendly materials in diverse applications.

J. Wizi · L. Ni (✉)
Key Laboratory of Integrated Regulation and Resource Development on Shallow Lakes, MOE, School of Environment, Hohai University, Nanjing, China
e-mail: nilixiao@hhu.edu.cn

J. Wizi
e-mail: wizijakpa@yahoo.com

B. Tawiah
Department of Textiles, Department of Industrial Art (Textiles and Fashion), Kwame Nkrumah University of Science and Technology, Kumasi, Ghana
e-mail: benuaby@yahoo.com

S. C. George and B. Tawiah (eds.), *Advanced Bionanocomposite Materials*, Advanced Structured Materials 241, https://doi.org/10.1007/978-981-96-8225-6_16

Keywords Bionanocomposites · Life cycle assessment · Ecological footprint · Sustainable materials · Biodegradable polymers

16.1 Introduction

Bionanocomposites (BNCs) are innovative hybrid composite materials. It consists of the integration of a continuous matrix phase usually organic in nature and a discontinuous inorganic phase, of which one should be a nanostructure enhancing their properties and functionalities, offering a sustainable alternative to conventional synthetic composites (Schmidt et al. 2023; Kaci et al. 2019). BNCs are made by combining biodegradable polymers, like polylactic acid (PLA), cellulose, chitosan, cellophane, and starch with nanofillers which are either organic or inorganic like clays, graphene, carbon nanotubes (CNTs) or cellulose nanocrystals (CNCs), silica, titanium dioxide to enhance mechanical, thermal, and gas barrier properties of the biopolymer matrix (Dhar et al. 2016; Nazrin et al. 2020). The biopolymers serve as the continuous phase, while the nanofillers as the dispersed phase and usually range from 1 to 100 nm in size. BNCs are at the forefront of sustainable materials, addressing global concerns on resource depletion and environmental pollution for being eco-friendly and derived from renewable resources (Sanni et al. 2022). These materials are sometimes labelled as nanocomposites, eco-friendly composite, bio-composite, nanocomposites, bio-hybrids or bio-based plastics, green composites, etc.

Bionanocomposites find application in various industries, including food packaging, biomedical devices, environmental remediation, construction, automotive and electronic components, and aerospace applications (Ali and Hoque 2022; Jayakumar et al. 2022, 2023) as shown in Fig. 16.1. For example, PLA-based nanocomposites have shown to enhance thermal and biodegradability stability, making them ideal for reducing dependency on fossil-fuel-derived plastics (Khoo and Chow 2022).

However, inconsistent findings and unresolved issues diminish BNCs environmental promise. For instance, while nanofillers like CNCs improve material durability, they can accidentally reduce biodegradation rates, compromising end-of-life benefits. Additionally, certain nanofillers, such as carbon nanotubes, create toxicity concerns if released during degradation, necessitating strict encapsulation techniques. Scalability further hinders their implementation, as industrial manufacturing faces barriers such variable nanoparticle dispersion and high energy demands. The dearth of established testing methodologies and research gaps on nanoparticles long-term environmental purpose necessitate thorough evaluation frameworks like life cycle assessment.

Life cycle assessment (LCA) is an essential technique for measuring the environmental impacts of BNCs, from raw material extraction, to production, to usage and disposal at end of life (Kokare et al. 2023). By examining factors such as energy consumption, greenhouse gas emissions, and biodegradability, LCA provides an

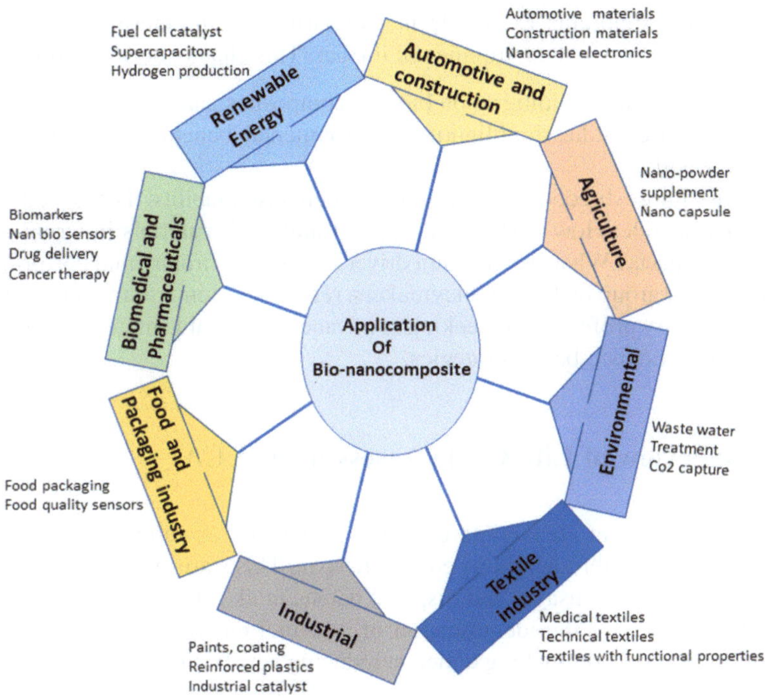

Fig. 16.1 Bionanocomposites application in various fields

overview of a material's ecological footprint, enabling informed sustainable design choice (Kupolati et al. 2022; Nwodo and Anumba 2020).

Aside LCA, there are other methods for environmental impact of products. Carbon footprint analysis (CFA), for instant, quantifies greenhouse gas emissions, particularly CO_2 equivalents (Mostafaei et al. 2023). CFA, while more limited than LCA, is more expedient and straightforward, rendering it appropriate for focused emissions reduction initiatives. The environmental impact assessment (EIA) method evaluates local project consequences such as waste disposal and land use, lacking the comprehensive perspective of LCA (Larrey-Lassalle 2017). Material flow analyses (MFA) enhance resource efficiency neglecting environmental consequences. The water footprint assessment (WFA) evaluates water consumption and contamination solely from a water-centric viewpoint. LCA is distinctive due to its cradle-to-grave methodology and comprehensive multi-impact analysis, encompassing global warming, resource depletion, and acidification (ISO 14040/14044). LCA incorporates these elements into a comprehensive evaluation, facilitating trade-off analysis and systemic enhancements in sustainability while CFA or WFA might concentrate on particular issue. A firm utilizing CFA may reduce emissions yet neglect water contamination, whereas LCA would emphasize both. LCA directs sustainable design from raw material selection to end-of-life disposal, whereas EIA guarantees project compliance.

This chapter highlights LCA role in circularity and in determining trade-offs between performance and environmental impact. This chapter seeks outline

(i) outline LCA methodologies on bionanocomposites analysis,
(ii) present case studies highlighting environmental benefits and challenges of BNCs, and
(iii) discuss future directions for optimizing their sustainability. The key addressed points include: How BNCs compare to traditional materials in terms of life-cycle impacts? What are the main drivers of their environmental performance? The target group includes policymakers, researchers, environmental engineers, and industry professionals seeking to advance sustainable material innovations through evidence-based strategies.

16.2 Overview of Life Cycle Assessment (LCA)

Life cycle assessment is a comprehensive system of examining environmental impacts of a product throughout its entire life span. This starts with the raw material extraction, production, usage, and disposal (Costa et al. 2019). By looking at each stage, LCA allows for the identification of potential environmental issues, facilitating informed decision-making aimed at reducing negative impacts and promoting sustainability.

LCA is key in sustainable development, as it bridges technical innovation with ecological responsibility. With BNCs, LCA assesses the trade-offs between biodegradability and energy demands in production, ensuring "green" claims are validated by quantifiable outcome (Pradhan 2022). An example would be, polylactic acid (PLA) nanocomposites which reduces dependence on fossil fuels, but growing the necessary feedstock requires water-intensive processes or land-use conflicts (Khoo and Chow 2022). LCA frameworks tackle these issues by comparing BNCs to traditional materials like petroleum-based plastics, considering carbon footprint, toxicity, and recyclability. Using LCA in material design help create bionanocomposites that balance performance, cost, and planetary boundaries, which aligns with global sustainability agendas like the UN Sustainable Development Goals (SDGs).

16.2.1 Significance of LCA in Achieving Sustainable Development

LCA is crucial as it:

a. Drives innovation: LCA acts as a driver for innovation by showing the adverse environmental impacts of current methods. In sectors like packaging or energy generation, LCA reveals the carbon footprint and resource inefficiencies linked to conventional materials (e.g., plastics, fossil fuels), encouraging the investigation of alternatives such as biodegradable polymers or renewable energy systems (Coltelli et al. 2023). Through the quantification of impacts, life cycle assessment encourages industries to redesign products, use cleaner technologies, and adopt concepts of the circular economy, such as recycling and remanufacturing.

b. Improves transparency: Using standardized life cycle assessment frameworks like ISO 14040/44, products and industries can be compared for environmental performance. Transparency builds consumer, investor, and regulator trust. Energy Star and EU Ecolabel certify items using life cycle assessment data, helping consumers make informed choices (Walker and Rothman 2020). The automobile industry uses life cycle assessment (LCA) to promote low-carbon electric vehicles to obtain a competitive edge.
c. Aids decision-making: LCA helps policymakers, businesses, and consumers prioritize sustainability by designing regulations like carbon taxes or renewable energy subsidies that target high-impact areas (Bartolo et al. 2021). Businesses use LCA to identify low-impact raw materials and optimize logistics to reduce emissions. Consumers use LCA-driven tools like carbon footprint labels to make sustainable choices.
d. Recognizing environmental hotspots: LCA analyses affect all of a product's life cycle (e.g., raw material extraction, production, disposal) to identify the most environmentally damaging stages. LCA may show that crop fertilizer use excessively pollutes water and emits GHG (Medel-Jiménez et al. 2024). Targeted actions like precision farming or organic alternatives can reduce consequences.

16.2.2 The Key Principles of LCA

The key principles of LCA include:

i. Life cycle perspective: LCA generally adheres to a cradle-to-grave framework, evaluating the effect on the environment from the extraction of raw materials (cradle) to the product's disposal (grave). This approach guarantees that all stages of a product's lifecycle are examined, facilitating a comprehensive assessment of its overall environmental footprint (Alaux et al. 2024).
ii. Holistic Evaluation: Focuses on factors such as energy use and greenhouse gas emissions to discern trade-offs and advantages. The approach assesses various environmental impact categories, including utilization of energy, greenhouse gas emissions, water consumption, and resource depletion. This thorough evaluation aids in recognizing trade-offs and synergies among various environmental aspects, enabling better informed decisions (Bishop et al. 2021).
iii. Data-driven analysis: Depending on reliable primary and secondary data sources for accurate assessment. LCA predominantly depends on quantitative data, which can be sourced from primary means (e.g., experiments, surveys) or secondary means (e.g., existing literature, databases). The precision of a life cycle assessment is dependent on the quality and thoroughness of the data utilized (Spierling et al. 2018).

iv. Standardization: Ensuring uniformity in all studies through international standards such as ISO 14040 and ISO 14044. Life cycle assessment methodologies are governed by these international standards, which offer frameworks for conducting evaluations. These standards guarantee consistency and comparability among various LCA studies (Rosenboom et al. 2022).

The Key Phases of LCA

LCA is a systematic method for assessing the environmental implications of items across their entire life cycle. Effective LCA requires a precise specification of objectives and the development of system boundaries to direct the assessment process. LCA typically encompasses four primary phases: goal and scope definition, data collection, impact assessment, and results interpretation. The initial step, defining the goal and scope of the assessment, is crucial and entails establishing objectives and defining system boundaries. The objective-setting phase is essential to life cycle assessment (LCA) and articulates the purpose and scope of the evaluation. In accordance with the ISO 14040 standard, the goal must elucidate the purpose of the LCA, its application, and the intended audience for the results (ISO 2006). This step entails identifying particular inquiries, such as comparing the environmental implications of various materials or evaluating a product's sustainability throughout its life cycle. System boundaries refer to the determination of which phases of a product's lifecycle will be examined, encompassing raw material extraction, production, usage, and disposal (cradle-to-grave) or a subset thereof (e.g., cradle-to-gate) (Guinee 2002). The boundaries must align with the study's purpose to ensure relevant impacts are considered (JRC 2010). For biofuels, system boundaries should encompass all life cycle stages (feedstock production, processing, use) and their impacts on ecosystem services (ES), such as carbon sequestration, water management, and habitats (Maia de Souza et al. 2018). The choice of boundaries significantly affects LCA results. For example, if disposal phase is excluded, important impacts related to waste management (such as landfill emissions or recycling benefits) might be missed, resulting in an incomplete assessment (Walker and Rothman 2020). Moreover, it is important that boundaries consider the geographical and temporal contexts, as these factors can influence data availability and certain relevance processes (Moutousidi and Kookos 2021). For example, the environmental impacts of crops can vary based on local farming practices and climatic conditions, which should be accounted for in LCA.

Inventory Analysis in LCA

Inventory analysis in LCA involves the collection and analysis of data on how much water, land, energy is used as well as the emissions of greenhouse gases and pollutants over the product's life cycle. For biofuels, it is crucial to account for direct and indirect interactions with ecosystem services to accurately assess sustainability (Maia de Souza et al. 2018). Data collection focuses on quantifying inputs like fertilizers, energy, and outputs like CO_2 emissions, and wastewater at each stage, from feedstock cultivation to end use. Key aspects include:

i. Resource Use: It is essential to measure the land, water consumption, and energy usage in agricultural activities like irrigation. For biofuels, a detailed data on crop yield, soil carbon sequestration is crucial, as they influence soil fertility and water management (Czyrnek-Delêtre et al. 2017).
ii. Emissions: Both direct emissions (e.g., NO_x from fertilizer application) and indirect emissions (e.g., CO_2 from land-use changes) need to be examined as they have impact on air quality and climate (Mirzaei et al. 2023).
iii. Contextual Factors: Differences across regional variability in soil type, climate, and land management practices require localized data to avoid oversimplification. For example, irrigation demands for corn-based biofuels differ significantly between arid and temperate regions (Makkawi 2023; McCarthy et al. 2021).

In data analysis stage, resource flows are combined and modeled, while uncertainties and gaps are addressed based on:

i. Allocation: When dealing with by-products like distillers' grains from ethanol production, it is essential to account for them within system boundaries. The ISO 14044 standard recommends either expanding the systems or partitioning based on physical or economic relationships (ISO 2006).
ii. Uncertainty and Variability: Ensuring data quality, such as its relevance to temporal and indicators geography are essential, because static assumptions can misrepresent the impact on ecosystem services (ES) (Bayer et al. 2021). Tools like input–output analysis help understand indirect effects, such as land-use changes impacting food crops (Koo et al. 2020; Geoghegan and O'Donoghue 2023).
iii. Integration with ES frameworks: Connecting inventory data to ES classification systems like CICES ensures that impacts on intermediate and final services like pollination and carbon storage are captured. For example, data on soil erosion can relate to habitat and nutrient cycling impact (Steinhoff-Knopp et al. 2021; Haines-Young and Potschin-Young 2018). In summary, LCA inventory analysis requires thorough data collection, resource examination, and emission taking into account various influencing factors and uncertainties. The process integrates information with the broader environmental frameworks to provide a comprehensive view of sustainability impacts.

16.2.3 Impact Assessment

Impact assessment in LCA translates information on resource use and emissions into potential environmental impacts, emphasizing on ecosystem services for thorough assessment, such as biofuel production (Kumari and Arora 2021; Harris et al. 2016). Key Considerations:

i. Non-linear ecosystem dynamics: LCA often assumes simple cause and effect between inputs (fertilizer use) and outputs (emissions), but ecosystems respond in a complex way. Expanding biofuel crops can disproportionately impact ES-like soil carbon sequestration or pollination due to limits in ecosystem resilience (Meyer and Leckert 2018). Tools like the Multiscale Integrated Model of Ecosystem Services (MIMES) simulates interactions between land management and ES provision (Noman et al. 2022).
ii. Spatial and temporal scale dependency: The effect of ES depends on the local and global scale. Locally, soil erosion from intensive farming affects water regulation and habitat quality, and globally, changes in carbon sequestration affect climate regulation (Noman et al. 2022). The cascade's model framework connects data on land use to ES impacts by linking biophysical structures like soil type to final benefits like crop yield.
iii. Integration of ES classification systems: The Common International Classification of Ecosystem Services (CICES) classifies ES to avoid double-counting and clarify trade-offs (Czúcz et al. 2018). For biofuels, this ensures that impacts on intermediate services (nutrient cycling) and final benefits (food security) are both distinct and yet interconnected.
iv. Midpoint to endpoint linkages: Initial data (midpoints) must connect to endpoint impacts like biodiversity loss and human health. Systems like Final Ecosystem Goods and Services Classification System (FEGS-CS) outlines path from land-use changes such as habitat fragmentation to endpoint damages (Landers et al. 2016). An example would be how pesticide emissions harms pollinator populations (midpoint), reducing crop pollination (endpoint) (Maia de Souza et al. 2018; Adhikari and Hansen 2018).

Impact Assessment Techniques

Impact assessment is an assessment process aimed at enhancing decision-making and to ensure that the development choices being evaluated are environmentally and socially feasible and sustainable. Impact assessment examines the potential impacts of a product on the environment, human health, and socioeconomic status. It guarantees that innovations meet sustainability goals, legal requirements, and public safety by weighing risks against advantages. Table 16.1 is a list of some assessment techniques and how they are conducted.

16.2.4 Interpretation

The interpretation phase of LCA merges inventory and impact assessment data to enhance sustainable biofuel production. This phase involves five steps discussed below with a careful consideration of complexities and contextual nuances.

Table 16.1 Impact assessment techniques, aims and how they are carried out

Impact assessment technique	Aim	How it is conducted
Life cycle assessment (LCA)	Quantify environmental impacts across a product's lifecycle	Inventory analysis of resources/emissions, impact modeling (e.g., carbon footprint, energy use)
Risk assessment (RA)	Identify hazards and quantify risks	Hazard identification, dose–response analysis, exposure assessment, risk characterization
Cost–benefit analysis (CBA)	Compare economic costs versus societal/environmental benefits	Monetize direct/indirect costs and benefits, calculate net present value
Social life cycle assessment (S-LCA)	Evaluate socio-economic impacts (e.g., labor rights, community well-being)	Stakeholder interviews, social indicator analysis (e.g., fair wages, health impacts)
Environmental impact assessment (EIA)	Evaluate environmental effects of proposed projects to prevent or mitigate negative impacts	Identify potential risks, predict impacts, propose mitigation strategies, prepare Environmental Impact Statement
Health impact assessment (HIA)	Assess health implications of policies or projects to promote positive health outcomes and minimize risks	Toxicological studies, exposure pathway analysis, epidemiological data review
Multi-criteria decision analysis (MCDA)	Rank alternatives using weighted criteria (e.g., safety, cost, sustainability)	Stakeholder surveys, criteria scoring (e.g., AHP, TOPSIS), aggregation of results
Policy impact assessment (PIA)	Analyze the economic, social, and environmental impacts of public policy measures aimed at enhancing transparency and decision-making	Define policy objectives, engage stakeholders, use indicators to measure alignment with policy aims
Climate impact assessment (CIA)	Assess greenhouse gas emissions and climate resiliency	Carbon foot printing, climate modeling, adaptation planning
Technology assessment (TA)	Analyze societal implications of technologies	Delphi methods, foresight workshops, ethical reviews
Land use impact assessment (LIA)	Manage land-use changes	GIS mapping, zoning reviews, stakeholder consultations

i. Identifying trade-offs and benefits: The cascade model is used to analyze how production impacts on ecosystem services (ES) across spatial–temporal scales (Wang et al. 2022). An increased in biofuel crop yield can boost energy provision (provisioning ES) but might degrade soil health (regulating ES). Tools like InVEST illustrates these trade-offs to prioritize balanced solutions (Ahmadi Mirghaed and Souri 2022). This approach might simplify reversible linkages and lacks clarity for micro-scale impacts like local biodiversity loss.

ii. Context-Specific Analysis: It is important to consider local factors like soil type and availability of water alongside farming practices like non-till farming. For example, cultivating irrigation-intensive biofuel crops in arid regions may exacerbate water shortages, necessitating shifts to drought-resistant crops due to economic dependencies (Ludovico and Lodovico 2020).
iii. Stakeholder engagement and impact valuation: ES impacts are evaluated through both monetary (System of Environmental-Economic Accounting-SEEA) and non-monetary (cultural significance) values to align with stakeholder priorities (Esen and Hein 2020; United Nations Statistics Division 2012). In rural communities for example, they would prioritize habitat preservation over marginal yield gains, guiding policy toward agroecological practices.
iv. Uncertainty and scenario analysis: Sensitivity analyses address data gaps (unmeasured biodiversity impacts) and model uncertainties. Scenario modeling compares land-use strategies like forest conservation versus expanding biofuel production to highlights risks and resilience pathways (Maia de Souza et al. 2018).
v. Policy and practice recommendations: Results inform strategies such as incorporating circular economy practices: Using agricultural residues for biofuels to minimize land competition (The International Resource Panel 2012). Policy alignment: Revising EU Renewable Energy Directive (EU-RED) criteria to account for ES impacts like pollination and carbon sequestration (European Council 2009).

Nuances in Interpretation

Interpreting results necessitates a balance between immediate advantages (biofuel yield) and long-term sustainability (soil health). For example, InVEST maps may prioritize energy production while obscuring biodiversity decline. Drought-resistant crops and other local water scarcity solutions may necessitate alterations in global supply chains. Additionally, power dynamics are crucial: marginalized stakeholders, such as smallholder farmers, typically possess minimal influence in valuation processes, which tend to favor dominant interests. Although beneficial, scenario models that overlook climate variability or cumulative threats risk obsolescence.

16.3 Bionanocomposite Materials

16.3.1 Composition and Characteristics

Bionanocomposite materials are composed of a biopolymer matrix reinforced with nanoscale fillers, combining renewable resources with nanotechnology to enhance material properties. The biopolymer matrix typically includes biodegradable polymers such as polylactic acid (PLA), cellulose, chitosan, starch, or polyhydroxyalkanoates (PHAs), sourced from natural plants, algae, or microbial fermentation (Sanni et al. 2022; Arrieta et al. 2018). Biopolymers are chosen for their eco-friendly characteristics, including biodegradability and biocompatibility. Nanofillers like silicon carbide (SiC), zinc oxide (ZnO), carbon nanotubes (CNTs), graphene, or clay nanoparticles, are incorporated to improve thermal, mechanical, and barrier properties (Sajna et al. 2017; Zia 2020). For example, the addition of SiC to starch increased thermal stability from 250 to 275 °C, while ZnO nanoparticles in PHB raised degradation temperatures by 24 °C, enhancing heat resistance (Díez-Pascual and Díez-Vicente 2014; Dash and Swain 2013).

BNCs exhibit unique properties like enhanced thermal stability. The addition of nanofillers like cellulose nanocrystals (CNC) to PLA increases crystallinity and thermal resistance (Khoo and Chow 2022). Improved mechanical strength is demonstrated through reinforcement with materials such as graphene or carbon nanotubes (CNTs), resulting in a 57% increase in tensile modulus for PLA/CNT composites (Cooper et al. 2019). The barrier properties are also enhanced, with nanoparticles such as Chitosan-$CaCO_3$, which reduced oxygen permeability by threefold (Dash and Swain 2013). The biodegradability and biocompatibility, of natural polymers like chitosan or PLA, make them suitable for medical and packaging application.

The performance of composites relies on effective dispersion of nanofiller (using techniques like sonication or surfactants) and processing methods (like melt-mixing or solution casting), ensuring a homogeneous integration and optimal interfacial interactions (Sanni et al. 2022). An example is the covalent bonding between chitosan and SiO_2 nanoparticles forms crosslinked networks, enhancing thermal stability (Chrissafis et al. 2008). The speciality of BNCs with improved properties makes them sustainable and high-performance choice for the packaging, biomedical, and automotive industries.

16.3.2 Applications of Bionanocomposites

Bionanocomposites are gaining attention as sustainable alternatives to traditional materials across various industries, for their enhanced properties and environmental benefits. Among other industries are the medical, packaging, and construction industries. Table 16.2 shows examples of BNCs composition and applications.

Table 16.2 Summary of bionanocomposites: composition, application, and properties

Bionanocomposite	Area of application	Properties	References
PLA-magnetic nanocomposites	Targeted drug delivery, magnetic resonance imaging	Magnetic targeting, imaging capabilities	Mohammadi-Zerankeshi and Alizadeh (2023)
PHA-silver nanocomposite	Antimicrobial application, medical implants	Antimicrobial activity, controlled drug release	Mukheem et al. (2018)
Starch-gold nanocomposite	Photothermal therapy, drug delivery	Targeted drug delivery, photothermal effect	Tian et al. (2012)
Starch-clay nanocomposites	Food packaging, controlled drug release	Improved mechanical strength, sustained release	Maity et al. (2021)
Cellulose-silver nanocomposite	Antimicrobial wound dressings, infection control	Enhanced antimicrobial properties, stability	Jeevanandam et al. (2023)
Alginate-chitosan nanocomposite	Controlled drug release, wound dressings	Sustained drug release, enhanced stability	Rezaei and Ehtesabi (2022)
Chitosan-carbon nanotube	Antibacterial coatings, biosensors, drug delivery	Improved conductivity, antibacterial activity	Farokh et al. (2023)
PLA-zinc oxide nanocomposite	UV-blocking, wound dressings, antibacterial applications	UV-blocking properties, improved antibacterial activity	Chong et al. (2023)
Collagen-silver nanocomposite	Antibacterial coatings, wound dressings	Antimicrobial activity, controlled release	Kalirajan and Palanisamy 2020)
Starch-mesoporous silica nanocomposites	Bioimaging, controlled drug release	Imaging capabilities, controlled release	Abukhadra et al. (2020)

(continued)

Table 16.2 (continued)

Bionanocomposite	Area of application	Properties	References
Chitosan-graphene oxide nanocomposite	Drug delivery, neural tissue engineering	Improved drug release, enhanced mechanical properties	Karthika et al. (2020)
Chitosan-nanohydroxyapatite	Dental applications, bone tissue engineering	Improved bone regeneration, bio-mineralization	Mohammadnia et al. (2022)
Pullulan-zinc oxide nanocomposite	Antibacterial applications, UV-blocking wound dressings	Improved antibacterial activity, UV-blocking properties	Bailore et al. (2021)
Cellulose-gold nanocomposites	Biosensing, photothermal therapy	Biosensing capabilities, photothermal effect	Kanaoujiya et al. (2022)
Gelatin-magnetic nanocomposite	Hyperthermia cancer treatment, targeted drug delivery	Magnetic targeting, controlled drug release	Araújo-Custódio et al. (2019)

In the packaging industry, PLA-bionanocomposites are often combine with cellulose nanocrystals (CNC). These materials are biodegradability and protect the products from oxygen and moisture, extending the shelf life of perishable goods. Starch-based nanocomposites with silicon carbide (SiC) offers superior thermal stability and lower oxygen permeability, making them suitable for high-temperature packaging applications (Dash and Swain 2013). Compared to conventional plastics, these materials help reduce reliance on fossil fuels and minimizing environmental pollution.

In the construction sector, bionanocomposites from chitosan-based incorporated with calcium carbonate ($CaCO_3$) or silica (SiO_2) nanoparticles enhanced thermal stability and mechanical strength. These composites find application as insulation and lightweight structural materials for their resistance to thermal degradation and eco-friendly nature (Sanni et al. 2022). For example, biobased-SiO_2 hybrids forms covalent bonding between polymer matrices and nanoparticles, enhancing stability high temperatures (Morales et al. 2023). These materials offer sustainable alternative to traditional concrete and steel, reducing carbon footprints in construction.

The automotive industry also benefits from bionanocomposites like polyetheretherketone (PEEK) reinforced with carbon nanotubes (CNTs) or graphene nanoplatelets (GNp), these provide high strength-to-weight ratios and thermal resistance (Swaminathan et al. 2020; Uddin et al. 2019). Bio-based poly (butylene succinate) (BioPBS) composites with pyrolyzed biomass and GNp have shown to improved tensile modulus and heat deflection temperatures, making them ideal for interior panels and engine components (Cooper et al. 2019). These materials reduce vehicle weight, enhancing fuel efficiency while maintaining durability advantages, which are unattainable with conventional metals or non-biodegradable polymers. BNCs help industries reduced environmental impact, enhanced performance, versatility, and cost-effectiveness compared to traditional materials, aligning with global sustainability goals.

16.3.3 Safety Concerns on Bionanocomposites Usage

Bionanocomposites are gaining attention across various sectors for their added advantages of being biodegradability, reduced toxicity, and renewability. However, a complete understanding is needed on the risk associated with the usage of BNCs in various sectors especially areas with close human contact. A number of concerns have been identified with the use of BNCs, hence the need for LCA on BNCs.

A major concern is the potential transfer of nanoparticles from packaging to food. The potential migration of NPs in nanoscale form remains uncertain; however, their unique characteristics relative to bigger particles may result in unforeseen health implications. Factors affecting migration comprise storage conditions (duration, temperature), material properties, food attributes (e.g., pH), and the interaction between migrants and packaging materials. For instance, acidic meals such as orange juice exhibit increased silver migration from nano-silver-coated screens in comparison with other foods (Metak et al. 2015).

The European Food Safety Authority (EFSA) has developed recommendations for analyzing nanoparticles in food chains. These include analyzing physicochemical features, exposure levels, hazard characterization, and toxicological hazards such as reproductive toxicity, chronic toxicity, immunotoxicity, neurotoxicity, and endocrine activity. Despite these attempts, defined limitations for nanomaterials in food contact applications remain lacking.

Improper disposal of BNCs may release nanomaterials into ecosystems, resulting in soil and water toxicity and disturbing both aquatic and terrestrial organisms. Research underscores the carbon footprint of specific coatings, an instant been soy protein isolate-cellulose nanocrystal, which resulted in a carbon footprint of about 0.33 kg carbon dioxide eq/kg of solution (Rajendran et al. 2024), stressing the urgent need of reconciling ecological impacts with economic scalability.

BNCs have significant potential in medicinal applications, although encounter obstacles in guaranteeing the safety and stability of nanoparticles during usage. The regulatory approval processes are prolonged due to apprehensions over toxicity and repeatability in large-scale production.

Addressing safety concerns necessitates extensive toxicological data, standardized testing protocols, and coordination among researchers, policymakers, and industry stakeholders to guarantee safe and sustainable application of BNCs in food packaging and other areas. All these concerns raises the need for LCA in BNCs.

16.4 Methodology for LCA of Bionanocomposites

16.4.1 Selection of Functional Unit

In LCA of bionanocomposites, being able to select the right functional unit is important. This unit aids in comparing the environmental impacts of various materials and processes, and must also accurately reflect the performance and purpose of BNCs, for fair comparisons across different contexts. Functional units are based on specific BNCs performance in various applications. Here some examples:

i. Mechanical Properties: In construction, the functional units might be "per unit of "per unit of tensile strength" or "per unit of flexural modulus" which focuses on material's capacity to withstand stress and pressure (Kumari and Arora 2021).
ii. Barrier Properties: In packaging usage, the units could be "the amount of material needed to preserve 1 kg of food for a specified time," emphasizing the material's ability to block oxygen and moisture to maintain food quality (Kupolati et al. 2022; Nizam et al. 2021).
iii. Thermal Properties: In construction, the units might be "the volume of insulation material needed to achieve a thermal resistance of 1 $m^2 \cdot K/W$" evaluating the material's effectiveness in heat retention or loss, with composites like chitosan-SiO_2 hybrids, offering enhanced thermal stability (Sanni et al. 2022).
iv. Weight Reduction: With automotive applications, the focus is on reducing components weight, such as using enough material to decrease a part's mass while maintaining mechanical strength as seen with polyetheretherketone (PEEK)–carbon nanotube composites (Xu et al. 2019; Liu et al. 2022) and similar results in biological application (Bashandeh et al. 2022).

Functional units must align with ISO 14040/44 standards, ensuring they are measurable, specific to the application and considers the entire lifecycle (production, use, disposal). An example, starch–SiC nanocomposites require units that accounts for their thermal stability and biodegradability to highlight their sustainability benefits (Dash and Swain 2013). The linking of functional unit to performance metrics enables LCA to capture the trade-offs between environmental impacts and the benefits of BNCs over conventional materials (Calvino et al. 2020).

16.4.2 Data Collection Techniques

Data collection is essential for research, especially when it comes to BNCs, which requires information accuracy in understanding material's performance, eco-friendliness, and health implication. In LCA, gathering data is foundation step in assessing the environmental impacts of a product. The two main data collection techniques are primary and secondary sources, with each presenting unique benefits and challenges. Accurate, relevant, and comprehensive data key for reliable of LCA results.

Primary data collection involves getting information directly from the source and is usually tailored to a specific product being studied. This approach is precise and reliable. Different techniques have been used for primary data collection including: On-site measurements, which is a direct data collecting from manufacturing facilities, such as energy consumption, raw material usage, and emissions. Zhang et al. (2015) measured energy usage and emissions at a bionanocomposite production site for their study. Surveys and interviews usually gather insights through questionnaires or interviews from key players like manufacturers, suppliers, and end-users. Kumari and Arora (2021) gathered data on raw material sourcing and transport on BNCs using surveys. Experiments and laboratory analysis undertaken measure characteristics like biodegradability, thermal stability, and mechanical strength. (Sharma et al. 2018) used this method to determine the biodegradation rates of nanocellulose-based composites. Moreover, process monitoring engages sensors and monitoring systems to collect real-time data on resource, emissions, and waste generation with IoT-based sensors monitored water and energy usage in a BNCs manufacturing site (Brandt et al. 2021; Tao et al. 2014; Asmara 2018). Primary data collection has its own set of challenges, including been time-consuming, labor-intensive, and financially demanding in complex supply chains (Okogwu et al. 2023). The cost involved is high covering for specialized equipment, trained personnel, and access to facilities (Saavedra-Rubio et al. 2022). Gaining access to proprietary or confidential data can be difficult (Atoum and Keshta 2021), and also, there can be inconsistency in data due to varying measurement techniques and operational conditions (Radnik et al. 2022; Gao and Lowry 2018).

Secondary data collection, on the other hand, involves using existing sources like databases, literature, industry reports to gather information and is used when primary data hard to obtain. LCA databases are a common secondary data source, with valuable information on materials, energy use, and emissions. An example is when Sanni et al. (2022) used Ecoinvent database for data on the environmental impact of polymer production related to bionanocomposites. Scientific literature, including journals and technical reports, often contains relevant data on material properties and processes. The following studies extracted information on energy consumption of nanocellulose production in their work (Gallo Stampino et al. 2021; Malovrh Rebec et al. 2024; Kane et al. 2023). Industry reports from associations, government agencies, and NGOs can provide insights into production processes and emissions. Data on global carbon nanotubes production reports was gathered by a study by Swaminathan et al. (2020). Government and public data, like national statistics and environmental reports, can provide context on resource usage and emissions and such a platform. Environmental Protection Agency was used by Khoo and Chow (2022) in their report on industrial processes emissions. Secondary data collection also presents several challenges, such as variation in data quality, impacting accuracy, and reliability of product or process being examined (Kupolati et al. 2022). Data gaps, with missing or incomplete information, limit the comprehensiveness of LCA (Kleinekorte et al. 2020). Additionally, data may be outdated and not relevant to some specific regions (Huang et al. 2022). Lastly, compatibility issues from differences in data formats, units, or system boundaries complicate secondary data incorporation into LCA framework (Furberg et al. 2022).

16.4.3 Software Tools for LCA

LCA software tools are used to assess environmental impacts' assessment and life cycle of products and processes. There is a variety of such tools with their distinct functionalities and applications. This section explores some known LCA software and their functionalities along with relevant case studies shown in Table 16.3.

16.5 Environmental Impact Categories

16.5.1 Global Warming Potential

Global Warming Potential (GWP) is a measure that indicates how different greenhouse gas (GHG) contribute to climate change by comparing their heat-trapping capacity relative to carbon dioxide (CO_2) over a period of 100 years (IPCC 2013a). This is especially vital for BNCs which are designed to be sustainable alternatives to traditional materials. Industries like packaging, automotive, and construction use GWP analysis to evaluate and enhance the environmental effect of these materials, aiding with the development of more sustainable technologies.

BNCs offer a lower carbon footprint compared to traditional petroleum-based materials because of their renewable nature and the ability to sequester carbon during the growth of the biomass (Thakur et al. 1411). However, their production involves emission from agricultural practices, uses energy-intensive processes, and decomposition in landfills which can release methane (CH_4), a potent GHG with a GWP 28–34 times higher than CO_2 over 100 years (IPCC 2013b; Herath et al. 2023).

A number of studies have looked into assessing the GWP of bionanocomposites. A study by Yadav et al. (2020) used LCA to examine the GWP of polyhydroxyalkanoate (PHA)-based nanocomposites in comparison with conventional petroleum-derived plastics. Their findings indicated that PHA nanocomposites exhibited lower GWP,

Table 16.3 LCA software tools and their functionalities

Software	Functionality	Case study	References
SimaPro	Comprehensive LCA with multiple impact assessments and extensive database integration	What are the main environmental impacts ad products contributing to the biodiversity footprint of EU consumption? A comparison of LCI impact assessment methods and models	Sanyé-Mengual et al. (2023)
GaBi	Industry focused LCA, product sustainability, and circular economy modeling	Life cycle economic and environmental implications of using nanocomposites in automobiles	Lloyd and Lave (2003)
OpenLCA	Flexible open-source, cost-effective for LCA integrated with multiple databases	LCA of engineered nanomaterials: state of the art and strategies to overcoming existing gaps	Hischier and Walser (2012)
BEES	Evaluates environmental and economic performance of building products	Measuring the life cycle environmental and economic performance of concrete: the BEES approach	Lippiatt and Ahmad (2004)
Umberto	Carbon footprint analysis and overall environmental impact assessments	Analysis of sugar cane production and transportation in Hoya del Rio Suárez from life cycle perspective	Sierra et al. (2022)
Brightway2	Open-source framework for complex analyses with focus on transparency	LCA of carbon capture and utilization as a negative emission technology: recommendations and case study	Duval-Dachary et al. (2024)
CMLCA	Educational and professional assessments; user-friendly interface	Thermodynamic resource indicators in LCA: a case study on the titania produced in Panzhihua city, southwest China	Liao et al. (2012)
iPoint	Suite of tools for sustainability assessments and regulatory compliance	Exergetic life cycle assessment for green hydrogen production	Gerhardt-m et al. (2024)

(continued)

Table 16.3 (continued)

Software	Functionality	Case study	References
Ecochain	Tailored for SMEs, circular economy projects and businesses for scalable solutions	Life cycle assessment for particleboards with recycled wood in an industrial context	Crociati et al. (2024)
LCA Calculator	Offers predefined templates, ideal for educators and small businesses	Development of LCA calculator to support community infrastructure co-design	Borrion et al. (2019)

primarily attributed to the carbon sequestration during the production phase of the biopolymer from waste streams. However, the study indicated that the addition of nanofillers and energy requirements significantly impacted the overall GHG emission profile. Various reviews noted that the environmental benefits of bionanocomposites are intricately linked to several factors including how raw materials are sourced, efficiency of manufacturing processes, and how waste is managed, highlighting the complexity in evaluating environmental impact of materials (Vasile and Cazacu 2013; Ren et al. 2018; Mohammed et al. 2024).

The incorporation of nanotechnology in BNCs adds complexity in GWP assessment. The production of nanofillers, like carbon nanotubes or graphene, often requires energy-intensive methods which counteract the carbon benefits from using biopolymers (Tripathy and Gupta 2024; Carroccio et al. 2022). Furthermore, the potential release of nanoparticles during BNCs degradation raises concerns about their long-term environmental impact, which is not fully captured by traditional GWP metrics (Mulla et al. 2021). To improve GWP of BNCs, experts propose strategies like optimizing agricultural practices to reduce emissions from biopolymer growth, adopting renewable energy sources for manufacturing, and establishing closed-loop recycling systems to minimize waste (Mujtaba et al. 2023). Additionally, advancements in green chemistry, usage of bio-based nanofillers and low-energy processing techniques, are expected to further reduce the carbon footprint of these materials (Formela et al. 2022). BNCs have the potential in reducing greenhouse gas emissions and mitigating climate change; however, consideration of their entire life cycle is essential to maximize their environmental benefits. Future research should aim at refining production processes, improving the sustainability of raw material sourcing, and developing effective waste management strategies to enhance the overall sustainability of BNCs.

16.5.2 Resource Depletion

Resource depletion occurs with the exhaustion of natural resources through excessive usage, creating both an environmental and economic challenge. With the rise in population and industrial development, there is increased consumption of natural resources, especially non-renewable ones like oil and minerals, raising concerns about their future availability. This section examines the utilization trends of natural resources and evaluates the sustainability of raw materials usage in various sectors, with a focus on BNCs. The global natural resource extraction has tripled in the past 40 years, with fossil fuels, minerals, and biomass being most heavily exploited (UNEP 2024). Non-renewable resources like petroleum, natural gas, and rare earth elements face particular depletion risks due to high industrial demand (Blass et al. 2017). Conventional materials production heavily relies on these finite resources. For example, petroleum-based plastics consume about 6% of global oil production annually, contributing to resource depletion and pollution. Similarly, rare earth element mining for electronics and renewable energy technologies raises sustainability concerns due to environmental impacts and geopolitical tensions (Saouter and Gibon 2024). The global material usage is projected at 190 billion tons by 2060, while greenhouse gas emissions could increase by 43% (Ekins and Zenghelis 2021; Modarress Fathi et al. 2023). This increased consumption poses challenges to achieving global climate, biodiversity, and economic prosperity goals (UNEP 2024).

Biopolymers, such as polylactic acid (PLA), starch, and cellulose, obtained from renewable resources like corn, sugarcane, and wood, make them sustainable than petroleum-based plastics (Brodin et al. 2017). However, their sustainability depends on factors like farming method, land use, and water consumption. For instance, growing crops for biopolymer production can deplete resources without proper management. Intensive farming practices result in soil degradation, water shortages, and biodiversity loss, undermining the environmental benefits of biopolymers (Yadav et al. 2020). Additionally, the use of fertilizers and pesticides in agriculture contributes to GHG emissions and water pollution, further complicating the sustainability assessment of biopolymer feedstocks (Mukherjee et al. 2019). The incorporation of nanofillers into bionanocomposites introduces further issues on resource depletion. Nanofillers like cellulose nanocrystals and clay nanoparticles are obtained from abundant natural resources, while carbon nanotubes and graphene require energy-intensive production processes and rare raw materials (Tan et al. 2022). The extraction and processing of nanomaterials can have significant environmental impacts, including high energy consumption and the generation of hazardous waste. To address these challenges, it is important for government, researchers, and industries to focus making raw materials used in bionanocomposites sustainable via: Sustainable Agriculture**:** It is crucial to adopt regenerative farming practices like crop rotation, organic farming, and precision agriculture, which can lessen environmental impact of biopolymer feedstocks while maintaining soil health and biodiversity (Stathatou et al. 2023). Circular Economy: Recycling and reusage of bionanocomposites can minimize waste and demand new materials. Chemical recycling techniques can break down biopolymers into their monomers for repolymerization, creating a circular lifecycle (Lizundia et al. 2022). Green Nanotechnology: Creating bio-based nanofillers and using low-energy production methods can decrease environmental impact of nanomaterials. The usage of use of agricultural waste for cellulose nanocrystals presents a sustainable alternative (Yadav et al. 2020). Policy and Certification: Governments can promote resource management through policies, regulations, and certification programs. The European Union's Circular Economy Action Plan encourages renewable materials usage and waste reduction (Souza et al. 2024; Polyakov et al. 2021).

16.5.3 Ecotoxicity

Ecotoxicity of a material refers to its harmful effects on the ecosystem and there is a need to understand these potential risks posed by BNCs on the environment and human health. BNCs usage is on the rise across various industries. Evaluating their ecotoxicity helps to understand their potential adverse effects through direct or indirect exposure. Even though bionanocomposites are marketed as environmentally friendly, it could still pose ecotoxicological risks due to the presence of nanofillers or degradation byproducts. For example, nanoparticles like carbon nanotubes, silver nanoparticles, or titanium dioxide, which are added to enhance product strength, can also be toxic to aquatic and terrestrial organisms (Turan et al. 2019), causing oxidative stress, DNA damage, and impaired growth and reproduction in species such as algae, fish, and invertebrates (Saxena et al. 2024; Hou et al. 2019). As BNCs degrade, they can also release micro- and nanoplastics, which are emerging pollutants. These particles can accumulate in water bodies and soil, leading to bioaccumulation in organisms and biomagnification through the food chain (Bolan et al. 2024). Studies have demonstrated microplastics adsorption of harmful chemicals like pesticides and heavy metals, increasing their toxicity to marine life (Vo and Pham 2021) and human health, as humans are part of the ecosystem and rely on it for food, water, and air. Exposure to toxic substances can occur through ingestion, inhalation, or dermal contact and with humans can cause cellular damage, inflammation, and even carcinogenic effects. This is particularly concerning for workers in manufacturing facilities who are highly exposed (Prata et al. 2020; Kauts 2023). Additionally, the presence of toxic additives like plasticizers or flame retardants can leach into water or food sources triggering health issues like endocrine disruption, developmental disorders, and other chronic health conditions in humans (Campanale et al. 2020; Kowalczyk et al. 2022). The long-term effects of low-dose exposure remains unclear, highlighting the need for further research.

To understand these risks, researchers employ laboratory tests, field studies, and computational modeling. Standardized toxicity tests by Organisation for Economic Co-operation and Development (OECD) assessed substances affect organisms like Daphnia magna (water fleas), zebrafish, and earthworms (Hund-Rinke et al. 2022; OECD 2011) focusing on mortality, growth inhibition, and reproductive success. Advanced techniques like omics technologies (e.g., genomics, proteomics, and metabolomics) aids in understanding the molecular level effect and identify biomarkers of exposure (Álvarez et al. 2023). Computational models, such as quantitative structure–activity relationship (QSAR) models, help predict the toxicity of new materials based on their chemical properties, reducing the need for extensive animal testing (Cronin et al. 2019). To minimize risk, strategies like regulatory framework, monitory and remediation, green chemistry, and LCA could be implemented.

16.5.4 Human Health Impacts

Bionanocomposites are used in the packaging, automotive, and biomedical applications for their environmental benefits; however, risks to human health may arise during production, use, and disposal. These risks arise from exposure to nanoparticles, chemical additives, and degradation byproducts, which can enter the human body through inhalation, ingestion, or dermal contact. This section discusses potential health risks with BNCs and highlights the need for comprehensive risk assessment and management. Their production involves synthesis and incorporation of nanofillers, such as carbon nanotubes, cellulose nanocrystals, or clay nanoparticles, into biopolymer matrices. This process can release nanoparticles into the air, posing inhalation risks to workers. Inhalation of nanoparticles is linked to respiratory and cardiovascular diseases, as well as systemic inflammation and oxidative stress (Oberdörster and Kuhlbusch 2018; Oberdörster and Graham 2018). Carbon nanotubes, which are similar to asbestos structurally, are shown to cause lung fibrosis and mesothelioma in animal studies (Barbarino and Giordano 2021; Donaldson et al. 2006). Additionally, workers exposure during production can lead to neurological disorders, endocrine disruption, and carcinogenic effects (Nel 2023). Occupational safety measures, such as the use of personal protective equipment (PPE) and proper ventilation systems, are essential to minimize these risks. In products like food packaging, textiles, and medical devices, concerns are raised about human exposure to nanoparticles and chemical additives. For instance, nanoparticles can migrate from food packaging into food under high temperatures or prolonged storage (Bott et al. 2018) which can cause gastrointestinal inflammation, disruption of the gut microbiome, and systemic toxicity in animal (McClements 2021). In biomedical applications, BNCs are used in drug delivery systems, wound dressings, and tissue engineering scaffolds with significant therapeutic benefits, and also the potential release of nanoparticles or degradation byproducts into the body can lead to adverse effects, such as immune responses, cytotoxicity, and organ damage (Akbar et al. 2020; Kuperkar et al. 2024). Hence, the need for strict testing and regulation is necessary to ensure the safety in medical applications.

The disposal of bionanocomposites through incineration or landfilling can release toxic substances into the environment, affecting health. Incineration can generate hazardous air pollutants, such as dioxins and furans, which causes cancer and reproductive disorders (Domingo et al. 2020). Landfilling, on the other hand, can lead to the leaching of toxic additives and nanoparticles into groundwater contaminating drinking water sources and posing risks to human health (Kumar et al. 2023). As BNCs break down in nature can also produce micro- and nanoplastics, which have been detected in food, water, and even the air. Human exposure to these particles has been associated with inflammatory responses, oxidative stress, and potential carcinogenicity (Le et al. 2023). The long-term health effects of long exposure to plastics are still not fully understood, necessitating further research and monitoring.

16.6 Case Studies

Case studies are necessary for translating theoretical concepts into practical applications across fields such as environmental science, engineering, and materials science. They enable researchers and practitioners to comprehend the implementation of theoretical principles, particularly in emerging areas like BNCs. Case studies promote critical thinking, problem-solving abilities, and innovation, allowing participants to evaluate both successful implementations and insights gained from failures. In sectors such as food packaging and healthcare, case studies demonstrate the environmental and financial implications of novel materials and technologies. They facilitate knowledge exchange among academia, industry, and policymakers, which is crucial for handling complex challenges and promoting sustainable development.

16.6.1 LCA of PLA-Based Nanocomposite Film for Food Packaging

This section examines a case study that illustrate LCA method for various BNCs, emphasizing their manufacture, application, and environmental performance. The case study examines the work by Pasha et al. (2023).

Case Study: New polylactic acid-based nanocomposites film for food packaging application

i. **Goal and scope definition**

The study examines the development of polylactic acid (PLA)-based nanocomposite films reinforced with nanoclays (C20A, C30B) and plasticized with triacetin (TA) to tackle the environmental drawbacks of conventional plastics (which rely on fossil fuel and are non-biodegradability) while improving mechanical, barrier, and thermal properties for food packaging. Although the study does not conduct a full LCA, its objectives align with LCA principles by emphasizing sustainability, biodegradability, and reduced environmental impact.

Key Scope Elements from the Study:

Functional Unit: Development of films with optimal tensile strength (TS), elongation-at-break (E_{ab}), water vapor permeability (WVP), and thermal stability.

System Boundaries: Cradle-to-gate (material production, film fabrication via solvent casting) but excludes downstream phases (transport, usage, end-of-life).

Environmental Motivation: Mitigating plastic pollution through biopolymers (PLA) and renewable additives (TA, nanoclays).

ii. **Inventory analysis (LCI)**

The study provides data on material inputs, energy use, and emissions during film production:

Material Inputs:

PLA: Derived from renewable resources (e.g., corn starch), biodegradable, but energy-intensive during polymerization.

Triacetin (TA): A bio-based plasticizer that improves flexibility but increases hydrophilicity and WVP.

Nanoclays (C20A, C30B): An organically modified montmorillonites improve barrier properties but require energy for surface modification.

Chloroform: A hazardous solvent used in solvent casting, posing toxicity risks.

Energy Use: Solvent evaporation at 50 °C for 12 h, homogenization, and ultrasonication. Nanoclay Modification: Hydrophobic treatment of MMT to enhance PLA compatibility.

Emissions/Waste: Chloroform emissions during solvent evaporation. Potential nanoclay nanoparticle release during degradation (not quantified).

iii. **Life Cycle Impact Assessment (LCIA)**

The study lacks quantitative environmental impact metrics; however, qualitative insights can be inferred:

Positive impacts: Reduced Fossil Fuel Dependency: PLA is synthesized from renewable biomass, avoiding petroleum-based plastics. Enhanced Biodegradability: PLA degrades faster than conventional plastics under industrial composting conditions.

Improved performance: Higher E_{ab} (78.5% with 20% TA) reduces film brittleness, minimizing packaging failure and food waste. Lower WVP (1.06×10^{-10} g/m s Pa with 3% C20A/10% TA) extends food shelf life. UV-blocking properties (reduced light transmission at 280 nm) protect photosensitive foods.

Negative impacts: Solvent use: Chloroform is toxic and volatile, requiring careful handling and recovery systems. Nanoclay Toxicity: Potential ecotoxicity of nanoclays (C20A/C30B) during disposal or degradation (not addressed in the study). Energy Intensity: PLA production and nanoclay modification are energy-intensive processes.

iv. **Interpretation and trade-offs**

The study demonstrates PLA-based nanocomposites balances performance and sustainability but highlights unresolved LCA challenges:

Strengths: Material efficiency: Optimal nanoclay loading (1–3 wt%) enhances TS and WVP without excessive material use. Waste reduction: Films with 20% TA indicate high ductility ($E_{ab} = 78.5\%$)), reducing breakage during transportation.

Limitations: End-of-life scenarios: No data on composability rates, nanoparticle leaching, or recycling feasibility. Toxicity: Chloroform and nanoclays may pose risks during production or disposal. Trade-offs: While TA improves flexibility, it increases moisture content (MC = 19.4% with 3% C30B/20% TA), potentially accelerating hydrolytic degradation.

v. **References to prior LCA-relevant studies**

PLA versus Petrochemical plastics: PLA's biodegradability and lower carbon footprint are highlighted.

Nanoclay Benefits: C20A reduces WVP by 72% compared to neat PLA, enhancing barrier properties and reducing food spoilage.

Thermal Stability: Nanocomposites show marginally improved thermal stability ($T_{\text{nitial}} = 338.7$ °C for C20A3/TA20), reducing energy use during high-temperature applications.

vi. **Recommendations for future LCA studies**

There is a need to quantify energy use and emissions during PLA synthesis, nanoclay modification, and solvent recovery; to assess end-of-life scenarios (composting, recycling, or incineration) for PLA/nanoclay films; and to evaluate nanoparticle toxicity and long-term environmental persistence.

This study does not conduct a full LCA, but offers critical insight into the environmental and functional advantages of PLA-based nanocomposites. Incorporating TA and nanoclays enhances mechanical and barrier properties, aligning with sustainability goals. However, there is a need to address unresolved issues like solvent toxicity, nanoparticle impacts, and end-of-life management. This case study illustrates BNCs potential to reduce plastic pollution while highlighting the necessity for holistic environmental assessments.

16.6.2 Comparative Analysis with Conventional Materials

This section provides a comparative analysis of the environmental impacts of BNCs and conventional materials, highlighting their advantages and disadvantages.

The analysis relies on essential environmental indicators such as global warming potential, decreasing resource reserves, usage of energy and impact of end-of-life (Table 16.4).

BNCs offer environmental benefits compared to traditional materials due to their diminished carbon footprint, dependence on renewable resources, and biodegradability. However, challenges including energy-intensive production, resource depletion from nanofillers, and potential ecotoxicity must be addressed to fully realize their sustainability potential. By optimizing production methods, utilizing renewable energy sources and developing safer nanofillers, BNCs will become a more sustainable alternative to traditional materials.

Table 16.4 Comparison of the environmental impacts of BNCs and conventional materials

Environmental indicator	Bionanocomposites	Conventional materials	Advantages of bionanocomposites	Disadvantages of bionanocomposites
Global warming potential (GWP)	Lower GWP due to carbon sequestration during biopolymer growth (e.g., PLA from corn, starch from potatoes)	Higher GWP due to fossil fuel extraction and processing (e.g., polyethylene, polypropylene)	Reduced GHG emissions and contribution to climate change	Energy-intensive production processes can partially offset GWP benefits
Resource depletion	Renewable raw materials (e.g., biopolymers from plants, cellulose from wood)	Non-renewable resources (e.g., petroleum for plastics, metals from mining)	Reduced reliance on finite resources and lower environmental degradation from extraction	Nanofiller production (e.g., silver nanoparticles, clay) can still contribute to resource depletion
Energy consumption	Moderate to high energy use during manufacturing (e.g., extrusion, injection molding)	High energy use in extraction, refining, and processing (e.g., petroleum refining, metal smelting)	Potential for lower energy use when renewable energy sources are adopted in production	Energy-intensive processes for nanofiller synthesis and composite production
End-of-life impacts	Biodegradable and compostable options available (e.g., PLA, starch-based composites)	Non-biodegradable; contributes to landfill waste and microplastic pollution (e.g., polyethylene, PVC)	Reduced long-term environmental impact and waste accumulation	Limited biodegradability in certain environments; potential release of nanoparticles
Toxicity	Low toxicity for most biopolymers; potential ecotoxicity from nanofillers (e.g., silver nanoparticles)	High toxicity from chemical additives (e.g., plasticizers, flame retardants) and heavy metals (e.g., lead)	Reduced exposure to harmful chemicals and additives	Potential ecotoxicity from nanofillers during degradation or incineration

16.7 Challenges in LCA of Bionanocomposites

LCA is necessary for understanding environmental impacts of materials, including BNCs. Conducting LCA for bionanocomposites is complicated due to its complex composition, unique production processes, and emerging applications. These factors can affect the accuracy, reliability and comprehensiveness of LCA results. The main challenges of LCA on BNCs are as follows:

i. Data availability and quality are limited, as trustworthy information on bionanocomposites is scarce due to their status as relatively new materials and the ongoing development of their production processes.
ii. Uncertainties in impact assessment, as well as variability, are intrinsic issues in LCA, especially noticeable for developing materials such as bionanocomposites.
iii. Lack of standardized methodologies for evaluating the environmental impacts of bionanocomposites presents a considerable challenge. Current LCA frameworks and guidelines, such as ISO 14040/14044, are not tailored for nanomaterials or multifunctional composites. Additionally, the inherent complexity of nanomaterials, coupled with the incorporation of nanoparticles into bionanocomposites, further complicates the LCA process.

Data limitations, impact assessment uncertainties, and the absence of established procedures make bionanocomposite LCAs complex. To address these problems, academics, industry stakeholders, and regulators must collaborate to improve data quality, develop nanomaterial-specific recommendations, and strengthen LCA frameworks. Overcoming these limitations, LCA can shed light on BNCs sustainability and promote their responsible use.

16.7.1 Data Availability and Quality

Reliable and precise data is essential for appropriately assessing the environmental impact of BNCs. Nonetheless, various challenges hinder data collection due to the novelty of these materials, the intricacy of their production processes and the absence of standardized data collection methodologies. A significant obstacle is the scarcity of data regarding bionanocomposites. Reliable data on the extraction of raw materials for new materials is usually scarce. Data regarding the environmental impacts of sourcing biopolymers such as corn and sugarcane, in conjunction with nanofillers, is inadequate. The energy and water demand for developing biopolymer feedstocks might differ markedly depending on agricultural methods (Mastropetros et al. 2022). Moreover, the data regarding energy consumption and emissions during production are from laboratory settings, which may not fully represent industrial processes (Lizundia et al. 2022), and also insufficient data regarding disposal, degradation, or recycling hinders evaluations (Lüftl and Visakh 2016; Biswas et al. 2022). The absence of comprehensive data creates uncertainties in LCA outcomes, compromising their reliability.

The variability of study data reporting is concerning. Different researchers use different methods, system boundaries, and functional units, complicating comparisons. Some studies omit lifecycle stages like production or disposal, leading to insufficient evaluations (Kumar and Mani 2022). Functional unit selection greatly impacts LCA results and uneven application hampers comparisons (Francart et al. 2021). Data from varied geographical or temporal contexts is heterogeneous and hampers comparison (Minx et al. 2021). Standardized reporting requirements improve LCA quality. Nanomaterials complicate data collection due to their unique properties that affect environmental impacts but are less understood. Nanoparticle behavior including durability and toxicity is difficult to quantify (Ahmad et al. 2022). Nanoparticles emitted during BNC use and disposal pose ecotoxicological risks but there are no recognized methods to measure them (Kim et al. 2024). Traditional LCA frameworks sometimes exclude nanomaterial lifecycle stages (Ejaz et al. 2023), limiting evaluations. High-quality databases with detailed lifetime inventory (LCI) data on production process inputs and outputs, energy consumption, emissions, and waste generation are needed for bionanocomposite LCA data availability and quality. Standardized nanomaterial impact assessment methods should be used. For the creation and preservation of bionanocomposite databases, researchers, industry stakeholders, and policymakers must collaborate on case studies on their manufacturing, use, and disposal (Kumar et al. 2023).

For accurate bionanocomposite life cycle assessments, data availability, and quality must be addressed. Unreliable data, reporting methodology differences, nanomaterial complexities, and temporal variations all contribute to LCA uncertainty. Standardized data collection and high-quality databases would improve LCA accuracy and reliability, enabling sustainable bionanocomposite use.

16.7.2 Uncertainties in Impact Assessment

Bionanocomposite impacts' evaluation is complicated by various uncertainties. LCA is used to understand these implications; however, methodology, data variability, and assumptions cause problems. These factors can reduce LCA reliability and interpretation, leading to incorrect conclusions. This section discusses impact assessment methodology uncertainties and their effects on bionanocomposite LCA studies.

Methodological choices taken during LCA setup and execution create uncertainty. Choosing impact categories carefully can affect outcomes; focusing on carbon footprint may neglect other important consequences like water or land use (Hardaker et al. 2022). Different characterization models like ReCiPe or TRACI can yield varying outcomes for the same impact category due to differing assumptions and data (Naeini et al. 2022). Also, assigning environmental impacts among co-products in multi-output processes introduces uncertainties, as varying allocation bases such as mass, economic value, or energy content can result in substantially different outcomes (Moretti 2022). These methodological decisions produce inconsistencies and impede comparability among LCA analysis.

Data-related uncertainties stem from the quality and comprehensiveness of data utilized in LCA. Challenges include data deficiencies, wherein insufficient information regarding certain lifecycle components, such as nanoparticle synthesis or end-of-life disposal, may result in inaccurate impact assessments (Schwirn et al. 2020). Variability in data collection periods and locations influences outcomes; for instance, carbon footprint of biopolymers might differ markedly due to agricultural techniques and regional climatic circumstances (Liu et al. 2017). Moreover, laboratory-scale data may not precisely reflect industrial-scale processes, resulting in uncertainties in evaluations (Neubauer et al. 2021). Data-related issues can undermine the accuracy and credibility of LCA outcomes.

The uncertainties in nanomaterial evaluation create complexity owing to the distinctive features of nanomaterials. The behavior of nanoparticles, including their mobility, aggregation in the environment and bioavailability, remains inadequately understood, misleading impact predictions (Markiewicz et al. 2018). The toxicological effects of nanoparticles depend on their sizes, morphology, and surface characteristics; nonetheless, standardized evaluation methodologies remain in development (Zhang et al. 2022). Additionally, the release possibilities of nanoparticles during the use and disposal of BNCs remain uncertain due to factors like material degradation and environmental conditions (Ates et al. 2020). These uncertainties make bionanocomposite environmental and health assessments difficult. Uncertainties in impact assessment approaches have major implications for decision-making. They might misinterpret LCA results and make poor choices. Overestimating bionanocomposites' benefits may lead to early adoption without considering potential risks. Sensitivity and uncertainty analysis can help reveal uncertainty sources and improve LCA results (Li et al. 2023). For LCA studies to be credible and reproducible, methodologies, data sources, and assumptions must be transparent (Müller et al. 2020).

In summary, impact assessment methodology uncertainties make BNC-LCAs difficult. Methods, data restrictions, nanomaterial complexities, and scenario assumptions cause uncertainty. Robust procedures, high-quality data and transparency are needed to address these issues. By enhancing LCA accuracy and reliability, researchers may better understand BNC environmental repercussions and influence decision-making.

16.7.3 Standardization Issues

The lack of standardization in LCA practices poses challenges that impact the comparability and credibility of results, particularly with evaluating bio-based and fossil-based polymers. The employment of diverse methodologies, data sources, and reporting formats results in inconsistencies that impact significant decision-making.

To start with, methodological variability presents a challenge as discrepancies in effect category selection, such as favoring carbon footprint above resource depletion and differing characterization methods like ReCiPe and TRACI, result in inconsistent conclusions (Pelton and Smith 2015; Drew et al. 2021; Arendt et al. 2020). Disregarding essential lifecycle steps, such as nanofiller manufacturing in BNCs, further impacts the data, complicating comparisons. The quality and consistency of data is an additional issue. Dependence on varied data sources, such as laboratory and industrial data, together with geographical and temporal fluctuations in farming practices complicates precise evaluations of biopolymer carbon footprints (Chopra et al. 2024; Nitkiewicz 2020). Gaps in data on nanoparticle synthesis heightens uncertainties.

In addition, documentation and transparency limitations compound these concerns. Inconsistent functional units such per kilogram versus per product and disclosed assumptions like recycling rates inhibit cross-study comparisons (Bjånesøy et al. 2023; Lam et al. 2020). Few studies conducted comprehensive uncertainty analyses, which raised credibility concerns (Hariri et al. 2019). Furthermore, problems inherent to nanomaterials introduce additional complexity. Uncertainties regarding nanoparticle behavior, such as aggregation, bioavailability, and disposal release, coupled with inadequately defined ecotoxicity assessment methodologies complicate the standardization of LCA (Glisovic et al. 2017; Saxe et al. 2022; Rauscher et al. 2017). The standardization gaps restrict policy development and decision-making as inconsistent methodologies and unclear reporting weaken trust in LCA outcomes (Bozeman et al. 2022; Dsilva et al. 2023; Meglin et al. 2022). Addressing these challenges necessitates cooperation among researchers, industry participants, and policymakers to develop standardized methodologies, data protocols, and reporting criteria. Increased transparency and specific guidelines for nanomaterials are essential for enhancing the reliability of LCA for bio-based nanocomposites and promoting sustainable material innovation.

16.8 Future Directions

16.8.1 Innovations in Bionanocomposite Materials

The study of BNCs is progressing rapidly, propelled by the demand for sustainable substitutes to traditional plastics and composites. Novel innovations are facilitating the creation of advanced BNCs with superior properties, minimized environmental repercussions, and expanded applications. Prominent trends include: Researchers are investigating sophisticated nanofillers, for example bio-based cellulose nanocrystals (CNCs), chitin nanofibers, and lignin nanoparticles, alongside functionalized nanofillers like graphene oxide and clay nanoparticles, which are modified to enhance compatibility with biopolymers, thereby improving mechanical and thermal properties while upholding sustainability (Akinnawo 2024; Jagadeesh et al. 2021; Olson et al. 2021). Hybrid nanofillers also attain synergistic effects and multifunctional properties (Wang and Lin 2021). Also, environmentally sustainable manufacturing methods, such as solvent-free procedures including melt blending, extrusion, and 3D printing, contribute to decreased energy usage and waste (Pacheco et al. 2023;

Youssef et al. 2022). Innovative energy-efficient production processes, such as microwave-assisted procedures, are being devised to reduce energy use and environmental repercussions (Naik et al. 2022). The integration of multifunctional properties like self-healing capabilities, antimicrobial properties, and stimuli-responsive behaviors in BNCs broaden its applications in food packaging, medical devices, and sensors (Vasile 2018; Chen et al. 2023; Adetunji et al. 2022). Introducing circularity in BNCs promotes recyclable and biodegradable products, as well as the upcycling of agricultural and industrial waste (Omran and Baek 2022).

Digital technology and modeling, comprising computational design, machine learning, and enhanced life cycle assessment tools, are transforming the industry by facilitating accurate material optimization and sustainability evaluation (Shen et al. 2023; Kusiak 2018; Rane 2023). These innovations are facilitating the development of sustainable, high-performance BNCs with many applications.

16.8.2 Advancements in LCA Methodologies

Conventional dynamic LCA incorporates temporal variations by utilizing time-specific data to reflect alterations in technology, energy sources, and the surroundings over time. LCA methodologies face challenges related to data limitations, uncertainties, and insufficient standardization. In response, researchers are formulating innovative methodologies and approaches to improve the accuracy, reliability, and applicability of LCA. This section highlights significant advancements in LCA methods.

Dynamic LCA incorporates temporal variations by utilizing time-specific data to reflect alterations in technology, energy sources, and climate conditions over time (Kouamé and Ghannadzadeh 2023). This approach provides precise evaluation of long-term effects, such as GWP, rendering it especially beneficial for BNCs that may deteriorate or emit nanoparticles over time. The incorporation of machine learning (ML) and big data is transforming LCA by improving data quality and analytical efficiency. ML algorithms can detect patterns, address data shortcoming, and forecast environmental consequences, thereby diminishing the necessity for extensive experimental investigations (Wang et al. 2023). This technique help lowers uncertainty and strengthens LCA for developing materials like bionanocomposites.

Hybrid LCA integrates process-based and input–output methodologies to encompass both direct and indirect impacts on the environment. This methodology reduces the necessity for extensive process data and is usable across diverse scales, rendering it appropriate for intricate systems such as BNCs with many life cycle phases (Yu and Wiedmann 2018; Finogenova et al. 2019). Frameworks specific to nanomaterials are being established to tackle the distinct properties and hazards associated with nanomaterials. These frameworks incorporate new impact categories such as nanoparticle release and ecotoxicity. Sophisticated models forecast the transport and behavior of nanoparticles in the environment, while standardized protocols improve the comparability of LCA studies (Jung 2021; Johnston et al. 2020).

Social LCA (S-LCA) examines the social implications of products and materials, utilizing metrics including labor rights and community health (Yadav et al. 2020; McMahon et al. 2021). Embracing stakeholder's views and integrating environmental LCA and S-LCA provide a more holistic assessment of sustainability, which is particularly relevant for BNCs marketed as socially responsible materials (Sánchez et al. 2019; Elnaggar 2024).

The incorporation of circular economy principles is an emerging trend, as life cycle assessment models increasingly encompass various end-of-life alternatives such as recycling and biodegradation (Dahiya et al. 2020). Integrating life cycle assessment (LCA) with Material Flow Analysis (MFA) monitors resource recovery and waste minimization, while eco-design techniques facilitate the creation of recyclable or biodegradable bio-based nanocomposites (BNCs) suited for circularity (Yousaf and Aqsa 2023; Timm et al. 2023). Advancements in LCA methodologies, including dynamic LCA, machine learning integration, hybrid LCA, nanomaterial-specific frameworks, social LCA, and circular economy integration, are improving the accuracy, reliability, and usefulness of LCA. By employing these methodologies, researchers can acquire profound insights into the sustainability of BNCs and facilitate informed decision-making.

16.8.3 Policy Implications and Recommendations

The development and commercialization and application of BNCs markedly improve sustainability across various sectors. Fully capitalizing on these prospects requires robust policy and regulatory frameworks. Presented herein are essential ideas for policymakers to promote sustainable behavior among BNCs:

i. Advocate for Research and Development (R&D): Policymakers must support and incentivize BNC R&D. This includes support for bio-based nanofillers and sustainable production methods, academia–industry–government collaboration, and pilot projects in packaging and automotive.
ii. Implement regulatory frameworks for nanomaterials: Establishing specific regulations to mitigate nanomaterial potential risks is essential. This involves implementing safety regulations for production and disposal, necessitating explicit labeling of nanomaterials and uniform risk assessment procedures.
iii. Promote Circular Economy Practices: Advocate for recycling and reusability through the implementation of policies that hold producers accountable for their products. Investing in recycling infrastructure for biodegradable composites and providing incentives for recyclables or reused materials.
iv. Advocate for Sustainable Manufacturing: Foster energy-efficient and low-emission processes via financial incentives, best practice guidelines for solvent-free techniques, and training initiatives for industry personnel.

v. Enhance Market Development and Consumer Awareness: Increase demand for sustainable materials via consumer education initiatives and certification programs for environmentally friendly products.
vi. Boost International Collaboration: Strive to harmonize worldwide norms for production and disposal, promote information exchange, and bolster global research networks to tackle common concerns.

By adopting these guidelines, governments can foster an environment conducive to the success of BNCs as sustainable substitutes for traditional materials, promoting resource efficiency and environmental stewardship.

16.8.4 Conclusions

Bionanocomposite materials represent a significant improvement in sustainable materials, providing alternatives to traditional petroleum-derived polymers and composites. This composition combines biopolymers with nanoscale fillers, improving mechanical, thermal, and barrier characteristics while adhering to global sustainability objectives, including lowered carbon footprints and resource efficiency. LCA assesses environmental implications throughout all lifespan phases, from raw material extraction to disposal.

This chapter emphasizes how BNCs can reduce greenhouse gas emissions, decrease dependence on non-renewable resources, and reduce waste. PLA-based nanocomposites exhibited reduced global warming potential relative to conventional polymers, while starch–clay composites provided lightweight alternatives for automotive applications. Nonetheless, issues such as the energy-intensive manufacture of nanofillers, ecotoxicity hazards associated with nanoparticle release, and discrepancies in life cycle assessment approaches underscore the necessity for optimization and standardization. To improve assessment reliability, it is essential to rectify data shortcoming, boost methodology transparency, and implement adaptable LCA frameworks that account for temporal and spatial variability. The advancement in green nanotechnology, circular economy initiatives, and policies such as uniform rules and incentives for sustainable manufacturing might further augment the potential of BNCs. Collaboration among academics, companies, and governments is essential to surmount present obstacles and ensure these materials significantly contribute to a resource-efficient, low-carbon future. In conclusion, BNCs offer potential advancements in sustainability across various industries, and their effective integration will necessitate a balanced evaluation of environmental advantages, technological obstacles, and ethical obligations. By enhancing LCA methods and promoting innovation, stakeholders utilize these resources to facilitate the transition to a circular economy and sustainable development.

References

Abukhadra MR, Refay NM, Nadeem A, El-Sherbeeny AM, Ibrahim KE (2020) Insight into the role of integrated carbohydrate polymers (starch, chitosan, and β-cyclodextrin) with mesoporous silica as carriers for ibuprofen drug; equilibrium and pharmacokinetic properties. Int J Biol Macromol 156. https://doi.org/10.1016/j.ijbiomac.2020.04.052

Adetunji CO et al (2022) Nanobiomaterials for food packaging sensor applications. In: Bio- and nano-sensing technologies for food processing and packaging. https://doi.org/10.1039/9781839167966-00167

Adhikari A, Hansen AJ (2018) Land use change and habitat fragmentation of wildland ecosystems of the North Central United States. Landsc Urban Plan 177. https://doi.org/10.1016/j.landurbplan.2018.04.014

Ahmad A, Imran M, Sharma N (2022) Precision nanotoxicology in drug development: current trends and challenges in safety and toxicity implications of customized multifunctional nanocarriers for drug-delivery applications. Pharmaceutics 14(11). https://doi.org/10.3390/pharmaceutics14112463

Ahmadi Mirghaed F, Souri B (2022) Effect of landscape fragmentation on soil quality and ecosystem services in land use and landform types. Environ Earth Sci 81(12). https://doi.org/10.1007/s12665-022-10454-1

Akbar MU et al (2020) Biomedical applications of bionanocomposites. In: Bionanocomposites: green synthesis and applications. https://doi.org/10.1016/B978-0-12-816751-9.00017-9

Akinnawo SO (2024) Advance nanocomposites from biopolymers and fillers: sources, characterization, and end-use applications. Polym Plast Technol Mater 63(5). https://doi.org/10.1080/25740881.2023.2296659

Alaux N, Vašatko H, Maierhofer D, Saade MRM, Stavric M, Passer A (2024) Environmental potential of fungal insulation: a prospective life cycle assessment of mycelium-based composites. Int J Life Cycle Assess 29(2). https://doi.org/10.1007/s11367-023-02243-0

Ali N, Hoque ME (2022) Bionanocomposites in the automotive and aerospace applications. https://doi.org/10.1007/978-981-16-8578-1_13

Álvarez M, Andrade MJ, Núñez F, Rodríguez M, Delgado J (2023) Proteomics as a new-generation tool for studying moulds related to food safety and quality. Int J Mol Sci 24(5). https://doi.org/10.3390/ijms24054709

Araújo-Custódio S et al (2019) Injectable and magnetic responsive hydrogels with bioinspired ordered structures. ACS Biomater Sci Eng 5(3). https://doi.org/10.1021/acsbiomaterials.8b01179

Arendt R, Bachmann TM, Motoshita M, Bach V, Finkbeiner M (2020) Comparison of different monetization methods in LCA: a review. Sustainability (Switzerland) 12(24). https://doi.org/10.3390/su122410493

Arrieta MP, López de Dicastillo C, Garrido L, Roa K, Galotto MJ (2018) Electrospun PVA fibers loaded with antioxidant fillers extracted from *Durvillaea antarctica* algae and their effect on plasticized PLA bionanocomposites. Eur Polym J 103. https://doi.org/10.1016/j.eurpolymj.2018.04.012

Asmara AP (2018) A preliminary study of investigating of compound group contained in ethanolic extract of Mahagony (*Swietenia mahagoni* L. Jacq.) seeds related to α-glucosidase inhibition. J Nat 18(2). https://doi.org/10.24815/jn.v18i2.9236

Ates B, Koytepe S, Ulu A, Gurses C, Thakur VK (2020) Chemistry, structures, and advanced applications of nanocomposites from biorenewable resources. Chem Rev 120(17). https://doi.org/10.1021/acs.chemrev.9b00553

Atoum IA, Keshta IM (2021) Big data management: security and privacy concerns. Int J Adv Appl Sci 8(5). https://doi.org/10.21833/ijaas.2021.05.009

Bailore NN, Balladka SK, Doddapaneni SJDS, Mudiyaru MS (2021) Fabrication of environmentally compatible biopolymer films of pullulan/Piscean collagen/ZnO nanocomposite and their antifungal activity. J Polym Environ 29(4). https://doi.org/10.1007/s10924-020-01953-y

Barbarino M, Giordano A (2021) Assessment of the carcinogenicity of carbon nanotubes in the respiratory system. Cancers 13(6). https://doi.org/10.3390/cancers13061318

Bashandeh K et al (2022) MXene-aromatic thermosetting copolyester nanocomposite as an extremely wear-resistant biocompatible implant material for osteoarthritis applications. Appl Surf Sci 600. https://doi.org/10.1016/j.apsusc.2022.154124

Bayer AD et al (2021) Diverging land-use projections cause large variability in their impacts on ecosystems and related indicators for ecosystem services. Earth Syst Dyn 12(1). https://doi.org/10.5194/esd-12-327-2021

Bishop G, Styles D, Lens PNL (2021) Environmental performance comparison of bioplastics and petrochemical plastics: a review of life cycle assessment (LCA) methodological decisions. Resour Conserv Recycl 168. https://doi.org/10.1016/j.resconrec.2021.105451

Biswas MC, Chowdhury A, Hossain MM, Hossain MK (2022) Applications, drawbacks, and future scope of nanoparticle-based polymer composites. In: Nanoparticle-based polymer composites. https://doi.org/10.1016/B978-0-12-824272-8.00002-6

Bjånesøy S, Kinnunen A, Einarsdóttir H, Heinonen J (2023) Carbon storage in the built environment: a review. Environ Res: Infrastruct Sustain 3(4). https://doi.org/10.1088/2634-4505/ad139f

Blass V, Chebach TC, Ashkenazy A (2017) Sustainable non-renewable materials management. In: Springer series in supply chain management, vol 4. https://doi.org/10.1007/978-3-319-29791-0_5

Bolan S et al (2024) The distribution, fate, and environmental impacts of food additive nanomaterials in soil and aquatic ecosystems. Sci Total Environ 916. https://doi.org/10.1016/j.scitotenv.2024.170013

Borrion A, Matsushita J, Austen K, Johnson C, Bell S (2019) Development of LCA Calculator to support community infrastructure co-design. Int J Life Cycle Assess 24(7). https://doi.org/10.1007/s11367-018-1492-2

Bott J, Störmer A, Albers P (2018) Investigation into the release of nanomaterials from can coatings into food. Food Packag Shelf Life 16. https://doi.org/10.1016/j.fpsl.2018.03.004

Bozeman JF, Nobler E, Nock D (2022) A path toward systemic equity in life cycle assessment and decision-making: standardizing sociodemographic data practices. Environ Eng Sci 39(9). https://doi.org/10.1089/ees.2021.0375

Brandt N, Ahlemann F, Reining S, Rehring K (2021) Internet of things in product lifecycle management—a review on value creation through product status data. In: 27th annual Americas conference on information systems, AMCIS 2021

Brodin M, Vallejos M, Opedal MT, Area MC, Chinga-Carrasco G (2017) Lignocellulosics as sustainable resources for production of bioplastics—a review. J Clean Prod 162. https://doi.org/10.1016/j.jclepro.2017.05.209

Calvino C, Macke N, Kato R, Rowan SJ (2020) Development, processing and applications of bio-sourced cellulose nanocrystal composites. Prog Polym Sci 103. https://doi.org/10.1016/j.progpolymsci.2020.101221

Campanale C, Massarelli C, Savino I, Locaputo V, Uricchio VF (2020) A detailed review study on potential effects of microplastics and additives of concern on human health. Int J Environ Res Public Health 17(4). https://doi.org/10.3390/ijerph17041212

Carroccio SC, Scarfato P, Bruno E, Aprea P, Dintcheva NT, Filippone G (2022) Impact of nanoparticles on the environmental sustainability of polymer nanocomposites based on bioplastics or recycled plastics—a review of life-cycle assessment studies. J Clean Prod 335. https://doi.org/10.1016/j.jclepro.2021.130322

Chen Z, Scharnagl N, Zheludkevich ML, Ying H, Yang W (2023) Micro/nanocontainer-based intelligent coatings: synthesis, performance and applications—a review. Chem Eng J 451. https://doi.org/10.1016/j.cej.2022.138582

Chong WJ et al (2023) Biodegradable PLA-ZnO nanocomposite biomaterials with antibacterial properties, tissue engineering viability, and enhanced biocompatibility. Smart Mater Manuf 1. https://doi.org/10.1016/j.smmf.2022.100004

Chopra J, Rangarajan V, Rathnasamy S, Dey P (2024) Life cycle assessment as a key decision tool for emerging pretreatment technologies of biomass-to-biofuel: unveiling challenges, advances, and future potential. Bioenergy Res. https://doi.org/10.1007/s12155-024-10741-8

Chrissafis K, Paraskevopoulos KM, Papageorgiou GZ, Bikiaris DN (2008) Thermal and dynamic mechanical behavior of bionanocomposites: Fumed silica nanoparticles dispersed in poly(vinyl pyrrolidone), chitosan, and poly(vinyl alcohol). J Appl Polym Sci 110(3). https://doi.org/10.1002/app.28818

Coltelli MB, Morganti P, Lazzeri A (2023) Sustainability assessment, environmental impact, and recycling strategies of biodegradable polymer nanocomposites. In: Biodegradable and biocompatible polymer nanocomposites: processing, characterization, and applications. https://doi.org/10.1016/B978-0-323-91696-7.00016-7

Cooper CJ, Abdelwahab MA, Mohanty AK, Misra M (2019) Hybrid green bionanocomposites of bio-based poly(butylene succinate) reinforced with pyrolyzed perennial grass microparticles and graphene nanoplatelets. ACS Omega 4(24). https://doi.org/10.1021/acsomega.9b01771

Costa D, Quinteiro P, Dias AC (2019) A systematic review of life cycle sustainability assessment: current state, methodological challenges, and implementation issues. Sci Total Environ 686. https://doi.org/10.1016/j.scitotenv.2019.05.435

Crociati L, Zambelli P, Gozzi D, Bolis V (2024) Life Cycle Assessment for particleboards with recycled wood in an industrial context Life Cycle Assessment for particleboards with recycled wood in an industrial context. https://doi.org/10.1088/1755-1315/1402/1/012029

Cronin MTD, Richarz AN, Schultz TW (2019) Identification and description of the uncertainty, variability, bias and influence in quantitative structure-activity relationships (QSARs) for toxicity prediction. Regul Toxicol Pharmacol 106. https://doi.org/10.1016/j.yrtph.2019.04.007

Czúcz B et al (2018) Where concepts meet the real world: a systematic review of ecosystem service indicators and their classification using CICES. Ecosyst Serv 29. https://doi.org/10.1016/j.ecoser.2017.11.018

Czyrnek-Delêtre MM, Smyth BM, Murphy JD (2017) Beyond carbon and energy: the challenge in setting guidelines for life cycle assessment of biofuel systems. Renew Energy 105. https://doi.org/10.1016/j.renene.2016.11.043

Dahiya S, Katakojwala R, Ramakrishna S, Mohan SV (2020) Biobased products and life cycle assessment in the context of circular economy and sustainability. Mater Circ Econ 2(1). https://doi.org/10.1007/s42824-020-00007-x

Dash S, Swain SK (2013) Synthesis of thermal and chemical resistant oxygen barrier starch with reinforcement of nano silicon carbide. Carbohydr Polym 97(2). https://doi.org/10.1016/j.carbpol.2013.05.061

de Souza VM, Fröhling M, Pigosso DCA (2024) A multi-level resource circularity index based in the European union's circular economy monitoring framework. Waste Biomass Valoriz 15(2). https://doi.org/10.1007/s12649-023-02193-6.

Dhar P, Bhasney SM, Kumar A, Katiyar V (2016) Acid functionalized cellulose nanocrystals and its effect on mechanical, thermal, crystallization and surfaces properties of poly (lactic acid) bionanocomposites films: a comprehensive study. Polymer (Guildf) 101. https://doi.org/10.1016/j.polymer.2016.08.028

Di Bartolo A, Infurna G, Dintcheva NT (2021) A review of bioplastics and their adoption in the circular economy. Polymers 13(8). https://doi.org/10.3390/polym13081229

Di Ludovico D, Di Lodovico L (2020) The Regional Management Risk Plan. Knowledge, scenarios and prevention projects in a regional context. Int J Disaster Risk Reduct 45. https://doi.org/10.1016/j.ijdrr.2019.101465

Díez-Pascual AM, Díez-Vicente AL (2014) Poly(3-hydroxybutyrate)/ZnO bionanocomposites with improved mechanical, barrier and antibacterial properties. Int J Mol Sci 15(6). https://doi.org/10.3390/ijms150610950

Domingo JL, Marquès M, Mari M, Schuhmacher M (2020) Adverse health effects for populations living near waste incinerators with special attention to hazardous waste incinerators. A review of the scientific literature. Environ Res 187. https://doi.org/10.1016/j.envres.2020.109631

Donaldson K et al (2006) Carbon nanotubes: a review of their properties in relation to pulmonary toxicology and workplace safety. Toxicol Sci 92(1). https://doi.org/10.1093/toxsci/kfj130

Drew J, Christie SD, Tyedmers P, Smith-Forrester J, Rainham D (2021) Operating in a climate crisis: a state-of-the-science review of life cycle assessment within surgical and anesthetic care. Environ Health Perspect 129(7). https://doi.org/10.1289/EHP8666

Dsilva J, Zarmukhambetova S, Locke J (2023) Assessment of building materials in the construction sector: a case study using life cycle assessment approach to achieve the circular economy. Heliyon 9(10). https://doi.org/10.1016/j.heliyon.2023.e20404

Duval-Dachary S, Lorne D, Beauchet S, Salou T, Hélias A (2024) Life cycle assessment of carbon capture and utilisation as a negative emission technology: Recommendations and case study. Int J Life Cycle Assess. https://doi.org/10.1007/s11367-024-02388-6

Ejaz M et al (2023) Nanotechnologies for environmental remediation and their ecotoxicological impacts. Environ Monit Assess 195(11). https://doi.org/10.1007/s10661-023-11661-4

Ekins P, Zenghelis D (2021) The costs and benefits of environmental sustainability. Sustain Sci 16(3). https://doi.org/10.1007/s11625-021-00910-5

Elnaggar A (2024) Nine principles of green heritage science: life cycle assessment as a tool enabling green transformation. Herit Sci 12(1). https://doi.org/10.1186/s40494-023-01114-z

Esen SE, Hein L (2020) Development of SEEA water accounts with a hydrological model. Sci Total Environ 737. https://doi.org/10.1016/j.scitotenv.2020.140168

European Council (2022) Directive 2009/48/EC of the European Parliament and of the Council of 18 June 2009 on the safety of toys (Latest Consolidated Version: 05/07/2022). Off J Eur Union

Farokh A, Pourmadadi M, Rashedi H, Yazdian F, Navaei-Nigjeh M (2023) Assessment of synthesized chitosan/halloysite nanocarrier modified by carbon nanotube for pH-sensitive delivery of curcumin to cancerous media. Int J Biol Macromol 237. https://doi.org/10.1016/j.ijbiomac.2023.123937

Finogenova N, Bach V, Berger M, Finkbeiner M (2019) Hybrid approach for the evaluation of organizational indirect impacts (AVOID): combining product-related, process-based, and monetary-based methods. Int J Life Cycle Assess 24(6). https://doi.org/10.1007/s11367-018-1544-7

Formela K, Kurańska M, Barczewski M (2022) Recent advances in development of waste-based polymer materials: a review. Polymers 14(5). https://doi.org/10.3390/polym14051050

Francart N, Widström T, Malmqvist T (2021) Influence of methodological choices on maintenance and replacement in building LCA. Int J Life Cycle Assess 26(11). https://doi.org/10.1007/s11367-021-01985-z

Furberg A, Arvidsson R, Molander S (2022) A practice-based framework for defining functional units in comparative life cycle assessments of materials. J Ind Ecol 26(3). https://doi.org/10.1111/jiec.13218

Gallo Stampino P, Riva L, Punta C, Elegir G, Bussini D, Dotelli G (2021) Comparative life cycle assessment of cellulose nanofibres production routes from virgin and recycled raw materials. Molecules 26(9). https://doi.org/10.3390/molecules26092558

Gao X, Lowry GV (2018) Progress towards standardized and validated characterizations for measuring physicochemical properties of manufactured nanomaterials relevant to nano health and safety risks. NanoImpact 9. https://doi.org/10.1016/j.impact.2017.09.002

Geoghegan C, O'Donoghue C (2023) An analysis of the social and private return to land use change from agriculture to renewable energy production in Ireland. J Clean Prod 385. https://doi.org/10.1016/j.jclepro.2022.135698

Gerhardt-m J, Incer-valverde J, Morosuk T, Minke C (2024). Exergetic life cycle assessment for green hydrogen production. Energy 299. https://doi.org/10.1016/j.energy.2024.131553

Glisovic S et al (2017) Emerging technologies and safety concerns: a condensed review of environmental life cycle risks in the nano-world. Int J Environ Sci Technol 14(10). https://doi.org/10.1007/s13762-017-1367-2

Guinee JB (2002) Handbook on life cycle assessment operational guide to the ISO standards. Int J Life Cycle Assess 7(5). https://doi.org/10.1007/bf02978897

Haines-Young R, Potschin-Young MB (2018) Common International Classification of Ecosystem Services (CICES) V5.1 and guidance on the application of the revised structure. One Ecosyst 3

Hardaker A, Styles D, Williams P, Chadwick D, Dandy N (2022) A framework for integrating ecosystem services as endpoint impacts in life cycle assessment. J Clean Prod 370. https://doi.org/10.1016/j.jclepro.2022.133450

Hariri RH, Fredericks EM, Bowers KM (2019) Uncertainty in big data analytics: survey, opportunities, and challenges. J Big Data 6(1). https://doi.org/10.1186/s40537-019-0206-3

Harris TM, Hottle TA, Soratana K, Klane J, Landis AE (2016) Life cycle assessment of sunflower cultivation on abandoned mine land for biodiesel production. J Clean Prod 112. https://doi.org/10.1016/j.jclepro.2015.09.057

Herath PL, Jayawardana D, Bandara N (2023) Quantification of methane and carbon dioxide emissions from an active landfill: study the effect of surface conditions on emissions. Environ Earth Sci 82(2). https://doi.org/10.1007/s12665-023-10745-1

Hischier R, Walser T (2012) Life cycle assessment of engineered nanomaterials: State of the art and strategies to overcome existing gaps. Sci Total Environ 425. https://doi.org/10.1016/j.scitotenv.2012.03.001

Hou J et al (2019) Toxicity and mechanisms of action of titanium dioxide nanoparticles in living organisms. J Environ Sci (China) 75. https://doi.org/10.1016/j.jes.2018.06.010

Huang X, Wang X, Chen B, Li F, Su S, Zhang T (2022) Temporal trend and regional disparity of agricultural green total factor productivity in china: data envelopment analysis with biennial environmental technology. Discret Dyn Nat Soc 2022. https://doi.org/10.1155/2022/2219779

Hund-Rinke K, Schlinkert R, Schlich K (2022) Testing particles using the algal growth inhibition test (OECD 201): the suitability of in vivo chlorophyll fluorescence measurements. Environ Sci Eur 34(1). https://doi.org/10.1186/s12302-022-00623-1

IPCC (2013a) Climate change 2013: the physical science basis summary for policymakers

IPCC, International Panel of Climate Change (2013b) Climate Change 2013 Phys. Sci. Basis. Contrib. Work. Gr. I to Fifth Assess. Rep. Intergov. Panel Clim. Chang.

ISO (2006) 14040: Environmental management—life cycle assessment—Principles and framework. Int. Organ. Stand.

Jagadeesh P, Puttegowda M, Mavinkere Rangappa S, Siengchin S (2021) Influence of nanofillers on biodegradable composites: a comprehensive review. Polym Compos 42(11). https://doi.org/10.1002/pc.26291

Jayakumar A et al (2023) Natural fiber-based bionanocomposites: thermal, morphological and mechanical properties. In: Epoxy-based biocomposites. https://doi.org/10.1201/9781003271017-10

Jayakumar A et al (2022) Recent innovations in bionanocomposites-based food packaging films—a comprehensive review. Food Packag Shelf Life 33. https://doi.org/10.1016/j.fpsl.2022.100877

Jeevanandam J, Rodrigues J, Pan S, Danquah MK (2023) Cellulose-based bionanocomposites: synthesis, properties, and applications. In: Advances in bionanocomposites: materials, applications, and life cycle. https://doi.org/10.1016/B978-0-323-91764-3.00011-5

Johnston LJ, Gonzalez-Rojano N, Wilkinson KJ, Xing B (2020) Key challenges for evaluation of the safety of engineered nanomaterials. NanoImpact 18. https://doi.org/10.1016/j.impact.2020.100219

JRC (2010) International Reference Life Cycle Data System (ILCD) Handbook: analysing of existing environmental impact assessment methodologies for use in life cycle assessment. European Commission

Jung F et al (2021) Predicting the environmental emissions arising from conventional and nanotechnology-related pharmaceutical drug products. Environ Res 192. https://doi.org/10.1016/j.envres.2020.110219

Kaci M, Aouat T, Devaux E, Lopez-Cuesta JM (2019) The effects of filler size and content on the fire behavior of melt-spun poly(lactic acid)/cellulose bionanocomposite fibers. AIP Conf Proc. https://doi.org/10.1063/1.5140290

Kalirajan C, Palanisamy T (2020) Hybrid collagen scaffold tethered with silver-catechin nanocomposite modulates angiogenesis and TGF-β toward scarless healing in chronic deep second degree infected burns. Adv Healthc Mater 9(12). https://doi.org/10.1002/adhm.202000247

Kanaoujiya R, Saroj SK, Srivastava S, Chaudhary MK (2022) Renewable polysaccharide and biomedical application of nanomaterials. J Nanomater 2022. https://doi.org/10.1155/2022/1050211

Kane S, Miller SA, Kurtis KE, Youngblood JP, Landis EN, Weiss WJ (2023) Harmonized life-cycle inventories of nanocellulose and its application in composites. Environ Sci Technol 57(48). https://doi.org/10.1021/acs.est.3c04814

Karthika V, AlSalhi MS, Devanesan S, Gopinath K, Arumugam A, Govindarajan M (2020) Chitosan overlaid Fe_3O_4/rGO nanocomposite for targeted drug delivery, imaging, and biomedical applications. Sci Rep 10(1). https://doi.org/10.1038/s41598-020-76015-3

Kauts S et al (2023) The evidence of in-vivo and in-vitro studies on microplastic and nano plastic toxicity in mammals: a possible threat for an upcoming generation? Phys Chem Earth 132. https://doi.org/10.1016/j.pce.2023.103511

Khoo RZ, Chow WS (2022) Thermal properties of the poly(lactic acid) bionanocomposites. https://doi.org/10.1007/978-981-16-8578-1_2

Kim M et al (2024) Human and environmental safety of carbon nanotubes across their life cycle. Nat Rev Mater 9(1). https://doi.org/10.1038/s41578-023-00611-8

Kleinekorte J et al (2020) Life Cycle Assessment for the design of chemical processes, products, and supply chains. Annu Rev Chem Biomol Eng 11. https://doi.org/10.1146/annurev-chembioeng-011520-075844

Kokare S, Oliveira JP, Godina R (2023) Life cycle assessment of additive manufacturing processes: a review. J Manuf Syst 68. https://doi.org/10.1016/j.jmsy.2023.05.007

Koo H, Kleemann J, Fürst C (2020) Integrating ecosystem services into land-use modeling to assess the effects of future land-use strategies in northern Ghana. Land 9(10). https://doi.org/10.3390/land9100379

Kouamé S, Ghannadzadeh A (2023) A comparative parametric study on dynamic biogenic carbon of harvested wood products: biomass rotation period vs. product lifetime. Energies 16(7). https://doi.org/10.3390/en16073163

Kowalczyk A, Wrzecińska M, Czerniawska-Piątkowska E, Araújo JP, Cwynar P (2022) Molecular consequences of the exposure to toxic substances for the endocrine system of females. Biomed Pharmacother 155. https://doi.org/10.1016/j.biopha.2022.113730

Kumar M, Mani M (2022) Sustainability assessment in manufacturing for effectiveness: challenges and opportunities. Front Sustain 3. https://doi.org/10.3389/frsus.2022.837016

Kumar V, Singh E, Singh S, Pandey A, Bhargava PC (2023) Micro- and nano-plastics (MNPs) as emerging pollutant in ground water: environmental impact, potential risks, limitations and way forward towards sustainable management. Chem Eng J 459. https://doi.org/10.1016/j.cej.2023.141568

Kumari N, Arora P (2021) Application of techno-ecological synergies in life cycle assessment (TES-LCA) to microalgal biofuel. In: European biomass conference and exhibition proceedings

Kuperkar K, Atanase LI, Bahadur A, Crivei IC, Bahadur P (2024) Degradable polymeric bio(nano)materials and their biomedical applications: a comprehensive overview and recent updates. Polymers 16(2). https://doi.org/10.3390/polym16020206

Kupolati WK, Sadiku ER, Eze AA, Ibrahim ID, Agboola O (2022) Life cycle assessment, recycling and re-use of the bionanocomposites. https://doi.org/10.1007/978-981-16-8578-1_11

Kusiak A (2018) Smart manufacturing. Int J Prod Res 56(1–2). https://doi.org/10.1080/00207543.2017.1351644

Lam KL, Zlatanović L, van der Hoek JP (2020) Life cycle assessment of nutrient recycling from wastewater: a critical review. Water Res 173. https://doi.org/10.1016/j.watres.2020.115519

Landers D, Nahil A, Rhodes CR (2016) The beneficiary perspective—benefits and beyond. In: Potschin M, Haines-Young R, Fish R, Turn RK (eds) Routledge handbook of ecosystem services. Routledge, London, New York, pp 74–88

Larrey-Lassalle P et al (2017) An innovative implementation of LCA within the EIA procedure: lessons learned from two wastewater treatment plant case studies. Environ Impact Assess Rev 63. https://doi.org/10.1016/j.eiar.2016.12.004

Le VG et al (2023) A comprehensive review of micro- and nano-plastics in the atmosphere: occurrence, fate, toxicity, and strategies for risk reduction. Sci Total Environ 904. https://doi.org/10.1016/j.scitotenv.2023.166649

Li J et al (2023) Identifying uncertainties in the whole life carbon assessment of buildings: sources, types, and potential actions. Build Environ 244. https://doi.org/10.1016/j.buildenv.2023.110779

Liao W, Heijungs R, Huppes G (2012) Thermodynamic resource indicators in LCA: a case study on the titania produced in Panzhihua city, southwest China. Int J Life Cycle Assess 17(8). https://doi.org/10.1007/s11367-012-0429-4

Lippiatt BC, Ahmad S (2004) Measuring the life-cycle environmental and economic performance of concrete: the bees approach

Liu C, Susilo YO, Karlström A (2017) Weather variability and travel behaviour—what we know and what we do not know. Transp Rev 37(6). https://doi.org/10.1080/01441647.2017.1293188

Liu Y et al (2022) Laser joining of carbon fiber reinforced plastics to titanium alloy via laser texturing. Zhongguo Jiguang/Chinese J Lasers 49(18). https://doi.org/10.3788/CJL202249.1803001

Lizundia E, Luzi F, Puglia D (2022) Organic waste valorisation towards circular and sustainable biocomposites. Green Chem 24(14). https://doi.org/10.1039/d2gc01668k

Lloyd SM, Lave LB (2003) Life cycle economic and environmental implications of using nanocomposites in automobiles. Environ Sci Technol 37(15). https://doi.org/10.1021/es026023q

Lüftl S, Visakh PM (2016) Polyethylene-based biocomposites and bionanocomposites: state-of-the-art, new challenges and opportunities. In: Polyethylene-based biocomposites and bionanocomposites. https://doi.org/10.1002/9781119038467.ch1

Maia de Souza D, Lopes GR, Hansson J, Hansen K (2018) Ecosystem services in life cycle assessment: A synthesis of knowledge and recommendations for biofuels. Ecosyst Serv 30. https://doi.org/10.1016/j.ecoser.2018.02.014

Maity S, Pandit P, Singha K (2021) Starch-based bionanocomposites in tissue engineering and regenerative medicines. In: Bionanocomposites in tissue engineering and regenerative medicine. https://doi.org/10.1016/B978-0-12-821280-6.00029-5

Makkawi Y et al (2023) A comparative analysis of second-generation biofuels and its potentials for large-scale production in arid and semi-arid regions. Fuel 343. https://doi.org/10.1016/j.fuel.2023.127893

Malovrh Rebec K, Turk J, Kunaver M (2024) Quantifying the environmental implication of cotton-fiber-based nanocrystalline cellulose: a life-cycle assessment. Sustain Prod Consum 45. https://doi.org/10.1016/j.spc.2023.12.026

Markiewicz M et al (2018) Changing environments and biomolecule coronas: consequences and challenges for the design of environmentally acceptable engineered nanoparticles. Green Chem 20(18). https://doi.org/10.1039/c8gc01171k

Mastropetros SG, Pispas K, Zagklis D, Ali SS, Kornaros M (2022) Biopolymers production from microalgae and cyanobacteria cultivated in wastewater: recent advances. Biotechnol Adv 60. https://doi.org/10.1016/j.biotechadv.2022.107999

McCarthy AC, Griffin TS, Srinivasan S, Peters CJ (2021) Regional variability in land and water use in fruit and vegetable production in the United States. Urban Agric Reg Food Syst 6(1). https://doi.org/10.1002/uar2.20020

McClements DJ (2021) Advances in edible nanoemulsions: digestion, bioavailability, and potential toxicity. Prog Lipid Res 81. https://doi.org/10.1016/j.plipres.2020.101081

McMahon K, Ryan-Fogarty Y, Fitzpatrick C (2021) Estimating job creation potential of compliant WEEE pre-treatment in Ireland. Resour Conserv Recycl 166. https://doi.org/10.1016/j.resconrec.2020.105230

Medel-Jiménez F, Krexner T, Gronauer A, Kral I (2024) Life cycle assessment of four different precision agriculture technologies and comparison with a conventional scheme. J Clean Prod 434. https://doi.org/10.1016/j.jclepro.2023.140198

Meglin R, Kytzia S, Habert G (2022) Uncertainty, variability, price changes and their implications on a regional building materials industry: the case of Swiss canton Argovia. J Clean Prod 330. https://doi.org/10.1016/j.jclepro.2021.129944

Metak AM, Nabhani F, Connolly SN (2015) Migration of engineered nanoparticles from packaging into food products. LWT 64(2). https://doi.org/10.1016/j.lwt.2015.06.001

Meyer MA, Leckert FS (2018) A systematic review of the conceptual differences of environmental assessment and ecosystem service studies of biofuel and bioenergy production. Biomass Bioenergy 114. https://doi.org/10.1016/j.biombioe.2017.05.003

Minx J et al (2021) A comprehensive dataset for global, regional and national greenhouse gas emissions by sector 1970–2019. Earth Syst Sci Data Discuss

Mirzaei J, Parashkoohi MG, Zamani DM, Afshari H (2023) Examining energy use efficiency and conducting an environmental life cycle assessment through the application of artificial intelligence: a case study on the production of cumin and fennel. Results Eng 20. https://doi.org/10.1016/j.rineng.2023.101522

Modarress Fathi B, Ansari A, Ansari A (2023) Green commercial aviation supply chain—a European path to environmental sustainability. Sustainability 15(8). https://doi.org/10.3390/su15086574

Mohammadi-Zerankeshi M, Alizadeh R (2023) 3D-printed PLA-Gr-Mg composite scaffolds for bone tissue engineering applications. J Mater Res Technol 22. https://doi.org/10.1016/j.jmrt.2022.12.108

Mohammadnia MS, Hemmati S, Fasihi-Ramandi M, Bahari M (2022) Synthesis of novel surface-modified nanohydroxyapatite containing chitosan-functionalized graphene oxide decorated with glycidyl methacrylate (GO–CS–GMA) via ATRP for biomedical application. Polym Bull 79(11). https://doi.org/10.1007/s00289-021-03968-4

Mohammed M, Oleiwi JK, Mohammed AM, Ja A (2024) A review on the advancement of renewable natural fiber hybrid composites: prospects, challenges, and industrial applications. https://doi.org/10.32604/jrm.2024.051201

Morales J, Michell RM, Sommer-Márquez A, Rodrigue D (2023) Effect of biobased SiO_2 on the morphological, thermal, mechanical, rheological, and permeability properties of PLLA/PEG/SiO_2 biocomposites. J Compos Sci 7(4). https://doi.org/10.3390/jcs7040150

Moretti C et al (2022) Using lignin from local biorefineries for asphalts: LCA case study for the Netherlands. J Clean Prod 343. https://doi.org/10.1016/j.jclepro.2022.131063

Mostafaei H, Keshavarz Z, Rostampour MA, Mostofinejad D, Wu C (2023) Sustainability evaluation of a concrete gravity dam: life cycle assessment, carbon footprint analysis, and life cycle costing. Structures 53. https://doi.org/10.1016/j.istruc.2023.04.058

Moutousidi ES, Kookos IK (2021) Life cycle assessment of biobased chemicals from different agricultural feedstocks. J Clean Prod 323. https://doi.org/10.1016/j.jclepro.2021.129201

Mujtaba M et al (2023) Lignocellulosic biomass from agricultural waste to the circular economy: a review with focus on biofuels, biocomposites and bioplastics. J Clean Prod 402. https://doi.org/10.1016/j.jclepro.2023.136815

Mukheem A et al (2018) Fabrication and characterization of an electrospun PHA/graphene silver nanocomposite scaffold for antibacterial applications. Materials (Basel) 11(9). https://doi.org/10.3390/ma11091673

Mukherjee A, Knoch S, G Chouinard, Tavares JR, Dumont MJ (2019) Use of bio-based polymers in agricultural exclusion nets: a perspective. Biosystems Eng 180. https://doi.org/10.1016/j.biosystemseng.2019.01.017

Mulla MZ, Rahman MRT, Marcos B, Tiwari B, Pathania S (2021) Poly lactic acid (PLA) nanocomposites: effect of inorganic nanoparticles reinforcement on its performance and food packaging applications. Molecules 26(7). https://doi.org/10.3390/molecules26071967

Müller LJ, Kätelhön A, Bachmann M, Zimmermann A, Sternberg A, Bardow A (2020) A guideline for life cycle assessment of carbon capture and utilization. Front Energy Res 8. https://doi.org/10.3389/fenrg.2020.00015

Naeini M, Cotton JS, Adams TA (2022) Dynamic life cycle assessment of solid oxide fuel cell system considering long-term degradation effects. Energy Convers Manag 255. https://doi.org/10.1016/j.enconman.2022.115336

Naik TP, Singh I, Sharma AK (2022) Processing of polymer matrix composites using microwave energy: a review. Compos Part A: Appl Sci Manuf 156. https://doi.org/10.1016/j.compositesa.2022.106870

Nazrin A, Sapuan SM, Zuhri MYM (2020) Mechanical, physical and thermal properties of sugar palm nanocellulose reinforced thermoplastic starch (TPS)/poly (lactic acid) (PLA) blend bionanocomposites. Polymers (Basel) 12(10). https://doi.org/10.3390/polym12102216

Nel A (2023) Carbon nanotube pathogenicity conforms to a unified theory for mesothelioma causation by elongate materials and fibers. Environ Res 230. https://doi.org/10.1016/j.envres.2022.114580

Neubauer P, Anane E, Junne S, Cruz Bournazou MN (2021) Potential of integrating model-based design of experiments approaches and process analytical technologies for bioprocess scale-down. Adv Biochem Eng Biotechnol 177. https://doi.org/10.1007/10_2020_154

Nitkiewicz T et al (2020) How sustainable are biopolymers? Findings from a life cycle assessment of polyhydroxyalkanoate production from rapeseed-oil derivatives. Sci Total Environ 749. https://doi.org/10.1016/j.scitotenv.2020.141279

Nizam NUM, Hanafiah MM, Woon KS (2021) A content review of life cycle assessment of nanomaterials: current practices, challenges, and future prospects. Nanomaterials 11(12). https://doi.org/10.3390/nano11123324

Noman M, Islam MN, Shoaib M (2022) Community-based ecosystem management (CbEM) of Arial Beel at Munshiganj District in Bangladesh by integrating MIMES model. Model Earth Syst Environ 8(1). https://doi.org/10.1007/s40808-021-01099-3

Nwodo MN, Anumba CJ (2020) Exergetic life cycle assessment: a review. Energies 13(11). https://doi.org/10.3390/en13112684

Oberdörster G, Graham U (2018) Predicting EMP hazard: lessons from studies with inhaled fibrous and non-fibrous nano- and micro-particles. Toxicol Appl Pharmacol 361. https://doi.org/10.1016/j.taap.2018.05.004

Oberdörster G, Kuhlbusch TAJ (2018) In vivo effects: methodologies and biokinetics of inhaled nanomaterials. NanoImpact 10. https://doi.org/10.1016/j.impact.2017.10.007

OECD (2011) Test Guideline No. 201. Freshwater Alga and Cyanobacteria, Growth Inhibition Test. Organization for Economic Cooperation and Development. Organ. Econ. Coop. Dev.

Okogwu C et al (2023) Exploring the integration of sustainable materials in supply chain management for environmental impact. Eng Sci Technol J 4(3). https://doi.org/10.51594/estj.v4i3.546

Olson E et al (2021) Self-assembly in biobased nanocomposites for multifunctionality and improved performance. Nanoscale Adv 3(15). https://doi.org/10.1039/d1na00391g

Omran BA, Baek KH (2022) Valorization of agro-industrial biowaste to green nanomaterials for wastewater treatment: approaching green chemistry and circular economy principles. J Environ Manag 311. https://doi.org/10.1016/j.jenvman.2022.114806

Pacheco A et al (2023) Polymeric materials obtained by extrusion and injection molding from lignocellulosic agroindustrial biomass. Polymers 15(20). https://doi.org/10.3390/polym15204046

Pasha HY, Mohtasebi SS, Taherimehr M, Tabatabaeekoloor R, Firouz MS, Javadi A (2023) New poly(lactic acid)-based nanocomposite films for food packaging applications. Iran Polym J (Engl Ed) 32(7):855–871. https://doi.org/10.1007/s13726-023-01170-z

Pelton REO, Smith TM (2015) Hotspot scenario analysis: comparative streamlined LCA approaches for green supply chain and procurement decision making Pelton and smith hotspot scenario analysis. J Ind Ecol 19(3). https://doi.org/10.1111/jiec.12191

Polyakov M, Khanin I, Bilozubenko V, Korneyev M, Shevchenko G (2021) Factors of uneven progress of the European Union countries towards a circular economy. Probl Perspect Manag 19(3). https://doi.org/10.21511/ppm.19(3).2021.27

Pradhan RA (2022) Life-cycle assessment and circular economy concepts for bionanocomposites. In: Bionanocomposites for food packaging applications. https://doi.org/10.1016/B978-0-323-88528-7.00002-2

Prata JC, da Costa JP, Lopes I, Duarte AC, Rocha-Santos T (2020) Environmental exposure to microplastics: an overview on possible human health effects. Sci Total Environ 702. https://doi.org/10.1016/j.scitotenv.2019.134455

Radnik J et al (2022) Automation and standardization—a coupled approach towards reproducible sample preparation protocols for nanomaterial analysis. Molecules 27(3). https://doi.org/10.3390/molecules27030985

Rajendran N, Runge T, Bergman RD, Nepal P, Pottackal NT, Rahman MM (2024) Economic and environmental analysis of producing soy protein-cellulose-based bionanocomposite fruit coating. Ind Crops Prod 211. https://doi.org/10.1016/j.indcrop.2024.118213

Rane NL (2023) Integrating leading-edge artificial intelligence (AI), internet of things (IoT), and big data technologies for smart and sustainable architecture, engineering and construction (AEC) industry: challenges and future directions. Int J Data Sci Big Data Anal 3(2). https://doi.org/10.51483/ijdsbda.3.2.2023.73-95

Rauscher H, Rasmussen K, Sokull-Klüttgen B (2017) Regulatory aspects of nanomaterials in the EU. Chemie-Ingenieur-Technik 89(3). https://doi.org/10.1002/cite.201600076

Ren J, Dang KM, Pollet E, Avérous L (2018) Preparation and characterization of thermoplastic potato starch/halloysite nano-biocomposites: effect of plasticizer nature and nanoclay content. Polymers (Basel) 10(8). https://doi.org/10.3390/polym10080808

Rezaei A, Ehtesabi H (2022) Fabrication of alginate/chitosan nanocomposite sponges using green synthesized carbon dots as potential wound dressing. Mater Today Chem 24. https://doi.org/10.1016/j.mtchem.2022.100910

Rosenboom JG, Langer R, Traverso G (2022) Bioplastics for a circular economy. Nat Rev Mater 7(2). https://doi.org/10.1038/s41578-021-00407-8

Saavedra-Rubio K, Thonemann N, Crenna E, Lemoine B, Caliandro P, Laurent A (2022) Stepwise guidance for data collection in the life cycle inventory (LCI) phase: building technology-related LCI blocks. J Clean Prod 366. https://doi.org/10.1016/j.jclepro.2022.132903

Sajna VP, Mohanty S, Nayak SK (2017) Influence of nanoclay and graft copolymer on the thermal and flammability properties of poly(lactic acid)/banana fiber biocomposites. J Vinyl Addit Technol 23. https://doi.org/10.1002/vnl.21556

Sánchez S, Cancio Y, Sánchez IR, Martirena JF, Rosa ER, Habert G (2019) Sustainability assessment in Cuban cement sector—a methodological approach. IOP Conf Ser: Earth Environ Sci. https://doi.org/10.1088/1755-1315/323/1/012128

Sanni SE, Sadiku ER, Okoro EE (2022) Thermal stabilities of bionanocomposites at elevated temperatures. https://doi.org/10.1007/978-981-16-8578-1_3

Sanyé-Mengual E, Biganzoli F, Valente A, Pfister S, Sala S (2023) What are the main environmental impacts and products contributing to the biodiversity footprint of EU consumption? A comparison of life cycle impact assessment methods and models. Int J Life Cycle Assess 28(9). https://doi.org/10.1007/s11367-023-02169-7

Saouter E, Gibon T (2024) A world full of energy. In: All you need to know about the next energy revolution. https://doi.org/10.1007/978-3-031-51332-9_1

Saxe JK, Hoffman L, Labib R (2022) Method to incorporate green chemistry principles in early-stage product design for sustainability: case studies with personal care products. Green Chem 24(12). https://doi.org/10.1039/d2gc00842d

Saxena P et al (2024) A critical review on fate, behavior, and ecotoxicological impact of zinc oxide nanoparticles on algae. Environ Sci Pollut Res 31(13). https://doi.org/10.1007/s11356-024-32439-2

Schmidt A, Bittmann-Hennes B, Montero B, Wetzel B, Barral L (2023) Green bionanocomposites based on polyhydroxybutyrate and filled with cellulose nanocrystals: melting processing and characterization. J Polym Environ 31(11). https://doi.org/10.1007/s10924-023-02835-9

Schwirn K, Voelker D, W Galert, J Quik, Tietjen L (2020) Environmental risk assessment of nanomaterials in the light of new obligations under the REACH regulation: which challenges remain and how to approach them? Integr Environ Assess Manag 16(5). https://doi.org/10.1002/ieam.4267

Sharma B, Malik P, Jain P (2018) Biopolymer reinforced nanocomposites: a comprehensive review. Mater Today Commun 16. https://doi.org/10.1016/j.mtcomm.2018.07.004

Shen SC, Khare E, Lee NA, Saad MK, Kaplan DL, Buehler MJ (2023) Computational design and manufacturing of sustainable materials through first-principles and materiomics. Chem Rev 123(5). https://doi.org/10.1021/acs.chemrev.2c00479

Sierra D, Cubillos-Varela A, Franco C (2022) Analysis of sugarcane production and transportation in Hoya del Río Suárez from a life cycle perspective. Clean Technol Environ Policy 24(10). https://doi.org/10.1007/s10098-022-02380-4

Spierling S et al (2018) Bio-based plastics—a review of environmental, social and economic impact assessments. J Clean Prod 185. https://doi.org/10.1016/j.jclepro.2018.03.014

Stathatou PM, Corbin L, Meredith JC, Garmulewicz A (2023) Biomaterials and regenerative agriculture: a methodological framework to enable circular transitions. Sustainability 15(19). https://doi.org/10.3390/su151914306

Steinhoff-Knopp B, Kuhn TK, Burkhard B (2021) The impact of soil erosion on soil-related ecosystem services: development and testing a scenario-based assessment approach. Environ Monit Assess 193. https://doi.org/10.1007/s10661-020-08814-0

Swaminathan PD, Uddin MN, Wooley P, Asmatulu R (2020) Fabrication and biological analysis of highly porous PEEK bionanocomposites incorporated with carbon and hydroxyapatite nanoparticles for biological applications. Molecules 25(16). https://doi.org/10.3390/molecules25163572

Tan SX, Andriyana A, Ong HC, Lim S, YL Pang, Ngoh GC (2022) A comprehensive review on the emerging roles of nanofillers and plasticizers towards sustainable starch-based bioplastic fabrication. Polymers 14(4). https://doi.org/10.3390/polym14040664

Tao F, Zuo Y, Da Xu L, Lv L, Zhang L (2014) Internet of things and BOM-based life cycle assessment of energy-saving and emission-reduction of products. IEEE Trans Ind Informatics 10(2). https://doi.org/10.1109/TII.2014.2306771

Thakur M, Chandel M, Rani A, Sharma A (2022) Introduction to biorenewable nanocomposite materials: methods of preparation, current developments, and future perspectives. In: ACS symposium series, vol 1411. https://doi.org/10.1021/bk-2022-1411.ch001

The International Resource Panel (2012) Responsible resource management for a sustainable world: findings from the International Resource Panel

Tian J, Huang J, Zhao Y, Zhao S (2012) Electrochemical immunosensor for prostate-specific antigen using a glassy carbon electrode modified with a nanocomposite containing gold nanoparticles supported with starch-functionalized multi-walled carbon nanotubes. Microchim Acta 178(1–2). https://doi.org/10.1007/s00604-012-0816-3

Timm JFG, Maciel VG, Passuello A (2023) Towards sustainable construction: a systematic review of circular economy strategies and ecodesign in the built environment. Buildings 13(8). https://doi.org/10.3390/buildings13082059

Tripathy DB, Gupta A (2024) Nanocomposites as sustainable smart materials: a review. J Reinf Plast Compos. https://doi.org/10.1177/07316844241233162

Turan NB, Erkan HS, Engin GO, Bilgili MS (2019) Nanoparticles in the aquatic environment: usage, properties, transformation and toxicity—a review. Process Saf Environ Prot 130. https://doi.org/10.1016/j.psep.2019.08.014

Uddin MN, Dhanasekaran PS, Asmatulu R (2019) Mechanical properties of highly porous PEEK bionanocomposites incorporated with carbon and hydroxyapatite nanoparticles for scaffold applications. Prog Biomater 8(3). https://doi.org/10.1007/s40204-019-00123-1

UN Environment Programme (2024) Global Resources Outlook 2024: bend the trend—pathways to a liveable planet as resource use spikes

United Nations Statistics Division (2012) System of Environmental-Economic Accounting for Water (SEEA-Water)

Vasile C (2018) Polymeric nanocomposites and nanocoatings for food packaging: a review. Materials 11(10). https://doi.org/10.3390/ma11101834

Vasile C, Cazacu G (2013) Biocomposites and nanocomposites containing lignin. In: Biopolymer nanocomposites. https://doi.org/10.1002/9781118609958.ch23

Vo HC, Pham MH (2021) Ecotoxicological effects of microplastics on aquatic organisms: a review. Environ Sci Pollut Res 28(33). https://doi.org/10.1007/s11356-021-14982-4

Walker S, Rothman R (2020) Life cycle assessment of bio-based and fossil-based plastic: a review. J Clean Prod 261. https://doi.org/10.1016/j.jclepro.2020.121158

Wang X, Lin Z (2021) Morphologic and synergistic effects of GNP/NS binary-filler-based multi-functional coatings with robust anti-corrosion and hydrophobic properties. Prog Org Coatings 157. https://doi.org/10.1016/j.porgcoat.2021.106286

Wang Z, Fu B, Zhang L, Wu X, Li Y (2022) Ecosystem service assessments across cascade levels: typology and an evidence map. Ecosyst Serv 57. https://doi.org/10.1016/j.ecoser.2022.101472

Wang H, Yang J, Chen G, Ren C, Zhang J (2023) Machine learning applications on air temperature prediction in the urban canopy layer: a critical review of 2011–2022. Urban Climate 49. https://doi.org/10.1016/j.uclim.2023.101499

Xu Z, Zhang M, Gao SH, Wang G, Zhang S, Luan J (2019) Study on mechanical properties of unidirectional continuous carbon fiber-reinforced PEEK composites fabricated by the wrapped yarn method. Polym Compos 40(1). https://doi.org/10.1002/pc.24600

Yadav B, Pandey A, Kumar LR, Tyagi RD (2020) Bioconversion of waste (water)/residues to bioplastics—a circular bioeconomy approach. Bioresource Technol 298. https://doi.org/10.1016/j.biortech.2019.122584

Yousaf MA, Aqsa R (2023) Integrating circular economy, SBTI, digital LCA, and ESG benchmarks for sustainable textile dyeing: a critical review of industrial textile practices. Glob Nest J 25(7). https://doi.org/10.30955/gnj.005145

Youssef SH, Abdella S, Garg S (2022) Development and validation of a novel tool for assessing the environmental impact of 3D printing technologies: a pharmaceutical perspective. Pharmaceutics 14(5). https://doi.org/10.3390/pharmaceutics14050933

Yu M, Wiedmann T (2018) Implementing hybrid LCA routines in an input–output virtual laboratory. J Econ Struct 7(1). https://doi.org/10.1186/s40008-018-0131-1

Zhang B, ZM Chen, Qiao H, Chen B, Hayat T, Alsaedi A (2015) China's non-CO_2 greenhouse gas emissions: inventory and input-output analysis. Ecol Inform 26(P1). https://doi.org/10.1016/j.ecoinf.2014.01.009

Zhang N, Xiong G, Liu Z (2022) Toxicity of metal-based nanoparticles: challenges in the nano era. Front Bioeng Biotechnol 10. https://doi.org/10.3389/fbioe.2022.1001572

Zia F et al (2020) Synthesis and characterization of graphene nanoplatelets-hydroxyethyl cellulose copolymer-based polyurethane bionanocomposite system. Int J Biol Macromol 165. https://doi.org/10.1016/j.ijbiomac.2020.10.069

Jakpa Wizi is a doctoral fellow at Hohai University in Nanjing, China, where he is affiliated with the School of Environment and the Key Laboratory of Integrated Regulation and Resource Development on Shallow Lakes. His research primarily explores environmental studies, focusing on controlling algal blooms, utilizing algae biomass, and waste valorization, all within the context of sustainability. Jakpa's work contributes to broader efforts aimed at leveraging natural resources to develop eco-friendly technologies.

Lixiao Ni is a Professor at Hohai University in Nanjing, China. She is affiliated with the School of Environment and focuses on ecological remediation, particularly algae bloom control. Ni leads several national and provincial natural science fund projects. Her research interests align with life sciences, though some sources also mention involvement in chemistry. She is actively engaged in environmental studies and contributes to projects related to clean water management under rapidly changing environments in China.

Benjamin Tawiah is a senior lecturer at Kwame Nkrumah University of Science and Technology, specializing in textile research and development. He holds a Ph.D. in Textile Technology from The Hong Kong Polytechnic University. His research focuses on high-performance polymers, flame retardants, and bio-inspired materials. Tawiah has worked as a postdoctoral fellow at Hong Kong Polytechnic University and has published numerous papers in reputable journals. He serves on the editorial board of the International Journal of Textile Science and Engineering and is recognized for his contributions to fire safety and polymer composites.

Chapter 17
Regulatory Framework and Standards on Bionanocomposites Manufacturing and Applications: Social Impact and Ethical Considerations

Bhawana Chand, Peeyush Phogat, Shreya, Sukhvir Singh, and Ranjana Jha

Abstract Bionanocomposites offer transformative solutions for sustainable development across diverse sectors, yet their adoption necessitates a robust regulatory framework and adherence to ethical standards. This chapter explores regulatory guidelines that shape the manufacturing, application, and lifecycle management of bionanocomposites, with a focus on compliance with international standards such as ISO and FDA. It addresses challenges in harmonizing regulations across jurisdictions to ensure product safety, quality, and environmental responsibility. The societal impact of bionanocomposites is evaluated through their contributions to healthcare, environmental sustainability, and industrial applications. This analysis includes risk assessments of human and ecological health, encompassing exposure pathways, biodegradability, and end-of-life considerations. Ethical concerns surrounding these materials, including transparency, equitable access, and public accountability, are discussed, proposing a framework grounded in precautionary principles and responsible innovation. Through an interdisciplinary approach, this chapter underscores the importance of establishing unified regulatory and ethical standards to support the safe and equitable integration of bionanocomposites into society, aligning technological progress with sustainable development goals.

Keywords Bionanocomposites · Regulatory framework · Ethical considerations · Sustainable development · Societal impacts

---

B. Chand · P. Phogat (✉) · Shreya · R. Jha
Research Lab for Energy Systems, Department of Physics, Netaji Subhas University of Technology, Dwarka, New Delhi, India
e-mail: peeyush.phogat@gmail.com

P. Phogat · Shreya
Council of Scientific and Industrial Research—National Institute of Science Communication and Policy Research (CSIR–NIScPR), Pusa, New Delhi, India

S. Singh
Department of Nanotechnology, Delhi Skill and Entrepreneurship University, New Delhi, India

S. C. George and B. Tawiah (eds.), *Advanced Bionanocomposite Materials*, Advanced Structured Materials 241, https://doi.org/10.1007/978-981-96-8225-6_17

17.1 Introduction

Bionanocomposites represent an innovative class of materials that combine the nanoscale precision of nanotechnology with bio-based components, offering vast potential in fields like healthcare, environmental remediation, and sustainable industrial applications (Puiggalí and Katsarava 2017). These materials are designed to be both functional and environmentally friendly, addressing critical global challenges such as resource depletion and pollution. However, the rapid development and integration of bionanocomposites into diverse industries necessitate comprehensive regulatory frameworks and ethical considerations to guide their manufacturing and application (Iroegbu and Ray 2022). Due to their unique composition, often involving biological and nanoscale elements, bionanocomposites pose complex challenges in terms of safety, efficacy, and environmental impact (Heidari et al. 2024a). Regulatory bodies, such as the International Organization for Standardization (ISO) (Sen et al. 2024) and the U.S. Food and Drug Administration (FDA) (Iqbal et al. 2017), are beginning to address these challenges, but existing regulations are largely inconsistent across regions. This lack of harmonization creates barriers to global market integration, complicates compliance for developers, and underscores the need for unified standards to ensure safe and effective use (Shatkin and Kim 2017).

In addition to regulatory concerns, the societal and ethical impacts of bionanocomposites are significant. As these materials become more integrated into consumer products, medical applications, and environmental technologies, questions about biocompatibility, environmental footprint, and public health implications become critical (Shatkin and Kim 2017). The potential environmental benefits of bionanocomposites, such as biodegradability and reduced carbon emissions, are frequently cited, but rigorous empirical research is essential to validate these claims and understand the full lifecycle impact from production to disposal. The ethical considerations surrounding bionanocomposites are equally important. Transparency and public awareness about the use and potential risks of these materials are necessary to maintain consumer trust, particularly in sectors like healthcare where safety is paramount (Shunin et al. 2018). Issues of equity and access also arise, as high costs and technical complexity may restrict the benefits of bionanocomposites to wealthier regions, potentially excluding lower-income areas that could benefit most from these sustainable technologies (Khan and Malik 2019). Addressing these ethical aspects is vital to ensuring that the development of bionanocomposites aligns with broader societal values and goals for sustainable development.

Manufacturing standards play a crucial role in ensuring the quality, safety, and environmental sustainability of bionanocomposites (Khan and Malik 2019). Given the complexity of combining bio-based and nanoscale materials, stringent quality control measures and standardized manufacturing practices are essential. This includes considerations for material sourcing, synthesis processes, and environmental safety standards (Bonwick et al. 2019). A comprehensive approach to lifecycle management is also necessary to prevent any adverse effects on ecosystems and human health, as bionanocomposites may present unique risks related to their biodegradability, bioaccumulation, and long-term environmental impact. By integrating rigorous regulatory frameworks, ethical standards, and responsible manufacturing practices, it is possible to foster the growth of bionanocomposites in a way that supports sustainable development goals. This chapter, therefore, examines the critical regulatory, societal, and ethical dimensions essential for the responsible innovation of bionanocomposites, underscoring the importance of a balanced approach that prioritizes both technological progress and public welfare.

17.2 Bionanocomposites: Applications and Impact

Bionanocomposites have emerged as a significant breakthrough in material science, offering solutions that integrate nanoscale precision and biologically derived materials. This unique blend of properties makes them highly applicable across healthcare, environmental management, and industrial sectors, each of which stands to benefit significantly from the versatility, functionality, and eco-friendliness these materials provide. With their growing use, bionanocomposites offer considerable advantages, but they also introduce new complexities regarding safety, efficacy, and environmental impact, demanding careful evaluation and a balanced approach to innovation.

17.2.1 Key Applications in Healthcare, Environment, and Industry

In healthcare, bionanocomposites are being extensively researched and applied to advance drug delivery, tissue engineering, and medical implants. There are many fields in healthcare where bionanocomposites are used for the treatment as shown in Fig. 17.1. These materials can be engineered to provide controlled, targeted drug release, enhancing the effectiveness of treatments while minimizing side effects. For instance, bionanocomposites designed with nanoparticles can carry therapeutic agents directly to specific cells or tissues, ensuring that drugs are delivered precisely where needed, such as in cancer treatments. This targeted approach can limit the systemic toxicity often associated with chemotherapy, improving patient outcomes and enhancing quality of life during treatment. Additionally, bionanocomposites in tissue engineering offer structures that mimic the extracellular matrix, a crucial element in promoting cell growth and regeneration. Materials such as collagen-based or chitosan-based bionanocomposites facilitate the repair of damaged tissues, from skin grafts to cartilage repair, by providing scaffolding that supports cell proliferation and differentiation, which is crucial for healing.

Another vital application in healthcare is in the development of biocompatible and durable implants. Traditional implants, made from metals or synthetic polymers, can

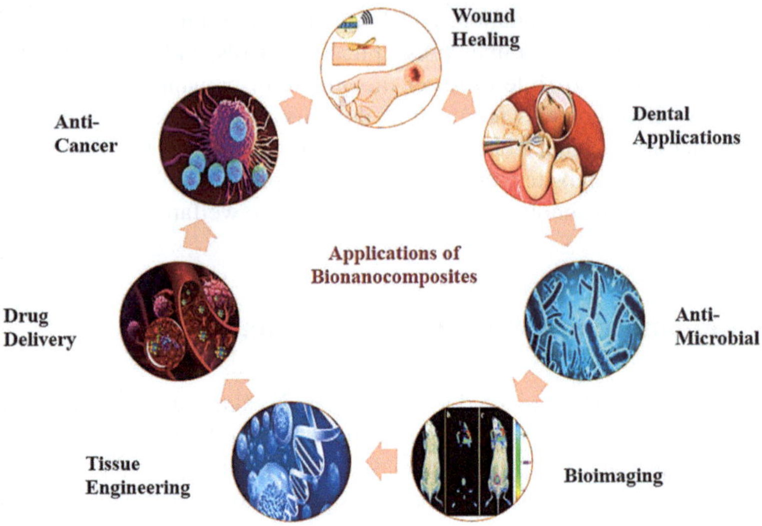

Fig. 17.1 Depict different healthcare application of bionanocomposites. Reproduced from Nandhini et al. (2024) copyright © 2024, The Author(s)

sometimes be rejected by the body or cause inflammatory responses. Bionanocomposites, however, can be engineered to be highly compatible with human tissue, reducing the likelihood of rejection and offering enhanced longevity. For example, hydroxyapatite-based bionanocomposites, which closely resemble natural bone, are increasingly used in orthopedics and dental implants. They provide the required mechanical strength while fostering the integration of the implant with surrounding bone tissue, thus improving implant success rates and reducing the need for future replacements. There are various bionanocomposites which are used for different healthcare treatment based on their characteristics, composition, shapes and sizes, some of them are listed in Table 17.1 with their composition and applications highlighting their roles in tissue engineering, drug delivery, wound healing, and other biomedical fields.

In environmental applications, bionanocomposites are harnessed for pollutant adsorption, water purification, and bioremediation efforts. Figure 17.2 highlights the role of bio-porous polymers, modified with cross-linkers and metal–organic frameworks, in water purification. These materials effectively adsorb contaminants such as oils, solvents, drugs, and dyes (Gupta et al. 2022). Their high surface area and chemical stability make them ideal for sustainable water treatment, contributing to cleaner water and reduced environmental impact. One of the major challenges in environmental management is the removal of toxic pollutants, including heavy metals, organic compounds, and pathogens, from water and soil. Bionanocomposites with high surface areas and tunable surface chemistries are well-suited for capturing and neutralizing these contaminants. For instance, cellulose-based bionanocomposites functionalized with nanoparticles are effective in removing heavy metals like lead

Table 17.1 Different types of bionanocomposites, their compositions, and their various medical applications

Bionanocomposite	Composition	Medical applications
Chitosan-based bionanocomposites	Chitosan, nanoparticles (e.g., silver, gold)	Wound healing, antimicrobial dressings, drug delivery systems, and tissue engineering scaffolds
Collagen-based bionanocomposites	Collagen, hydroxyapatite, or bioactive glass	Bone regeneration, cartilage repair, and skin tissue engineering
Gelatin-based bionanocomposites	Gelatin, nano clay, graphene oxide	Controlled drug delivery, wound dressing materials, and scaffolds for regenerative medicine
Cellulose-based bionanocomposites	Cellulose nanocrystals, silver or gold nanoparticles	Antibacterial coatings, tissue engineering scaffolds, and drug delivery systems
Alginate-based bionanocomposites	Alginate, carbon nanotubes, or magnetic nanoparticles	Drug encapsulation, tissue regeneration, and cancer therapy
Polymer-ceramic bionanocomposites	Polylactic acid (PLA) or polycaprolactone (PCL) with hydroxyapatite	Bone tissue engineering and dental implants
Silk fibroin bionanocomposites	Silk fibroin, nanoparticles (e.g., gold, silica)	Wound healing, drug delivery, and scaffolds for nerve regeneration
Polyhydroxyalkanoate (PHA)-based bionanocomposites	PHA, carbon nanotubes, or nano clays	Biodegradable implants, sutures, and drug delivery systems
Starch-based bionanocomposites	Starch, graphene oxide, or montmorillonite	Biodegradable wound dressings, controlled drug release, and antimicrobial packaging for medical devices
Hyaluronic acid bionanocomposites	Hyaluronic acid, silica or graphene nanoparticles	Injectable hydrogels for tissue repair, anti-aging products, and targeted drug delivery
Magnesium-based bionanocomposites	Magnesium oxide nanoparticles, biodegradable polymers	Biodegradable orthopedic implants, stents, and bone fixation devices
Zinc-based bionanocomposites	Zinc oxide nanoparticles, biopolymers	Antimicrobial coatings, wound dressings, and skin tissue regeneration
Protein-based bionanocomposites	Soy protein, keratin, or casein with nanoparticles	Scaffolds for tissue engineering, drug delivery, and biodegradable sutures
Graphene-based bionanocomposites	Graphene oxide, biopolymers	Biosensors, cancer therapy, and high-strength scaffolds for tissue engineering
Carbon nanotube bionanocomposites	Carbon nanotubes, polymers	Neural regeneration, biosensors, and high-strength medical implants

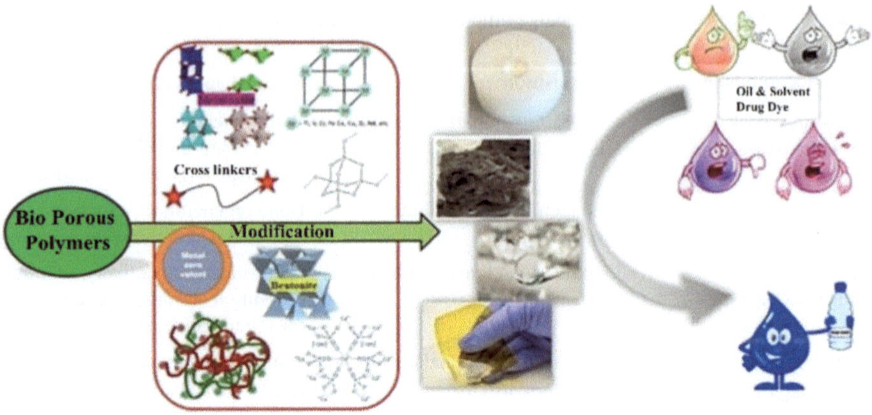

Fig. 17.2 Illustration of bio-porous polymers modified with cross-linkers and metal frameworks for efficient removal of contaminants such as oil, solvents, drugs, and dyes, contributing to sustainable water purification systems. Reproduced from Tadayoni et al. (2024) copyright © 2024 by the authors

and mercury from water, providing a low-cost, sustainable solution for clean water access in regions affected by industrial pollution (Darmenbayeva et al. 2024). Similarly, bionanocomposites composed of algae-derived materials and silver nanoparticles have shown efficacy in eradicating bacterial contaminants in water, combining both adsorption and antimicrobial functions (Anjaneyulu et al. 2024).

In industry, bionanocomposites are being adopted as sustainable alternatives in sectors such as packaging, electronics, and construction shown in Fig. 17.3. Traditional materials, particularly in packaging, often rely on fossil-based plastics that are neither biodegradable nor environmentally sustainable. Bionanocomposites, such as those based on polylactic acid (PLA) and cellulose, offer biodegradable options that can replace single-use plastics without sacrificing durability as illustrated in Fig. 17.3a (Tripathi et al. 2021). These materials can be designed to be robust, flexible, and resistant to moisture, making them ideal for packaging applications that maintain product safety while reducing plastic waste. Similarly, in electronics, bionanocomposites are beginning to replace synthetic materials in applications like electrodes as shown in Fig. 17.3b, printed electronics, and flexible displays, where biocompatibility and biodegradability are becoming increasingly important (Nawafleh and Al-Oqla 2024). In construction, bionanocomposites based on fibers like hemp and flax are used as reinforcements in building materials depicted in Fig. 17.3c, offering lightweight, strong, and eco-friendly alternatives to traditional construction materials, thereby reducing the carbon footprint associated with construction (Satdive et al. 2022).

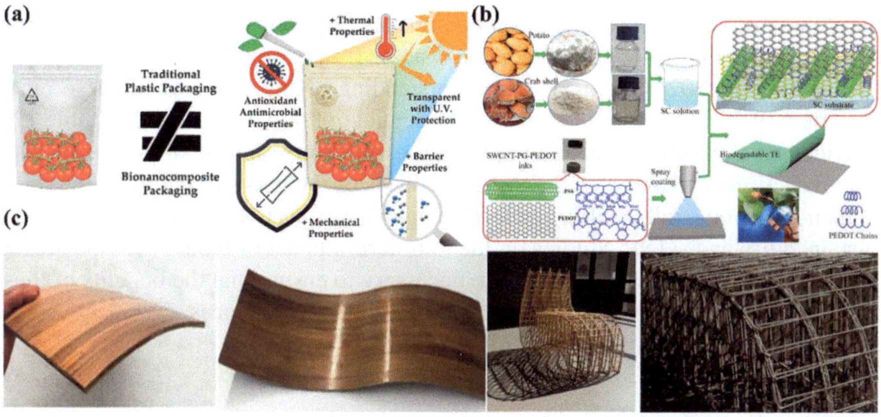

Fig. 17.3 **a** Replacing traditional plastic packaging with enhanced properties such as antimicrobial and UV protection. Reproduced from Pires et al. (2023) copyright © 2023 by the authors, **b** fabrication process for biodegradable and flexible 3D interconnected SWCNT-PG-PEDOT-based thermoelectric materials (TE), showcasing the use of natural resources and innovative coating techniques for sustainable electronics. Reproduced from Miao et al. (2018) copyright © 2018 American Chemical Society, and **c** structural applications in construction owing to their superior mechanical and environmental benefits. Reproduced from Morales et al. (2017) copyright © 2017 by the authors Dahy (2019) copyright © 2019 by the author

17.2.2 Potential Benefits and Risks of Bionanocomposites

The potential benefits of bionanocomposites are substantial, especially concerning sustainability, biocompatibility, and multifunctionality. By incorporating biodegradable and renewable components, bionanocomposites help reduce the reliance on nonrenewable resources, addressing critical environmental issues like plastic pollution and greenhouse gas emissions (Kausar 2021). In healthcare, the tailored bioactivity and biocompatibility of bionanocomposites make them highly effective in promoting tissue growth, delivering drugs with precision, and supporting implants that integrate seamlessly with human tissue. These materials are also relatively easy to customize, allowing researchers to design them for specific applications, whether in medical, environmental, or industrial contexts. This level of customization is valuable because it means that bionanocomposites can be engineered to meet specific functional, mechanical, and safety requirements, thereby expanding their applicability across a diverse range of fields (Xiong et al. 2018).

However, the risks associated with bionanocomposites must also be acknowledged and managed. In healthcare, despite the benefits of controlled drug delivery and biocompatible implants, there is a risk of unforeseen reactions between bionanocomposites and biological tissues, which could lead to toxicity or immune responses. For instance, while some nanoparticles can enhance the therapeutic effects of a drug, they may also pose a risk if they accumulate in certain organs, leading to potential toxicity over time (Jiang et al. 2019). Moreover, in environmental applications, there is a risk that nanoscale components in bionanocomposites may contribute to bioaccumulation or release harmful by-products into ecosystems if they do not degrade as expected (Olawore et al. 2024). A thorough understanding of the lifecycle of these materials from their production to disposal is essential to mitigate these risks. Table 17.2 provides an overview of bionano applications in diverse fields, summarizing their materials, uses, and associated risks. It emphasizes the need for careful assessment of their environmental, health, and technological impacts to balance innovation with sustainability and safety. Studies on the long-term impact of bionanocomposites on human health and the environment are necessary to ensure that the benefits of these materials are realized without unintended negative consequences.

17.2.3 Case Studies on Bionanocomposites Applications

Several case studies illustrate the transformative impact of bionanocomposites across various fields. In healthcare, the chitosan-$AgIO_3$ bionanocomposite demonstrates significant potential for healthcare applications due to its excellent antibacterial properties illustrated in Fig. 17.4. Effective against a wide range of pathogenic microorganisms, including Pseudomonas aeruginosa, Klebsiella pneumoniae, and Staphylococcus aureus, this material outperforms commercial drugs like penicillin and silver sulfadiazine in some cases (Ahghari et al. 2022). Its bactericidal mechanism, driven by reactive oxygen species (ROS) production, makes it a promising candidate for use in infection control, wound healing, and sterilization.

In environmental applications, the use of biocompatible nanosilica for soil stabilization is a promising approach to enhance soil properties in an environmentally friendly way. Nanosilica, a fine silica nanoparticle, improves the mechanical properties of soil by interacting with its particles, leading to better compaction, strength, and durability. Unlike traditional stabilizers like cement or lime, which can have negative environmental impacts, nanosilica is biocompatible and non-toxic, making it an eco-friendly alternative for soil improvement. When mixed with water, nanosilica forms a gel-like matrix that fills the voids between soil particles, resulting in increased cohesion and reducing soil plasticity. This mechanism leads to improved shear strength and reduced expansiveness, which is particularly beneficial in clayey soils prone to instability. The addition of nanosilica also reduces soil permeability, making it more resistant to water-induced erosion and enhancing its overall stability. Figure 17.5a displays a microscopic image showcasing the uniform dispersion of silica nanoparticles within the soil matrix, highlighting the effective incorporation of bionanocomposites for enhancing subgrade soil stabilization. Figure 17.5b presents a graph illustrating the relationship between the percentage of silica nanoparticles and the stabilization effect, as represented by the weight percentage (wt%). The graph indicates a consistent increase in stabilization properties with rising silica nanoparticle content, reaching a peak at 3%. This combined analysis demonstrates the significant role of silica nanoparticles in improving the structural and mechanical properties of subgrade soil, contributing to enhanced load-bearing capacity and durability. The potential

Table 17.2 Risks and effects associated with bionano applications across various fields, highlighting potential hazards and their implications

No.	Field	Bionano material	Application	Observed risk/effect	References
1	Medicine	Antibacterial coatings	Antibacterial coatings	Cytotoxicity at high concentrations; potential to harm healthy cells	You et al. (2012)
2	Agriculture	Zinc oxide nanoparticles	Fertilizer and pesticide additives	Accumulation in soil affecting microbial ecosystems	Al Jabri et al. (2022)
3	Water treatment	Algae-derived bionanocomposites	Removal of bacterial contaminants	Effective antibacterial activity but potential for nanoparticle leaching	Li et al. (2022)
4	Food packaging	Chitosan-based bionanocomposites	Biodegradable food packaging materials	Migration of nanoparticles into food at elevated temperatures	Kishore et al. (2023)
5	Energy storage	Carbon nanotube composites	Supercapacitors	Limited recyclability and concerns about respiratory toxicity during manufacturing	Sousa et al. 2020)
6	Environmental remediation	Iron oxide nanoparticles	Heavy metal adsorption in soil	Unintended release of nanoparticles potentially contaminating water sources	Kaleem et al. (2024)
7	Heating	Metal oxide nanoparticles	Thermal insulation coatings	Release of harmful byproducts at elevated temperatures	Meierhofer and Fritsching (2021)

applications of nanosilica-treated soil are numerous, especially in construction and geotechnical engineering. It can be used for road construction, foundation work, and land reclamation, providing a cost-effective and durable solution for areas with weak or unstable soil. Furthermore, the environmentally friendly nature of nanosilica makes it an ideal choice for sustainable development in construction projects, where reducing the environmental footprint is a priority.

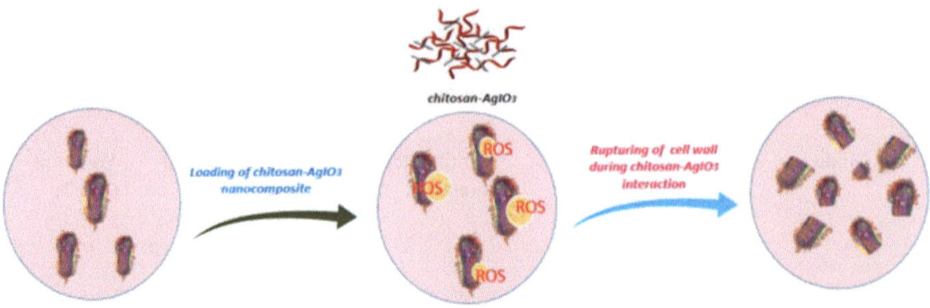

Fig. 17.4 Mechanism of action of chitosan-$AgIO_3$ nanocomposite against microbial cells. The nanocomposite interacts with the cell wall, generating reactive oxygen species (ROS) that lead to cell wall rupture and microbial cell destruction. Reproduced from Ahghari et al. (2022) copyright © 2022, The Author(s)

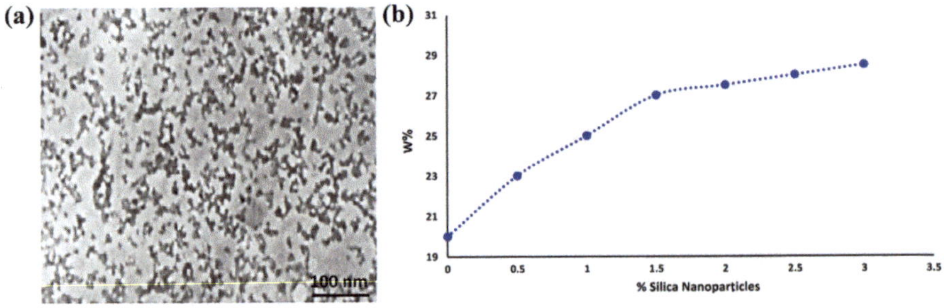

Fig. 17.5 **a** TEM image of silica nanoparticles biosynthesized using wheat husk ash. **b** The influence of green silica nanoparticles on the ideal moisture content of subgrade soil. Reproduce from Buazar (2019) copyright © 2019, The Author(s)

In the industrial sector, (PLA) and graphene oxide-based bionanocomposites are gaining traction in the development of biodegradable packaging. These materials combine the durability and flexibility of traditional plastics with the added advantage of being biodegradable, offering a promising solution to the plastic waste crisis (Xu et al. 2016). PLA-based bionanocomposites are particularly suitable for food packaging, where they maintain the necessary barrier properties to protect products from moisture and contaminants while providing an eco-friendly alternative to conventional plastics. The use of these materials in the packaging industry underscores the potential of bionanocomposites to drive sustainable practices, reducing plastic waste and lowering environmental impact.

These case studies highlight not only the versatility of bionanocomposites but also the critical need for responsible innovation. While these materials offer significant benefits, their lifecycle impacts, safety profiles, and regulatory considerations must be carefully managed to ensure that their integration into society does not lead to unintended consequences. Ensuring that bionanocomposites are developed with attention to both their advantages and potential risks will be crucial to maximizing their positive impact across healthcare, environmental management, and industry.

17.3 Existing Regulatory Frameworks and Standards for Bionanocomposites

As the application of bionanocomposites expands across fields like healthcare, environmental management, and industry, there is an increasing need for regulatory frameworks and standards to ensure their safe and effective use. Bionanocomposites, which incorporate nanoscale and biological materials, pose unique challenges in terms of health, safety, and environmental impact. While international organizations like the ISO and the U.S. FDA have begun establishing standards, the regulatory landscape for bionanocomposites remains complex and often fragmented across regions (Rehm and Fata Moradali 2020). This variation in regulations creates challenges for companies and researchers aiming to develop and market bionanocomposites globally, highlighting the need for a more harmonized approach.

17.3.1 International Standards (e.g., ISO, FDA)

International standards play a foundational role in regulating bionanocomposites. The ISO, a globally recognized body, develops standards to address the specific technical and safety aspects of nanotechnology and biotechnology. ISO standards, such as ISO/TS 80004-1 for nanotechnology terminology (Chemical Analysis | Medical Device | Smithers 2024) and ISO 10993 for biocompatibility testing of medical devices, are particularly relevant to bionanocomposites used in healthcare (Zhou et al. 2024). These standards guide the characterization, testing, and risk assessment of nanomaterials, offering a common framework for evaluating safety and performance. Additionally, ISO standards promote the use of scientifically validated methods to assess potential risks, enhancing the credibility and safety profile of bionanocomposites in various applications.

In the United States, the FDA regulates bionanocomposites under the categories of medical devices, drugs, or food additives, depending on their application. The FDA has established guidelines for nanomaterials, focusing on safety, effectiveness, and quality control (Dorato 2018), and it frequently collaborates with the ISO to align its regulations with international standards. In healthcare, bionanocomposites must comply with the FDA's rigorous testing protocols to ensure biocompatibility, as well as efficacy and safety, particularly if used in drug delivery systems or implants. For instance, any bionanocomposite intended for drug delivery would undergo extensive clinical trials to assess not only its therapeutic efficacy but also its potential long-term interactions with biological tissues (Bhat et al. 2015). Through these measures, the

FDA aims to prevent adverse reactions and ensure that products meet strict safety standards before they reach consumers.

The European Union (EU) has also developed specific guidelines under its European Medicines Agency (EMA) and European Food Safety Authority (EFSA) frameworks, particularly for bionanocomposites used in pharmaceuticals and food products (Trotta et al. 2023). The EU employs a "precautionary principle" in its regulatory approach, which requires that products demonstrate a high level of safety before approval. This principle often leads to stricter regulatory requirements in Europe compared to other regions, especially concerning environmental impact and biodegradability (De Smedt and Vos 2022). Additionally, the European Commission's REACH regulation (Registration, Evaluation, Authorization, and Restriction of Chemicals) requires comprehensive safety data on chemicals used in bionanocomposites, which impacts both their development and marketability within the EU. Through REACH, manufacturers are held accountable for providing scientific evidence that their products do not pose risks to human health or the environment, aligning with the EU's sustainability and safety goals (Lacourt et al. 2024).

17.3.2 Regional Regulations and Variations

Although international bodies like ISO and FDA provide overarching guidelines, regional regulations for bionanocomposites differ significantly due to variations in policy priorities, scientific approaches, and economic conditions. In North America, the United States and Canada have regulatory frameworks with some common principles but also notable differences. While the U.S. FDA enforces rigorous testing and approval procedures, Health Canada places additional emphasis on environmental safety, particularly concerning the long-term environmental impact of nanomaterials. Health Canada requires companies to demonstrate that their bionanocomposites are not only safe for human use but also environmentally sustainable, with minimal impact on ecosystems during their lifecycle. This dual focus on human health and environmental safety reflects Canada's commitment to sustainable development and risk minimization (Health and the Environment—Canada.ca 2024).

In Asia, countries like Japan, South Korea, and China are rapidly advancing in bionanocomposite research and development, but their regulatory frameworks vary. Japan, for example, follows ISO guidelines closely but also incorporates additional safety assessments specific to its domestic market, particularly in the areas of cosmetics and consumer products (Dorato 2018). South Korea has stringent regulations regarding the environmental impact of nanomaterials, ensuring that all bionanocomposites undergo thorough environmental risk assessments before market entry (Hwang et al. 2012). China, one of the largest producers and consumers of nanomaterials, has made significant progress in establishing regulations for nanotechnology applications. However, Chinese regulations are often seen as less strict in terms of environmental impact, focusing more on fostering rapid technological advancement and commercialization. Table 17.3 highlights country-specific regulations governing bionanocomposites across various regions, outlining their objectives to ensure safety, environmental sustainability, and effective use. This regional variation creates a regulatory asymmetry that affects how bionanocomposites are marketed, tested, and used in different parts of the world.

Table 17.3 Country-specific regulations on bionanocomposites

No.	Region/ country	Regulation	Objective	References
1	European Union (EU)	EU Regulation (EC) No 1223/2009 on Cosmetic Products	Ensure the safety of bionano materials in cosmetics, particularly in nano-formulations	Regulation-1223/ 2009-EN-Cosmetic Products Regulation-EUR-Lex
2	China	China's 2018 National Bio-safety Regulations	Regulate the use of bionanocomposites in biotechnology and food applications	Ministry of Ecology and Environment of the People's Republic of China, National Biosafety Regulations (2018)
3	India	Food Safety and Standards (Food Products Standards and Food Additives) Regulations, 2011	Oversee the use of nanomaterials in food packaging and food contact materials	FSSAI (2025)
4	Japan	Act on the Evaluation of Chemical Substances and Regulation of Their Manufacture, etc.	Safety and environmental assessment of nanomaterials used in industrial applications	Japanese Law Translation (2025)
5	Singapore	Health Sciences Authority (HSA) Guidelines for Nano-based Products	Ensure the safe use of nano-based products in pharmaceuticals and medical devices	Health Sciences Authority (HSA) (2025)
6	United States (USA)	Toxic Substances Control Act (TSCA)—Nanomaterials Reporting Rule	Requires reporting and testing of new chemical substances including nanomaterials	Control of Nanoscale Materials under the Toxic Substances Control Act I US EPA (2025)
7	Australia	Australian Industrial Chemicals Introduction Scheme (AICIS)	Chemicals, including bionanocomposites, in new products	AICIS (2025)
8	South Africa	Management Act (NEMA)	Guidelines for the environmental impact of nanomaterials used in industrial sectors	Reddy and Singh (2022)

17.3.3 Challenges in Harmonizing Regulations Globally

Harmonizing regulations for bionanocomposites on a global scale face several challenges. First, there are scientific and technical complexities inherent in the characterization and risk assessment of bionanocomposites, especially due to the varying properties and behaviors of nanomaterials (Jamwal and Mittal 2024). Each bionanocomposite may require specific testing and risk analysis, which complicates the creation of universal standards. Additionally, the rapid pace of technological advancements in this field means that regulatory frameworks often struggle to keep up with new developments, leading to gaps in oversight. This is particularly challenging when bionanocomposites with unique properties or applications are developed, as existing standards may not cover these novel aspects adequately.

Another major challenge in harmonizing regulations is the differing policy priorities among regions. In the European Union, for example, there is a strong emphasis on environmental sustainability and the precautionary principle, leading to stricter regulations compared to the United States, which often prioritizes innovation and market access (Risk Regulation in Europe: Assessing the Application of the Precautionary... - Jale Tosun - Google Books 2024). These differing priorities can hinder the creation of unified standards, as regulatory bodies in different regions may have conflicting objectives. Moreover, there are economic considerations; countries with emerging economies may prioritize the rapid commercialization of bionanocomposites to drive growth, potentially at the expense of stringent safety and environmental standards. Economic and logistical factors also play a role in the harmonization challenge. Testing and compliance costs can be prohibitively high, especially for smaller companies or those in developing countries. If global regulations were to require costly and extensive testing, this could create a barrier to entry for these businesses, limiting the potential of bionanocomposites to be adopted globally (The Impacts of Nanotechnology on Companies 2010). Therefore, any harmonization efforts must balance the need for comprehensive safety assessments with considerations of accessibility and affordability for companies of all sizes and regions.

Despite these challenges, progress is being made toward regulatory harmonization. Collaborative efforts, such as joint research initiatives and information sharing between regulatory agencies, are helping to bridge some of the gaps. The OECD's Working Party on Manufactured Nanomaterials (WPMN), for example, has been working on developing standardized testing protocols and risk assessment frameworks for nanomaterials, which could serve as a foundation for bionanocomposite regulations (Maria et al. 2019). Furthermore, bilateral and multilateral trade agreements increasingly include provisions for regulatory alignment in areas like nanotechnology, fostering a more cooperative approach to standard-setting across borders.

17.4 Manufacturing Standards for Bionanocomposites

The production of bionanocomposites has introduced a range of challenges and opportunities in the field of materials science. By integrating nanoscale particles and biologically derived materials, these composites offer remarkable properties, making them valuable in diverse applications from healthcare and electronics to environmental management and packaging. However, manufacturing these materials requires stringent standards to ensure product quality, environmental safety, and human health protection (Jeevanandam et al. 2022). The emerging field of bionanocomposites demands specialized approaches to quality control, health and environmental safety, and sustainable manufacturing practices. This section explores current standards, guidelines, and evolving frameworks necessary to address the complex demands of bionanocomposite production.

17.4.1 Quality Control and Safety Standards

Quality control is essential in manufacturing bionanocomposites, as it ensures that products meet consistent standards in terms of structural integrity, functionality, and safety. Given the combination of nanoscale and biocompatible materials in bionanocomposites, manufacturers must adopt precise characterization techniques to maintain control over factors such as particle size, distribution, surface area, and chemical compositions (Heidari et al. 2024b). ISO and the American Society for Testing and Materials (ASTM) are among the primary organizations establishing quality standards for nanomaterials, including those used in bionanocomposites (Botlhoko et al. 2024; Machado et al. 2015a).

ISO's series of standards related to nanotechnology (e.g., ISO/TS 80004 for terminology and ISO/TR 13121 for risk assessment) are crucial in providing guidelines for quality control in nanomaterial production (Haydon 2015). These standards ensure that bionanocomposites are produced with well-characterized nanoscale components, helping manufacturers maintain control over the specific properties that give these materials their unique functionalities. For example, ISO 10993, a standard for biocompatibility in medical devices, is particularly relevant for bionanocomposites in healthcare applications (Sharma and Luthra 2023; Khan and Chauhan 2022). These standard outlines rigorous testing protocols to ensure that materials are safe and compatible with biological systems, addressing potential issues like cytotoxicity and inflammatory responses.

Moreover, advanced testing technologies such as X-ray diffraction (XRD), scanning electron microscopy (SEM), and dynamic light scattering (DLS) are often required to ensure the stability and structural integrity of bionanocomposites (Dmour 2022). These techniques allow for precise control over parameters like particle uniformity and dispersion within the composite matrix, which are critical for performance consistency. For example, in drug delivery applications, the uniformity of nanoparticle size is crucial for achieving controlled release rates, and these testing standards help manufacturers achieve desired consistency.

In addition to ISO and ASTM, national regulatory bodies such as the U.S. FDA and the EMA enforce quality control standards for bionanocomposites used in healthcare. These agencies require comprehensive testing to ensure that bionanocomposites meet rigorous safety, efficacy, and stability requirements, particularly for applications like medical implants, wound dressings, and drug delivery systems. The FDA, for instance, mandates stability studies to assess the durability and shelf-life of products containing bionanocomposites. Meeting these standards is essential to prevent product variability and potential risks associated with inconsistent quality in these high-stakes applications.

17.4.2 Environmental and Health Safety Guidelines

The inclusion of nanoscale components in bionanocomposites introduces unique health and environmental safety concerns that demand specialized regulatory attention. Manufacturing and handling these materials can result in nanoparticle exposure, raising concerns about worker safety and potential environmental contamination. Due to the small size of nanoparticles, they can easily penetrate biological membranes and interact with cellular structures in unpredictable ways, which can lead to toxicological risks if not properly managed (Störmer et al. 2017). Regulatory agencies, including the Occupational Safety and Health Administration (OSHA), the Environmental Protection Agency (EPA), and the European Chemicals Agency (ECHA), are actively developing safety guidelines specific to the handling and disposal of nanomaterials within bionanocomposites (A. A.-E.-F. S. S. for Natural 2024).

OSHA, for example, provides detailed recommendations on occupational exposure limits for nanoparticles, emphasizing the importance of personal protective equipment (PPE) and ventilation systems in manufacturing facilities. These guidelines aim to protect workers from potential inhalation, ingestion, or dermal exposure to nanoparticles, which could lead to respiratory, cardiovascular, or dermal health issues (Howard and Murashov 2009). Ensuring that facilities adhere to OSHA guidelines is crucial in safeguarding workers' health in environments where nanoparticles are manipulated. The EPA and ECHA also provide guidance on environmental safety regarding the lifecycle of bionanocomposites, from production to disposal. The EPA's Toxic Substances Control Act (TSCA) requires that companies assess the environmental impact of nanomaterials used in their products. This includes evaluating the potential for nanoparticle release into air, water, or soil, and ensuring that materials do not adversely affect ecosystems (Miettinen 2019). In Europe, REACH regulations require manufacturers to provide safety data and risk assessments before marketing products containing nanoparticles (Donchevskaya 2009; Hristozov et al. 2012). These regulations address concerns over bioaccumulation and potential ecological impacts, requiring companies to demonstrate that their materials are environmentally safe over their entire lifecycle.

In healthcare applications, additional safety protocols are often necessary to evaluate the potential toxicity of bionanocomposites when they interact with biological systems (Perera et al. 2024). Regulatory bodies mandate that manufacturers conduct biocompatibility tests, including cytotoxicity, genotoxicity, and allergenicity assessments, to minimize risks associated with bionanocomposites in contact with human tissues. These protocols are essential for products like implants or drug delivery systems, where direct interaction with human cells requires stringent safety validation. By enforcing these health and environmental safety guidelines, regulatory bodies help ensure that bionanocomposites are manufactured, handled, and disposed of in ways that minimize risks to both human health and the environment.

17.4.3 Emerging Standards for Sustainable Manufacturing

As the global emphasis on sustainability intensifies, manufacturers of bionanocomposites are increasingly pressured to adopt environmentally responsible production methods. Sustainability in bionanocomposite manufacturing involves minimizing waste, reducing carbon emissions, and prioritizing the use of renewable resources. Emerging standards and frameworks from organizations like the Global Reporting Initiative (GRI) and the United Nations Sustainable Development Goals (SDGs) provide guidelines to incorporate sustainability into manufacturing practices (Tsalis et al. 2020).

In recent years, ISO has introduced standards aimed at promoting sustainable manufacturing practices, such as ISO 14001 for environmental management systems (Salim et al. 2018) and ISO 14067 for carbon footprint assessment (Suer et al. 2021). These standards encourage manufacturers to measure and reduce their environmental impact, which is particularly relevant in bionanocomposite production where resource-intensive processes, including the synthesis of nanomaterials, can contribute to pollution and high energy consumption. By adopting ISO 14001, manufacturers can establish an environmental management system that helps monitor and mitigate emissions, resource usage, and waste production (Massoud et al. 2010). Additionally, ISO 14067 provides a framework for assessing the carbon footprint of products, allowing manufacturers to identify and reduce greenhouse gas emissions associated with their processes. Life cycle assessment (LCA) is another emerging tool to assess the environmental sustainability of bionanocomposites. By evaluating the entire lifecycle of a product from raw material extraction to disposal LCA provides a comprehensive view of its environmental impact. LCA helps manufacturers understand the ecological footprint of their products and make informed decisions to improve resource efficiency (Sharma et al. 2023). For example, bionanocomposites produced with biodegradable or bio-based materials, such as cellulose or chitosan, generally exhibit lower environmental impacts than those derived from synthetic components. An LCA can highlight these benefits, encouraging manufacturers to prioritize sustainable material choices (Cruz et al. 2024).

The circular economy model is increasingly influencing standards for bionanocomposite manufacturing. This model promotes the recycling and reuse of materials to minimize waste and environmental harm (Coltelli et al. 2023). In the context of bionanocomposites, implementing circular economy principles involves designing materials that are either biodegradable or recyclable. For instance, bionanocomposites used in packaging could be engineered to decompose naturally or to be recycled into new products. Standards like ISO 18604, which outlines requirements for material recycling, are relevant in this area as they provide guidelines for designing products with recyclability in mind (Luu et al. 2022). Moreover, as industries prioritize sustainable practices, green chemistry principles are becoming central to bionanocomposite manufacturing. Green chemistry emphasizes reducing the use of hazardous substances, increasing energy efficiency, and using renewable feedstocks. Bionanocomposite manufacturers are encouraged to select synthesis methods that generate minimal waste and avoid toxic solvents or reagents, thus making production more sustainable and less harmful. The implementation of green chemistry aligns with global sustainability initiatives, such as the UN SDGs, which call for responsible consumption and production practices to protect the environment (Anastas and Zimmerman 2018).

17.4.4 Societal Impacts of Bionanocomposites

Bionanocomposites, which combine nanoscale materials with bio-derived compounds, have demonstrated transformative potential across fields including healthcare, environmental management, and manufacturing. However, as the utilization of bionanocomposites expands, understanding their broader societal impacts becomes crucial. These materials, while advancing innovative applications, also prompt discussions around their environmental footprint, public health implications, and economic significance. Here, we explore the societal impacts of bionanocomposites through the lenses of environmental sustainability, public health, and economic contributions.

17.4.5 Environmental Footprint and Sustainability

The environmental sustainability of bionanocomposites depends largely on their life cycle, from raw material sourcing to disposal or recyclability. One of the key advantages of bionanocomposites lies in their potential to replace traditional, petroleum-based materials, which are associated with high carbon emissions and limited biodegradability (Singha et al. 2024). Many bionanocomposites use renewable resources such as cellulose, starch, or chitosan, derived from plants, algae, or crustaceans, thereby reducing reliance on non-renewable resources. These bio-based materials are often biodegradable or compostable, which can significantly lower waste accumulation in landfills and reduce environmental pollution, particularly in applications like packaging where disposability is a factor (Singh et al. 2023). However, the environmental benefits of bionanocomposites are tempered by the energy-intensive processes sometimes required to synthesize nanomaterials, which may offset gains in sustainability. For instance, the production of nanoparticles like carbon nanotubes or metal oxides can be resource-intensive, involving substantial energy consumption, and potentially generating hazardous byproducts. To address this, researchers and manufacturers are increasingly focused on green synthesis methods that minimize energy consumption and reduce or eliminate toxic solvents. Techniques such as microwave-assisted synthesis or enzymatic reactions are being explored as sustainable alternatives for manufacturing bionanocomposites with lower environmental impact (Azadi et al. 2024).

Another aspect of sustainability for bionanocomposites relates to their end-of-life options. While many bio-based components are biodegradable, the inclusion of inorganic nanomaterials like metal oxides can complicate disposal. Such materials may persist in the environment, raising concerns about bioaccumulation and ecotoxicity. Figure 17.6 illustrates the biodegradation behavior of biopolymers (PHB, PBAT, PLA, PBS) under four environmental conditions: home composting, industrial composting, soil, and marine environments. In home composting at 28 °C, PHB, PBAT, and PLA achieve 90% biodegradation within 12 months, while PBS does not degrade. Under industrial composting at 28 °C, all biopolymers degrade efficiently, with PHB and PLA completing 90% biodegradation within 6 months. In soil at 25 °C, only PHB and PBAT show biodegradation, achieving 90% degradation in 2 years, whereas PLA and PBS remain non-biodegradable. Notably, in marine environments, PHB is the only biopolymer that degrades, with 90% biodegradation occurring in 12 months. This comparison highlights the suitability of each biopolymer for specific environmental applications based on degradation parameters (Yu and Flury 2024). Effective waste management practices, along with LCAs, are essential to ensure that bionanocomposites do not inadvertently contribute to environmental degradation (Shaik and Franchetti 2022). LCAs provide a comprehensive evaluation of the environmental impact throughout a product's life cycle, offering insights into how materials can be optimized to improve their ecological footprint. By promoting LCAs and supporting recycling infrastructure, industries can help ensure that bionanocomposites contribute to a sustainable economy (Sarkar et al. 2020).

17.4.6 Public Health Considerations

The public health implications of bionanocomposites span potential benefits and concerns. In healthcare, bionanocomposites have shown promise in applications such as drug delivery, wound healing, and tissue engineering, where they can enhance

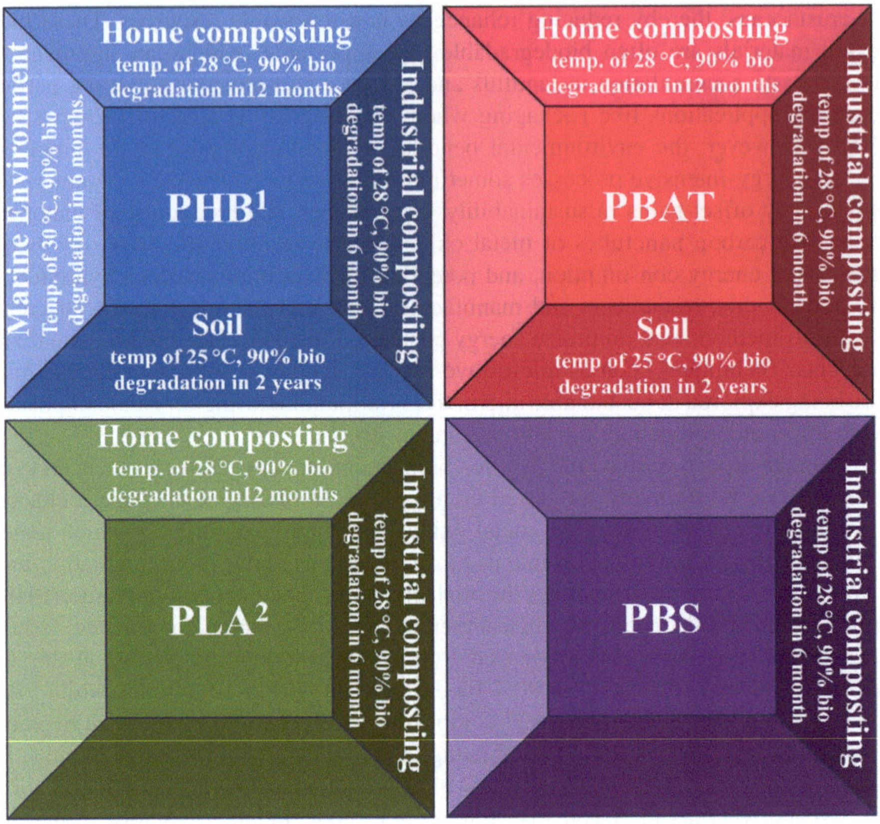

Fig. 17.6 Comparison of biodegradation rates of biopolymers under different environmental conditions

treatment efficacy and improve patient outcomes (Öztürk et al. 2024). For instance, bionanocomposites with antimicrobial properties, such as those incorporating silver nanoparticles, are used in medical dressings to prevent infections and accelerate healing. Similarly, bio-based nanoparticles can be tailored to target specific cells, providing efficient drug delivery systems that reduce side effects and improve therapeutic outcomes (Erduran et al. 2024). These innovations demonstrate the potential of bionanocomposites to make healthcare more effective, affordable, and accessible.

Despite these advantages, the integration of nanomaterials into bionanocomposites raises safety concerns related to human exposure, especially as these materials become more prevalent in consumer products. Nanoparticles, due to their small size and high surface reactivity, can penetrate biological barriers and enter cells, potentially triggering toxicological responses (Ganguly et al. 2018). For instance, inhaled nanoparticles may reach the bloodstream and accumulate in organs, posing respiratory and cardiovascular risks. Studies on specific nanoparticles, like silver or titanium dioxide, have shown potential cytotoxic and genotoxic effects at certain concentrations, though research is ongoing to fully understand these risks (Yamani et al. 2017). Given these concerns, regulatory bodies such as the U.S. FDA and the EMA require rigorous testing and safety assessments for bionanocomposites in healthcare settings (Rehm and Fata Moradali 2021).

In non-medical applications, such as packaging or textiles, public exposure to bionanocomposites can occur through prolonged skin contact or incidental ingestion (Harun-Ur-Rashid et al. 2022). Ensuring that these materials are safe under various exposure conditions is crucial for public health. Standards from organizations like the ISO help establish guidelines for evaluating biocompatibility and toxicity in consumer products (Connolly et al. 2019). Manufacturers are increasingly adopting these standards, conducting extensive biocompatibility tests to address public safety concerns. Furthermore, transparent labeling practices can empower consumers to make informed decisions about products containing bionanocomposites, enhancing public trust in these innovative materials.

17.4.7 Economic and Industrial Implications

Bionanocomposites offer significant economic and industrial advantages, positioning themselves as key drivers of innovation in multiple sectors. In the packaging industry, for example, the demand for sustainable alternatives has spurred the adoption of bionanocomposites. These materials not only meet consumer demand for environmentally friendly products but also enhance product functionality by providing properties like antimicrobial activity and barrier protection. As companies shift toward biodegradable or recyclable packaging solutions, bionanocomposites offer a competitive edge, potentially reducing costs associated with waste disposal and enhancing brand reputation through sustainable practices (Sabet 2024).

In the automotive and construction industries, bionanocomposites are used to create lightweight, durable components that improve energy efficiency and reduce emissions (Ali and Hoque 2022). For instance, natural fiber-reinforced composites are increasingly used in automotive parts, replacing heavier, non-renewable materials like metals and synthetic polymers as depicted in Fig. 17.7. By reducing vehicle weight, these composites contribute to fuel efficiency, which not only lowers operational costs for consumers but also aligns with global initiatives to reduce carbon footprints. Similarly, bionanocomposites with enhanced mechanical properties are utilized in building materials, where they provide long-lasting, eco-friendly alternatives to conventional construction materials. As industries prioritize sustainability, the economic benefits of integrating bionanocomposites become increasingly evident, fostering job growth and encouraging innovation across manufacturing sectors.

However, the production and commercialization of bionanocomposites also present challenges. Manufacturing processes for nanoscale components can be costly, and the specialized equipment and expertise required for quality control add to production expenses (Charitidis et al. 2014). For smaller companies, these barriers

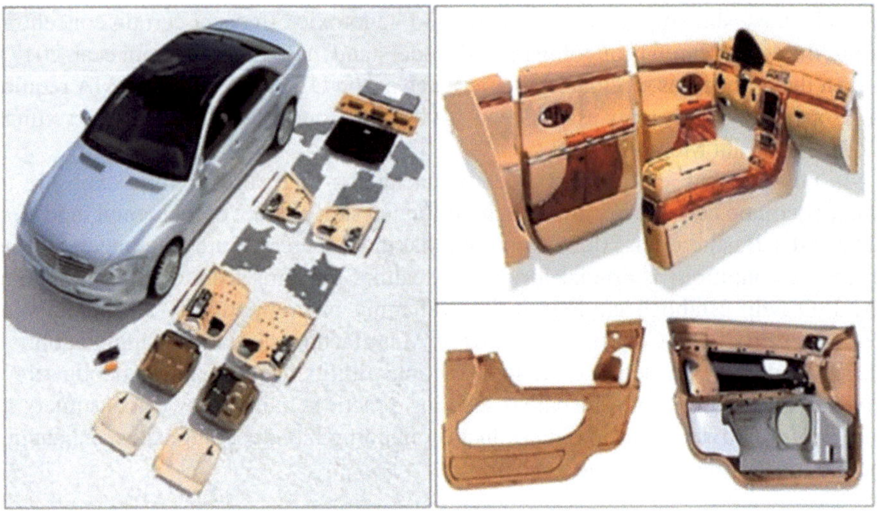

Fig. 17.7 Various parts of bionanocomposites used in the automobile industry for interior panels, dashboards, and other components, showcasing their versatility, lightweight, and sustainable design advantages. Reproduced from Teles et al. (2015) copyright 2015 SciELO Brazil

can limit their ability to enter the market, leading to a concentration of manufacturing capabilities among larger corporations. Additionally, due to varying regulations across regions, companies face compliance costs and logistical hurdles, which may slow down market growth and innovation in certain regions. Addressing these challenges will require coordinated efforts to harmonize regulations, reduce production costs, and foster cross-sector collaboration.

Economic policies, including subsidies and incentives for sustainable manufacturing, can also play a pivotal role in supporting the bionanocomposite industry (Maluin 2024). Governments in regions like the European Union and Asia–Pacific have implemented initiatives to promote green technology, encouraging investments in research and development for sustainable materials (Khanna 2020). By providing funding opportunities and establishing favorable policies, these regions are setting the stage for a robust bionanocomposite market that aligns with sustainability goals. As the industry matures, increased competition and economies of scale are likely to drive down production costs, making bionanocomposites more accessible and economically viable.

17.5 Ethical Considerations in Bionanocomposites

As bionanocomposites become increasingly prevalent in fields ranging from healthcare to environmental management, it is essential to examine the ethical implications that accompany their development and deployment. The rapid advancement

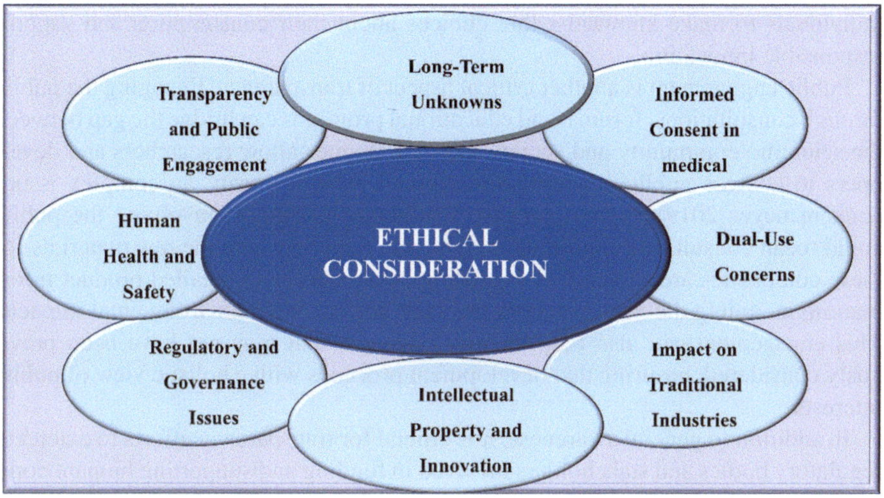

Fig. 17.8 Shows different types of ethical consideration while dealing with bionanocomposites

of bionanocomposite technologies offers numerous societal benefits but raises questions about transparency, equitable access, and environmental responsibility. The ethical considerations associated with bionanocomposites encompass a range of critical issues that influence their development, application, and societal acceptance as illustrated in Fig. 17.8, these considerations include transparency and public engagement, human health and safety, regulatory and governance challenges, intellectual property rights, and the potential impact on traditional industries. Other important aspects include the ethical implications of dual-use concerns, the need for informed consent in medical applications, and the long-term unknowns related to environmental and societal impacts. These considerations are critical in guiding the responsible development and use of bionanocomposites (Srinivas et al. 2025). This section will discuss ethical concerns in terms of transparency and public engagement, equity and accessibility, and long-term environmental responsibility.

17.5.1 Transparency and Public Engagement

Transparency in the development and application of bionanocomposites is essential to build trust and foster informed public engagement (Voorde and Jeswani 2021). As these materials enter the market, there is a significant need for clear, accessible information about their properties, benefits, and potential risks. Ensuring that consumers, researchers, and stakeholders understand how bionanocomposites are made, used, and disposed of can help avoid public mistrust and promote informed decision-making (Frewer et al. 2011). For instance, explaining the types of nanomaterials used in these composites and any associated environmental or health risks allows individuals to make knowledgeable choices about their consumption and support responsible innovation.

Public engagement is another critical aspect of transparency. Engaging the public through consultations, forums, and educational programs can bridge the gap between the scientific community and society. Open dialogues allow researchers and developers to address public concerns directly and provide clarity on complex issues (Lukensmeyer 2019). In the context of bionanocomposites, involving the public could mean consulting community members in regions where the raw materials for these composites are sourced, or providing consumers with detailed product information, including the nanomaterials involved and potential environmental impacts. This engagement can also help identify concerns that may not have been previously considered, ensuring that development proceeds with a holistic view of public interests.

In addition to general awareness, it is crucial for transparency efforts to extend to regulatory bodies and stakeholders involved in funding and supporting bionanocomposite research. As bionanocomposites gain traction, there may be a temptation to downplay potential risks in favor of commercial success. To counter this, institutions and manufacturers should establish clear communication channels, disclosing relevant information on environmental impact, safety data, and regulatory compliance. The scientific community can play a leading role in setting standards for transparency, advocating for open-access publications, and prioritizing studies that explore both the benefits and drawbacks of bionanocomposites. By fostering a transparent environment, we can ensure that advancements in bionanocomposite technology align with societal values and public trust.

17.5.2 Equity and Accessibility

Equity and accessibility in bionanocomposites pertain to ensuring that the benefits of these materials are distributed fairly across different communities and socioeconomic groups (Ramsden 2018). While bionanocomposites hold promise for numerous applications, there is a risk that their benefits could become accessible only to certain regions or demographics, exacerbating existing inequalities (Anwer et al. 2023). The development of equitable frameworks for bionanocomposite use requires considering accessibility for both high-income and low-income populations, especially as these materials are increasingly used in healthcare, agriculture, and consumer products. One example of this challenge is in healthcare, where advanced bionanocomposite-based drug delivery systems or wound dressings may offer significant therapeutic advantages but may remain unaffordable for under-resourced healthcare systems or low-income patients. If these materials are only accessible to wealthy populations or countries, the benefits of bionanocomposites may contribute to health disparities. To address this, government subsidies, industry partnerships, and non-profit initiatives could help bridge the gap, making advanced bionanocomposite technologies available to disadvantaged communities (Programmes Initiatives | Department of

Science Technology 2024). Developing policies that promote equitable distribution, perhaps through tiered pricing models or subsidized distribution, can ensure that innovations in bionanocomposites do not widen health and economic disparities but instead contribute to societal well-being on a broad scale.

In addition to financial barriers, geographical disparities in access to bionanocomposites are also a concern. Many of the raw materials for these composites are sourced from specific regions, often in low- and middle-income countries, where workers may not benefit from the final products they help produce. Ethical sourcing practices are essential to ensure fair compensation for these workers and to avoid exploitative labor practices (Raigrodski 2016). Furthermore, developers of bionanocomposites should consider the potential environmental impact on communities near extraction sites, as they may be disproportionately affected by pollution or resource depletion (Ho et al. 2022). Emphasizing corporate social responsibility and supporting local economies can help address these disparities, ensuring that bionanocomposites contribute positively to both local and global communities.

17.5.3 Long-Term Environmental Responsibility

The use of bionanocomposites raises significant questions about long-term environmental responsibility, particularly concerning their sourcing, production, and disposal. Given that one of the primary motivations for developing bionanocomposites is to create sustainable alternatives to traditional materials, it is essential that their life cycle aligns with environmental goals. This includes sourcing raw materials responsibly, minimizing environmental impact during production, and implementing effective disposal or recycling methods to avoid environmental harm at the end of their lifecycle (Atodariya et al. 2024).

Sourcing bio-based materials for bionanocomposites often involves harvesting renewable resources, such as plant fibers or chitosan, which, when managed sustainably, can provide a low-impact alternative to non-renewable resources (Iroegbu and Ray 2022). However, unsustainable sourcing practices can have detrimental effects, such as deforestation, habitat loss, and biodiversity reduction (Prajapati et al. 2024). To address this, manufacturers can prioritize sustainably sourced bio-based materials, supporting responsible agricultural and forestry practices, and ensuring that their supply chains meet environmental and social standards. Implementing certifications such as Forest Stewardship Council (FSC) for wood-based composites or other eco-labels can assure consumers that these products are developed with environmental integrity (Bowers 2019).

An additional consideration in environmental responsibility is the energy required to produce bionanocomposites, particularly when incorporating nanomaterials. High energy consumption can diminish the environmental benefits of bionanocomposites, and resource-intensive processes may not be sustainable in the long term. Advances in green synthesis methods, such as using low-energy production techniques and

Table 17.4 Key environmental responsibility initiatives in the production of bionanocomposites management strategies to promote sustainability and reduce environmental impact

Environmental responsibility initiatives	Key objectives
Sustainable sourcing of bio-based raw materials	Reduce resource depletion, promote biodiversity
Green synthesis methods	Lower energy use, reduce carbon footprint
Recycling and waste management strategies	Minimize end-of-life environmental impact

renewable energy sources, are essential for minimizing the carbon footprint associated with these materials. Furthermore, adopting production standards focused on environmental performance, such as ISO 14001 for environmental management, can help companies align their manufacturing processes with sustainability goals (Bravi et al. 2020). A major challenge in bionanocomposite environmental responsibility is end-of-life management. While some bio-based materials in bionanocomposites are biodegradable, the inclusion of non-biodegradable nanoparticles can complicate disposal. These particles may persist in the environment and have unknown ecological effects, potentially accumulating in soil or water and impacting ecosystems. As such, it is crucial to develop waste management strategies for bionanocomposites, such as recycling programs or decomposition methods that mitigate environmental harm (Tripathi et al. 2023). Designing for recyclability or creating standardized disposal protocols can help address this challenge, reducing the environmental footprint of bionanocomposites and ensuring that they remain sustainable over time. To promote environmental sustainability in the production of bionanocomposites, several key environmental responsibility initiatives are being implemented. As outlined in Table 17.4, these initiatives focus on sustainable sourcing of bio-based raw materials, green synthesis methods, and effective recycling and waste management strategies. The objectives behind these initiatives are to reduce resource depletion, promote biodiversity, lower energy use, and minimize the environmental impact at the end of the product's lifecycle.

17.6 Risk Assessment and Lifecycle Management

As the field of bionanocomposites expands, the importance of comprehensive risk assessment and lifecycle management has become paramount. To maximize the benefits while minimizing potential adverse impacts, it is essential to consider each phase of a bionanocomposite's lifecycle, from development and use to disposal. Proper risk assessment models, biodegradability strategies, and lifecycle management protocols contribute to the sustainable and safe integration of these materials into various industries (Biodegradable and Biocompatible Polymer Composites: Processing, Properties... Google Books 2024). This section will explore risk assessment methodologies, end-of-life management with a focus on biodegradability, and strategies for ensuring a safe lifecycle for bionanocomposites.

17.6.1 Risk Assessment Models and Methodologies

Risk assessment is a critical process in evaluating the potential hazards associated with bionanocomposites across their lifecycle. Effective models and methodologies can predict possible risks to human health, environmental safety, and even economic sustainability. Given the unique properties of bionanocomposites such as their nanostructures, bio-based components, and the potential for degradation or transformation over time risk assessments must consider a broad spectrum of factors. Several models are commonly used in risk assessment, including quantitative, qualitative, and probabilistic approaches. Quantitative models focus on numerical data to calculate precise risk levels, taking into account exposure, toxicity, and environmental impact. For example, when assessing a bionanocomposite used in food packaging, a quantitative model would analyze the migration rates of nanoparticles into food products over time (Jiménez and Ruseckaite 2012). Qualitative models, on the other hand, provide a descriptive analysis that can be useful for understanding risks when data is limited or when dealing with new materials. These models are often used in preliminary assessments or to guide regulatory compliance.

Probabilistic risk assessment models add an additional layer by incorporating probabilities and uncertainties to provide a more comprehensive analysis of potential outcomes. These models are particularly useful in the case of bionanocomposites where the long-term effects on health and the environment are not yet fully understood. They allow researchers and manufacturers to evaluate the likelihood of different scenarios, such as the potential for nanoparticles to bioaccumulate in soil or water, posing risks to ecosystems. In addition to these models, various methodologies like the LCA and Environmental Impact Assessment (EIA) provide holistic frameworks for evaluating the environmental and health impacts of bionanocomposites (Ramesh et al. 2022). LCAs, in particular, assess the entire lifecycle from production to disposal, identifying the stages with the highest environmental footprint. These assessments are crucial for manufacturers who aim to minimize waste, reduce energy consumption, and lower greenhouse gas emissions during the production and disposal of bionanocomposites.

17.6.2 End-of-Life Management and Biodegradability

End-of-life management is a vital consideration in the lifecycle of bionanocomposites, given the environmental impact of improper disposal. Unlike traditional materials, bionanocomposites offer the potential for biodegradability, which can significantly reduce their long-term environmental footprint. However, not all bionanocomposites are biodegradable, particularly those that incorporate non-biodegradable nanoparticles. Therefore, tailored end-of-life strategies are necessary to ensure that these materials do not contribute to environmental pollution (Sagha Zadeh et al. 2018). Biodegradability is determined by several factors, including the composition of the bio-based materials, the presence of nanomaterials, and environmental conditions. Some bionanocomposites are designed to break down in composting environments, where microorganisms can degrade the bio-based components into harmless byproducts. For instance, PLA-based bionanocomposites are known for their compostability, as they can decompose in industrial composting facilities within a few months under optimal conditions. However, biodegradability does not mean that a material will automatically degrade in all environments, such as in marine or anaerobic conditions (Calmon-Decriaud et al. 1998).

In cases where bionanocomposites contain non-biodegradable elements, such as certain metal oxide nanoparticles, recycling may be a more feasible end-of-life strategy than biodegradation. Recycling initiatives that focus on recovering valuable nanoparticles or other components can help reduce waste, conserve resources, and mitigate environmental harm. Closed-loop recycling systems, where materials are continually reused in new products, can be an effective strategy for materials that do not degrade easily. This approach aligns with circular economy principles, which aim to extend the lifecycle of materials and minimize waste generation. Regulatory frameworks also play an essential role in end-of-life management. Guidelines on biodegradability standards, waste treatment processes, and disposal protocols help ensure that bionanocomposites are managed responsibly throughout their lifecycle. For example, the European Union's Waste Framework Directive (WFD) sets out obligations for producers and handlers to ensure proper disposal and recovery of biodegradable materials, encouraging environmentally friendly end-of-life practices for biocomposites and bionanocomposites alike (Briassoulis et al. 2021). ASTM and ISO standards for biodegradability testing of plastics under various conditions, including aerobic (soil, compost, marine) and anaerobic (landfill, digestion) listed in Table 17.5. Key parameters such as CO_2, CH_4, and disintegration are monitored, ensuring environmental safety, regulatory compliance, and promoting sustainable material development across global industries.

Table 17.5 Various ISO and ASTM standards for biodegradation of bio-based plastics (Machado et al. 2015b)

Standard	Title	Conditions/ parameters monitored	Environment	Relevance to bionanocomposites
ISO 14851 (1999)	Determination of the Ultimate Aerobic Biodegradability of Plastic Materials in an Aqueous Medium	Oxygen demand (BOD)	Liquid (aerobic)	Applicable for assessing polymer matrix biodegradation
ISO 14852 (1999)	Determination of the Ultimate Aerobic Biodegradability of Plastic Materials in an Aqueous Medium	CO_2 evolution	Liquid (aerobic)	Useful for monitoring carbon dioxide evolution
ISO 17556 (2012)	Determination of the Ultimate Aerobic Biodegradability of Plastic Materials in Soil	Oxygen demand or CO_2 evolution	Soil (aerobic)	Ideal for bionanocomposites intended for soil applications
ASTM D5338-11	Standard Test Method for Determining Aerobic Biodegradation of Plastic Materials Under Controlled Composting Conditions	CO_2 evolution	Compost (Aerobic)	Evaluates biodegradability in compost environments
ISO 14855-1 (2012)	Aerobic Biodegradability Under Controlled Composting Conditions—CO_2 Analysis (General Method)	CO_2 evolution	Compost (aerobic)	Measures biodegradation in controlled compost systems
ISO 14855-2 (2007)	Aerobic Biodegradability Under Controlled Composting Conditions—Gravimetric CO_2 Analysis	CO_2 evolution	Compost (aerobic)	Precision analysis for compostable bionanocomposites
ISO 15985 (2004)	Ultimate Anaerobic Biodegradation Under High-Solids Anaerobic-Digestion Conditions	Biogas (CO_2, CH_4)	Solid (anaerobic)	Suitable for materials in anaerobic digestion setups
ASTM D5511-12	Anaerobic Biodegradation Under High-Solids Anaerobic Digestion Conditions	CO_2, CH_4 production	Solid (anaerobic)	Assesses biogas production for anaerobic systems
ASTM D5526-12	Anaerobic Biodegradation Under Accelerated Landfill Conditions	CO_2, CH_4 production	Solid (anaerobic)	Evaluate landfill biodegradation potential

(continued)

Table 17.5 (continued)

Standard	Title	Conditions/ parameters monitored	Environment	Relevance to bionanocomposites
ISO 16929 (2013)	Degree of Disintegration of Plastic Materials Under Defined Composting Conditions (Pilot-Scale Test)	Physical disintegration	Compost (aerobic)	Ideal for studying the structural breakdown of composites
ASTM D6691-09	Aerobic Biodegradation of Plastic Materials in the Marine Environment	CO_2 evolution	Marine (aerobic)	Suitable for bionanocomposites exposed to marine settings
ISO 20200 (2004)	Disintegration of Plastic Materials Under Simulated Composting Conditions (Lab-Scale Test)	Physical disintegration	Compost (aerobic)	Laboratory-scale analysis for controlled biodegradation

17.6.3 *Strategies for Safe Lifecycle Management*

To ensure that bionanocomposites are managed safely and responsibly from production to disposal, it is essential to implement strategies that encompass every phase of the lifecycle. Safe lifecycle management involves establishing guidelines for quality control, health and safety, environmental impact mitigation, and regulatory compliance. One effective strategy for safe lifecycle management is to adhere to existing quality and safety standards, such as those developed by the ISO or the EPA. ISO standards provide comprehensive guidelines for the safe handling, processing, and disposal of composite materials, including bionanocomposites. For example, ISO 10993 offers standards for biocompatibility testing, which is essential for bionanocomposites used in medical devices or pharmaceuticals (Rehm and Fata Moradali 2021).

Safety guidelines for production processes, such as containment measures to prevent nanoparticle exposure, are also crucial. These guidelines can reduce occupational risks for workers and help prevent accidental release of nanoparticles into the environment. Implementing these measures requires cooperation between manufacturers, regulatory bodies, and safety organizations to create standardized procedures that prioritize human and environmental health. In addition to these standards, several emerging strategies focus on promoting sustainable manufacturing practices. Green synthesis methods, which use low-energy processes and eco-friendly solvents, can reduce the carbon footprint associated with the production of bionanocomposites. For instance, replacing traditional solvents with water-based solutions or using renewable energy sources during production can make manufacturing more sustainable.

Another critical aspect of lifecycle management is continuous monitoring and assessment. Tracking the performance and environmental impact of bionanocomposites throughout their lifecycle allows manufacturers and regulatory bodies to make data-driven decisions and adjust practices as needed. Implementing digital tools such as blockchain technology for transparent data tracking can enhance accountability and ensure compliance with environmental and safety regulations (Allena 2020). Blockchain can enable real-time tracking of materials, from raw sourcing to end-of-life, providing a transparent record that manufacturers, regulators, and consumers can access.

Finally, promoting public awareness and education about safe handling and disposal of bionanocomposites is essential for lifecycle management. Providing accessible information to consumers, industries, and communities helps prevent improper disposal and encourages responsible usage. Educational campaigns on recycling or composting options, for example, can empower consumers to make environmentally friendly choices at the end of a product's life.

17.7 Future Directions and Recommendations

As the field of bionanocomposites continues to evolve and expand, it is essential to anticipate future directions that will shape their development and integration into various industries. A key challenge lies in creating unified global standards that can harmonize regulations across different countries and regions. Additionally, fostering ethical frameworks in research and development, alongside encouraging responsible innovation, will play a crucial role in ensuring that bionanocomposites contribute positively to society while minimizing any potential harm. This section outlines the need for global standards, ethical frameworks, and responsible innovation, providing recommendations for advancing these areas.

17.7.1 Need for Unified Global Standards

One of the primary challenges in the bionanocomposite sector is the lack of uniformity in regulatory and safety standards worldwide (Idumah and Hassan 2016). Given the global nature of the supply chain for these materials and the potential for widespread use, particularly in healthcare, packaging, and electronics, it is crucial to establish cohesive and harmonized standards that are accepted internationally. At present, different regions often have varying approaches to the testing, regulation, and use of bionanocomposites, which can create barriers to market entry, inconsistency in product quality, and delays in innovation.

International organizations such as the ISO, the European Union, and the U.S. FDA have developed some frameworks for nanomaterials, but these regulations are often fragmented and lack coherence when applied across borders (Allan et al. 2021). The need for global alignment is pressing, as the rapid advancement in nanotechnology and its integration into bionanocomposites requires swift adaptation of regulatory frameworks. The creation of universal standards would ensure consistency in product quality, safety, and environmental impact, allowing for seamless international trade and the safe use of these materials. A unified approach to regulation would also streamline the approval process for new bionanocomposite-based products, ensuring that manufacturers and researchers comply with consistent safety protocols and environmental guidelines. By developing a comprehensive global regulatory framework, policymakers can provide a more predictable and transparent environment for both manufacturers and consumers (Berry and Rondinelli 1998). Additionally, regulatory bodies should work in close collaboration with industry stakeholders to ensure that these standards are reflective of the latest scientific advancements and adequately address emerging risks associated with the materials' production, use, and disposal.

17.7.2 Proposals for Ethical Frameworks in Research and Development

As bionanocomposites emerge as a transformative technology with the potential to revolutionize a range of industries, the ethical implications of their use and development must be carefully considered. Research and development (R&D) in this area should not solely focus on technological innovation but also prioritize ethical concerns related to human health, environmental sustainability, and social equity (Bryden and Gezelius 2017). Researchers, manufacturers, and regulatory bodies should work together to establish ethical guidelines that guide the responsible development and application of bionanocomposites.

One key ethical concern is ensuring transparency in the research process (Mertz and Strech 2014). Scientists and developers must disclose the full range of materials used in the creation of bionanocomposites, their potential risks, and their environmental impact. Public trust in new technologies can only be achieved through transparent practices and clear communication with both the scientific community and the general public. Moreover, open-access publication of research findings related to the safety and efficacy of bionanocomposites can foster accountability and provide a foundation for informed decision-making by policymakers. Another ethical consideration is equity in the development and distribution of bionanocomposites. Access to these advanced materials and the benefits they offer should not be restricted to certain groups or regions. Researchers and policymakers must prioritize inclusivity and work toward ensuring that the benefits of bionanocomposites are shared equitably (Singh et al. 2024). For instance, sustainable packaging solutions developed from bionanocomposites should be made accessible to industries in developing regions that may struggle to adopt such technologies due to financial or infrastructural barriers. The consideration of long-term environmental impacts is also essential in ethical R&D practices. Bionanocomposites should not be developed in isolation from the larger environmental context. Researchers must evaluate the full lifecycle of these materials, including production, usage, and disposal, to understand the broader ecological consequences (Finnveden et al. 2009). Ethical frameworks should emphasize the importance of minimizing toxicity, environmental degradation, and resource depletion during the development and deployment of bionanocomposites. This approach aligns with the principles of sustainable development, ensuring that future generations can benefit from these innovations without compromising their own ability to thrive.

17.7.3 Encouraging Responsible Innovation

The rapid pace of technological innovation in bionanocomposites presents both immense opportunities and significant risks. To ensure that these materials are developed and used responsibly, there is a need for a paradigm shift toward responsible innovation. Responsible innovation involves the anticipation and mitigation of potential risks, while fostering an inclusive, transparent, and ethically sound approach to technological advancement (Dreyer et al. 2017).

Key to responsible innovation is the involvement of diverse stakeholders, including researchers, policymakers, manufacturers, environmental organizations, and the public. A collaborative approach ensures that all perspectives are considered, and potential societal impacts are identified early in the innovation process (Secco et al. 2019). For example, engaging communities that may be affected by the widespread use of bionanocomposites, such as agricultural workers exposed to nanoparticle-based products or consumers of food packaged in these materials, will help ensure that their concerns are addressed. Public consultation and stakeholder engagement are vital in shaping technologies that are socially responsible and ethically sound.

Furthermore, responsible innovation in bionanocomposites requires a proactive approach to risk management. Instead of addressing risks reactively after the fact, researchers and manufacturers should anticipate potential issues and implement safeguards early in the development process. This could involve conducting comprehensive risk assessments and employing sustainability metrics throughout the design and manufacturing phases. Manufacturers should also prioritize the development of recyclable or biodegradable bionanocomposite products to reduce waste and minimize environmental impact. By embedding responsibility into the design phase, rather than as an afterthought, companies can significantly enhance the sustainability and safety of these materials (Antón et al. 2015). Regulatory bodies and standard-setting organizations also play a crucial role in encouraging responsible innovation. These bodies must stay ahead of technological developments by continuously updating standards to reflect the evolving nature of bionanocomposites. Policies should be designed to incentivize the development of safer, more sustainable materials, and research funding should be directed toward projects that prioritize ethical considerations, sustainability, and environmental stewardship.

17.8 Conclusion

Bionanocomposites, as a transformative material, hold tremendous promise in addressing global challenges in healthcare, the environment, and various industries, offering a wide range of applications due to their unique properties. However, the successful and responsible integration of these materials into commercial products and real-world applications requires addressing a spectrum of challenges that span regulatory, ethical, environmental, and societal considerations. This chapter has emphasized the importance of developing robust regulatory frameworks, creating clear ethical guidelines, ensuring sustainable manufacturing practices, and addressing the potential societal impacts of bionanocomposites.

One of the central conclusions of this work is the urgent need for unified global standards for bionanocomposites. Despite the growing recognition of their potential, regulatory frameworks for these materials remain fragmented across regions. Currently, different regions follow divergent standards, creating inconsistencies in safety protocols, product quality, and market accessibility. This regulatory fragmentation can lead to delays in the market introduction of innovative bionanocomposite products and create barriers for manufacturers seeking to enter international markets. Thus, there is an imperative to align regulatory standards globally, ensuring that bionanocomposites are tested, certified, and used safely and effectively worldwide. Harmonizing these standards not only facilitates trade and innovation but also ensures that bionanocomposites meet consistent safety and environmental criteria, minimizing potential risks associated with their use.

Furthermore, the ethical considerations surrounding bionanocomposites cannot be overlooked. As these materials have the potential to revolutionize numerous industries, the development of clear ethical frameworks to guide their research, development, and application is paramount. Issues such as transparency in the composition of materials, the potential risks associated with nanomaterials, and the accessibility of these innovations to underserved populations must be considered in every stage of product development. It is essential that ethical guidelines prioritize public health and safety while also promoting equity and inclusivity, ensuring that the benefits of bionanocomposite technologies are accessible to all, particularly in developing regions. By fostering an open, transparent, and equitable approach to innovation, we can build public trust in these materials and encourage broader acceptance.

In addition to regulatory and ethical frameworks, sustainable manufacturing practices are crucial for the future of bionanocomposites. These materials, by their very nature, can offer significant environmental benefits over conventional composites, such as biodegradability, reduced environmental impact, and resource efficiency. However, without careful consideration of the environmental footprint of their production, bionanocomposites could contribute to resource depletion or environmental degradation. Manufacturers must adopt green manufacturing techniques that emphasize sustainability, such as the use of renewable resources, energy-efficient processes, and the minimization of waste. Additionally, the lifecycle of bionanocomposites, including their disposal and biodegradability, should be considered from the outset to ensure that these materials do not pose long-term environmental hazards. A circular economy approach—where materials are reused, recycled, or biodegraded—should be prioritized to minimize the ecological impact of bionanocomposites and contribute to a more sustainable material economy.

The societal impacts of bionanocomposites also merit significant attention. While these materials have the potential to offer solutions for pressing global issues, including sustainable packaging, medical advancements, and pollution reduction, their introduction could also bring unintended consequences. Concerns about the potential health risks associated with nanoparticles, their impact on ecosystems, and their role in exacerbating environmental inequalities must be carefully evaluated. Public health considerations, environmental risks, and economic implications must be considered in the development of new applications for bionanocomposites. It is essential that manufacturers, researchers, and policymakers work together to address these concerns proactively, ensuring that any potential risks are mitigated before these materials are widely deployed.

Acknowledgements I wish to express my deepest gratitude to the esteemed faculty members and PhD Scholars of Netaji Subhas University of Technology for their invaluable guidance, insightful suggestions, and constant support throughout the preparation of this book chapter. Their expertise and mentorship have been instrumental in shaping my understanding of the subject and refining the quality of my work. I am especially thankful to Prof. Anand Srivastava, Honourable Vice Chancellor of Netaji Subhas University of Technology, for fostering an exceptional academic and research environment. His commitment to excellence and provision of essential resources have greatly contributed to the successful completion of this chapter. I am sincerely appreciative of the faculty's unwavering encouragement and intellectual input, which have been a cornerstone of this.

References

A. A.-E.-F. S. S. for Natural (2024) Aging products. books.google.com [Online]. Available at: https://books.google.com/books?hl=en&lr=&id=fwg1EQAAQBAJ&oi=fnd&pg=PA335&dq=Occupational+Safety+and+Health+Administration+(OSHA),+the+Environmental+Protection+Agency+(EPA),+and+the+European+Chemicals+Agency+(ECHA),+are+actively+developing+safety+guidelines+specific+to+the+handling+and+disposal+of+nanomaterials+within+bionanocomposites.&ots=p5HBEIQZAW&sig=Bo5idlEsBTu7ceGfTNIJJoEz5p8. Accessed 28 Dec 2024

Act on the Evaluation of Chemical Substances and Regulation of Their Manufacture, etc.—English—Japanese Law Translation [Online]. Available at: https://www.japaneselawtranslation.go.jp/en/laws/view/63/en. Accessed 27 Jan 2025

Ahghari MA, Ahghari MR, Kamalzare M, Maleki A (2022) Design, synthesis, and characterization of novel eco-friendly chitosan-$AgIO_3$ bionanocomposite and study its antibacterial activity. Sci Rep 12(1):1–13. https://doi.org/10.1038/s41598-022-14501-6

AICIS | Australian Industrial Chemicals Introduction Scheme (AICIS) [Online]. Available at: https://www.industrialchemicals.gov.au/. Accessed 27 Jan 2025

Al Jabri H, Saleem MH, Rizwan M, Hussain I, Usman K, Alsafran M (2022) Zinc oxide nanoparticles and their biosynthesis: overview. Life 12:594. https://doi.org/10.3390/LIFE12040594

Ali N, Hoque ME (2022) Bionanocomposites in the automotive and aerospace applications, pp 237–253. https://doi.org/10.1007/978-981-16-8578-1_13

Allan J, Belz S, Hoeveler A, Hugas M, Okuda H et al (2021) Regulatory landscape of nanotechnology and nanoplastics from a global perspective. Elsevier [Online]. Available at: https://www.sciencedirect.com/science/article/pii/S0273230021000258. Accessed 29 Dec 2024

Allena M (2020) Blockchain technology for environmental compliance. Environ Law. JSTOR [Online]. Available at: https://www.jstor.org/stable/27010194. Accessed 29 Dec 2024

Anastas PT, Zimmerman JB (2018) The United Nations sustainability goals: how can sustainable chemistry contribute? Curr Opin Green Sustain Chem 13:150–153. https://doi.org/10.1016/J.COGSC.2018.04.017

Anjaneyulu B, Mittal C, Chauhan V, Tiwari A, Bhathiwal AS (2024) Wastewater treatment with sustainable bionanocomposites: a comprehensive review. Nanotechnol Environ eng 9(3):483–518. https://doi.org/10.1007/S41204-024-00380-X

Antón PS, Silberglitt R, Schneider J (2001) The global technology revolution: bio/nano/materials trends and their synergies with information technology by 2015 [Online]. Available at: https://books.google.com/books?hl=en&lr=&id=nwqlU7VPwQcC&oi=fnd&pg=PR3&dq=By+embedding+responsibility+into+the+design+phase,+rather+than+as+an+afterthought,+companies+can+significantly+enhance+the+sustainability+and+safety+of+bionano+composite+materials.&ots=q18aw3D5fS&sig=IflbZKRwat-j93OD-kB48umwBl0. Accessed 29 Dec 2024

Anwer AH et al (2023) State-of-the-art advances in nanocomposite and bio-nanocomposite polymeric materials: a comprehensive review. Adv Colloid Interface Sci 318:102955. https://doi.org/10.1016/J.CIS.2023.102955

Atodariya J, Agrawal M, Singh A, Patni N (2024) Advancements in green nanocomposites: a comprehensive review on cellulose-based materials in biocomposites and bionanocomposites. Adv Struct Mater 210:213–239. https://doi.org/10.1007/978-981-10-2978-3_9

Azadi E, Dinari M, Derakhshani M, Reid K, Karimi B (2024) Sources and extraction of biopolymers and manufacturing of bio-based nanocomposites for different applications. Molecules. mdpi.com [Online]. Available at: https://www.mdpi.com/1420-3049/29/18/4406. Accessed 28 Dec 2024

Berry MA, Rondinelli DA (1998) Proactive corporate environmental management: a new industrial revolution. Acad Manag Exec 12(2):38–50. https://doi.org/10.5465/AME.1998.650515

Bhat MR, Payghan SA, Kumar Batra A, Chimkode RM, Bhandari A (2015) Bionanocomposites: technique towards enhancement of solubility, drug release and bioavailability. J Med Pharmaceut Innov 2(9) [Online]. Available at: https://www.jmedpharm.com/index.php/home/article/view/53. Accessed 27 Dec 2024

Biodegradable and biocompatible polymer composites: processing, properties... Google Books [Online]. Available at: https://books.google.co.in/books?hl=en&lr=&id=8ixHDgAAQBAJ&oi=fnd&pg=PA379&dq=+FDA+%2Bbionanocomposite+&ots=JasV01op9C&sig=AN-EV_2IvVQp1TFOwBSXDE7Vzc8&redir_esc=y#v=onepage&q=FDA%20%2Bbionanocomposite&f=false. Accessed 27 Dec 2024

Bonwick G, Bradley E, Lock I, Romero R (2019) Bio-based materials for use in food contact applications. images.chemycal.com [Online]. Available at: https://images.chemycal.com/Media/Files/bio-based-materials-for-use-in-food-contact-applications.pdf. Accessed 08 Nov 2024

Botlhoko OJ, Lekalakala R, Muniyasamy S (2024) Bioplastics and its bio-nanocomposites for a circular economy. Biodegradable polymers, blends and biocomposites, pp 257–270. https://doi.org/10.1201/9781003304142-12

Bowers C (2019) Next generation of residential construction: adoption of green building programs, environmentally certified wood products and the transparency of environmental … [Online]. Available at: https://digital.lib.washington.edu/researchworks/handle/1773/45202. Accessed 29 Dec 2024

Bravi L, Santos G, Pagano A, Murmura F (2020) Environmental management system according to ISO 14001:2015 as a driver to sustainable development. Corp Soc Responsib Environ Manag 27(6):2599–2614. https://doi.org/10.1002/CSR.1985

Briassoulis D, Pikasi A, Hiskakis M (2021) Organic recycling of post-consumer/industrial bio-based plastics through industrial aerobic composting and anaerobic digestion—techno-economic

sustainability criteria and indicators. Polym Degrad Stab 190:109642. https://doi.org/10.1016/J.POLYMDEGRADSTAB.2021.109642
Bryden J, Gezelius SS (2017) Innovation as if people mattered: the ethics of innovation for sustainable development. Innov Dev 7(1):101–118. https://doi.org/10.1080/2157930X.2017.1281208
Buazar F (2019) Impact of biocompatible nanosilica on green stabilization of subgrade soil. Sci Rep 9(1):1–9. https://doi.org/10.1038/s41598-019-51663-2
Calmon-Decriaud A, Bellon-Maurel V, Silvestre F (1998) Standard methods for testing the aerobic biodegradation of polymeric materials. Review and perspectives. Adv Polym Sci 135:207–226. https://doi.org/10.1007/3-540-69191-X_3
Charitidis CA, Georgiou P, Koklioti MA, Trompeta AF, Markakis V (2014) Manufacturing nanomaterials: from research to industry. Manuf Rev (Les Ulis) 1:11. https://doi.org/10.1051/MFREVIEW/2014009
Chemical Analysis | Medical Device | Smithers [Online]. Available at: https://www.smithers.com/industries/life-science/medical-device/chemical-analysis?gad_source=1&gclid=Cj0KCQiAvbm7BhC5ARIsAFjwNHtB4qrxabpAUEejXyVxUqcskLcsE-4pWAaCW-cQxs1D2Jukud0Bl4gaAjoWEALw_wcB. Accessed 27 Dec 2024
Coltelli MB, Morganti P, Lazzeri A (2023) Sustainability assessment, environmental impact, and recycling strategies of biodegradable polymer nanocomposites. Biodegradable and biocompatible polymer nanocomposites: processing, characterization, and applications, pp 699–737. https://doi.org/10.1016/B978-0-323-91696-7.00016-7
Connolly M et al (2019) Novel polylactic acid (PLA)-organoclay nanocomposite bio-packaging for the cosmetic industry; migration studies and in vitro assessment of the dermal toxicity of migration extracts. Polym Degrad Stab 168:108938. https://doi.org/10.1016/J.POLYMDEGRADSTAB.2019.108938
Control of Nanoscale Materials under the Toxic Substances Control Act | US EPA [Online]. Available at: https://www.epa.gov/reviewing-new-chemicals-under-toxic-substances-control-act-tsca/control-nanoscale-materials-under. Accessed 27 Jan 2025
da Cruz TT, Las-Casas B, Dias IKR, Arantes V (2024) Nanocelluloses as sustainable emerging technologies: state of the art and future challenges based on life cycle assessment. Sustain Mater Technol 41:e01010. https://doi.org/10.1016/J.SUSMAT.2024.E01010
Dahy H (2019) Natural Fibre-Reinforced Polymer Composites (NFRP) fabricated from lignocellulosic fibres for future sustainable architectural applications, case studies: segmented-shell construction, acoustic panels, and furniture. Sensors 19:738. https://doi.org/10.3390/S19030738
Darmenbayeva A et al (2024) Cellulose-based sorbents: a comprehensive review of current advances in water remediation and future prospects. Molecules 29:5969. https://doi.org/10.3390/MOLECULES29245969
De Smedt K, Vos E (2022) The application of the precautionary principle in the EU. The responsibility of science. library.oapen.org [Online]. Available at: https://library.oapen.org/bitstream/handle/20.500.12657/57314/1/978-3-030-91597-1.pdf#page=168. Accessed 28 Dec 2024
Dmour I (2022) Physicochemical characterization of nanobiocomposites. Biocomposites—recent advances [Online]. Available at: https://www.intechopen.com/chapters/85357. Accessed 28 Dec 2024
Donchevskaya E (2009) REACH (Registration, Evaluation, Authorization and Restriction of Chemicals) [Online]. Available at: https://essuir.sumdu.edu.ua/bitstream/123456789/13426/1/Donchevskaya.pdf. Accessed 28 Dec 2024
Dorato S (2018) General concepts: current legislation on cosmetics in various countries. Analysis of cosmetic products, 2nd edn., pp 3–37. https://doi.org/10.1016/B978-0-444-63508-2.00001-1
Dreyer M et al (2017) Responsible innovation: a complementary view from industry with proposals for bridging different perspectives. Sustainability 9:1719. https://doi.org/10.3390/SU9101719
El Yamani N et al (2017) In vitro genotoxicity testing of four reference metal nanomaterials, titanium dioxide, zinc oxide, cerium oxide and silver: towards reliable hazard assessment. Mutagenesis 32(1):117–126. https://doi.org/10.1093/MUTAGE/GEW060

Erduran M, Çankaya N, Yalcin S (2024) Biobased nanomaterials in drug delivery. Biobased nanomaterials, pp. 173–222. https://doi.org/10.1007/978-981-97-0542-9_7

Finnveden G et al (2009) Recent developments in Life Cycle Assessment. J Environ Manag 91(1):1–21. https://doi.org/10.1016/J.JENVMAN.2009.06.018

Frewer LJ, Norde W, Fischer A, Kampers F (2011) Nanotechnology in the agri-food sector: implications for the future. https://doi.org/10.1002/9783527634798

FSSAI [Online]. Available at: https://fssai.gov.in/cms/food-safety-and-standards-regulations.php. Accessed 27 Jan 2025

Ganguly P, Breen A, Pillai SC (2018) Toxicity of nanomaterials: exposure, pathways, assessment, and recent advances. ACS Biomater Sci Eng 4(7):2237–2275. https://doi.org/10.1021/acsbiomaterials.8b00068

Gupta P, Rana N, Dash S, Gupta N, Singh M, Kaushik S (2022) Bionanocomposite membranes and adsorbents for water and wastewater treatment. Bionanotechnology towards sustainable management of environmental pollution, pp 141–166. https://doi.org/10.1201/9781003270959-6

Harun-Ur-Rashid M, Bin Imran A, Susan MABH (2022) Green polymer nanocomposites in automotive and packaging industries. Curr Pharm Biotechnol 24(1):145–163. https://doi.org/10.2174/1389201023666220506111027

Haydon B (2015) Nanoengineering: a toolbox of standards for health and safety. Nanoengineering: global approaches to health and safety issues, pp 557–580. https://doi.org/10.1016/B978-0-444-62747-6.00017-8

Health and the Environment—Canada.ca [Online]. Available at: https://www.canada.ca/en/health-canada/services/science-research/emerging-technology/biotechnology/health-environment.html?utm_source=chatgpt.com. Accessed 28 Dec 2024

Health Sciences Authority (HSA) [Online]. Available at: https://www.hsa.gov.sg/. Accessed 27 Jan 2025

Heidari M, Al Hasani S, Thangavel S, Kumar A (2024) Challenges and perspectives of polymeric nanocomposites for biomedical applications. books.google.com [Online]. Available at: https://books.google.com/books?hl=en&lr=&id=__AhEQAAQBAJ&oi=fnd&pg=PA317&dq=bionano+composites+pose+complex+challenges+in+terms+of+safety,+efficacy,+and+environmental+impact.+&ots=aLTSVG4XAx&sig=v4dIdyDBBdRtGJN0L90TXetUNfk. Accessed 08 Nov 2024

Heidari M, Al Hasani S, Thangavel S, Kumar A (2024b) Challenges and perspectives of polymeric nanocomposites for biomedical applications. Nanomanufacturing techniques in sustainable healthcare applications, pp 317–339. https://doi.org/10.1201/9781003470311-16

Ho J, Shalishali M, (Bill) Tseng T-L, Ang D (2022) Opportunities in green supply chain management. Coastal Bus J 8(1) [Online]. Available at: https://digitalcommons.coastal.edu/cbj/vol8/iss1/2. Accessed 29 Dec 2024

Howard J, Murashov V (2009) National nanotechnology partnership to protect workers. J Nanopart Res 11(7):1673–1683. https://doi.org/10.1007/S11051-009-9682-2/METRICS

Hristozov DR, Gottardo S, Critto A, Marcomini A (2012) Risk assessment of engineered nanomaterials: a review of available data and approaches from a regulatory perspective. Nanotoxicology 6(8):880–898. https://doi.org/10.3109/17435390.2011.626534

Hwang M et al (2012) Risk assessment principle for engineered nanotechnology in food and drug. Toxicol Res 28(2):73–79. https://doi.org/10.5487/TR.2012.28.2.073

Idumah CI, Hassan A (2016) Emerging trends in flame retardancy of biofibers, biopolymers, biocomposites, and bionanocomposites. Rev Chem Eng 32(1):115–148. https://doi.org/10.1515/REVCE-2015-0017/ASSET/GRAPHIC/J_REVCE-2015-0017_CV_002.JPG

Iqbal HM, Keshavarz T (2017) The challenge of biocompatibility evaluation of biocomposites. Elsevier [Online]. Available at: https://www.sciencedirect.com/science/article/pii/B9780081007525000147. Accessed 08 Nov 2024

Iroegbu AOC, Ray SS (2022) Recent developments and future perspectives of biorenewable nanocomposites for advanced applications. Nanotechnol Rev 11(1):1696–1721. https://doi.org/10.1515/NTREV-2022-0105/HTML

Iroegbu AOC, Ray SS (2022) Recent developments and future perspectives of biorenewable nanocomposites for advanced applications. Nanotechnol Rev 11(1):1696–1721. https://doi.org/10.1515/NTREV-2022-0105/ASSET/GRAPHIC/J_NTREV-2022-0105_FIG_004.JPG

Jamwal V, Mittal A (2024) Recent progresses in nanocomposite films for food-packaging applications: synthesis strategies, technological advancements, potential risks and challenges. Food Rev Int. https://doi.org/10.1080/87559129.2024.2368065

Jeevanandam J, Vadanasundari V, Pan S, Barhoum A, Danquah MK (2022) Bionanotechnology and bionanomaterials: emerging applications, market, and commercialization. Bionanotechnology: emerging applications of bionanomaterials, pp 3–44. https://doi.org/10.1016/B978-0-12-823915-5.00009-5

Jiang Z et al (2019) Toxic effects of magnetic nanoparticles on normal cells and organs. Life Sci 220:156–161. https://doi.org/10.1016/J.LFS.2019.01.056

Jiménez A, Ruseckaite RA (2012) Nano-biocomposites for food packaging. Green Energy Technol 50:393–408. https://doi.org/10.1007/978-1-4471-4108-2_15

Kaleem M et al (2024) Biogenic synthesis of iron oxide nanoparticles and experimental modeling studies on the removal of heavy metals from wastewater. J Saudi Chem Soc 28(1):101777. https://doi.org/10.1016/J.JSCS.2023.101777

Kausar A (2021) Progress in green nanocomposites for high-performance applications. Mater Res Innov 25(1):53–65. https://doi.org/10.1080/14328917.2020.1728489

Khan T, Chauhan A (2022) Polymer-based bionanomaterials for biomedical applications. Bionanotechnology: emerging applications of bionanomaterials, pp 187–225. https://doi.org/10.1016/B978-0-12-823915-5.00006-X

Khan ST, Malik A (2019) Engineered nanomaterials for water decontamination and purification: from lab to products. Elsevier [Online]. Available at: https://www.sciencedirect.com/science/article/pii/S0304389418308859. Accessed 08 Nov 2024

Khanna M (2020) Growing green business investments in Asia and the Pacific: trends and opportunities [Online]. Available at: https://www.adb.org/sites/default/files/publication/664446/sdwp-072-green-business-investments-asia-pacific.pdf. Accessed 29 Dec 2024

Kishore A, Mithul Aravind S, Singh A (2023) Bionanocomposites for active and smart food packaging: a review on its application, safety, and health aspects. J Food Process Eng 46(5):e14320. https://doi.org/10.1111/JFPE.14320

Lacourt C, Mukherjee K, Garthoff J, O'Sullivan A, Meunier L, Fattori V (2024) Recent and emerging food packaging alternatives: chemical safety risks, current regulations, and analytical challenges. Compr Rev Food Sci Food Saf 23(6):e70059. https://doi.org/10.1111/1541-4337.70059

Li M, Zamyadi A, Zhang W, Dumée LF, Gao L (2022) Algae-based water treatment: a promising and sustainable approach. J Water Process Eng 46:102630. https://doi.org/10.1016/J.JWPE.2022.102630

Lukensmeyer CJ (2019) Public deliberation: a manager's guide to citizen engagement

Luu D-N et al (2022) Recycling of post-use bioprocessing plastic containers—mechanical recycling technical feasibility. mdpi.com. https://doi.org/10.3390/su142315557

Machado AV, Araújo AIS, Oliveira M (2015) Assessment of polymer-based nanocomposites biodegradability. Biodegradable polymers. Volume 1: Advancement in biodegradation study and applications, pp 209–224 [Online]. Available at: https://repositorium.sdum.uminho.pt/handle/1822/39292. Accessed 28 Dec 2024

Machado A, Araújo A, Oliveira M (2015b) Assessment of polymer-based nanocomposites biodegradability [Online]. Available at: https://repositorium.sdum.uminho.pt/handle/1822/39292. Accessed 29 Dec 2024

Maluin FN (2024) Enhancing chitosan nanofilm with agricultural waste fillers for sustainable and safe functional food packaging. ACS Agric Sci Technol. https://doi.org/10.1021/ACSAGSCITECH.4C00398/ASSET/IMAGES/MEDIUM/AS4C00398_0007.GIF

Maria L, Moses JA, Anandharamakrishnan C (2019) Ethical and regulatory issues in applications of nanotechnology in food. Food nanotechnology, pp 67–92. https://doi.org/10.1201/9781315153872-5

Massoud MA, Fayad R, Kamleh R, El-Fadel M (2010) Environmental management system (ISO 14001) certification in developing countries: challenges and implementation strategies. Environ Sci Technol 44(6):1884–1887. https://doi.org/10.1021/ES902714U

Meierhofer F, Fritsching U (2021) Synthesis of metal oxide nanoparticles in flame sprays: review on process technology, modeling, and diagnostics. Energy Fuels 35(7):5495–5537. https://doi.org/10.1021/ACS.ENERGYFUELS.0C04054/ASSET/IMAGES/MEDIUM/EF0C04054_0009.GIF

Mertz M, Strech D (2014) Systematic and transparent inclusion of ethical issues and recommendations in clinical practice guidelines: a six-step approach. Implement Sci 9(1):184. https://doi.org/10.1186/S13012-014-0184-Y/FIGURES/5

Miao J, Liu H, Li Y, Zhang X (2018) Biodegradable transparent substrate based on edible starch-chitosan embedded with nature-inspired three-dimensionally interconnected conductive nanocomposites for wearable green electronics. ACS Appl Mater Interfaces 10(27):23037–23047. https://doi.org/10.1021/ACSAMI.8B04291/SUPPL_FILE/AM8B04291_SI_001.PDF

Miettinen M (2019) Empirical analysis of regulative risk assessment processes of nanomaterials under the Toxic Substances Control Act (TSCA) and European Union regulation concerning the Registration, Evaluation, Authorization and restriction of Chemicals (REACH). J Phys: Conf Ser. iopscience.iop.org [Online]. Available at: https://doi.org/10.1088/1742-6596/1323/1/012023/meta. Accessed 28 Dec 2024

Ministry of Ecology and Environment of the People's Republic of China, National Biosafety Regulations (2018) Google Search [Online]. Available at: https://www.google.com/search?q=Ministry+of+Ecology+and+Environment+of+the+People%E2%80%99s+Republic+of+China%2C+National+Biosafety+Regulations%2C+2018.&rlz=1C1CHBF_enIN1130IN1130&oq=Ministry+of+Ecology+and+Environment+of+the+People%E2%80%99s+Republic+of+China%2C+National+Biosafety+Regulations%2C+2018.&gs_lcrp=EgZjaHJvbWUyBggAEEUYOTIHCAEQIRiPAjIHCAIQIRiPAjIHCAMQIRiPAtIBCTE3MTVqMGoxNagCCLACAQ&sourceid=chrome&ie=UTF-8. Accessed 27 Jan 2025

Pozo Morales A, Güemes A, Fernandez-Lopez A, Carcelen Valero V, de La Rosa Llano S (2017) Bamboo–polylactic acid (PLA) composite material for structural applications. Materials 10:1286. https://doi.org/10.3390/MA10111286

Nandhini J, Karthikeyan E, Rajeshkumar S (2024) Eco-friendly bio-nanocomposites: pioneering sustainable biomedical advancements in engineering. Discover Nano 19(1):1–31. https://doi.org/10.1186/S11671-024-04007-7

Nawafleh N, Al-Oqla FM (2024) Applications of bio-composites in electronics. Phys Sci Rev 9(12):3601–3620. https://doi.org/10.1515/PSR-2024-0076/MACHINEREADABLECITATION/RIS

Olawore O, Ogunmola M, Desai S (2024) Engineered nanomaterial coatings for food packaging: design, manufacturing, regulatory, and sustainability implications. Micromachines 15:245. https://doi.org/10.3390/MI15020245

Öztürk A, Alarçin E, Yaşayan G, Avci-Adali M, Khosravi A, Zarepour A, Iravani S, Zarrabi A (2024) Innovative approaches in skin therapy: bionanocomposites for skin tissue repair and regeneration. pubs.rsc.org [Online]. Available at: https://pubs.rsc.org/en/content/articlehtml/2024/ma/d4ma00384e. Accessed 28 Dec 2024

Perera KY, Hopkins M, Jaiswal AK, Jaiswal S (2024) Nanoclays-containing bio-based packaging materials: properties, applications, safety, and regulatory issues. J Nanostruct Chem 14(1):71–93. https://doi.org/10.1007/S40097-023-00525-5/TABLES/5

Pires JRA, Rodrigues C, Coelhoso I, Fernando AL, Souza VGL (2023) Current applications of bionanocomposites in food processing and packaging. Polymers 15:2336. https://doi.org/10.3390/POLYM15102336

Prajapati D et al (2024) Bionanotechnology: a paradigm for advancing environmental sustainability. Indian J Microbiol 2024:1–27. https://doi.org/10.1007/S12088-024-01389-1
Programmes & Initiatives | Department of Science & Technology [Online]. Available at: https://dst.gov.in/programmes-innitiatives?utm_source=chatgpt.com. Accessed 29 Dec 2024
Puiggalí J, Katsarava R (2017) Bionanocomposites. Clay-polymer nanocomposites, pp 239–272. https://doi.org/10.1016/B978-0-323-46153-5.00007-0
Raigrodski D (2016) Creative capitalism and human trafficking: a business approach to eliminate forced labor and human trafficking from global supply chains. Wm Mary Bus L Rev. HeinOnline [Online]. Available at: https://heinonline.org/hol-cgi-bin/get_pdf.cgi?handle=hein.journals/wmaybur8&section=5. Accessed 29 Dec 2024
Ramesh M, Deepa C, Kumar LR, Sanjay MR, Siengchin S (2022) Life-cycle and environmental impact assessments on processing of plant fibres and its bio-composites: a critical review. J Ind Text 51(4):5518S-5542S. https://doi.org/10.1177/1528083720924730/ASSET/IMAGES/LARGE/10.1177_1528083720924730-FIG5.JPEG
Ramsden J (2018) Applied nanotechnology: the conversion of research results to products [Online]. Available at: https://books.google.com/books?hl=en&lr=&id=aB8wDwAAQBAJ&oi=fnd&pg=PP1&dq=Equity+and+accessibility+in+bionanocomposites+pertain+to+ensuring+that+the+benefits+of+these+materials+are+distributed+fairly+across+different+communities+and+socioeconomic+groups.+&ots=5Sai8ara6X&sig=DZebdTPlgczMg-0E3dennCqbNuc. Accessed 29 Dec 2024
Reddy K, Singh S (2022) Regulation of nanotechnology and nano-enabled products: the implications of the Consumer Protection Act 68 of 2008 in South Africa. Afr J Inter/multidiscip Stud 4(1):325–337. https://doi.org/10.51415/AJIMS.V4I1.956
Regulation-1223/2009-EN-Cosmetic Products Regulation-EUR-Lex [Online]. Available at: https://eur-lex.europa.eu/legal-content/EN/TXT/?uri=CELEX%3A32009R1223. Accessed 27 Jan 2025
Rehm BHA, Fata Moradali M (2020) Advances in biopolymers for biomedical and biotechnological applications [Online]. Available at: https://books.google.de/books?id=qVwhtAEACAAJ. Accessed 27 Dec 2024
Rehm B, Fata Moradali M (2021) Biopolymers for biomedical and biotechnological applications. https://doi.org/10.1002/9783527818310
Risk Regulation in Europe: Assessing the Application of the Precautionary... - Jale Tosun - Google Books [Online]. Available at: https://books.google.co.in/books?hl=en&lr=&id=K6OqfTE7IvgC&oi=fnd&pg=PR3&dq=In+the+European+Union,+there+is+a+strong+emphasis+on+environmental+sustainability+and+the+precautionary+principle,+leading+to+stricter+regulations+compared+to+the+United+States,+which+often+prioritizes+innovation+and+market+access.&ots=M6p3MoRCAl&sig=N-4UhLFGm4D7ZflIkNWe7YW8Z4o&redir_esc=y#v=onepage&q&f=false. Accessed 28 Dec 2024
Sabet M (2024) Exploring biodegradable polymer composites for sustainable packaging: a review on properties, manufacturing techniques, and environmental impacts. Iranian Polym J (Engl Ed) 1–20. https://doi.org/10.1007/S13726-024-01365-Y/TABLES/7
Sagha Zadeh R, Eshelman P, Setla J, Kennedy L, Hon E, Basara A (2018) Environmental design for end-of-life care: an integrative review on improving the quality of life and managing symptoms for patients in institutional settings. J Pain Symptom Manag 55(3):1018–1034. https://doi.org/10.1016/J.JPAINSYMMAN.2017.09.011
Salim HK et al (2018) Global trends in environmental management system and ISO14001 research. J Clean Prod 170:645–653. https://doi.org/10.1016/J.JCLEPRO.2017.09.017
Sarkar S, Gulati K, Poluri KM (2020) Life cycle assessment and future perspectives of green polymeric nanocomposites. Green polymeric nanocomposites, pp 251–280. https://doi.org/10.1201/9781351045155-9
Satdive A, Tayde S, Tonde S, Hazra C, Kundu D, Chatterjee A (2022) Bionanocomposites in the construction and building applications, pp 293–310. https://doi.org/10.1007/978-981-16-8578-1_16

Secco L et al (2019) Towards a method of evaluating social innovation in forest-dependent rural communities: first suggestions from a science-stakeholder collaboration. For Policy Econ 104:9–22. https://doi.org/10.1016/J.FORPOL.2019.03.011

Sen A, Kurian MS, Narayanan DP, Abraham A, Jayalakshmi PS, Nair SG, Joseph C, Unnikrishnan NV, Simon SM (2024) Bio-nanocomposites: innovative solutions for addressing issues in health, agriculture, energy and environmental domains. Elsevier [Online]. Available at: https://www.sciencedirect.com/science/article/pii/S2352507X24001811. Accessed 08 Nov 2024

Shaik S, Franchetti MJ (2022) Recycling and sustainability of nanocomposites. Applications of nanocomposites, pp 236–260. https://doi.org/10.1201/9781003247074-11

Sharma B, Thomas S, Bajpai P, Ghosal K, Shekhar S (2024) Advances in bionanocomposites: materials, applications, and life cycle [Online]. Available at: https://books.google.com/books?hl=en&lr=&id=KrLBEAAAQBAJ&oi=fnd&pg=PP1&dq=Life+cycle+assessment+(LCA)+is+another+emerging+tool+to+assess+the+environmental+sustainability+of+bionanocomposites&ots=1NhpBu5ZFA&sig=CSQ4sTR93N4TRR_L6hz5prw28O4. Accessed 28 Dec 2024

Sharma A, Luthra G (2023) Significance of ISO 10993 standards in ensuring biocompatibility of medical devices: a review. J Pharm Res Int 35(8):23–34. https://doi.org/10.9734/JPRI/2023/V35I87342

Shatkin JA, Kim B (2017) Environmental health and safety of cellulose nanomaterials and composites. Handbook of nanocellulose and cellulose nanocomposites, pp 683–729. https://doi.org/10.1002/9783527689972.CH21

Shunin Y, Bellucci S, Gruodis A, Lobanova-Shunina T (2018) Nanotechnology application challenges: nanomanagement, nanorisks and consumer behaviour, pp 337–395. https://doi.org/10.1007/978-3-319-69167-1_11

Singh NB, Kumar B, Usman UL, Susan MABH (2024) Nano revolution: exploring the frontiers of nanomaterials in science, technology, and society. Nano-Struct Nano-Objects 39:101299. https://doi.org/10.1016/J.NANOSO.2024.101299

Singh M, Changmai M, Ghosh T, Karwa A (2023) Natural resource-based nanobiomaterials. Advanced materials and manufacturing techniques for biomedical applications, pp 61–101. https://doi.org/10.1002/9781394166985.CH4

Singh S, Bala N, Kaur A (2024) Green sustainable bionanocomposites materials: short review. J Manag Eng Sci 1:98–106. https://doi.org/10.61552/JMES.2024.03.003

Sousa SPB, Peixoto T, Santos RM, Lopes A, Paiva MC, Marques AT (2020) Health and safety concerns related to CNT and graphene products, and related composites. J Compos Sci 4:106. https://doi.org/10.3390/JCS4030106

Srinivas A, Chandran KR, Nehra S, Kumar D (2025) Future aspects in the field of bio-nanomaterials toward environmental assessment. Bio-nanomaterials in environmental remediation, pp 295–317. https://doi.org/10.1002/9783527848546.CH12

Störmer A, Bott J, Kemmer D, Franz R (2017) Critical review of the migration potential of nanoparticles in food contact plastics. Trends Food Sci Technol [Online]. Available at: https://www.sciencedirect.com/science/article/pii/S0924224416303843. Accessed 26 Dec 2024

Suer J, Traverso M, Ahrenhold F (2021) Carbon footprint of scenarios towards climate-neutral steel according to ISO 14067. Elsevier [Online]. Available at: https://www.sciencedirect.com/science/article/pii/S0959652621027931. Accessed 28 Dec 2024

Tadayoni NS, Dinari M, Roy A, Karimi Abdolmaleki M (2024) Recent advances in porous biopolymer composites for the remediation of organic pollutants. Polymers 16:1543. https://doi.org/10.3390/POLYM16111543

Teles MCA et al (2015) Evaluation of the diameter influence on the tensile strength of pineapple leaf fibers (PALF) by Weibull method. Mater Res 18:185–192. https://doi.org/10.1590/1516-1439.362514

The Impacts of Nanotechnology on Companies (2010). https://doi.org/10.1787/9789264094635-EN

Tripathi N, Misra M, Mohanty AK (2021) Durable polylactic acid (PLA)-based sustainable engineered blends and biocomposites: recent developments, challenges, and opportunities. ACS Eng Au 1(1):7–38. https://doi.org/10.1021/ACSENGINEERINGAU.1C00011

Tripathi A et al (2023) Substantial utilization of food wastes for existence of nanocomposite polymers in sustainable development: a review. Environ Dev Sustain 26(10):24727–24753. https://doi.org/10.1007/S10668-023-03756-2/FIGURES/4

Trotta F et al (2023) Silver bionanocomposites as active food packaging: recent advances & future trends tackling the food waste crisis. Polymers 15(21):4243. https://doi.org/10.3390/POLYM15214243

Tsalis TA, Malamateniou KE, Koulouriotis D, Nikolaou IE (2020) New challenges for corporate sustainability reporting: United Nations' 2030 Agenda for sustainable development and the sustainable development goals. Corp Soc Responsib Environ Manag 27(4):1617–1629. https://doi.org/10.1002/CSR.1910

Van de Voorde M, Jeswani G (2021) Ethics in nanotechnology: social sciences and philosophical aspects, vol II, pp 1–364. https://doi.org/10.1515/9783110719932/PDF

Xiong R, Grant AM, Ma R, Zhang S, Tsukruk VV (2018) Naturally-derived biopolymer nanocomposites: interfacial design, properties and emerging applications. Mater Sci Eng R Rep 125:1–41. https://doi.org/10.1016/J.MSER.2018.01.002

Xu H, Adolfsson KH, Xie L, Hassanzadeh S, Pettersson RN, Hakkarainen M (2016) Zero-dimensional and highly oxygenated graphene oxide for multifunctional poly (lactic acid) bionanocomposites. ACS Sustain Chem Eng 4(10):5618–5631. https://doi.org/10.1021/acssuschemeng.6b01524

You C et al (2012) The progress of silver nanoparticles in the antibacterial mechanism, clinical application and cytotoxicity. Mol Biol Rep 39(9):9193–9201. https://doi.org/10.1007/S11033-012-1792-8/FIGURES/1

Yu Y, Flury M (2024) Unlocking the potentials of biodegradable plastics with proper management and evaluation at environmentally relevant concentrations. npj Mater Sustain 2(1):1–7. https://doi.org/10.1038/s44296-024-00012-0

Zhou Z, Wang Y, Liu H, D'avenio G, Daniele C, Grigioni M (2024) Nanostructured medical devices: regulatory perspective and current applications. Materials 17(8):1787. https://doi.org/10.3390/MA17081787

MIX
Papier aus verantwortungsvollen Quellen
Paper from responsible sources
FSC® C105338

If you have any concerns about our products,
you can contact us on
ProductSafety@springernature.com

In case Publisher is established outside the EU,
the EU authorized representative is:
Springer Nature Customer Service Center GmbH
Europaplatz 3, 69115 Heidelberg, Germany

Printed by Libri Plureos GmbH
in Hamburg, Germany